BIOMEDICAL ENGINEERING

Bridging Medicine and Technology

This is an ideal text for an introduction to biomedical engineering. The book presents the basic science knowledge used by biomedical engineers at a level accessible to all students and illustrates the first steps in applying this knowledge to solve problems in human medicine.

Biomedical engineering now encompasses a range of fields of specialization including bioinstrumentation, bioimaging, biomechanics, biomaterials, and biomolecular engineering. This introduction to bioengineering assembles foundational resources from molecular and cellular biology and physiology and relates them to various subspecialties of biomedical engineering.

The first two parts of the book present basic information in molecular/cellular biology and human physiology; quantitative concepts are stressed in these sections. Comprehension of these basic life science principles provides the context in which biomedical engineers interact. The third part of the book introduces the subspecialties in biomedical engineering and emphasizes – through examples and profiles of people in the field – the types of problems biomedical engineers solve.

W. Mark Saltzman is the Goizueta Foundation Professor of Chemical and Biomedical Engineering at Yale University. His research interests include materials for controlled drug delivery, drug delivery to the brain, and tissue engineering. He has taught at Johns Hopkins University and Cornell University and, after joining the Yale faculty in 2002, was named the first Chair of the Department of Biomedical Engineering.

Professor Saltzman has published more than 150 research papers, 3 authored books, and 2 edited books, and he is an inventor on more than 10 patents. His many honors and awards include a Camille and Henry Dreyfus Foundation Teacher-Scholar Award (1990), the Allan C. Davis Medal as Maryland's Outstanding Young Engineer (1995), the Controlled Release Society Young Investigator Award (1996), Fellow of the American Institute of Biological and Medical Engineers (1997), the Professional Progress in Engineering Award from Iowa State University (2000), Britton Chance Distinguished Lecturer in Engineering and Medicine at the University of Pennsylvania (2000), and Distinguished Lecturer of the Biomedical Engineering Society (2004).

CAMBRIDGE TEXTS IN BIOMEDICAL ENGINEERING

The *Cambridge Texts in Biomedical Engineering* series provides a forum for high-quality, accessible textbooks targeted at undergraduate and graduate courses in biomedical engineering. It covers a broad range of biomedical engineering topics from introductory texts to advanced topics, including, but not limited to, biomechanics, physiology, biomedical instrumentation, imaging, signals and systems, cell engineering, and bioinformatics. The series blends theory and practice and is aimed primarily at biomedical engineering students, but it is suitable for broader courses in engineering, life science, and medicine.

BIOMEDICAL ENGINEERING

Bridging Medicine and Technology

W. Mark Saltzman

Yale University

CAMBRIDGE UNIVERSITY PRESS
Cambridge, New York, Melbourne, Madrid, Cape Town, Singapore, São Paulo, Delhi

Cambridge University Press
32 Avenue of the Americas, New York, NY 10013-2473, USA

www.cambridge.org
Information on this title: www.cambridge.org/9780521840996

First published 2009

Printed in the United States of America

A catalog record for this publication is available from the British Library.

Library of Congress Cataloging in Publication data

Saltzman, W. Mark.
Biomedical engineering : bridging medicine and technology / W. Mark Saltzman.
p. cm. – (Cambridge texts in biomedical engineering)
Includes bibliographical references and index.
ISBN 978-0-521-84099-6 (hardback)
1. Biomedical engineering. I. Title. II. Series.
R856.S25 2009
610.28 – dc22 2008044766

ISBN 978-0-521-84099-6 hardback

To Zach and Alex
There is no luckier, happier father on earth than I.

Contents

PART 3. BIOMEDICAL ENGINEERING

Preface

The field of biomedical engineering has expanded markedly in the past ten years. This growth is supported by advances in biological science, which have created new opportunities for development of tools for diagnosis of and therapy for human disease. This book is designed as a textbook for an introductory course in biomedical engineering. The text was written to be accessible for most entering college students. In short, the book presents some of the basic science knowledge used by biomedical engineers and illustrates the first steps in applying this knowledge to solve problems in human medicine.

Biomedical engineering now encompasses a range of fields of specialization including bioinstrumentation, bioimaging, biomechanics, biomaterials, and biomolecular engineering. Most undergraduate students majoring in biomedical engineering are faced with a decision, early in their program of study, regarding the field in which they would like to specialize. Each chosen specialty has a specific set of course requirements and is supplemented by wise selection of elective and supporting coursework. Also, many young students of biomedical engineering use independent research projects as a source of inspiration and preparation but have difficulty identifying research areas that are right for them. Therefore, a second goal of this book is to link knowledge of basic science and engineering to fields of specialization and current research.

As a general introduction to the field, this textbook assembles foundational resources from molecular and cellular biology and physiology and relates this science to various subspecialties of biomedical engineering. The first two parts of the book present basic information in molecular/cellular biology and human physiology; quantitative concepts are stressed in these sections. Comprehension of these basic life science principles provides the context in which biomedical engineers interact. The third part of the book introduces the subspecialties in biomedical engineering and emphasizes—through examples and profiles of people in the field—the types of problems biomedical engineers solve. Organization of the chapters into these three major parts allows course instructors and students to customize their usage of some or all of the chapters depending on the background of the students and the availability of other course offerings in the curriculum.

WHICH STUDENTS PROFIT FROM THIS BOOK?

A significant number of students come to college with a clear idea of pursuing a career in biomedical engineering. Of course, these students benefit tremendously from a rigorous overview of the field, ideally provided in their first year. Most of these students leave the course even more certain about their choice of career. Many of them jump right into independent study or research projects: This overview of the diverse applications of biomedical engineering provides them with the information that they need to select research projects—or future courses—that will move them in the right direction.

I have also found this material to be interesting to engineering students who are trying to decide which of the engineering degree programs is right for them. The material in this textbook might also be used to introduce undeclared or undecided engineering majors to the field of biomedical engineering. Students enter college with varying degrees of competence in science and math. Some do not know what biomedical engineering encompasses and whether they have the adequate secondary education training to succeed. Exposure to the topics presented here may inspire some of these students to further their studies in biomedical engineering.

Also, I encourage instructors to make their course accessible to students who are not likely to become engineering majors; biomedical technology is increasingly important to the life of all educated citizens. I have taught courses in this subject to freshmen at three different universities over the past 20 years; students with a variety of intended majors always enroll in the course (mathematics, history, economics, English, fine arts, and anthropology majors have participated in the past few years). In fact, it is these students who appear to be most changed by the experience.

TO THE INSTRUCTOR

Teachers of courses directed to early undergraduates in biomedical engineering struggle against competing forces: The diverse backgrounds of the students pull you to start from first principles, and the rapid progress of the field pushes you to cover more and more topics. To address this, I have presented more material than I am capable of covering in a one-semester course for freshmen students. In a typical 13-week semester, I find that only 12–13 of the 16 chapters can be covered comfortably. Assuming that this will be true for your situation as well, I recommend that you assess the level of experience of your students and decide which chapters are most valuable in creating a coherent and satisfying course. Many students arrive at college with a sophisticated understanding of cellular and molecular biology; therefore, I do not cover Part 1 (Chapters 2–5) in detail. Condensing this early material allows me to include almost all of the other chapters. Part 1 is still available to the student, of course, and most of them profit from reading these chapters, as they need as the course progresses, even if the details are not covered in lecture. In courses that emphasize biomedical engineering, and not the biological sciences, the instructor might want to cover only Part 3 of the book and use the previous parts as reference material.

Some examples of approaches for arranging the chapters into semester-long courses that emphasize different aspects of biomedical engineering are presented in the following table.

Modular approaches to teaching an introductory course in biomedical engineering using this text

Week of the course	Comprehensive approach	Applications emphasis	Physiology emphasis	Cellular engineering emphasis
1	Chap. 1 and 2	Chap. 1	Chap. 1	Chap. 1
2	Chap. 3 and 4	Chap. 2–5 (selected)	Chap. 2–4 (selected)	Chap. 2 and 3
3	Chap. 5	Chap. 10	Chap. 5	Chap. 4
4	Chap. 7	Chap. 10 and 11	Chap. 6	Chap. 5
5	Chap. 8	Chap. 11	Chap. 7	Chap. 6–9 (selected)
6	Chap. 9	Chap. 12	Chap. 8	Chap. 10
7	Midterm review	Midterm review	Midterm review	Midterm review
8	Chap. 10	Chap. 2–5 (selected)	Chap. 9	Chap. 11
9	Chap. 11	Chap. 13	Chap. 10	Chap. 12
10	Chap. 12	Chap. 14	Chap. 11 and 12	Chap. 13
11	Chap. 13 and 14	Chap. 15	Chap. 13	Chap. 14
12	Chap. 15	Chap. 16	Chap. 14	Chap. 15
13	Chap. 16	Chap. 16	Chap. 15	Chap. 16

Acknowledgments

I have many people to thank, for encouragement and direct participation. It is a long list, and undoubtedly incomplete. For the past seven years, I have been immersed in a milieu rich in inspiration, creation, and succor. So I profited from brushes and asides, from long conversations and wisdom overheard.

I thank the Whitaker Foundation for their generous financial support, which made it possible for me to transform notes and notions into text. I am particularly grateful to Jack Linehan, who has been a steady source of inspiration and advice to me over the past decade.

I thank Peter Gordon of Cambridge University Press, who has been the most stalwart supporter of this project. Peter is everything one could hope for in an editor: He is wise, generous with praise, and direct (yet kind) with criticism. Thankfully, he is also patient. I thank Michelle Carey for her brilliant support. What a pleasure, to be an author for Cambridge University Press.

I thank Veronique Tran for her help in the inception of this project, her critical assistance in overall organization of the book, and her work on early versions of Chapters 2, 6, and 11. It was Veronique who urged this project forward at the start, and it would not have happened without her effort and enthusiasm. I thank Lawrence Staib, who co-authored Chapter 12 and shaped it into one of my favorite chapters in the book. I thank Rachael Sirianni, who continues to amaze me with the breadth of her talents: Rachael's photography enhances every chapter.

I burdened generous friends; each of them gave one of the chapters a careful reading and provided thoughtful edits and suggestions, which made each chapter better, more readable. I thank Ian Suydam (Chapter 2), Kim Woodrow (Chapter 3), Michael Caplan (Chapter 5), Michelle Kelly (Chapter 7), Peter Aronson (Chapter 9), Deepak Vashishith (Chapter 10), and Themis Kyriakides (Chapter 15).

I am grateful to my co-instructors in Physiological Systems (BENG 350) at Yale, who have been exceptional colleagues and patient, enthusiastic teachers of physiology. The influence of Michael Caplan, Walter Boron, Emile Boulpaep, and Peter Aronson can be felt in Chapters 5, 7, 8, and 9, respectively. I have profited from their examples as teachers.

A number of people contributed essential administrative and research support—tracking down papers and facts, producing figures, proofreading, and creating and solving homework problems. I thank Tiffanee Green, Michael Henry, Kofi Buaku Atsina, Florence Kwo, and Salvador Joel Nunez Gastelum. Two special people did this and much more: Audrey Lin and Jennifer Saucier-Sawyer proofread, edited, pursued figures (and permissions for figures), and managed to keep binders, drafts, and sticky notes organized. More than this, they smiled at every obstacle, accommodated every idea, and remained positive as I missed deadlines. Without Audrey's expert help in the final push—and her never-say-no generosity—this text would still be in binders.

I thank Caroline, for letting me be me, as she is she. It is a marvel, isn't it, this unexpected shower, rescuing a late summer afternoon? Thanks, Caroline, for sharing it with me.

Abbreviations and Acronyms

3D-	three-dimensional
3DCRT	three-dimensional conformal radiation therapy
Ab	antibody
ADA	adenosine deaminase deficiency
ADH	anti-diuretic hormone
ADP	adenosine diphosphate
AIDS	acquired immune deficiency syndrome
AML	acute myeloid leukemia
APC	antigen presenting cell
ATP	adenosine-5′-triphosphate
AV	atrioventricular
BBB	blood-brain barrier
BCG	Bacillus Calmette-Guérin
BME	biomedical engineering
BMR	basal metabolic rate
BSA	bovine serum albumin
CABG	coronary artery bypass graft
CLL	chronic lymphocytic leukemia
CT	computed tomography
DAG	diacylglycerol
DNA	deoxyribonucleic acid
EBRT	external beam radiation therapy
ECF	extracellular fluid
ECG	electrocardiography, electrocardiogram
ECM	extracellular matrix
EGF	epidermal growth factor
EGFR	epidermal growth factor receptor
EVAc	poly(ethylene-co-vinyl acetate)
FBR	foreign body response
FDA	U.S. Food and Drug Administration
fMRI	functional magnetic resonance imaging
GFR	glomerular filtration rate
GFP	green fluorescent protein

HIV	human immunodeficiency virus
HPV	human papillomavirus
HSC	hematopoietic stem cells
HUVEC	human umbilical vein endothelial cells
ICAM	intercellular adhesion molecule
Ig	immunoglobulin
IL-2	interleukin 2
IMRT	intensity-modulated radiation therapy
IR	infrared
IRS	insulin receptor substrate
ISF	interstitial fluid
LDL	low-density lipoprotein
mAbs	monoclonal antibodies
MHC	major histocompatibility complex
MRI	magnetic resonance imaging
MW	molecular weight
NHL	non-Hodgkin's lymphoma
NMR	nuclear magnetic resonance
PAH	para-aminohippuric acid
PAN	polyacrylonitrile
PCR	polymerase chain reaction
PDMS	polydimethylsiloxane
PE	polyethylene
PEG	poly(ethylene glycol)
PET	positron emission tomography *or* poly(ethylene terephthalate)
PEU	polyurethane
pHEMA	poly(2-hydroxymethacrylate)
PIP3	phosphatidylinositol 3,4,5-trisphosphate
PKB	protein kinase B
PLGA	poly(lactide-co-glycolide)
pMMA	poly(methyl methacrylate)
PP	polypropylene
PS	polystyrene
PSA	prostate specific antigen
PSu	polysulphone
PTFE	poly(tetrafluoroethylene)
PVC	poly(vinyl chloride)
PVP	poly(vinyl pyrrolidone)
RBC	red blood cell
RF	radio frequency
RGD	three peptide sequence of arginine (R), glycine (G), and aspartic acid (D)
RNA	ribonucleic acid
RPF	renal plasma flow
rRNA	ribosomal RNA

RSV	respiratory syncytial virus
RTK	receptor tyrosine kinase
SA	sinoatrial
SARS	Severe Acute Respiratory Syndrome
SGOT	serum glutamic oxaloacetic transaminase
siRNA	small interfering RNA
sMRI	structural magnetic resonance imaging
SPECT	single photon emission computed tomography
TIL	tumor-infiltrating lymphocytes
tRNA	transfer RNA
UV-VIS	ultraviolet-visible spectroscopy
VEGF	vascular endothelial cell growth factor
WBC	white blood cells
WHO	World Health Organization

1 Introduction: What Is Biomedical Engineering?

LEARNING OBJECTIVES

After reading this chapter, you should:

- Be familiar with how changes in medicine have enhanced life span and quality of life.
- Understand a few examples of the role of engineering in defining medical treatments.
- Have developed your own definition of biomedical engineering.
- Understand some of the subdisciplines that are included in biomedical engineering.
- Understand the relationship between the study of biomedical engineering and the study of human physiology.
- Be familiar with the structure of this book, and have developed a plan for using it that fits your needs.

1.1 Prelude

The practice of medicine has changed dramatically since you were born. Consider a few of these changes, some of which have undoubtedly affected your own life: Couples can test for pregnancy in their homes, a new vaccine is available for chicken pox, inexpensive contact lenses provide clear vision, artificial hips allow recipients to walk and run, ultrasound imaging follows the progress of pregnancy, and small reliable pumps administer insulin continuously for diabetics. For your parents, the changes have been even more sweeping. Overall life expectancy—that is, the span of years that people born in a given year are expected to live—increased from 50 in 1900 to almost 80 by 2000 (Figure 1.1). You can expect to live 30 years longer than your great-grandparents; you can also expect to be healthier and more active during all the years of your life.

How has this happened? One answer is obvious. People are living longer because they are not dying in situations that were previously fatal, such as childbirth and bacterial infections. The growth of biomedical engineering is a major factor in this extension of life and improvement of health. Biomedical engineers have contributed to every field of medicine—from radiology to obstetrics to cancer treatment—but in the next few paragraphs this growth is illustrated with examples from emergency medicine.

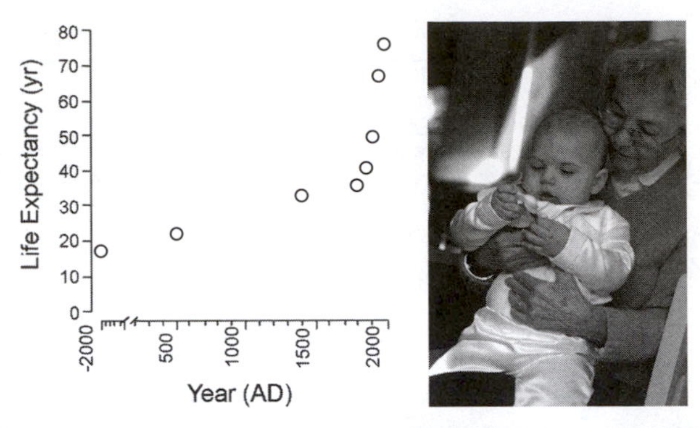

Figure 1.1 **Human life expectancy.** Human life expectancy has increased dramatically in the past 200 years.

Accidents and trauma are major causes of death and disability around the world. In the United States, it is overwhelmingly the leading cause of death among people of college age, and it is ranked fifth among causes of death for all ages (1). Automobile accidents account for many of these deaths: 42,116 people were killed in automobile accidents in the United States in 2001. Victims of trauma often have internal injuries, which are life threatening but not easy to diagnose by visual observation. Many accident victims are rushed to emergency rooms for treatment, and actions performed in the first few minutes after arrival can often mean the difference between life and death. Emergency room treatment has improved enormously over the past few decades, chiefly due to advances in the technology for looking inside of people quickly and accurately (Figure 1.2). Ultrasound imaging, which can provide pictures of internal bleeding within seconds, has replaced exploratory surgery and other slower, more invasive approaches for localization of internal injuries. Old ultrasound imaging machines weighed hundreds of pounds, but new instruments are smaller and lighter—some weighing only a few pounds, making it possible to get them to the patient faster. Other imaging technologies have also improved: Helical computed tomography (CT) scanners produce rapid three-dimensional internal images of the whole body, and new magnetic resonance imaging (MRI) techniques can reveal the chemistry, not just the shape, of internal structures. As a result of faster and better diagnosis of internal injuries, more accident victims are saved today.

In the near future, emergency medicine providers will probably use ultrasound imagers that are small enough to be carried in a pocket and inexpensive enough for every physician to own, like a stethoscope is today. Reduction in size and cost will surely save the lives of more accident victims. A pill-sized sensor is already available that patients can swallow; it continuously reports internal temperature as it passes through the intestinal tract. In the future, similar devices will probably be used to report other internal conditions such as sites of bleeding or abnormal cells. Further in the future, these small devices will be guided to specific locations in the body, where they can initiate repair of disease that is deep within the body.

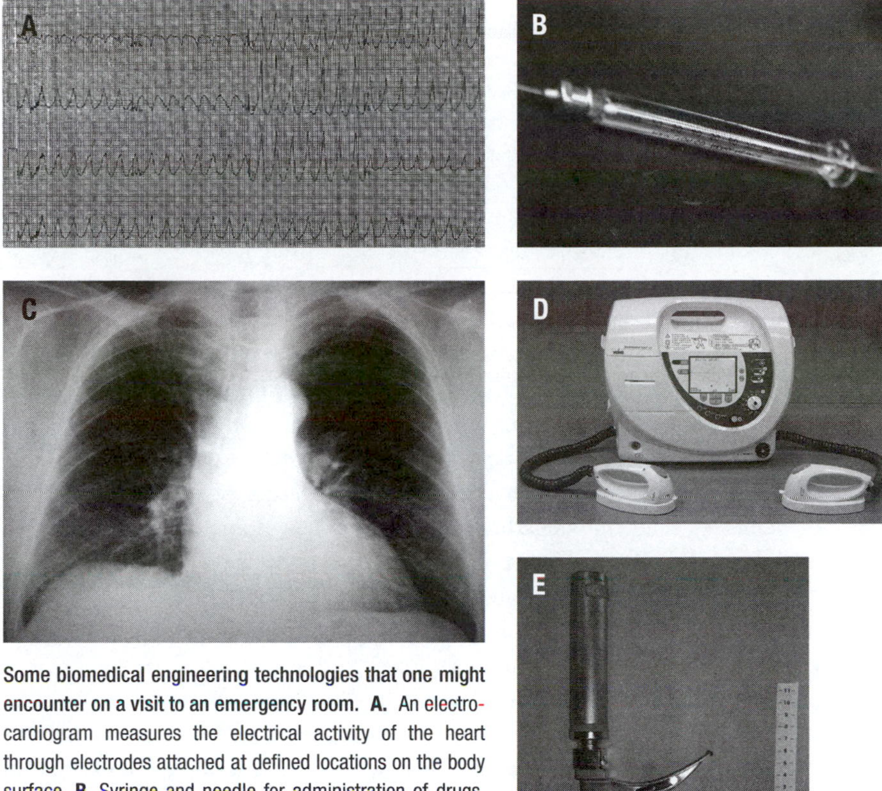

Figure 1.2 **Some biomedical engineering technologies that one might encounter on a visit to an emergency room. A.** An electrocardiogram measures the electrical activity of the heart through electrodes attached at defined locations on the body surface. **B.** Syringe and needle for administration of drugs. **C.** Chest x-rays are used to screen for lung diseases such as tuberculosis. **D.** Defibrillator for restoring normal heart rhythm. Photo courtesy of Dr. Yury Masloboev. **E.** Laryngoscope for intubation to provide breathing. Photo courtesy of Abinoam Praxedes Marques Junior.

The trends in emergency medicine are not unique. Innovations produced by biomedical engineers are saving lives once lost to kidney failure, improving eyesight lost to disease and aging, and producing artificial hips, knees, and hearts.

Do you want to be a part of this story or similar stories that are changing the conduct of medicine in operating rooms, doctors' offices, emergency vehicles, and homes? Then you want to be a biomedical engineer. This book will introduce you to the field of biomedical engineering and show how your knowledge of math, chemistry, physics, and biology can be used to understand how the human body works. It will show you how biomedical engineers work to develop new methods to diagnose problems with the human machine and new approaches to treat disease efficiently and inexpensively. This book will also show you how biomedical engineering and medicine will grow in the future, and point you in some directions that you can pursue to be part of this future. Biomedical engineering has been performed under different titles throughout history (Box 1.1); this book will help put this development into perspective, so that you can focus on how biomedical engineers will contribute to the future.

Box 1.1 Too many names?

As you read about the subject of biomedical engineering, you will encounter a variety of names that sound similar: bioengineering, biological engineering, biotechnology, biosystems engineering, bioprocess engineering, biomolecular engineering, and biochemical engineering. Some of the differences between these names are important, but unfortunately the terminology is not used consistently. Therefore, students of biomedical engineering need to approach the terminology with care (and without assuming that the person using the terminology has the same definition that they do!).

Biomedical engineering and bioengineering are often used interchangeably [e.g., see ref. (2)]. This is certainly true in the case of names of academic departments at universities. Some departments are called Department of Biomedical Engineering and others Department of Bioengineering, but in most cases the educational mission and research programs associated with these departments are similar. Still, it is wise for prospective students to look closely at the classes that are offered at each university and to decide if the emphasis of the department is the right one for them.

Some of the terms represent subsets of the larger discipline of biomedical engineering. Biomolecular engineering, for example, is now used to describe the contributions of chemical engineering to the larger field of biomedical engineering. In that sense, biochemical engineering and bioprocess engineering, which have historically been used to indicate the use of chemical engineering tools in the development of industrial processing methods for biological systems, are now embraced by the larger subdiscipline of biomolecular engineering. One could argue that all of these are subsets of the larger field of bioengineering or biomedical engineering.

Biotechnology is a trickier term to characterize because it has been used in a variety of different contexts over the past few decades. To many people, biotechnology is the end result of DNA manipulation: for example, transgenic animals, recombinant proteins, and gene therapy. Some common definitions are "the application of the principles of engineering and technology to the life sciences"; "the application of science and engineering to the direct or indirect use of living organisms, or parts or products of living organisms, in their natural or modified forms"; or "the use of biological processes to solve problems or make useful products." Again, one could argue that these definitions are equivalent to biomedical engineering. Even the technologies most commonly associated with biotechnology (e.g., production of recombinant proteins as pharmaceuticals) are examples of biomedical engineering. They are treated as such in this textbook, and are discussed in Chapters 13 and 14.

1.2 Engineering in modern medicine

Our experience of the world is shaped by engineering and technology. Because of the work of engineers, we can move easily from place to place, communicate with people at distant sites (even on the moon!), live and work in buildings that are safe from natural elements, and obtain affordable and diverse foods. It is widely, although not universally, accepted that the quality of life on our planet has improved as a result of the proliferation of technology that occurred during the 20th century. There is little doubt that the presence of technology creates constraints on the way that we live, and that the daily choices we make are shaped by the technologies that have infiltrated widely (think about the ways that television, computers, airplanes, cell phones, and ATMs have influenced your progress through this past day). Choices that we make in the future, and maybe even historical trajectories, will be influenced by future technologies such as (perhaps) nanomachines, efficient fuel cells, and small, inexpensive global positioning devices. It is the work of engineers to make technology possible, and then to make that technology reliable and inexpensive enough to influence people throughout the world.

Medical technology is one of the most visible aspects of the modern world; it is impossible to avoid and uniquely compelling. People from all walks of life are eager to hear about new machines, new medicines, and new devices that will uncover hidden disease, treat previously untreatable ailments, and mend weary or broken organs. Evidence for this high interest is everywhere; for example, new medical technology appears routinely on the covers of news magazines such as *Time* and *Newsweek* and in daily newspaper reports. We know that modern medicine is built on steady progress in science, but it is just as heavily dependent on innovations in engineering. It is engineers who transfer scientific knowledge into useful products, devices, and methods; therefore, progress in biomedical engineering is arguably more central to our experience of modern medicine than are advances in science. Some of the most fascinating stories of the 20th century involved the development of new medical technologies (Figure 1.3). Whole-organ transplantation, such as the first heart transplant in 1967, could not occur until there were machines to sustain life during the operation, tools for the surgeons to operate with and repair the wounds they created, and methods for preserving organs during transport. Thousands of transplants are performed annually in the United States today, but the need for organs far exceeds the supply. Biomedical engineers have been working for many decades to create an artificial heart, and there is no doubt that this work will continue until it is successful (see Chapter 15). Clinical testing of the Salk polio vaccine, in which millions of doses were administered to children, could not happen without the engineering methods to cheaply produce the vaccine in large quantity (see Chapter 14). The Human Genome Project would have not been possible without automated machines for deoxyribonucleic acid (DNA) sequencing.

Medical technology has also invaded our homes in surprising and influential ways. Every home has a thermometer, specially designed to permit the recording of body temperature. But we can now also test for pregnancy at home, so that one of the most life-changing medical discoveries can be done in privacy. Blood glucose tests,which are essential for proper treatment of diabetes, have advanced rapidly and now are commonly done at home. Your home can be easily equipped to be a screening center for high blood pressure, high cholesterol, glucose monitoring, and ovulation prediction.

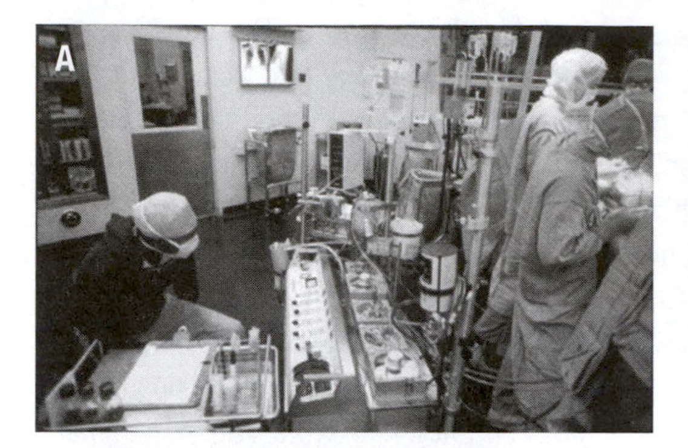

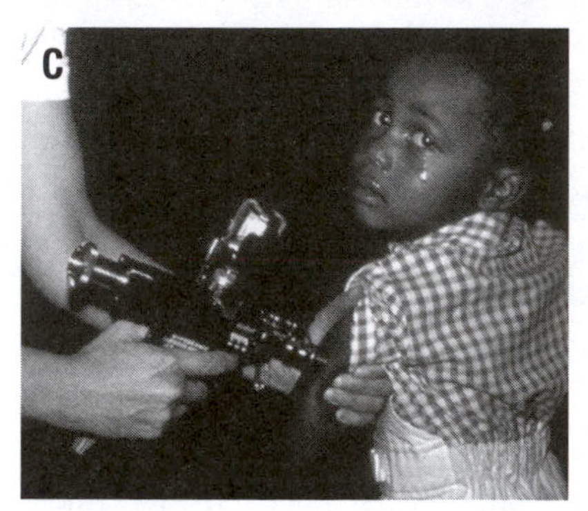
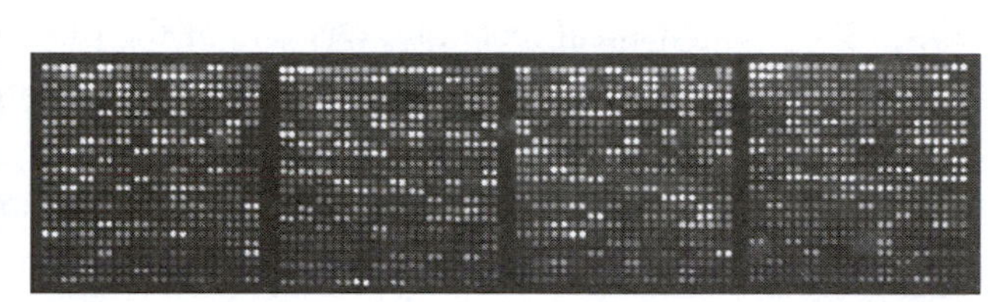

Figure 1.3 **Examples of new technology that permitted medical advances. A.** Heart–lung machine that permits heart transplantation and surgery. Photo courtesy of National Institutes of Health. **B.** Jet airplanes are used for rapid transport of a preserved organ to a distant operating room. **C.** An injector for vaccine delivery. Photo courtesy of The Centers For Disease Control and Prevention. **D.** DNA microarrays can be used to measure the expression of genes in cells and tissues. Photo courtesy of the W.M. Keck Foundation at Yale University. (See color plate.)

In addition, medical technologies have entered our bodies. Many people now elect to use contact lenses instead of eyeglasses; this change has resulted from the development of materials that can remain in contact with the eye for extended periods without causing damage. Artificial joints and limbs are common, as are artificial heart valves; synthetic components, usually metals and polymers, are fashioned into implantable devices that can replace the function of the human skeleton. We are not yet able to reanimate dead tissue (as Shelley predicted in *Frankenstein*), but we are close to the technology required for a 6 million dollar man.

This book supplies an introduction to biomedical engineering, the most rapidly growing of the engineering disciplines. Biomedical engineers invent, design, and build new technologies for diagnosis, treatment, and study of human disease. Usually, they work as a part of a team of engineers, scientists, and physicians, but the role of the engineer is essential. It is the engineer who is responsible for converting new knowledge into a useful form.

1.3 What is biomedical engineering?

New students to the field of biomedical engineering ask versions of this question: "What is biomedical engineering?" Often, they ask the question directly but, just as often, they ask it in indirect and interesting ways. Some of the forms of this question that I have heard in the past few years are:

- Do biomedical engineers all work in hospitals?
- Do you have to have an MD degree to be a biomedical engineer?
- How can I learn enough biology to understand biomedical engineering and enough engineering to be a real engineer?
- Is biomedical engineering the same as genetic engineering?
- How much of biomedical engineering is biology, chemistry, physics, and mathematics?

Some versions of the question are easy to answer. For example, most biomedical engineers do not work in hospitals and do not hold MD degrees. Other questions can inspire answers that take up whole books (such as this book), and still be incomplete. All of the chapters in this book are designed to address these questions from different perspectives. In this introduction, the overall question is examined from several different angles.

1.3.1 We can learn something about biomedical engineering from standard definitions

Our working definition of biomedical engineering can start in an obvious place. According to the Merriam-Webster Dictionary:

> engineering *noun*: a) the application of science and mathematics by which the properties of matter and the sources of energy in nature are made useful to people; b) the design and manufacture of complex products.

Biomedical engineering is engineering that is applied to human health. Because human health is multifaceted—involving not only our physical bodies but also the things that we put in our bodies (such as foods, pharmaceuticals, and medical devices) and the things that we put on our bodies (such as protective clothing and contact lenses)—biomedical engineers are interested in a wide range of problems. The breadth of modern biomedical engineering is reflected in the table of contents for this book (shown in diagrammatic form in Figure 1.4).

The work of engineers is often hidden from view of the general public, occurring in laboratories, office buildings, construction sites, pilot plants, and testing facilities. This is true for biomedical engineering as well as civil engineering and other engineering disciplines. Although the work might be hidden, the end result is often visible and important (e.g., the Brooklyn Bridge or the artificial heart; see Figure 1.5). Because of this, society has huge expectations for engineers, and engineers have large goals for themselves.

The importance of engineers to human progress is worthy of celebration. Consider this quote about the role of engineers from the president of the American

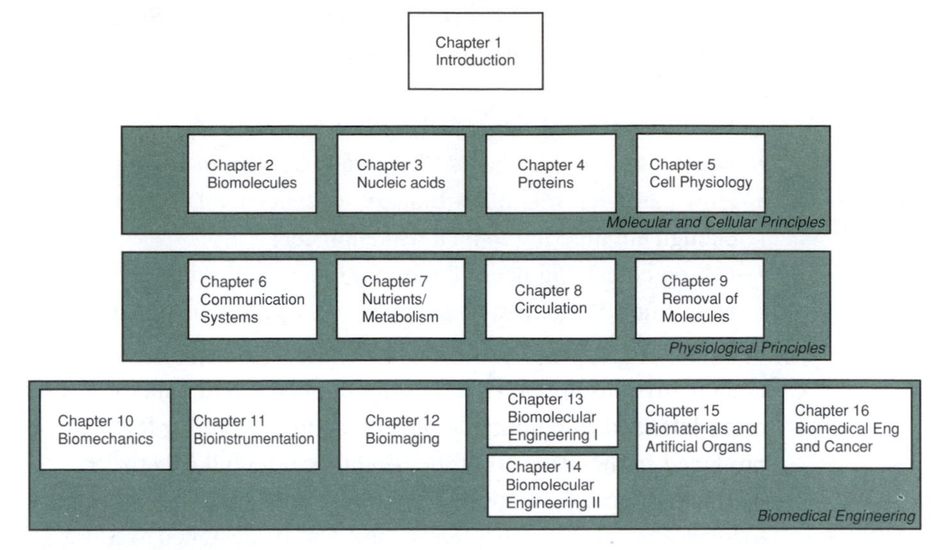

Figure 1.4 **Organization of this book.**

Society of Civil Engineers, Robert Moore. Mr. Moore, in a speech to the society in May 1902, said [from ref. (3)]:

> And in the future, even more than in the present, will the secrets of power be in his keeping, and more and more will he be a leader and benefactor of men. That his place in the esteem of his fellows and of the world will keep pace with his growing capacity and widening achievement is as certain as that effect follow cause.

Mr. Moore was speaking in the shadow of incredible engineering achievements: The work of engineers to build bridges over large spans of water changed the flow of society, for example. The substance of this quote—although not its selection of pronouns—is relevant today. Engineers of today have ambitious visions for their profession, and they are still called upon to be worthy inheritors of the engineering

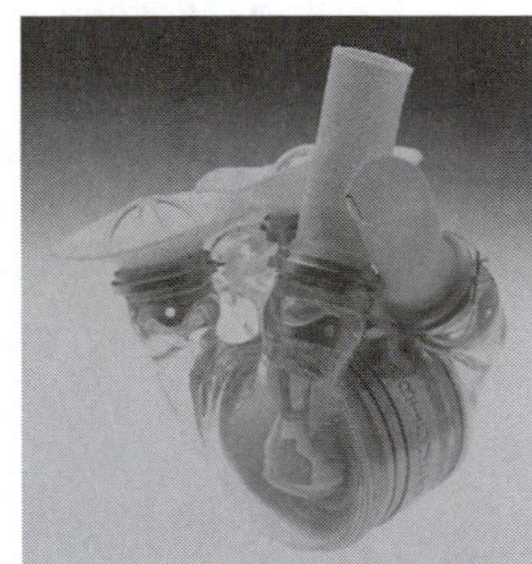

Figure 1.5 **Examples of engineering on the heroic scale. A.** Brooklyn Bridge (opened May 24,1883). **B.** AbioCor™ artificial heart (reprinted with permission from Jewish Hospital & St. Mary's HealthCare and the University of Louisville).

tradition to do good works. Imagine the confidence in your profession that is required to suggest that you can build a machine to replace the human heart, which is one of the most durable, reliable, and complex of machines. As we will see in Chapters 13 and 15, biomedical engineers now imagine that the creation of reliable replacement tissues and organs such as the heart is achievable. The success of the Abiomed artificial heart (called AbioCor™, Figure 1.5, which had been implanted into 10 patients as of March 2003), is an example of progress in this heroic effort.

A simple definition of engineering might be this: Engineering is the art of making practical application of the knowledge of pure science (3). Engineering is a creative discipline (like sculpture, poetry, and dance), but the end result is often intended to be durable, useful, abundant, and safe. Engineering art is not produced for museums, but intended to infiltrate the world.

Technology is a broader and more comprehensive term than engineering; in general, technology is the end result of a practical application of knowledge in a particular area. Anyone can produce technology, but engineers—because their training is focused on providing the knowledge tools needed to produce technology—have had the dominant role.

1.3.2 Biomedical engineers seek to understand human physiology and to build devices to improve or repair it

Other textbooks and review articles have described the origins of biomedical engineering, which can be identified even in ancient sources (2). Rather than reviewing this history in detail, we instead offer a schematic, speculative view of progress in biomedical engineering (Figure 1.6). Early humans learned that tools could improve the quality of their life; one might argue that the first engineers were the clever individuals who either recognized the value of wheels, levers, and sharpened rocks or figured out new ways to use these tools. As humans used tools, and as a result found new leisure time for other activities, some curious individuals probably began to use these implements to study themselves. As people learned more about the structure and function of their own bodies (that is, as they learned more about human anatomy and physiology), they were able to apply this knowledge to the creation of improved tools for repair of function (such as splints and sutures).

Observed in this way, the history of biomedical engineering involves a sequential and iterative process of discovery and invention: new tools for studying the human body leading to a deeper understanding of body function leading to the invention of improved tools for repair and study of the human body, and so forth. The dual nature of biomedical engineering is alive today; some biomedical engineers are concerned with careful analysis and study of the operation of body systems, others are concerned with the development of new techniques for the study and repair of the body, and still others do a bit of each.

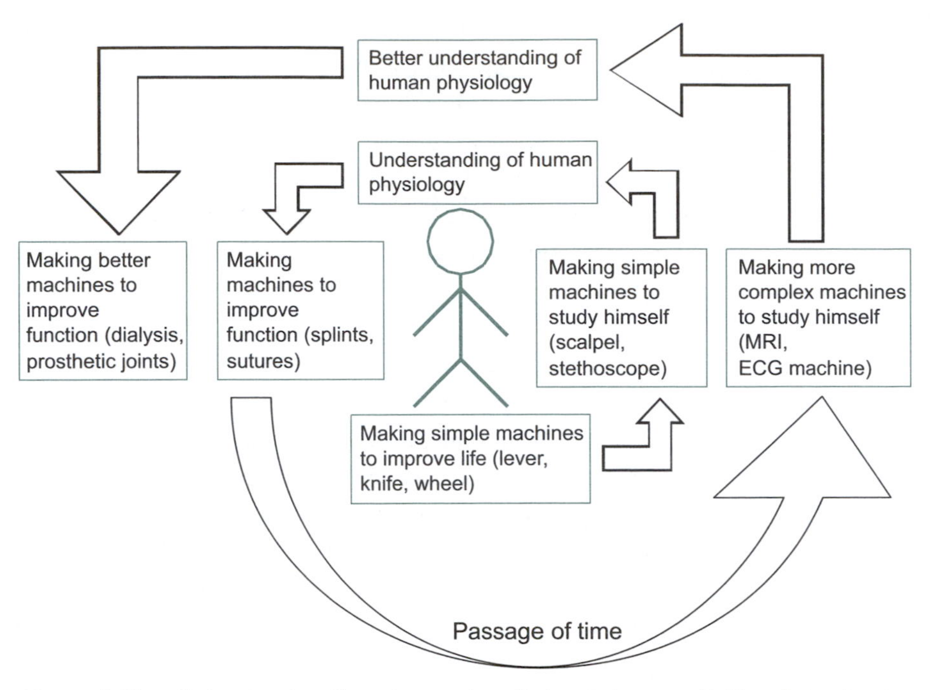

Figure 1.6

Advances in biomedical engineering. Figure shows a schematic view of advancement in biomedical engineering, which relies on sequential development of improved tools for studying physiology and the subsequent increased understanding of physiology that results. MRI, magnetic resonance imaging; ECG, electrocardiogram.

This speculative view of biomedical engineering can be confirmed through history; consider the timeline provided in Table 1.1, which shows highlights in the development of contact lenses. Many vision problems can now be corrected in humans; how did we get to this state? It was probably recognized very early in human history that the eye was involved in human vision; placing a hand in front of the eye blocks vision, and injuries to the eye destroy it. This knowledge of the source of vision was eventually translated into efforts to repair faulty vision with lenses (by Bacon in 1249 and Nicholas of Cusa in 1451); these developments could not happen until people (we would later call them engineers) had developed sufficient experience with the optical properties of materials and the construction of the lens. Leonardo da Vinci suggested a lens that was directly applied to the human eye early in the 15th century, but the technical skill required to make polished glass lenses shaped like the eye did not appear until 1887, in the hands of the German glassblower F. E. Muller. These early lenses were difficult to make (and therefore expensive), and they were not well tolerated by the eye. Study of the response of the eye to the presence of these materials revealed new aspects of eye physiology, such as the eye's nonspherical geometry and the circulation pathway for tears. New materials were developed especially for lenses; plastics were particularly valuable (Figure 1.7). Long wear contact lenses required an understanding of the cornea's need for oxygen (which is a wonderful engineering problem that illustrates an aspect of physiology, see Problem 15 in Chapter 2). Today, lenses are

Table 1.1 Development of contact lenses

Year	Event
1249	Roger Bacon writes about convex lens eyeglasses for treating farsightedness
1451	Nicholas of Cusa invents concave lens spectacles to treat nearsightedness
1508	Leonardo da Vinci conceives of a water-filled hemisphere that could be worn directly on the eye
1636	Rene Descartes proposes placing a lens at one end of a water-filled tube with the other end placed on cornea of the eye
1801	English scientist Thomas Young develops a model based on the theories of da Vinci and Descartes
1827	Sir John Herschel studies how to mold lenses for accurate fitting over the cornea
1884	The development of anesthesia allows for molding the cornea
1887	Glassblower F.E. Muller produces the first glass contact lens to protect a diseased eye
1888	A.E. Fick makes the first glass contact lens to correct vision
1889	August Muller creates lenses to correct his myopia by molding the human eye
1936	W. Feinbloom is the first to use plastic in contact lenses
1938	Obrig and Mullen produce the first all-plastic scleral contact lens using Poly(methyl methacrylate) (PMMA). (Scleral lenses covered the entire eye, including the white part of the eye.)
1947	Tuohy develops an all-plastic corneal contact lens, a “hard” lens
1960	Wichterle, Lim, and Dreifus begin to work on making “soft” lenses with hydroxyethyl methacrylate (HEMA) hydrogels
1971	Bausch & Lomb receive U.S. Food and Drug Administration (FDA) approval to sell soft lenses developed by Wichterle
1980s	FDA approves contact lenses for extended wear
1983	CIBA Vision introduces BiSoft, the first FDA-approved soft bifocal contact lens
1987	First disposable extended wear lenses are introduced
1988	Vistakon invents the first soft disposable contact lenses, the ACUVUE brand
1995	Johnson & Johnson launches the first daily disposable contact lens, 1-DAY ACUVUE
2001	FDA approves 30-night continuous wear contact lenses developed by CIBA Vision (Focus Day and Night)
2002	CIBA Vision launches FOCUS Dailies Toric, the first daily disposable contact lens for astigmatism
2002	FDA approves contact lenses for corrective refractive therapy (Paragon CRT) that reshape the cornea during sleep and temporarily correct vision. The lenses are only worn during sleep and provide clear vision when removed the following morning! (http://www.paragoncrt.com)

manufactured from synthetic oxygen-permeable materials using computer-aided techniques; the manufacturing process is inexpensive and reliable enough to render the lenses disposable.

This example also demonstrates the kinds of science that a biomedical engineer must master: physics (e.g., light refraction and mechanics); anatomy; physiology (e.g., tear production and circulation); materials science; immunology (e.g., the body’s response to foreign materials); and mathematics (e.g., evaluation of oxygen diffusion).

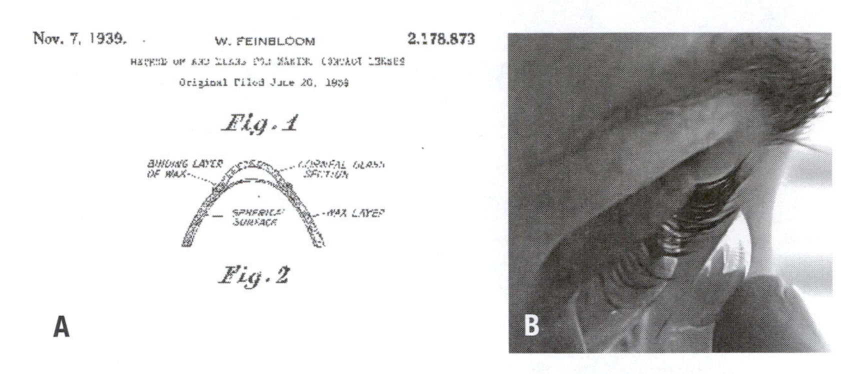

Figure 1.7 Contact lenses. **A.** An early plastic contact lens by Feinbloom. **B.** A modern hydrogel contact lens.

1.3.3 Biomedical engineering has been taught in universities for many decades, but is growing at present

The academic discipline of engineering became defined around the mid-1800s. The teaching of engineering in the United States began in the years before the Civil War. For example, Yale University began offering a course on civil engineering in 1852. In 1863, Yale awarded the first PhD in engineering in the United States to J. Willard Gibbs (whose "free energy" you will encounter in Chapter 2). President Abraham Lincoln signed into law the Land Grant Colleges Act in 1862, which provided land and perpetual endowments to each state for the support of colleges of agriculture and mechanical arts. Some of these land grant colleges became major engineering schools that thrive today including Iowa State University, Massachusetts Institute of Technology, Washington State University, Michigan State University, Texas A & M, Cornell University, California Polytechnic State University, and Purdue University. Engineering education and engineering practice both accelerated during World War II.

Much has been written recently about the history of biomedical engineering; for example, a history of accomplishments of the past 50 years is available (4). A history of the development of the academic field was written by Peter Katona, president of the Whitaker Foundation (5). According to Katona, biomedical engineering began to appear as a subject of study at universities and colleges in the late 1950s and early 1960s. The number of university programs and students has expanded tremendously over the past few decades (Figure 1.8).

Why this recent increase in student interest in biomedical engineering? Is something new happening in the field that contributes to this increase? Biomedical engineering has been a productive area of study for decades, and many life-saving products have emerged from this study including heart pacemakers, kidney dialysis machines, and artificial joints, but our understanding of biology and human medicine has expanded at an explosive rate in the past decade. The Human Genome Project is just one example of newly acquired riches of biological information. Biology has been transformed from a descriptive science

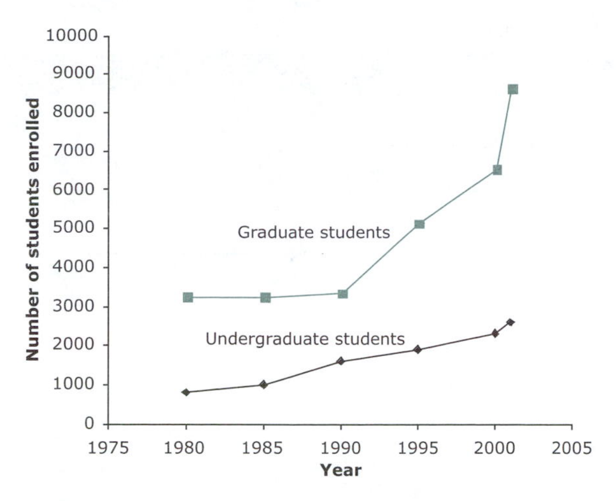

Figure 1.8 Numbers of students enrolled in biomedical engineering degree programs.

into a quantitative science. As such, it now provides an easier point of entry for engineering.

The current growth of biomedical engineering might be similar to expansions that occurred in civil, mechanical, and electrical engineering in the past. As we learned about mechanics and properties of materials during the 20th century, engineers acquired new tools that allowed them to build bridges, buildings, and other physical structures that were unimaginable in 1900. Similarly, the growth of basic knowledge in physics during the 20th century enabled new accomplishments (such as transistors, computers, and telecommunications systems) by electrical engineers. The increase of biological knowledge is already leading to new potentials, but it will be the work of this new and expanded group of biomedical engineers (including you) to convert that knowledge into safe and affordable new products that improve life for future generations.

1.3.4 Biomedical engineering can be divided into subdisciplines

Biomedical engineers study a variety of different kinds of problems related to human health (Table 1.2). Some concepts are important to all biomedical engineers; quantitative human physiology and mathematical analysis are essential to all of biomedical engineering, but some biomedical engineers study systems that are best approached through an understanding of electrical signals and circuits, or mechanics, or chemistry. For this reason, biomedical engineering is conveniently divided into subdivisions that reflect the kinds of tools that are best used to approach the problem of interest and the nature of the problem itself. Understanding each of these subdivisions—and recognizing the similarities and the differences between them—is another way of seeing what biomedical engineering is about. The next few paragraphs describe briefly the subdivisions that we will explore in this book.

Table 1.2 Subdisciplines of biomedical engineering

Subspecialty	Examples	
Systems biology and Bioinformatics	Modeling of cellular networks DNA sequence analysis Microarray technology	Chapters 3 and 16
Physiological modeling	Physiology of excitable cells Dynamics of the microcirculation Models of cellular mechanics Pharmacokinetic models of chemotherapy drugs	Chapters 6–9
Biomechanics	Gait analysis Prosthetic joints and limbs Cellular mechanics	Chapter 10
Biomedical instrumentation and Biomedical sensors	Electrocardiogram Cardiac pacemaker Glucose sensor O_2 sensor pH sensor	Chapter 11
Biomedical imaging	Radiographic imaging Ultrasound imaging Magnetic resonance imaging Optical imaging	Chapter 12
Biomolecular engineering and Biotechnology	Drug-delivery systems Artificial skin (tissue engineering) Protein engineering Chromatography and other separation methods Vaccines	Chapters 13 and 14
Artificial organs	Biomaterials Hemodialysis Artificial heart	Chapter 15

PHYSIOLOGICAL MODELING

Biomedical engineers are experts in physiology and mathematics, so it is not surprising that they have been pioneers in the development of physiological models. Often, biomedical engineers make mathematical models of the systems that they are working on to help them understand and predict system behavior. For example, biomedical engineers that are designing prosthetic hips use mathematical models of hip mechanics to predict the stresses and strains that their artificial hip must endure (Figure 1.9).

Physiological modeling often has long-lasting influences. Mathematical models of blood flow in small vessels is one of the most important results of biomedical engineering, and still guides the development of tissue-engineered blood vessels and cardiovascular biomaterials such as stents. Low velocity flow through a cylindrical conduit is called Hagen-Poiseuille flow in honor of Jean Leonard Marie Poiseuille and Gotthilf Heinrich Ludwig Hagen; Poiseuille (a physiologist) and Hagen (an engineer) independently published the first systematic measurements of pressure drop within flowing fluids in simple tubes in 1839 and 1840

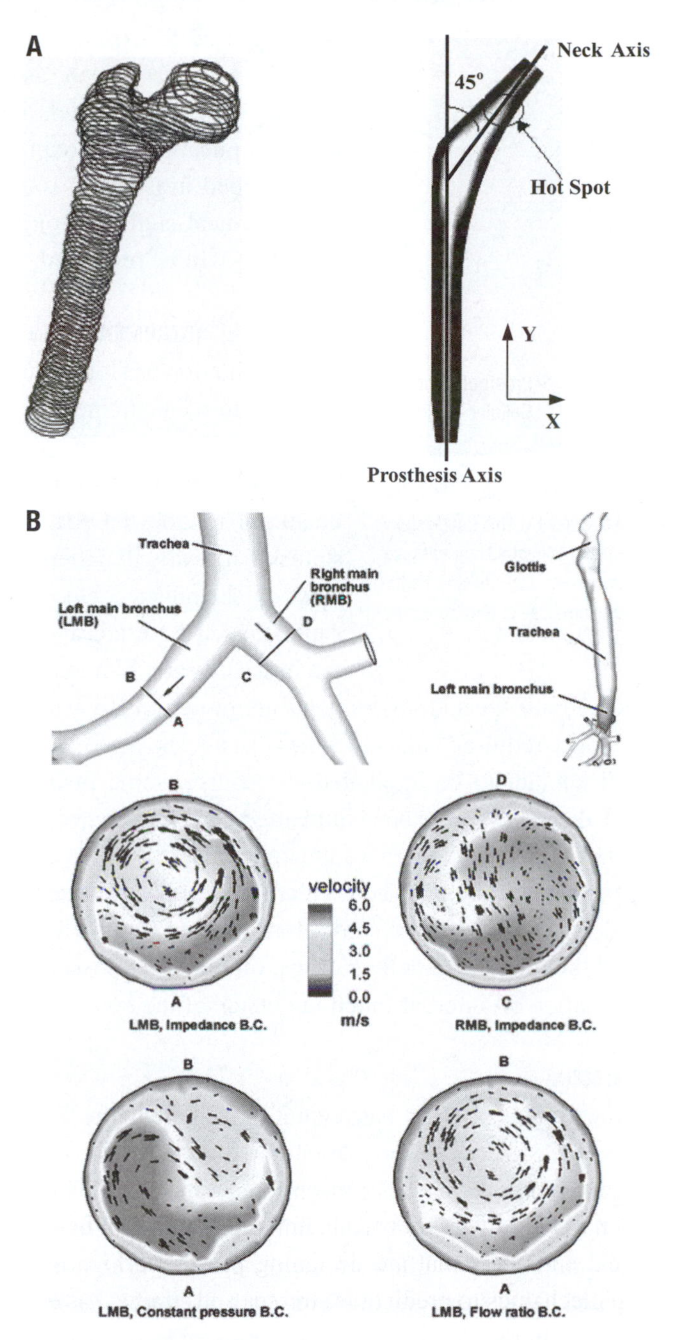

Figure 1.9 **Examples of physiological modeling. A.** Modeling of forces acting on a human hip, from reference (6) with permission. **B.** Modeling of air flows in the left main bronchus and right main bronchus under different conditions within the lung. Models such as this one are useful in evaluating breathing patterns for patients with different lung diseases, and quantifying the extent of disease in a particular patient, from reference (7) with permission. (See color plate.)

(this work is reviewed in more detail in Chapter 8). As we will see in Chapter 16, physiological modeling is becoming even more important in understanding biology; models of the networks of chemical reactions that occur within cells is an essential ingredient of systems biology and bioinformatics. Now, mathematics

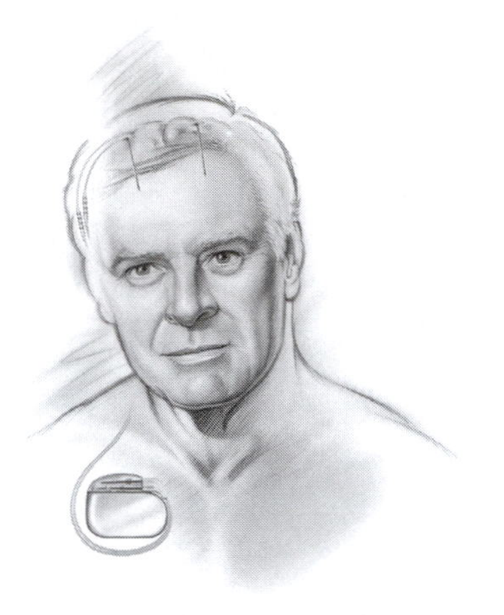

Figure 1.10 **Biomedical instrumentation.** This field is described in detail in Chapter 11. This image shows a deep-brain stimulator (developed by Medtronic), which is now used to treat patients with Parkinson's disease. Image courtesy of Medtronic, Inc.

is being used to explore complex physiological responses, such as angiogenesis (Figure 1.9); these models may have profound implications for treatment of cancer, as described in Chapter 16. The influence of biomedical engineers on the biology of the future will be profound.

BIOMEDICAL INSTRUMENTATION

Instrumentation has long been an important component of medicine. Hospitals depend on electronic instruments, such as heart and blood pressure monitors, to provide continuous and reliable feedback on the health status of critically ill patients. As the microelectronics industry has developed more sophisticated materials and techniques over the past 20 years, biomedical instrumentation has become multifaceted. Instruments are more sensitive and smaller; many functions that once required large machines can now be performed on microchips that are small enough to be implanted. Of course, some instruments, such as cardiac pacemakers, have long been implanted. Pacemakers, which are now multifunctional and highly reliable, have created a major advance in human health. Similar devices are now being used to treat Parkinson's disease and other neurological disorders by delivery of electrical stimulation (Figure 1.10). Implantable instruments of the future will deliver drugs, monitor local tissue states, and send detailed information on internal function outside of the body.

BIOMEDICAL IMAGING

Biomedical imaging technology has revolutionized medicine. Physicians all over the world now have access to reliable and safe methods for collecting medical images using technologies such as conventional radiographic imaging, CT, MRI, and ultrasound imaging. The wide availability and safety of these techniques have improved our standards for following the progress of common conditions such as pregnancy and life-threatening ailments such as cancer. Biomedical engineers have been leaders in the design and construction of new imaging machines, the creation of medical imaging approaches using these machines, and the analysis of image data that are acquired from patients. Each year the state of the art in imaging improves; each improvement in image resolution and quality means more accurate diagnosis of disease and improved health care. Most current imaging methods provide information on tissue anatomy, but imaging techniques of the future will also provide information on the function of tissue (Figure 1.11), opening new doors for the study of disease progress and creating opportunities for development of new treatments.

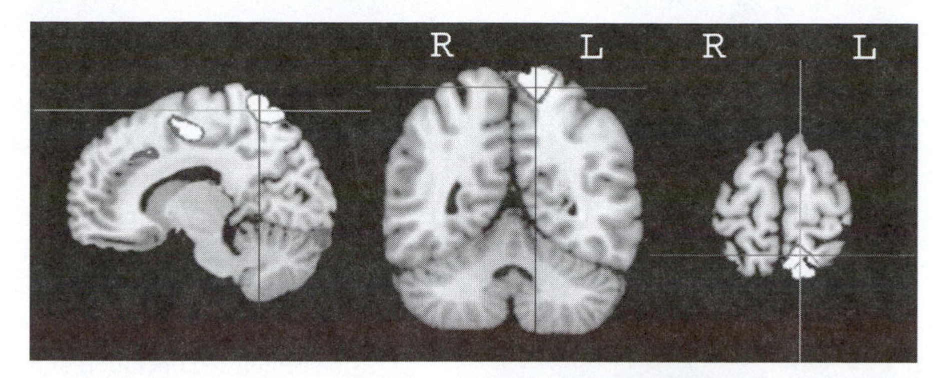

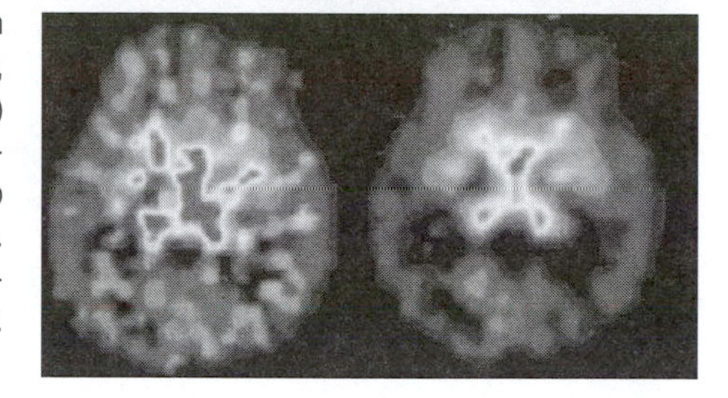

Figure 1.11

Biomedical imaging. This field is described in Chapter 12. Functional magnetic resonance (fMRI, top) and positron emission tomography (PET, right) are now providing functional and metabolic information in addition to anatomical information. Top panel provided by Todd Constable, Yale University. Bottom panel reprinted by permission from Macmillan Publishers Ltd: *J Cereb Blood Flow Metab.* 2003;23:1096–1112. (See color plate.)

BIOMECHANICS

Humans live in a physical world. Biomedical engineers have long studied the performance of humans as mechanical objects and determined the role of mechanical forces on human function. Some engineers who are interested in biomechanics study the role of forces (produced by exercise, work conditions, or the activities of normal life) on tissue physiology or human performance. Others are interested in the consequences of mechanical injury and the design of better ways to protect humans from mechanical forces by the design of seat belts or helmets, for example. Still others examine the ways that diseases affect the mechanical performance of tissues such as the heart or the ability of humans to move after loss of mechanical function in their bones or control of muscles. Of course, biomedical engineers have been leaders in the design of mechanical replacements for hips, joints, heart valves, and organs (Figure 1.12).

As biology advances, the role of biomechanical analysis is expanding. Biomedical engineers are studying the mechanical function of cells, for example, by studying the mechanics of cell movement through circulation or cell motility through tissues. Structural proteins within cells, such as the proteins that make muscle cells contract or the proteins that regulate cell division, also operate as mechanical objects, and biomedical engineers are leading the effort to understand how these systems work.

BIOMOLECULAR ENGINEERING

Drugs are chemical substances used to improve health, but they often have unwanted, even deadly, side effects. A major advance in drug therapy over the

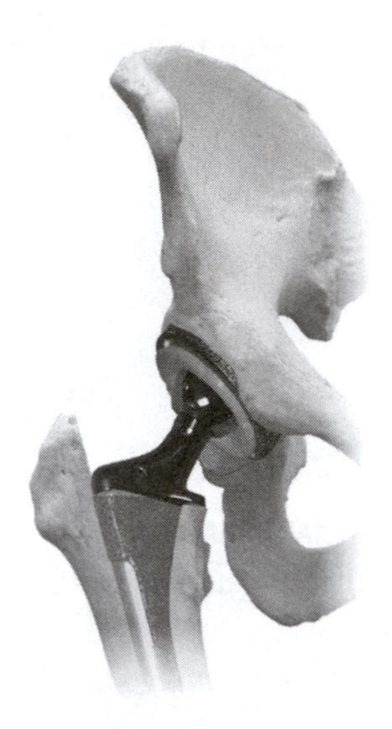

Figure 1.12 **Biomechanics.** This field is described in detail in Chapter 10. This image shows a metal hip implant, with a polyethylene cup to lubricate. Image courtesy of Zimmer, Inc.

past century is the development of pharmacokinetic analysis, which predicts patterns of drug absorption and metabolism (often based on mathematical models), thereby providing tools that can be used to give drugs safely and effectively. Biomedical engineers, especially those trained in chemical engineering, have been pioneers in this area. Similar mathematical tools applied to the design and operation of biological reactors have been enormously important in the large-scale production of drugs. Biomolecular engineering is the branch of biomedical engineering that emphasizes the use of chemical engineering principles for design and analysis.

Today, biomolecular engineers are using similar approaches to design drug-delivery systems and to create new treatments using tissue and cellular engineering (Figure 1.13). Many areas of interest to biomedical engineers can be approached using the tools of chemical engineering such as biomaterials design, nanobiotechnology, and genomic analysis. For this reason, biomolecular engineering is growing and developing as a subdiscipline of biomedical engineering. In fact, it is growing so rapidly that we have divided our

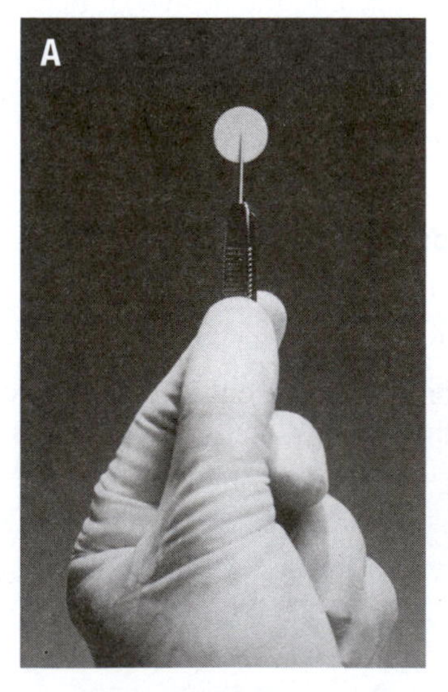

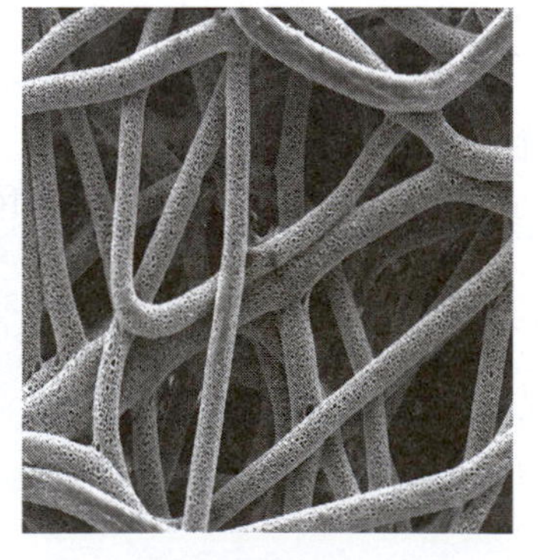

Figure 1.13 **Biomolecular engineering.** This field is described in detail in Chapters 13 and 14. **A.** The GLIADEL® drug-delivery system for treating brain cancer allows surgeons to give a long-lasting dose of chemotherapy directly at the site of the tumor. Photo used with permission. **B.** Polymer scaffolds built by biomolecular engineers are being used for tissue engineering. Each polymer fiber in this electron microscopic image is ~10 microns in diameter.

description of biomolecular engineering into three sections: a general introduction (Chapter 13), a more specialized treatment concerning applications in the immune system such as vaccines (Chapter 14), and a more detailed description of treatments for cancer (Chapter 16), which combines biomolecular engineering, radiation physics, and imaging.

Figure 1.14 **Artificial organs.** This field is described in detail in Chapter 15. The development of synthetic heart valves, which have provided reliable mechanical function for decades, is a union of artificial organ design and biomechanics analysis. This image shows a bileaflet tilting disk mechanical heart valve (Photo courtesy of St. Jude Medical, Inc.). From Schoen and Padera (8).

ARTIFICIAL ORGANS

Synthetic materials can be combined with biological components to produce devices that function like tissues and organs. The use of natural materials—often derived from animal tissues—to repair tissues was described in ancient times. But the development of synthetic materials (metals, ceramics, and polymers) has provided biomedical engineers with tools to expand and improve the design of artificial organs. For example, polymeric materials are routinely used in vascular grafts, and combinations of synthetic polymers and living cells may someday lead to implantable replacement cartilage, liver, or nervous tissue (see Chapter 15).

Synthetic materials are critical components in extracorporeal systems for blood purification (i.e., systems that treat blood by taking it out of the body). Willem Kolff, a Dutch physician, developed the first successful kidney dialysis unit in 1943, using cellophane to remove urea from the blood of diabetics; further work by biomedical engineers has made hemodialysis a life-saving procedure that is widely available. The addition of cells to a dialysis-like machine can make it function as an artificial liver or pancreas. Biomedical engineers design artificial hearts and heart components, such as valves (Figure 1.14), and also design the machines that keep patients alive during cardiac surgery.

SYSTEMS BIOLOGY

Systems biology is a frontier area for biomedical engineers. Engineers, of course, are specialists in the analysis of all kinds of systems; engineers are trained to develop models of complex systems, to learn how to control these systems, modify them, or replicate them in alternate forms. Systems analysis may be the quintessential engineering exercise, and the revolution in modern molecular biology has placed engineers in a position to apply these analytical tools to deep and fundamental biological problems.

Systems biology requires contributions in many areas of strength for biomedical engineers. Of course, the development of models of biological function (usually at the cellular and molecular level) is a key component, as is the development

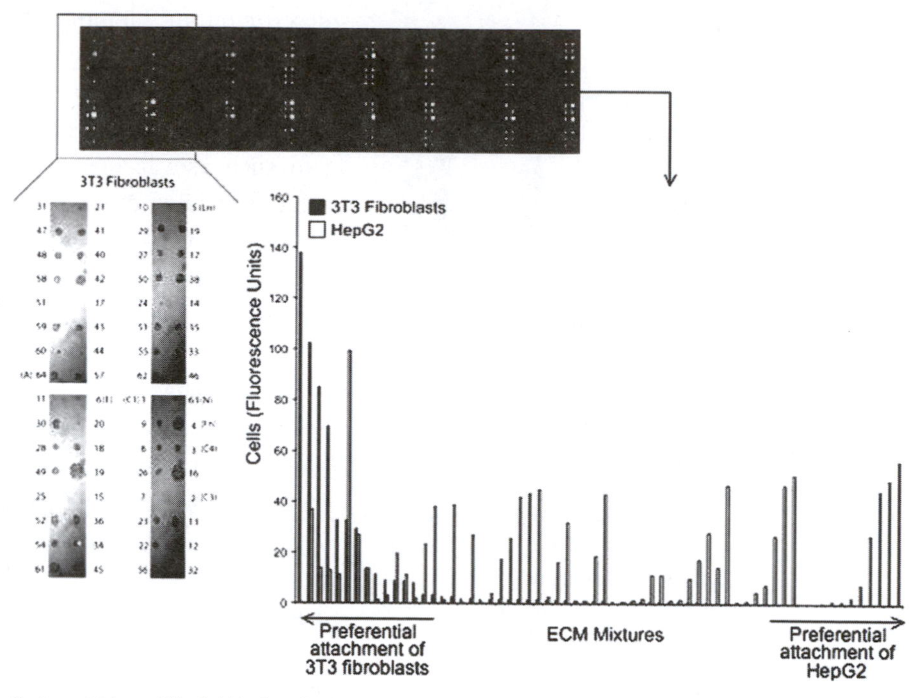

Figure 1.15 **Systems biology.** This field is described in detail in Chapter 16. Here, a protein microarray is used to probe the relative strength of adhesion of hepatocytes and fibroblasts to small spots of differing protein composition (Woodrow K. and Saltzman W.M., unpublished data, 2008).

of efficient computer methods for examining biological databases (such as gene or protein databases) to find or sort new biological information. Biomedical engineers have also created new methods to manipulate cells through genetic and cellular engineering. All of these advances will require new methods for measuring the state of function of individual cells. Biomedical engineers are already contributing to this effort by analyzing the protein composition of living cells (proteomics), creating methods for simultaneous measurement of thousands of genes and proteins within a cell (array technologies), creating arrays of cells and tissues for diagnostic purposes (Figure 1.15), and designing devices that can physically interface with cells and proteins (biomicroelectromechanical systems, see Chapter 11).

1.4 Biomedical engineering in the future

There have been enormous advances in human health care over the past 100 years, and our life expectancy has increased dramatically during this period (remember Figure 1.1). Much of this progress is because of success in the battle with infectious diseases. In London, in 1665, 93% of deaths were the result of infectious disease, whereas in the United States, only 4% of deaths were the result of infectious disease in 1997. Engineers contributed significantly to this effort by developing sanitation methods for cities, large-scale processes for manufacture of vaccines and antibiotics, and delivery methods for drugs.

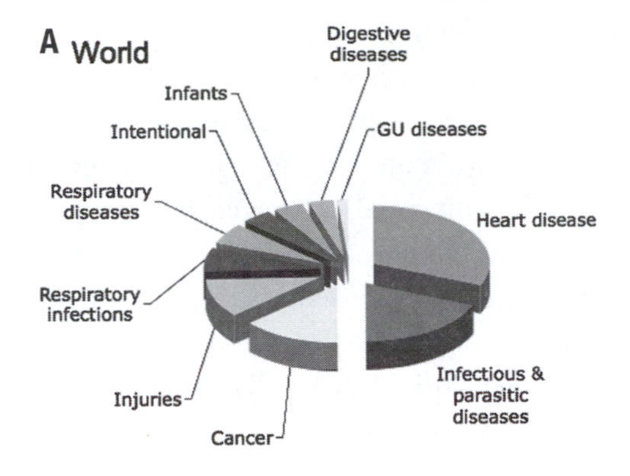

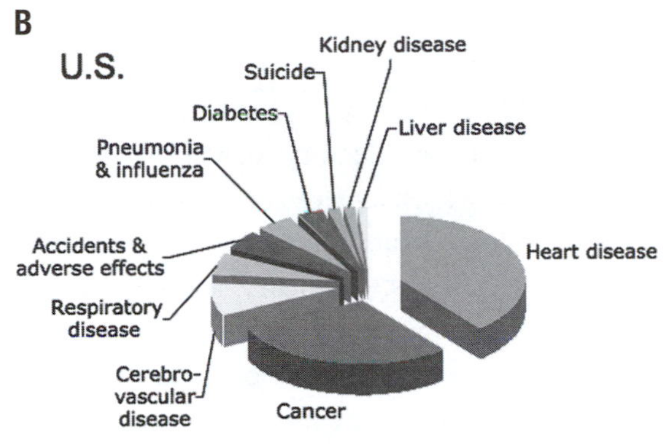

Figure 1.16

Causes of death in the world and the United States, 1997. GU, genitourinary.

We have made progress, but the problems are not solved (Figure 1.16). Infectious diseases are still the second leading cause of death in the world and the most important cause of premature death in many developing countries. Current vaccines are not perfect (in fact, some people get chicken pox even after getting an expensive and painful vaccine!). Biomedical engineers can do a better job in making technology that can also be translated to people with limited resources; we need to make vaccines less expensive, easier to administer, and easier to transport. And of course, there are established infectious diseases (such as the ones caused by human immunodeficiency virus [HIV] or hepatitis C) and emerging diseases (such as Severe Acute Respiratory Syndrome [SARS] and West Nile virus infection) that are important health problems in all areas of the world. We will not achieve control of these diseases without the work of biomedical engineers.

Most people die from non-infectious diseases such as cancer and heart disease (Figure 1.16). Chronic diseases such as rheumatoid arthritis, diabetes, and Alzheimer's disease impair quality of life. In the same way that past engineers revolutionized treatment of kidney disease, biomedical engineers are now studying imaging systems, biomaterials, biomechanics, and all of the subdisciplines described earlier to help patients with these incurable ailments. The future of medicine will be shaped by these inventive minds.

Engineers work to improve the human condition, but they must also be aware of the context and potential consequences of their work. We can learn something from the history of the use of technology, but the lessons are not always clear nor can they be easily translated from one field to another. Mechanical engineers have given us the marvel of commercial air travel, but no one could predict that commercial airliners would one day be used as weapons. Today, there is legitimate concern about the possible consequences of stem cell engineering, cloning of organisms, and other technologies on which biomedical engineers will work. Each of us is obliged to look carefully at the relationship between engineering, ethics, and impact on society.

1.5 How to use this book

This book is intended for use with introductory courses in biomedical engineering, for students who have no previous experience with engineering and limited experience with biology. We do assume that most students will be taking the standard freshman courses in physics, chemistry, and mathematics. Most of the mathematical analysis, however, is separated from the main part of the text within colored boxes. All of the necessary biological background material is included in the chapters of the book (Figure 1.4); surveys of molecular and cellular principles (Part 1) and physiological principles (Part 2) are provided. Students and instructors can use these parts of the book as introductory surveys to the biological foundations of biomedical engineering, or they can use these chapters as reference materials, to be consulted as they focus on the major subdivisions of biomedical engineering practice, which is contained in Part 3.

TO THE STUDENT

Within each chapter, I have highlighted two types of information. First, mathematical analysis is a central part of biomedical engineering, but I realize that students taking a first course in biomedical engineering might still be taking their first course in college calculus. Therefore, I have illustrated the use of mathematics in biomedical engineering in a way that (I hope) complements, but does not interfere, with the reading of the main text. Mathematical concepts are presented in boxes that appear alongside the main text. In addition, the problems at the end of each chapter are arranged roughly in order of the level of mathematical sophistication that is required for their solution. I also provide guidance, throughout the book, on developing an approach to problem solving. Box 1.2 contains some initial advice.

Second, I have profited enormously from my personal associations with other biomedical engineers and I realize that many students are searching for a career path that matches their personal strengths with their professional ambitions. To help with this process, I have included profiles of other individuals in biomedical engineering, including pioneers of the field as well as recent graduates. In each

Box 1.2 Solving engineering problems

Good engineers are skilled at making estimates and solving problems. The problems in Chapter 1 were designed to exercise your skill at estimation. As you solve these problems, and those throughout the book, develop a systematic approach for thinking about and presenting your work on each problem.

Here are some guidelines to help in developing and presenting your solutions to engineering problems.

General features

- Use fine-lined paper, preferably lined in grids (like graph paper) to facilitate drawing of diagrams and graphs.
- Include—in the top right hand corner of the first page (and perhaps on every page)—a title block that includes your name, the date of submission, the class number, and so forth.
- Always be clear about the units. Break complex units down into their components (length, mass, time, etc.), using the unit conversion table in Appendix C.

For each problem

- Include a concise statement of the problem you are solving.
- Describe clearly what things you are trying to determine: What is your goal?
- Start each problem by writing down all of your assumptions; leave room so that you can add assumptions as you work through the problem.
- Draw a diagram to help in defining the important variables and clarifying the problem statement.
- Define the key variables in the problem and assign an appropriate symbol for each.
- As you develop equations that relate the variables in the problem, use the symbols for as long as possible, substituting known values only near the end.
- If you estimate values for any of the variables, justify your choices as clearly as you can; if you find values in reference tables, indicate the source clearly.
- Look at your answer and challenge it. Does the answer you found make sense, given what you know?

profile I summarize briefly the career of the individual and then, where possible, I have asked each person to describe his or her own feelings about biomedical engineering, how she or he got started in the field, and how it has impacted her or his life.

Summary

- Life expectancy and quality of life have increased for people in most nations of the world during the last century; the development of reliable, safe, and inexpensive medical technology by biomedical engineers has played an important role in this enhancement.

Profile of the Author: W. Mark Saltzman

I was born in Des Moines, Iowa, and spent my childhood in Iowa and Illinois. I was always interested in biology and medicine, and I started my college career as a premedical student at Drake University in Des Moines. During my freshmen year of college, I discovered physics, math, chemistry, and the pleasure of using quantitative tools to find answers to complex problems. I transferred to the College of Engineering at Iowa State University, where I earned a Bachelor of Science degree in chemical engineering in 1981.

During my senior year at Iowa State, I listened to a lecture on using chemical engineering tools to solve problems in biomedical engineering; Richard Seagrave, one of my Iowa State professors, delivered the lecture elegantly and at the chalkboard. I listened, and it felt as if a door was opening to a previously hidden world that combined my interest in medicine with my new skills as a chemical engineer. I followed this path to the Massachusetts Institute of Technology (MIT), where I entered the MIT/Harvard Division of Health Science and Technology graduate program in medical engineering. At MIT, I earned a master's degree in chemical engineering in 1984 and a doctorate in medical engineering in 1987. I discovered many things during my years at MIT, including the joy of teaching, and I decided to pursue a career that combined biomedical engineering research and teaching. My mentor at MIT was Robert Langer, who is profiled in Chapter 13; Bob has been a bottomless source of inspiration for me for the past 20 years.

After graduating from MIT, I was fortunate to join the faculty at Johns Hopkins University, where I started my own research program and developed new classes at the intersection of chemical and biomedical engineering, including a class for freshmen called "Introduction to Biotechnology." My research program developed quickly—thanks to outstanding students and terrific collaborators at Hopkins such as Henry Brem and Richard Cone—to include projects involving controlled drug delivery to the brain, polymers for supplementing or stimulating the immune system, cell interactions with polymer materials, and tissue engineering. In 1996, I moved to Cornell University, which allowed me to expand my research program into new areas of biomaterials and nanotechnology. I continued developing my course for freshmen students, extending its scope and enhancing its focus on human physiology to become "Introduction to Biomedical Engineering." I joined the faculty of engineering at Yale University, as the Goizueta Foundation Professor of Chemical and Biomedical Engineering, in July of 2002 and became the first chair of Yale's Department of Biomedical Engineering in 2003. I continue to teach my course for freshmen at Yale, which is now called "Frontiers in Biomedical Engineering" and covers the material presented in this book.

I currently live happily in New Haven, Connecticut. I have two sons, Alexander and Zachary, who (so far) are mute on the role of biomedical engineering in their futures.

- Emergency rooms, hospitals, doctors' offices, and homes contain medical instruments and products that resulted from 20th century biomedical engineering.
- Biomedical engineering is the application of science, mathematics, and engineering design principles to improve human health.

- Human physiology is the foundational science that distinguishes biomedical engineering from other forms of engineering; throughout history, advances in our understanding of physiology have led to new biomedical engineering technology.
- Biomedical engineering is growing in interest among students, and opportunities for biomedical engineers to find productive work and contribute to society are increasing rapidly.
- Biomedical engineers often specialize in a variety of subdisciplines or fields such as physiological modeling, biomedical instrumentation, biomedical imaging, biomechanics, biomolecular engineering, artificial organs, and systems biology.
- Emerging human diseases and new discoveries in physiology and human health promise to present new problems for biomedical engineers of the future.

REFERENCES

1. Table 1. Deaths, percent of total deaths, and death rates for the 10 leading causes of death in the selected age groups, by race and sex: United States, 2000. National Vital Statistics Report. 2002;50(16):13–48.
2. Enderle J, Blanchard S, Bronzino J, eds. *Introduction to Biomedical Engineering*. San Diego: Academic Press; second edition, 2005.
3. Florman SC. *The Existential Pleasures of Engineering*. New York: St. Martins Press, Inc.; 1976.
4. Nebeker F. Golden accomplishments in biomedical engineering. *IEEE Eng Med Biol Mag*. 2002;21(3):17–47.
5. Katona PG. The Whitaker Foundation: The end will be just the beginning. *IEEE Trans Med Imaging*. 2002;21(8):845–849.
6. Li C, Granger, C, Schutte, H, Biggers, SB, Kennedy, JM, Latour, RA. Failure analysis of composite femoral components for hip arthroplasty. *J Rehabil Res Dev*. 2003;40(2):131–146.
7. Ma B, Lutchen KR. An anatomically based hybrid computational model of the human lung and its application to low frequency oscillatory mechanics. *Ann Biomed Eng*. 2006;34(11):1691–1704.
8. Schoen FJ, Padera RF. Cardiac surgical pathology. In: Cohn LH, Edmunds LHJ, eds. *Cardiac Surgery in the Adult*. New York: McGraw-Hill; 2003:119–185.

FURTHER READING

Schwan HP. The development of biomedical engineering: Historical comments and early developments. *IEEE Trans Biomed Eng*. 1984;31:730–736.

Schwan HP, ed. The history of biomedical engineering. *IEEE Eng Med Biol Mag*. 1991;10:24–50.

USEFUL LINKS ON THE WORLD WIDE WEB

http://www.bmenet.org/BMEnet/

This site contains a wide collection of information on Biomedical Engineering including information on all of the academic programs that are available in the field, links to research on latest advances, and information on jobs.

http://www.whitaker.org

The Whitaker Foundation is a private philanthropic organization that has been an influential participant in the growth of biomedical engineering as an academic discipline. Their Web site contains information on academic programs in biomedical engineering and reports of research that is supported by the foundation.

http://www.bmes.org/

This is the official site for the Biomedical Engineering Society (BMES), with information about the professional society for biomedical engineers. Information about the BMES annual meeting is also provided here.

http://www.asee.org/precollege/

This site has a guide for precollege students who are interested in careers in engineering. Presented by ASEE (The American Society for Engineering Education), 1818 N Street, N.W., Suite 600, Washington, DC, 20036.

http://www.nibib1.nih.gov/

The National Institute of Biomedical Imaging and Bioengineering (NIBIB) is the newest member of the National Institutes of Health (NIH) in the United States. The mission of the NIBIB is to improve health by promoting fundamental discoveries, design and development, and translation and assessment of technological capabilities. The NIBIB Web site contains information on all aspects of the NIBIB including organization and mission, research and training grant opportunities and related information, breaking news, publications, events, bioimaging and bioengineering information of general interest, interagency activities, and the NIH Bioengineering Consortium (BECON), which is administered by the NIBIB.

http://www.greatachievements.org/greatachievements/

The greatest engineering achievements of the 20th century were assembled by the U.S. National Academy of Engineering. For each broad category of achievement (i.e., #1 is electrification, #2 is the automobile), a history and a timeline of significant events during the century is provided. Biomedical engineers directly impacted both imaging (#14) and health technologies (#16) on the list of top 20 achievements.

http://www.uh.edu/engines

This is the Web site that accompanies the Engines of Our Ingenuity *radio program. Professor John Lienhard provides historical (and entertaining) information on the origins of technological innovation and their impact on society.*

QUESTIONS

1. Write a definition (1–2 sentences) of biomedical engineering in your own words (test yourself by not looking back at any of the definitions in the text when you write your own definition).
2. Make two lists (of at least 10 items each) in response to the following two questions:
   a. What products of biomedical engineering have you personally encountered? Pick three of these products and write a description of what you think is good, and what could be improved in that product.
   b. What products of biomedical engineering do you expect to encounter in the next 50 years?
3. Pick a faculty member who teaches or does research in biomedical engineering from your department for this exercise.
   a. Perform a Medline search with your selected faculty member as the author to find a list of articles that he or she has written in the past two years.
   b. Select one of the articles and find a copy of it in your library (or online if it is available in that format). Read it and write a brief review of the findings for a general audience. In which subdiscipline of biomedical engineering does this research work belong?
4. Interview an older family member (parent, grandparent, aunt, uncle) about an advance in medicine that they remember. Why was this advance memorable to them? How did they find out about it?

PROBLEMS

1. Drugs are often administered in capsules. Some capsules act as containers, which hold many smaller particles that contain the active agent. Administration of the drug is improved by the capsule; when the capsule breaks down in the intestine and the particles are freed from the container, the large surface area of the drug particles allows rapid dissolution of the drug.
   a. Assume that a capsule is approximately 1 cm long and 3 mm in diameter. Calculate the surface-to-volume ratio of the capsule.
   b. Assume that the capsule is filled with particles that are 0.4 mm in diameter. How many of these particles will fit into one capsule?
   c. What is the total surface area of the particles within the capsule?
2. Using only the information provided in Figure 1.1, estimate the following:
   a. Human life expectancy in the year 1250.
   b. Human life expectancy in the year 2050.
3. From Figure 1.8, estimate the fraction of biomedical engineering students who were undergraduates in 1980, 1985, 1990, 1995, and 2000.

4. Figure 1.13 shows a surgeon holding a GLIADEL® wafer. From this photographic evidence only, estimate the following:
   a. The dimensions of the wafer.
   b. The dose of drug that it contains, if the loading of drug is 3.85% by mass (i.e., 3.85% of the wafer mass is due to the drug).
5. Figure 1.14 shows a mechanical artificial heart valve. If this valve opens and closes once during each cycle, or beat, of the heart, how many times will the valve need to open and close during 10 years of continuous use? Measure your own heart rate (describe how you do it) and use this rate in answering this question.
6. Figure 1.15 shows an image of a microscope slide, onto which small spots of protein have been printed. The slide was used to identify protein compositions that provide for good adhesion of cells. If each spot of protein is 300 microns (micrometers) in diameter, and each cell is 15 microns in diameter, what is the maximum number of cells that can fit into each spot?

PART 1

MOLECULAR AND CELLULAR PRINCIPLES

2 Biomolecular Principles

LEARNING OBJECTIVES

After reading this chapter, you should:

- Understand the types of chemical bonds that hold atoms together in molecules.
- Understand the difference between polar and nonpolar molecules, and the important role that polarity plays in interactions of biological molecules.
- Understand the basic concepts of biochemical energetics, including the role of adenosine-5′-triphosphate (ATP) in the transformation of energy into biochemical work.
- Understand the concepts of acids, bases, pH, and buffering.
- Know the major classes of biological polymers: proteins, polysaccharides, and nucleic acids.
- Understand the chemical structure of polysaccharides as polymers of monosaccharides, including the simple sugars glucose, galactose, and fructose.
- Understand the basic structure of nucleic acids as polymers of nucleotides and how that structure is different in deoxyribonucleic acid (DNA) and ribonucleic acid (RNA) polymers.
- Understand the basic structure of proteins, which are polymers of amino acids, and how the diversity of amino acid structure influences protein three-dimensional structure and function.
- Understand how the chemical structure of phospholipids contributes to the properties of biological membranes.
- Understand the basic features of biological membranes, which are lipid bilayers that are decorated with proteins and carbohydrates.
- Understand the mechanisms of diffusion and osmotic pressure generation.

2.1 Prelude

Biomedical engineers are engaged in a great diversity of activities: Chapter 1 described many of the fields in which biomedical engineers make significant contributions. This chapter, together with Chapters 3 and 4, reviews fundamental chemistry concepts that are important for understanding human physiology and biomedical engineering (BME). These chapters introduce several families of

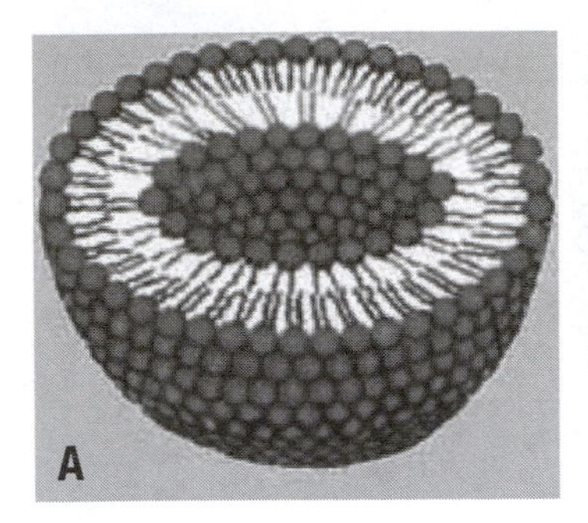

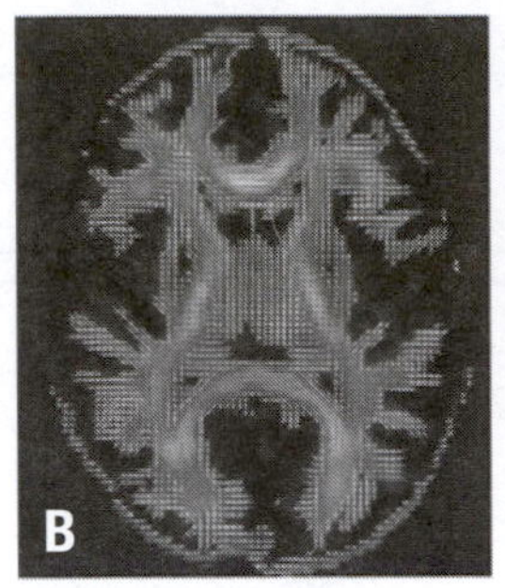

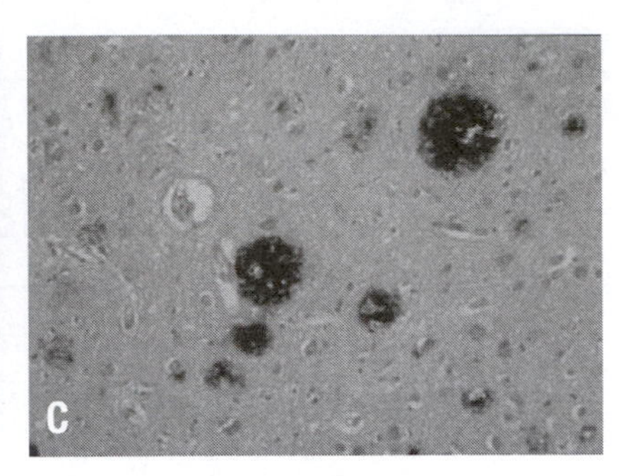

Figure 2.1 **Examples of chemistry in biomedical engineering. A**. Liposomes are synthetic structures produced by assembly of lipids into small sacs or vesicles. **B**. Diffusion tensor imaging (DTI), which maps the diffusion of water in the brain using magnetic resonance imaging (MRI). Photo courtesy of A Brock and L Staib, Yale University. **C**. Collection of proteins that accumulate in the brains of patients with Alzheimer's disease, called an Alzheimer's plaque. Photo credit: BRACE-Alzheimer's Research Registered UK Charity No. 297965. (See color plate.)

biological molecules—proteins, nucleic acids, carbohydrates, and lipids—that will be explored in more detail throughout the rest of the book.

Why should biomedical engineers understand chemistry? Knowing how molecules interact with each other and with their environments helps biomedical engineers to manipulate these molecules to create new tools for treating disease. For example, biomedical engineers have developed methods to synthesize lipid molecules into **liposomes** (Figure 2.1). Liposomes have already found many uses in human health—as carriers of the anticancer drug doxorubicin and as alternate vehicles for gene therapy that do not require the use of viruses. Similarly, biomedical engineers have used their skills in mathematical modeling and their understanding of molecular interactions to understand the formation of molecular complexes, such as the extracellular clumps of protein-rich materials (plaques) that form in Alzheimer's disease (Figure 2.1). A better understanding of the properties of the proteins that form plaque may someday lead to treatments for Alzheimer's disease.

Understanding basic chemical concepts is important in almost every aspect of BME. Artificial hips are made of synthetic materials, usually metals and polymers. Early efforts in creating artificial devices sometimes failed because of unwanted interactions between molecules of the artificial device and molecules of the body. Magnetic resonance imaging (MRI), one of the most powerful methods for non-invasive imaging of the internal structure of humans, is derived from a method that has been used for decades by chemists to understand molecules and their interactions. Even projects that appear to be dominated by physics and mechanics, such as the design of imaging systems, artificial hips, and many others, are based on a deep understanding of molecules and their interactions.

As Chapter 12 will describe in more detail, the images that biomedical engineers create using MRI (Figure 2.1) are based on the chemistry and interactions of water within tissues in the body (Figure 2.2). To understand MR images, it is

Figure 2.2 **Water. A**. Structure of liquid water. Photo courtesy of Anders Nilsson, Stanford University. **B**. Water is an indispensable part of human health.

helpful to understand the properties of water and how water interacts with other molecules in cells and tissues.

This chapter begins with a brief introduction to bonding in atoms and molecules and builds to include descriptions of proteins and nucleic acids and other large biological molecules. Every chapter in this book attempts to relate chemical, biological, and physiological facts to engineering analysis; an introduction to engineering analysis is provided in Box 2.1.

2.2 Bonding between atoms and molecules

All basic life processes that allow us to digest food, move, and grow involve chemical reactions: reactions that yield energy, build new molecules, or break down unneeded molecules. The molecules in our body are involved in thousands of chemical reactions. Before learning about the function of biological molecules, it is useful to examine the ways they can interact with one another by reviewing key concepts in chemistry.

2.2.1 Atomic bonding

There are two types of bonds that can be formed *between atoms*: ionic and covalent bonds. **Ions** are molecules with a net charge, either positive or negative. **Ionic bonds** are formed when electrons are transferred from one atom to another (e.g., Na^+Cl^-). This transfer results in two ions: a positively charged molecule, or **cation**, caused by the loss of electrons, and a negatively charged molecule, or **anion**, caused by the gain of electrons. **Covalent bonds** result from the sharing of electrons (e.g., H_2). Covalently bonded molecules can further be classified as **polar** or **nonpolar**. Molecules are called "polar" because they have partially negative and partially positive charges at the poles of the molecule. This polarity of charge is caused by unequal sharing of electrons between atoms within a molecule. For example, water (H_2O) is a polar molecule because the oxygen atom within the molecule is slightly negative, whereas the hydrogen atoms are

Box 2.1 Engineering analysis and boxes in this book

Biomedical engineers make extensive use of engineering tools and mathematical models to describe the systems that they study. One of the most valuable and far-reaching aspects of an education in engineering is the development of tools and techniques for solving real-world problems. Box 1.2 provides you with some general techniques for approaching and solving engineering problems. This box, and the others like it that appear throughout the rest of this book, provide more information on engineering analysis. Careful study of these boxes, in conjunction with the main text of the chapters, will provide a thorough introduction to the science and technology of biomedical engineering.

Engineering analysis invariably begins by defining the system under study. The system might be a supporting beam in a bridge, a set of components on an integrated circuit board, the human body, or an individual cell or organ in the body. In describing a system, it is important to pay careful attention to the **system boundaries**, i.e., the physical sites of intersection between the system under study and the rest of the world (Figure Box 2.1).

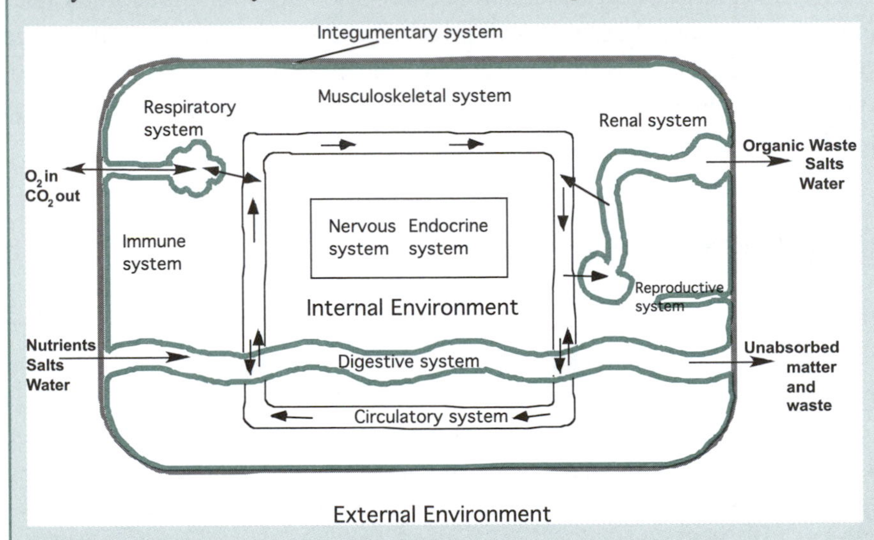

Figure Box 2.1 Schematic diagram of body systems. Notice two methods for defining the system under study: The colored boundary does not include the space inside of the intestinal, reproductive, and urinary systems; the grey boundary does include them. Notice also the arrows, which represent the movement of mass across system boundaries.

Often there are multiple choices for the system boundaries, which lead to alternate descriptions of the system under study. For example, if the system is the human body, the boundary might be defined as the skin and the physical openings between the body and the environment (such as the mouth, nostrils, and urethra). Alternately, the boundary could be the skin and the mucus epithelial surfaces that line the intestinal, respiratory, and reproductive tracts. The engineer is often free to select the system boundary that makes the problem either easier to solve or more interesting. In the first case, the contents of the intestines and lungs would be part of the system; in the second case, they would not be in the system, but instead part of the environment in which the system resides. The boundary selected might depend on the question that the engineer is trying to answer.

This chapter is primarily concerned with chemistry and chemical reactions. When analyzing chemical systems, it is often convenient to define the system as 1 mole of the material of interest. For example, in Box 2.2, the formation of water from atomic hydrogen and oxygen is described. In this example, a convenient system to consider is 1 mole of hydrogen and 1 mole of oxygen.

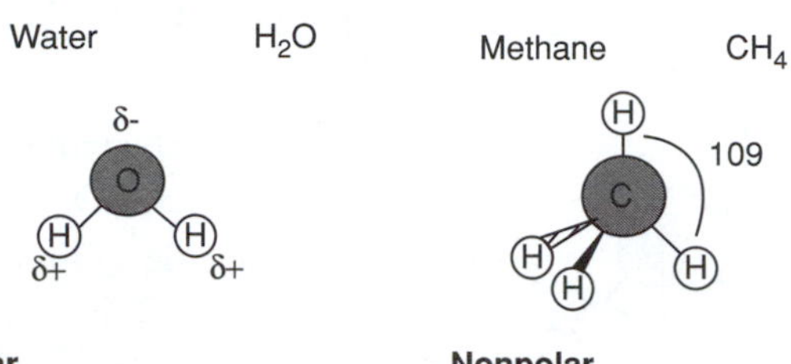

Polar
Unequal sharing of electrons results in polar distribution of charges

Nonpolar
Charges are distributed symmetrically

Figure 2.3 **Polar and nonpolar molecules.** Water is an example of a polar molecule. The unequal sharing of electrons between oxygen and hydrogen atoms creates a distribution of charge, which creates electrical polarity. The oxygen atom has a partial negative charge (δ−), whereas the two hydrogen atoms are partially positive (δ+). Methane is a nonpolar molecule because the charges are distributed equally; the hydrogens are arranged with equal three-dimensional spacing, each separated by an angle of 109°. The symbol δ indicates a partial charge.

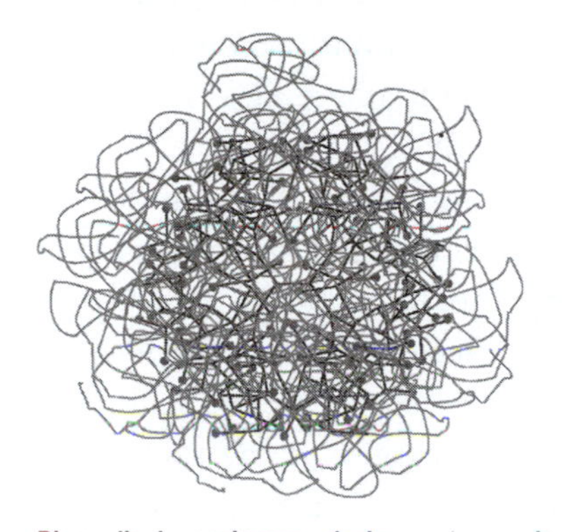

Figure 2.4 **Biomedical engineers design new molecules.** This diagram shows the structure of a polymer built for drug delivery and targeting. A polyamidoamine dendritic polymer (black) with poly(ethylene glycol) arms (blue) attached to the dendrimer via covalent linkages (red). Reproduced from (1) with permission, copyright 2002 American Chemical Society. (See color plate.)

slightly positive (Figure 2.3). This charge difference occurs because oxygen atoms hold on to shared electrons more tightly than do hydrogen atoms. In contrast, a nonpolar molecule, such as methane (CH_4), has uniform and symmetrical charge distribution within the molecule (Figure 2.3).

Biomedical engineers are frequently involved in synthesis of new molecules for medical applications. One example is dendrimers for **gene delivery** (Figure 2.4), which is the act of transferring foreign DNA into a cell. These new agents are created by the formation of new covalent bonds between simple precursor molecules. In the example shown in Figure 2.4, a complex gene delivery molecule is constructed by covalent assembly of a number of simpler starting ingredients. The resulting covalent complex has new properties—such as the ability to bind reversibly to DNA molecules and protect them during entry into a cell—not found in the simpler starting materials.

2.2.2 Molecular bonding

Other types of bonding can occur between molecules (and sometimes between small segments of large molecules, as we will see in Chapter 3, Section 3.3 and Chapter 4, Section 4.2). Two molecules can be weakly attracted to one another through intermolecular forces. These forces may include van der Waals interactions and hydrogen bonding.

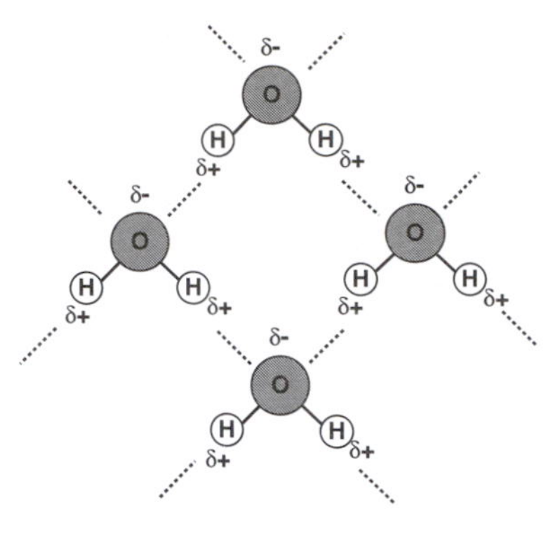

Figure 2.5 **Hydrogen bonding.** Polar water molecules can form hydrogen bonds with each other. Partial positive hydrogen atoms in one water molecule are weakly attracted to partial negative oxygen atoms in another water molecule.

Figure 2.6 **Spider silk.** Spider silk is composed predominantly of proteins. Some spiders make many different kinds of silk, each with a specialized protein composition and unique physical properties. Some spider silks are amazingly strong—stronger than the best man-made materials and more resilient. Hydrogen bonding between parts of the protein contributes to the unusual mechanical properties.

Hydrogen bonding occurs when a partially positive hydrogen atom in a polar molecule is attracted to a slightly negative atom (usually O, N, or F) in a neighboring molecule (Figure 2.5). Hydrogen bonds (1–5 kcal/mol) are much weaker than covalent bonds (50–300 kcal/mol). However, the additive effect of thousands of these weak interactions makes hydrogen bonding an effective glue, which holds molecules—particularly large molecules—together. Chapters 3 and 4 describe the importance of hydrogen bonds in the formation of large molecules—often called **macromolecules** because of their size and their construction from repeated smaller molecular units—such as double-stranded DNA and proteins. Hydrogen bonds are also involved in the formation of **molecular complexes**, or collections of molecules held together often by multiple weak bonds. One such complex occurs in the binding of some molecules to specialized proteins, called **enzymes**, which speed up their chemical conversion to another form (more about enzymes in Chapter 4.4). Spiders are able to manufacture special proteins that form exceptionally strong fibers (Figure 2.6); some forms of spider silk are stronger than steel, but also elastic. Hydrogen bonding interactions between different segments of the spider silk protein appear to be important in creating its unique material properties.

Van der Waals interactions are also weak noncovalent attractions but they are due to temporary and unequal electron distributions around atoms rather than the permanent dipole found in hydrogen-bonded atoms.

2.3 Water: The medium of life

The chemical reactions that drive life occur predominantly in aqueous—or water-rich—environments. For this reason, water is often called "the source of life," but it is also the molecular product of a chemical reaction. Three atoms—two

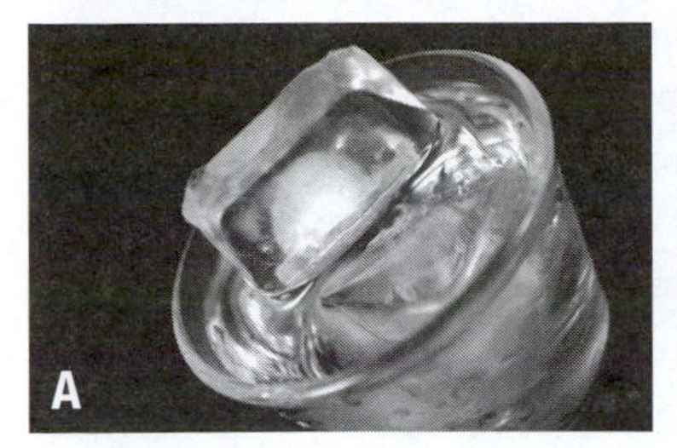

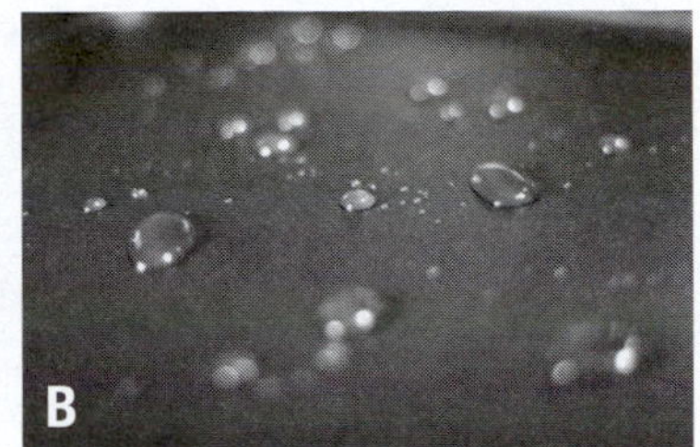

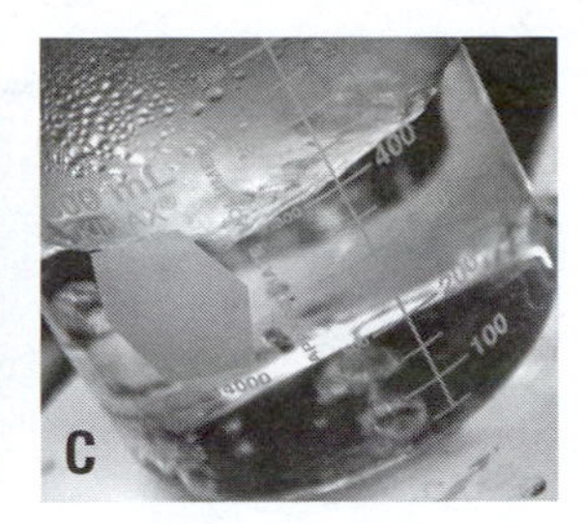

Figure 2.7

The unique properties of water. **A**. Because solid water (ice) is less dense than liquid water, ice floats on the top of water in the ocean, lakes, and streams. **B**. High surface tension of water causes it to bead up on many surfaces (and allows some creatures to walk on water). **C**. Water absorbs a tremendous quantity of heat before it eventually boils.

hydrogens and one oxygen—are held together by covalent bonds to form water. In addition, water can be a product or a reactant in other chemical reactions.

Because the human body is approximately 70% water, it is the ideal environment for these reactions. The extensive hydrogen bonding network that can form between water molecules gives rise to its unique properties. These properties include high melting and boiling temperatures, high surface tension, and a higher density than ice (Figure 2.7).

The ability to form hydrogen bonds also makes water an excellent **solvent**. Water can easily dissolve ions or other polar molecules that are capable of forming hydrogen bonds (Figure 2.5). These water-soluble molecules are referred to as **hydrophilic** ("water loving"). Nonpolar molecules are not easily dissolved in water and are called **hydrophobic** ("water fearing"). Hydrophobic molecules aggregate together to exclude water as best they can: This behavior is often described as the **hydrophobic effect**. As described in Chapter 4, the hydrophobic effect is an important driving force in **protein folding**, the process by which a long macromolecule of amino acids forms a three-dimensional, biologically active protein molecule (Figure 2.8).

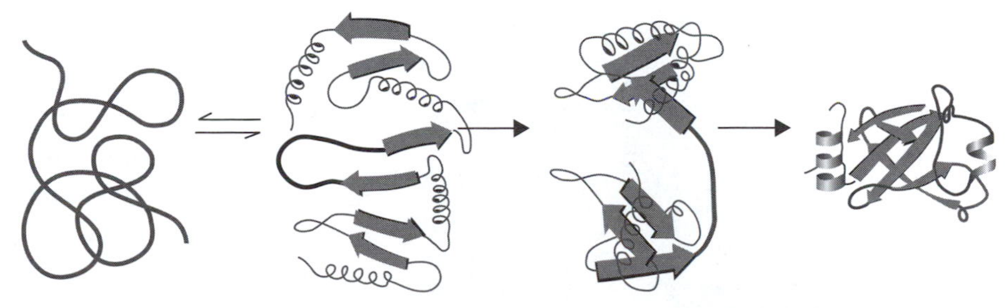

Figure 2.8

Protein folding. A polypeptide, which is a macromolecule composed of amino acids, is converted into an active protein by a process known as protein folding, in which the chemical properties of the amino acids within the polymer interact with other amino acids on the chain. The polypeptide is flexible, allowing it to fold into complex three-dimensional shapes. The ultimate shape of the protein depends on interactions between amino acids and their linear sequence along the molecule. Hydrophobic interactions are important in formation of the three-dimensional structure of proteins.

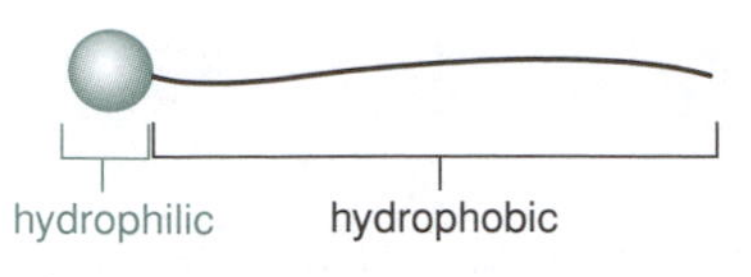

Figure 2.9 **Amphiphilic molecule.** An amphiphilic molecule has both a hydrophilic and a hydrophobic portion within the same molecule.

Molecules that contain both hydrophilic and hydrophobic groups are called **amphiphilic**. For example, phospholipids are amphiphilic molecules (Figure 2.9) that form the plasma membrane surrounding cells (Section 2.7.1). Bioengineers use these kinds of molecules to form complex structures such as the liposomes described earlier (Figure 2.1).

2.4 Biochemical energetics

Humans gain energy through the food that they eat. This energy is stored or expended to sustain life. Running, jumping, and breathing require the activity of muscles; energy must be expended to power these muscles (Figure 2.10). Humans can store energy in the form of molecules such as glycogen and triglycerides within the body. These storage molecules are digested in reactions that release energy when it is needed.

All of the chemical reactions in our bodies result in utilization or accumulation of energy. The factors that determine whether a particular reaction will release energy or require energy are discussed in Box 2.2. It is important to separate the possibility of a reaction occurring (which is considered in Box 2.2) from the rate at which the reaction will proceed (which is considered in Box 2.3). These concepts are related but distinct. Later, in Chapter 4, the rate of biochemical reactions will be considered in more detail. In addition, the role of **enzymes**, which are proteins that are specialized to serve as biological catalysts, or agents that speed up biochemical reactions, will be described.

Figure 2.10 **Human activities require expenditure of energy.** Photo credit: Steve Woltmann, courtesy of Augustana College.

One of the most important chemical reactions for energy utilization in cells involves another molecule, ATP (Figure 2.11). ATP is similar to the nucleotides that make up DNA and RNA, but ATP has three phosphate groups linked together (whereas the nucleotide adenosine of DNA has only one phosphate). The covalent bonds between the phosphate groups are said to be "high energy bonds" because chemical reactions that break these bonds release substantial amounts of energy. A **hydrolysis** reaction is any chemical reaction in which water is a reactant. The hydrolysis of ATP—in which ATP reacts with water to form adenosine

Box 2.2 Thermodynamics of chemical reactions

Water is formed by the reaction of hydrogen atoms and oxygen atoms. The chemical reaction describing the formation of water can be written as:

$$2H_2 + O_2 \rightarrow 2H_2O \qquad \text{(Equation 1)}$$

In this reaction, some covalent bonds are broken and new ones are formed: For example, the bonds that hold the two hydrogen atoms in the stable molecular form of hydrogen (H_2) are broken, and new bonds between O and H are formed. Suppose that a system is defined as 2 moles of hydrogen and 1 mole of oxygen. What is the change in energy of the system, if those 3 moles of hydrogen and oxygen are converted into 2 moles of water? Energy is always released upon formation of a chemical bond, and required for their formation. Therefore, the total energy change for a reaction, such as the reaction in Equation 1, depends on the net energy change for all of the bonds broken and formed. If net energy is released, then the change in bond energy is converted into heat within the system.

Enthalpy is a thermodynamic property of chemicals; the enthalpy of a compound is a measure of the amount of internal energy of the compound.* Therefore, the amount of heat associated with a chemical reaction is equal to the change of internal energy on reaction and depends on the sum of energies released and consumed as bonds are broken and formed in the overall reaction.

The reaction of Equation 1 is an example of a **formation reaction**, in that it describes the formation of the molecule of water from the most stable form of its component atoms. The overall heat of formation—or **enthalpy** change of formation reaction—is a measure of the amount of energy that is either consumed or released when water is formed and is called ΔH_f° (the superscript "o" means that the energy is measured at standard conditions, 25°C and 1 atm). Appendix B contains a table with heats of formation for some compounds: The ΔH_f° is equal to −242 kJ/mole for gaseous water and −286 kJ/mole for liquid water. A negative value of a heat of formation indicates that heat is released when the reaction occurs (the enthalpy of the products is less than the enthalpy of the reactants). The heat of formation for gaseous water indicates that less heat is released to form a gas; this makes sense because some of the heat released by the formation reaction is used to convert the liquid water to the gaseous state.

Heats of formation can be used to calculate the enthalpy change for other kinds of reactions. Consider the more general chemical reaction:

$$\mathrm{aA} + \mathrm{bB} \rightarrow \mathrm{cC} + \mathrm{dD} \qquad \text{(Equation 2)}$$

where compounds A and B react to form compounds C and D. The coefficients a, b, c, and d are stoichiometric coefficients, indicating the number of moles of each reactant (or product) consumed (or produced) during the reaction. The overall energy change that occurs in this reaction can be determined from heats of formation (ΔH_f°) (see Figure Box 2.1):

$$\Delta H^\circ = c\Delta H_{\mathrm{f,c}}^\circ + d\Delta H_{\mathrm{f,D}}^\circ - \left(a\Delta H_{\mathrm{f,A}}^\circ + b\Delta H_{\mathrm{f,B}}^\circ\right) \qquad \text{(Equation 3)}$$

(continued)

Box 2.2 *(continued)*

The heats of formation ΔH_f° for each of the reactants (A, B) and products (C, D) can be found in tables, such as the one in Appendix B. A negative ΔH° indicates an **exothermic** reaction, in which heat is released as the reaction proceeds. A positive ΔH° means that the reaction absorbs heat, or is **endothermic**.

The **entropy** (S) of a system is a measure of disorder in a system or the amount of energy in a system that cannot be used to do **work**. For any change in state of a system, a change in entropy, or ΔS, can be calculated. The standard entropy change, ΔS°, can be calculated similarly to ΔH° by using standard entropy tables.

The energetics of biochemical reactions are often described in terms of **Gibbs free energy**, G, which is related to both the enthalpy and entropy:

$$\Delta G = \Delta H - T\Delta S \qquad \text{(Equation 4)}$$

The value of ΔG can be used to predict whether a reaction is favorable under the given conditions:

If ΔG is < 0, the reaction is spontaneous and will proceed.

If ΔG is > 0, the reaction is not favorable and will not proceed without some input of energy.

Consider, for example, a combustion reaction, which is strongly exothermic (i.e., it has a large negative ΔH). Unless $T\Delta S$ is also large and negative—which is unlikely because combustion reactions tend to increase disorder, so ΔS will be positive—the combustion reaction will proceed spontaneously.

Most biological reactions have a positive ΔG, so they do not occur spontaneously. How do the unfavorable reactions proceed? They require an input of energy, which most often comes from the breaking of a high-energy phosphate bond found in a special biochemical called adenosine triphosphate (ATP). ATP hydrolysis, the breakdown by addition of water occurs as follows:

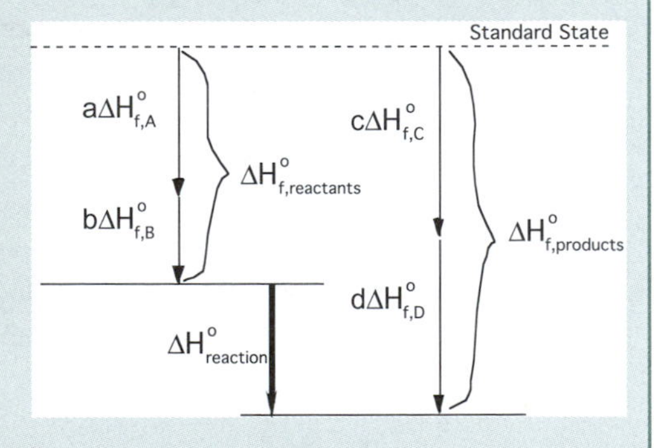

$$ATP + H_2O \rightarrow ADP + P_i \qquad \text{(Equation 5)}$$

This reaction is energetically favorable, with: $\Delta G^\circ = -7.3$ kcal/mol for the reaction in Equation 5. Unfavorable reactions can occur in combination with the ATP hydrolysis reaction (Equation 5); the overall free-energy change is the sum of the ΔG for two reactions. Thus, the negative free energy of ATP hydrolysis can be used to drive a reaction that is energetically unfavorable on its own.

* Enthalpy is actually equal to the internal energy of the system plus the product of volume and pressure, which is a measure of the mechanical **work** performed on the systems by the surroundings. For our purposes, it is reasonable to think of enthalpy as a measure of the internal energy.

Box 2.3 Kinetics of chemical reactions

Box 2.2 introduced some concepts regarding biochemical reactions, including the amount of heat released or absorbed during reaction, and some methods for determining whether reactions will occur spontaneously. Those concepts are related to the **thermodynamic** behavior of compounds: Thermodynamics is the branch of chemistry that is mainly concerned with states of matter and the ways that they change with pressure, temperature, and volume. The thermodynamic properties described in Box 2.2 indicate changes of state that are possible, how the energy will change as the reaction proceeds, and if a reaction will proceed under certain conditions, but these properties do not indicate the speed or rate of a reaction. Study of rates of reactions or rates of change in chemical systems is called **chemical kinetics**.

To illustrate chemical kinetics, consider the important biochemical reaction between water and carbon dioxide:

$$H_2O + CO_2 \underset{k_r}{\overset{k_f}{\rightleftarrows}} H_2CO_3 \qquad \text{(Equation 1)}$$

Notice that, in Equation 1, a forward reaction (in which carbon dioxide and water react to form carbonic acid) and a reverse reaction (in which carbonic acid dissociates into carbon dioxide and water) are both shown. This notation illustrates that the reaction is **reversible**: It can proceed in either the forward or the reverse direction. In a real system—for example, a closed vessel that was filled with liquid water that contained dissolved carbon dioxide and carbonic acid—both the forward and the reverse reaction would be happening at all times. The net amount of change in the concentration of any of the chemical species would depend on the rate of the forward reaction (which is consuming water and carbon dioxide and generating carbonic acid) and the rate of the reverse reaction (which is doing the opposite).

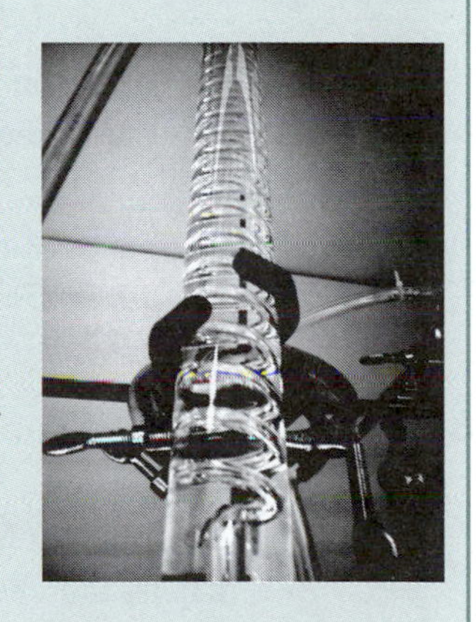

Rates of chemical reactions are described mathematically by rate constants, usually designated by the symbol k.* In the example of Equation 1, the rate constant for the forward reaction is k_f and the rate constant for the reverse reaction is k_r. The rate of formation of carbonic acid by the forward reaction is determined by the rate constant and the concentrations of the reacting species:

$$Rate_f = k_f\,[H_2O]\,[CO_2] \qquad \text{(Equation 2)}$$

where [X] indicates the concentration of component X. Likewise, the rate of the reverse reaction, which is the rate of formation of water or carbon dioxide by the reverse reaction, can be written:

$$Rate_r = k_r\,[H_2CO_3] \qquad \text{(Equation 3)}$$

To determine the net rate of production of carbonic acid at any given time, one would need to consider both the rate of generation by the forward reaction and the rate of consumption by the reverse reaction:

$$\text{Net Rate of } H_2CO_3 \text{ production} = k_f\,[H_2O]\,[CO_2] - k_r[H_2CO_3] \qquad \text{(Equation 4)}$$

(continued)

Box 2.3 *(continued)*

A reaction reaches **chemical equilibrium** when the forward and reverse reactions are occurring at the same rate; therefore, the concentration of reactants and products stop changing. This phenomenon can be written mathematically; the rate of the forward reaction (Equation 2) is equal to the rate of the reverse reaction (Equation 3), so that $k_f[H_2O][CO_2] = k_r[H_2CO_3]$ or:

$$K = \frac{k_f}{k_r} = \frac{[H_2CO_3]}{[H_2O]\,[CO_2]} \qquad \text{(Equation 5)}$$

This new quantity, the equilibrium constant K, is equal to the ratio of the forward and reverse rate constants. Although this discussion has considered chemical kinetic properties and rates of change, the equilibrium constant is a thermodynamic property: It is a state of matter that represents a stable equilibrium condition. In fact, it can be shown that K is related to the standard free-energy change, ΔG°, for the reaction, which was defined in Box 2.2:

$$\Delta G^\circ = -RT\ln(K) \qquad \text{(Equation 6)}$$

The expression for equilibrium can be written more generally, for the reaction shown in Equation 2 of Box 2.2:

$$K = \frac{k_f}{k_r} = \frac{[C]^c\,[D]^d}{[A]^a\,[B]^b} \qquad \text{(Equation 7)}$$

The equilibrium constant for this more general reaction is still related to the standard free-energy change for the reaction, as described by Equation 6.

* In this example, the reaction rate is assumed to be a **first order reaction**, which means that all of the reaction rates are equal to the rate constant multiplied by the concentration. This assumption is true for many reactions, but not all. The interested reader is referred to other books that discuss reaction rates and rate constants more broadly. When the kinetics of enzyme-catalyzed reactions is considered in Chapter 4, Michaelis Menton kinetics, which is not first order, will be described.

diphosphate (ADP) and an inorganic phosphate—releases 30.5 kJ/mole of energy; that is, the change in Gibbs free energy (defined in Box 2.2) for the reaction shown in Figure 2.11 is $\Delta G^\circ = -30.5$ kJ/mole.

ATP has a number of properties that make it an important and active participant in biochemical energetics. It can be generated efficiently from other energy-rich

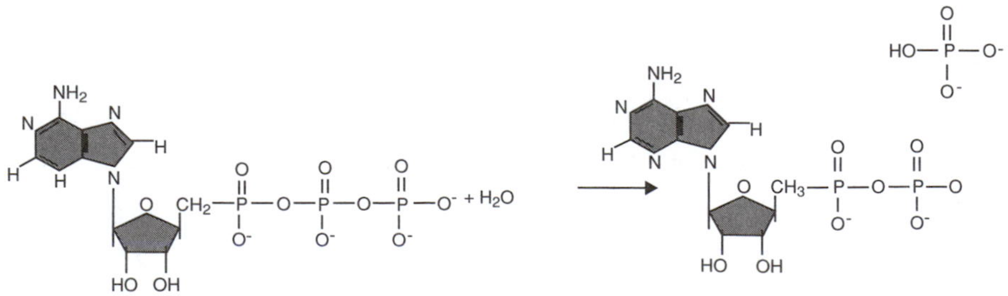

NH2 N N H H N O CH2—P—O—P—O—P—O- + H2O HO OH → NH2 N N H N N O CH3—P—O—P—O HO OH HO—P—O-

Figure 2.11 Hydrolysis of adenosine-5′-triphosphate (ATP).

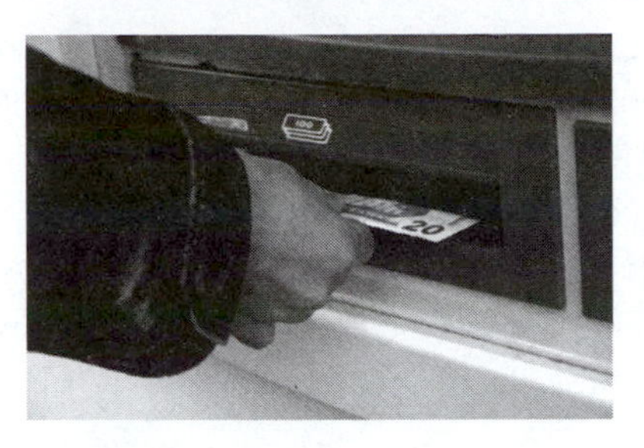

Figure 2.12 **Adenosine-5′-triphosphate (ATP) as currency.** It might be helpful to thinking of ATP as the body's instant energy currency. Like cash in your pocket, ATP is ready to be converted into other forms of energy.

substances to serve as a temporary store of energy, and rapidly hydrolyzed to release the stored energy. This ease of conversion makes ATP an excellent currency for efficient, immediate energy exchange. Some people like to think about ATP in analogy to monetary currency, in the sense that the energy stored in macromolecules such as glycogen is similar to money in the bank, whereas ATP is similar to money in your pocket (Figure 2.12).

The generation of inorganic phosphate during ATP hydrolysis is also convenient, in that many biochemical reactions can be coupled to ATP hydrolysis, allowing the energy that is released by ATP conversion to be used to drive an otherwise unfavorable reaction (Figure 2.13). For example, the conversion of glutamic acid to glutamine, an amino acid that is necessary for protein synthesis, does not occur spontaneously:

$$\text{Glutamic acid} + \text{Ammonia} \rightarrow \text{Glutamine} \tag{2.1}$$

because the change in free energy for the reaction is positive (see Box 2.2). However, if glutamic acid is first converted to an intermediate, glutamyl phosphate, and the phosphate group is provided by an ATP molecule, the conversion of glutamic acid to glutamine can occur in two sequential reactions, both of which do occur spontaneously:

$$\begin{array}{c} \text{Glutamic acid} + \text{ATP} \rightarrow \text{Glutamyl phosphate} + \text{ADP} \\ \text{Glutamyl phosphate} + \text{ammonia} \rightarrow \text{Glutamine} + \text{P}_\text{i} \\ \hline \text{Glutamic acid} + \text{ATP} + \text{ammonia} \rightarrow \text{Glutamine} + \text{ADP} + \text{P}_\text{i} \end{array} \tag{2.2}$$

Equation 2.2 is an example of coupled reactions, in which the energy released by the conversion of ATP to ADP is used to make an energetically unfavorable

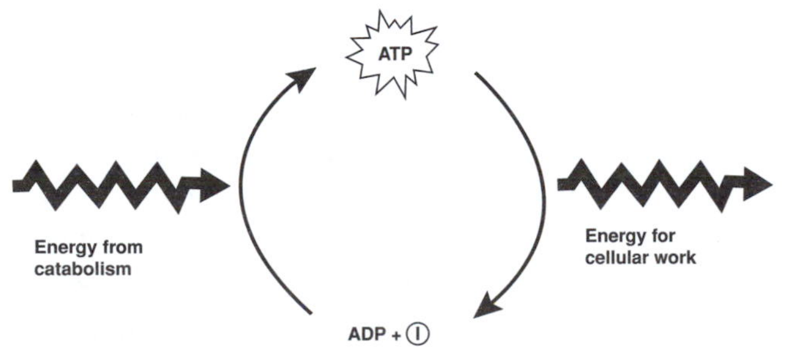

Figure 2.13 **Adenosine-5′-triphosphate (ATP) as an intermediate in metabolism.** ADP, adenosine diphosphate. From http://Fig.cox.miami.edu/~cmallery/150/metab/ATPcycle.jpg.

reaction (one from which the changes in free energy are positive) occur spontaneously. In the rest of this book, there are other examples of reactions that are driven by the hydrolysis of ATP (see Chapter 6, in particular Section 6.2).

For this type of coupling to occur, it is necessary that small amounts of ATP be constantly available in cells throughout the body. In this way, ATP serves as an intermediary in the overall process of conversion of energy from foods to the activities of life. Chapter 7 will describe the generation of ATP from other energy sources derived from food.

SELF-TEST 1 When carbon dioxide dissolves in water, it can react to form carbonic acid according to the reaction: $H_2O(l) + CO_2(aq) \rightarrow H_2CO_3(aq)$. According to the table of standard heats of formation in Appendix B, Table B.2, ΔH_f° for H_2O, CO_2, and H_2CO_3 are -285.8, -393.5, and -699.7 kJ/mole, respectively. Calculate the standard change of enthalpy for this reaction.

ANSWER: $-699.7 - (-285.8 - 393.5)$ kJ/mole $= -20.4$ kJ/mole

2.5 Importance of pH

2.5.1 Hydrogen ions and water

Water dissociates—or breaks down into components—in solution. Dissociation of water produces a hydrogen ion (H^+, which is often called a **proton**, because each hydrogen atom, H, consists of only one proton and one electron, so that H^+—hydrogen without an electron—is a proton), and a hydroxide ion (OH^-):

$$H_2O \rightarrow H^+ + OH^- \tag{2.3}$$

However, free hydrogen ions do not exist in solution; because they are so small and positively charged, protons associate with water (the positively charged proton is attracted to the negative pole of water at the oxygen atom; see Figure 2.5). Equation 2.3a is therefore written:

$$2H_2O \rightarrow H_3O^+ + OH^-, \tag{2.3a}$$

where H_3O^+ is the hydronium ion.

The fraction of water molecules that undergo dissociation is very small. In fact, the concentrations of the dissociated H^+ and OH^- in pure water at 25°C are, respectively, 10^{-7} M and 10^{-7} M. Of course, in pure water, H^+ and OH^- concentration must be the same, because one molecule of each is produced with each water molecule that dissociates. One way to describe the extent of dissociation is to report the number of ions in solution: 10^{-7} M each of H^+ and OH^- in this case. But the use of such small numbers is cumbersome, so an alternate method was developed to report the concentration of H^+ in a solution. A new variable, called pH for "potential of hydrogen," is defined as the negative

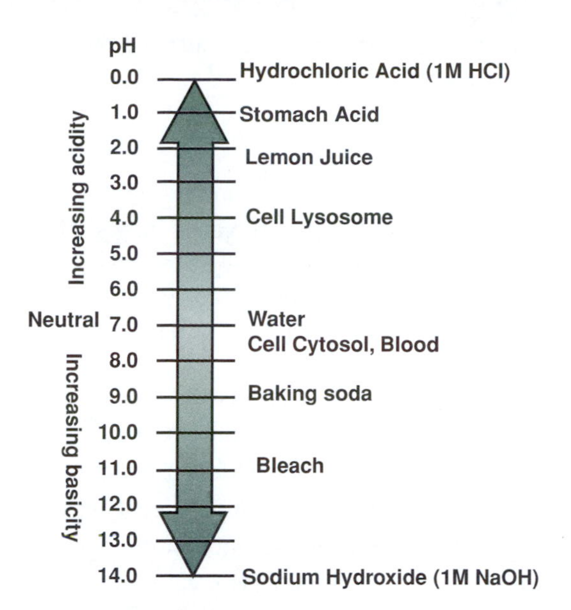

Figure 2.14 Scale of pH. The pH of water is neutral (pH = 7.0). Substances with pH less then 7.0 are more acidic, and those with values greater than 7.0 are more basic. Physiological pH is about 7.4, which is the pH of cytosol—fluid inside of cells (see Chapter 5, Section 5.2)—and blood. Digestive enzymes in the stomach function in an acidic pH = 1.0.

logarithm of the hydrogen ion concentration:

$$\mathrm{pH} = -\log[\mathrm{H_3O^+}] \quad (2.4)$$

Thus, for pure water, with 10^{-7} M of protons, the pH is equal to 7. A solution in this state is called **neutral**, meaning that the solution has an equal number of H^+ and OH^- ions.

2.5.2 pH change in disease

Deviation of a patient's blood pH from its normal value—which is 7.4, near neutral—is always a sign of serious illness. An understanding of acid/base chemistry is important in medicine because many of the body's processes occur best at a specific pH value; therefore, control of pH is critical to the function of cells, organs, and the whole body. For example, most **enzymes** function in or near the neutral physiological pH of 7.4, whereas some enzymes function in acidic environments, such as the digestive enzymes in the stomach. Variations in pH are found in different parts of the body as well as within cellular compartments. Figure 2.14 lists some common pH values for physiological and common substances.

2.5.3 Acids and bases

Following the definition in Box 2.3, the equilibrium constant for the reaction in Equation 2.3 is:

$$K = \frac{[\mathrm{H_3O^+}][\mathrm{OH^-}]}{[\mathrm{H_2O}]} \quad (2.5)$$

SELF-TEST 2 The density of pure water is approximately 1 g/mL. What is the concentration of water in units of moles/L?

ANSWER: 55.5 mole/L

Because the concentration of water, $[H_2O]$, varies only slightly, Equation 2.5 is often rewritten:

$$K_w = K[H_2O] = [H_3O^+][OH^-] \tag{2.6}$$

where K_w (called the ion product of water) is defined as the product of the hydrogen ion concentration and hydroxyl ion concentration (at 25°C).

Acids are molecules that release H^+ when added to aqueous solutions, and **bases** are molecules that release OH^- when added to aqueous solutions.[1] Therefore, when an acid is added to an aqueous solution, it dissociates, forming more H^+ ions and decreasing the pH: This process creates an acidic solution with $pH < 7$. When a base is added to an aqueous solution, it dissociates to form OH^- ions, thereby lowering $[H_3O^+]$ and increasing the pH to create a basic or alkaline solution with $pH > 7$. Note that, because K_w is a constant, $[H_3O^+]$ and $[OH^-]$ cannot vary independently from one another. Therefore, the existence of either acids or bases in solution will act to alter the pH. Likewise, multiple acids and bases present in solution will act in synergy to create a pH of the solution that reflects the presence of all of the components.

STRONG ACIDS AND BASES

Consider first the contribution of strong acids and bases to pH. Strong acids and bases, such as hydrochloric acid (HCl) and sodium hydroxide (NaOH) dissociate completely in solution:

$$\begin{aligned} HCl &\rightarrow H^+ + Cl^- \\ NaOH &\rightarrow Na^+ + OH^- \end{aligned} \tag{2.7}$$

When strong acids are added to an aqueous solution, they alter the pH directly by the addition of H^+ and OH^- to the solution.

WEAK ACIDS AND BASES

In addition to these strong acids and bases, there are many weak acids and bases, in which the dissociation and release of H^+ or OH^- ions is not complete:

$$HA \rightleftarrows H^+ + A^-, \tag{2.8}$$

[1] NOTE: This is one definition of acids and bases, which was first proposed by Svante Arrhenius, a Swedish scientist. This definition is not perfect (it does not explain why the bicarbonate ion HCO_3^- acts as a base in solution, for example), but it is adequate for the purposes of this book. Later definitions by the Dutch scientist Johannes Brønsted and the English scientist Thomas Lowry, refined the Arrhenius definition, making it applicable to more situations.

Box 2.4 Acid–base equilibrium

When an acid (HA) is placed in water, it dissociates to release hydrogen ions and the acid's deprotonated form (A^-), which is also called the **conjugate base**:

$$HA \rightleftarrows H^+ + A^- \quad \text{(Equation 1)}$$

For a weak acid, the dissociation is not complete; the arrows in both directions in Equation 1 indicate that the reaction is reversible, as described in Box 2.3. Because the reaction is reversible, equilibrium exists between the products and the reactants. Therefore, as described in Box 2.3, an equilibrium constant can be defined for this acid:

$$K_A = \frac{[H^+][A^-]}{[HA]} \quad \text{(Equation 2)}$$

By rearranging Equation 3 and taking the logarithm of both sides, we get a relationship between the equilibrium constant and pH:

$$-\log[H^+] = -\log K_A + \log\frac{[A^-]}{[HA]} \quad \text{(Equation 3)}$$

Similar to the definition of pH, a new variable called pK_A can be defined as the $-\log[K_A]$. Substitution of this definition, together with the definition for pH, yields a useful expression, often called the Henderson–Hasselbalch equation:

$$pH = pK_A + \log\frac{[A^-]}{[HA]} \quad \text{(Equation 4)}$$

The Henderson–Hasselbalch equation shows that when $[A^-] = [HA]$, the pH equals the pK_A. Thus, the pK_A for a weak acid or base is the pH at which half of the molecules are dissociated and half are neutral. Equation 4 can be used to determine the degree of dissociation, i.e., the ratio $\frac{[A^-]}{[HA]}$, of the weak acid.

Some molecules have multiple regions that can exhibit acid/base behavior. For example, vitamin C has two groups that behave as weak acids. Therefore, the compound has two pK_A values: one that describes the behavior of each of the acidic groups: $pK_A^1 = 4.2$ and $pK_A^2 = 11.6$.

where HA is the weak acid and A^- is called the **conjugate base**. The degree of dissociation of an acid can be characterized by the pK_A value, which is defined in Box 2.4: The lower the pK_A, the stronger the acid. A table of pK_A values is located in Appendix B (Table B.1).

Many molecules are weak acids and bases, including molecules that are produced by sustained muscular activity such as lactic acid, molecules in foods such as citric acid, vitamins such as ascorbic acid (vitamin C), and proteins such as albumin (a major component of blood). Large molecules, such as proteins, often contain multiple independent chemical groups within them that behave as weak acids and bases. The contribution of weak acids and bases to pH is described quantitatively in Box 2.4.

Weak acids and bases are often used as buffers in solution to resist changes in pH caused by addition of other acids or bases. The bicarbonate buffering system is one mechanism for the body to maintain a constant, near-neutral pH of blood and intracellular and extracellular fluid. When bicarbonate is present, as when any weak acid is present as a buffer in solution, the anion (A^-) can act as a conjugate base to bind free H^+ ions, neutralizing the protons so that the pH does not change appreciably.

SELF-TEST 3 Classify each of these compounds as an acid or a base: HBr, NO_2^-, HCO_3^-

ANSWER: Acid, Base, Base

CARBONIC ACID

The bicarbonate buffering system in the blood consists of a mixture of carbonic acid (H_2CO_3) and sodium bicarbonate ($NaHCO_3$). These substances act to buffer the blood from extreme changes in pH; consider what happens when either a strong acid or base is added to a solution containing both carbonic acid and bicarbonate:

$$\begin{aligned} \mathrm{HCl} + \mathrm{NaHCO_3} &\rightarrow \mathrm{H_2CO_3} + \mathrm{NaCl} \\ \mathrm{NaOH} + \mathrm{H_2CO_3} &\rightarrow \mathrm{NaHCO_3} + \mathrm{H_2O} \end{aligned} \tag{2.9}$$

The H^+ ion from hydrochloric acid is taken up by $NaHCO_3$ to form carbonic acid. The OH^- ion from sodium hydroxide combines with the H^+ from carbonic acid to form water. Hence, the presence of carbonic acid and bicarbonate, in equilibrium, tends to neutralize the solution. Another way to say this is that the presence of carbonic acid and bicarbonate, both at the same time in the solution, buffers the solution.

THE HENDERSON–HASSELBALCH EQUATION

The Henderson–Hasselbalch equation (Box 2.4) allows a quantitative description of the effect of buffers in solution. The pK_A of carbonic acid is 6.1; therefore, for the bicarbonate system, the Henderson–Hasselbalch equation is:

$$\mathrm{pH} = 6.1 + \log \frac{[\mathrm{HCO_3^-}]}{[\mathrm{H_2CO_3}]} \tag{2.10}$$

The body has mechanisms to maintain the ratio of $\frac{[HCO_3^-]}{[H_2CO_3]}$ near 20, which gives a pH of 7.4. The effectiveness of the bicarbonate system at buffering in the blood is caused by 1) the pK_A of carbonic acid/bicarbonate, which is near-neutral pH

Table 2.1 Size of selected molecules, monomers, and polymers

Molecule	Size (nm)	Molecular weight (Daltons)
Oxygen (O_2)	0.2	32
Water (H_2O)	0.3	18
Glucose ($C_6H_{12}O_6$), monosaccharide	0.7	180
Tyrosine ($C_9H_{11}NO_3$), amino acid	0.8	181
Adenosine ($C_{10}H_{13}N_5O_3$), nucleotide	1.0	267
DNA, double helix, diameter	2	–
Albumin, protein	7	68,000
Cell membrane, thickness	3–9	–
DNA, human, extended	9 cm	–

and 2) the fact the kidneys can control excretion of HCO_3^- and the lungs can control H_2CO_3 via respiration to increase or decrease CO_2.

The bicarbonate system will come up again later in the book, when respiratory physiology is discussed in Chapter 7. There, the relationships among carbon dioxide, carbonic acid, and bicarbonate will be developed more completely.

Bicarbonate buffering is an example of a control system that the body uses to maintain **homeostasis** or a constant internal environment. For example, a decrease in blood pH (**acidosis**) caused by a problem in metabolism triggers the increased ventilation through the lungs and increased bicarbonate ion secretion from the kidneys. As the Henderson–Hasselbalch equation shows, these two responses help to return the pH to the normal value.

SELF-TEST 4 As described previously, a solution containing $H_2CO_3/NaHCO_3$ is a buffered solution. Which of the following solutions would act as buffers: a) KH_2PO_4/H_3PO_4, b) $NaClO_4/HClO_4$, c) C_5H_5N/C_5H_5NHCl, d) NaBr/HBr.

ANSWER: Yes, No, Yes, No

2.6 Macromolecules: Polymers of biological importance

Polysaccharides, proteins, nucleic acids, and lipids are key classes of molecules, which are tremendously important in the biology of all forms of life. These classes of large molecules are constructed by linking together smaller molecules. Nucleic acids, proteins, and polysaccharides are members of a more general class of chemicals called **polymers**, which are large molecules formed by the bonding of many smaller chemicals, called **monomers**, into one long molecule (Table 2.1). Because of their large size, these polymers are sometimes called macromolecules. Monomers are usually small molecules that share certain chemical features, which allow them to be linked together through covalent bonds. Not every molecule can

$$n\ \mathrm{H_2C{=}CH_2}$$

ethylene

↓ polymerization

$$\mathrm{R{-}O{-}CH_2{-}CH_2{-}(CH_2{-}CH_2){-}CH_2{-}CH_2{-}CH_2{-}CH_2{-}(CH_2{-}CH_2){-}CH_2{-}CH_2{-}O{-}R}$$

or more simply $\left(\mathrm{CH_2{-}CH_2}\right)_n$ n = a very large integer

polyethylene

Figure 2.15 Polyethylene is formed by reactions that couple many ethylene molecules into one long chain.

be a monomer for formation of polymers: Monomers must contain the proper reactive groups to enable the sequential chemical reactions that lead to a polymer (Figure 2.15). The chemical process of making a polymer from a collection of monomers is called **polymerization**.

2.6.1 Introduction to biological polymers

Polymers are an inescapable part of the modern world. Many of the polymers that we encounter in daily life are synthetic, or man-made. Polyethylene, for example, is a polymer that is formed by coupling together many monomers of ethylene to form long molecules that are high in molecular weight (Figure 2.15). Polyethylene is manufactured and used in a wide range of products including plastic bags, tubing, packaging materials, furniture, toys, and medical implants (Figure 2.16). Other synthetic polymers—such as polypropylene, poly(ethylene terephthalate), and silicone rubber—are used in myriad applications, including medical devices such as artificial organs (see Chapter 15).

MONOMER DIVERSITY

The naturally occurring, biological polymers considered in this section are more complex than polyethylene. In most biological polymers, there is not a single

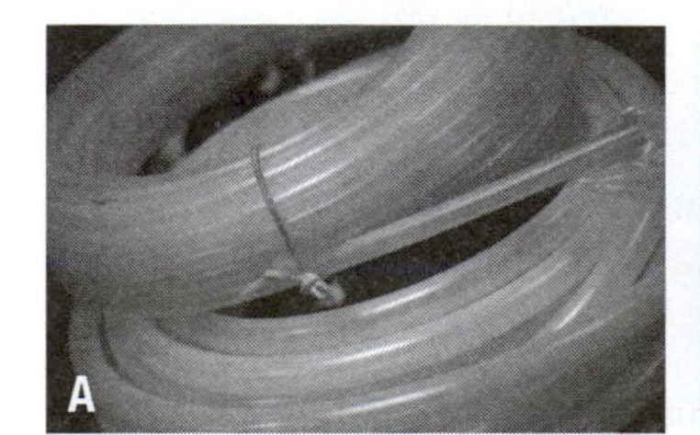

Figure 2.16 Examples of polyethylene. **A.** Tubing. **B.** Bubble wrap. **C.** Acetabular cups, which are a critically important component of artificial hips. Photo courtesy of Biomet Orthopedics, Inc.

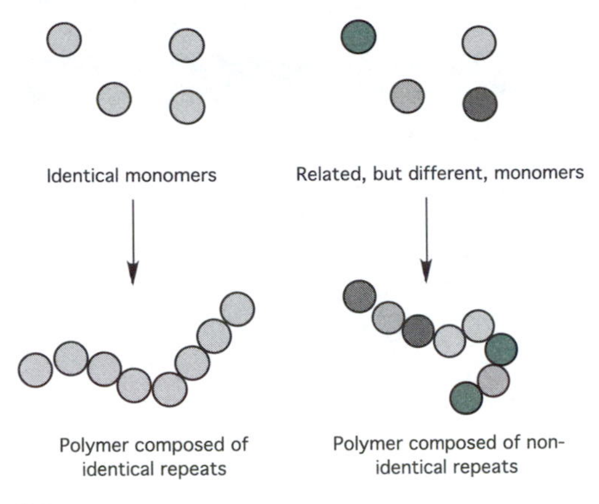

Figure 2.17 Polymers.

monomer but a collection of monomers that are chemically different but share some chemical characteristics. This sharing of some characteristic is necessary to allow the monomers to be polymerized into a single molecule (Figure 2.17).

The biological polymers of great interest to biomedical engineers are **polysaccharides**, **proteins**, and **nucleic acids**. Polysaccharides are made up of small sugars called monosaccharides, proteins are formed using amino acids, and nucleic acids are composed of nucleotides. These polymeric biomolecules contain hundreds (carbohydrates and proteins) to millions (nucleic acids) of monomers (Table 2.2).

LIPIDS

Chapters 3 and 4 will discuss the structure and function of proteins and nucleic acids in greater detail. The rest of this chapter will present a brief description of each of these classes of polymers and their functions. A discussion of lipids is also included. **Lipids** are not polymers, but they are large molecules built from smaller units to create a diversity of structures. Lipids include **triglycerides**, which are used for energy storage, and **phospholipids**, which assemble into cell membranes. The properties of lipids are introduced in the next section, after the discussion of biological polymers.

Table 2.2 Classes of biological polymers

Monomer	Diversity of monomer in macromolecule	Biological macromolecule
Amino acids	~20	Protein
Nucleotides	4	DNA or RNA
Monosaccharides	Extensive	Polysaccharides

Notes. Macromolecules are made up of smaller components. Sugar molecules combine to form carbohydrates such as glycogen, which is a storage form of energy for the body. Genetic information is stored in nucleic acids (DNA and RNA), which are polymers of nucleotides. Proteins are polymers of amino acids and can serve myriad functions, including as structural support for cells or catalysts for chemical reactions.

Table 2.3 Functional groups found in biomolecules

Functional group	Structure	Properties	Examples
Aliphatic	H H H / R-C-C-C-H / H H H	Uncharged, nonpolar	Hydrocarbon tails of lipids
Hydroxyl (Alcohols)	R-C-O-H	Uncharged, polar	Hydroxyl groups in carbohydrates
Carboxylic	O ‖ R-C-OH	Weak acid, polar	Carboxylic acid groups in proteins
Amine	H / R-N-H	Weak base, polar	Amine groups in proteins
Phosphate	O ‖ R-C-O-P-O⁻ / O	Negative charge; high energy transfer	Phosphate groups in phospholipids and nucleic acids, ATP

Notes. Aliphatic groups are uncharged and nonpolar, which render molecules that contain them hydrophobic. Hydroxyl, carboxylic, and amine groups are all polar and thereby hydrophilic. Carboxylic and amine groups can become deprotonated or protonated and may be negatively charged ($R\text{-}COO^-$) or positively charged ($R\text{-}NH_3^+$) depending on the pH. Negatively charged phosphate groups are found in phospholipids, which make up the plasma membrane and the backbone of nucleic acids (DNA and RNA). Breakage of the high-energy phosphate bond in ATP is used as an energy source for many reactions in the cell.

FUNCTIONAL GROUPS IN MONOMERS

One of the features of biological polymers that is important to their function, but makes them difficult to understand, is the great diversity of chemical structures that can be created by linking many similar monomers. To make it easier to understand, it is helpful to consider the chemical properties of small elements of the larger polymer; some of the properties of the large molecule can be predicted from the sum of all of the smaller chemical features that it contains. These small chemical features are often referred to as **functional groups**. An example of this concept was encountered earlier in amphiphilic molecules: These molecules have regions that are hydrophobic and other regions that are hydrophilic (Figure 2.9). Another way to describe this structure is that amphiphilic molecules contain two different functional groups: one that is hydrophobic and one that is hydrophilic.

The amino acids, nucleotides, and monosaccharides that make up biological macromolecules contain a diverse set of functional groups that contribute to their chemical behavior. As shown in Table 2.3, the functional groups convey properties specific to the molecules that contain them. In many cases, the functional groups are also reactive—that is, capable of participating in chemical reactions—which makes it possible for other molecules to react with polymers, a process that can lead to permanent changes in either the reacting molecule or the polymer. Figure 2.18 shows a biological molecule with a number of different functional groups to illustrate the concept.

Certain kinds of chemical reactions occur commonly in the synthesis and breakdown of the polymers listed in Table 2.2. Of particular note are reactions involving the removal and addition of water (Figure 2.19). Synthesis involves a

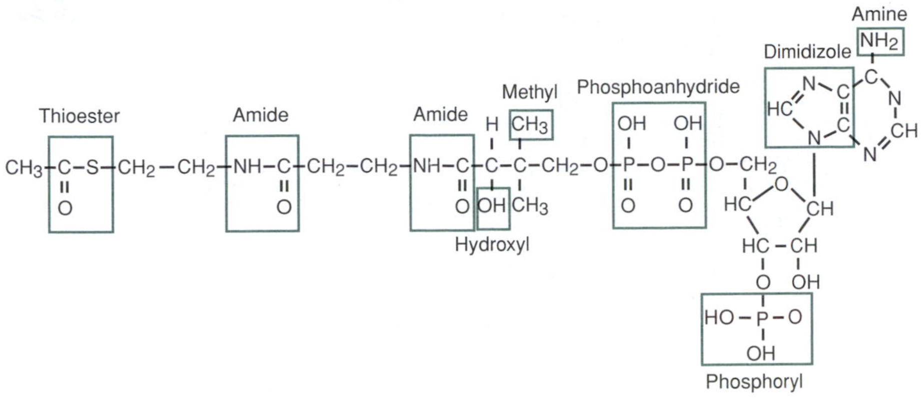

Amine
Dimidizole
Methyl
Phosphoanhydride
Thioester
Amide
Amide
Hydroxyl
Phosphoryl

Figure 2.18

Functional groups. Some of the important functional groups are illustrated on acetyl-coenzyme A.

condensation reaction in which a water molecule is removed and a new bond is formed between the two subunits. The overall chemistry of the reaction—and the chemical structure of the bond—differs among the various biomolecules; this will be described more completely in Chapters 3 and 4. Metabolism or breakdown of a biomolecule into its subunits requires the addition of water—a hydrolysis reaction.

2.6.2 Carbohydrates

Carbohydrates are abundant in nature and a major source of energy in the human diet. Carbohydrates are divided into groups according to their size, such as **monosaccharides**, **disaccharides**, and **polysaccharides**. These groups are related: A disaccharide is created by coupling together two monosaccharides, and a polysaccharide is formed by linking together many monosaccharides.

Carbohydrates are usually described by the formula $(CH_2O)_n$, where n is the number of carbon atoms in the molecule, although there are exceptions to this rule. Monosaccharides typically form cyclic structures in aqueous solutions; the bonding of atoms within the molecule creates a ring. Among the most common monosaccharides are the five-carbon sugars **ribose** and **deoxyribose** (which are

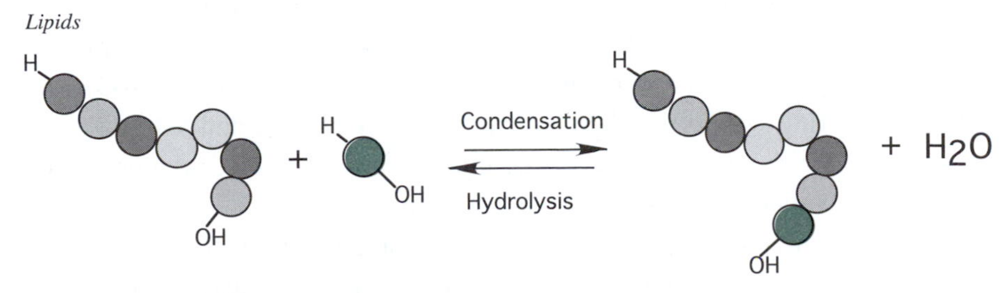

Figure 2.19

Condensation and hydrolysis reactions. Biomolecules can be formed from their subunits by a condensation reaction in which a hydroxyl group (OH) from one monomer combines with a hydrogen (H) atom from another to yield water. A new bond is formed, which holds the two monomers together. In the reverse reaction, water is added to the polymer, and the bond is broken to yield the two subunits. Both reactions are catalyzed by enzymes.

Glucose Galactose

Fructose

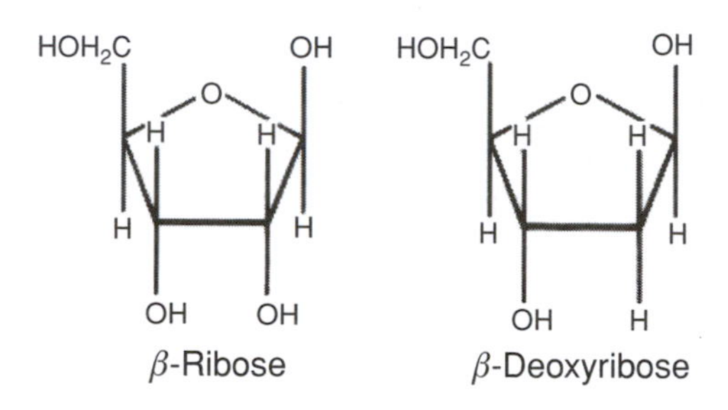

Figure 2.20 **Monosaccharides.** The chemical structures for three common six-carbon sugars and two common five-carbon sugars are shown.

a part of the structure of nucleotides, as described in the next section) and the six-carbon sugars **glucose**, **fructose**, and **galactose** (Figure 2.20).

Sugar is a common term used to refer to certain forms of carbohydrate in the diet. Sugars are carbohydrates that form white crystals or powders, dissolve easily in water, and usually have a sweet taste, making them a popular ingredient in foods. Common table sugar is **sucrose**, a disaccharide of glucose and fructose, which is derived from sugar cane, sugar beets, maple syrup, honey, and other natural sources. **Lactose**, or milk sugar, provides the sweetness in milk; it is also a disaccharide, but of glucose and galactose. When people use the term "blood sugar" they are referring to glucose, which is the form of simple carbohydrate that circulates in the blood.

Monosaccharides are small molecules, but their structure is complex. The ring structure of the molecule introduces one source of complexity: For example, glucose and fructose have the same chemical formula, $(CH_2O)_6$, but the size of the ring is different (there are six vertices in glucose and five in fructose). The cyclic, or ringed, structure of the molecules creates other opportunities for variation, which appear subtle in diagrams, but lead to important chemical and biological differences. Glucose and galactose appear quite similar—the only difference is the position of the –OH group on the carbon on the left-hand side of the structures in Figure 2.19, which is either above (galactose) or below (glucose) the ring. Our bodies easily recognize these differences. Glucose is the most important source of energy for the brain, whereas galactose is not usable by the brain.

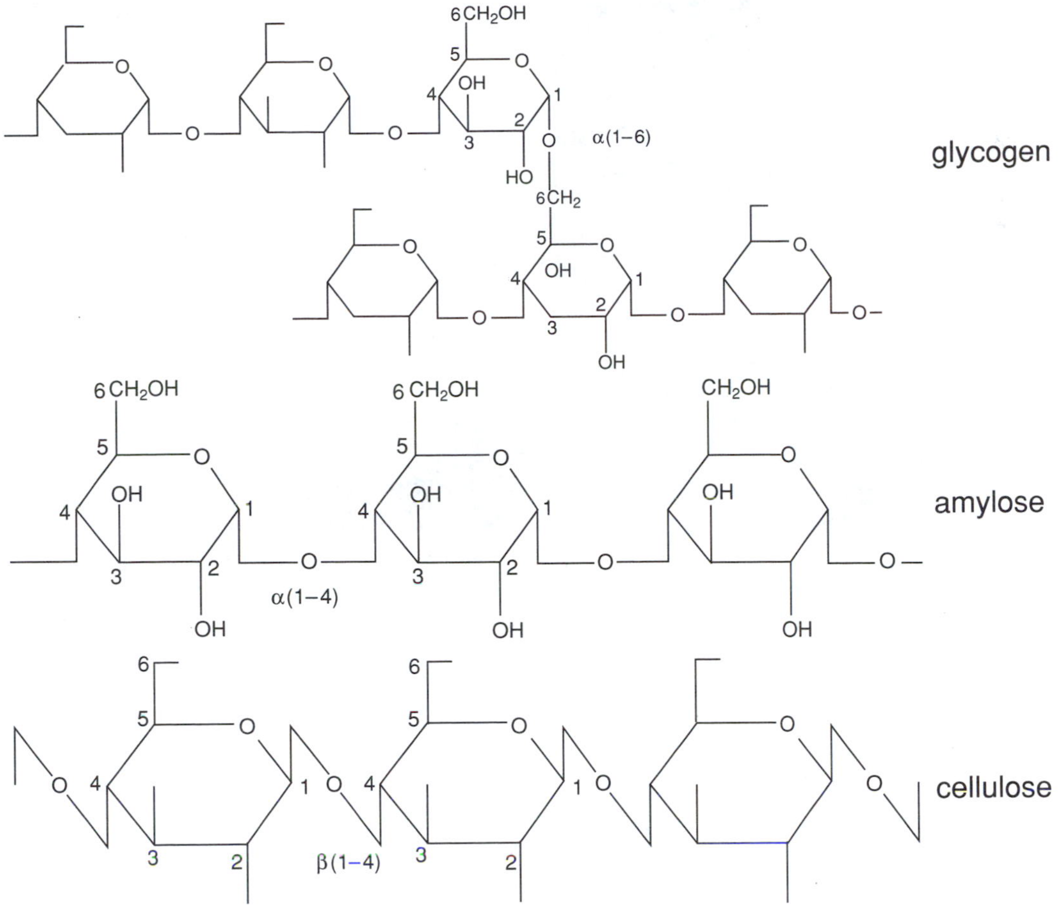

Figure 2.21 **Polysaccharides.** Polysaccharides are formed from polymerization of monosaccharides or disaccharides. Unlike proteins and nucleic acids, which are also linear, carbohydrates can be linear or branched. Some common examples of carbohydrates are glycogen (top), starch (middle), and cellulose (bottom). Glycogen is a major storage form of carbohydrates in humans; it is a polymer of glucose and is synthesized in the liver when fuel is abundant. Starch is a major form of polysaccharide in the diet. Cellulose is also present in the diet, but is not digestible by humans. Notice that the polysaccharides differ in the type of monosaccharides present and in the form of the linkage between the monosaccharides.

Polysaccharides are large molecules formed by polymerization of monosaccharides (Figure 2.21). Glycogen is a polymer of glucose; it can be hydrolyzed to provide fuel when needed. Starch is used for energy storage in plants and is a source of fuel in humans when it is broken down to its monomers by the digestive system. Plants also use polysaccharides such as cellulose to provide structural support for cell walls (Figure 2.22).

2.6.3 Nucleic acids

Nucleic acids are polymers of nucleotides that have a special function and significance in biology. As Chapter 3 will describe in more detail, nucleic acids provide a mechanism for storage of **genetic information** within cells. In

Figure 2.22

Cellulose in the diet. Cellulose is a major component of edible plants but cannot be digested by humans. Although cellulose and starch are both polymers of glucose, the linkage between the glucose monomers differs. Humans lack enzymes that are necessary for speeding up the reaction that breaks down the polymer cellulose into glucose, whereas humans have abundant enzymes for breakdown of starch. Cellulose adds bulk to the diet, but cannot be absorbed. Cellulose is a major component of the insoluble part of dietary fiber.

addition, the unique chemical properties of nucleic acids provide for a mechanism of transmission of genetic information from parents to offspring.

NUCLEOTIDES

The monomer for nucleic acid polymers is called a **nucleotide**. The complete chemical structure of nucleotides is described in Chapter 3; as an introduction, structures of the monomers and polymers are illustrated schematically in Figures 2.23 and 2.24. Nucleotides are composed of three different segments—a five-carbon sugar molecule (or **pentose**), a phosphate group, and an organic unit—that are covalently coupled together into a single molecule. Within one nucleic acid, each nucleotide unit contains identical pentose and phosphate regions, but the organic unit can differ. The most important nucleic acids are DNA and RNA. All DNA molecules use deoxyribose as the pentose; all RNA molecules use ribose as the pentose (the structures of these pentoses are illustrated in Figure 2.20).

The "organic unit" of the nucleotide deserves special attention. It is this part of the molecule that provides chemical distinction to the nucleotide. The organic units are members of a larger family of molecules called **nitrogenous bases**; they behave as bases because of the presence of a nitrogen atom within the ring structure of the unit.

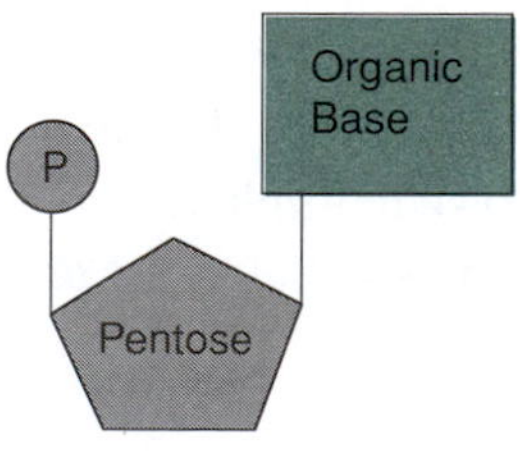

Figure 2.23

Nucleotides. Nucleotides are composed of a pentose, a phosphate group (P), and an organic unit, which has basic properties. The type of pentose differs in deoxyribonucleic acid (DNA) and ribonucleic acid (RNA). DNA contains deoxyribose sugars whereas RNA contains ribose sugars.

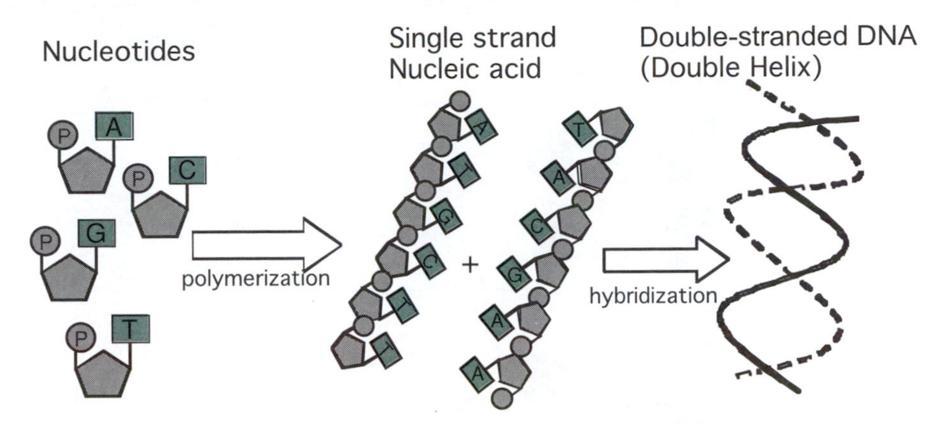

Figure 2.24 **Nucleic acids.** Nucleic acids are polymers of nucleotides, which are composed of a pentose, a phosphate group, and an organic unit, which is either a purine or a pyrimidine. The type of pentose differs in deoxyribonucleic acid (DNA) and ribonucleic acid (RNA). DNA contains deoxyribose sugars whereas RNA contains ribose sugars. A single nucleic acid chain can combine with another complementary chain to form a double helix.

Ammonia is not one of the units used in nucleotides; it is a simple nitrogenous base and can be used to illustrate the behavior of this class of molecules as bases:

$$NH_3 + H_2O \rightarrow NH_4^+ + OH^- \tag{2.11}$$

There are five nitrogenous bases important in nucleic acids: **adenine**, **guanine**, **cytosine**, **thymine**, and **uracil**. The chemical structures of these molecules will be revealed in Chapter 3; for the present discussion, these bases will be identified simply as A, G, C, T, and U. These organic bases are important because their ordering on the nucleic acid polymer is the basis for information storage.

NUCLEIC ACIDS

Nucleic acids are polymers of nucleotides, in which the phosphate region of one nucleotide is covalently coupled to the pentose of the adjacent nucleotide. Polymerization to produce the nucleic acid polymer always occurs in an orderly fashion, with the phosphate of each new nucleotide attaching to the pentose of the previous nucleotide. The resulting nucleic acid is a linear polymer (Figure 2.24 and Figure 2.25). (More details on the synthesis of nucleic acids are in Chapter 3, Section 3.3.) Although the polymer chains are linear, they are not perfectly straight: The bonding between the phosphate and the pentose creates a light twist in the chain that gives it a helical shape.

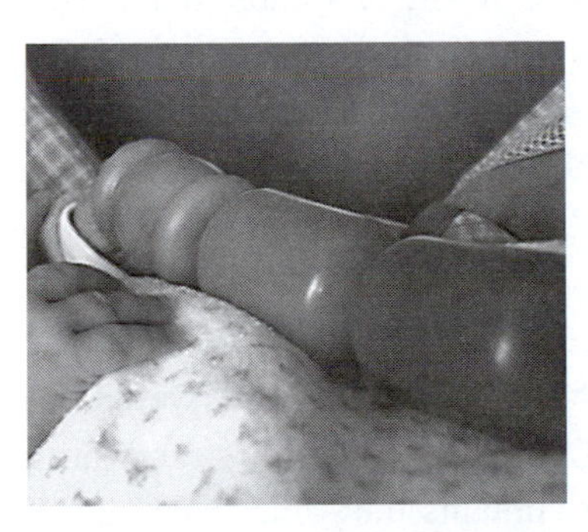

Figure 2.25 **Linear polymers.** Linear polymers have no cross-links or branching. Most often, they are produced by polymerization of monomers, in which each monomer has two reactive groups, one on each end, so that the monomers add sequentially to the growing polymer chain. The growth of the polymer—one block at a time—is similar to the infant toy in which colored beads are connected to form a chain.

NUCLEIC ACID HYBRIDIZATION

Nucleic acids can exist as single-stranded linear polymer chains or they can assemble

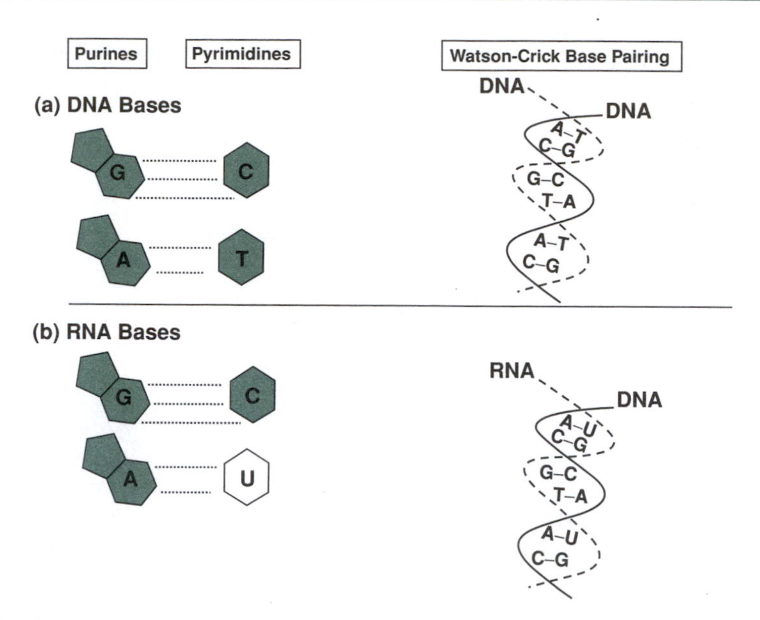

Figure 2.26 **Nucleotide bases.** Of the five organic bases that occur in deoxyribonucleic acid (DNA) and ribonucleic acid (RNA), two are double-ring structures (called purines) and three are single-ring structures (called pyrimidines). Each base pair is composed of a purine and a pyrimidine capable of forming hydrogen bonds with each other. Note that guanine and cytosine can form three hydrogen bonds, whereas adenine and thymine or uracil can only form two hydrogen bonds indicated by the dotted lines. This hydrogen bonding capability of the bases determines the base-pairing rules: G–C, A–T, A–U and vice versa.

further to create double-stranded molecules, such as the famous double-stranded double helix of DNA. The double helix structure consists of two intertwining nucleic acid polymers that assemble, through noncovalent interactions, into a double-stranded structure (Figure 2.24). The process of two single strands of DNA assembling into double-stranded DNA is called **hybridization**.

Not every pair of DNA strands can form a double helix: Hybridization can only occur if the two strands have **complementary** sequences. What does it mean to have a complementary sequence? Remember that the information on a nucleic acid polymer is contained in the order of bases on each linear strand. For example, the sequence of nucleotides within a region of a DNA single strand might be ATGCTT (Figure 2.24). Notice that the DNA polymer has a direction: The nucleotides, which are connected to each other from one phosphate group to the next pentose, are not symmetrical, so the molecule would be different if the order was turned around (TTCGTA). Complementary strands are often described as "mirror images": each strand contains the same information (although the strands are not identical), but they are mirror images prepared in a special way. First, the complementary strand is pointed in the opposite direction: If one strand is arranged phosphate to pentose, phosphate to pentose facing upward, then the complementary strand is arranged pentose to phosphate, pentose to phosphate. Second, the ordering of the nucleotides on the complementary strand is perfectly predictable. The bases on the complementary strands match a particular pattern: A goes with T, C goes with G, G goes with C, and T goes with A (Figure 2.26).

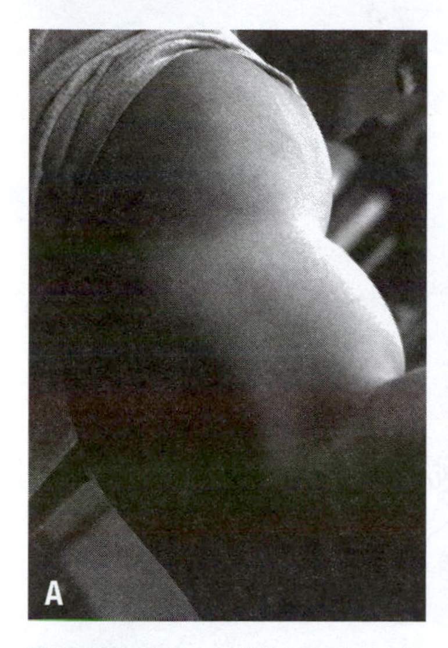

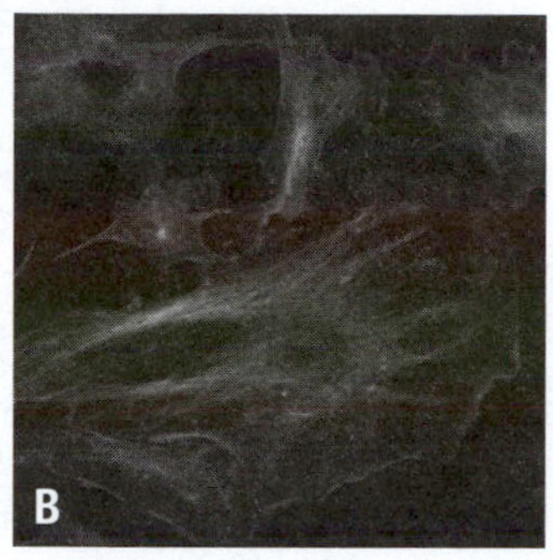

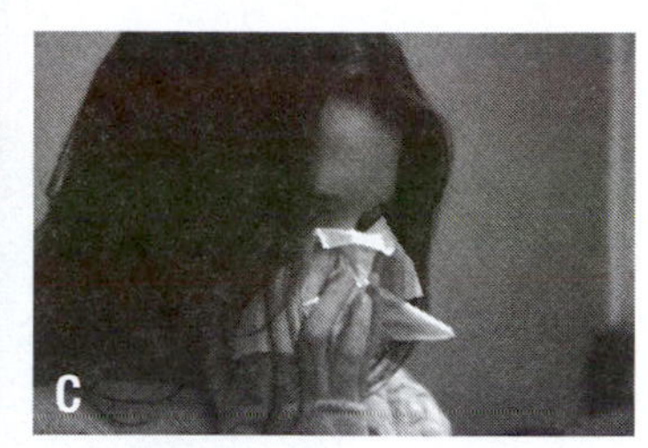

Figure 2.27 **Proteins are the working molecules of cells and tissues.** **A.** Protein assemblies do the work of muscle contraction. **B.** Proteins within cells form a cytoskeleton, which determines cell shape. **C.** Proteins, such as antibodies, protect people from infectious diseases.

As Chapter 3 explains, these matches—which are often called base pairings—are determined by hydrogen bonding interactions between the nucleotides. It is the hydrogen bonding of complementary base-pair matching that holds the two DNA strands together in a stable double helix.

2.6.4 Proteins

Proteins are the working molecules of cells, performing thousands of functions that are essential for the life of the cell and the organism. Proteins do the work of muscle contraction, speed up the chemical reactions required to break down food to usable forms, convert light into electrical signals that our brains can use for vision, and protect us from infectious diseases by neutralizing infectious agents (Figure 2.27). All of these functions of proteins are revealed in later chapters. For now, the basic structure of proteins is sketched at the first level of detail.

Proteins are produced by chemical reactions that are directed by DNA. One of the main functions of the DNA in our cells is to provide the information blueprint for synthesis of the proteins that our cells will need. The chemical reactions that lead to protein synthesis are divided into two main groups, which are called **transcription** and **translation**. During transcription, the sequence of nucleotides that contains the information important for making a protein is transcribed from its location on DNA within the cell into another nucleic acid polymer, called **messenger RNA** (mRNA). During translation, the information on the mRNA is translated, through a series of chemical reactions, into a linear sequence of amino acids that will become a protein. This sequence of events is often called the central dogma of molecular biology (Figure 2.28).

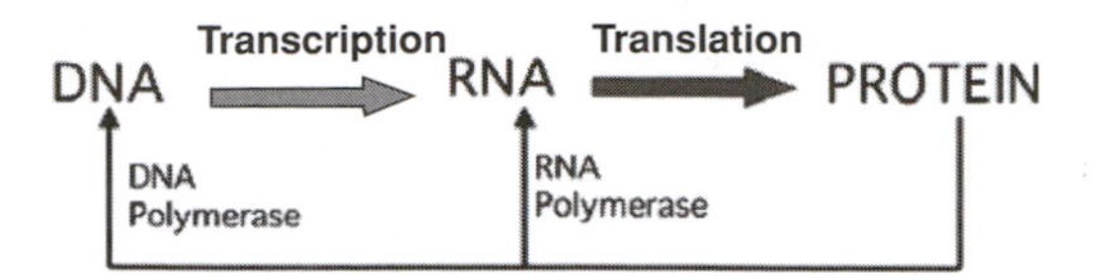

Figure 2.28 **Central dogma of molecular biology.** The processes of translation and transcription that produce messenger ribonucleic acid (mRNA) and proteins require coordinated chemical reactions, which are catalyzed by enzymes such as deoxyribonucleic acid (DNA) polymerase and RNA polymerase.

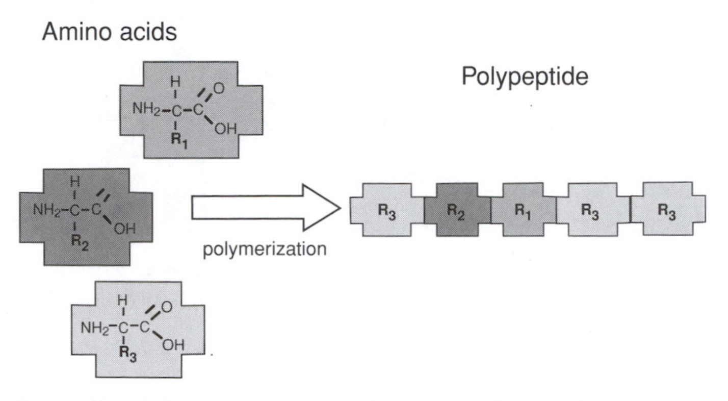

Figure 2.29 **Polypeptides.** Polypeptides are polymers of amino acids. There are 20 amino acids with varying side groups (R). The properties of the side chains determine the three-dimensional structure of the polypeptide. Proteins are made up of one or more polypeptide chains. Many proteins are enzymes catalyzing chemical reactions; others provide structural support inside and outside of the cell.

Proteins are made up of one or more **polypeptide** chains. Polypeptides are polymers of amino acids, which have the general chemical structure shown in Figure 2.29.

Each amino acid has three functional groups, an amine (NH_2), a carboxylic acid (COOH), and a side group, R. The amine and carboxylic acid groups are identical on every amino acid, but the side group is distinct for each different amino acid and confers unique chemical properties to each. There are 20 amino acids, which can be grouped into categories: nonpolar, polar, acidic, and basic (Table 2.4). Many of the side groups behave like weak acids or bases; therefore, their state (i.e., whether they have released or absorbed a proton) depends on the pH.

All double-stranded DNA molecules have the same three-dimensional structure.[2] Proteins, in contrast, exhibit a tremendous diversity of three-dimensional structures. It is well known that the three-dimensional structure of a protein is both 1) critical to its function and 2) dependent on the linear sequence of amino acids that comprises the protein. It is not yet possible to predict the three-dimensional structure of a protein from its sequence of amino acids, although interactions between amino acids and water in the aqueous environment contribute to the structure. The kinds of interaction forces that were discussed earlier appear to

[2] Chapter 3 will show us that long strand of DNA double helix can be arranged into compact units for packing into the cell nucleus.

Table 2.4 Amino acid groups

Group	Example	Chemical structure
Nonpolar	Alanine	H_3C–CH(NH_2)–COOH
Polar	Serine	HO–CH_2–CH(NH_2)–COOH
Acidic	Aspartic acid	HOOC–CH_2–CH(NH_2)–COOH
Basic	Lysine	H_2N–$(CH_2)_4$–CH(NH_2)–COOH

Notes. The 20 amino acids can be grouped into one of four categories: nonpolar, polar, acidic, and basic. The R groups are highlighted in the shaded boxes. Alanine is an example of a nonpolar amino acid, whereas serine is a polar amino acid. Aspartic acid has an acidic side group, and lysine has a basic one. The protonation state of these groups is dependent on pH and the p*Ka* values of the various amino acids. At pH = 7.4, almost all acidic side groups are deprotonated (COO^- for aspartic acid) and basic side groups are protonated (NH_3^+ for lysine) so the charged form of the amino acids is predominant. Refer to Figure 4.8 for a complete list of all amino acids.

be important, including hydrophobic/hydrophilic interactions, hydrogen bonding, and electrostatic interactions (recall Figure 2.8).

Knowing that proteins are the working molecules of the body, biomedical engineers have converted them into devices that perform useful work to improve health. One approach is to purify proteins and immobilize them to synthetic materials. This procedure permits the action of proteins to be controlled at a certain location. One popular example of this approach is home pregnancy test kits, in which proteins that are immobilized on test strips are used to detect a change in the properties of urine that occur only in pregnancy (Figure 2.30).

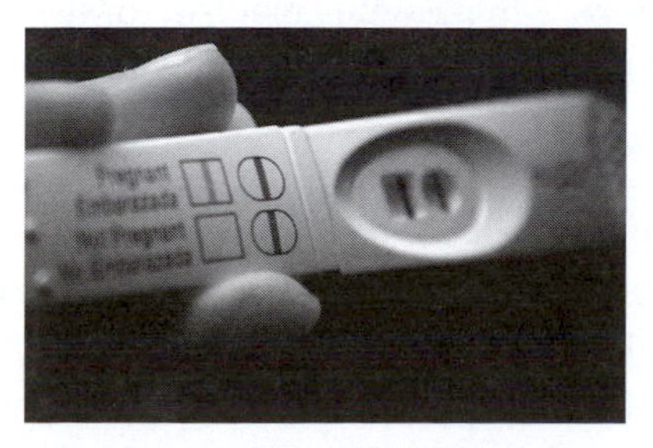

Figure 2.30 **Home pregnancy test.** Home pregnancy tests are based on antibodies that produce color changes in the presence of hormones related to pregnancy.

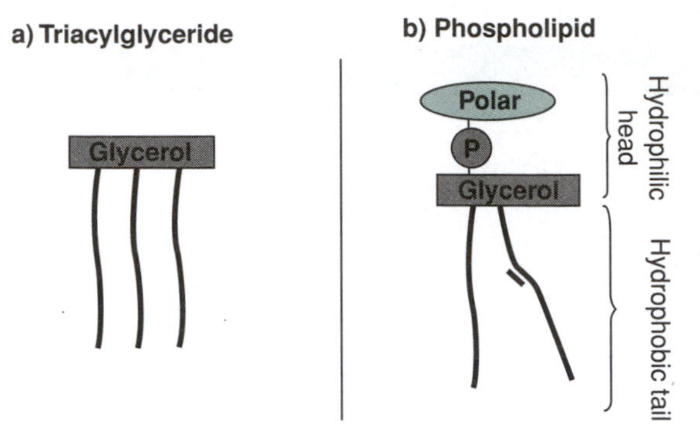

Figure 2.31 **Lipid structure.** There are two major classes of lipids. **A.** Triglycerides (or triacylglycerides) are composed of three fatty acid molecules (thick wavy lines) connected to a glycerol. The body uses triglycerides for energy storage in the form of fat. **B.** Phospholipids are amphiphilic molecules that have two fatty acid chains (thick lines), which make up the hydrophobic tail and a hydrophilic head consisting of a glycerol molecule, a phosphate, and a polar molecule. Double bonds within the fatty acid chains (i.e., unsaturation of the chain) cause the fatty acid chains to bend slightly.

2.7 Lipids

In contrast to the biological macromolecules described so far, lipids are not polymers, but they are fairly large molecules built from a combination of other simpler units. Lipids do not dissolve in water and include two major classes: triglycerides and phospholipids. Steroids are also classified as lipids and are involved in cell communication (steroids are discussed in Chapter 6, Section 6.4.3).

2.7.1 Chemical structure of lipids

Triglycerides are the molecules that make what is commonly known as fat. Some fat is useful: It is a major form of energy storage in the body. As shown in Figure 2.31, triglycerides are composed of three fatty acid chains and a glycerol molecule. Fatty acids are long hydrocarbon chains of varying lengths with a carboxylic acid functional group, $CH_3(CH_2)_nCOOH$; they can be saturated or unsaturated. Unsaturated fatty acids have double bonds in their hydrocarbon chains, resulting in a slight kink or bend in their structure. Triglyceride molecules added to water will aggregate into globs that separate from the water. The molecules are hydrophobic, so they try to get away from water by grouping together, just as oil separates from water.

Phospholipids are composed of two fatty acid chains, a glycerol molecule, a phosphate, and a polar group attached via the phosphate (Figure 2.31). Thus, phospholipids are amphiphilic molecules with a polar head group and a nonpolar tail. These features enable lipids to self-assemble in aqueous environments into complex structures, such as a lipid bilayer, in which the hydrophobic tails are facing each other in the middle of the bilayer and the polar head groups are exposed to the aqueous environment on either side of the bilayer (Figure 2.32).

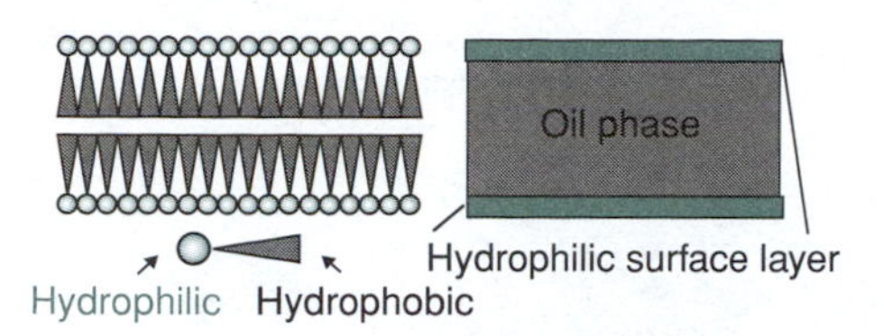

Figure 2.32

Cell membrane structure. The plasma membrane is described as a fluid mosaic composed of lipids, carbohydrates, and proteins. **A.** A lipid bilayer, in which amphiphilic phospholipids are arranged with their hydrophobic tails pointing toward the interior of the membrane and their hydrophilic heads exposed to the adjacent water phases, serves as the main element of the membrane. **B.** The lipid bilayer behaves as a thin oil phase, which does not permit the penetration of molecules that do not dissolve in oil, such as charged molecules and hydrophilic molecules.

This structure forms naturally in an aqueous environment because the polar heads tend to associate with water molecules in the solvent, whereas hydrophobic nonpolar tails tend to bury themselves in the interior of the structure.

Biomedical engineers have also used lipids to construct devices (Figure 2.1). Liposomes are spherical bilayers that can be used as attractive drug carriers because they are natural (because they are made from naturally occurring phospholipids), protect drugs from degradation (because they can encapsulate the drug and protect it from degradation by enzymes), and can potentially be targeted to a particular tissue (because they can be chemically modified on their outer surface). Many anticancer drugs are hydrophobic, but must be injected to be useful. Injection of these compounds is difficult, because they do not dissolve easily in water, but sometimes these compounds can be entrapped in liposomes, which can then be suspended in water (because they are hydrophilic on the exterior surface) and injected intravenously.

2.7.2 Structure of the cell membrane

Cells have a major problem. The cell interior is largely water, with high concentrations of proteins and organelles—formed substructures in cells—swimming in a water-rich fluid called **cytosol**. Cells also exist in an extracellular environment that is also water-rich. How can a cell keep its contents in place, when it is floating in a sea of water?

Lipids are a major component of the **cell membrane**, which separates the cytosol from the extracellular environment. The main structure of the cell membrane is a lipid bilayer formed of various phospholipids (Figure 2.32). The thickness of the lipid bilayer is approximately 3 nm. In addition to phospholipids, the cell membrane contains proteins that provide the membrane with intelligence, that is, the abilities to interact with molecules on the outside of the cell and to move molecules (that would otherwise not dissolve and be excluded) across the membrane (Figure 2.33). These accessory functions are described in later chapters. The external surface of the plasma membrane also carries a carbohydrate-rich coat called the **glycocalyx**, which is made up of the exposed carbohydrate groups of glycolipids and transmembrane glycoproteins. Charged groups in the glycocalyx cause the surface to be negatively charged. On average, the plasma

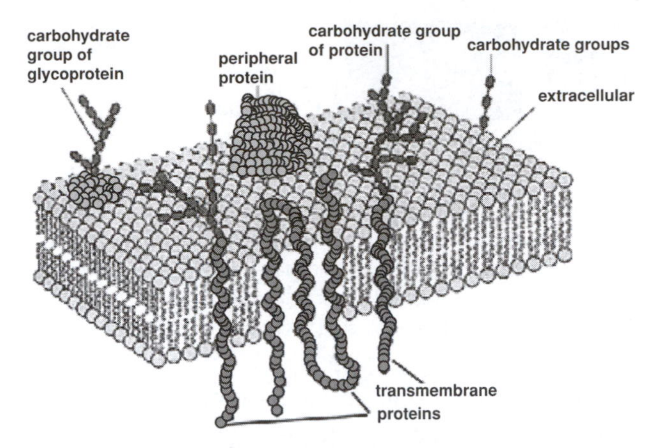

Figure 2.33

Proteins and polysaccharides decorate the plasma membrane. Some of the lipids and proteins are free to move laterally within the two-dimensional plane of the bilayer. Types of proteins include transmembrane proteins, which function as receptors for cell signaling and channels, which help transport molecules across the membrane. Carbohydrate-modified proteins (glycoproteins) and lipids (glycolipids) are also found on the outside of the membrane. These carbohydrate groups make up the negatively charged glycocalyx.

membrane of human cells contains 50% protein, 45% lipid, and 5% carbohydrate. The membrane is described as a "fluid mosaic," which means that proteins and lipids easily move laterally within the membrane, as if they are floating within a two-dimensional membrane sea.

2.7.3 Transport across the cell membrane

The plasma membrane separates the cell cytosol from the extracellular environment. However, molecules such as nutrients and waste products must be able to cross this barrier. How are these molecules transported across the membrane? There are three major modes of transport: passive diffusion, facilitated diffusion, and active transport (Figure 2.33). **Diffusion** is the spontaneous movement of particles from an area of high concentration to an area of low concentration (Box 2.5). **Passive diffusion** is the process by which water and small uncharged molecules such as oxygen (O_2) and carbon dioxide (CO_2) pass through the plasma membrane.

For the special case of water, the diffusion of water down its concentration gradient is called **osmosis**. Thus, in osmosis, water flows from an area of high water concentration to an area of low water concentration. Box 2.6 reviews important concepts regarding osmosis and **osmotic pressure**.

The lipid bilayer is a thin, oil-like barrier (Figure 2.34). Molecules that are charged or polar cannot dissolve in the lipid bilayer and, therefore, will not diffuse across it. Thus, the cell membrane is not permeable to ions, glucose, or most macromolecules, but the cell often needs these kinds of molecules. To allow for transport into the cell, there are specialized proteins embedded in the plasma membrane that function as channels to facilitate movement of the molecules from one side of the membrane to the other. As described in later chapters,

Box 2.5 Diffusion

Molecules are in a constant state of motion. Einstein showed that the average speed of molecules depends on temperature; for example, albumin, the most abundant protein in blood, is a large molecule, but its average velocity because of thermal energy is 600 cm/s. Even though they move with high speed, molecules like albumin do not move between different locations in our body very quickly because they are constantly colliding with other molecules and changing direction. Instead of motion in a straight path, molecules in liquids drift from one location to another, moving at high speed but constantly colliding and changing direction in a statistical process called a random walk. If we did not have a circulatory system, which moves molecules in a directed flow, it would take a very long time for a molecule to move from your brain to your toes (many years, in fact), and its time of arrival would depend as much on chance as it does on molecular speed. Even though this rapid, temperature-induced molecular motion is undirected, molecules that are initially placed in one location tend to spread out over time. This process of spreading is called **diffusion**, and one of the important characteristics of diffusion is that molecules tend to spread from regions of high to low concentration.

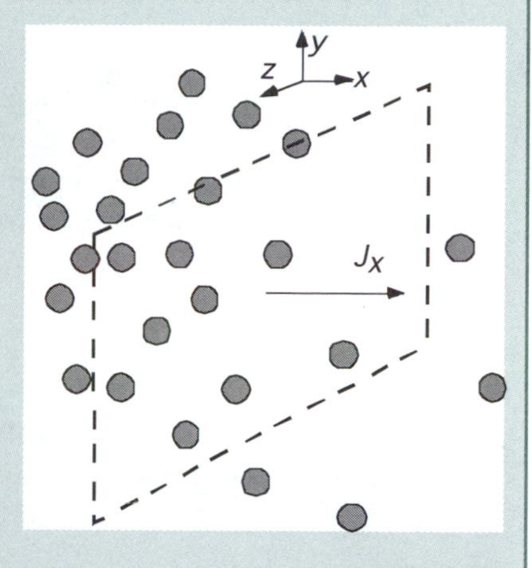

Diffusion of neutral molecules can be described using Fick's First Law of Diffusion (Equation 1), which states that the rate of diffusion or **flux**, J, is proportional to the **concentration gradient**, which is determined mathematically by the first derivative of the concentration, c. Flux is a vector quantity; that is, it has both a direction and a magnitude. As illustrated in the upper diagram at the right, the flux of molecules in the x-direction, J_x, is equal to the number of molecules (or mass or moles) that pass through an imaginary plane, which is perpendicular to the x-axis, per time per surface area. This flux is given by Fick's Law of Diffusion:

$$J_x = -D\frac{dc}{dx} \qquad \text{(Equation 1)}$$

The proportionality constant in this case is the diffusion coefficient, D, which has units of cm^2/s. For diffusion in water, D is $\sim 10^{-5}$ cm^2/s for small molecules (such as oxygen) and $\sim 10^{-7}$ cm^2/s for large molecules (such as proteins and DNA).

Consider the biological system illustrated in the lower diagram at the right: A membrane of thickness, Δx, is maintained under steady-state conditions, in which the concentration on one side of the membrane, C_1, is higher than the concentration on the other side of the membrane, C_2. **Steady state** means that, although there is a movement of molecules from one side of the membrane to the other, the concentration at any location in the system does not change with time. Think for yourself: How can this happen? Do steady-state conditions apply for biological systems?

(continued)

Box 2.5 *(continued)*

Assuming steady state, and the conditions illustrated in the diagram, show that the following statements are true:

1. The flux, J_x, is a constant value at all positions in the membrane (that is, for all x between $x = 0$ and $x = \Delta x$).
2. The concentration varies linearly between $x = 0$ and $x = \Delta x$ (that is, c varies according to: $c = Ax + B$, where A and B are constants).
3. The actual concentration variation, or concentration profile is:

$$c = (C_2 - C_1)\frac{x}{\Delta x} + C_1$$

4. The constant flux through the membrane is equal to: $J_x = D(C_1 - C_2)/\Delta x$.
5. Therefore, in this situation, one can estimate the diffusion flux as:

$$J_x = -D\frac{dc}{dx} = \text{constant} \qquad \text{(Equation 2)}$$

As a review, write down all of the assumptions that we made in going from Equation 1 to Equation 2. As you will see in later chapters (such as Chapters 8 and 15), an understanding of diffusion and the equations that describe diffusion allows biomedical engineers to model transport, transport of gases across alveoli, nutrients across capillaries, and ions across membranes.

these channels can facilitate diffusion, regulate transport, and provide selective transport. In some cell types, water channels called **aquaporins** facilitate rapid transport of water. Both passive diffusion and facilitated diffusion through channels do not require energy as long as molecules are moving down their concentration gradient from high to low concentration (Box 2.5). Sometimes an ion or molecule needs to be pumped *against* its concentration gradient. **Active transport** is the process of moving a molecule from an area of low concentration on one

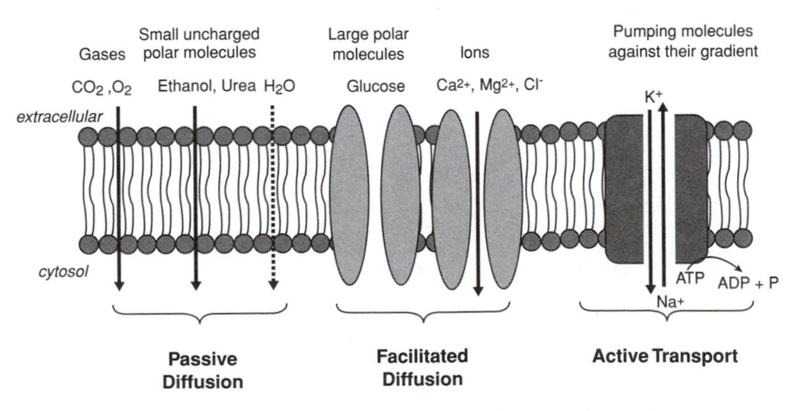

Figure 2.34 **Modes of transport across the cell membrane.** Gases, small uncharged molecules, and water can diffuse across the plasma membrane passively (passive diffusion). Large polar molecules and ions require channels to help them diffuse through the hydrophobic lipid bilayer (facilitated diffusion). The membrane is slightly permeable to water so, for bulk water movement, specialized water channels called aquaporins are used. Active transport is used to move molecules against their concentration gradient and requires a transporter and the input of energy from hydrolysis of adenosine-5′-triphosphate (ATP), which yields adenosine diphosphate (ADP) and inorganic phosphate.

Box 2.6 Osmosis and osmotic pressure

Bulk movement of water in and out of the cell—and across other tissue barriers—is important in all organs but particularly in the small intestine and the kidneys. **Osmosis** is the movement of water across a semipermeable membrane from an area of high water concentration to an area of low water concentration. For osmosis to occur, the membrane must be permeable to water but not **solute** (molecules or ions). Therefore, in osmosis, the water moves from an area of low *solute* concentration to an area of high *solute* concentration. Water is migrating to try to equalize the solute concentration gradient; water must move because the solute (which is not able to diffuse through the membrane) cannot.

Consider a U-tube show below. Compartment B is filled with a solution containing a solute, such as sugar. Compartment A is filled with pure water. The two compartments are separated by a semipermeable membrane, which allows water but not sugar molecules to pass through. At time zero, the volume of the two compartments is equal. As time passes, water will flow from A to B because of osmosis; this flow can be measured by noticing the increase in volume in compartment B compared to compartment A.

The van't Hoff equation provides the **osmotic pressure**, Π, of an ideal solution:

$$\Pi = RT\Delta C \qquad \text{(Equation 1)}$$

$$\Pi = RT(C_A - C_B), \qquad \text{(Equation 2)}$$

where R is the ideal gas constant, T is the absolute temperature (K), and C is the concentration of the solute in the solution (mol/L). Why is this called osmotic *pressure*? The process of osmosis, that is, the movement of water from A to B, can be stopped by applying pressure to the solution in compartment B. Application of a pressure exactly equal to the osmotic pressure will stop osmosis. Alternately, if no pressure is applied, osmosis will occur and the fluid level in B will rise above the level in A; this increases the pressure in B and the additional pressure on compartment B is equal to the difference in fluid height. Osmosis in this system will stop when the hydrostatic pressure caused by the difference in the fluid levels is equal to the osmotic pressure. In experiments for measuring osmotic pressure for a solution in compartment B, the solution in compartment A is pure water, so C_A would be zero. For a cell, the solute may be present both outside and inside

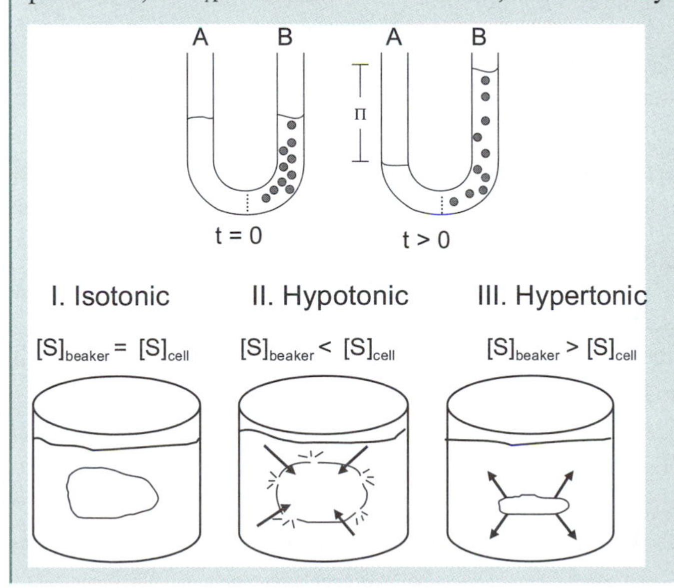

(continued)

Box 2.6 *(continued)*

the cell. Thus, osmotic pressure is equal to the hydrostatic pressure that must be applied to B to prevent net water flow across the membrane.

To illustrate the importance of osmosis in the life of a cell, we can observe what happens to a cell when it is placed in solutions with differing solute concentrations. An isotonic solution (I) contains the same concentration of solute, [S], as the cell, so no net movement of water occurs. If the cell is placed in a hypotonic solution (II), the concentration of solutes is higher in the cell than in the beaker; as a result, water will diffuse into the cell potentially causing it to swell and burst. In contrast, a cell placed in a hypertonic solution (III) will shrink because of the movement of water from inside the cell to outside.

side of the membrane to an area of high concentration on the other. Active transport requires energy input supplied by ATP through a specialized transmembrane active transport protein. An example of an active transporter is the sodium potassium pump or Na^+/K^+ ATPase, which transports three Na^+ ions out of the cell for every two K^+ ions pumped into the cell. This pump allows the cell to maintain a high extracellular $[Na^+]$ outside the cell and a high intracellular $[K^+]$ in the cytosol. The difference in ion concentrations is important for maintaining cell volume by osmotic pressure (Box 2.6), but also for generating a voltage gradient across the membrane, in which the inside is more negative than the outside. More information about the physiological importance of the Na^+ K^+ pump is provided in Chapter 6 (Section 6.3.1).

Summary

This chapter reviewed biochemical concepts that are important in understanding the interaction between molecules, between molecules and their solvents, and between molecules and the cell membrane.

- Atoms can form ionic or covalent bonds with one another.
- Hydrogen bonds and van der Waals interactions are weak bonds between molecules.
- Most chemical reactions in the body take place in an aqueous environment. The type of molecule—polar, nonpolar, acidic, basic—affects how it behaves in an aqueous environment.
- Hydrogen bonding is important in water chemistry as well as in the assembly of macromolecules, such as nucleic acids and proteins.
- Buffer systems in the blood help to maintain near-neutral pH, which is critical to the function of many enzymes.
- The body maintains homeostasis through negative feedback mechanisms, such as the bicarbonate buffering system, that detect a change and act to reduce the magnitude of the change.
- Biomolecules contain various functional groups that confer different properties.

- Carbohydrates, nucleic acids, and proteins are polymers of small subunits.
- The cell is isolated from its extracellular environment by the phospholipid bilayer membrane. Phospholipids are amphiphilic molecules that make up the bilayer along with proteins and glycolipids.
- Diffusion is the movement of a solute from an area of high concentration to low concentration.
- Molecules can cross the membrane via passive or facilitated diffusion depending on their permeability.
- Active transport allows molecules to be transported against their concentration gradient by requiring energy input in the form of ATP.
- Diffusion of water through a semipermeable membrane is called osmosis.

REFERENCE

1. Luo D, Haverstick K, Belcheva N, Han E, Saltzman WM. Poly(ethylene glycol)-conjugated PAMAM dendrimer for biocompatible high-efficiency DNA delivery. *Macromolecules*. 2002;35:3456–3462.

FURTHER READING

GENERAL CHEMISTRY

Chang R. *General Chemistry*, Third Edition. New York: McGraw-Hill; 2003.

Ebbing DD, Wrighton MS. *General Chemistry*. Boston: Houghton Mifflin Company; 1990.

Jones L, Atkins P. *Chemical Principles: The Quest for Insight*. New York: W. H. Freeman; 2004.

MOLECULAR BIOLOGY

Alberts B, Bray D, Lewis J, Raff M, Roberts K, Watson JD. *Molecular Biology of the Cell*. 3rd ed. New York and London: Garland Publishing; 1994.

Lodish H, Berk A, Zipursky SL, Matsudaira PL, Baltimore D, Darnell JE. *Molecular Cell Biology*. New York: W. H. Freeman; 2000.

BIOCHEMISTRY

Berg JM, Tymoczko JL, Stryer L. *Biochemistry*. New York: W. H. Freeman; 2002.

Matthews CK, van Holde KE. *Biochemistry*. New York: Benjamin(Cummings Publishing Co.; 1990.

Voet D, Voet JG. *Biochemistry*. Hoboken, NJ: John Wiley and Sons; 2004.

PHYSIOLOGY

Guyton AC. *Textbook of Medical Physiology*. Philadelphia: WB Saunders; 1996.

DIFFUSION

Berg HC. *Random Walks in Biology*. Princeton, NJ: Princeton University Press. 1993.

Profiles in BME: Veronique V. Tran

I immigrated to the United States with my family in 1975 after the Fall of Saigon at the end of the Vietnam War. We eventually settled in the Houston area. My interest in math and biology started in high school. I fulfilled pre-med requirements in biology and chemistry while majoring in Chemical Engineering at the University of Houston. Dr. Richard Willson was the first person to expose me to the possibility of combining engineering and biology through a course in the fundamentals of biochemical engineering. We kept in touch through the years and he remains a close friend and mentor. After graduation in 1991, I worked at Shell Oil Company in the production division as a facilities engineer. At first, I found my job challenging and exciting—I was applying what I learned in college by designing large-scale production facilities. Even though this on-the-job training was a great learning environment, as the years passed, I felt that the projects were routine and lacked meaning—focused on improving the bottom line. I began debating whether to quit my job and go back to school. But it wasn't until January 1995, when my uncle passed away due to complications during open-heart bypass surgery that I finally acted on my desire for a career change. My uncle's surgeon mentioned that one possibility for why my uncle developed adult respiratory distress syndrome was that his blood reacted unfavorably to the surfaces lining the heart–lung machine. That was when I realized that there are many problems in modern medicine yet to be solved, and that biomedical engineers would be on the front lines working with basic scientists and physicians to develop and test new tools and therapies. I wanted to be a part of that team.

Later that year, I enrolled in the Joint Program in Biomedical Engineering at the University of Texas Southwestern (UTSW) Medical Center at Dallas and the University of Texas at Arlington. I completed graduate coursework in both the Biomedical Engineering Program and the Biochemistry and Molecular Biology Program. Compared to more than four years earlier when I was an undergraduate student, the pace of scientific discovery and understanding was astounding. Through laboratory rotations, I saw firsthand how biomedical engineers contributed to the areas of drug discovery, bioinformatics, and artificial organs. An internship at Johnson & Johnson further reinforced that I made the right decision to pursue BME. Dr. Christopher B. Newgard mentored my doctoral research in his Diabetes Research Laboratory. My research focused on engineering beta cells for resistance to small molecule mediators of the immune system. The lessons learned from these and other studies can be applied to engineer surrogate beta cells, which someday could be transplanted to serve as a bioartificial pancreas for treatment of Juvenile Diabetes. Dr. Robert C. Eberhart was one of my early mentors in BME and encouraged me to pursue a career in academia.

After receiving my PhD degree, I was interested in learning more about the area of drug delivery and tissue engineering so I joined Dr. W. Mark Saltzman's lab at Yale University for my postdoctoral training. I continued to do research but more importantly, inspired by Dr. Saltzman, I discovered a love for teaching and mentoring students. There are many problems yet to be solved, especially in the area of "translational research" or "bench-to-bedside." I find this the most exciting and challenging part of being a biomedical engineer—scientists, engineers, and physicians working together to solve problems in the laboratory and translating those discoveries to practical applications for diagnosis and treatment of disease. As biomedical engineers, we have an opportunity to apply our knowledge and skills to develop new tools and therapies that can alleviate disease and improve the quality of life of others. I think that I finally found the "meaning" I was looking for in my second career in academia. I was a tenure-track Assistant Professor of Biomedical Engineering at the University of Houston for two years—helping to build the undergraduate program. From that experience, I realized that my true passion lies in helping others and I have been working for the past years in administration, first at Rice University and now at University of Houston. I look forward to many years of service in my role as administrator and mentor.

With my husband, Tim Vu, I also look forward to continuing the adventures of parenting our two young sons, Ethaniel Hiep and Lucas Liem.

USEFUL LINKS ON THE WORLD WIDE WEB

http://www.ncbi.nlm.nih.gov/entrez/query.fcgi?db=Books
Searchable Electronic Textbooks, National Center for Biotechnology Information (NCBI)

http://www.biology.arizona.edu/
The Biology Project—University of Arizona; online interactive resource for learning Biology and Biochemistry (including acid-base chemistry)

http://www.mhhe.com/physsci/chemistry/essentialchemistry/flash/acid13.swf

http://www.mhhe.com/physsci/chemistry/essentialchemistry/flash/buffer12.swf
Flash Animations for Acid Ionization and Buffers: Raymond Chang, Essential Chemistry, McGraw-Hill

http://arbl.cvmbs.colostate.edu/hbooks/cmb/cells/pmemb/hydrosim.html
Osmosis and Hydrostatic Pressure Simulator, Colorado State University

http://sun.science.wayne.edu/~bio340/Applets/
Membrane Potential Tutorial, Dr. Robert Stephenson, Wayne State University

http://www.physiologyeducation.org/
Various Freeware including Membrane Transport and Membrane Potential; Physiology Educational Research Consortium

http://cr.middlebury.edu/biology/labbook/diffusion/
An excellent primer on diffusion, osmosis, diffusion potentials, and the Nernst equation available from Dr. Joseph Patlak and Dr. Chris Watters of the University of Vermont and Middlebury College.

KEY CONCEPTS AND DEFINITIONS

acid a compound that can donate a proton (H^+). The carboxyl and phosphate groups are the primary acidic groups in biological molecules.

acidosis excess of acid in the body fluids, as may occur in kidney disease or diabetes

active transport the transport of molecules in an energetically unfavorable direction across a membrane coupled to the hydrolysis of ATP or other source of energy

adenine a compound that is one of four constituent bases of nucleic acid that is a purine derivative and hybridizes thymine in double-stranded DNA

adenosine 5′-triphosphate (ATP) a nucleotide that is the most important molecule for capturing and transferring free energy in cells. Hydrolysis of each of the two high-energy phosphoanhydride bonds in ATP is accompanied by a large *free-energy change* (ΔG) of 7 kcal/mol.

amino acid monomeric building block of proteins, consisting of a carbon atom bound to a carboxyl group, an amino group, a hydrogen atom, and a distinctive side chain

amphiphilic a molecule that has both hydrophobic and hydrophilic regions

anion negatively charged ion (e.g., Cl^-)

aquaporin a water channel protein that allows water molecules to cross the cell membrane much more rapidly than through the phospholipid bilayer

base a compound, usually containing nitrogen, that can accept a proton (H^+)

buffer a solution of the acid (HA) and base (A) form of a compound that undergoes little change in pH when small quantities of strong acid or strong base are added

carbohydrate a large group of organic compounds containing hydrogen and oxygen molecules in the ratio 2:1 that can be found in foods and living tissues including sugars, starch, and cellulose and can be broken down to release energy

cation positively charged ion (e.g., Na^+)

cell membrane a semipermeable membrane surrounding the cytoplasm of the cell that regulates transport of material from the external environment into the cell and vice versa

complementary sequence a sequence the nucleotide bases of which match or hybridize (mirror images) with the original sequence

concentration gradient difference in concentration between adjacent regions

condensation the opposite of a hydrolysis reaction: a reaction that gives off a by-product (usually water) during the reaction of two molecules to form one

conjugate acid species formed by the addition of a proton (H^+) to a base. CH_3COOH (aq) is the conjugate acid of CH_3COO^- (aq)

conjugate base species formed by the loss of a proton (H^+) of an acid. CH_3COO^- (aq) is the conjugate base of CH_3COOH (aq)

covalent bond stable chemical force that holds the atoms in molecules together by sharing of one or more pairs of electrons. Such a bond has strength of 50–200 kcal/mol.

cytosine a constituent of nucleic acid that is a pyrimidine derivative and pairs with guanine

diffusion net movement of molecules from high concentration to low concentration

disaccharide a sugar that is formed by coupling together two monosaccharides (or simple sugars)

dissociation constant the equilibrium constant for the decomposition of a complex ion into its components in solution. The smaller the value of K, the lesser the dissociation of the species in solution. This value varies with temperature, ionic strength, and the nature of the solvent.

electrochemical gradient a difference in chemical concentration and electric potential across a membrane

endothermic referring to a chemical reaction that absorbs heat. Such a reaction has a positive change in enthalpy.

enthalpy change (ΔH) heat; in a chemical reaction, the enthalpy of the reactants or products is equal to their total bond energies

entropy change (ΔS) a measure of the degree of disorder or randomness in a system; the higher the entropy, the greater the disorder

enzyme a protein that catalyzes a chemical reaction

equilibrium constant ratio of forward and reverse rate constants for a reaction. For a binding reaction, $A + B \rightarrow AB$, it equals the association constant, K_A; the higher the K_A, the tighter the binding between A and B. The reciprocal of the K_A is the dissociation constant, K_D; the higher the K_D, the weaker the binding between A and B.

exothermic referring to a chemical reaction that releases heat. Such a reaction has a negative change in enthalpy.

facilitated diffusion protein-aided transport of molecules across a membrane down its concentration gradient at a rate greater than that obtained by passive diffusion

flux the rate of transfer of fluid, particles, or energy across a given surface

free-energy change (ΔG) the difference in the free energy of the product molecules and of the reactants in a chemical reaction. A large negative value of ΔG indicates that a reaction has a strong tendency to occur; that is, at chemical equilibrium the concentration of products will be much greater than the concentration of reactants.

fructose a hexose sugar especially found in honey and fruits

functional groups specific and variable chemical groups that give an organic compound its characteristic properties

galactose a hexose sugar that is a constituent of lactose and many other polysaccharides

gene delivery the process of transfer of foreign DNA into a cell

genetic information useful hereditary information that is carried by nucleic acids present in the gene

Gibbs free energy (G) a measure of the potential energy of a system, which is a function of the enthalpy (H) and entropy (S)

glucose a simple sugar that is an important energy source. It has the formula $C_6H_{12}O_6$.

glycocalyx a carbohydrate coat covering the cell surface

glycolipid a lipid consisting of two hydrocarbon chains linked to a polar head group containing carbohydrates

glycoprotein a protein linked to oligosaccharides

guanine a constituent of nucleic acid that belongs to a class called purines and hybridizes with cytosine in double-stranded DNA

homeostasis the tendency of physiological systems to maintain a stable internal environment

hybridization the process of two single strands of DNA assembling into double-stranded DNA

hydrogen bond a *noncovalent bond* between an electronegative atom (commonly oxygen or nitrogen) and a hydrogen atom covalently bonded to another electronegative atom; particularly important in stabilizing the three-dimensional structure of proteins and formation of base pairs in nucleic acids

hydrolysis the opposite of a condensation reaction; a reaction in which a covalent bond is cleaved with addition of an H from water to one product of the cleavage and of an OH from water to the other

hydrophilic literally "water-loving"; compounds that have an affinity for water because of an ability to form hydrogen bonds

hydrophobic literally "water-fearing"; compounds that do not form hydrogen bonds and, therefore, do not dissolve easily in water

hydrophobic effect the property of nonpolar molecules to self-aggregate or cluster together to shield themselves from aqueous molecules

hypertonic referring to an external solution the solute concentration of which is high enough to cause water to move out of cells because of *osmosis*

hypotonic referring to an external solution the solute concentration of which is low enough to cause water to move into cells because of *osmosis*

ion an atom or group of atoms that carries a positive or negative electric charge as a result of having lost or gained one or more electrons, respectively

ionic bond a *noncovalent bond* between a positively charged ion (cation) and negatively charged ion (anion)

isotonic referring to a solution the solute concentration of which is such that it causes no net movement of water in or out of cells

lactose a disaccharide made of glucose and galactose units, usually found in milk

lipids small biological molecules that do not dissolve in water, including fatty acids and steroids

liposome a spherical *phospholipid bilayer* structure with an aqueous interior that forms in vitro from phospholipids and may contain proteins

macromolecule a large molecule formed by the repeated coupling of smaller units, called monomers

messenger RNA (mRNA) a form of RNA in which genetic information transcribed from RNA is transferred to the ribosome

molecular complex two or more molecules that are held together in an assembly by multiple weak noncovalent bonds, such as hydrogen bonds

monomer any small molecule that can be linked with others of the same type to form a *polymer*. Examples include amino acids, nucleotides, and monosaccharides.

monosaccharide any simple sugar with the formula $(CH_2O)n$ where $n = 3$ to 7

neutral neither acidic nor basic: pH ~ 7

nitrogenous base nitrogen-containing base that makes up the nucleotides that form nucleic acids

noncovalent bond any relatively weak chemical bond that does not involve an intimate sharing of electrons. Multiple noncovalent bonds often stabilize the conformation of macromolecules and mediate highly specific interactions between molecules.

nonpolar referring to a molecule or structure that lacks any net electric charge or asymmetric distribution of positive and negative charges. Nonpolar molecules generally are insoluble in water.

nucleic acid a polymer of *nucleotides* linked by phosphorous-containing bonds. DNA and RNA are the primary nucleic acids in cells.

nucleotide a *nucleoside* with one or more phosphate groups linked to the sugar moiety. DNA and RNA are polymers of nucleotides.

osmosis net movement of water across a semipermeable membrane from a solution of lesser to one of greater solute concentration. The membrane must be permeable to water but not to solute molecules.

osmotic pressure hydrostatic pressure that must be applied to the more concentrated solution to stop the net flow of water across a semipermeable membrane separating solutions of different concentrations

passive diffusion the process by which water and small uncharged molecules such as oxygen (O_2) and carbon dioxide (CO_2) pass through the plasma membrane

pentose a five-carbon sugar

peptide a small polymer usually containing fewer than 30 amino acids connected by peptide bonds

pH a measure of the acidity or alkalinity of a solution defined as the negative logarithm of the hydrogen ion concentration in moles per liter: pH $= \log[H^+]$. Neutrality is equivalent to a pH of 7; values below this are acidic and those above are alkaline.

phospholipid the principal components of cell membranes, consisting of two hydrocarbon chains (usually fatty acids) joined to a polar head group containing phosphate

polar referring to a molecule or structure with a net electric charge or asymmetric distribution of positive and negative charges. Polar molecules are usually soluble in water.

polymer any large molecule composed of multiple identical or similar units (monomers) linked by covalent bonds

polymerization the chemical process of making a polymer from a collection of monomers

polypeptide a linear organic polymer consisting of a large number of amino acid residues bonded together in a chain

polysaccharide a biological macromolecule composed of monosaccharide subunits

protein a biological macromolecule composed of amino acid subunits

protein folding a process by which a polypeptide (or collection of polypeptides) assumes its functional shape or conformation; the process occurs through attractive and repulsive interactions of the polypeptide subunits

proton a positively charged subatomic particle, equivalent to a hydrogen atom without an electron, H^+

purine a class of nitrogenous compounds containing two fused heterocyclic rings. Two purines, adenine and guanine, commonly are found in DNA and RNA.

pyrimidine a class of nitrogenous compounds containing one heterocyclic ring; two pyrimidines, cytosine and thymine, commonly are found in DNA; in RNA, uracil replaces thymine

solute a substance that is dissolved to form a solution

steady state a stable condition that does not change over time or in which change in one direction is continually balanced by change in another

sucrose a disaccharide made up of one glucose molecule and one fructose molecule ($C_{12}H_{22}O_{11}$)

thymine a constituent of DNA that belongs to a class of nitrogenous bases called pyrimidines and hybridizes with adenine in double-stranded DNA.

transcription the process whereby genetic information contained represented as DNA nucleotides is copied into newly synthesized strands of RNA using the DNA as a template

translation the process whereby a sequence of nucleotide triplets in mRNA gives rise to a specific sequence of amino acids during protein synthesis

triglycerides a storage form of fat consisting of a glycerol molecule linked to three fatty acids

uracil a constituent base of RNA; hybridizes with adenine

van der Waals interaction a weak noncovalent attraction due to small, transient asymmetric electron distributions around atoms (dipoles)

NOMENCLATURE

a, b, c, d	Stoichiometric coefficients	
G	Gibb's free energy	J/mole
H	Enthalpy	J/mole
K	Chemical reaction rate constant	1/s (for first-order reaction); L/mole-s (for second-order reaction), etc.
K	Equilibrium constant	$(\text{mole/L})^x$, where x = number of product molecules—number of reactant molecules in balanced reaction equation
K_w	Ion product of water	$(\text{mole/L})^2$
R	Gas constant	J/(mole-K)
S	Entropy	J/(mole-K)

SUBSCRIPTS

A, B, C, D, etc.	Chemical compound A, B, C, D, etc.
F	Formation reaction
A	Association reaction or acid-base equilibrium
D	Dissociation reaction

SUPERSCRIPTS

$^{\circ}$	Related to the standard state

QUESTIONS

1. Describe the properties of acids and bases. It might be helpful to look in a chemistry book, to find information beyond that available in this chapter.
2. Why are polar molecules hydrophilic and nonpolar molecules hydrophobic?
3. If hydrogen bonds are much weaker than covalent bonds, why do you think hydrogen bonds are used to hold biomolecules together?
4. Does entropy increase or decrease during a polymerization reaction? Why?
5. Explain the difference between passive and active transport. Why is active transport necessary for some ions?
6. Normal saline solution (0.9% NaCl by mass) is used for intravenous administration or for lubrication of dry eyes. Do you think that this solution is isotonic, hypertonic, or hypotonic compared to the body fluids? Why?
7. If you are on a deserted island, why must you find water from a stream or well rather than drink the seawater? Explain your answer in terms of osmotic pressure.
8. How does hyperventilation—that is, very rapid deep breathing—disturb the HCO_3^-/H_2CO_3 equilibrium? Does it result in acidosis or alkalosis?
9. Do some research in the library or on the internet, using reliable sources. Cystic fibrosis is a genetic disease. What is the defect in cystic fibrosis patients, and how does that defect manifest into the symptoms for the disease?

PROBLEMS

1. Write the condensation reactions involved in the synthesis of a disaccharide from two monosaccharides, a dipeptide from two amino acids, and a dinucleotide from two nucleotides.
2. For each of the following compounds, classify it as an acid or a base: a) NH_3, b) H_3PO_4, c) LiOH, d) HCOOH (formic acid), e) H_2SO_4, f) HF, g) $Ba(OH)_2$.

3. For the following substances, draw the chemical structure and determine whether the substance is polar or nonpolar. If it is polar, indicate the partial negative and positive charges on the appropriate atoms.
   a. Carbon dioxide (CO_2)
   b. Carbon tetrachloride (CCl_4)
   c. Hydrochloric acid (HCl)
   d. Ammonia (NH_3)
   e. Oxygen (O_2)
4. Tyrosine, serine, and threonine are amino acids, which can be modified by phosphorylation (addition of phosphate group). As you will see, this is an important mechanism for turning enzymes on or off. (a) Find the chemical structures for tyrosine, serine, and threonine and draw them (see Appendix B, Table B.1). For each of the structures, (b) identify each functional group in the molecule, and (c) determine whether the molecule can undergo hydrogen bonding. Mark the partial charges on the appropriate atoms.
5. Identify the acid and conjugate base in each reaction. Calculate the p*Ka* for each acid. List them in order from the strongest to weakest acid. The acid-ionization constants, K_a, at 25°C are listed for each.
   a. $HC_2H_3O_2 + H_2O \leftrightarrow H_3O^+ + C_2H_3O_2$-acetic acid, $K_A = 1.7 \times 10^{-5}$
   b. $HC_7H_5O_2 + H_2O \leftrightarrow H_2O^+ + C_7H_5O_2$-benzoic acid, $K_A = 6.3 \times 10^{-5}$
   c. $HC_6H_4NO_2 + H_2O \leftrightarrow H_3O^+ + C_6H_4NO_2$-nicotinic acid, $K_A = 1.4 \times 10^{-5}$
6. Calculate the $[H^+]$ of stomach acid and blood. Which has a higher $[H^+]$? What generalization can you make regarding the relationship between $[H^+]$ and pH?
7. A solution contains 0.45 M hydrofluoric acid (HF; $K_A = 6.8 \times 10^{-4}$). Write the dissociation reaction. Determine the degree of ionization and the pH of the solution.
8. The pH of a 0.1 M acetic acid solution is 2.885. What is the dissociation constant of acetic acid?
9. What is the pH of a buffer solution that is 0.20 M proprionic acid ($HC_3H_4O_2$) and 0.1 M sodium proprionate ($NaC_3H_4O_2$)? The K_A of proprionic acid is 1.3×10^{-5}.
10. Carbohydrates in foods are a source of energy. The combustion of glucose ($C_2H_{12}O_6$) is:

$$C_6H_{12}O_6(s) + 6O_2(g) \rightarrow 6CO_2(g) + 6H_2O$$

   a. Calculate the standard enthalpy of the reaction. HINT: Use heats of formation from Appendix B, Table B.2.
   b. Is this an exothermic or endothermic process? How much heat (kcal) is generated for each gram of glucose that is burned?
   c. Calculate the value of $\Delta G°$ at 37°C if $\Delta S°$ is 212 J/(K-mol). Is this a favorable reaction?

   [Note: 1 cal = 4.184 J]

11. A U-tube apparatus (as in Box 2.6) is separated by a membrane permeable to water, but not to sodium chloride (NaCl). NaCl (8 g) is dissolved in 0.5 L of water and placed on one side of a semipermeable membrane with pure water on the other side of the membrane. Draw a diagram of the beaker. Which direction will the water flow? If the temperature of the water is constant at 25°C, what is the osmotic pressure? If compartment A and B begin with equal volumes, what will be the difference in the height of the fluid columns at equilibrium?
12. The first step in glycolysis (breakdown of sugar) is to convert glucose to glucose-6-phosphate. Calculate the equilibrium constant for the reaction at 25°C. Is this reaction favorable or not? If it is not favorable, what can drive the reaction to proceed as written?

$$\text{Glucose} + \text{Phosphate} \rightarrow \text{Glucose-6-phosphate} + H_2O$$

$$\Delta G^\circ = 3.3 \text{ kcal/mol}$$

13. In vitro experiments are conducted at pH = 7.4 to simulate physiological conditions. A phosphate buffer system is often used.

$$H_2PO_4^- \rightarrow H_2PO_4^{2-} + H^+ \quad pK_A = 7.2$$

   a. What must be the ratio of the concentrations of HPO_4^{2-} to $H_2PO_4^-$ ions?
   b. What mass of NaH_2PO_4 must be added to 500.0 mL of 0.10 M Na_2HPO_4 (aq) in the preparation of the buffered solution?
14. Estimate the flux (mg/cm^2/s) by diffusion of a steroid through a lipid bilayer membrane. Assume the diffusion coefficient for steroid in the lipid bilayer is 10^{-6} cm^2/s, and that the concentration is 1 ng/mL on the outside of the membrane and 0 on the inside.
15. For the membrane of thickness Δx shown in the Box 2.5:
   a. Draw a graph of the concentration of solute as a function of x at steady state.
   b. Estimate the concentration profiles that you expect during the approach to steady state. That is, assume that the membrane is initially saturated with solute at concentration, C_2, and then the concentration on the left boundary (at $x = 0$) is suddenly increased to C_1. Sketch the concentration profile immediately after the increase to C_1. Sketch the concentration profile a little later, but before steady state is achieved.
16. A solution of 1 M glucose is separated by a selectively permeable membrane from a solution of 0.2 M fructose and 0.7 M sucrose. The membrane is not permeable to any of the sugar molecules. Indicate which side of the membrane is initially hypertonic, which is hypotonic, and the direction of water movement.
17. Consider a U-shaped tube (as illustrated in Box 2.6) in which the arms of the U-tube are separated by a membrane that is permeable to water and glucose but not sucrose. The left side (side A) is filled with a solution of 2.0 M sucrose

and 1.0 M glucose. The right side (side B) is filled with 1.0 M sucrose and 2.0 M glucose.

a. What changes would you observe, as the system moves toward equilibrium?

b. During the period from initial filling to equilibrium, which molecule(s) will show net movement through the membrane?

18. One of the components in the head of "strike-anywhere" matches is tetraphosphorus trisulfide, P_4S_3. The combustion is shown below.

a. Calculate the standard enthalpy of the reaction.

b. Draw a graphical representation of the standard enthalpy change for this reaction.

c. Is this an exothermic or endothermic process? Explain your answer.

$$P_4S_3(s) + 8O_2(g) \rightarrow P_4O_{10}(s) + 3SO_2(g)$$

$$\Delta H_f^\circ\,(P_4S_3) = -155\,\text{kJ/mol}$$

$$\Delta H_f^\circ\,(O_2) = 0\,\text{kJ/mol}$$

$$\Delta H_f^\circ\,(P_4O_{10}) = -2942\,\text{kJ/mol}$$

$$\Delta H_f^\circ\,(SO_2) = -296.8\,\text{kJ/mol}$$

19. The decomposition of calcium carbonate is shown below along with the standard enthalpy and entropy values.

a. Calculate the ΔH_f° for the reaction.

b. Calculate the ΔS for the reaction.

c. What is the standard Gibb's free-energy change expression for the reaction?

d. Is the reaction spontaneous at 25°C? Is the reaction spontaneous at 1000°C? Explain your answers.

e. Calculate the equilibrium constant at 25°C and 1000°C.

$$CaCO_3(s) \rightarrow CaO(s) + CO_2(g)$$

	$CaCO_3(s)$	$CaO(s)$	$CO_2(g)$
ΔH_f° (kJ/mol)	−1206.9	−635.1	−393.5
ΔS (kJ/mol)	92.9	38.2	213.7

20. The reaction in which urea is formed from NH_3 and CO_2 is shown below. The standard free-energy change ΔG at 25°C is −13.6 kJ/mol.

$$2NH_3(g) + CO_2(g) \rightarrow NH_2CONH_2(aq) + H_2O(l)$$

a. Write an expression for the equilibrium constant, K, in terms of the molar concentrations of the reactants and products.

b. Write an expression for the equilibrium constant, K, as a function of ΔG° and temperature.

c. Determine the value of the equilibrium constant, K, for this reaction at 25°C.

21. Acetic acid, CH_3COOH, is a typical weak acid. It is an ingredient in vinegar.

a. Acetic acid partially ionizes in water. Write a balanced chemical reaction for the dissociation of acetic acid into its conjugate base and hydrogen ion.

b. Write an expression for the equilibrium constant for acetic acid.

c. The equilibrium concentrations are $[CH_3COOH] = 0.15M$, $[CH_3COOH] = 0.15M$, and $[CH_3COO\text{-}] = 1.63mM$. What is the equilibrium constant of ionization, K_A?

d. Calculate the pK_A of acetic acid.

22. A solution initially contains 42 mM formic acid ($HCHO_2$, $pK_A = 3.76$).

a. Formic acid is a weak acid and partially ionizes in water. Write a balanced chemical reaction for its dissociation.

b. Determine the conjugate base and H+ concentration at equilibrium.

c. Calculate the percentage of ionization.

23. For the dissociation reaction of a weak acid shown below, begin with defining the K_a and show all the steps for the derivation of the Henderson–Hasselbalch equation.

$$HA \Leftrightarrow H^+ + A^-$$

The Henderson–Hasselbalch equation for the blood bicarbonate system is shown as follows:

$$pH = 6.1 + \frac{[HCO_3^-]}{[H_2CO_3]}$$

a. Calculate the $\frac{[HCO_3^-]}{[H_2CO_3]}$ ratio for a blood pH of 5.8.

b. Is this patient experiencing acidosis or alkalosis? Why?

c. What can the body due to restore the blood pH to normal?

3 Biomolecular Principles: Nucleic Acids

LEARNING OBJECTIVES

After reading this chapter, you should:

- Understand the importance of deoxyribonucleic acid (DNA) in storing genetic information in cells.
- Know the chemical structures of DNA and ribonucleic acid (RNA), and how these chemical structures are related to the functions of these biological macromolecules.
- Understand the mechanism of DNA replication and its importance in cell division.
- Understand the central dogma of molecular biology and the concepts of biological transcription and translation.
- Understand that RNA exists in different forms in the cell, with each form contributing uniquely to the processes of transcription and translation.
- Recognize the importance of gene cloning and how recombinant DNA technology has revolutionized biology and biomedical engineering (BME).
- Understand the technique of the polymerase chain reaction (PCR) and how it is used to synthesize DNA.
- Know the common gene delivery vectors that are used in human cells, as well as their advantages and disadvantages.

3.1 Prelude

One of the most fascinating and well-known stories in science is that of the discovery of the structure of DNA, which was accomplished by James Watson and Francis Crick in 1953, when both were young men working at Cavendish Laboratory in Cambridge, England. Watson's autobiographical book, *The Double Helix*, describes that period of accomplishment, but it retains its popularity because it deals directly with a more general theme. It might be the best description for modern readers of the magical quality of science and its appeal for young people seeking adventure, mystery, and fame. In this way, the story of DNA, beginning with its unveiling, has been linked to romance, celebrity, and power.

DNA, the most famous of the family of biological polymers called nucleic acids, is worthy of this glamour. Nucleic acids are the key information storage

Table 3.1 Diseases resulting from a single gene defect

Disease	Location of Defect	Incidence	Treatment
	Enzyme involved		
Lesch–Nyhan syndrome	Hypoxanthine–guanine phosphoribosyltransferase	1 in 100,000–380,000 (males only)	
Adenosine deaminase deficiency	Adenosine deaminase		Enzyme infusion; gene therapy
Gaucher's disease	Lysosomal glucocerebrosidase		Enzyme infusion
Phenylketonuria	Phenylalanine hydroxylase		Diet
Sickle cell anemia	Hemoglobin	0.4% of African-American males	
Hemophilia	Factor VIII	1 in 10,000 males	Blood transfusion; protein infusion
	Membrane protein involved		
Cystic fibrosis	Cystic fibrosis transmembrane conductance receptor (CFTR)	1 in 2000 live births in Caucasians	Gene therapy not yet successful
Disaccharidase deficiency (lactose intolerance)	Lactase	Frequent	Diet
Duchenne's muscular dystrophy	Dystrophin gene product	1 in 3500 males	None
	Protein involved		
Huntington's disease	Huntington		None

Notes: Information was derived from (1, 2). These diseases are caused by a defect in a single gene, which results in the deficiency of an enzyme, membrane proteins, or other gene product.

molecules of life. Genetic information is encoded within the nucleic acids of almost every cell in our body, where it is capable of being inherited from one cell to the next, generation after generation. The study of nucleic acids is multifaceted: It involves examination of the structure and function of nucleic acid polymers, the capacity of these molecules to hold information, the changes in information content that happen upon modifications of nucleic acids, the diverse roles of nucleic acids in the life of a cell, and the characteristics that make them useful tools in biomedical engineering. In fact, as in many other areas of science and technology, the use of the nucleic acids DNA and RNA is becoming central to the work of biomedical engineers. The biological polymers called nucleic acids were introduced in Chapter 2; this chapter will expand on that introduction, showing how nucleic acids work to store genetic information and perform other essential biochemical functions.

DNA is intimately involved in human health. Many diseases result from failures at the DNA level. These failures can arise from defects in genes themselves (causing genetic diseases) or in the regulatory regions of genes (causing cancer). Some diseases are the result of a defect in a single gene. Although many of these diseases are rare, some—such as cystic fibrosis and muscular dystrophy—are relatively common (Table 3.1). Many biomedical engineers are now involved in

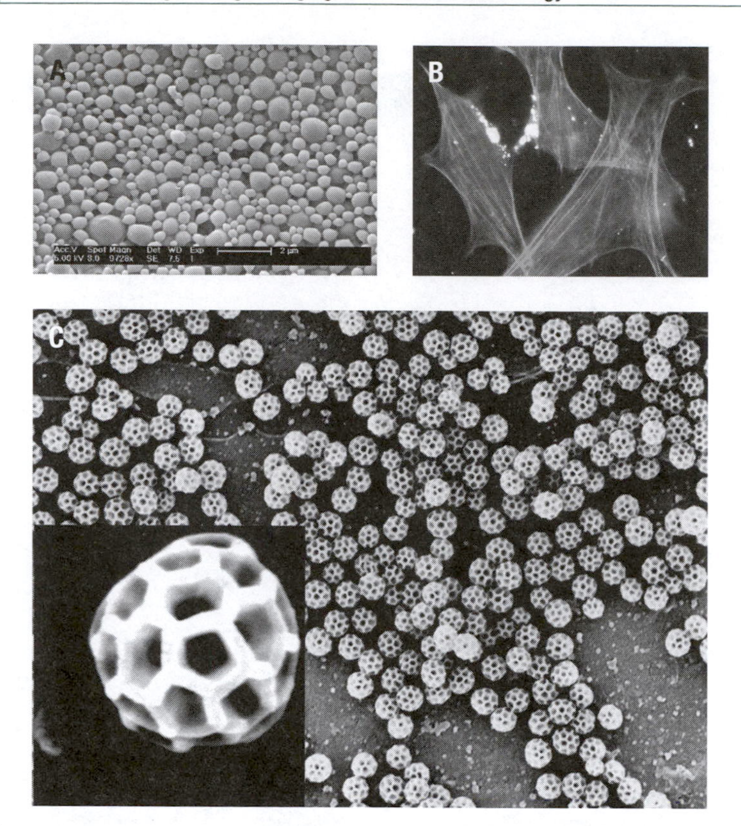

Figure 3.1

Gene therapy vectors and designer deoxyribonucleic acid (DNA). A. Biomedical engineers are working to develop new gene therapy vectors such as plasmid-loaded biodegradable particles and also to understand how these vectors interact with human cells. Photo courtesy of Jeremy Blum. **B.** Polyoma virus-like particles associating with actin fibers in cultured cells. (See color plate.) Reprinted with permission from Macmillan Publishers Ltd: *Gene Therapy*. 2006;7(24):2122–2131. **C.** A Buckyball-like structure constructed entirely of DNA. Photo courtesy of Dan Luo, Cornell University.

the search for safe and effective methods for gene therapy in humans, in the hope that the defective genes can be replaced with new functional genes, which will cure the disease (Figure 3.1).

Other diseases appear to be caused by changes in more than one gene. These diseases are called polygenic and include some common disorders including Type II diabetes and heart disease. The genetic origin of these diseases—and the relationship of changes at the genetic level to the manifestations of the disease itself—is often complex and not yet completely understood. In many cases, a person inherits a predisposition to a polygenic disease, but development of the disease is dependent on environmental or behavioral factors. Biomedical engineers are working with biologists and clinicians to understand these diseases, often using computer models to predict the consequence of genetic mutations on cell function. Moreover, the whole subspecialty of **systems biology** (Figure 3.2)—which will employ thousands of biomedical engineers in the near future—has arisen primarily from the Human Genome Project and its effort to sequence all 3 billion bases of the human genome. A **genome** is the total of all the genetic

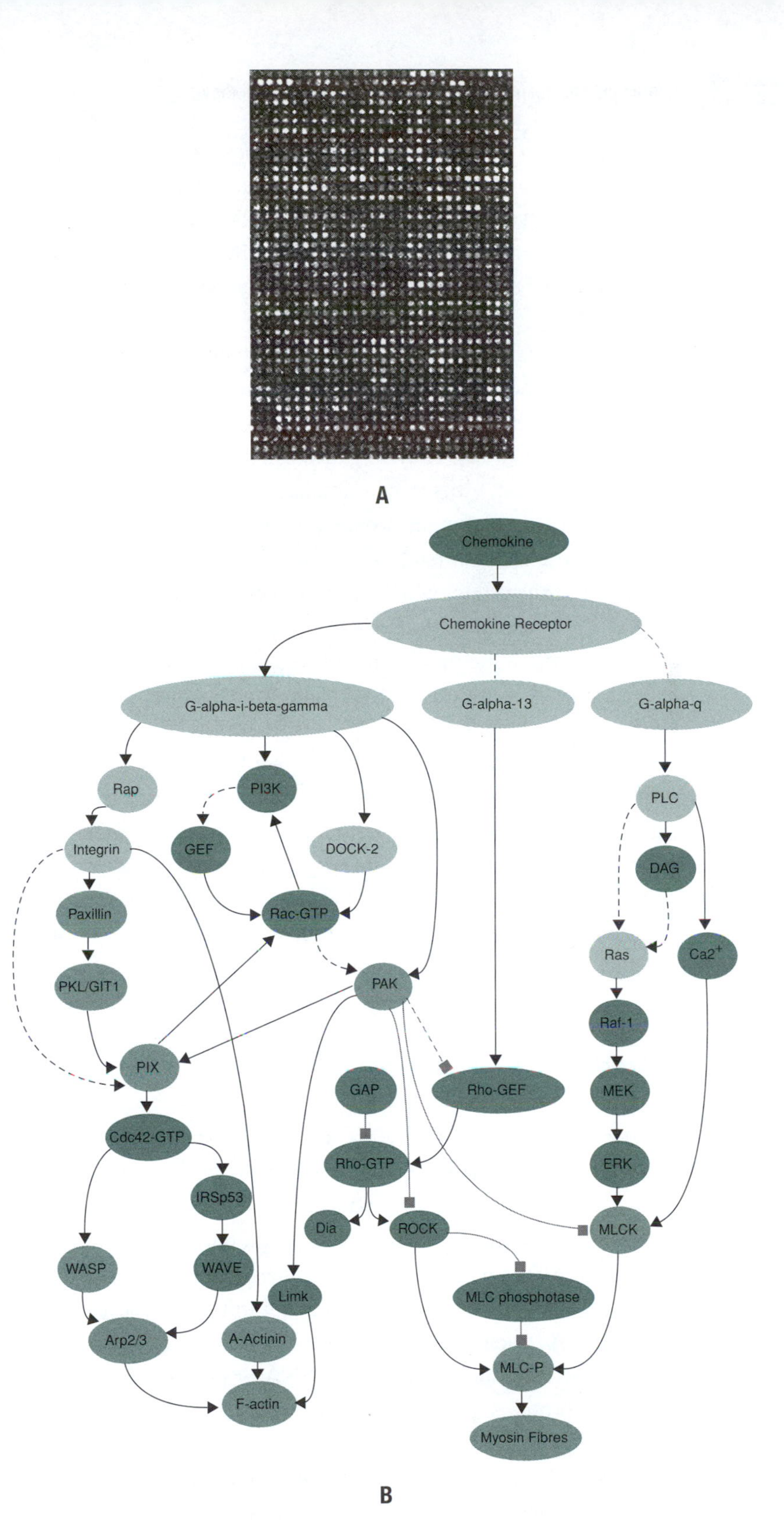

Figure 3.2 **Systems biology is one of the frontier areas for biomedical engineers. A.** Deoxyribonucleic (DNA) microarray, in which DNA probes are attached to a solid surface, with each dot on the surface containing many copies of one particular known DNA sequence. This microarray allows researchers to screen for the presence of many individual genes in a small sample of tissue or cells. Photo credit: Dr. M. A. Saghai Maroof, Virginia Polytechnic Institute and State University. (See color plate.) **B.** An example of the signaling pathway for a cytokine.

materials within an organism's chromosomes. Biomedical engineers working in systems biology helped to design automated equipment to sequence the human genome as well as computer software tools to analyze the massive amount of data that was generated. The ability to synthesize and clone DNA also led to the development of microarray chips for the analysis of gene expression in hundreds of samples.

From the origins of recombinant DNA technology in the 1970s, our society has quickly gained the capability to manipulate and use DNA as a tool for understanding and treating disease. DNA technology has also changed the way that we manufacture drugs. Human genes can now be inserted safely and efficiently into bacteria, yeast, viruses, or animal cells. This capability has already led to the production of recombinant proteins as therapeutic drugs; diabetics around the world now use recombinant human insulin that is safer and less expensive than previous insulin drugs, which were harvested from animal tissues. DNA technology is also changing forensic science, agriculture, and other aspects of contemporary life. Some bioengineers are even learning how to use DNA as a tool, deploying DNA molecules as molecular Tinker Toys® for building tiny but well-defined objects (Figure 3.1).

This chapter's discussion of nucleic acids begins with an introduction to the science of genetics and ends with descriptions of some of the ways that DNA is used in BME. Along the path from genetics to bioengineering, the structure of nucleic acids, which was introduced in Chapter 2, is described in more detail, and the basic concepts of molecular biology are discussed. Finally, the applications of molecular biology, in the form of recombinant DNA technology, are presented together with their abundant connections to BME.

3.2 Overview: Genetics and inheritance

Genetics is the study of heredity and hereditary variations. Each human has a unique assembly of **traits**. Some traits are easily recognizable, such as brown eye color, a cleft in the chin, or a widow's peak hairline; other traits are not easily visible, such as blood type or a predisposition to diabetes (Figure 3.3). **Heredity** is the transmission of traits from one generation to subsequent generations. DNA molecules provide the basic units for heredity because genes are stored in the form of DNA. Each gene is a unique sequence of polynucleotides that contains the information to make a particular RNA or protein. The vast majority of genes encode for proteins. For example, every human cell has a gene that **encodes** the protein insulin.

Genes are located on **chromosomes**, which are strands of DNA packed within the nucleus of most cells. In human cells, nearly 2 m of DNA is compacted into a nucleus that is 5–10 μm in diameter. DNA is folded and packaged into chromosomes to achieve this high density of DNA storage in the nucleus

Figure 3.3 **There are many examples of genetic variation.** Transfused blood must be matched to a person's type, to account for genetic variations in proteins on red blood cells. Physical characteristics, such as eye color, a cleft in the chin, and shape of the nose are other indications of genetic variation among individuals.

(Figure 3.4). Individual strands of DNA form a double helical structure, which would ordinarily be an extended, long molecule. But within the nucleus, this double-stranded DNA is packaged with proteins called **histones** to form a material called **chromatin**. During cell division, chromatin undergoes further condensation to form the familiar "X"-shaped chromosome structure.

The chromosomes from a cell can be isolated and examined with a microscope. When pictures of a set of chromosomes are arranged in a specific order, the resulting overall image is called a **karyotype**. In a karyotype obtained from a typical human cell, there are 46 chromosomes arranged into 23 pairs (Figure 3.5). Human cells that contain 46 chromosomes are called **diploid**, which means that they contain two copies of each chromosome. (Notice that the "X" and "Y" chromosomes of a human male, which together make chromosome pair 23, are not actually a matched pair.)

The paired chromosomes are called **homologous chromosomes**: One is inherited from each parent (Figure 3.5). The number of chromosomes in a cell is often a defining characteristic of that cell. Sperm cells and egg cells are the reproductive cells and are often called **gametes**. These cells participate in biological reproduction. Their function is to each contribute half of the genetic material to an offspring (Figure 3.6); therefore, they contain only 23 chromosomes each, one for each pair of the 23 pairs of chromosomes in the offspring. Gametes are **haploid** cells, because they contain only one chromosome for each chromosome pair (Figure 3.7). The gametes are examples of **germ cells**, or the cells within the body that are normally haploid. Only the gametes and some of their precursors are germ cells; all other cells in the body are called **somatic** cells. Most somatic cells are diploid, but there are some exceptions. Red blood cells do not contain a nucleus; such cells are called **nulliploid**. Other cells, such as cells within the regenerating liver, may contain more than two copies of the chromosomes; such cells are called **polyploid**.

How is this arrangement of chromosomes—and the joining of haploid chromosomes from egg and sperm—related to the inheritance of genes? To understand this, consider how genes on the paired chromosomes lead to traits, which were defined earlier. The physical location of a gene on a particular chromosome

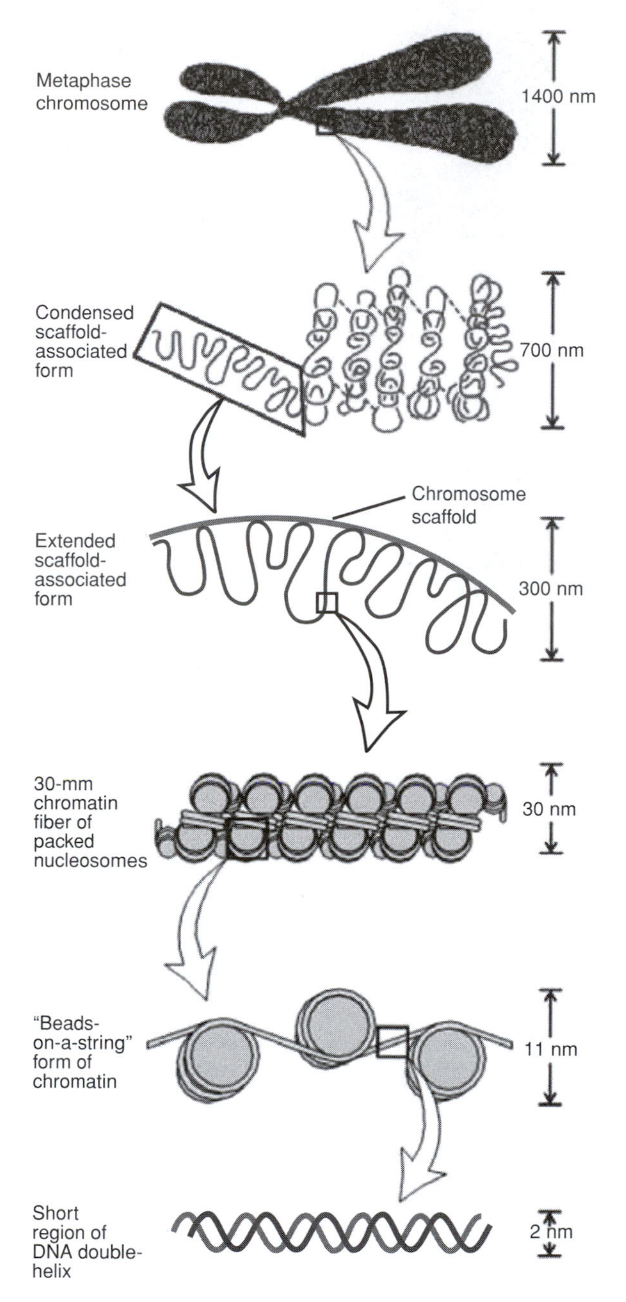

Figure 3.4 **Deoxyribonucleic acid (DNA) packing in the nucleus.** The negatively charged DNA double helix is wrapped around basic proteins called histones to form nucleosomes. Nucleosomes are further packed together to form a 30-nm chromatin fiber. Additional levels of packing result in a highly condensed chromosome structure.

is called the **gene locus** (the plural form is *gene loci*). Each heritable feature, such as eye color, is called a **character**; a variant of character, such as blue eye color, is called a **trait**. The homologous chromosomes contain an identical set of genes, one set from each parent. The DNA sequence in the genes can vary from one chromosome to another; each alternate version of the gene is called an **allele**. Therefore, homologous chromosomes possess either identical alleles or

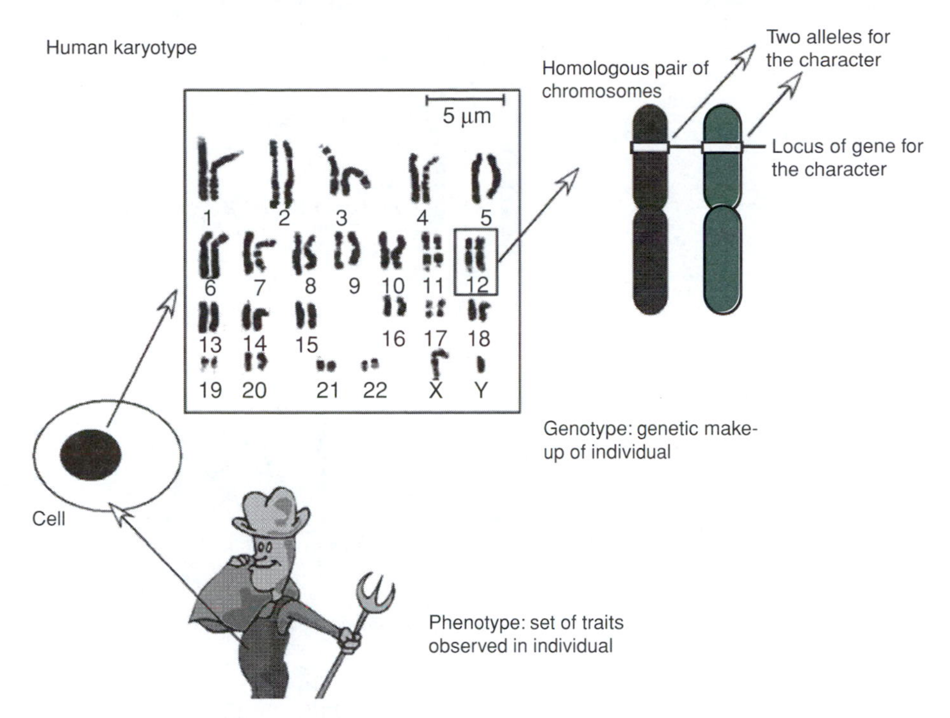

Figure 3.5 **A human karyotype.** Chromosomes can be isolated from a cell and displayed in a karyotype as 23 pairs of homologous chromosomes. The nuclei of all somatic cells contain 22 identical pairs of chromosomes plus a pair of sex chromosomes (XX in females; XY in males). Chromosomal aberrations can be detected by karyotype analysis; for example, Down syndrome is caused by the presence of three copies of the 21st chromosome.

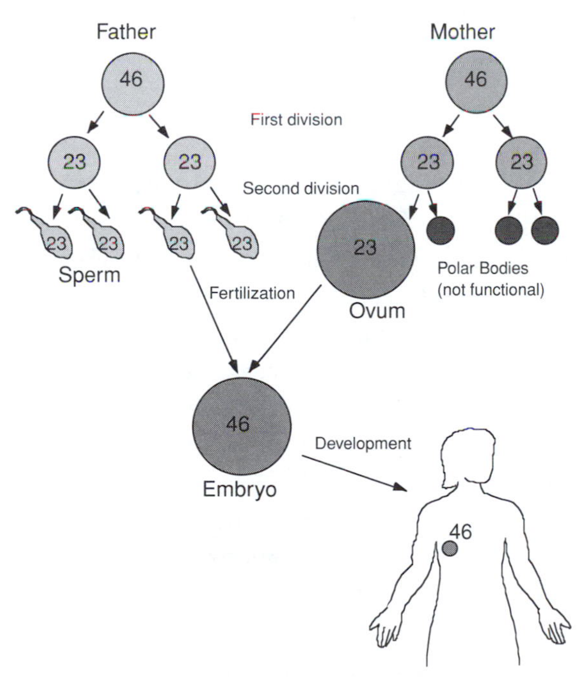

Figure 3.6 **Inheritance of genes from gametes to the embryo.** Meiosis involves two cell-division steps in succession, and results in haploid cells (called sperm in men and ova in women). Human sperm and ova have 23 single (unpaired) chromosomes. These cells fuse, and their chromosomes are combined, to form an embryo. The genetic composition of the embryo is duplicated in all of the somatic cells of the individual it becomes.

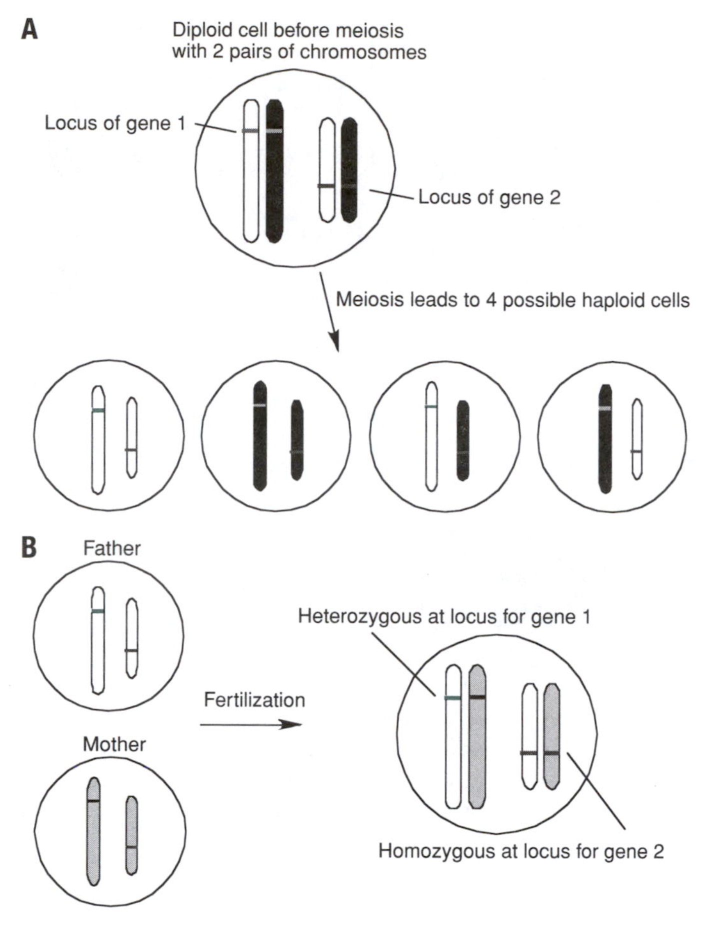

Figure 3.7 **Meiosis and fertilization.**

two different alleles. When the alleles are identical, the individual is **homozygous** for the gene controlling that character; when the alleles differ, the individual is **heterozygous**. In individuals with two different alleles (i.e., heterozygotes), one of the alleles may be **dominant**, or fully expressed in the individual, whereas the other is **recessive**, or without a noticeable effect. The presence of one recessive allele is not obvious from the physical appearance of an individual because the **phenotype** (or set of traits exhibited) may be similar whether the person possesses a homozygous dominant pair or a heterozygous pair of alleles. Therefore, it is not possible to infer the **genotype** (or genetic makeup) of an individual from the **phenotype** (or set of traits exhibited) unless the individual exhibits the recessive trait because of the presence of a homozygous recessive pair. Box 3.1 describes some ways to think about the inheritance of traits in populations.

SELF TEST 1 Using the Punnett square approach outlined in Box 3.1, calculate the probabilities for the outcomes of mating a homozygous male (AA) with a heterozygous female (Aa).

ANSWER: 50% AA and 50% Aa

Box 3.1 Allelic and genotypic frequencies

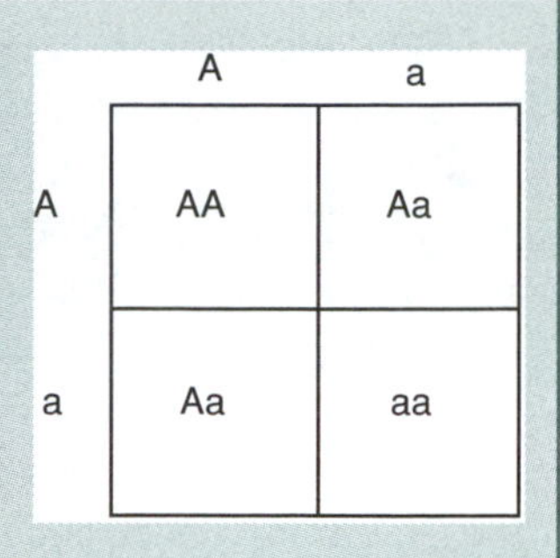

Consider a gene with two alleles **A** and **a**. The possible genotypes resulting from the union of a sperm and egg can be visualized using a Punnett square, which is illustrated in the figure at the right. In this simple square, it is assumed that each parent is heterozygous for the gene. Therefore, the sperm and egg each contain either allele A or a. The resulting offspring will be AA, Aa, or aa. The probability of obtaining each of these combinations in the offspring can be calculated from the Punnett square. There are two occurrences of Aa, one of AA, and one of aa: Therefore, there is a 25% chance of AA, 25% chance of aa, and 50% chance of Aa in the offspring from this mating.

You might have wondered something like the following: If the allele for brown hair is dominant over the allele for blonde hair, how does the blonde trait remain in the population? This problem was considered 100 years ago independently by two scientists, the mathematician G.H. Hardy in England and the biologist physician W. Weinberg in Germany. They developed a mathematical model, illustrating what has come to be known as the **Hardy–Weinberg** equilibrium. Their model allows us to identify the conditions under which a trait remains stable in a population.

Assume that there is a population that contains two alleles **A** and **a**, such that the frequency of allele **A** in the population is p and the frequency of allele **a** is q. Because the population contains only two alleles:

$$p + q = 1 \qquad \text{(Equation 1)}$$

	A (p = 0.8)	a (q = 0.2)
A (p = 0.8)	AA (pp)	Aa (pq)
a (q = 0.2)	Aa (pq)	aa (qq)

Assume that the frequency of **A** and **a** in the population at some time is $p = 0.8$ and $q = 0.2$ (i.e., 80% **A** and 20% **a**). A Punnett square for the population is shown to the right (as opposed to the Punnett square for an individual mating, which is above). From this square, with $p = 0.8$ and $q = 0.2$, you can calculate that the probability of AA in the new population is $0.8 \times 0.8 = 0.64$; the probability of aa is $0.2 \times 0.2 = 0.04$; and the probability of Aa is $2 \times 0.8 \times 0.2 = 0.32$ (notice that these probabilities still add up to 1).

This process can be repeated for the more general case to reveal the Hardy–Weinberg principle:

$$p^2 + 2pq + q^2 = 1 \qquad \text{(Equation 2)}$$

What is the frequency of the alleles in this new generation? The frequency of AA is p^2, the frequency of aa is q^2, and the frequency of Aa is $2pq$. Therefore, in terms of the individual alleles, the new generation gets two A alleles from each AA individual and one from each Aa individual. Assuming that there are N individuals in the new generation, the frequency of each allele is:

$$f_A = \frac{N \times 2 \times p^2 + N \times 2pq}{2N} = p^2 + pq$$
$$f_a = \frac{N \times 2 \times q^2 + N \times 2pq}{2N} = q^2 + pq \qquad \text{(Equation 3)}$$

(continued)

Box 3.1 *(continued)*

In the case we were considering here, the frequencies in the starting population were $p = 0.8$ **A** and $q = 0.2$ **a**. According to Equation 3, the frequencies in the next generation are $f_A = 0.64 + 0.16 = 0.8$ and $f_a = 0.04 + 0.16 = 0.2$. Notice that the frequencies of the alleles stay constant through the generation! This is the **Hardy–Weinberg principle**: The frequencies of the alleles stay constant from generation to generation as long as the frequencies are governed by the statistics of random mating (and no other forces).

What if the frequency of an allele is found to change with time? Deviations from the Hardy–Weinberg frequencies indicate that external factors have affected the inheritance pattern of the gene in that population. Some other factor—such as gene mutations, gene flow into the population, natural selection, or selective breeding among a subpopulation—must be significant.

The relationship between genotype and phenotype is not always as simple as the previous discussion suggests because the correspondence between genes and traits is not one-to-one. Most characteristics are influenced by several genes (or **polygenic**). Also, each gene can influence a variety of traits in a feature called **pleiotropy**. In addition, some genes have **incomplete dominance**, in which the presence of multiple alleles in heterozygotes can lead to a trait that appears intermediate to the homozygous phenotypes. Most genes exist in multiple forms; that is, there are more than two alleles in the population. For example, there are various eye colors.

Understanding inheritability of diseases helps scientists and biomedical engineers to develop appropriate diagnostic tests and treatment strategies. There are several genetic diseases that arise because of alterations in one gene (e.g., cystic fibrosis or Huntington's disease; see Table 3.1). For these single-gene diseases, gene therapy to restore the defective gene is being investigated as a possible cure. However, most diseases are polygenic or caused by alterations in multiple genes. Most diseases also have additional environmental factors that influence their progression. Thus, prevention and effective treatment of cancer, cardiovascular disease, and Type II diabetes remain the most challenging goals of scientists and biomedical engineers.

Many biomedical engineers are now involved in work related to reproductive health (Figure 3.8). Some of their efforts are aimed at diagnostic tools: Biomedical engineers have long been leaders in the development of ultrasound systems, which are now routinely used to follow fetal development in utero. Home pregnancy test kits were created with the help of biomedical engineers. Biomedical engineers have been involved in the development of contraceptives, including the NuvaRing® and other barrier methods. In one of the newest developments, biomedical engineers are using tissue-engineering approaches to preserve oocytes (the precursor cells to ova) outside of the body. The role of biomedical engineers in ultrasound imaging is described more completely in Chapter 12 and drug delivery and tissue engineering in Chapter 13.

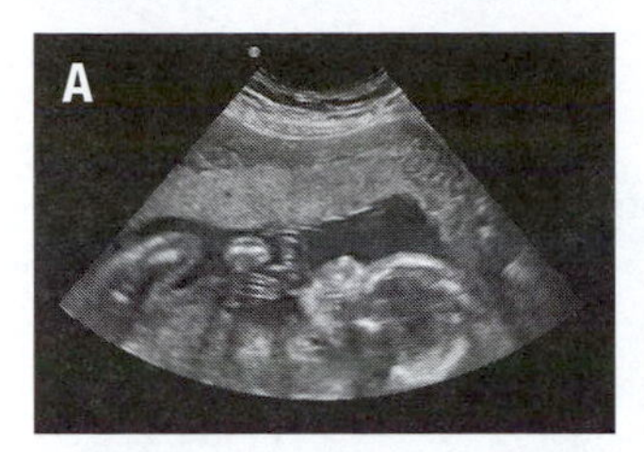

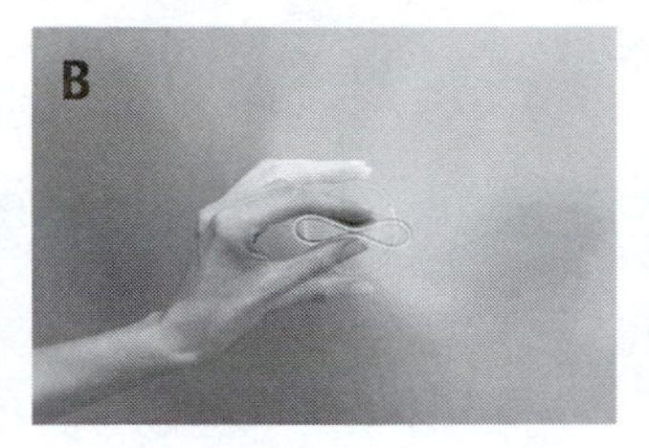

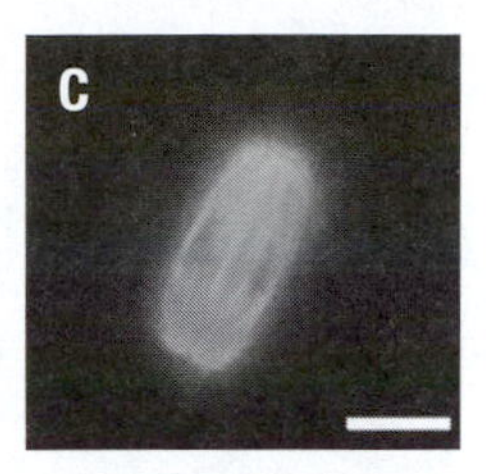

Figure 3.8

Biomedical engineers and reproductive health. A. Image of a 22-week-old fetus obtained by ultrasound. Photo courtesy of Jenica Chung. **B.** NuvaRing® contraceptive is a soft polymer vaginal ring that releases hormonal contraceptives over a period of several weeks. Lower doses of hormones can be used, because they are released directly into the reproductive tissues. Photo used with permission. **C.** An oocyte that was maintained in a three-dimensional tissue engineering matrix retains its ability to undergo mitosis. Reprinted from (3) with permission from Elsevier. (See color plate.)

3.3 Molecular basis of genetics

3.3.1 DNA structure

Nucleic acids (DNA or RNA) are linear polymers of nucleotides (recall Chapter 2). DNA of a typical human cell contains 3 billion nucleotides, which make up its **genome**. A gene is a segment of nucleotides that codes for one polypeptide chain.[1] Each nucleotide is composed of three components: a pentose (five-carbon sugar), a phosphate group, and an organic base. The pentose differs between DNA and RNA: DNA nucleotides contain deoxyribose and RNA nucleotides contain ribose. In both DNA and RNA, the five carbons in the pentose are numbered from $1'$ to $5'$ (Figure 3.9 has more detail than the images in Chapter 2). DNA and RNA have completely different functions in the cell, so it is startling to recognize that their structures differ in such subtle ways.

The chemical structures of the bases found in DNA and RNA are shown in Figure 3.10. Adenine (A) and guanine (G) are purines (double-ringed nitrogen-containing bases), whereas thymine (T), cytosine (C), and uracil (U) are pyrimidines (single-ring nitrogen-containing bases). In addition to the difference in the structure of the pentose, RNA and DNA differ in the use of uracil and thymine: RNA uses U, whereas DNA uses T. One of the most important discoveries of Watson and Crick was that of the rules for association, or pairing, of these bases during the formation of double-stranded DNA: The pairs are C-G, A-T (in DNA), and C-G, A-U (in RNA). The maximum number of hydrogen bonds results from pairing a pyrimidine with a purine. This pairing also leads to the same distance between the nitrogens of each respective base pair, giving the α-helix a regular diameter along its length. For example, A and T form two hydrogen bonds between them, whereas G and C form three. Because these hydrogen bonding matches are so predictable, the sequence of a complementary nucleic acid strand

[1] Proteins can be composed of more than one polypeptide chain: Insulin, for example, contains two chains. Chapter 4 describes this in more detail. This can occur by several mechanisms: One long polypeptide, encoded by a single gene, can be broken into two parts, or polypeptides encoded separately by different genes can combine after translation to form a protein.

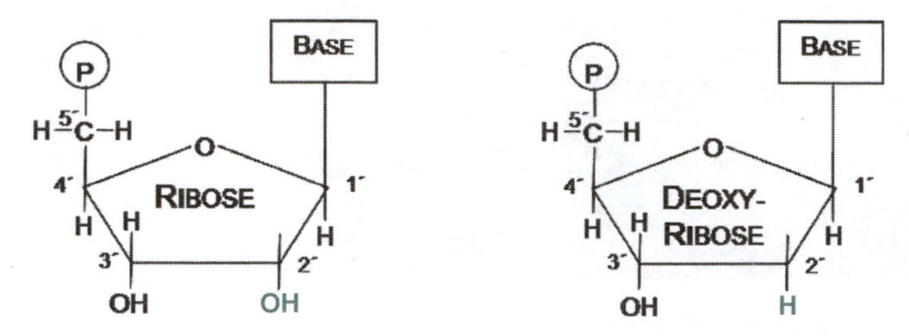

Figure 3.9 **Nucleotides.** Nucleotides are the monomers of nucleic acids. Each nucleotide is composed of a five-carbon sugar (ribose in ribonucleic acid [RNA] or deoxyribose in deoxyribonucleic acid [DNA]), a phosphate group and a nitrogen-containing base. The 3′ hydroxyl and 5′ phosphate groups are used to form covalent bonds between nucleotides.

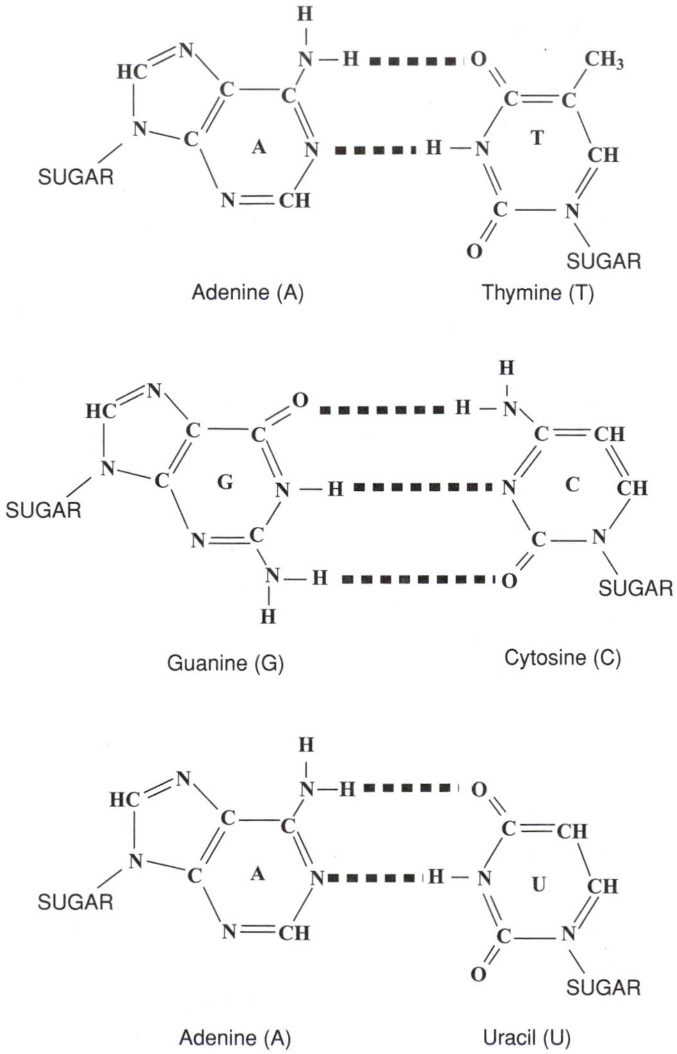

Adenine (A) Thymine (T)

Guanine (G) Cytosine (C)

Adenine (A) Uracil (U)

Figure 3.10 **Nucleotide bases and base pairing.** The bases make predictable hydrogen bonds with each other as indicated by the Watson–Crick base pairing rules. In deoxyribonucleic acid (DNA), adenine (A) forms two hydrogen bonds with thymine (T), and guanine (G) forms three hydrogen bonds with cytosine (C). In ribonucleic acid (RNA), thymine (T) is replaced by uracil (U).

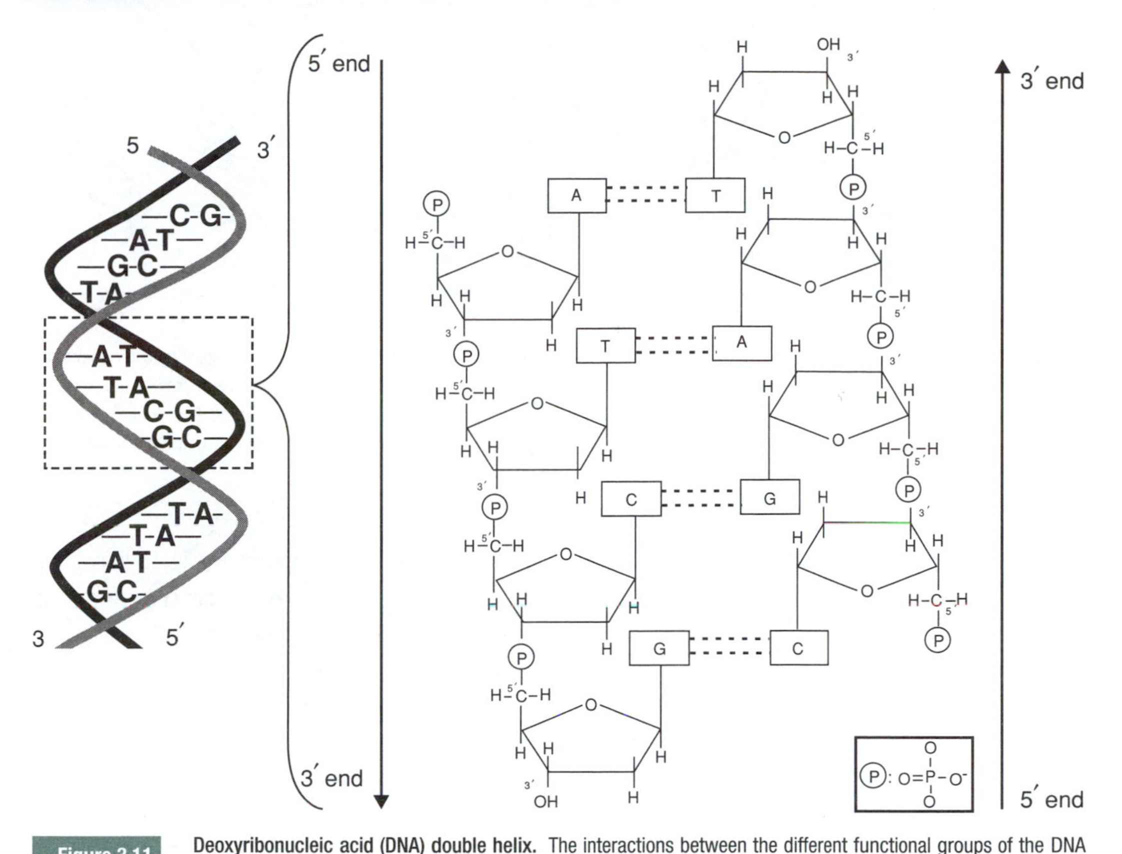

Figure 3.11 **Deoxyribonucleic acid (DNA) double helix.** The interactions between the different functional groups of the DNA bases cause the two strands to adopt a double helical structure. Two complementary strands of DNA align such that the sugar–phosphate backbone of each strand is joined together by the Watson–Crick paired bases.

can be determined by applying Watson and Crick base-pairing rules to a parent, or single, strand.

The molecular interactions between the two strands of double-stranded DNA cause them to adopt a double helical structure, with two sugar phosphate backbones and the bases bridging the sides to form the steps of a ladder (Figure 3.11). As first indicated in Chapter 2, it is important to note the opposite orientation of the two strands that make up the backbone of the double helix. To describe the orientation of each strand, it is common to refer to the orientation of the two carbon atoms that are coupled together by the phosphate bond between nucleotides. On one nucleotide, the phosphate is coupled to carbon 3′ on the pentose; on the other nucleotide, the phosphate is linked to carbon 5′. Therefore, one of the nucleic acid strands runs from the 5′ phosphate to the 3′ hydroxyl ($5' \rightarrow 3'$), and the complementary strand is oriented in an antiparallel fashion from the 3′ hydroxyl to the 5′ phosphate ($3' \rightarrow 5'$) (i.e., they are aligned in opposite directions). This orientation is not easy to describe, or to visualize from words, but careful study of Figure 3.11 may help make it clear.

The two strands of DNA in a double helix are held together by hydrogen bonds. Because hydrogen bonding is weak, compared to the covalent bonds that hold the nucleotides together in each single strand, it is possible to separate the strands reversibly (Figure 3.12). The hydrogen bonds can be **denatured** (disrupted)

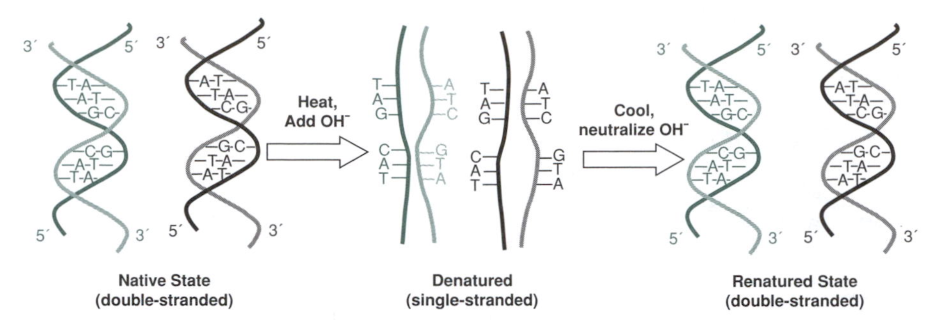

Figure 3.12 **Nucleic acid hybridization.** The hydrogen bonds between bases can be disrupted (denatured) by the addition of heat or hydroxide ions to form single-stranded nucleic acids. When heat is removed, or the hydroxide neutralized, the two strands hybridize with each other because of their complementary bases and assume their double-stranded structures. Hybridization can occur between two complementary deoxyribonucleic acid (DNA) strands or one DNA and one ribonucleic acid (RNA) strand.

by heat or by alkaline (basic) solutions. When the temperature is lowered or hydroxide ion concentration is decreased, the strands **renature** (reassociate) according to their Watson and Crick base-pairing rules. This process of renaturing of complementary single strands to form a double strand is called **nucleic acid hybridization**. The ability of DNA to hybridize so faithfully—that is, the property of DNA to bind only to strands with its exact complementary sequence—is extremely valuable. Box 3.2 illustrates how DNA hybridization is used to perform DNA fingerprinting.

3.3.2 DNA replication

When a cell reproduces itself, it must duplicate its contents and divide into two identical daughter cells. The process by which this occurs is called **meiosis** in germ cells, which was described briefly, and **mitosis** in somatic cells. Mitosis is described in Chapter 5. Cell division is critical in the developing embryo: The single cell produced by the union of sperm and egg divides repeatedly to form a newborn with trillions of cell. But division is also important in adults. For example, cells in the skin must continually divide to replenish the outer layer of skin, which contains dead cells that slough off from the surface. Lymphocytes, specialized immune cells in the blood, proliferate as part of the immune response to a foreign pathogen.

How is double-stranded DNA copied to ensure that daughter cells each receive an exact version of the parent DNA? Replication of the DNA double helix begins at **replication origins**, specific sequences of nucleotides, at which the two parental DNA strands are separated to form a **replication fork**. A tiny biological machine containing many proteins (called the replication machinery) performs the work of DNA replication. This machine first recognizes the replication origin, and then opens it, to create the replication fork (Figure 3.13). Replication proceeds bidirectionally with the two forks moving in opposite directions from each origin on the very long DNA strand (a human chromosome contains between 51 and

Box 3.2 DNA fingerprinting

The sequence of nucleotides within the chromosomes is unique to each person. In addition, most of the cells of the body contain copies of the chromosome. As people move through the world, they leave unique evidence of presence behind: within saliva on a drinking glass or in hair on a brush or comb. Because the DNA sequence is so uniquely paired with an individual, it can be used to match an individual to an unknown sample. This technique has become a staple of law enforcement, earning the popular name "DNA fingerprinting" because it is analogous to the long-used practice of linking people to a crime scene by matching the fingerprints they leave.

The most common method of DNA fingerprinting involves a technique called restriction fragment length polymorphism (RFLP). To perform this test, the DNA from the sample (or from a small volume of blood drawn from a known person) is isolated. RFLP uses special enzymes, called **restriction enzymes**, to cut the long DNA strands from the chromosome into small fragments. Restriction enzymes, as described in Section 3.6.1, are like smart and reliable molecular scissors: They cut the DNA at every spot on the strand that contains a certain base sequence. For example, the restriction enzyme called *Eco*RI cuts double-stranded DNA at every site that contains the sequence GAATTC. Because this sequence (or any sequence of ~6 bases) occurs many times in any person's DNA, use of the enzyme results in the production of many DNA fragments of different length. Because the DNA of each person is unique, cutting with the enzyme produces a set of fragments of different length in each person. To create a fingerprint, however, a method for visualizing the set of fragments made from each distinct DNA sample is needed.

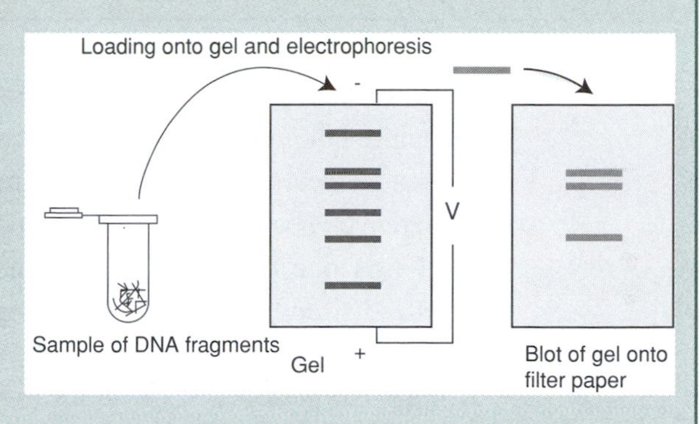

The first step in visualization is to sort the fragments according to size. This is accomplished by a technique called gel electrophoresis, in which the DNA fragments are loaded into the top of a gel (a soft material similar to soft contact lenses or gelatin) and a voltage applied to pull the fragments (remember that DNA is a charged molecule because of the many phosphates) through the gel. For a given length of treatment, smaller fragments are pulled farther into the gel. This process produces a pattern of "bands" on the gel: Each band represents DNA fragments of a certain size. Each sample, from a different person, will have a different distribution of fragment size and, therefore, a different pattern of bands.

These bands are DNA molecules within a gel, so how can they be seen? The most common method involves two steps. First, the gel is pressed against a piece of filter paper to create a replica, or blot, of the gel. Second, the filter paper is soaked in a solution of a radioactive DNA probe (shown as grey in the figure). This probe is a piece of DNA that contains a sequence that will hybridize with its complementary sequence on the sample DNA (which is now spread out on the filter). Here, finally, the fidelity of DNA **hybridization** is put to practical use. Every fragment of DNA on the filter that contains the complement to the probe will get radioactively

(continued)

Box 3.2 *(continued)*

labeled because it hybridizes with the probe. (Notice, in the figure, that not all of the bands are labeled by the grey probe. Why?) The radioactive bands can be photographed with x-ray film, making a picture of the DNA fingerprint.

This approach to DNA fingerprinting is conceptually simple. In practice, there are many variables that can be used to refine the approach: The properties of the restriction enzyme (or enzymes) and the probes have a large impact on the resulting pattern. Scientists and engineers modify this basic approach to develop tests that have a high probability for making a unique match between samples from the same person.

245 million base pairs, so the picture in Figure 3.13 shows only a small portion of the entire DNA strand). The replication machinery has mechanisms to ensure that DNA is accurately duplicated before every cell division; in this way, genetic information can be passed onto the two daughter cells without error. Each daughter cell gets one of the original strands from the parent cell and one newly synthesized strand: Therefore, DNA replication is said to be **semiconservative** (Figure 3.14).

Among the proteins of the replication machinery is an enzyme called **DNA polymerase**: This enzyme catalyzes the reaction that adds a new nucleotide to a growing strand of DNA (Figure 3.13). Recall the discussion of biochemical

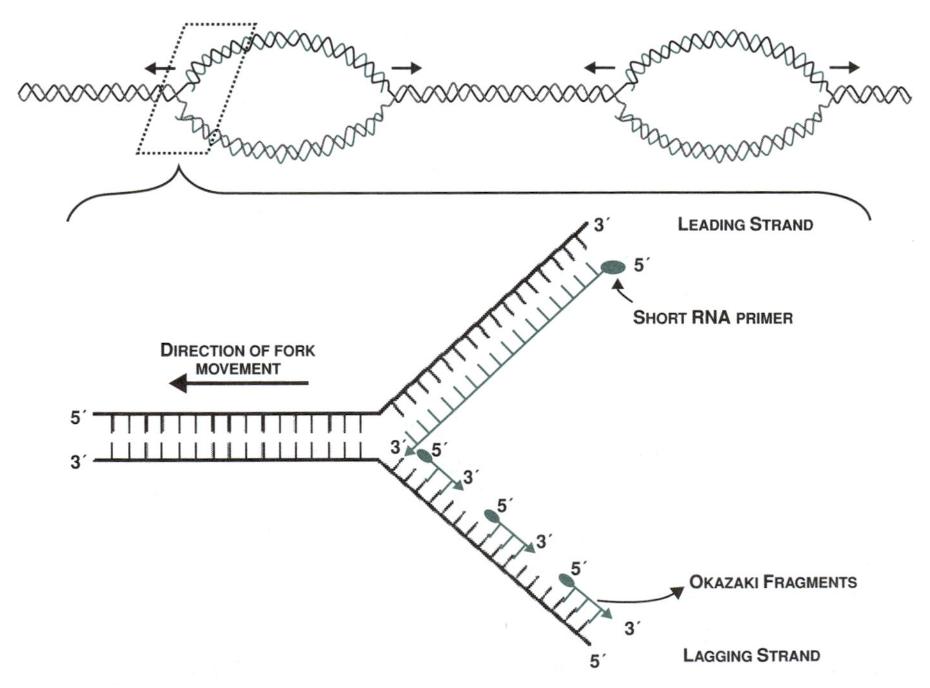

13 **Deoxyribonucleic acid (DNA) replication.** DNA replication is initiated at sites called replication forks. DNA polymerase requires a primer or short segment of nucleotides to initiate polymerization. This primer is provided by ribonucleic acid (RNA) primase. Because DNA synthesis can only occur in the 5′ to 3′ direction, the two template strands are replicated in different ways. The leading strand can be replicated continuously, in the 5′ to 3′ direction. The lagging strand can only be replicated discontinuously to form Okazaki fragments, which must then be joined together to form a continuous strand.

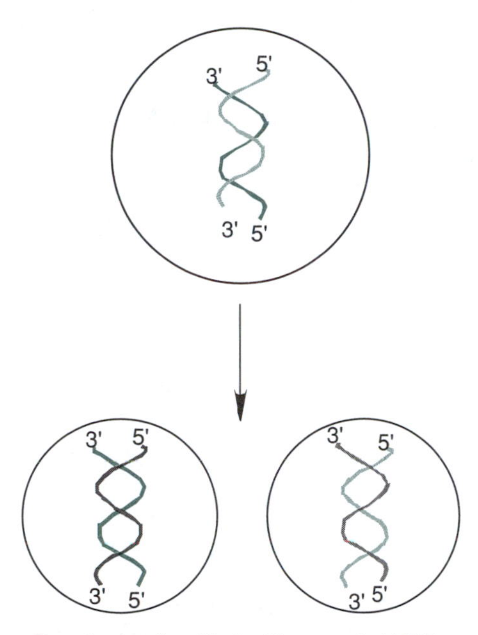

Figure 3.14 **Deoxyribonucleic acid (DNA) replication during cell division.** DNA replication is semiconservative: Each of the new double-stranded units contains one strand from the parent.

energetics and the role of adenosine-5′-triphosphate (ATP) from Chapter 2. Because DNA polymerization is energetically unfavorable, energy is required for the reaction to proceed. In this case, that energy comes by coupling the polymerization reaction to another reaction, in which a triphosphate group that is attached to incoming nucleotide is hydrolyzed: A nucleotide containing the base A arrives in the form of a nucleoside triphosphate, called deoxyribose ATP (dATP).[2] C, G, and T arrive in similar forms of the other nucleotide: dCTP, dGTP, and dTTP. Hydrolysis of a triphosphate (dATP, dCTP, dGTP, or dTTP) releases pyrophosphate (PPi); this reaction is energetically favorable and drives the polymerization reaction (Figure 3.15). The 5′ phosphate of the new nucleotide is attached at the 3′ hydroxyl position of the old nucleotide to form a **phosphodiester bond**. Thus, DNA polymerization is said to proceed only in the 5′ to 3′ direction: New nucleotides are added to the 3′ end of the growing chain. DNA replication has other requirements: DNA polymerase requires a template to match complementary base pairs as well as a short segment of a few nucleotides (called a **primer**) to start the polymerization process.

Because DNA polymerization can only proceed in the 5′ to 3′ direction, the two strands are replicated differently: Replication of the two strands is asymmetrical (Figure 3.13). One of the strands, called the **leading strand**, is replicated in a

[2] dATP has the same structure as ATP (shown in Figure 2.11), except that the pentose ribose is replaced with deoxyribose. The other deoxynucleotides (dCTP, dGTP, dTTP) have analogous structures.

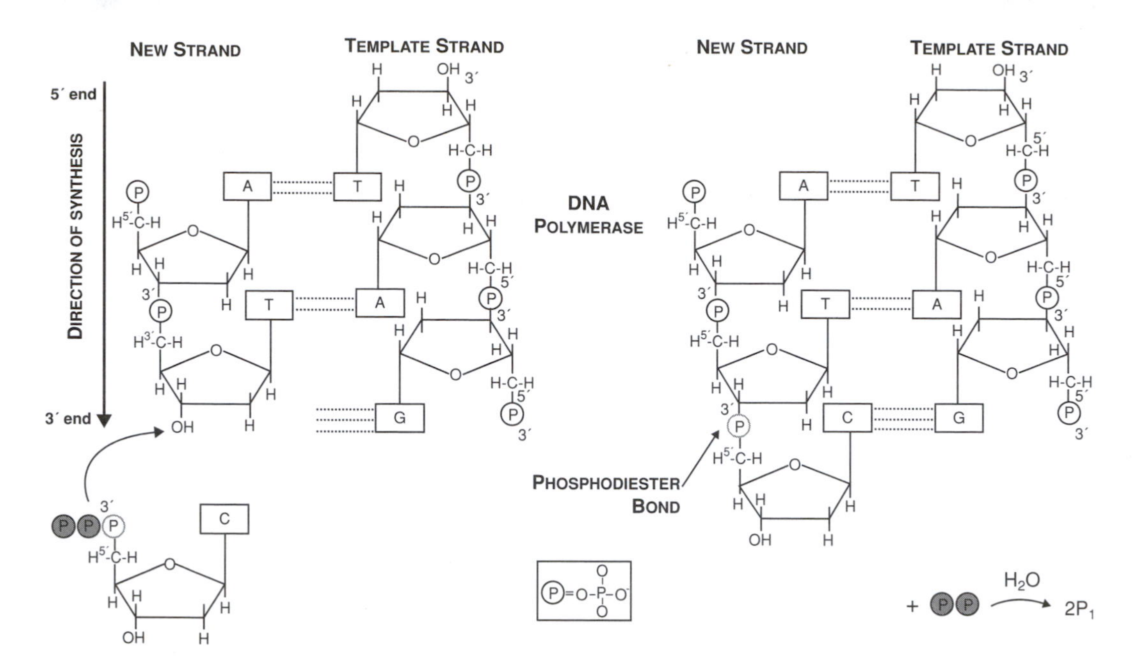

Figure 3.15 **Deoxyribonucleic acid (DNA) polymerization.** A polynucleotide has a 5′ phosphate end and a 3′ hydroxyl end. The polymerization reaction for adding nucleotides to a growing polynucleotide strand is catalyzed by DNA polymerase. A nucleoside triphosphate with a base complementary to the template strand base is added to the 3′ end of the new strand. Hydrolysis of the pyrophosphate (PPi) from the 5′ triphosphate of the incoming nucleotide provides the energy to drive the polymerization reaction. The newly formed covalent bond is called a phosphodiester bond. DNA polymerization can only occur in the 5′ to 3′ direction. The released pyrophosphate can be further hydrolyzed to form two inorganic phosphate molecules.

continuous manner because its replication occurs in the natural 5′ to 3′ direction. The other strand cannot be replicated continuously because of its reverse orientation. This strand, called the **lagging strand**, must be replicated discontinuously. On the lagging strand, short fragments of newly synthesized DNA are created, named **Okazaki fragments** after the husband and wife team Reiji and Tsuneko Okazaki of Japan. Eventually, all of these short strands must be joined together: This is accomplished by an enzyme called **DNA ligase** that catalyzes formation of the phosphate bond.

DNA polymerases proofread the newly synthesized DNA by going back and cutting out incorrect bases (similar to the function of the "backspace" key on a keyboard). Errors still occur, although usually less than one incorrect base per 10^8 base pairs incorporated. Spontaneous errors in DNA replication give rise to mutations. Mutations can also be induced by external factors such as ultraviolet light and chemical **carcinogens** (cancer-causing chemicals). Depending on where the mutations occur, these errors at the DNA level can result in production of abnormal proteins or overexpression or suppression of proteins. Any alterations in DNA in the germ cells are passed on to the offspring; mutations in somatic cells influence only the individual. Many types of cancer result from accumulated mutations in somatic cells (see Chapter 16).

In addition to DNA polymerase, the replication machinery contains several other proteins, each with a unique function (Figure 3.16). A **helicase** helps to

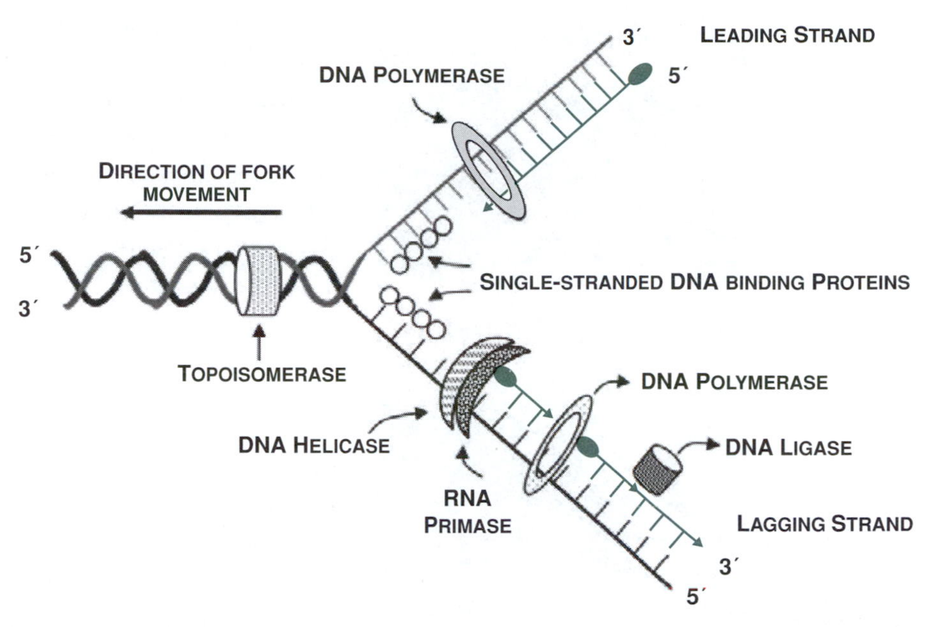

Figure 3.16 **Deoxyribonucleic acid (DNA) replication machinery.** Several enzymes work together to replicate the two template strands. DNA helicase unwinds the double helix so that both strands can be replicated. Ribonucleic acid (RNA) primase makes the short oligonucleotide primers needed by DNA polymerase to begin synthesis. Single-stranded DNA binding proteins help to keep the single strands from re-annealing. DNA ligase joins the Okazaki fragments together on the lagging strand. Topoisomerase prevents twisting of the chromosome.

unwind the double helix ahead of the replication fork; **DNA binding proteins** help to stabilize the single-stranded DNA so that it can be copied. Specialized proteins help to load DNA polymerase and keep it on the strand. **RNA primase** makes the initial primers needed to initiate DNA polymerization. These RNA primer bases are subsequently deleted by exonucleases. DNA ligase joins the Okazaki fragments together on the lagging strand. **Topoisomerases** prevent twisting of the chromosome by introducing nicks (single-stranded cuts) in the helix and rejoining them. Each of the components of the replication machinery is critical for DNA polymerization to proceed with speed and precision.

The importance of one of these enzymes can be illustrated by the effectiveness of a chemotherapeutic drug, **camptothecin** (Figure 3.17). Camptothecin is an agent that, when present in a cell, inhibits the action of the protein topoisomerase. By inhibiting topoisomerase, camptothecin prevents DNA synthesis and, therefore, stops the cell from dividing. Through its action on just one enzyme in this complex machine, the drug causes dividing tumor cells to fail.

3.4 The central dogma: Transcription and translation

How is the sequence of nucleotide bases in a gene used to synthesize a protein? Proteins are made by an orderly and sequential process, which is outlined in the "central dogma of molecular biology," introduced in Chapter 2 (Figure 3.18). In short, the instructions contained in DNA must first be transcribed to RNA and then translated to the corresponding amino acid sequence that forms the polypeptide

Figure 3.17 **Camptothecin is derived from the Chinese plant *Camptotheca acuminata Decne*.** Photo courtesy of James Manhart, Department of Biology, Texas A&M University, College Station, TX.

chain. It is unfortunate that the words used for these events—"transcription" and "translation"—are so similar, but students (and teachers) use a number of memory aids to remember the differences. For example, the process of converting DNA to RNA involves converting one nucleic acid sequence to another, similar to the copying of text from a page onto another page, a clerical operation that we often call transcription. Conversion of RNA to proteins, in contrast, involves conversion from nucleic acid to amino acid and involves a "handbook," which is called the **genetic code**. This process is similar to the linguistic operation of "translation" from one language to another.

The entire sequence of events that occurs in converting the DNA sequence, which resides somewhere on one of the chromosomes inside a cell, into a protein is often called **gene expression**. When a cell manufactures a certain protein, using its internal machinery to accomplish this synthesis, the cell is said to "express" the protein.

3.4.1 Gene transcription: RNA synthesis

The first step in gene expression is transcription of the DNA sequence into a corresponding RNA molecule. Thus, gene transcription is also called RNA

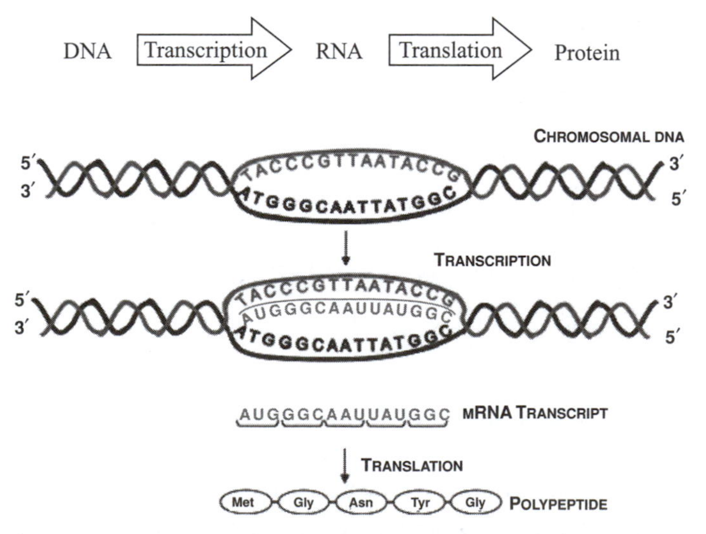

Figure 3.18 **Central dogma.** A segment of deoxyribonucleic acid (DNA) encoding for a gene is transcribed into ribonucleic acid (RNA) by RNA polymerase. The resulting RNA transcript is translated into a polypeptide chain based on the genetic code, which relates three-letter codons to amino acids.

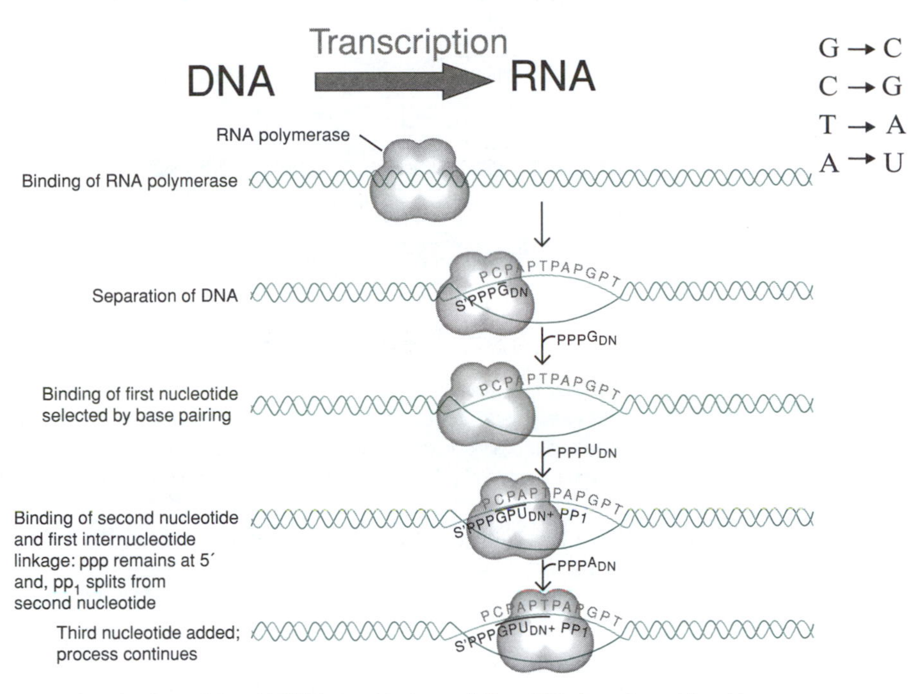

Figure 3.19 **Production of a ribonucleic acid (RNA) strand by transcription.** DNA, deoxyribonucleic acid.

synthesis. The result of transcription is a special RNA molecule called **messenger RNA** or mRNA. Gene transcription is similar to the process of DNA replication, except that only a small segment (i.e., one gene) of one strand instead of the entire genome is copied, as in DNA replication.

As in DNA replication, the work is done by a microscopic, protein-rich machine. The transcription machinery consists of several proteins, referred to as the **general transcription factors**, and **RNA polymerase**, the enzyme catalyzing the synthesis of a complementary RNA sequence. RNA polymerase requires a DNA template; it follows that template one base at a time, using it to decide which complementary base to add to the growing RNA strand. Like DNA synthesis, RNA synthesis also proceeds exclusively in the 5′ to 3′ direction. However, unlike DNA polymerase, RNA polymerase does not require a primer to initiate RNA synthesis.

RNA polymerase binds to a region of the DNA template, called the **promoter** region, located just upstream (to the left in Figure 3.19) of the start site of the gene to be transcribed. The DNA helix is unwound to expose the DNA template strand. The reaction catalyzed by RNA polymerase is similar to that of DNA polymerase in which nucleotides are added to the 3′ hydroxyl end of the growing RNA transcript (the reaction, which forms a new phosphodiester bond, is similar to the reaction shown in Figure 3.15). Complementary nucleotides are added to the growing RNA strand according to base-pairing rules (C→G, G→C, T→A, and A→U). There is a difference in base pairing in RNA compared to DNA synthesis; thymine is replaced by uracil in the synthesis of RNA. Transcription

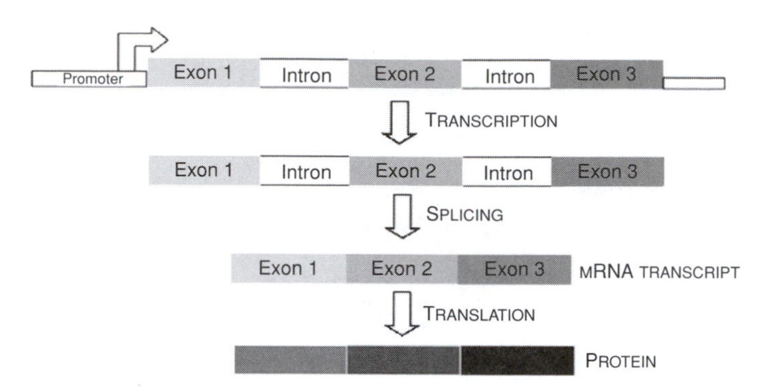

Figure 3.20 **Eukaryotic gene structure.** A typical eukaryotic gene is composed of an upstream promoter region that contains binding sites for the general transcription enzymes as well as other transcription factors. Encoding regions (exons) are separated by noncoding regions (introns). The introns are spliced or removed to yield a messenger ribonucleic acid (mRNA) transcript consisting of only exons. The resulting mRNA is then translated into a protein.

proceeds along the gene until a special sequence called the **termination codon** (see Section 3.4.3) is encountered.

3.4.2 RNA processing

A closer look at a typical eukaryotic gene structure—such as a gene found in humans—shows that the gene contains more information than just the base sequences necessary for describing the polypeptide (Figure 3.20). Of course, there are sequences that specify, or encode, the protein; these sequences are called **exons**. There are also sequences involved in regulation (such as the **promoter** described previously), and there are noncoding gene sequences (**introns**). In human cells (as in cells from all eukaryotes), the primary RNA transcript undergoes a process called **splicing** to remove noncoding regions called **introns**. The function of these regions is still not known, although some appear to regulate the processes of gene expression and RNA splicing.

Because both exons and introns are transcribed, the interspersed introns must be removed after RNA synthesis to form an mRNA that can be correctly translated into a protein. Post-transcriptional processing leads to the formation of a mature mRNA that consists only of exons and also a 5′ “cap” and about 200 adenylic residues at the 3′ end, referred to as a poly A tail. The 5′ cap is involved in translation, where it is recognized by the ribosome, and the 3′ poly A tail determines the stability of the mRNA. Simple organisms called prokaryotes—such as the bacterium *Escherichia coli*—have genes that do not contain introns, so splicing after RNA synthesis is not necessary. Therefore, in humans and other eukaryotes, a revised “central dogma” that includes the RNA processing step is appropriate (Figure 3.21).

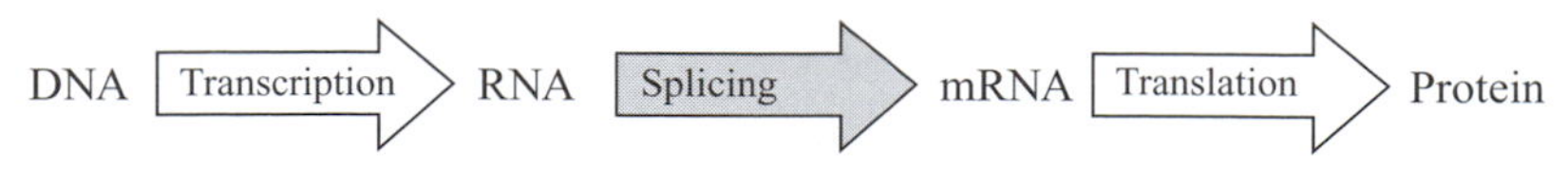

Figure 3.21 **Revised central dogma.** DNA, deoxyribonucleic acid, RNA, ribonucleic acid, mRNA, messenger RNA.

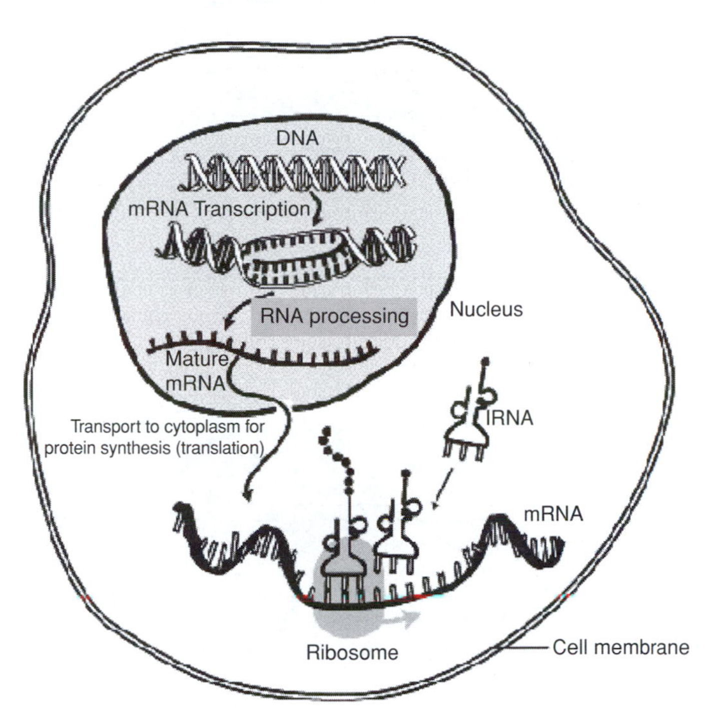

Figure 3.22 **Overview of protein production.** DNA, deoxyribonucleic acid, RNA, ribonucleic acid, mRNA, messenger RNA.

3.4.3 mRNA translation: Protein synthesis

All of the events described previously—transcription and RNA processing—occur in the nucleus of eukaryotic cells.[3] The newly synthesized mRNA moves out of the nucleus into the cytoplasm and is translated into a polypeptide chain (Figure 3.22). Translation of mRNA is the final step of protein synthesis, and is sometimes referred to simply as protein synthesis. Protein synthesis takes place on specialized organelles called **ribosomes**, which are examples of tiny, multicomponent machines, in this case composed of proteins and specialized packages of nucleic acids called **ribosomal RNAs** (rRNAs).

The challenge of protein synthesis is to convert a linear sequence of nucleotides (mRNA) into a linear sequence of amino acids (polypeptide). As in the previous steps, cells solve this problem in an orderly way. Starting at the 5′ end of the transcript, the mRNA bases are read by the ribosome, and a corresponding polypeptide chain is synthesized. Reading is accomplished in steps of three bases: Each set of three bases along the mRNA chain (called a triplet) defines a distinct element—a word, using the analogy of reading. This three-base word in the mRNA transcript is called a **codon**. Each codon that is encountered is translated into a specific amino acid: The rules for converting from the codon to the amino acid are given by the genetic code (Table 3.2). The genetic code contains 61

[3] In prokaryotes, transcription and translation both occur in the cytoplasm, where ribosomes bind the newly synthesized mRNA often before it is completely transcribed. As such, the process of transcription and translation are said to be coupled in prokaryotes.

Table 3.2 The genetic code

FIRST LETTER	SECOND LETTER: U	C	A	G	THIRD LETTER
U	UUU Phe	UCU Ser	UAU Tyr	UGU Cys	U
	UUC Phe	UCC Ser	UAC Tyr	UGC Cys	C
	UUA Leu	UCA Ser	UAA Stop	UGA Stop	A
	UUG Leu	UCG Ser	UAG Stop	UGG Trp	G
C	CUU Leu	CCU Pro	CAU His	CGU Pro	U
	CUC Leu	CCC Pro	CAC His	CGC Pro	C
	CUA Leu	CCA Pro	CAA Gin	CGA Pro	A
	CUG Leu	CCG Pro	CAG Gin	CGG Pro	G
A	AUU Ile	ACU Thr	AAU Asn	AGU Ser	U
	AUC Ile	ACC Thr	AAC Asn	AGC Ser	C
	AUA Ile	ACA Thr	AAA Lys	AGA Arg	A
	AUG Met	ACG Thr	AAG Lys	AGG Arg	G
G	GUU Val	GCU Ala	GAU Asp	GGU Gly	U
	GUC Val	GCC Ala	GAC Asp	GGC Gly	C
	GUA Val	GCA Ala	GAA Glu	GGA Gly	A
	GUG Val	GCG Ala	GAG Glu	GGG Gly	G

Notes: Three-letter messenger RNA (mRNA) bases form codons that correspond to one of the 20 amino acids. There are also three codons (UAA, UAG, and UGA) that code for STOP signals; AUG, which codes for methionine, also serves as a START codon. Note that there are 61 codons that encode for 20 amino acids. The genetic code is degenerate because more than one codon can encode the same amino acid.

codons for 20 amino acids with three stop codons (or termination signals, which were mentioned earlier). The start codon, AUG, which codes for the amino acid methionine, begins almost every polypeptide chain in eukaryotic and prokaryotic cells.[4] Because more than one codon can encode for the same amino acid (notice in Table 3.2 that both UGU and UGC encode for the amino acid cysteine), the genetic code is said to be **degenerate**. Each mRNA polymer translates to a unique polypeptide sequence, but any given polypeptide could have been specified by many different mRNA sequences.

The work of translation is done by a hybrid molecule called **transfer RNA** (tRNA). Transfer RNA is a hybrid of an amino acid and a short sequence of RNA, which serves an adapter molecule. The tRNA recognizes the codon and carries with it the appropriate amino acid to add to the growing polypeptide chain (Figure 3.23). One end of each tRNA carries an anticodon complementary to the bases of the codon on the mRNA transcript. Attached, at the other end of the tRNA, is the corresponding amino acid. New amino acids are added to the carboxyl end of the previous amino acid to form a peptide bond.

After the polypeptide chains are synthesized, they undergo folding and processing reactions, which convert them into functioning proteins. More on this topic is described in Chapter 4. As mentioned earlier, many proteins are made up of more than one polypeptide chain. In eukaryotic cells, newly synthesized polypeptides undergo post-translational modifications such as cleavage of the chain (**proteolysis**), attachment of carbohydrates (**glycosylation**), or attachment of fatty

[4] Actually, formylmethionine is used in prokaryotes, which has consequences for expressing recombinant mammalian proteins.

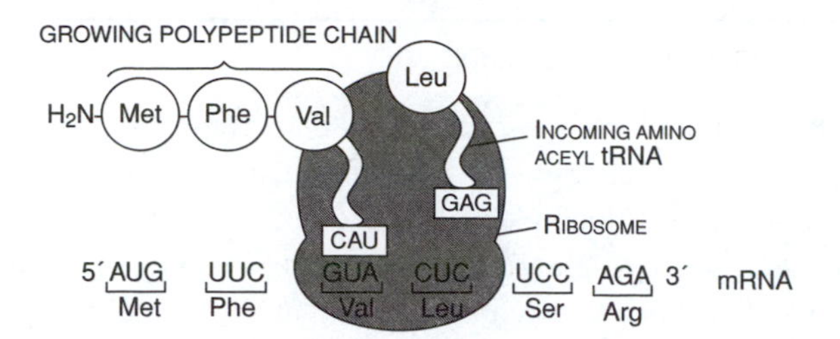

Figure 3.23

Polypeptide synthesis. Amino acids are added to a growing polypeptide chain by transfer ribonucleic acids (tRNAs), which serve as adapter molecules. Each tRNA has an anticodon on one end and the corresponding amino acid on the other. Three messenger RNA (mRNA) bases (a codon) are read at a time by an incoming tRNA. If the anticodon on the tRNA is complementary to the codon in the mRNA, then the amino acid from that tRNA is added to the polypeptide chain.

acids (**N-myristoylation**) or lipids (**prenylation**). These modifications are critical for the proper function of the protein.

Prokaryotic cells are not capable of post-translational modification. This limitation creates a potential problem for engineers who want to make human proteins in bacterial cells. Sometimes this problem can be solved, as it was by the biologists and engineers who learned how to produce insulin in bacterial cells (Box 3.3). Often, however, to produce correctly modified proteins, genetically engineered proteins are made in eukaryotic cells because they have the necessary machinery to make the modifications (see Section 3.6.2).

3.5 Control of gene expression

All of the cells of the human body (except red blood cells) contain the same genetic information, but only a fraction of the genes in a particular cell type are expressed. For example, only certain cells of the pancreas make insulin, although every cell with a nucleus has the gene for insulin. Cells of the brain, liver, skin, and heart each express a different set of genes, which gives rise to their unique properties (Figure 3.24). A muscle cell and a skin cell contain the same genetic information and, therefore, have the capacity to make all of the proteins of the body, but each expresses different genes. How is gene expression controlled in individual cells?

Gene expression can be regulated at each step in the pathway of converting DNA into protein (Figure 3.25). At the DNA level, the promoter regions of genes can bind specific **transcription factors**, which can either enhance or inhibit transcription of the gene. Steroid hormone receptors are examples of transcription factors; they act to increase transcription of target genes in response to the presence of a hormone (this is discussed in more detail in Chapter 6, Section 6.4.3). Another level of control occurs during the RNA processing step: RNA transcripts can sometimes be spliced in alternate ways to yield different mRNA. The stability of the mRNA transcript can be controlled by increasing degradation before translation. After the mRNA is translated, the resulting proteins can be inactivated or compartmentalized until they are needed.

Box 3.3 Making human insulin in bacteria

The human disease diabetes mellitus—or more commonly, diabetes—has been known since ancient times. In diabetes, the body produces an insufficient quantity of the protein hormone insulin, which is important in regulating blood glucose levels (see Chapter 6). Diabetic patients often produce large quantities of glucose-rich urine; diabetes mellitus literally means "to flow honey." For hundreds of years, diabetes was diagnosed in patients by "water tasters" who specialized in detecting sweetness in urine.

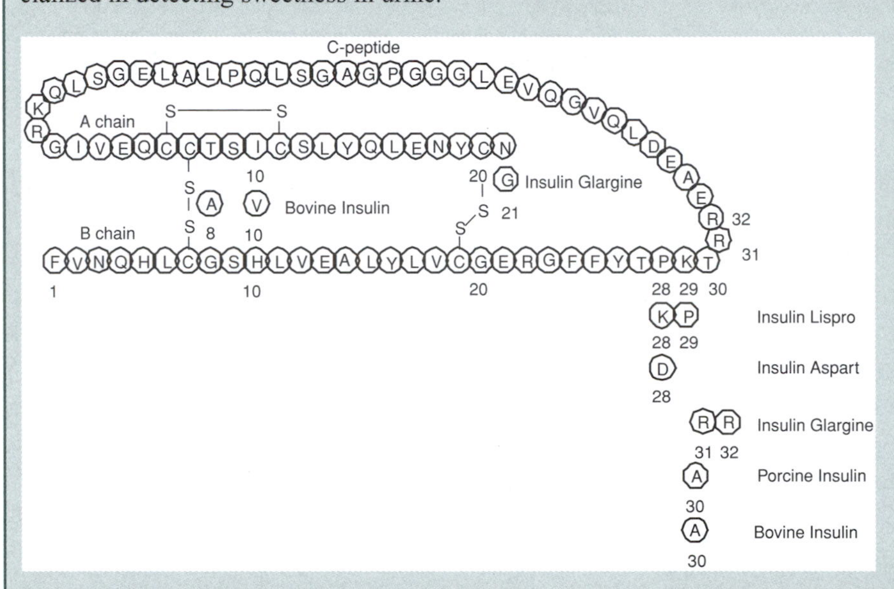

Amino acid sequence of human proinsulin and the various commercially available insulin analogs. Insulin itself is composed of two chains, A and B, which are produced by cleavage of the proinsulin polypeptide to remove the C-peptide. Species differences in insulin are also noted. From (6).

The Egyptian physician Hesy-Ra described symptoms of the disease in 1552 BCE. For a long time, the only useful treatment for diabetes was severe control of diet. Eventually a series of experimental studies linked the pancreas to diabetes. Following this line of research, in 1921, Canadian surgeon Frederick Banting persuaded Professor John MacLeod, the new head of the department of physiology at the University of Toronto, to give him laboratory space and animals to test an idea that he had for purifying the substance from the pancreas that was responsible for diabetes. Working with an assistant, Charles Best, Banting showed that pancreatic extracts from one dog could lower the blood glucose in another. In 1922 they tested this approach in a 14-year-old diabetic Canadian boy, Leonard Thompson, who responded remarkably to the treatments. Banting and Macleod were awarded the Nobel Prize for this discovery in 1923 (and shared the money from the prize with Best and others). Over the next decades, the active ingredient in these extracts was identified as insulin; safe, reliable methods for preparing insulin from animals for treatment of human disease were developed.

Prior to the era of DNA technology, insulin that was used to treat diabetics was extracted from animals, primarily cows and pigs. Bovine (cow) and porcine (pig) insulin are similar to human insulin, but not identical. As the figure illustrates, bovine insulin differs from human insulin by three amino acids (positions 8 and 10 of the A chain and position 30 of the B chain), whereas porcine insulin differs by one amino acid (position 30 of the B chain). Because of these differences, the biological activity of animal insulin is not identical to that of human insulin. More importantly, because these are foreign proteins (i.e., they are not identical to human insulin), some patients develop antibodies that neutralize insulin action.

To create recombinant human insulin, the gene for insulin was inserted into a plasmid, pBR322 (Figure 3.30), using the restriction enzyme *Bam*HI. The plasmid was then introduced into *Escherichia coli* for production of the protein. By inserting the insulin gene into the tetracycline-resistance gene of pBR322, it was possible to identify bacteria with the recombinant gene by growth of the cells in tetracycline. (Why will cells with the recombinant plasmid grow in tetracycline? How do you identify these cells from cells that do not receive any plasmid?)

Notice that there are at least two ways to make insulin: 1) make each polypeptide chain (A and B) in separate bacteria and join them together subsequently, or 2) make proinsulin and chemically cut out the C-peptide. Although both approaches work, the second method is generally used for making recombinant human insulin.

Scientists and engineers use various methods to block expression of specific genes. In the laboratory, this is often done to study the function of genes and their products, but this approach is moving quickly toward new therapies for disease. There are several methods for blocking gene expression. In one approach, **antisense** oligonucleotides can bind to complementary base pairs in DNA or RNA to block transcription or translation, respectively (Figure 3.26). However, these single stranded, antisense oligonucleotides are short-lived. More recently, **small interfering RNAs** (siRNA) have been used to target degradation of specific mRNA transcripts. siRNAs are duplex strands of less than 30 bp that are stable and can persist for several weeks. When a duplex siRNA binds to the complementary bases in the target mRNA transcript, the cell's machinery causes degradation of that transcript. This method of selectively turning off the expression of specific proteins is being used in cell culture experiments as well as in vivo animal experiments. Both antisense and siRNAs have great therapeutic potential, but a

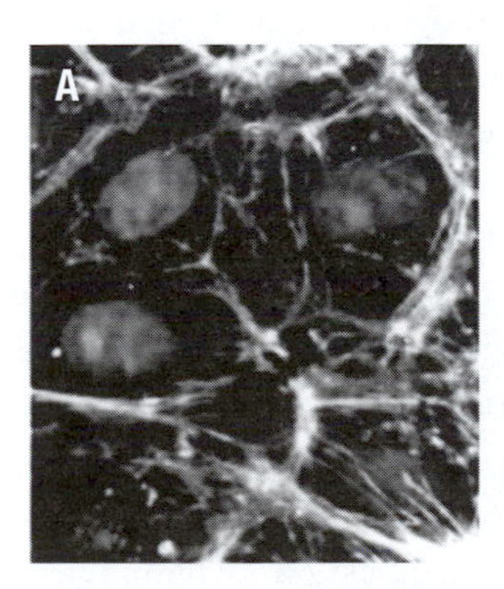

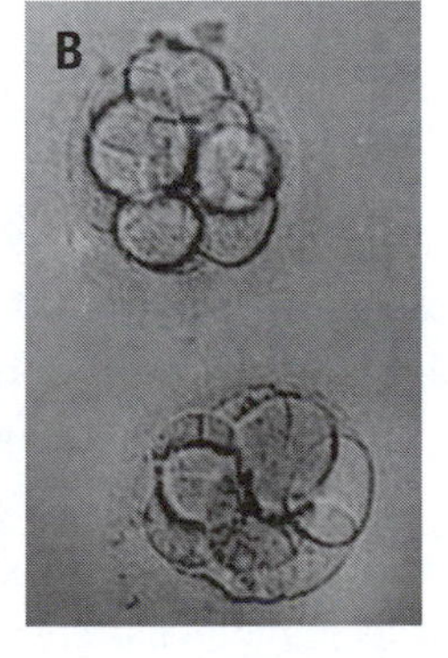

Figure 3.24 **Images of different types of cells. A.** This fluorescent image shows endothelial cells, the cells that form the lining of the cardiovascular system. The red fluorescence shows the nucleus of the cells, and the green shows the cytoskeleton. Figure from (4), with permission copyright (2001) National Academy of Sciences, USA. **B.** An embryo at the 12-cell stage. Photo courtesy of Dianne G.

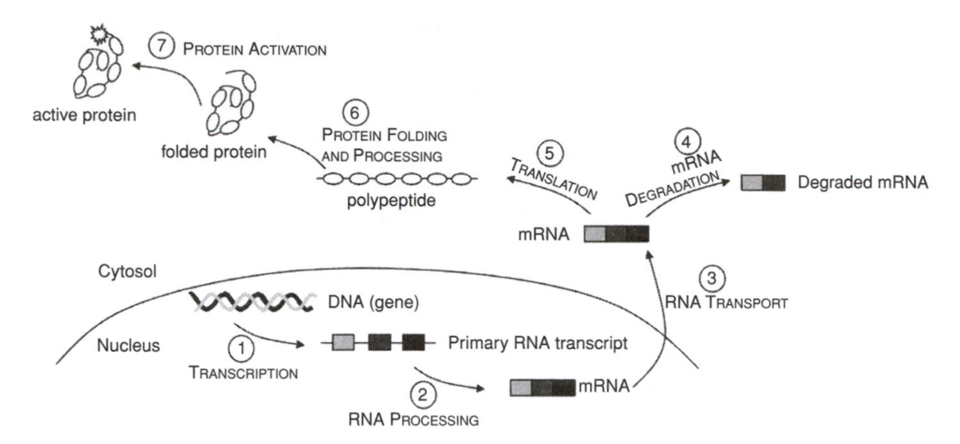

Figure 3.25 **Control of gene expression.** The expression of a particular gene can be controlled at different stages. (1) The initiation of transcription can be enhanced or inhibited by transcription factors that bind to regulatory sequences outside of the coding region. (2) Once the primary transcript is made, it can be alternatively spliced to yield different messenger ribonucleic acids (mRNAs) during the RNA-processing step. (3) The translocation of the transcript out of the nucleus and into the cytoplasm where the translation machinery is located can be affected. (4) The presence of the mRNA in the cytoplasm can be controlled by degradation. (5) The rate of translation can be altered. (6) After being formed, the polypeptide may undergo post-translational modification and processing that may affect where the protein is expressed (i.e., cytoplasm, membrane, or secreted). (7) A protein can remain inactive until a signaling event causes it to become activated.

key problem is adequate delivery of these agents through the phospholipid bilayer of the cell membrane and into the cytoplasm. Biomedical engineers are helping to design better drug-delivery systems not only to improve efficiency and duration of nucleic acids for siRNA, but also for gene therapy strategies; some of these methods are discussed in Chapter 13.

3.6 Recombinant DNA technology

The previous sections describe some of the molecular events that occur in cells, allowing them to reproduce their genes and make proteins from them. In the relatively short period from 1953, when Watson and Crick discovered the DNA double helix, to now, humans have learned how to manipulate genes, to replicate them efficiently, to move genes between cells and organisms, and to control their expression. **Recombinant DNA technology** refers to a set of techniques that enables scientists to transfer genetic information from one organism to another. It is one of the major technological achievements of the past 100 years.

This section reviews some of the basic concepts underlying recombinant DNA technology. This technology has emerged quickly from basic science to commercial and medical applications (Table 3.3). Just as discoveries in physics led to increased work for electrical and mechanical engineers, these new discoveries in molecular biology have created extraordinary opportunities for biomedical

Table 3.3 Recombinant proteins

Protein	Use
Erythropoietin	Treats anemia
Granulocyte-colony stimulating factor	Treats blood disorders
Growth hormone	Promotes growth
Insulin	Lowers blood glucose (diabetes treatment)
Interferons	Anti-viral, anti-tumor agent
Nerve growth factor	Promotes nerve damage repair
Tissue plasminogen activator	Thrombolytic agent

Notes: Some of the human proteins produced by recombinant DNA technology are used as drugs to treat various conditions and diseases. Adapted from (8).

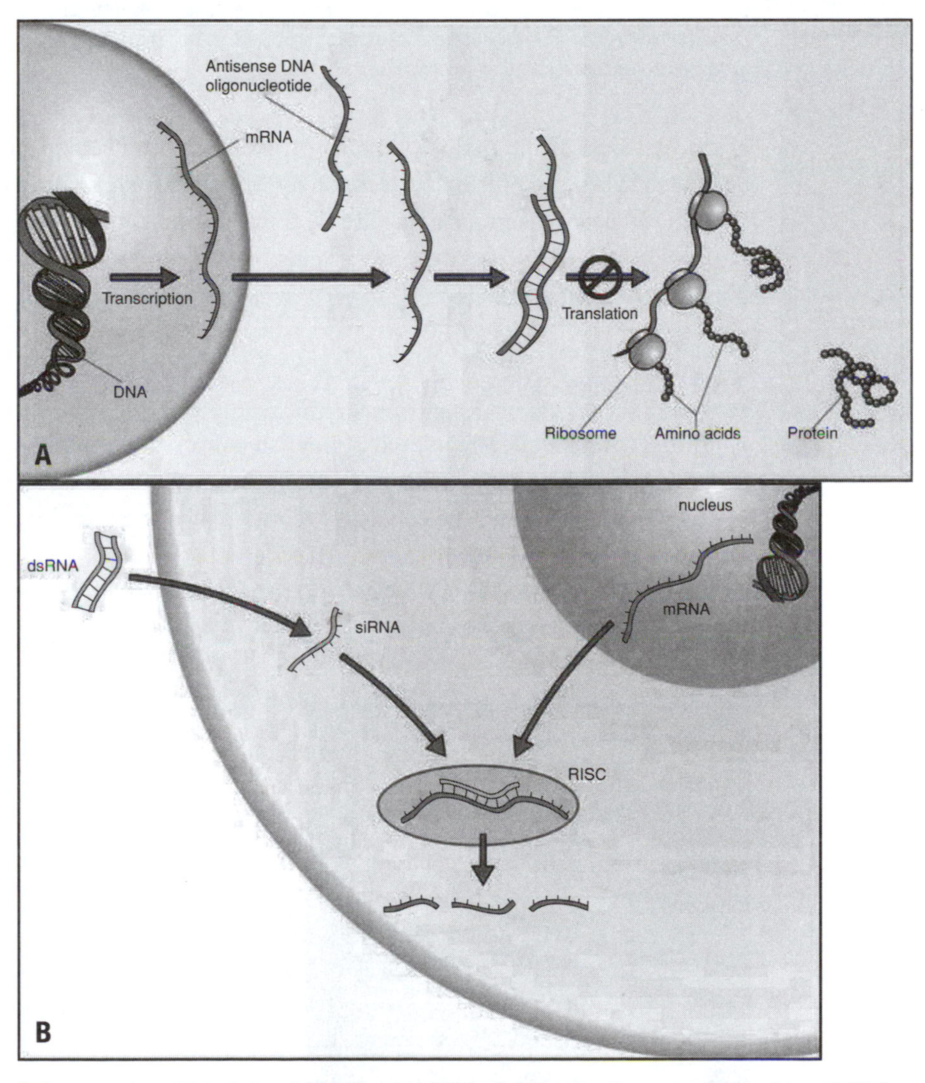

Figure 3.26

Antisense and small interfering ribonucleic acid (siRNA). A. Action of antisense deoxyribonucleic acid (DNA) or RNA, which binds to messenger RNA (mRNA) and prevents protein synthesis. Adapted from (5). **B.** siRNA activates a cellular process of mRNA degradation. Adapted from (5). dsRNA, double-stranded RNA; RISC, RNA-induced silencing complex.

Figure 3.27 **Biomolecular engineering seeks to invent new approaches to human health through the basic science of molecular biology.** Engineers use approaches based on stem cells, genetic engineering, and molecular interactions to design new approaches for therapy of disease.

engineers. One of the major subfields of BME—biomolecular engineering—is dedicated to bringing forth new technology from this basic science (Figure 3.27). Biomolecular engineers of today are working on smart biomaterials, agents for molecular imaging, gene therapy systems, and stem cells (among many other things; see Chapters 13–16).

3.6.1 Molecular cloning

Cloning is now frequently uttered in English conversations. The word originates from the Greek *klon*, which means twig. Because of the biological ramifications of cloning, the word and the concept it represents have entered our culture, with influence beyond biological science (Figure 3.28). Indeed, even the biological consequences of cloning have been more profound than predicted when the techniques were first discovered.

Figure 3.28 In the 20th century, cloning became an important cultural concept.

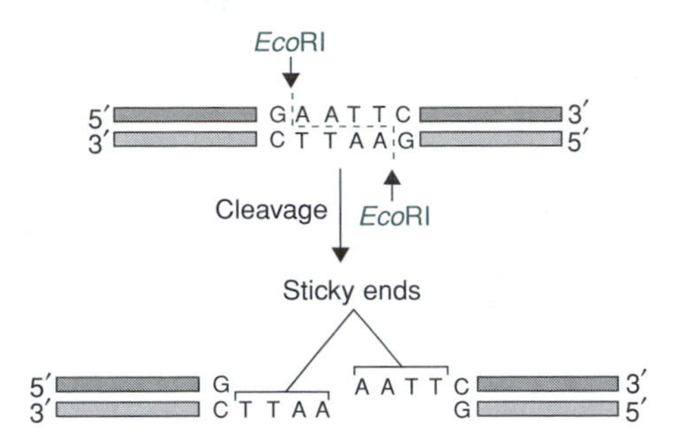

Figure 3.29

Action of a restriction enzyme on double-stranded deoxyribonucleic acid (DNA). This diagram shows the action of the particular enzyme called ***Eco* RI**.

WHAT IS MOLECULAR CLONING?

Cloning, at the molecular level, is the crux of recombinant DNA technology. All of the applications listed in Table 3.3 require some type of molecular cloning. The technology of cloning grew out of the desire of biologists to study genes. To study a particular gene, it is necessary to produce large quantities of it. Genes are polymers of DNA, and every gene, because it is just a reordering of the same four monomers, has roughly the same chemical properties. It is possible to separate genes based on size (see Box 3.2 for an example), but how can a favorite gene be separated from a mixture of thousands of other genes that are the same size? As an alternate for separation, biologists discovered that they could make millions of copies of a gene of interest. By making many copies of the gene they wanted, they could avoid the problem of separating rare genes from a mixture of similar molecules.

Cloning means making identical copies. A good photocopy machine clones documents. In molecular biology, cloning usually refers to genes. Therefore, to clone a gene means to make many identical copies of a particular length of DNA. When a gene is cloned, it is also sometimes said that the gene is amplified. Cells and organisms can be cloned as well. When a cell is cloned, it is reproduced under conditions such that all of the new cells are progeny of a single cell and, therefore, are identical. This topic is considered further in Chapter 5. A cloned organism is genetically identical to its parent, as discussed later in this chapter.

HOW IS A GENE AMPLIFIED OR CLONED?

Molecular cloning begins with the use of specialized enzymes called **restriction endonucleases**, also called **restriction enzymes**, which cut the DNA at specific sites. These enzymes are introduced in Box 3.2, as they are used in DNA fingerprinting. Often, the restriction enzymes create fragments that have unpaired bases or "sticky ends" (Figure 3.29). These fragments of DNA can then be inserted into a **cloning vector** (such as a **plasmid**), where it pairs with the corresponding sticky ends in the DNA of the vector (Figure 3.30). This pairing is another example of the importance of DNA hybridization: The gene finds the correct place to insert into the plasmid because of hybridization, which is driven by base-pair–matching interactions. The enzyme DNA ligase is then used to covalently link the

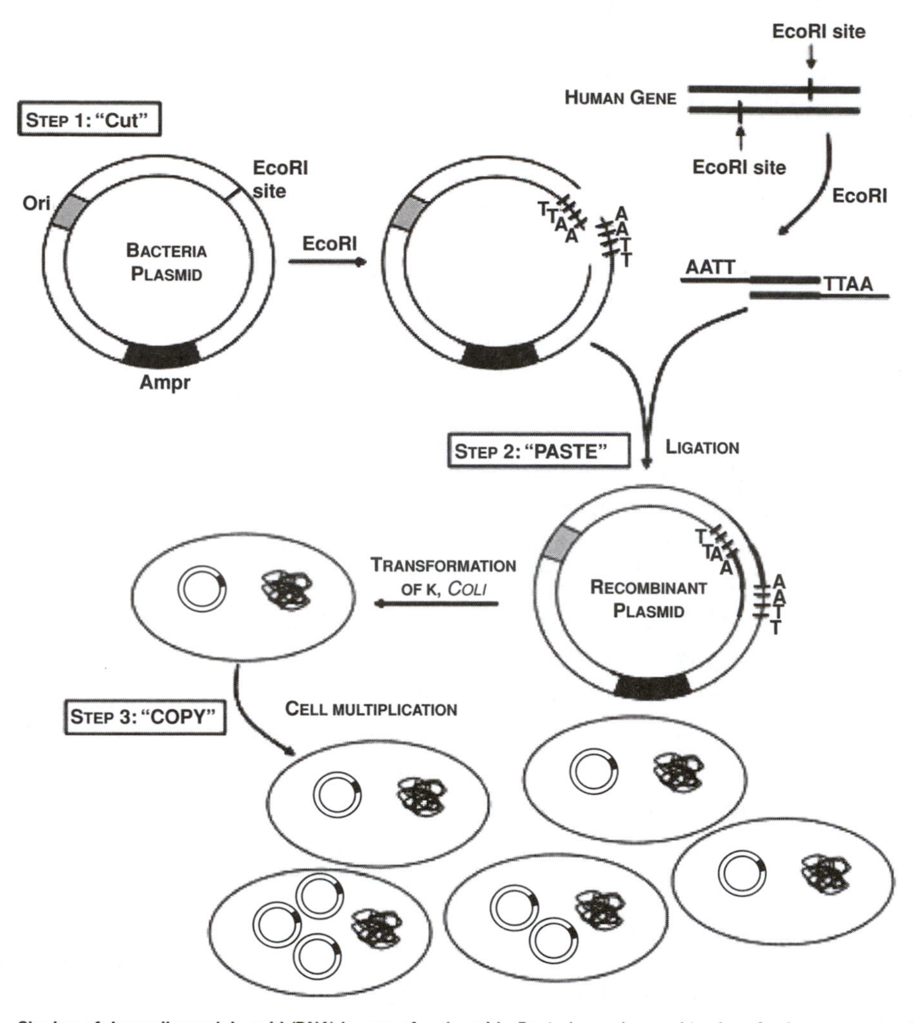

Figure 3.30 **Cloning of deoxyribonucleic acid (DNA) by use of a plasmid.** Bacteria can be used to clone foreign segments of DNA. First, the cloning plasmid and the gene of interest must be cut with the same restriction enzyme to create sticky ends. The bases at the ends of the gene segment are complementary to those in the newly cut plasmid. Next, for the "paste" step, DNA ligase catalyzes the formation of a phosphodiester bond between two nucleotides to seal the gap. The recombinant plasmid containing the new gene segment is used to transform bacteria. The bacteria are grown on an agar plate containing nutrients and antibiotics. Only those bacteria that have been transformed will survive because they carry the antibiotic-resistance gene in their foreign plasmids. Finally, the transformed bacteria cells divide and make copies of the plasmid DNA at each cell division.

foreign DNA with the plasmid DNA. In 1972, Paul Berg of Stanford University was the first to use restriction enzymes and ligases to produce a recombinant DNA molecule—that is, a DNA molecule that contained DNA from two different organisms. Berg later won the Nobel Prize in chemistry for that discovery.

To clone a gene from a plasmid, the recombinant vector must be inserted into an organism that can serve as a host for its replication. Stanley Cohen of Stanford University and Herbert Boyer of the University of California San Francisco described the first molecular cloning in 1973. The technique has ramifications far beyond the copying function described here. Because DNA is chemically identical in all cells, this technique can be used to move genes from one organism to another. Genentech, a company started by Herbert Boyer and Robert Swanson in 1976, is representative of the larger biotechnology industry, which is based largely on the concept of expression of human proteins in alternate organisms. Recombinant human insulin, for example, is made from *E. coli* cells that contain the genes for human insulin (Box 3.3): Genentech produced the technology for **recombinant** human insulin, which was licensed to the pharmaceutical company Eli Lilly and approved for use in treating human diabetes in 1983. Several other recombinant proteins are now available for treating human disease (Table 3.3).

Cloning depends on specialized vehicles, or vectors, that are matched with a cellular host to produce abundant replication of the vehicle (and therefore the gene of interest that is riding within it). **Plasmids** are popular vectors for gene cloning; plasmids are small circular DNA molecules that replicate independently of chromosomal DNA in bacteria. Plasmids and restriction endonucleases occur naturally in microorganisms such as bacteria. Plasmids naturally function to move genes between individual microorganisms; restriction enzymes are part of a microorganism's natural defense mechanisms, allowing them to cut up or destroy genes from invaders that enter their cytoplasm.

Most commonly used cloning plasmids contain an origin of replication, sites for cutting by restriction enzymes, and at least one antibiotic-resistance gene. A map for a typical plasmid, called pBR322, is shown in Figure 3.31; the restriction enzyme cutting sites are indicated by the abbreviation for the restriction enzyme and its position on the plasmid (such as *Eco*RV 185, which indicates that the enzyme *Eco*RV—is at position 185). The restriction enzymes are named for the organism in which they were found: Thus, *Eco*R is strain R of *E. coli*, and V indicates that it is the fifth restriction enzyme found in that strain.

Most restriction enzymes recognize symmetric sequences that are 4, 5, or 6 base pairs in length. A symmetrical sequence is one in which the two strands of the double-stranded DNA will be identical when each is read in the 3′ to 5′ direction. Because they are complementary, they will look like mirror images in their double-stranded positions. For example, the enzyme *Sa*lI recognizes the sequence GTCGAC. A few restriction enzymes do recognize longer sequences, and some recognize slightly asymmetrical sequences, but these are in the minority. The enzymes usually cut symmetrically, although the symmetrical cut can leave overhanging single-stranded tails on the 3′ or the 5′ ends. These are called "sticky ends," and are useful in inserting a gene of interest and re-closing the plasmid, as discussed above.

SELF TEST 2 Show that *Sa*lI is a symmetrical sequence.

ANSWER: Single strand—GTCGAC; complementary strand—CAGCTG, which is the reverse of the single strand.

A gene of interest is inserted into this cloning vector, using restriction enzymes to open the plasmid at known locations. Cloning plasmids usually contain a gene that will be expressed in the host, conferring resistance to an antibiotic such as

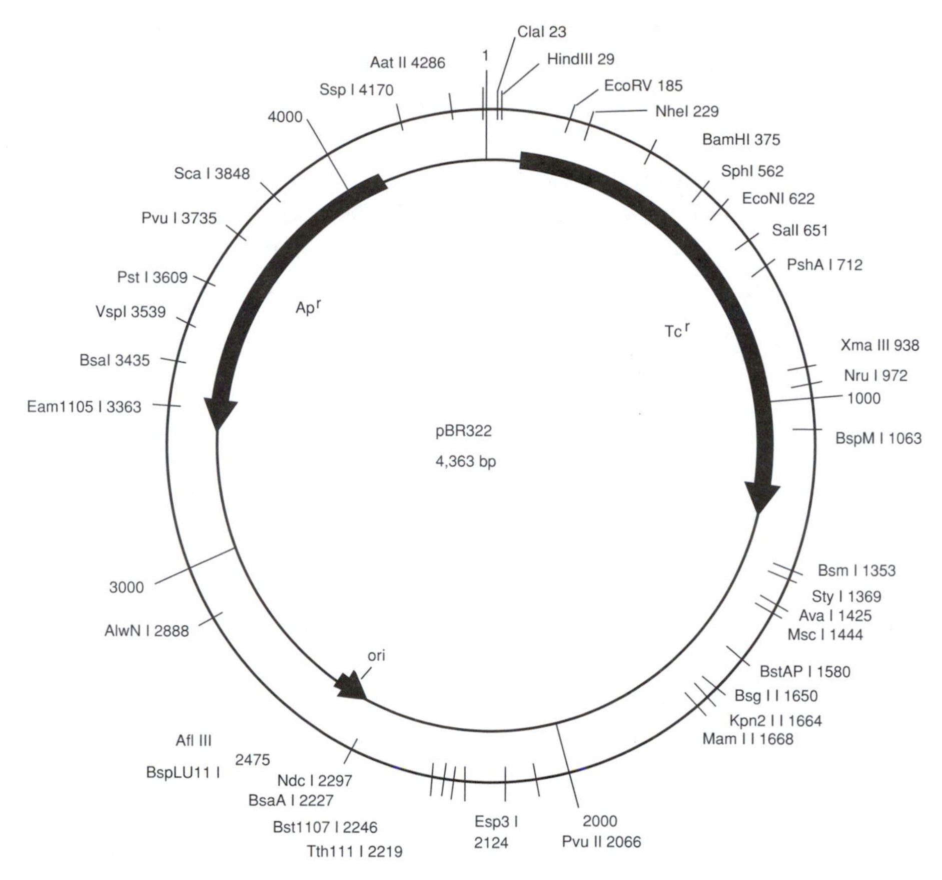

Figure 3.31 **Typical plasmid pBR322 (available commercially from Invitrogen).** This vector contains an origin of replication (ori) and two antibiotic-resistance genes: Ap (resistance to ampicillin) and Tc (resistance to tetracycline). The plasmid contains three unique restriction endonuclease recognition sites within the b-lactamase (AmpR) gene and eight unique sites within the TetR gene. The plasmid, purified from DH10B^{TM} *Escherichia coli*, also has 14 nonselectable unique restriction endonuclease recognition sites. Unique cleavage sites for *Hin*dIII and *Cla*I are found in the promoter of the tetracycline resistance gene. Insertion of deoxyribonucleic acid (DNA) at either of these two sites usually results in loss of tetracycline resistance.

ampicillin or tetracycline. This gene provides an easy method for finding cells that contain the plasmid: If the cells will grow in ampicillin, they must be resistant to it, and they probably obtained that resistance from a functional plasmid that resides within their cytoplasm.

To express the foreign gene on the plasmid, the plasmid is introduced into host cells. There are plasmids that can be used in either bacteria or mammalian cells: It is important to select a plasmid that matches the host. The process of inserting a plasmid into bacterial cells is called **transformation**. Because the plasmid is a large molecule, it does not enter the cell easily. Therefore, to perform transformation, the bacteria are usually exposed to chemical agents that break down the cell membrane barrier temporarily or the cell membrane is mechanically disrupted, using direct microinjection or electroporation (where the cell is exposed to a high-voltage electrical pulse). In animal cells, the introduction of a plasmid into the cells is called **transfection**. Again, the plasmids do not

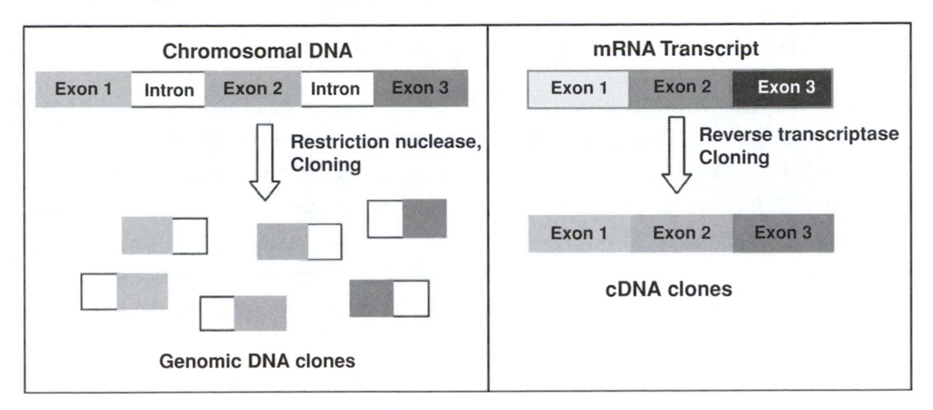

Figure 3.32 **Genomic versus complementary deoxyribonucleic acid (cDNA) clones.** Genomic clones are made from chromosomal DNA so both introns and exons are present. cDNA clones are made from messenger ribonucleic acid (mRNA) transcripts that have already been spliced to remove introns. Thus, cDNA clones only contain gene encoding exons. The abundance of mRNA transcripts in the source cell will determine the relative amount of cDNA produced. Thus, certain transcripts could result in a high number of cDNA clones, whereas genomic DNA clones represent an equal sampling of all sequences regardless of the level of tissue-specific transcription.

enter the cell naturally, but can be encouraged to enter by chemical or mechanical means. Of course, these chemical and mechanical treatments often kill the cells that they intend to transfect; as a result, many biomedical engineers have worked on methods that can cause transfection with fewer toxic effects on cells (for a review of this progress, see [7]).

Genes can also be introduced into animal cells using viruses. This process is discussed in Box 3.4. When genes are introduced into cells by viral methods, the process is called **transduction**. Thus, bacterial cells are transformed, and animal cells are either transfected or transduced.

Figure 3.30 illustrates the process of cloning by transformation of bacterial cells. To create DNA clones, the plasmid containing the foreign DNA fragment is used to transform bacteria. The bacterial cells are grown on an agar plate that contains an antibiotic; only those cells that have been transformed will grow on the plate because they contain a plasmid with the antibiotic-resistance gene. This method is called **antibiotic selection** because an antibiotic is being used to provide conditions that allow only the cells of interest (the selected cells) to grow. After this antibiotic selection step, the transformed bacteria can be collected from the plate and grown in a culture flask to produce many copies. When adequate quantities of transformed bacteria have been produced, the plasmid DNA can be purified from the bacterial cell components and genomic DNA. The resulting copies of recombinant DNA can be isolated from the plasmid (using restriction enzymes) and used or further analyzed.

HOW IS RNA CLONED?

RNA can also be cloned, but mRNA must first be converted to its **complementary DNA (cDNA)** by a special enzyme called **reverse transcriptase** (see Figure 3.32). Because it is derived from mRNA, cDNA contains only the coding sequences

Box 3.4 Gene therapy and viral vectors

Gene therapy generally refers to the introduction of an exogenous wild-type transgene to correct for a defective gene. This type of therapy is limited to diseases with single-gene mutations (Table 3.1). Other applications of gene therapy include: 1) introduction of toxic genes to kill tumor cells, and 2) introduction of genes that encode for therapeutic proteins such as insulin or growth factors.

Current gene therapy is performed in somatic cells and not reproductive germ line cells (egg and sperm). Therefore, the effects of somatic gene therapy cannot be passed on to progeny. A major obstacle in somatic gene therapy is transfection efficiency of the desired cells with the gene of interest. Because delivering naked DNA is inefficient, gene delivery vectors are used to help the vector with DNA get into the cell. These vectors can also be used for ex vivo gene therapy—which involves infecting cells outside the body and then introducing the modified cells back into the patient.

What are viruses and how can they be used as expression vectors? Let us take a closer look at the life cycle of a **retrovirus** in the figure below. Retroviruses are single-stranded RNA viruses that infect animal cells and use the host transcription and translation machinery to produce viral proteins. Human immunodeficiency virus (HIV) is an example of a retrovirus that specifically attacks immune cells called helper T cells. The retrovirus binds to molecules located on the plasma membrane and enters the host cell by fusion of the virus outer coat with the membrane. When inside the host cell, the **capsid** breaks down, and reverse transcriptase carried within the virus synthesizes a DNA copy of the viral RNA. A second DNA strand is also synthesized to yield a double-stranded DNA molecule. This double-stranded viral DNA is eventually randomly integrated into the host genome. Thus, the viral genes are transcribed and expressed using the host machinery. The viral genes direct production of viral proteins, such as components of the capsid and envelope, and the enzyme reverse transcriptase. These components are assembled to form new virus particles, which all contain a copy of the original viral RNA sequence. To leave the cell, the plasma membrane folds around the capsid to form buds, which pinch off to form new virus particles. Depending on the type of retrovirus, infection can either have no pathological consequence for the host cell or can damage the infected cell (such as HIV).

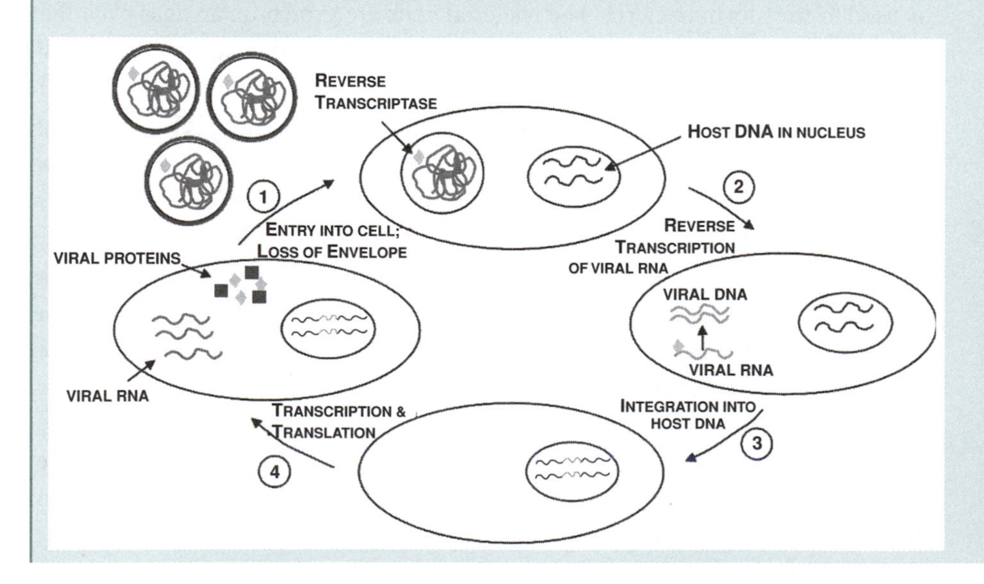

Retrovirus life cycle

Retroviruses that have been rendered replication defective can be used as gene delivery vectors. In this case, genes that code for viral proteins are replaced with complementary DNA encoding for the therapeutic protein (or other protein of interest). Because the retroviral genome integrates into the host genome, expression of the therapeutic gene is stable. However, the integration site within the host genome is random, so there is the potential that insertional mutagenesis may occur (i.e., an important gene may be disrupted at the site of integration). Retroviruses can only transduce dividing cells because they require cell division for the integration step. Another disadvantage of retroviral vectors is that they can only harbor genes that are smaller than ~7 kb.

Other viruses are also used to deliver genes, such as **adenoviruses**, which are self-replicating DNA viruses. Genes transferred via adenovirus transduction remain as extrachromosomal DNA called episomes, which are only transiently expressed. An adenovirus naturally infects the epithelial cells of the respiratory track. Therefore, adenoviruses containing an inserted wild-type CFTR gene have been delivered via a nose spray for gene therapy treatment of cystic fibrosis. The normal CFTR gene encodes for an ion channel, which allows the release of Cl^- and other ions. Unfortunately, protein expression in the transduced cells is transient because the DNA is not integrated into the chromosome of the target gene. Expression lasts about 2–3 weeks, so repeated transductions are necessary to maintain stable expression of a gene. Adenoviral vectors can transduce nondividing cells—which is advantageous because most of the cells in an adult are quiescent.

So far, none of the current vectors—either viral or nonviral—possess all of the features of an ideal gene delivery vector: 1) the ability to incorporate large genes, 2) the absence of immunogenicity, 3) the ability to target a specific cell type, 4) high transduction efficiency, 5) the ability to turn gene expression on and off, and 6) the ability to regulate gene expression in response to exogenous or endogenous hormonal signals.

Chapter 13 describes an alternate approach for gene therapy, using nonviral or synthetic systems.

(i.e., the parts of the DNA found in exons, often called the **structural gene**), so the amino acid sequence of the original protein can be deduced.

CLONING BY PCR

For small DNA fragments (<2 kb), an alternate approach to cloning can be used. **PCR** will amplify any intervening DNA sequences between two primer **oligonucleotides** (Figure 3.33). Recall that DNA polymerase requires both a template and a primer. The primers in this case are chemically synthesized and are complementary to the sequences flanking the gene to be amplified. The template is often genomic DNA isolated from a cell. The resulting PCR clone can be analyzed for mutations in genes (genetic testing) or matching patterns between two samples (paternity or forensic testing).

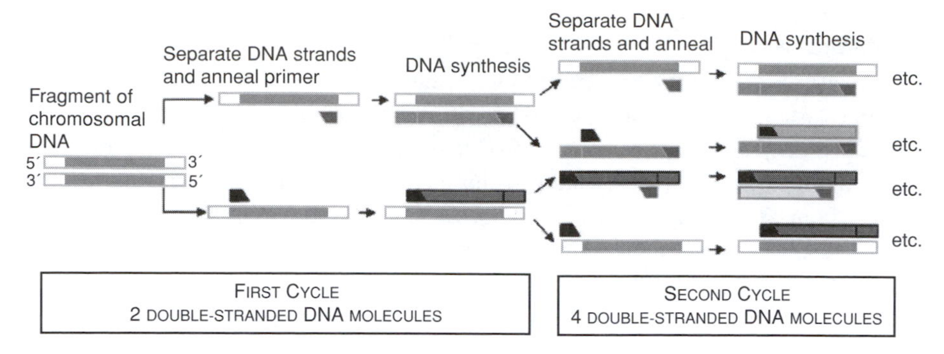

Figure 3.33 **Polymerase chain reaction (PCR).** As described in the text, PCR works by sequential separation, annealing of primer, and amplification steps. DNA, deoxyribonucleic acid.

3.6.2 Applications of recombinant DNA technology

Many of the important applications of recombinant DNA technology are presented in the previous sections. The use of DNA technology in forensic science is described in Box 3.2; similar techniques are used to diagnose genetic diseases. These applications share some common techniques, particularly the use of restriction enzymes to cut DNA and specially designed probes, which hybridize with the cut segments, to identify sequences within the fragments. The use of DNA technology to express proteins is illustrated in Box 3.3 for human insulin. DNA technology in human gene therapy is described in Box 3.4. In all of these applications, biomedical engineers are working to learn how to apply DNA technology safely and efficiently.

Often the work of engineers is to convert these technologies from the laboratory bench into the larger scale needed for commercial applications (Figure 3.34). Since the 1980s, for example, engineers have worked side by side with biologists at Genentech to convert their newly expressed proteins into commercial products.

The following section briefly describes some other applications of recombinant DNA technology (Table 3.4).

GENETICALLY MODIFIED ORGANISMS

Complex organisms, such as animals and plants, can be modified using recombinant DNA techniques to produce **genetically modified organisms**. Often, genetic modification is attempted to confer desirable traits such as resistance to insect pest or viral infection, resistance to environmental damage, and production of specific nutrients.

Recently, genetically engineered rice was developed that could produce provitamin A and iron in its grain (Figure 3.35). This "Golden Rice" is helping to reduce the vitamin A deficiency and anemia in developing countries. By using

Table 3.4 Applications of recombinant DNA technology

Application	Example
Forensic science	DNA fingerprinting (Box 3.2)
Production of therapeutic proteins	Human insulin in bacterial cells (Box 3.3)
Genetically modified foods	Rice containing vitamin A and iron; fruits that resist frost damage
Human gene therapy	Adenosine deaminase (ADA deficiency) (Box 3.4)
Genetic testing	Paternity testing; detection of inherited diseases

Figure 3.34 **Bioreactors for producing proteins.** Photo credit: Harry Turner, courtesy of National Research Council of Canada.

Figure 3.35 **Golden rice.** This image shows ordinary rice; "Golden Rice" would look similar, but with an unusual yellow color because of the presence of beta-carotene. The gene for beta-carotene, which is found in abundance in carrots and other vegetables, is also important in the health of people in regions of the world where rice is easier to grow than carrots. Beta-carotene is a precursor to vitamin A, which has several functions in the body, but is most famously involved in vision.

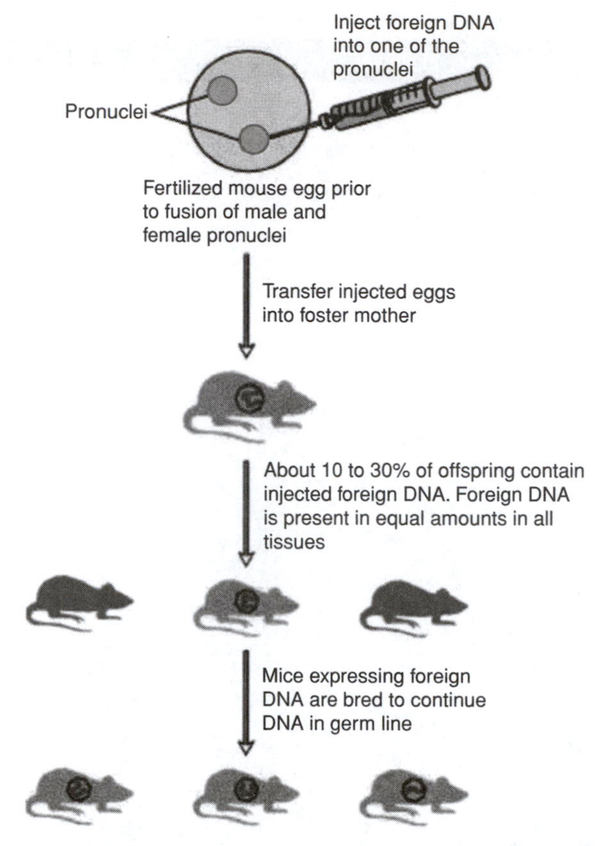

Figure 3.36 **Transgenic mice.** Transgenic mice are produced by injecting foreign deoxyribonucleic acid (DNA) into the pronuclei of a freshly fertilized egg. The manipulated egg is transferred into a female mouse. A fraction of these eggs will develop into mice, of which a fraction will express the foreign DNA.

recombinant DNA approaches, the genomes of crops and animals can be directly modified without laborious breeding regimens of the past.

Future developments in agricultural biotechnology may include the use of genetically modified plants and animals to produce medicinals or protein pharmaceuticals. For example, therapeutic proteins can already be expressed in sheep or cow milk. The protein can be purified from the milk obtained from the genetically modified animals. In addition, it is possible to produce proteins and vaccines in edible plants, such as corn and potatoes. Although much of this is now possible in the laboratory—and is ready for translation to a larger scale by bioengineers—there are still concerns about the long-term safety of these methods that should be answered before widespread use.

ANIMAL MODELS OF HUMAN DISEASES

The genome of rodents (mouse and rat) and other animals can be altered by recombinant DNA techniques to produce phenotypes that mimic human diseases. **Transgenic** animals have foreign DNA incorporated into their genome at the embryonic stage by microinjection (Figure 3.36). Specific genes can also be disrupted to create **knockout** animals, or animals in which a specific gene has been permanently deleted. Those manipulations that result in phenotypes similar to human diseases can be used for testing the efficacy of potential therapeutics prior to testing in patients. Both transgenic and knockout models may also be used to study the function of unknown proteins. Moreover, tissue-specific knockouts or overexpression of the gene of interest can be analyzed for additional insight into gene function.

MOLECULAR BIOLOGY RESEARCH TOOL

Much of what we know about the function of proteins was learned using the tools of recombinant DNA technology. For example, a gene that is linked to a particular disease via inheritance studies can be expressed by cloning; the function of the protein it encodes can then be determined. The amino acid sequence can be deduced by sequencing the cDNA clone. The amino acid sequence of the unknown protein can be compared to other known proteins in a database to predict a possible function. In addition, the cDNA can be inserted into an expression vector to produce sufficient quantities of proteins for biochemical and functional tests or assays.

Conversely, the DNA sequence of a gene encoding for a protein of unknown function can be determined. Similar to the "knockout" animals, the sequences in the gene can be modified to produce a **mutant** protein with abnormal function. By assessing the phenotype of the mutant protein, insight into the function of the normal protein can often be deduced.

GENE SEQUENCING

Recombinant DNA technology has enabled the sequencing of whole genomes of pathogens, plants, and animals. These sequence maps enable genetic engineering of these plants and animals for nutritional and therapeutic purposes. The rough draft of the human genome was completed—revealing that approximately 30,000 human genes are encoded in 3 billion base pairs. The next challenge is to find which of the sequences are most useful and to determine the function of the RNA and proteins encoded by these sequences. Biomedical engineers working in the bioinformatics field are helping to develop tools to allow discoveries in these new areas of genomics and proteomics (Chapter 16).

GENETIC TESTING

Various laboratory methods can be used to test for particular mutations in a gene or to verify identity or paternity. The methods used for genetic testing include Southern blotting, DNA fingerprinting, or restriction fragment length polymorphism (RFLP) analysis. This type of testing is already available for some diseases. Parents-to-be can be tested to see if they carry recessive alleles for a genetic disease (Figure 3.37). The fetus or newborn can be tested to determine the presence of a mutant allele so that an appropriate course of therapy can be administered.

Figure 3.37 **Prenatal genetic testing.** Photo courtesy of Sean Dreilinger/durak.org.

Summary

- Nucleic acids are linear polymers made up of monomer units called nucleotides. Each nucleotide is composed of a pentose sugar, a phosphate group, and a nitrogenous base. DNA nucleotides have deoxyribose sugars, whereas RNA nucleotides have ribose sugars.
- DNA is the main information storage molecule of the cell. Information is stored in the form of a nucleotide base sequence. A gene is a sequence of DNA bases that code for RNA or a polypeptide chain.
- DNA is a double helix with two sugar phosphate backbones and nucleotide bases bridging the two chains.
- Base pairing is accomplished by hydrogen bonding between purines and pyrimidines. Watson–Crick base pairing dictates the base pairs in DNA are A-T and C-G. In RNA, the thymine is replaced by a uracil, so base pairs in RNA are A-U and C-G.
- DNA replication occurs every time a cell divides, and is semiconservative, with each daughter cell receiving one parental strand and one newly synthesized strand. DNA polymerization only occurs in the 5′ to 3′ direction.
- Errors in replication or environmental factors can give rise to mutations in genes.
- The central dogma of molecular biology dictates that DNA must first be transcribed into RNA, which is then translated into protein.
- Genes are transcribed into RNA by RNA polymerase. In eukaryotic cells, transcribed RNA must undergo processing to form an mRNA transcript.

- mRNA is translated into amino acids using the genetic code. Ribosomes and tRNA are involved in translation of codons to corresponding amino acids. The newly synthesized polymer of amino acids makes up a polypeptide chain. Proteins can be composed of more than one polypeptide chain.
- Recombinant DNA technology enables researchers to manipulate DNA for various purposes. Genes can be made to express in other species.
- PCR is used to amplify short segments of DNA.
- Plasmids and viruses are used as cloning vectors and expression vectors.
- Both nonviral and viral methods can be used to introduce a foreign gene into eukaryotic cells.

REFERENCES

1. Sege-Peterson K, Nyhan WL, Page T. Lesch-Nyhan disease and HPRT deficiency. In: Rosenberg RN, Prusiner SB, Dimauro S, Barchi RL, Kunkel LM, eds. *The Molecular and Genetic Basis of Neurological Disease*. Boston: Butterworth-Heinemann; 1993:241–259.
2. Devlin TM, ed. *Textbook of Biochemistry with Clinical Correlations*. New York: Wiley-Liss; 1997.
3. Kreeger PK, Deck JW, Woodruff TK, Shea LD. The in vitro regulation of ovarian follicle development using alginate-extracellular matrix gels. *Biomaterials*. 2006;27:714–723.
4. García-Cardeña G, Commander J, Anderson KR, Blackman BR, Gimbrone MA Jr. Biomechanical activation of vascular endothelium as a determinant of its functional phenotype. *Proc Natl Acad Sci U S A*. 2001;98(8):4478–4485.
5. Robinson R. RNAi therapeutics: How likely, how soon. *PLoS Biol*. 2004;2(1):18–20.
6. Tofade TS, Liles EA. Intentional overdose with insulin glargine and insulin aspart. *Pharmacotherapy*. 2004;24(10):1412–1418.
7. Luo D, Saltzman WM. Synthetic DNA delivery systems. *Nat Biotechnol*. 2000;18:33–39.
8. Glick BR, Pasternak JJ. *Molecular Biotechnology*. Washington DC: ASM Press; 1994.

FURTHER READING

Watson J. *The Double Helix, a personal account of the discovery of the structure of DNA*. New York: Touchstone; 2001.

For more detail on the basic biology of nucleic acids and genes, please see Sections 4.1–4.5 and Sections 7.1–7.2 of:

Lodish H, Berk A, Zipursky SL, Matsudaira P, Baltimore D, Darnell JE. *Molecular Cell Biology*, 4th ed. New York: W. H. Freeman; 1999. This book is available free online at the National Institutes of Health (NIH) book collection Web site: http://www.ncbi.nlm.nih.gov:80/books/bv.fcgi?call=bv.View.ShowTOC&rid=mcb.TOC

Profiles in BME: Tiffanee M. Green

I was born in Washington, DC in 1981 and have always lived in Maryland. I can actually remember the day I was first introduced to engineering: I was a Girl Scout in the second grade. A guest speaker told my troop that if we love science and math, we should consider engineering. I remember flipping through the Yellow Pages and seeing categories for various engineering disciplines. (When I think about it today, the Yellow Pages was a poor choice of reference!) I enjoyed science and math, so I decided to become an engineer when I "grew up." Although I learned about engineering at that early age, it wasn't until high school that I discovered BME. My brother (who is 15 years older than I) had a set of World Book encyclopedias—the 1974 edition. For some reason, I was browsing through the pages of the B volume and stumbled across "biomedical engineering." The two-page description captured my attention and, that day, I decided that I would study BME in college as I really enjoyed math and science, particularly biology.

That same brother dared me to apply to Yale University during my senior year of high school "just to see if I could get in." I used two essays for all of my applications and one confidently discussed my determination to study BME. Well, I was accepted and, in the fall of 1999, I entered. Unlike most freshmen, I already knew my major and I never considered any other major, although I was exposed to other interesting subjects. During my freshman and sophomore years, I took the prerequisites for the BME major, which were a mixture of math (e.g., multivariable calculus and differential equations) and science (e.g., chemistry, biology, physiology, physics, and labs). While taking those courses, it was sometimes difficult to see the relevance to BME. The courses in the major, however, applied those basic concepts to biomedical problems and emphasized logical and analytical thinking rather than memorization.

Although I find BME fascinating, I did not want to be a traditional biomedical engineer. In the summer of 1999 before I entered Yale, I conducted research on biosensors at the Naval Research Laboratory's Center for Bio/Molecular Science and Engineering in Washington, DC. I found the research interesting, but I learned that I did not enjoy *doing* the research. A few months earlier, my high school government teacher suggested that I might want to consider patent law. She, like most of my high school teachers, thought I would be an excellent lawyer because I was actively involved in mock trials. I agree (except that food and drug law, not patent law, is most suitable for me).

Engineers, in general, develop and possess skills that enable them to do nearly anything. I have been taught to think logically, analytically, and critically. I doubt that I will use any of the equations from my math classes, but I will use the skills that I acquired from those exercises. For food and drug law (and patent law), a technical background is an asset. I was fortunate to attend Yale, where I was required to take courses in subjects other than math and science. I could explore my interest in law in classes such as "Genetics, Ethics, and the Law," "Science and Public Policy," and "Medical Imaging and Its Impact Since 1895." In May 2003, I graduated with a Bachelor of Science degree in Biomedical Engineering. A year later, I received a Master of Engineering degree in Engineering & Applied Science, through a special program offered by Yale Engineering, which has a business component in the curriculum.

I believe that my BME background will enable me to pursue nearly any career path. I can do research in a lab. I can work in regulatory affairs at a major pharmaceutical company. I can grant patents at the U.S. Patent and Trademark Office. I can commercialize research from universities into technology in industry. I can market pharmaceutical products to the public. I can advise companies at a management or consulting firm. The opportunities are truly infinite, and anyone can pave his or her own path as I have.

For more current information, consult the fifth edition of this book, which was published in 2003, or the fourth edition of a similar book, *Molecular Biology of the Cell*, by Alberts et al. (Garland Publishing, 2002).

USEFUL LINKS ON THE WORLD WIDE WEB

http://www.ncbi.nlm.nih.gov/
National Center for Biotechnology Information

http://www.kumc.edu/gec/
Genetics Education Center, University of Kansas Medical Center, which includes listing of Web sites related to genetics and the Human Genome Project

http://www.accessexcellence.org/AE/AEPC/NIH/
Understanding Gene *Testing brochure—National Cancer Institute, National Institutes of Health, U.S. Department of Health and Human Services*

http://www.nbii.gov/portal/server.pt
National Biological Information Infrastructure

http://www.ornl.gov/sci/techresources/Human_Genome/
http://www.ornl.gov/sci/techresources/Human_Genome/posters/chromosome/chooser.shtml
Human Genome Project information

http://www.genome.gov/
National Human Genome Research Institute

KEY CONCEPTS AND DEFINITIONS

adenovirus a group of DNA-containing viruses that cause respiratory disease, including one form of the common cold. Adenoviruses can also be genetically modified and used in gene therapy to treat cystic fibrosis, cancer, and potentially other diseases.

allele one of two or more alternative forms of a gene located at the corresponding site (locus) on homologous chromosomes. Different alleles produce variation in inherited characteristics such as hair color or blood type. In an individual, one form of the allele (the dominant one) may be expressed more than another form (the recessive one).

anneal similar to hybridize, to bring complementary strands of nucleic acids together

antibiotic selection a method used to ascertain whether a particular plasmid has been successfully integrated into the DNA sequence of a bacterial cell

antiparallel arranged in an opposite but parallel manner

antisense RNA RNA, with sequence complementary to a specific RNA transcript or mRNA, the binding of which prevents processing of the transcript or translation of the mRNA

bioinformatics the science of managing and analyzing biological data using advanced computing techniques. Especially important in analyzing genomic research data.

bioremediation the use of biological organisms such as plants or microbes to aid in removing hazardous substances from an area

camptothecin a chemotherapeutic agent that can inhibit protein synthesis by inhibiting the action of topoisomerase

carcinogen a substance capable of causing cancer in living tissues

character a feature that is transmissible from parent to offspring

chromatin a complex of DNA and proteins called histones

chromosome the self-replicating genetic structure of cells containing the cellular DNA that bears in its nucleotide sequence the linear array of genes. In prokaryotes, chromosomal DNA is circular, and the entire genome is carried on one chromosome. Eukaryotic genomes consist of a number of chromosomes whose DNA is associated with different kinds of proteins.

clone to create an exact copy made of biological material such as a DNA segment (e.g., a gene or other region), a whole cell, or a complete organism

cloning vector DNA molecule originating from a virus, a plasmid, or the cell of a higher organism into which another DNA fragment of appropriate size can be integrated without loss of the vector's capacity for self-replication; vectors introduce foreign DNA into host cells, where the DNA can be reproduced in large quantities

codon sequence of three nucleotides in DNA or mRNA that specifies a particular amino acid during protein synthesis; also called *triplet*. Of the 64 possible codons, three are stop codons, which do not specify amino acids

complementary DNA (cDNA) DNA that is synthesized in the laboratory from a messenger RNA (mRNA) template

daughter cells two cells resulting from division of a single parental cell

denaturation alteration in the conformation of a protein or nucleic acid caused by disruption of various noncovalent bonds caused by heating or exposure to certain chemicals; usually results in loss of biological function

deoxyribonucleic acid (DNA) long linear polymer, composed of four kinds of deoxyribose nucleotides, that is the carrier of genetic information. In its native state, DNA is a double helix of two antiparallel strands held together by hydrogen bonds between complementary purine and pyrimidine bases.

diploid a full set of genetic material consisting of paired chromosomes, one from each parental set. Most animal cells except the gametes have a diploid set of chromosomes. The diploid human genome has 46 chromosomes.

DNA binding proteins protein molecules that help to stabilize the single-stranded DNA during DNA replication

DNA ligase an enzyme that links together the 3′ end of one DNA strand with the 5′ end of another forming a continuous strand

DNA polymerase an enzyme that copies one strand of DNA (the template strand) to make the complementary strand, forming a new double-stranded DNA molecule. All DNA polymerases add deoxyribonucleotides one at a time in the 5′ to 3′ direction to a short pre-existing primer strand of DNA or RNA.

dominant an allele that is almost always expressed, even if only one copy is present

encode in general use, to put into code; in genetics, to specify the genetic code for

exon the protein-coding DNA sequence of a gene

expression vector a modified plasmid or virus that carries a gene or cDNA into a suitable host cell and there directs synthesis of the encoded protein. Some expression vectors are designed for screening DNA libraries for a gene of interest; others, for producing large amounts of a protein from its cloned gene.

gamete mature male or female reproductive cell (sperm or ovum) with a haploid set of chromosomes (23 for humans)

gene the fundamental physical and functional unit of heredity. A gene is an ordered sequence of nucleotides located in a particular position on a particular chromosome that encodes a specific functional product (i.e., a protein or RNA molecule).

genetic code the nucleotide triplets of DNA and RNA molecules, which carry genetic information in living tissue

genetic expression the entire sequence that occurs in converting a DNA sequence into a protein

genetically modified organism plant or animal that has been modified using recombinant DNA technology

genetics the study of inheritance patterns of specific traits

genome a complete set of genes or genetic material present in an organism's chromosome

genotype entire genetic constitution of an individual cell or organism

germ cells sperm and egg cells and their precursors. Germ cells are haploid and have only one set of chromosomes (23 in all), whereas most other cells in humans have two copies (46 in all).

glycosylation the addition of carbohydrates to proteins

haploid a single set of chromosomes (half the full set of genetic material) present in the egg and sperm cells of animals and in the egg and pollen cells of plants. Human beings have 23 chromosomes in their reproductive cells.

helicase an enzyme that moves along a DNA duplex using the energy released by ATP hydrolysis to separate (unwind) the two strands; required for the replication and transcription of DNA

heredity the genetic transmission of characteristics from parent to offspring

heterozygous referring to a diploid cell or organism having two different alleles of a particular gene

histone a group of proteins packaged with DNA to form chromatin that plays a role in gene regulation

homologous corresponding or similar in structure (or function)

homozygous referring to a diploid cell or organism having two identical alleles of a particular gene

incomplete dominance a phenomenon whereby the presence of multiple alleles in a heterozygote leads to a trait that appears intermediate to the homozygous phenotypes

intron DNA sequence that interrupts the protein-coding sequence of a gene; an intron is transcribed into RNA but is cut out of the message before it is translated into protein

karyotype a photomicrograph of an individual's chromosomes arranged in a standard format showing the number, size, and shape of each chromosome type; used in low-resolution physical mapping to correlate gross chromosomal abnormalities with the characteristics of specific diseases

knockout inactivation of specific genes. Knockouts are often created in laboratory organisms such as yeast or mice so that scientists can study the knockout organism as a model for a particular disease.

lagging strand the DNA template of a double-stranded DNA molecule opposite the leading strand that codes in the direction 5′ to 3′

leading strand the DNA template of a double-stranded DNA molecule opposite the lagging strand that codes in the direction 3′ to 5′

liposome spherical phospholipid bilayer structure with an aqueous interior that forms in vitro from phospholipids and may contain protein

locus the specific site of a gene on a chromosome. All the alleles of a particular gene occupy the same locus

meiosis the process of two consecutive cell divisions in the diploid progenitors of sex cells. Meiosis results in four rather than two daughter cells, each with a haploid set of chromosomes.

messenger RNA (mRNA) RNA that serves as a template for protein synthesis

mitosis the process of nuclear division in cells that produces daughter cells that are genetically identical to each other and to the parent cell

mutagenesis development of a mutation (change in DNA base)

mutant individual, organism, or new genetic character arising or resulting from mutation

mutation alteration or change in DNA base(s)

N-myristoylation the addition of fatty acids to proteins

nucleic acid hybridization the process of joining two complementary strands of DNA or one each of DNA and RNA to form a double-stranded molecule

nulliploid devoid of a nucleus or nuclei

Okazaki fragments short DNA fragments that are joined to form the lagging strand of DNA

oligonucleotide a molecule usually composed of 25 or fewer nucleotides; used as a DNA synthesis primer

phenotype the physical characteristics of an organism or the presence of a disease that may or may not be genetic

phosphodiester bond a bond between the 5′-phosphate of one nucleotide and the 3′-hydroxyl of another

plasmid autonomously replicating extrachromosomal circular DNA molecule, distinct from the normal bacterial genome and nonessential for cell survival under nonselective conditions. Some plasmids are capable of integrating into the host genome. A number of artificially constructed plasmids are used as cloning vectors.

pleiotropy one gene that causes many different physical traits such as multiple disease symptoms

polygenic determined by more than one gene

polymerase chain reaction (PCR) a method for amplifying a region of DNA by repeated cycles of DNA synthesis in vitro

polyploid having multiple chromosome sets as a result of a genetic event that is abnormal (e.g., constitutional or mosaic triploidy, tetraploidy) or programmed (e.g., some plants and certain human body cells are naturally polyploid)

post-translational modification chemical change in the polypeptide chain after translation

prenylation the addition of lipids to proteins

primer short preexisting polynucleotide chain to which new deoxyribonucleotides can be added by DNA polymerase

promoter a DNA site to which RNA polymerase will bind and initiate transcription

proteolysis the directed degradation of proteins or peptides by the actions of enzymes

recessive gene a gene that will be expressed only if there are two identical copies or, for a male, if one copy is present on the X chromosome

recombinant DNA technology procedure used to join together DNA segments in a cell-free system (an environment outside a cell or organism). Under appropriate conditions, a recombinant DNA molecule can enter a cell and replicate there, either autonomously or after it has become integrated into a cellular chromosome.

renaturation the process by which proteins or complementary strands of nucleic acids re-form their native conformations

replication fork a structure that forms when two parental DNA strands separate

replication origin a site for the initiation of DNA replication

restriction endonuclease an enzyme that cleaves DNA at a specific sequence

retrovirus a type of virus that contains RNA as its genetic material. The RNA of the virus is translated into DNA, which inserts itself into an infected cell's own DNA. Retroviruses can cause many diseases, including some cancers and AIDS.

reverse transcriptase an enzyme used by retroviruses to form a complementary DNA (cDNA) sequence from their RNA. The resulting DNA is then inserted into the chromosome of the host cell.

ribonucleic acid (RNA) a chemical similar to a single strand of DNA. In RNA, the base uracil (U) is substituted for thymine (T). RNA delivers DNA's genetic message to the cytoplasm of a cell where proteins are made.

ribosome a small molecule containing RNA and associated proteins, present within the cytoplasm of living cells, that is involved in protein synthesis by binding to mRNA and tRNA.

ribosomal RNA a multicomponent system of RNA synthesized in the nucleolus of cells. It constitutes the central component of ribosomes.

RNA interference (RNAi) short RNA sequences that are complementary to mRNA sequences and may interfere with translation

RNA polymerase an enzyme that catalyzes the synthesis of RNA

RNA primase an enzyme that catalyzes the synthesis of RNA

semiconservative describes a form of replication whereby the newly formed DNA in the daughter cell is made up of an original parental strand and one newly synthesized strand

single nucleotide polymorphism (SNP) DNA sequence variations that occur when a single nucleotide (A, T, C, or G) in the genome sequence is altered

small interfering RNA (siRNA) a class of 20- to 25-nucleotide-long RNA molecules that play a number of roles in biology including RNA interference

somatic cell any cell in the body except gametes and their precursors

Southern blotting transfer by absorption of DNA fragments separated in electrophoretic gels to membrane filters for detection of specific base sequences by radiolabeled complementary probes

splice the process of removing introns in an RNA transcript

structural gene any gene that codes for the amino acid sequences in a protein except the regulatory protein (nonregulatory gene)

systems biology the study of interactions between the components of a biological system and how these interactions give rise to the function and behavior of that system

termination codon any of three mRNA sequences (UGA, UAG, UAA) that do not code for an amino acid and therefore signal the end of protein synthesis

topoisomerase an enzyme that catalyzes the reversible breakage and rejoining of DNA strands

trait a variation of a character

transcription the synthesis of an RNA copy from a sequence of DNA (a gene); the first step in gene expression

transcription factor a protein that binds to regulatory regions and helps control gene expression

transduction the introduction of genetic materials into cells using viral methods

transfection the introduction of genetic material into eukaryotic cells

transfer RNA (tRNA) a class of RNA having structures with triplet nucleotide sequences that are complementary to the triplet nucleotide coding sequences of mRNA. The role of tRNAs in protein synthesis is to bond with amino acids and transfer them to the ribosomes, where proteins are assembled according to the genetic code carried by mRNA.

transform (bacteria) the genetic alteration of a cell caused by the uptake and expression of foreign DNA

transgenic being an organism the genome of which has been altered by the transfer of a gene or genes from another species or breed

translation the process in which the genetic code carried by mRNA directs the synthesis of proteins from amino acids

NOMENCLATURE

A, a, C, c	Gene Alleles A, a, C, c, etc . . .
f_A, f_a	Frequencies of gene alleles A and a
N	Number of individuals in a generation
p, q	Frequencies of alleles

QUESTIONS

1. Explain why DNA and RNA synthesis only occurs in the 5′ to 3′ direction.
2. What is a plasmid vector and what is its significance?
3. Are the following sequences located in a DNA or an RNA molecule? How can you tell?
   a. . . . CGCAGAAGGCAA . . . (sequence 1)
   b. . . . CGCTCTTCG . . . (sequence 2)
4. How many amino acids (just the number, not the identity) are specified in each of the sequences in Question 3?
5. The following sequence is from the transcribed strand of a DNA molecule. What is the sequence on the untranscribed strand?

   . . . GGGGATGCGAAA . . .

6. Explain the method behind gel electrophoresis and how it works to separate DNA into fragments.
7. Why was there interest in creating a gene map of the human genome? How could this information be used?
8. Explain how (in certain situations) the same gene can be spliced differently, and state an advantage of alternative splicing.
9. DNA replication takes place in which direction? What are Okazaki fragments, why are they formed, and how are the corrected?

PROBLEMS

1. Fill in the missing letters (*):

DNA	Untranscribed	T	G	T	*	*	*	*	G	T
DNA	Transcribed	*	*	A	C	*	*	G	*	*
mRNA		U	*	*	*	C	A	*	*	*

Answer:

DNA	Untranscribed	T	G	T	G	C	A	C	G	T
DNA	Transcribed	A	C	A	C	G	T	G	C	A
mRNA		U	G	U	G	C	A	C	G	U

2. What will be the amino acid sequence specified by the following gene?

3′ C A G G T A C T T G C C A A A T C G G T A C A T C 5′

3. A given strand of DNA undergoes a mutation because of an error in replication. The given DNA strand (before mutation) has the following sequence:

3′ –C-T-T-A-C-A-T-G-C-G-A-T-G-T-C-C-G-G-G-T-A-C-T-G-5′

Upon mutation, the G nucleotide that is eight bases from the 3′ end is deleted, resulting in the following sequence:

3′ –C-T-T-A-C-A-T-G-C-G-A-T-G-T-C-C-G-G-T-A-C-T-G-5′

Write out mRNA formed in transcription and the corresponding sequence of amino acids for both the normal and mutated sequence, highlighting the specific change that comes about from this mutation.

4. To be able to properly analyze a sample of DNA, you need to have at least 1 μg of the DNA of interest. *Assumptions:* PCR is run with 30 cycles; molecular weight per nucleotide in DNA is about 660 g/bp (330 g/bp for single-stranded DNA or RNA); bp = base pair.
   a. Given that you can use PCR to amplify DNA from the original sample and the size of the fragment to be amplified is 10,000 nucleotides long, estimate how many DNA molecules you need to obtain the minimum amount required to analyze the sample.
   b. Suggest a way to increase your sensitivity. In other words, amplify the DNA starting from even fewer molecules of template DNA.
5. Humans have a large amount of DNA that is uncoded compared to yeast and bacteria. The human genome contains about 3 billion bases of DNA; this is much more than the 3 million bases in a typical bacterium. The genome of a yeast cell (the simplest eukaryote) is about 14 megabases. Using the following table, calculate the density of genes per unit length of genome in each case. Also, for each cell type, state the fraction of genome that is used to code for proteins (i.e., contains genetic information). Comment on the reasons for the changes in complexity of proteins as well as the percentage of coded genome that is coded in higher organisms.

	Human	Yeast	Bacteria
Number of genes	100,000	6000	3500
Size of genes	3 kb	1.2 kb	1.1 kb

6. For the following sequence:

   5′–G-C-A-U-G-G-A-C-C-C-C-G-U-U-A-U-U-A-A-A-C-A-C-A-C-3′

   a. What is the sequence of amino acids in the peptide encoded for by the following stretch of mRNA?
   b. Give the sequence of bases in the strand of DNA that resulted in this sequence of mRNA.
7. If a typical chromosome is 100 million bases long, and each replication fork moves at 200 bases per second, how many origins of replication are required to replicate a chromosome in 8 hours? (Hint: Be careful! Think about exactly what happens during replication.)
8. The tRNA for phenylalanine has the anticodon triplet AAA. What is the mRNA for phenylalanine?
9. If a triplet on the untranscribed strand is CAT, what is the corresponding anticodon on tRNA?
10. Not all of the bases are used in the translation of mRNA. Using the genetic code, determine the amino acid composition of the polypeptide specified by the following DNA sequence. (Remember that AUG is the start codon as well as the methionine codon.)

    TGCTACGTGGATATTGGG

11. A virus with a circular double-stranded DNA chromosome contains approximately 10,000 bp. You want to begin characterizing this chromosome by making a map of the cleavage sites of three restriction endonucleases: *Eco*RI, *Hind*III, and *Bam*HI. You digest the viral DNA under conditions that allow the endonuclease reactions to go to completion and then subject the digested DNA to electrophoresis on agarose to determine the lengths of the restriction fragments produced in each reaction. Based on the resulting data, draw a map of the viral chromosome indicating the relative positions of the cleavage sites for these restriction endonucleases:

Endonuclease	Length of fragments (kb)
*Eco*RI	6.9, 3.1
*Hind*III	5.1, 4.4, 0.5
*Bam*HI	10.0
*Eco*RI + *Hind*III	3.6, 3.3, 1.5, 1.1, 0.5
*Eco*RI + *Bam*HI	5.1, 3.1, 1.8
*Hind*III + *Bam*HI	4.4, 3.3, 1.8, 0.5
*Eco*RI + *Hind*III + *Bam*HI	3.3, 1.8, 1.5, 1.1, 0.5

12. A circular piece of DNA (e.g., a plasmid) was cleaved using three different restriction enzymes. The cut sizes of the fragments of DNA were detected by gel electrophoresis. The results of the restriction enzyme treatments are shown here (note that 1 kb = 1000 base pairs).

Endonuclease	Size of resulting DNA fragments
*Eco*RI	13 kb, 2 kb
*Bam*HI	15 kb
*Hind*III	8 kb, 7 kb
*Eco*RI + *Bam*HI	8 kb, 5 kb, 2 kb
*Eco*RI + *Hind*III	7 kb, 5 kb, 2 kb, 1 kb
*Bam*HI + *Hind*III	8 kb, 4 kb, 3 kb

   a. What is the length (in kb) of the circular DNA?
   b. Draw a map of the circular DNA (including the restriction sites for *Hind*III, *Eco*RI, *Bam*HI) and determine the distance (in kb) between all the restriction sites on the circular DNA.

13. Researchers studying blood types in a population found the following genotypic distribution about the people sampled: 1050 MM, 1423 MN, and 502 NN.
   a. Calculate the allele frequencies of M and N.
   b. Assuming random mating, determine the three genotypic frequencies.

14. One in 1700 U.S. Caucasian newborns has cystic fibrosis. The allele C is normal and dominant over allele c for cystic fibrosis.
   a. What percentage of the population has cystic fibrosis?
   b. Draw the Punnett square for this gene.
   c. From your answer in (a), determine the expected frequencies of all of the genotypes assuming that the population is in Hardy–Weinberg equilibrium. *Hint: calculate q first.*
   d. How many of the 1700 are homozygous normal?
   e. How many in the 1700 are carriers of the cystic fibrosis gene (heterozygous)?

15. A population of 1000 individuals in Hardy–Weinberg equilibrium contains 250 individuals who exhibit a recessive phenotype. Find the frequency of both alleles: A (dominant) and a (recessive).

16. Figure 3.31 shows the plasmid pBR322, which is commercially available from Invitrogen. The plasmid is 4363 bases in size; the sequence of bases within the plasmid is shown in the following table. Use only the information in this chapter to answer the following questions.
   a. The restriction enzyme *Hind*III cuts this plasmid at only one site, at position 29. What sequence does *Hind*III recognize?
   b. Identify a restriction enzyme that cuts the plasmid in only one location and cuts the gene for resistance to tetracycline (Tc^r) but not the gene for resistance to ampicillin (Ap^r). What sequence does this enzyme

recognize? What would be the value of using this restriction enzyme for cloning?

c. Use Figure 3.30 and the sequence below to find the amino acid sequence of the Tc^r protein. Note the general location of the gene from the plasmid map. Also note that polypeptide sequences start with a start codon (AUG) and end with a termination codon (UAA, UGA, or UAG).

TTCTCATGTT TGACAGCTTA TCATCGATAA GCTTTAATGC 60
GGTAGTTTAT CACAGTTAAA

TTGCTAACGC AGTCAGGCAC CGTGTATGAA ATCTAACAAT 120
GCGCTCATCG TCATCCTCGG

CACCGTCACC CTGGATGCTG TAGGCATAGG CTTGGTTATG 180
CCGGTACTGC CGGGCCTCTT

GCGGGATATC GTCCATTCCG ACAGCATCGC CAGTCACTAT 240
GGCGTGCTGC TAGCGCTATA

TGCGTTGATG CAATTTCTAT GCGCACCCGT TCTCGGAGCA 300
CTGTCCGACC GCTTTGGCCG

CCGCCCAGTC CTGCTCGCTT CGCTACTTGG AGCCACTATC 360
GACTACGCGA TCATGGCGAC

CACACCCGTC CTGTGGATCC TCTACGCCGG ACGCATCGTG 420
GCCGGCATCA CCGGCGCCAC

AGGTGCGGTT GCTGGCGCCT ATATCGCCGA CATCACCGAT 480
GGGGAAGATC GGGCTCGCCA

CTTCGGGCTC ATGAGCGCTT GTTTCGGCGT GGGTATGGTG 540
GCAGGCCCCG TGGCCGGGGG

ACTGTTGGGC GCCATCTCCT TGCATGCACC ATTCCTTGCG 600
GCGGCGGTGC TCAACGGCCT

CAACCTACTA CTGGGCTGCT TCCTAATGCA GGAGTCGCAT 660
AAGGGAGAGC GTCGACCGAT

GCCCTTGAGA GCCTTCAACC CAGTCAGCTC CTTCCGGTGG 720
GCGCGGGGCA TGACTATCGT

CGCCGCACTT ATGACTGTCT TCTTTATCAT GCAACTCGTA 780
GGACAGGTGC CGGCAGCGCT

CTGGGTCATT TTCGGCGAGG ACCGCTTTCG CTGGAGCGCG 840
ACGATGATCG GCCTGTCGCT

TGCGGTATTC GGAATCTTGC ACGCCCTCGC TCAAGCCTTC 900
GTCACTGGTC CCGCCACCAA

ACGTTTCGGC GAGAAGCAGG CCATTATCGC CGGCATGGCG 960
GCCGACGCGC TGGGCTACGT

CTTGCTGGCG TTCGCGACGC GAGGCTGGAT GGCCTTCCCC 1020
ATTATGATTC TTCTCGCTTC

CGGCGGCATC GGGATGCCCG CGTTGCAGGC CATGCTGTCC AGGCAGGTAG ATGACGACCA 1080
TCAGGGACAG CTTCAAGGAT CGCTCGCGGC TCTTACCAGC CTAACTTCGA TCACTGGACC 1140
GCTGATCGTC ACGGCGATTT ATGCCGCCTC GGCGAGCACA TGGAACGGGT TGGCATGGAT 1200
TGTAGGCGCC GCCCTATACC TTGTCTGCCT CCCCGCGTTG CGTCGCGGTG CATGGAGCCG 1260
GGCCACCTCG ACCTGAATGG AAGCCGGCGG CACCTCGCTA ACGGATTCAC CACTCCAAGA 1320
ATTGGAGCCA ATCAATTCTT GCGGAGAACT GTGAATGCGC AAACCAACCC TTGGCAGAAC 1380
ATATCCATCG CGTCCGCCAT CTCCAGCAGC CGCACGCGGC GCATCTCGGG CAGCGTTGGG 1440
TCCTGGCCAC GGGTGCGCAT GATCGTGCTC CTGTCGTTGA GGACCCGGCT AGGCTGGCGG 1500
GGTTGCCTTA CTGGTTAGCA GAATGAATCA CCGATACGCG AGCGAACGTG AAGCGACTGC 1560
TGCTGCAAAA CGTCTGCGAC CTGAGCAACA ACATGAATGG TCTTCGGTTT CCGTGTTTCG 1620
TAAAGTCTGG AAACGCGGAA GTCAGCGCCC TGCACCATTA TGTTCCGGAT CTGCATCGCA 1680
GGATGCTGCT GGCTACCCTG TGGAACACCT ACATCTGTAT TAACGAAGCG CTGGCATTGA 1740
CCCTGAGTGA TTTTTCTCTG GTCCCGCCGC ATCCATACCG CCAGTTGTTT ACCCTCACAA 1800
CGTTCCAGTA ACCGGGCATG TTCATCATCA GTAACCCGTA TCGTGAGCAT CCTCTCTCGT 1860
TTCATCGGTA TCATTACCCC CATGAACAGA AATTCCCCCT TACACGGAGG CATCAAGTGA 1920
CCAAACAGGA AAAAACCGCC CTTAACATGG CCCGCTTTAT CAGAAGCCAG ACATTAACGC 1980
TTCTGGAGAA ACTCAACGAG CTGGACGCGG ATGAACAGGC AGACATCTGT GAATCGCTTC 2040
ACGACCACGC TGATGAGCTT TACCGCAGCT GCCTCGCGCG TTTCGGTGAT GACGGTGAAA 2100
ACCTCTGACA CATGCAGCTC CCGGAGACGG TCACAGCTTG TCTGTAAGCG GATGCCGGGA 2160
GCAGACAAGC CCGTCAGGGC GCGTCAGCGG GTGTTGGCGG GTGTCGGGGC GCAGCCATGA 2220
CCCAGTCACG TAGCGATAGC GGAGTGTATA CTGGCTTAAC TATGCGGCAT CAGAGCAGAT 2280
TGTACTGAGA GTGCACCATA TGCGGTGTGA AATACCGCAC AGATGCGTAA GGAGAAAATA 2340

CCGCATCAGG CGCTCTTCCG CTTCCTCGCT CACTGACTCG CTGCGCTCGG TCGTTCGGCT 2400
GCGGCGAGCG GTATCAGCTC ACTCAAAGGC GGTAATACGG TTATCCACAG AATCAGGGGA 2460
TAACGCAGGA AAGAACATGT GAGCAAAAGG CCAGCAAAAG GCCAGGAACC GTAAAAAGGC 2520
CGCGTTGCTG GCGTTTTTCC ATAGGCTCCG CCCCCCTGAC GAGCATCACA AAAATCGACG 2580
CTCAAGTCAG AGGTGGCGAA ACCCGACAGG ACTATAAAGA TACCAGGCGT TTCCCCCTGG 2640
AAGCTCCCTC GTGCGCTCTC CTGTTCCGAC CCTGCCGCTT ACCGGATACC TGTCCGCCTT 2700
TCTCCCTTCG GGAAGCGTGG CGCTTTCTCA TAGCTCACGC TGTAGGTATC TCAGTTCGGT 2760
GTAGGTCGTT CGCTCCAAGC TGGGCTGTGT GCACGAACCC CCCGTTCAGC CCGACCGCTG 2820
CGCCTTATCC GGTAACTATC GTCTTGAGTC CAACCCGGTA AGACACGACT TATCGCCACT 2880
GGCAGCAGCC ACTGGTAACA GGATTAGCAG AGCGAGGTAT GTAGGCGGTG CTACAGAGTT 2940
CTTGAAGTGG TGGCCTAACT ACGGCTACAC TAGAAGGACA GTATTTGGTA TCTGCGCTCT 3000
GCTGAAGCCA GTTACCTTCG GAAAAAGAGT TGGTAGCTCT TGATCCGGCA AACAAACCAC 3060
CGCTGGTAGC GGTGGTTTTT TTGTTTGCAA GCAGCAGATT ACGCGCAGAA AAAAAGGATC 3120
TCAAGAAGAT CCTTTGATCT TTTCTACGGG GTCTGACGCT CAGTGGAACG AAAACTCACG 3180
TTAAGGGATT TTGGTCATGA GATTATCAAA AAGGATCTTC ACCTAGATCC TTTTAAATTA 3240
AAAATGAAGT TTTAAATCAA TCTAAAGTAT ATATGAGTAA ACTTGGTCTG ACAGTTACCA 3300
ATGCTTAATC AGTGAGGCAC CTATCTCAGC GATCTGTCTA TTTCGTTCAT CCATAGTTGC 3360
CTGACTCCCC GTCGTGTAGA TAACTACGAT ACGGGAGGGC TTACCATCTG GCCCCAGTGC 3420
TGCAATGATA CCGCGAGACC CACGCTCACC GGCTCCAGAT TTATCAGCAA TAAACCAGCC 3480
AGCCGGAAGG GCCGAGCGCA GAAGTGGTCC TGCAACTTTA TCCGCCTCCA TCCAGTCTAT 3540
TAATTGTTGC CGGGAAGCTA GAGTAAGTAG TTCGCCAGTT AATAGTTTGC GCAACGTTGT 3600
TGCCATTGCT GCAGGCATCG TGGTGTCACG CTCGTCGTTT GGTATGGCTT CATTCAGCTC 3660

CGGTTCCCAA CGATCAAGGC GAGTTACATG ATCCCCCATG TTGTGCAAAA AAGCGGTTAG 3720
CTCCTTCGGT CCTCCGATCG TTGTCAGAAG TAAGTTGGCC GCAGTGTTAT CACTCATGGT 3780
TATGGCAGCA CTGCATAATT CTCTTACTGT CATGCCATCC GTAAGATGCT TTTCTGTGAC 3840
TGGTGAGTAC TCAACCAAGT CATTCTGAGA ATAGTGTATG CGGCGACCGA GTTGCTCTTG 3900
CCCGGCGTCA ACACGGGATA ATACCGCGCC ACATAGCAGA ACTTTAAAAG TGCTCATCAT 3960
TGGAAAACGT TCTTCGGGGC GAAAACTCTC AAGGATCTTA CCGCTGTTGA GATCCAGTTC 4020
GATGTAACCC ACTCGTGCAC CCAACTGATC TTCAGCATCT TTTACTTTCA CCAGCGTTTC 4080
TGGGTGAGCA AAAACAGGAA GGCAAAATGC CGCAAAAAAG GGAATAAGGG CGACACGGAA 4140
ATGTTGAATA CTCATACTCT TCCTTTTTCA ATATTATTGA AGCATTTATC AGGGTTATTG 4200
TCTCATGAGC GGATACATAT TTGAATGTAT TTAGAAAAAT AAACAAATAG GGGTTCCGCG 4260
CACATTTCCC CGAAAAGTGC CACCTGACGT CTAAGAAACC ATTATTATCA TGACATTAAC 4320
CTATAAAAAT AGGCGTATCA CGAGGCCCTT TCGTCTTCAA GAA 4363

4 Biomolecular Principles: Proteins

LEARNING OBJECTIVES

After reading this chapter, you should:

- Understand the concepts of primary, secondary, tertiary, and quaternary structure in proteins.
- Understand the contribution of amino acid ionization to the structure of proteins.
- Understand the role of disulfide bonds in stabilizing protein structure.
- Recognize some of the methods used to determine the structure of proteins.
- Understand how post-translational modifications such as glycosylation and myristoylation contribute to protein structure and function.
- Understand the kinetics of enzyme action.

4.1 Prelude

Proteins are the workhorses of the cell (Figure 4.1): They provide structural support in the cytoskeleton, facilitate communication with other cells by acting as receptors, neutralize foreign pathogens, generate contraction forces in muscle, and most ubiquitously catalyze chemical reactions. Proteins are abundant in biological systems, such as eggs (Figure 4.2). Proteins are one of the major macronutrients in the human diet (Figure 4.3).

Some recombinant proteins now serve as therapeutic drugs for treatment or prevention of disease. Biomedical engineers also use recombinant proteins, such as growth factors, to promote growth and differentiation of cells in engineered tissues. Some biomedical engineers have been using techniques of protein engineering to design new biomaterials for use in tissue engineering, drug-delivery systems, or other medical applications.

This chapter describes the structure and function of proteins and also includes a brief introduction to some of the techniques used to determine protein structure, chiefly nuclear magnetic resonance (NMR) and x-ray crystallography. Researchers in the pharmaceutical industry use these protein structures in structure-guided drug design. Chemicals that interfere with protein function have

Figure 4.1

Proteins are often described as the workhorses of the cell and of the body. Actual workhorses are less frequently encountered in the United States in the 21st century, but played an important role in many earlier societies. Photo credit: Ralf Roletschek.

long been used as drugs, but traditionally these chemical–protein interactions have been discovered empirically (or by accident). Now, chemical agents can be designed rationally, based on a detailed knowledge of the protein's structure, to interact with enzymes or receptors and enhance or inhibit their function. Because the structure of a protein determines its function, and because chemicals that interact with specific structural units within proteins can be useful as drugs, biomedical engineers and computer scientists are developing computer programs that will help predict the three-dimensional structure of a protein based on its primary amino acid sequence.

Genomic deoxyribonucleic acid (DNA) is the DNA resident within the chromosomes of cells within an organism. The goal of the Human Genome Project was to map these genes in humans; that project revealed that humans possess 30,000 genes that encode proteins, many with function yet unknown. By using bioinformatics tools, scientists can compare unknown amino acid sequences with

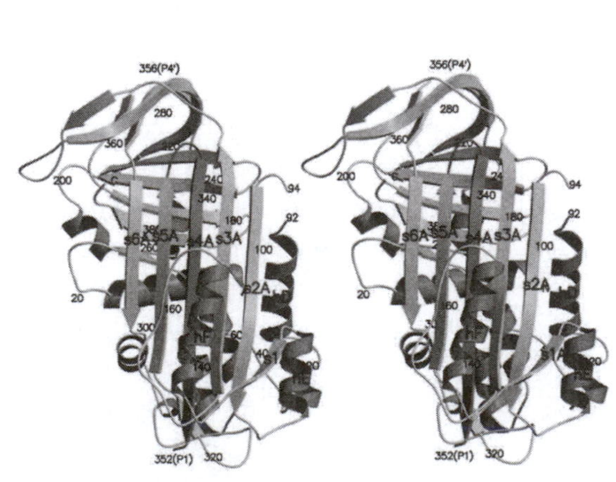

Figure 4.2

Eggs and egg proteins. A. Eggs are an important source of protein in the diet. **B**. Ovalbumin is one of the most abundant proteins in eggs. Its three-dimensional structure is shown here.

Figure 4.3 **Protein in the diet.** The diet of a typical American has abundant protein, but many are now pursuing "high protein" diets to lose weight. As a result, there are many commercial sources of ready protein, including packaged protein bars.

databases of sequences from known proteins to infer function: This approach is known as sequence homology.

Now that the genome is better understood, the next logical step is to uncover the **proteome**, or the complete set of proteins present within a cell at any given time. This quest is called **proteomics**. Unlike the genome, which is relatively constant in individuals throughout their body and over their life span, the proteome of a cell changes with time and environmental conditions. Even more importantly, the proteome within different cells in the same organism can differ dramatically: The set of proteins present in our brain cells is different from the set of proteins present in our kidney cells. Because of this added complexity, study of the proteome requires new tools for detecting the presence of myriad proteins in very small volumes of samples and new methods for organizing and understanding the vast quantity of data generated in this search. Researchers are developing these tools and using them to gather information about the differences in protein expression patterns in normal versus diseased cells and in cells subjected to different environmental conditions.

4.2 Protein structure

Proteins can be composed of one or more polypeptide chains. Chapters 2 and 3 describe the general structure of polypeptide chains, which are linear polymers of amino acids. During protein synthesis, the carboxyl group of one amino acid is covalently linked to the amino group of the next amino acid to form a peptide bond (Figure 4.4); formation of a peptide bond is an example of a condensation reaction (Chapter 2, Section 2.6.1). When synthesized, the linear polypeptide chain must properly fold into its correct three-dimensional shape to carry out its function. Because the sequence of amino acids within each polypeptide chain is the most important determinant of the three-dimensional structure of the protein molecule, the linear structure (i.e., the amino acid sequence) of a protein is often said to determine its function.

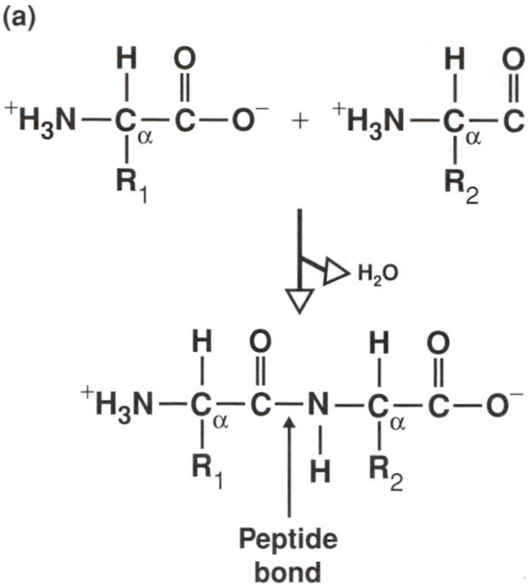

(a)

$$^{+}H_3N-C_{\alpha}H(R_1)-C(=O)-O^{-} + {}^{+}H_3N-C_{\alpha}H(R_2)-C(=O)-O^{-}$$

$$\downarrow \; H_2O$$

$$^{+}H_3N-C_{\alpha}H(R_1)-C(=O)-N(H)-C_{\alpha}H(R_2)-C(=O)-O^{-}$$

Peptide bond

Figure 4.4 **Peptide bond.** The amino terminus of one amino acid is added to the carboxyl terminus of another amino acid to form a covalent peptide bond. The condensation reaction results in the formation of water. Synthesis of a polypeptide chain proceeds from the amino terminus to the carboxyl terminus.

4.2.1 Primary, secondary, tertiary, and quaternary structure

Proteins have varying levels of structure, which are illustrated in Figure 4.5 and Figure 4.6. The amino acid sequence of a polypeptide is called the primary structure. The linear amino acid sequence has directionality: The end of the chain that has a free amino group is called the amino terminus and the end with the carboxyl group is called the carboxyl terminus. In describing the primary

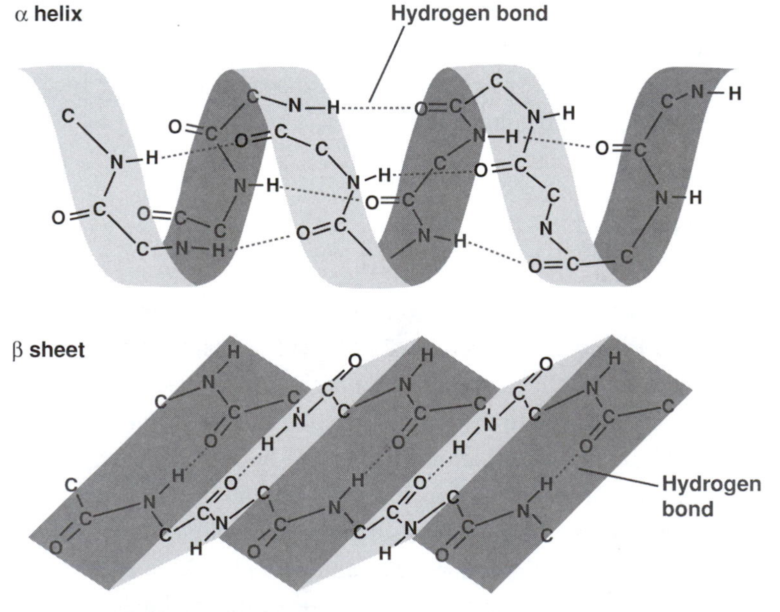

Figure 4.5 **Examples of secondary structure in proteins.** The α helix (top) and the β sheet (bottom) are examples of secondary structures found in proteins. Both of these structures occur in some regions of a polypeptide and are stabilized by hydrogen bonds between amino acids.

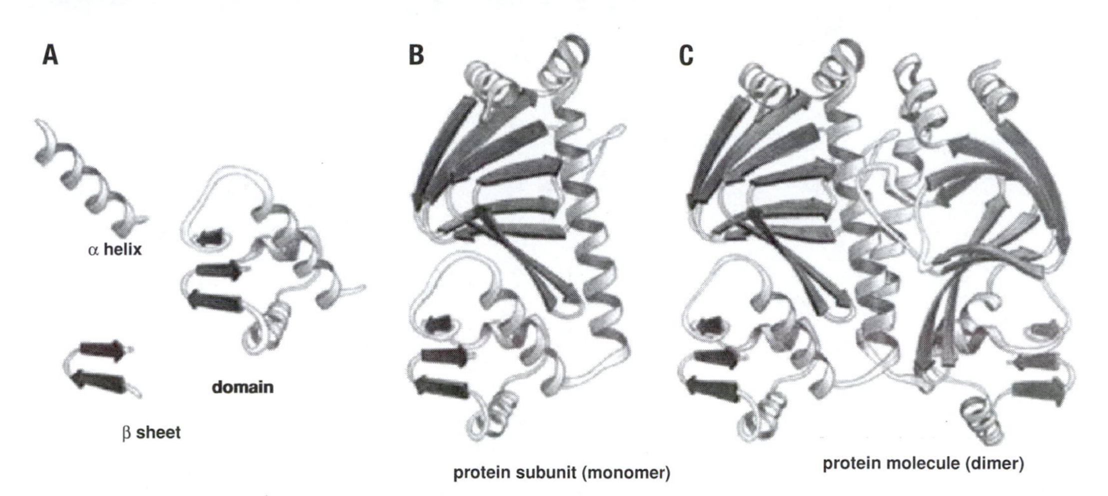

Figure 4.6 **Secondary, tertiary, and quaternary protein structure.** **A.** Secondary structure consists of structures such as the α helix and β sheet, which are found in many polypeptides. **B.** Tertiary structure refers to the three-dimensional arrangement of these secondary structures and the intervening domains within one polypeptide chain (which is also called a protein subunit if it is one polypeptide within a multipolypeptide protein). **C.** Quaternary structure consists of the structural relationships between all of the polypeptides within a complete protein that contains multiple subunits.

structure, the sequence of amino acids is usually recited in order from the amino to the carboxyl terminus.

Secondary structure refers to structure in local regions of the polypeptide chain, which often fold into repeated motifs, such as the α-helix or the β-sheet. These secondary structures form as a result of hydrogen bonding between the N–H and the C=O groups of peptide bonds in neighboring amino acids (Figure 4.5). The α-helix is shaped like a single helical rod and results from hydrogen bonding between every fourth amino acid. The β-sheet results from hydrogen bonding between two segments of the polypeptide chain that are adjacent to one another.

The tertiary structure of a polypeptide refers to its overall conformation—that is, the three-dimensional structure of the complete molecule. One of the most important efforts in modern biology has been the attempt to deduce the overall three-dimensional structure of a polypeptide from its amino acid sequence (this is often called the "protein-folding problem"). The first clues that three-dimensional protein structure is determined by amino acid sequence were uncovered in the 1950s by researchers working on polypeptides such as ribonuclease. Polypeptides can be denatured by chemical treatments, similar to the treatments that lead to denaturation DNA (which is discussed in Chapter 3, Section 3.3.1). Denaturation of a polypeptide usually involves the addition of heat or high urea concentration to disrupt the nonpeptide bonds within the molecule (reducing agents may also be needed to break disulfide interactions, which are described near the end of this section). As a result of this treatment, the denatured polypeptides "unfold"; they lose their secondary and tertiary structure, but retain their primary structure. These early researchers found that denatured, unfolded ribonuclease could be

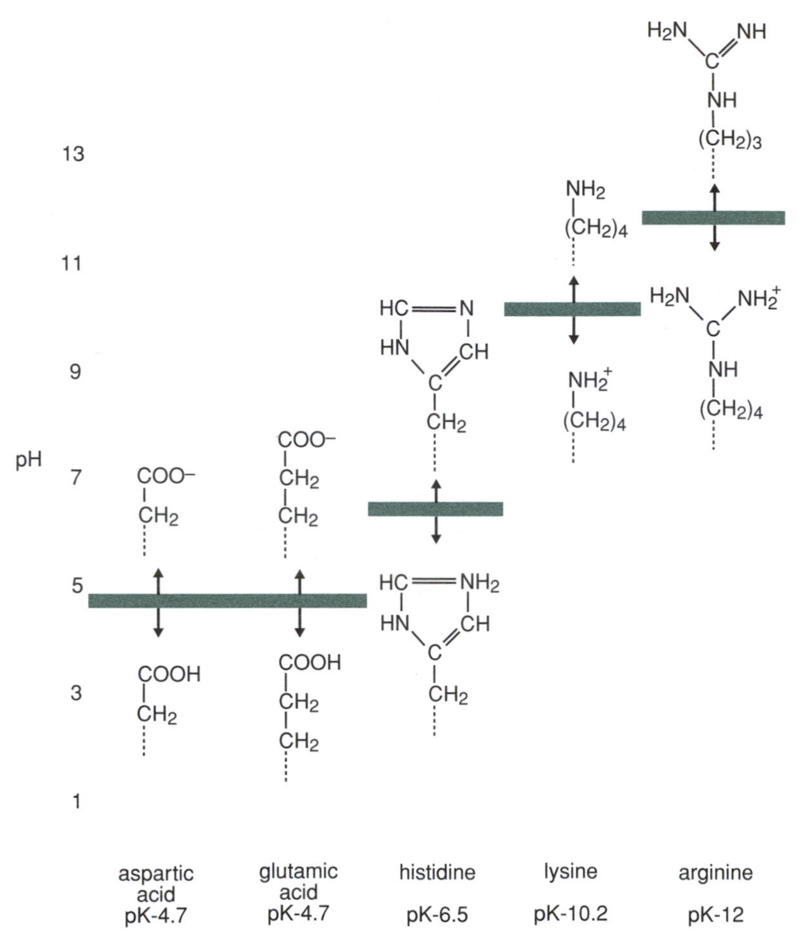

Figure 4.7 **Ionization state of amino acids.** Because many of the amino acids have side groups that are weak acids, the side groups can become ionized at high pH. The pH at which a particular amino acid becomes ionized is called the pK, which is illustrated here for some representative amino acid side groups.

renatured: If placed in its natural environment, it would spontaneously recover its normal secondary and tertiary structure. Therefore, the information needed to "encode" the three-dimensional structure must be present within the linear primary structure.

The ionization state of the amino and carboxyl terminals and the basic and acidic side chains is one key determinant of local protein structure. Ionization of an amino acid can be determined using the Henderson–Hasselbalch equation (see Box 2.4). Recall that pK_A is equal to the pH at which half of the weak acid molecules are ionized. Figure 4.7 shows the ionization states of the various basic and acidic side chains at various pH values. At physiological pH (values near 7.4), most acidic groups are deprotonated (i.e., they contain one less proton, and are therefore negatively charged) and most basic groups are protonated (i.e., they contain one more proton and hence positively charged). Electrostatic interactions between charged residues help to stabilize local secondary structures and domains.

SELF TEST 1 Of the amino acids listed in Figure 4.7, which are positively charged at pH 7?

ANSWER: Lysine and arginine

How does the three-dimensional structure arise from the linear amino acid sequence of a polypeptide? Can people learn how to predict the three-dimensional structure of a polypeptide from its linear sequence? The answers to these questions are not yet complete. In addition to electrostatic forces, intermolecular forces such as hydrogen bonding and van der Waals interactions (which were introduced in Chapter 2) create interactions between segments of a polypeptide; these interactions can be predicted by examining the chemical structure of each participating amino acid. Side chains confer particular chemical properties to each of the 20 amino acids; these side chains can be nonpolar, polar, acidic, or basic (Figure 4.8). The chemical interactions of amino acids with water molecules from the solution also contribute to the formation of the tertiary structure. Polypeptides fold into stable structures to reduce their overall free energy in solution; therefore, folding of a polypeptide chain into its three-dimensional configuration must be an energetically favorable process ($\Delta G < 0$) and should, theoretically, be describable by thermodynamic modeling. The problem is complex, however: Polypeptides are large, flexible molecules, with many interacting units and many possible three-dimensional configurations. Molecular models of these interactions now make it possible to predict the three-dimensional structure of certain simple polypeptides from their primary sequence; more sophisticated models will expand our ability to make predictions of more complex proteins in the future.

For some polypeptides, the primary structure is not sufficient to produce a properly folded, functional three-dimensional molecule. Protein folding is often aided by molecular chaperones, which are proteins that help **nascent proteins** to fold into the correct conformation or three-dimensional structure. (A nascent protein is one that has not yet achieved its functional, final folding pattern.) Although the three-dimensional structure of proteins is based mostly on noncovalent interactions, there is one important covalent bond to consider. The amino acid cysteine reacts with other cysteines in polypeptide chains to form **disulfide bonds** (–S–S–) between two parts of a polypeptide. Disulfide bonds are found mainly in extracellular proteins and help to stabilize their structures. To denature a protein that is stabilized by disulfide bonds, the covalent disulfide bonds must be broken using reducing agents.

A quaternary structure is formed when two or more polypeptide chains combine to form a protein. Noncovalent forces (such as hydrogen-bonding forces) hold the polypeptide chains, which are now called **subunits** of the larger protein, together. Quaternary structures can involve a few polypeptides, as when two identical polypeptides combine to form a two-chain protein (called a **dimer**; see Figure 4.6), or many, as when multiple polypeptide subunits combine to form a protein complex (such as the protease core complex).

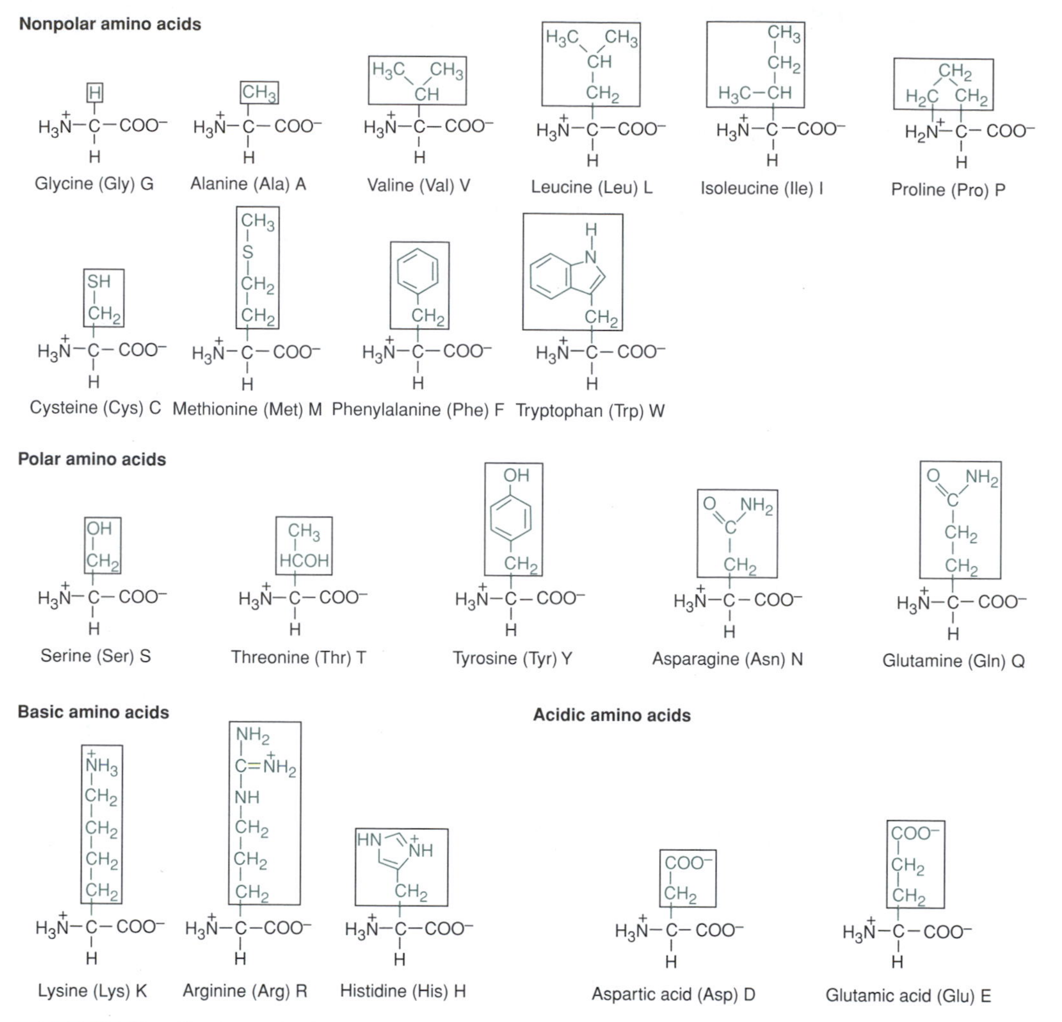

Figure 4.8 **Structures of amino acids.** The structures of the 20 most common amino acids in human proteins are shown. The amino acids are often separated into groups according to their chemical properties. The nonpolar amino acids have side chains that do not dissolve easily in water, whereas the polar amino acids have side chains that do associate with water. The basic amino acids have pK values in the basic range (pK > 7), whereas the acidic amino acids have pK values in the acidic range (pK < 7).

SELF TEST 2 Recall the structure of insulin and its synthesis from proinsulin, which was described in Box 3.3. Does insulin have a quaternary structure?

ANSWER: No. Quaternary structure consists of a specific noncovalent association of subunits having their own tertiary structures. Insulin is a multichain protein covalently joined by a cystine bond.

4.2.2 Determination of protein structure

NMR spectroscopy and x-ray crystallography are commonly used for the experimental determination of protein structure. NMR spectroscopy is most often used to find the structure of small proteins in solution. In x-ray crystallography, the protein of interest must first be crystallized. A crystal is a solid in which all of the molecules are arranged in a regular repeating pattern. For proteins, this special arrangement is often difficult to achieve, but methods for production of crystals of many different proteins (and even protein–protein complexes) are now known. To determine the structures of the molecules within the crystal, it is bombarded with x-rays; x-rays are used because their wavelength is similar to the distances between atoms. Because of this correspondence between the size of the structure and the wavelength, the atoms within the crystal act as a **diffraction grating**. (A diffraction grating is a tool that physicists use to characterize light. It is a material with a series of bands or grooves of a certain dimension. Because these bands have different properties for transmission of light waves, light that strikes the grating gets dispersed. The characteristics of light can be determined by observing its behavior when illuminated on a grating of known properties (i.e., a grating in which the dimension of the bands is known. Alternately, the characteristics of the grating can be inferred from the dispersion pattern that is obtained with light of known properties.)

To fully determine the molecular structure, the crystals are exposed to x-rays from many different angles: Each exposure produces a diffraction pattern, which can be converted into a map of electron density, and further converted (eventually, by comparing maps at different angles using computer algorithms) into the three-dimensional spatial coordinates for the whole molecule.

New methods, such as cryo-electron microscopy, are emerging for structural determination of large proteins that are difficult to crystallize, such as membrane proteins. To complement these experimental methods, biomedical engineers and scientists in the field of bioinformatics use computer models that will allow one to predict the three-dimensional structure based on the primary structure, as described earlier in this chapter.

4.2.3 Protein diversity and protein function

The chemical diversity of proteins is immense. Because there are 20 amino acids, there are 20,300 possible primary sequences for a polypeptide with 300 amino acids. The mean molecular weight (MW) of an amino acid residue is ~110; thus, a protein with 300 amino acids probably has an MW of ~33,000 daltons (or 33 kilodaltons [kd]). The mass of a protein is usually expressed in kilodaltons; one dalton is equal to one atomic mass unit. In addition, each of these proteins will assume one of a wide variety of sizes and shapes, depending on the sequence of amino acids that is present (Figure 4.9).

Proteins are functionally diverse as well. Different proteins serve as structural components of cells and tissues, conduits for transport molecules through biological barriers, elements in communication systems, recognition elements in the defense against infection, and chemical catalysts. Many of these functions of proteins are discussed in other parts of this book. Structural proteins are found in the

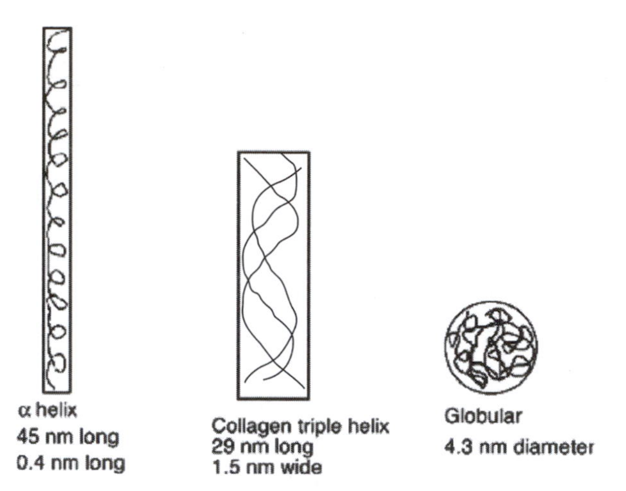

Figure 4.9 Examples of diversity of overall three-dimensional size and shape of 300-amino-acid proteins.

cell cytoskeleton (recall Figure 2.27) as well as the protein-rich gel that surrounds cells in tissues (this gel is called **extracellular matrix**). **Ion channels** are transmembrane proteins that help transport ions, which are otherwise impermeable. **Receptors** also span the membrane but have extracellular domains that bind to molecules called **ligands**. Ligand binding to the extracellular domain triggers a signal transduction pathway that is transmitted into the cell. **Antibodies** are specialized immune proteins, which help to fight infection by neutralizing pathogens and tagging them for destruction. Antibodies have also become an important tool for engineers and scientists working in the area of biomolecular engineering. One of the most important functions of proteins is to speed up biochemical reactions by acting as biochemical catalysts. More detail on **enzyme** function is provided later in this chapter. The proteins involved in muscle contraction have fascinating structures that are highly evolved for their function: These proteins work like tiny machines to perform mechanical work (Box 4.1).

Cloning of genes, and determination of the primary structure of the encoded protein, can lead to important information on the structure or function of an unknown gene. Proteins with similar functions share regions of homologous or similar amino acid sequences. Thus, clues about the function of an unknown protein can be gained by comparing the primary amino acid sequence of the unknown protein with a database of known proteins to determine if there are any matches. This type of analysis is called **sequence homology analysis**.

4.3 Modification and processing of polypeptides

Most polypeptide chains undergo chemical modifications within the cell after synthesis of the linear amino acid sequence, or translation, is complete. **Post-translational modification** is the name often given to this set of chemical modifications (see Table 4.1). These modifications often provide improved functionality or new capability to the protein. For example, proteins that are anchored to the

Box 4.1 Muscle protein mechanics

Muscle cells are rich in protein; these proteins are essential for contraction, and generation of force and movement, which is the main function of muscle. Muscle cells are unusually large, multinucleated cells called **myofibrils**. Myofibrils contain regions of light and dark staining, which are called **sarcomeres** (Panel A of figure); these regions repeat with a fixed pattern along the axis of the cell (i.e., stacked up in the direction of contraction). They named these regions **A bands** and **I bands** (for anisotropic and isotropic, corresponding to their appearance under polarized light). The line that divided the sarcomere in half was called the **M line**, and the line between adjacent sarcomeres was called the **Z line**. They also noticed that during muscle contraction the I band shortened, but the A band did not.

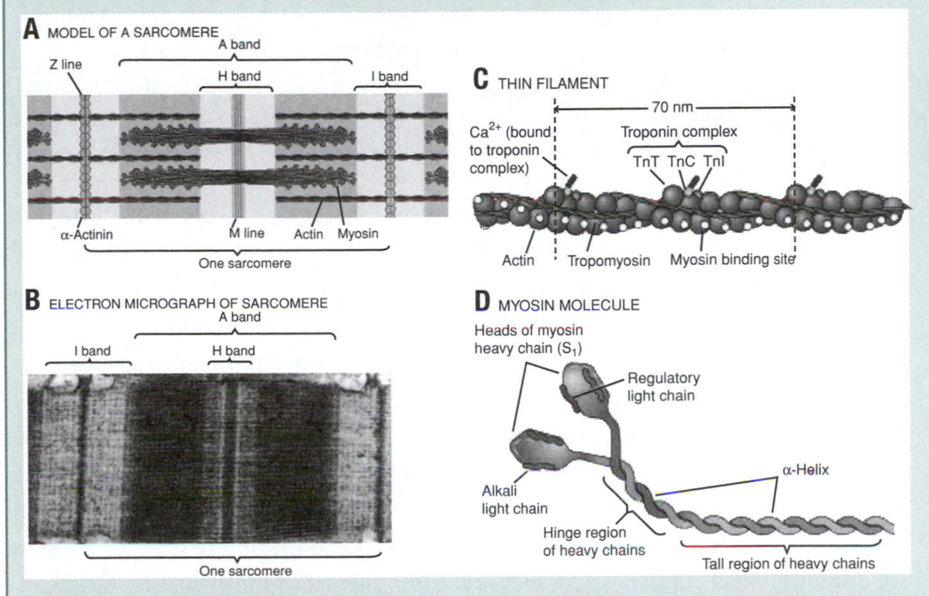

Actin and myosin proteins in muscle. A. The functional unit of a muscle cell (myofibril) is the sarcomere, which was first identified by microscopy as regions of dark and light staining (bottom). These regions are now known to correspond to regions of the cell that are occupied by thick (myosin-rich) and thin (actin-rich) filaments. **B.** Actin and myosin proteins have a characteristic shape and corresponding mechanical properties.

Further research since that time has shown that the sarcomere is composed of highly organized protein assemblies called **thin filaments** and **thick filaments** (Panel B of figure). The thin filaments are rich in the protein actin, and the thick filaments are rich in the protein myosin. The structure of these proteins and their ability to form multiprotein complexes underlie the function of muscle. The thin filaments are elongated "polymers" of actin: The individual actin proteins are stacked in a regular pattern to form a helical, elongated filament that is 12 to 18 nm in diameter and 1.6 μm in length. These thin filaments are attached at one end to the Z lines that form the boundary of the sarcomere. Thick filaments, in contrast, are bundles of myosin protein, which are tethered at the M line in the middle of the sarcomere. Myosin has an interesting structure: It is composed of six polypeptide chains (four smaller, light chains and two larger, heavy chains) that are noncovalently associated to produce an elongated, hockey stick–shaped molecule. These filaments are arranged side by side in a pattern that produces the A and I bands visible to microscopy: The darker staining A band contains both thin and thick filaments (except in the center, a region called the H band, which contains only thick filaments), whereas the light staining I band contains only thin filaments.

(continued)

Box 4.1 *(continued)*

Muscle contraction happens when the sarcomere shortens. Shortening of the sarcomere is driven by interactions between the thin and thick filaments. A more detailed description of the molecular contraction process is in Chapter 10. Briefly, contraction requires three elements: a mechanism for binding between thin and thick filaments, a mechanism for "sliding" or movement of the bound filaments relative to each other, and a mechanism for regulating the binding event so that binding can be turned on and off. All of these functions are provided by proteins. The head region of myosin (the part of the "hockey stick" that strikes the puck) binds to actin in the thin filament, forming a **cross-bridge**. Relative movement is created by a change in shape of the myosin molecule; the head region pivots in response to adenosine-5′-triphosphate (ATP) hydrolysis. Regulation is provided by accessory proteins—troponin and tropomysin—which are sensitive to Ca^{++} ions and interfere with or allow myosin binding to the thin filament.

The relationship between the properties of thick and thin filaments and the generation of work in muscle is described in Chapter 10.

plasma membrane are improved by addition of a hydrophobic tail via **myristoylation**. Secreted proteins and some membrane proteins are modified with carbohydrate groups by **glycosylation**, which can aid the protein in folding or improve protein stability in tissues. **Histones**, proteins in the cell nucleus that help maintain chromatin structure, are modified by **acetylation**. Many of the enzymes involved in cell communication are regulated by the addition of phosphate groups, a chemical process called **phosphorylation**. These post-translational modifications, which provide a finer level of control over the ultimate chemical structure of the protein, add to the versatility of proteins as working molecules in the cell.

Some proteins undergo a post-translational processing step in which a segment of amino acids is clipped off, or cleaved, from the rest of the structure. Often, the protein is inactive until a peptide segment is cleaved from it. For example, many digestive enzymes can be safely stored in the cells of the digestive organs because they only become active upon cleavage after secretion into the intestinal lumen. Cleavage of a polypeptide chain is usually catalyzed by specialized enzymes.

Table 4.1 Examples of post-translational modifications

Phosphorylation	The addition of a phosphate group, usually to a serine, tyrosine, threonine, or histidine residue
Acetylation	The addition of an acetyl group, $-COCH_3$, usually at the amino terminus of a polypeptide
Alkylation	The addition of an alkyl group (methyl, ethyl, etc.). Methylation usually occurs on lysine or arginine residues.
Isoprenylation	The addition of an isoprenoid group, which is a member of a large family of lipid molecules. This modification increases the polypeptide's ability to interact with membranes.
Glycosylation	The addition of a glycosyl group to asparagine, hydroxylysine, serine, or threonine residues
Ubiquitination	The coupling of the protein ubiquitin to a protein, which usually leads to the protein's destruction
Myristoylation	The addition of a myristoyl group, $CH_3\text{-}(CH_2)_{12}-$, to a glycine residue at the N terminal end of a polypeptide

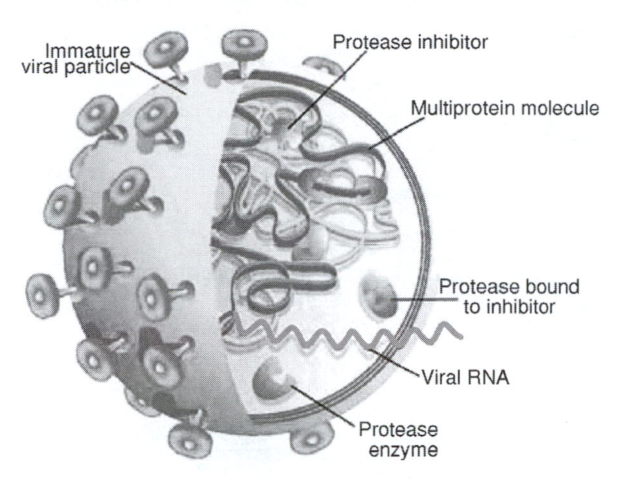

Figure 4.10 **Protease proteins are essential for formation of mature human immunodeficiency virus (HIV).** HIV particles contain a protease that is essential for cleavage of a polypeptide into smaller units. These smaller units are necessary for maturation of the virus into a mature, active form. RNA, ribonucleic acid. Protease inhibitors can be used to inhibit maturation of the virus. Reprinted by permission from Macmillan Publishers Ltd: *Nature Medicine* 9(7), 867–873 © 2003.

Because their function is to break down proteins, these enzymes are called **proteases**. Viruses sometimes use proteases to make active proteins from precursors. The HIV protease, for example, converts a multidomain precursor protein into active units (Figure 4.10); inhibition of this protease can eliminate the infectivity of the virus. Thus, drugs that are **protease inhibitors** are now a key component in the treatment of acquired immune deficiency syndrome (AIDS).

Post-translational processing can also be a key step in the synthesis of some proteins. Insulin is synthesized (i.e., produced from messenger ribonucleic acid [mRNA] by the biological process of translation) as a long single polypeptide chain called proinsulin. The active insulin protein is produced by cleavage of a large segment in the middle of the polypeptide, leaving the disulfide coupled A- and B-chains in their correct three-dimensional configuration (Figure 4.4, Figure 4.11, and Figure 4.12).

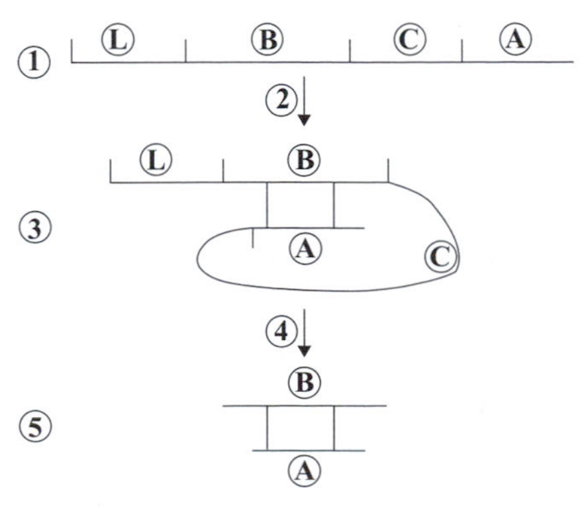

Figure 4.11 **Synthesis of insulin.** The preproinsulin molecule (1) folds spontaneously (2) into a structure that is stabilized by the addition of two disulfide linkages (3). Cleavage of the leader sequence (L) and protein C result in the active insulin molecule (4).

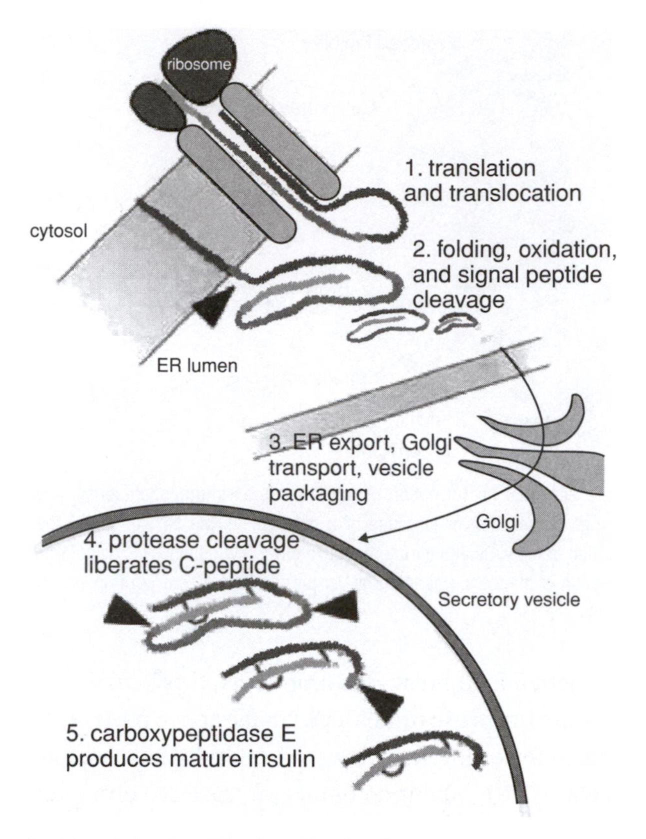

Figure 4.12 **Post-translational modification of insulin.** The proinsulin polypeptide is produced by translation and transported through a series of organelles—the endoplasmic reticulum (ER) and the Golgi apparatus—into a secretory vesicle. Within the secretory vesicle, proinsulin is cleaved by proteases into the insulin protein.

Some proteins have signaling peptides located at their amino-terminal end. Proteins that are water-soluble, or destined to be membrane-bound or secreted, contain **signal peptides** that are used by the cell to direct the proteins into a compartment called the **endoplasmic reticulum**. Other proteins that are destined for functions in the nucleus contain signaling peptides called **nuclear localization signals**.

4.4 Enzymes

Enzymes catalyze the thousands of reactions that occur in the cell. A **catalyst** is a molecule that speeds up a chemical reaction, but is not consumed or generated in the reaction. A variety of enzymes that are essential for DNA and RNA polymerization and protein synthesis, as well as the use of restriction enzymes for creation of recombinant DNA, are discussed in Chapter 3. Enzymes can speed up the rate of the reaction by a million or more (10^6 to 10^{12}) times more than an uncatalyzed reaction. Reactions that would take thousands of years to complete under enzyme-free conditions can occur within seconds or minutes in the presence of the right enzyme.

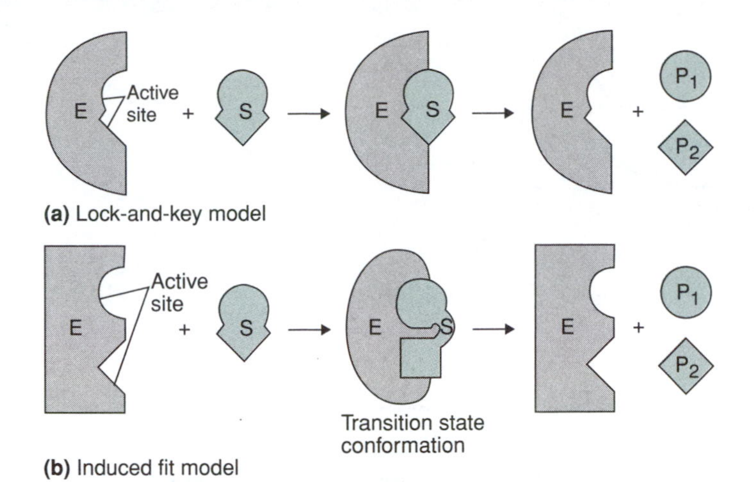

Figure 4.13 **Models for mechanism of enzyme action.** Enzymes work by physical association with substrates. The physical interaction changes the chemical environment of the substrate in such a way that the energy barrier associated with the conversion of the substrate to products is reduced (recall the discussion of transition states and energy barriers in Box 2.2). The mechanism of enzyme–substrate interaction used to be described as analogous to a lock and key, implying that the physical structure of the enzyme was the correct geometric shape and size to fit the substrate. An induced fit model—which allows the enzyme to mold around a substrate with the proper chemical properties—may be more accurate.

Enzymes are also incredibly specific, acting to speed up only certain reactions. They achieve specificity by only recognizing their **substrates** (the reactants in the reaction that they catalyze), and not other potential substrates. The region of the enzyme that binds to the substrate is called the **active site**. The *lock-and-key model* was initially used to describe the specificity of an enzyme binding to its substrate (Figure 4.13): The three-dimensional configuration and chemical conditions within the active site provide a proper fit for only one substrate (or maybe a small range of substrates). It is now believed that an *induced fit model* may be more appropriate. In this model, both enzyme and substrate undergo a conformation change. Perhaps the change in conformation of the substrate is the first step in its conversion into the reaction product.

A reaction in which an enzyme (E) catalyzes the conversion of substrate (S) to product (P) can be represented as shown as follows:

$$E + S \leftrightarrow ES \rightarrow E + P \tag{4.1}$$

Enzymes stabilize the **transition state** (ES) of the reaction. The transition state has the highest free energy; formation of the transition state is the rate-limiting step in the overall sequence of reactions in Equation 4.1. The change in Gibbs free energy between the transition state and the substrate is called the **activation energy** ($\Delta G^{\ddagger}$). Enzymes are able to lower the activation energy of a reaction and thereby speed up the reaction (Figure 4.14). Enzymes do not affect the overall change in Gibbs free energy (ΔG) of the reaction. Therefore, enzymes can only catalyze reactions that are already energetically favorable ($\Delta G < 0$; see Chapter 2, Box 2.2). A single enzyme can catalyze the conversion of many

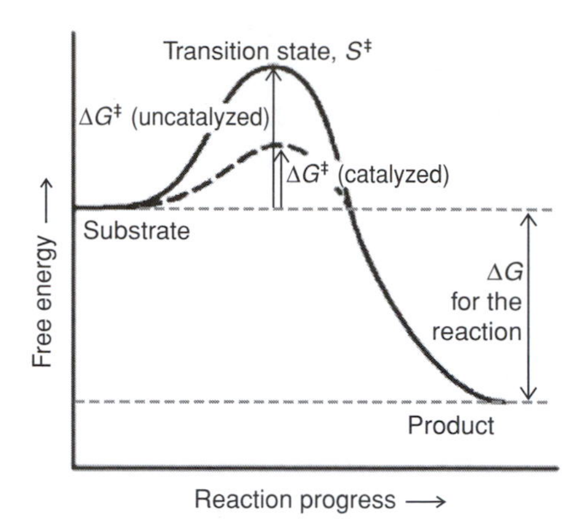

Figure 4.14 Illustration of enzyme action by reduction of energy for the transition state.

substrate molecules because the enzyme itself is not consumed in the reaction (Figure 4.13).

The kinetic properties of many enzymes can be described using the Michaelis–Menten equation:

$$V_0 = \frac{V_{\max}[S]}{[S] + K_{\mathrm{m}}}, \tag{4.2}$$

where V_0 is the rate or velocity of the chemical reaction, [S] is the concentration of substrate, and $V_{\max}$ and K_{m} are two constants that characterize the particular reaction (see Box 4.2). This equation matches the behavior of many enzymes (Figure 4.15). At low substrate concentrations, the rate increases linearly with substrate concentrations. As the substrate concentration increases, the active sites in the enzyme become saturated and the maximum reaction rate $V_{\max}$ is reached.

The substrate concentration at half-maximal velocity ($V_{\max}/2$) is equal to the K_{m} of the enzyme. The K_{m} value for a particular enzyme–substrate combination is often called the **affinity** of that enzyme for that substrate. When the K_{m} value

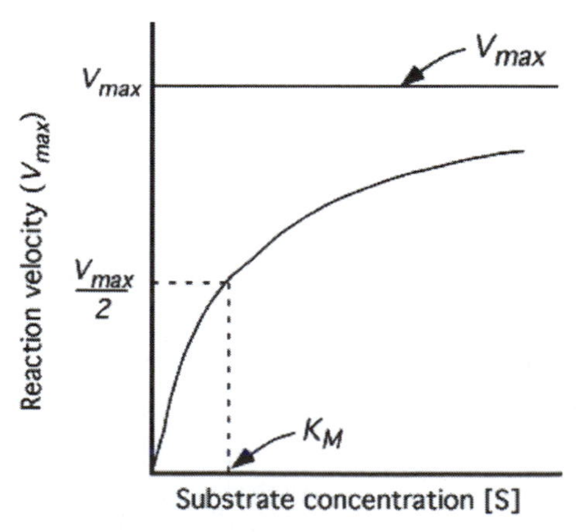

Figure 4.15 **Kinetics of a typical enzyme-catalyzed reaction.** The curved line represents Equation 4.2, the equation for Michaelis–Menten kinetics, which is first order in [S] at low concentrations.

Box 4.2 Enzyme kinetics

The rate of reaction in the presence of an enzyme can be analyzed by viewing the enzyme-catalyzed process as a series of simpler steps. The steps for a common model are shown in Figure 4.13. In the first step, the substrate molecule (or reactant) becomes associated with the enzyme to form an enzyme–substrate complex. In the second step, the substrate is converted into a product, while remaining associated with the enzyme. Finally, the product molecule is released from the enzyme–product complex.

$$E + S \underset{k_{-1}}{\overset{k_1}{\rightleftarrows}} ES \xrightarrow{k_2} EP \xrightarrow{k_3} E + P \qquad \text{(Equation 1)}$$

It is common to assume that the first step is at equilibrium, that is, the rate of substrate molecule association with the active site is the same as the rate of dissociation. The next two steps—conversion of the substrate to product and release of the product—therefore determine the rate of product formation. If the rate of release of product is rapid compared to the rate of chemical conversion ($k_3 \gg k_2$), the reaction equation can be simplified:

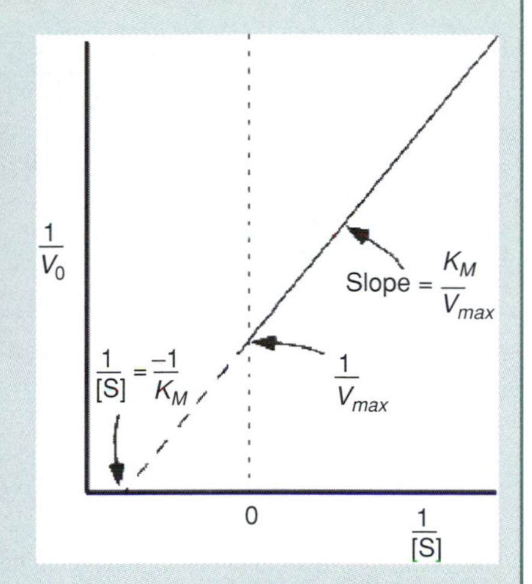

$$E + S \underset{k_{-1}}{\overset{k_1}{\rightleftarrows}} ES \xrightarrow{k_{\text{cat}}} E + P \qquad \text{(Equation 2)}$$

The overall reaction rate (or rate of product formation) is then simply determined by:

$$V_0 = \frac{d[P]}{dt} = k_{\text{cat}}[ES]. \qquad \text{(Equation 3)}$$

Often, the overall reaction rate is called the reaction velocity (V_0). At steady state, the concentration of ES is assumed constant (i.e., the rate of formation of ES is equal to the rate of dissociation), which can be expressed:

$$k_1[S][E] = k_{-1}[ES] + k_{\text{cat}}[ES]. \qquad \text{(Equation 4)}$$

Enzyme molecules are conserved in the reaction, so that the total number of enzyme molecules is constant, $[E]_{\text{TOT}}$:

$$[E_{\text{TOT}}] = [E] + [ES]. \qquad \text{(Equation 5)}$$

Equation 4 can be rewritten as

$$[E] = \frac{(k_{-1} + k_{\text{cat}})[ES]}{k_1[S]}, \qquad \text{(Equation 6)}$$

(continued)

Box 4.2 *(continued)*

which can be simplified by defining a constant, K_m, called the **Michaelis constant**:

$$K_m = \frac{k_{-1} + k_{cat}}{k_1}. \quad \text{(Equation 7)}$$

Using the definition of K_m in Equation 6 yields:

$$[E] = \frac{K_m[ES]}{[S]} \quad \text{(Equation 8)}$$

Rearranging Equation 5:

$$[E_{TOT}] = K_m\frac{[ES]}{[S]} + [ES]. \quad \text{(Equation 9)}$$

Equation 9 can be solved for [ES] and the result substituted into Equation 3 to yield the final expression:

$$V_0 = \frac{k_{cat}[E_{TOT}][S]}{[S] + K_m} \quad \text{(Equation 10)}$$

When [S] is much greater than K_m, and all the catalytic sites are saturated with substrate, Equation 10 reduces to:

$$V_{max} = k_{cat}\,[E_{TOT}] \quad \text{(Equation 11)}$$

Substitution of Equation 11 into Equation 10 yields the **Michaelis–Menten equation:**

$$V_0 = \frac{V_{max}\,[S]}{[S] + K_m}, \quad \text{(Equation 12)}$$

which is discussed in the text accompanying Figure 4.15 Often a **Lineweaver–Burke plot** (or double-reciprocal) can be used to determine the values of V_{max} and K_m from experimental data. Taking the reciprocal of both sides results in a linear expression for $1/V_0$ versus 1/[S]:

$$\frac{1}{V_0} = \left(\frac{K_m}{V_{max}}\right)\left(\frac{1}{[S]}\right) + \frac{1}{V_{max}} \quad \text{(Equation 13)}$$

The K_m/V_{max} and V_{max} can be determined graphically from this double-reciprocal plot as shown in the figure.

is low, that corresponds to a rapid reaction with low concentrations of substrate or high affinity of the enzyme for the substrate. In contrast, a higher value of K_m corresponds to a substrate that must be present at higher concentrations to achieve a rapid reaction. The enzyme has a lower affinity for this substrate.

For a given number of enzyme molecules (or a fixed concentration of enzyme in a solution), the enzyme-catalyzed reaction has a maximal rate that represents full occupation of all of the enzyme's active sites. The turnover number of an enzyme (k_{cat}) is defined as the number of substrate molecules converted into

Table 4.2 Turnover times and reaction times for some enzymes

Enzyme	Turnover number (s^{-1})	Reaction time (ms)
Carbonic anhydrase	600,000	0.0017
3-Ketosteroid isomerase	280,000	0.0036
Acetylcholinesterase	25,000	0.040
Penicillinase	2000	0.5
Lactate dehydrogenase	1000	1
Chymotrypsin	100	10
DNA polymerase I	15	67
Tryptophan synthetase	2	500
Lysozyme	0.5	2000

product per unit time when the enzyme is fully saturated with substrate. This number is related to V_{max}:

$$V_{max} = k_{cat}\,[E_{TOT}]\,, \tag{4.3}$$

where $[E_{TOT}]$ is the concentration of enzyme present.

The reaction time is the reciprocal of the turnover number. Table 4.2 lists some turnover numbers and reaction times for various enzymes. Another useful criterion for comparing the catalytic efficiency of enzymes and their substrates (in nonsaturating conditions normally found in the cell) is the ratio k_{cat}/K_M.

SELF-TEST 3 Experimentally, how would you determine the turnover number (k_{cat}) for an enzyme?

ANSWER: The turnover number can be determined by measuring the maximum reaction velocity, V_{max}. This maximum velocity can be found by measuring the reaction rate for different substrate concentrations [S], increasing the concentration until it is high enough so that subsequent increases no longer increase the reaction rate. The turnover number is then determined by dividing V_{max} by the enzyme concentration (which was maintained at a constant value).

Summary

- Linear polymers of amino acids are polypeptide chains. A protein can be composed of one or more polypeptide chains.
- There are 20 different amino acids, each with unique chemical properties conferred by the side chain. The amino acids can be placed into broad categories: polar, nonpolar, acidic, and basic.
- There are four levels of structure for proteins. The amino acid sequence is the primary structure, the local domains are the secondary structure, the

overall three-dimensional shape is the tertiary structure, and the formation of a complex with other polypeptide chains is the quaternary structure.
- All the information necessary for a protein to fold properly into its tertiary structure is contained in the primary amino acid sequence.
- Noncovalent interactions, such as hydrogen bonding or the hydrophobic/hydrophilic forces, hold a protein in its native form. Covalent disulfide bonds can be formed between two cysteine amino acids located at different regions of the polypeptide chain.
- The three-dimensional protein structure can be determined experimentally using NMR spectroscopy or x-ray crystallography.
- Proteins are often modified after translation by the addition of chemical groups.
- Proteins can also be processed via proteolytic cleavage of the polypeptide chain into smaller segments.
- Proteins have diverse functions such as maintaining cell structure, transporting small molecules, facilitating cell communication, protecting against foreign invaders, and catalyzing chemical reactions.
- Enzymes catalyze chemical reactions by lowering the activation energy of a reaction.

FURTHER READING

Although some of the basic principles of biochemistry are covered in this overview material, you are encouraged to find additional reading material on subjects that are unfamiliar to you. There are many good sources of introductory material. Two books that are highly recommended—and available for free reading online—are listed below:

Berg JM, Tymoczko JL, Stryer L. *Biochemistry*. New York: W. H. Freeman; 2002.

Lodish H, Berk A, Zipursky SL, Matsudaira P, Baltimore D, Darnell JE. *Molecular Cell Biology*. 4th ed. New York: W. H. Freeman; 1999.

Both books are available online at http://www.ncbi.nlm.nih.gov/entrez/query.fcgi?db/Books

USEFUL LINKS ON THE WORLD WIDE WEB

http://www.proteomicworld.org/DatabasePage.html
Proteomics Databases: Proteomic World—Resources for Proteomics and Protein Expression

http://ca.expasy.org/
ExPASy (Expert Protein Analysis System) Proteomics Server of the Swiss Institute of Bioinformatics

http://www.rcsb.org/pdb/home/home.do
RCSB Protein Data Bank

http://www.hprd.org/
Human Protein Reference Database

Profiles in BME: Brenda K. Mann

I grew up in southeastern Iowa, and have always had a strong interest in math and science. However, I have also always found the need to have outside interests, including sports and music. In high school, I was drawn toward chemistry, but was not sure I wanted to major in it in college. My dad, an engineer, steered me toward chemical engineering. Following in the footsteps of many family members, I went to Iowa State University, where I obtained my BS in chemical engineering and was a member of the swim team. While at ISU I worked on bioseparations with Dr. Charles Glatz, a professor in chemical engineering. I found that I loved research, and I found biological applications of chemical engineering more interesting than the petroleum applications that were common at that time. Thus, Dr. Glatz encouraged me to go to graduate school to develop more experience in bioengineering.

I started graduate school in Spring 1992 at Rice University, and I got my PhD in chemical engineering in 1997. My thesis work involved using in vivo nuclear magnetic resonance (NMR) spectroscopy to study cellular metabolism in yeast and bacteria. This was different than what I had done for my research previously, and is very different than what I have done since. However, I learned a lot about research and gained invaluable advice from my advisor, Dr. Jackie Shanks. As the only female professor in the department at that time, she had an interesting perspective to provide a female graduate student.

Upon finishing my PhD, I was not sure what kind of career path I wanted. Despite my love of research, I began working as a technical advisor for Arnold, White and Durkee, an Intellectual Property (IP) law firm in Houston. I initially found it very interesting and learned about patent law. I passed the patent exam and became a registered patent agent. However, I missed research and did not want to go to law school. I decided to return to research and to Rice University, this time as a postdoc in the new Department of Bioengineering. My research focused on using a poly(ethylene glycol)–based hydrogel to create a tissue-engineered vascular graft. My postdoctoral advisor was Dr. Jennifer West, who was extremely supportive of my return to research, albeit in a different field. Dr. West opened my eyes a little wider to the academic world and gave me yet another female perspective in academic life.

Following my postdoc, I became a founding faculty member of the Keck Graduate Institute (KGI) of Applied Life Sciences. While there, I helped to develop the bioengineering curriculum for this innovative school that integrates the life sciences and business. I had my own research lab examining hydrogels for peripheral nerve regeneration and taught courses in bioengineering (covering topics such as bioreactors, biomaterials, tissue engineering, and medical devices), most of which included industrial aspects. This experience helped foster an entrepreneurial spirit I never knew I had, and I managed to absorb a lot of information about the life science industry.

After my experience at KGI, I became Director of Pre-Clinical Research & Development at Sentrx Surgical, a small start-up biomaterials company in Salt Lake City. When the company secured its initial funding after a year, the venture capital firm moved the company (now called Carbylan BioSurgery) to Palo Alto. I then helped co-found a new company, Sentrx

(continued)

Profiles in BME *(continued)*

Animal Care, to use the same biomaterials for veterinary applications, and am currently the Chief Operating Officer. Because the company is very small, my job encompasses many different aspects: overseeing development, testing, production, packaging, and shipping of the materials, as well as assisting with business aspects such as marketing and regulation. I am also a Research Assistant Professor in the Department of Bioengineering at the University of Utah, splitting my time between the two.

My outside interests include coaching my kids in swimming and rediscovering the violin. Having these other activities helps keep me grounded and gives me time to think about something other than work. My family also keeps me on my toes: My husband is an Assistant Professor in the Physics Department at the University of Utah (biophysics research), and we have a 5-year-old daughter, Marina, and 1-year-old son, Kai.

KEY CONCEPTS AND DEFINITIONS

acetylation the addition of one or more acetyl groups to a protein; the formation of an acetyl derivative

activation energy the input of energy required to (overcome the barrier to) initiate a chemical reaction. By reducing the activation energy, an enzyme increases the rate of a reaction.

active site the region of an enzyme that binds substrates and catalyzes an enzymatic reaction

acylation the addition of one or more acyl groups to a protein

affinity a measure of the degree to which an substance tends to bind to another

alpha (α) helix a common secondary structure of proteins in which the linear sequence of amino acids is folded into a right-handed spiral stabilized by hydrogen bonds between carboxyl and amide groups in the backbone; a coiled secondary structure of a polypeptide chain formed by hydrogen bonding between amino acids separated by four residues

antibody a specialized immune protein, which helps fight infection by neutralizing pathogens and tagging them for destruction

beta (β) sheet a planar secondary structure of proteins that is created by hydrogen bonding between the backbone atoms in two different polypeptide chains or segments of a single folded chain; a sheetlike secondary structure of a polypeptide chain, formed by hydrogen bonding between amino acids located in different regions of the polypeptide

catalyst a substance that increases the rate of a reaction without itself undergoing any chemical change

dimer a compound formed by two molecules of a simpler compound; a polymer formed from two molecules of a monomer

disulfide bond (–S–S–) a covalent linkage formed between two sulfhydryl groups on cysteines. For extracellular proteins, disulfide bonding is a common way of joining two proteins together or linking different parts of the same protein. A disulfide bond is formed in the endoplasmic reticulum of eukaryotic cells.

enzyme a biological substance produced by living organisms to increase the rate of a biochemical reaction

glycosylation the addition of one or more sugars to a protein or lipid molecule

hydrophobic effect the tendency of nonpolar groups to cluster so as to shield themselves from contact with an aqueous environment

ion channel a transmembrane protein that transports ions, which are otherwise impermeable to the cells

ligand any molecule, other than an enzyme substrate, that binds tightly and specifically to a macromolecule, usually a protein, forming a macromolecule-ligand complex

Lineweaver–Burke plot a graphical model used to determine the maximum reaction rate, V_{max}, and Michaelis constant, K_m. The linear graph is obtained by taking the reciprocal of both sides of the Michaelis–Menten equation.

Michaelis constant (K_m) the value equal to the substrate concentration at which the enzyme reaction proceeds at half of the maximum velocity

Michaelis–Menten equation describes the velocity of a given reaction developed from a simple model of enzyme substrate kinetics

myristoylation the addition of myristic acid (a 14-carbon fatty acid) to the N-terminal glycine residue of a polypeptide chain

nascent proteins proteins that have not yet achieved their functional final folding pattern

nuclear magnetic resonance (NMR) spectroscopy a technique in which the magnetic nucleus of an atom is aligned with an external magnetic field and disturbs this alignment via the use of an electromagnetic field. The response of this disturbance is interpreted and gives information concerning physical, chemical, electronic, and structural information of compounds.

peptide bond a covalent bond that links adjacent amino acid residues in proteins; formed by a condensation reaction between the amino group of one amino acid and the carboxyl group of another with the release of a water molecule

phosphorylation a reaction in which a phosphate group becomes covalently coupled to another molecule

post-translational modification the enzyme-catalyzed change to a protein made after it is synthesized

protease an enzyme such as trypsin that degrades proteins by hydrolyzing some of their peptide bonds

protease inhibitor a substance that functions by inhibiting the actions of a protease

proteolysis the degradation of polypeptide chains

proteome the complete set of proteins present in the cell

proteomics a branch of molecular biology concerned with determining the proteome

receptor transmembrane protein a transmembrane protein that has an extracellular domain that binds to molecules called ligands

sequence homology analysis the comparison of an unknown primary amino acid sequence to a database of known primary amino acid sequences in an attempt to identify functional capabilities of the unknown amino acid

substrate the molecule on which an enzyme acts

transition state the structure that forms transiently in the course of a chemical reaction and has the highest free energy of any reaction intermediate. Its formation is a rate-limiting step in the reaction.

turnover number (k_{cat}) a rate constant that is equal to the number of substrate molecules processed per enzyme molecule each second

x-ray crystallography a technique for determining the three-dimensional structure of macromolecules (particularly proteins and nucleic acids) by passing x-rays through a crystal of the purified molecules and analyzing the diffraction pattern of discrete spots that results; a method in which the diffraction pattern of x-rays is used to determine the arrangement of individual atoms within a molecule

NOMENCLATURE

E_{TOT}	Total enzyme concentration	Mol/L
k	Rate of reaction	1/s
k_{cat}	Turnover number	1/s
K_M	Michaelis constant	
pK_A	Negative logarithm of acid dissociation constant	
V_0	Reaction velocity	Mol/L-s
V_{max}	Maximum reaction velocity	Mol/L-s

SUBSCRIPTS

1,2,3, etc. . .	Forward reaction steps in enzyme-catalyzed reaction
–1, etc. . .	Reverse reaction step for reaction step 1.

QUESTIONS

1. Describe the varying levels of structure for proteins. How do scientists experimentally determine the protein structure?
2. What is remarkable about post-translational modification of proteins?
3. Describe the hydrophobic effect and how it contributes to an energetically favorable protein folding reaction.
4. Describe the difference between the lock-and-key model and the induced fit model, then give a reason why the induced fit model may be more appropriate in describing enzyme mechanism.
5. Describe some ways that drugs might act as enzyme inhibitors.
6. Using a biochemistry textbook, or online resources, determine which of the following proteins has quaternary structure: α-chymotrypsin, hemoglobin, insulin, myoglobin, and trypsin.

PROBLEMS

1. Cells have surface receptors that can recognize and bind to the tripeptide RGD (Arg-Gly-Asp). Tissue engineers sometimes use this adhesive peptide to render synthetic biomaterials attractive to cells.
   a. Draw the chemical structure of the RGD peptide.
   b. Write the acid ionization chemical equation for each ionizable group. Use the Henderson–Hasselbalch equation to determine the ionization state of each ionizable group at pH 7.4. Be sure to consider ALL the ionizable groups: α-COOH ($pK_A = 2.0$), α-NH_2 ($pK_A = 9.0$), the Arg -NH ($pK_A = 12.5$), Asp –COOH ($pK_A = 3.9$).
   c. What is the net charge of the RGD peptide at physiological pH?
2. Select one of the enzymes involved in DNA replication (see Chapter 2).
   a. Describe the function of the enzyme.
   b. Search the National Center for Biotechnology Information (NCBI) STRUCTURE database for a three-dimensional structure of the enzyme. Print a copy of the enzyme structure and attach it to your homework set. (http://www.ncbi.nlm.nih.gov/Structure/).
   c. How do you think the structure of this enzyme facilitates its function?
   d. Note: To print the file, open the structure file in Cn3D and select the view that you want to save. Select "File/Export PNG" and it will ask you for a file name and folder location that you want to save. When it is saved, you can go to the saved folder and double click on the file name; it should open up in a default image viewer. When the image is visible then <right click> and select "print."
3. The concentration of a product, P, is measured by ultraviolet-visible spectroscopy (UV-VIS) as a function of time. The data are tabulated here. Estimate the initial reaction rate, V_0.

Time (h)	[P] (moles/L)
0	0
1	0.24
2	0.47
3	0.70
4	0.91

4. Penicillin is hydrolyzed and thereby rendered inactive by penicillinase (also known as β-lactamase), an enzyme present in some resistant bacteria. The mass of this enzyme in *Staphylococcus aureus* is 29.6 kd. The amount of penicillin hydrolyzed in 1 minute in a 10-mL solution containing 10^{-9} g of purified penicillinase was measured as a function of the concentration of penicillin. Assume that the concentration of penicillin does not change appreciably during the assay. [Note that kilodalton is the unit for MW for biological molecules. A dalton is another name for atomic mass unit. For example, a protein with a mass of 15 kd means that the MW is 15,000 g/mol.]

[Penicillin] (micromolar)	Amount hydrolyzed (nanomoles)
1	0.11
3	0.25
5	0.34
10	0.45
30	0.58
50	0.61

a. Plot V_0 versus [S] for these data. Does penicillinase appear to obey Michaelis–Menten kinetics? If so, what is the value of K_m?
b. What is the value of V_{max}? Indicate on the plot how you determined K_m and V_{max}.
c. Write the Michaelis–Menten equation and show the derivation of the Lineweaver–Burke equation.
d. Plot $1/V_0$ versus 1/[S].
e. Determine K_m and V_{max} using the Lineweaver–Burke plot. Indicate on the plot how you determined K_m and V_{max}.
f. Assume there is one active site per enzyme molecule. What is the turnover number of penicillinase under these experimental conditions?
g. How much time does it take for the penicillinase to hydrolyze one penicillin molecule?

5. Glutamic acid (1 of the 20 amino acids) has a side-chain carboxyl group (COOH, $pK_A = 4.3$) as shown in Figures 4.7 and 4.8.
   a. Write the chemical equation for the dissociation of the side-chain COOH. Label the weak acid and the conjugate base.
   b. The Henderson–Hasselbalch equation can be used to determine the ionization status of a weak acid: $pH = pK_A + \log \frac{[base]}{[acid]}$. Use the Henderson–Hasselbalch equation to determine whether the glutamic acid side-chain carboxyl group is protonated or deprotonated at physiological pH.
6. Enzymes function as catalysts in biochemical reactions.
   a. How does an enzyme speed up a reaction?
   b. Is the Gibbs free energy of a reaction affected in the presence of an enzyme? Explain.
7. The Michaelis–Menten equation (Equation 4.2) provides information about enzyme kinetics.
   a. Draw a schematic of a plot of the Michaelis–Menten equation for a particular enzyme substrate pair. Be sure to label the *x*- and *y*-axes. Indicate the V_{max} and K_m values on the graph.
   b. On your plot, also show the kinetic curve (V_0 vs. [S]) for the same enzyme with a substrate for which the enzyme has a lower affinity than the substrate in a.

c. On your plot, also show the kinetic curve for the situation described in (a) but with an increased quantity of enzyme. Assume that the enzyme concentration is two times higher than in (a).

8. The value of K_m was determined for three enzymes as shown below. Which of the enzymes has high affinity for its substrate and why?

Enzyme	K_m (M)
Chymotrypsin	0.0150
Pepsin	0.0003
Ribonuclease	0.0079

9. Carbonic anhydrase is an enzyme important to the management of CO_2 (carbon dioxide). About 11% of the blood's CO_2 is transported by hemoglobin. Most of the CO_2 that enters the erythrocytes (red blood cells) dissolves in the cytoplasm (cellular fluid). It then combines with water molecules to form carbonic acid, which immediately disassociates into hydrogen ions and bicarbonate. This reaction is sped up about 250 times by the enzyme carbonic anhydrase ($K_m = 1000\ \mu M$) that is located in the erythrocytes. Most of the CO_2 is converted into bicarbonate as soon as it enters red blood cells, thus keeping the CO_2 level in the cell lower than that of the interstitial fluid. This lowering of intracellular level is important as the concentration gradient between the interstitial fluid and the surrounding tissue increases the diffusion efficiency of CO_2, allowing it to be removed quickly from tissues. In a reaction vessel (which simulates body conditions) it was found that, in the presence of carbonic anhydrase, 5% of the initial 0.840993 M of CO_2 was converted after 2 seconds. How much CO_2 will be converted after 10, 30, and 60 seconds? State any simplifying assumptions.

5 Cellular Principles

LEARNING OBJECTIVES

After reading this chapter, you should:

- Understand the basic components of eukaryotic cells and the differences between eukaryotic and prokaryotic cells.
- Understand the basic role of the cytoskeleton, ribosomes, endoplasmic reticulum (ER), Golgi apparatus, mitochondria, lysosomes, and genomic deoxyribonucleic acid (DNA) in cell function.
- Understand the structure of extracellular matrix (ECM) and its role in tissue function.
- Understand the role of membrane proteins in regulating transport through cell membranes and regulating cell adhesion.
- Understand the cell cycle and cell division by mitosis and meiosis.
- Understand the basic principles of stem cells and differentiation.
- Understand the basic elements of cell culture and its importance in modern biomedical science and engineering.

5.1 Prelude

The cell is the basic functional unit in the body. The human body is composed of more than 200 different types of cells (Figure 5.1). Each cell of an individual is genetically the same: They all share the same genetic information, but cell types within an individual differ with respect to size, shape, and constituent molecules (Figure 5.2); therefore, they have different properties. For example, liver cells have abundant enzymes for detoxification of chemicals whereas red blood cells instead have abundant hemoglobin for oxygen transport. These differences are important to the function of the cell in the context of the organ in which it resides.

Despite this diversity of cell composition and function, the trillions of cells in each person (most estimates range from 50 to 200 trillion cells in an average person) share common properties. In addition to their identical genetic material, most cells in the body have the same structural organization. Cells in humans are all surrounded by a lipid bilayer membrane and contain many of the same structural features. For that reason, biologists can discuss general principles of

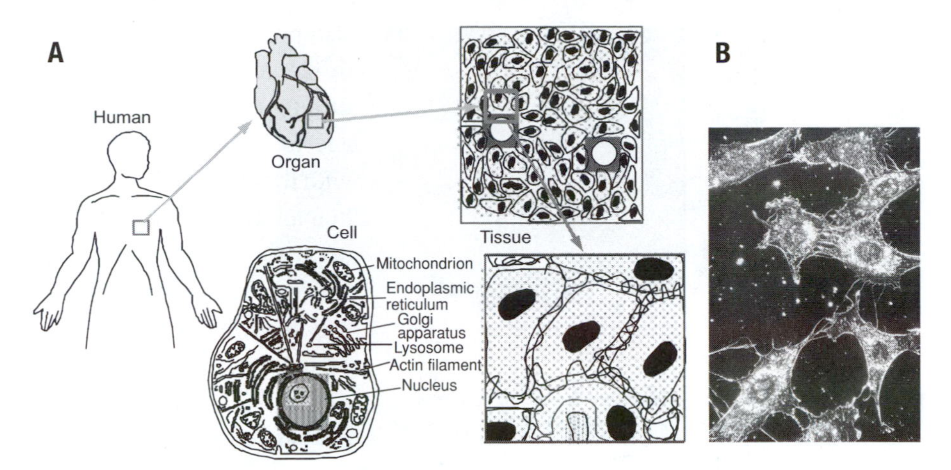

Figure 5.1

Cells are the basic functional unit of the human body. A. Cells are organized into tissues, which are collections of similar cells that perform a specific function. Organs are collections of two or more primary tissues, united for a particular function. The heart, for example, contains muscular, nervous, and connective tissues. **B**. Human connective tissue cells in culture. Photo credit: Dr. Cecil Fox, courtesy of National Cancer Institute.

the biology of the cell. Of course, specific types of cells exhibit behaviors and properties that are specific to that cell type.

Much of what is known about cells has been learned from the study of cells in culture (Figure 5.1). **Cell culture**, or the maintenance and growth of cells outside of the body, was first discovered in the early 1900s and is now routine. The ability to maintain and study cells in a controlled and reproducible fashion has enabled scientists to dissect and classify many of the functions of human cells. Experiments in cell culture are often called in vitro experiments—from the Latin phrase meaning "in glass"—because they occur in an artificial environment, outside of the living organism. It is sometimes difficult to determine if the observations made from in vitro experiments are also applicable for cells in the body, which are often referred to as cells in vivo—also from the Latin meaning "in life." Compounding this difficulty, often the cell culture studies are performed on cells from another species, making the experiments easier to accomplish and

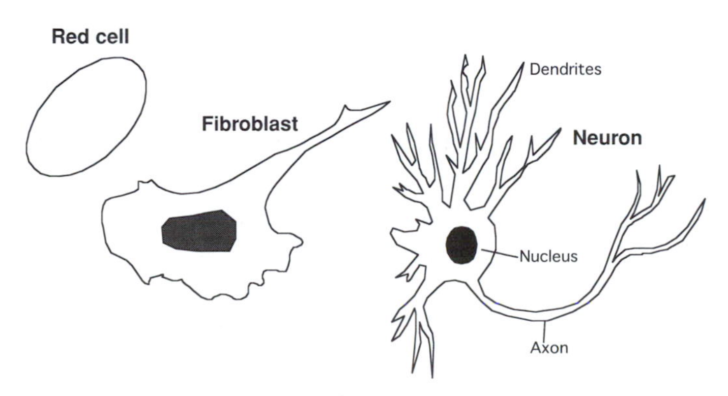

Figure 5.2

Differences in shape among human cells. The differences in shape are illustrated by the shapes of the red blood cell, which has no nucleus; the fibroblast, a connective tissue cell; and the neuron.

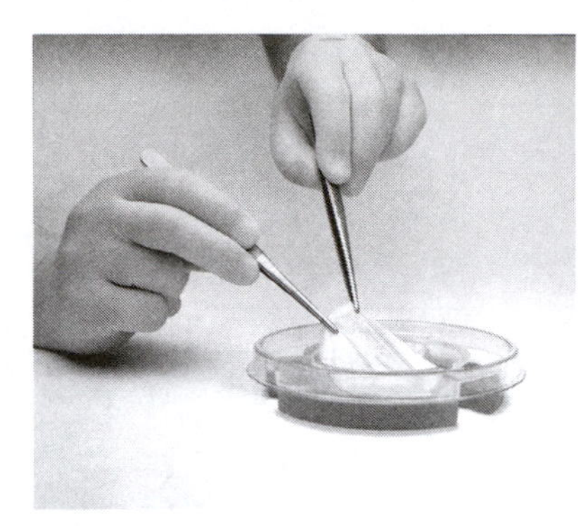

Figure 5.3 **Cells can be used in tissue engineering.** Apligraf® (Organogenesis Inc) is a skin substitute containing living cells. The skin substitute is constructed by first culturing human dermal cells within a gel of collagen protein. After culturing for 6 days, a layer of epidermal cells is added to the top of the collagen/dermal cell gel. This layered structure can be used to treat skin wounds that are difficult to heal, such as skin ulcers that occur in diabetics. Photo used with permission. (See color plate.)

also providing valuable information on the differences in function among cells from different species, but making the linkage to human biology unclear. Still, even with these difficulties, our understanding of human biology has advanced dramatically because of our facility with cell culture.

This chapter reviews basic cell structure and function as well as different aspects of the cell life cycle including proliferation, differentiation, and death. Biomedical engineers use cultured cells as experimental models to study basic biological functions, such as cell migration or cell adhesion, or to perform preliminary screening tests for new therapeutic strategies. For example, biomedical engineers commonly use cell cultures as an initial step in the production of new biomaterials, to test that the material is not toxic to living cells. Tissue engineers are now using cells as components in artificial skin, cartilage replacements, and blood vessels (Figure 5.3). All of these developments are further enhanced by the ability to genetically engineer cells that either stimulate or inhibit expression of specific genes. For example, tissue engineers are now able to use cells that are genetically modified to improve their survival in the body or to avoid reactions with the immune system.

5.2 Cell structure and function

Some basic elements of cell structure were introduced in Chapter 2. For example, all cells are surrounded by a lipid bilayer membrane, which separates the intracellular space from the extracellular space. This **plasma membrane** restricts the movement of water and ions into and out of the cell. Cells can be divided into two main classes: **prokaryotic** and **eukaryotic** (Table 5.1 and Figure 5.4).

Prokaryotic cells (e.g., bacteria, cyanobacteria, archaebacteria) are simple, lacking a nucleus, cytoplasmic organelles, and a cytoskeleton. These cells maintain their shape because of the presence of a rigid **cell wall**. In bacteria, the cell wall is composed of **peptidoglycan**, a polysaccharide polymer that is cross-linked by amino acids for additional stability. Cell wall structures differ between the two major classes of bacteria. Gram-positive bacteria have a predominantly peptidoglycan wall, whereas Gram-negative bacteria have two walls: a thinner inner wall, containing peptidoglycan, and an outer **lipopolysaccharide** layer.

Table 5.1 Comparison of prokaryotic and eukaryotic cells

Characteristic	Prokaryote	Eukaryote
Nucleus	Absent	Present
Diameter of a typical cell	1 μm	10–100 μm
Cytoskeleton	Absent	Present
Cytoplasmic organelles	Absent	Present
DNA content (base pairs)	1×10^6 to 5×10^6	1.5×10^7 to 5×10^9
Chromosomes	Single circular DNA molecule	Multiple linear DNA molecules

Despite their small size (<2 μm) and simplicity, prokaryotic cells are biochemically diverse. Their rapid doubling time—doubling in some cells can occur in as little as 30 minutes—allows them to multiply rapidly and to adapt quickly to changes in their environment. There are species of bacteria known that can use almost every type of organic molecule as food (including sugars, amino acids, fats, polypeptides, polysaccharides, and hydrocarbons). Biomedical engineers often use bacterial cells for production of recombinant proteins, as discussed in Chapter 2.

Eukaryotic cells (e.g., cells of fungi, algae, protozoa, plants, and animals) are more complex in structure. The outer surface of the cell, the plasma membrane, is a deformable fluid sheet that completely surrounds the internal contents of the cell, or cytoplasm. Plant cells have cell walls to give them structure. In animal cells, however, there is no rigid cell wall to provide shape or structure for the cell.

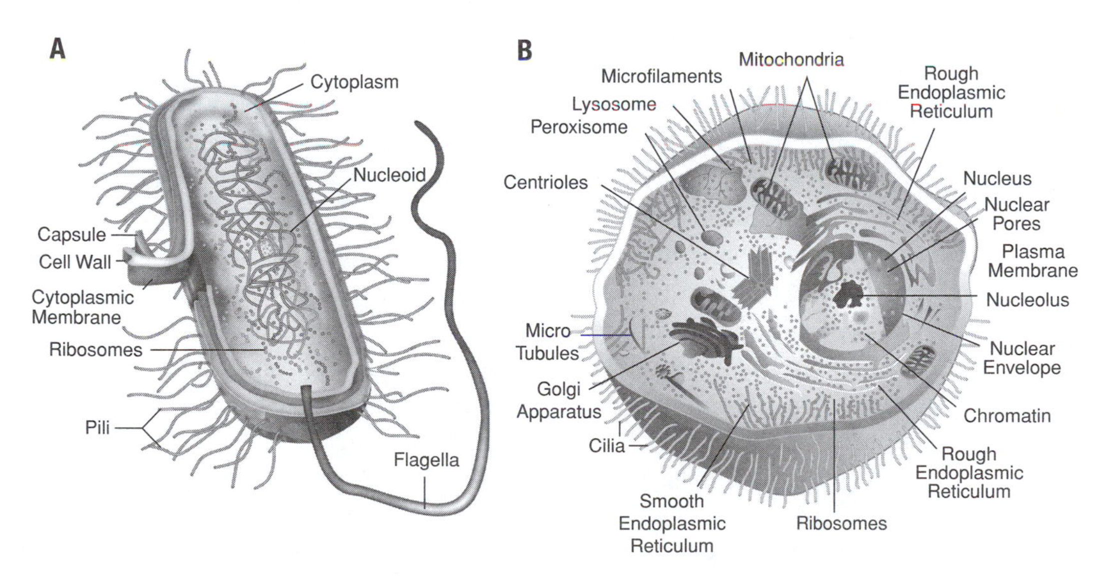

Figure 5.4 **Structure of prokaryotic and eukaryotic cells. A.** Prokaryotic cells are typically small (~1 micron) and have a rigid cell wall that provides their shape. The cytoplasm is unorganized, with the chromosomal deoxyribonucleic acid (DNA) and other cell components mixing throughout. **B.** Eukaryotic cells are larger (typically 10 microns) and possess no cell wall. Structure is provided instead by an intracellular assembly of rigid proteins, called the cytoskeleton. The intracellular space is organized into organelles, such as the nucleus, mitochondria, and lysosomes.

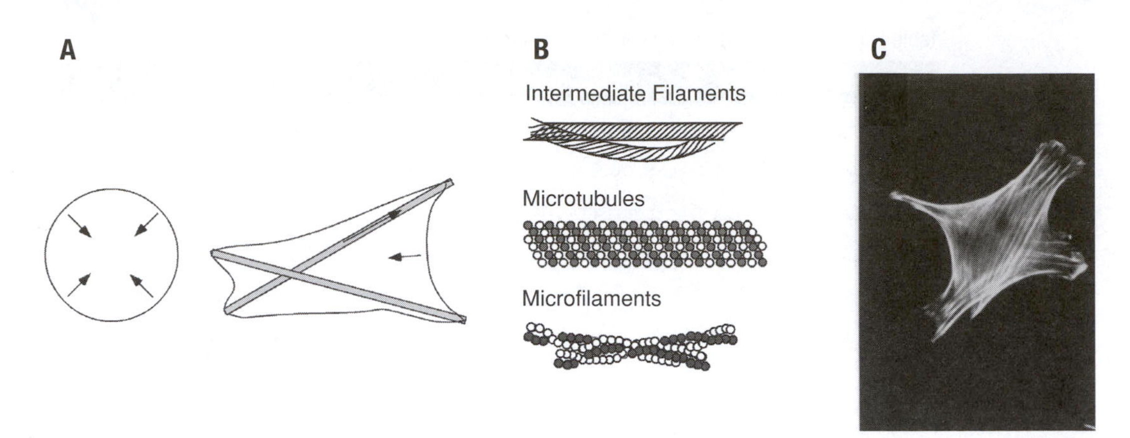

Figure 5.5

Cytoskeleton. A. The tension of the cell membrane would pull inward, tending to make each cell a small sphere. If a solid network of supports were present in the cell, the arrangement of the supports would determine the shape of the cell. **B.** Intermediate filaments, microtubules, and microfilaments are present in differing amounts in most cells. **C.** A fibroblast in culture is shown with the actin-rich microfilaments stained with a fluorescent dye. Reproduced from Saltzman WM. *Tissue Engineering*, New York: Oxford University Press, 2004 by permission of Oxford University Press, Inc.

If the cell contents were completely homogeneous and liquid (i.e., if the cell were a bag full of water)—and if the membrane were under tension (i.e., if the bag were elastic like a balloon)—then the overall cell shape would be spherical. In contrast, if this same membrane-bound liquid bag enclosed a number of incompressible sticks, then its overall shape would depend on both the tension in the membrane and the stiffness of the sticks (Figure 5.5a). For this imaginary cell, morphology depends on the properties of the membrane (e.g., tension and deformability) and the characteristics of the internal sticks (e.g., their size, number, position, and mechanical properties).

Cells are considerably more complicated than balloons filled with sticks, but they appear to share some common characteristics. The elastic and fluid properties of the membrane have already been mentioned. The cytoplasm also contains a wide spectrum of molecules that probably serve as molecular struts (Figure 5.5b). **Microfilaments**, **microtubules**, and **intermediate filaments** are nanometer-scale fibers that extend throughout the cytoplasm, creating interconnected and entangled networks. These protein filaments together form the **cytoskeleton**. The cytoskeleton is an internal skeleton that gives rise to cell shape and coordinates cell division and cell movement. Some cytoskeletal proteins serve as tracks to help transport substances from one organelle to another.

Organelles are specialized compartments surrounded by membranes but located inside the cell. Figure 5.4b shows the major organelles of a eukaryotic cell. Each organelle has a specific function, which is summarized in Table 5.2. The **nucleus** is the largest organelle in most cells and contains the chromosomes. Genomic DNA is contained within the nucleus. DNA is transcribed in the nucleus, and the messenger ribonucleic acid (mRNA) transcript moves out of the nucleus to the **ribosomes** for translation into a polypeptide chain. Ribosomes bound to the **rough ER** (rER) synthesize proteins that will be incorporated into the membranes

Table 5.2 Functions of the major organelles of eukaryotic cells

Organelle	Function
Nucleus	Control center for cell metabolism and reproduction; stores genetic information
Ribosomes	Synthesize proteins
Endoplasmic reticulum (rough and smooth)	Transports proteins and lipids
Golgi apparatus	Packages and processes proteins
Lysosomes	Digests food particles, old organelles, invaders
Cell (plasma) membrane	Controls what comes into and out of the cell, regulates cellular homeostasis
Mitochondria	Performs respiration for all eukaryotic cells—takes glucose and oxygen and converts it to adenosine-5′-triphosphate (ATP) energy; has role in maintaining cell life and death

of cells or secreted. The rER contains chaperone proteins that help properly fold the newly synthesized polypeptide chains as well as enzymes that assist in post-translational modifications of proteins (see Chapter 4, Section 4.3). Membrane or secreted proteins are transported in vesicles from the rER to the **Golgi apparatus** for further processing and sorting. The **smooth ER** is not associated with ribosomes and is the site of lipid synthesis (such as the phospholipids discussed in Chapter 2). The **mitochondria** is often referred to as the "powerhouse" of the cell because of the energy it produces in the form of adenosine-5′-triphosphate (ATP) from oxidative metabolism that exploits the energy released from the breakdown of carbohydrates and fatty acids. **Peroxisomes** contain about 50 different types of enzymes (such as catalase and other oxidative enzymes) that are involved in a variety of reactions involved in detoxification. **Lysosomes** contain digestive enzymes—proteases, lipases, nucleases, and carbohydrases—that break down proteins or other cellular components.

Humans are collections of eukaryotic cells. Within the body, a collection of cells of a similar type is called a **tissue**. There are four general types of tissues: muscle, nervous, epithelial, and connective tissues. **Muscle tissue** is specialized for movement; the cells within muscle tissue are rich in specialized proteins (described in Box 4.1) that enable the cells to contract. **Nervous tissue** is capable of initiating signals and transmitting them from cell to cell, in a coordinated fashion. **Epithelial tissues** are organized into sheets that cover body surfaces: They separate the body from its environment (skin and intestinal epithelium) and they line glands to enable regional secretion of specialized fluids. **Connective tissue** is rich in extracellular material that provides mechanical strength and anchors adjacent tissues. All of the various types of circulating blood cells are considered to be specialized forms of connective tissues. An **organ** is a collection of two or more primary tissues, which are organized into a functional unit. Some organs contain all four tissue types; for example, the stomach has an epithelial lining, a muscular layer for mixing of stomach contents, a nervous tissue for coordination of muscular action, and a connective tissue for binding the other tissues into a unit.

5.3 ECM

What holds cells together and allows them to form tissues? The extracellular space contains a three-dimensional array of protein fibers and filaments, which are embedded in a hydrated gel of high-molecular-weight, carbohydrate-rich molecules (Figure 5.6). These components are collectively called the ECM. ECM is produced by cells: Specialized structural molecules are secreted locally by cells and assemble to form a scaffold that supports cell attachment, spreading, proliferation, migration, and differentiation. Cells influence the chemistry of their surrounding ECM by direct secretion of molecules, but they also modify the physical characteristics of the matrix locally by release of enzymes, which can digest or stabilize the gel matrix, or by application of physical forces, which can physically rearrange the gel components.

The composition and organization of macromolecules in the ECM helps determine the tissue structure and physical properties. Examples of different types of ECM include the soft cartilage, which is found in the nose and ear, the basal lamina sheet underlying epithelial cells in the intestine, and tendons, which attach muscles to bone. Bone contains a mineral-rich ECM.

The specialized molecules of the ECM are predominantly of three classes: proteins, **glycosaminoglycans** (GAGs), and **proteoglycans**. The gel that fills the extracellular space is composed of GAGs, which are unbranched polysaccharide chains. Most GAGs are covalently attached to a protein core to form **proteoglycans**. There are two categories of proteins in the ECM: structural and adhesive. Structural proteins include **collagen** and **elastin**. Collagen provides strength whereas elastin provides elasticity. Collagen is the most abundant protein in the human body and is found, in various forms, in almost every tissue. The different types of collagen have different chemical and physical properties: Some form strong fibers (type I) and others form a meshlike network (type IV). Elastin is particularly abundant in certain tissues in which stretch and recovery are important for their function, such as the vessel wall of aorta.

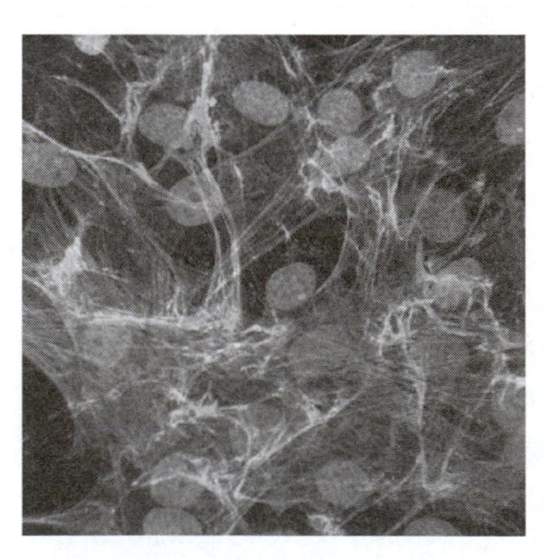

Figure 5.6 **Fibroblast cells are surrounded by an extracellular matrix.** Image by Dr. Boris Hinz, Laboratory of Cell Biophysics, Ecole Polytechnique Fédérale de Lausanne, Lausanne, Switzerland. (See color plate.)

Other adhesive proteins such as **fibronectin** and **laminin** help to bind the other matrix components together and facilitate attachment of cells to the ECM. Fibronectin is a glycoprotein found mainly in connective tissue and serves as a crosslinker, or intermediate, in the ECM by binding to both collagen and GAGs. Fibronectin also contains binding sites for adhesion receptors found on the surface of cells. Thus, cells can adhere to the ECM via specialized adhesion receptors. Laminin also

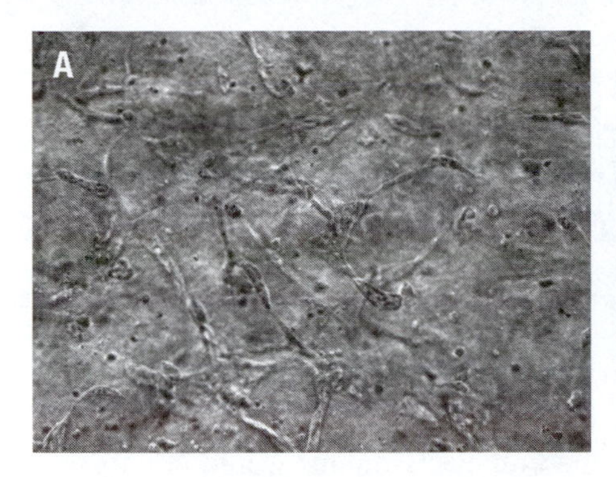

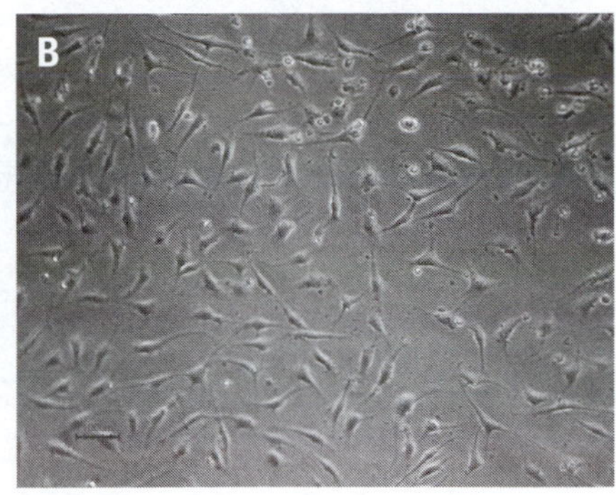

Figure 5.7

Biomedical engineering uses of extracellular matrix A. Human endothelial cells are suspended in a gel of collagen, to help determine the effectiveness of delivering a growth factor to the cells. Photo credit: Steven Jay. **B**. Fibroblasts are cultured on a peptide-based hydrogel, which might be useful for tissue engineering. Reprinted in part with permission from Salick DA, Kretsinger JK, Pochan DJ and Schneider JP, Inherent antibacterial activity of a peptide-based b-hairpin hydrogel. *JACS* 2007, 129:14793–14799. Copyright 2007, American Chemical Society.

contains binding sites for cell attachment and binding sites that promote neurite outgrowth. Laminin and collagen type IV are the major components of the special ECM called **basal lamina**, which is secreted by epithelial cells.

Biomedical engineers are using many of these biological principles to create artificial scaffolds that mimic selected properties of ECM (Figure 5.7). For example, using cell attachment to fibronectin as an example, synthetic tissue-engineering scaffolds are often grafted with either fibronectin or, as an alternate, the short peptide sequence Arg-Gly-Asp (or RGD). RGD works by binding to integrins, proteins in the plasma membrane of cells that mediate binding to ECM components (see Section 5.4.2). This RGD peptide is much more stable and easier to work with than the full fibronectin molecule, but this small sequence occurs naturally in the fibronectin protein (and other adhesive proteins) and it can be used as a mimic for the effect of the whole protein.

5.4 Molecules in the cell membrane

The plasma membrane, which separates the interior of the cell from the rest of its environment, is a phospholipid bilayer that supports a wealth of embedded and coupled membrane proteins (see Chapter 2, Sections 2.7.1 and 2.7.2). The proteins associated with the membrane provide a variety of functions including cell recognition and signaling. This section reviews the role of membrane proteins in the transport of molecules and adhesion to the ECM.

All cells in the human body require proteins to facilitate and regulate the transport of molecules in and out of the cell. In fact, cells cannot live for long without the function of transport proteins. Glucose, the primary fuel for most cells, cannot permeate through a lipid bilayer and is only available to the cell by movement across the membrane within a specialized transporter, which is discussed later. Proteins for membrane transport are also major contributors to the physiology of

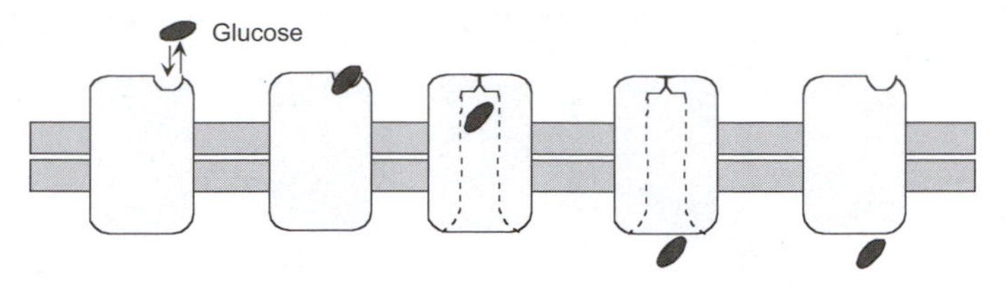

Figure 5.8 **Illustration of the mechanism of action of the glucose transport protein.** Glucose binds to the extracellular side of the glucose transport protein, which is capable of conformational change. The change includes opening of a channel that allows glucose to move through the membrane.

the digestive system and the kidneys: Humans would not absorb nutrients from their diet or regulate the concentration of sodium and potassium ions in their body fluids without the work of specialized transporters. Biomedical engineers are now beginning to learn how to incorporate these specialized molecules into devices, such as biosensors, that exploit the biological properties of recognition and selective transport of these molecules.

5.4.1 Membrane proteins that regulate transport

Many molecules do not diffuse through lipid bilayers. Polar molecules—such as sugars, ions, and most amino acids—are not able to dissolve easily into the lipid-rich inner region of the bilayer (recall Figures 2.32 and 2.33). One of the most important functions of accessory molecules in the membrane is regulation of transport of molecules that do not pass freely through the lipid bilayer. Several classes of accessory molecules are engaged in membrane transport, as described in the sections that follow.

FACILITATED TRANSPORT VIA TRANSPORTERS

Although the extracellular concentration of glucose is usually higher than the intracellular concentration, the low permeability of the lipid bilayer to glucose prevents passive transport of sufficient molecules of glucose to support metabolism. Glucose transport proteins in the cell membrane solve this problem by providing aqueous pathways that shuttle glucose through the hydrophobic bilayer. The glucose transporter facilitates glucose permeation by periodic changes in conformation: In one conformation, a glucose binding site is exposed on the extracellular face, whereas in another conformation, the binding site is exposed to the intracellular face (Figure 5.8). Conformational changes occur because of natural thermal fluctuations in the membrane; conformational changes occur whether the glucose binding site is occupied or vacant. As a result of this periodic change in structure, the transport protein permits the passage of glucose (or another molecule that is able to bind to the transporter binding site), without the addition of any additional energy. Glucose molecules can move in either direction across the bilayer; the net flux will occur from the region of high to low concentration.

Facilitated transport proteins, such as the glucose transporter, are present in the cell membrane in limited number. As the concentration of the solute increases at the external membrane surface, the transporter binding sites become saturated,

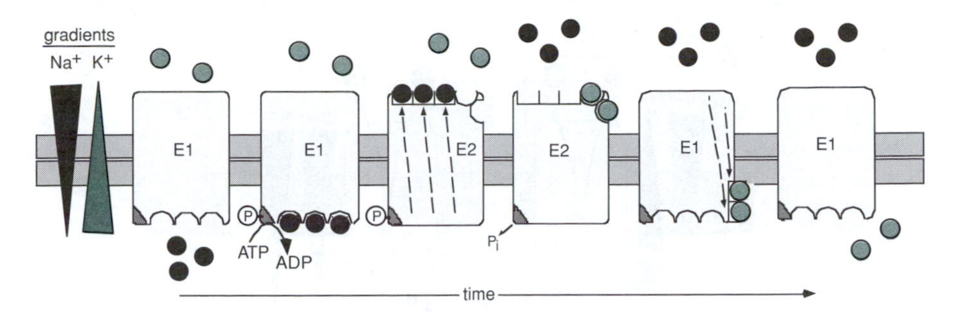

Figure 5.9

Na-K pump moves ions across the plasma membrane. Energy-dependent, adenosine-5′-triphosphate (ATP)-powered ion pumps such as the Na^+/K^+ exchange ATPase allow transport of molecules against their concentration gradient. Na^+ molecules are indicated by black circles; K^+ molecules are indicated by colored circles; gradients are illustrated by the triangles, with the width of the triangle representing concentration. The conformational change is triggered by ATPase activity, which phosphorylates an aspartate residue of the protein. Formation of a high-energy phosphate bond drives the E1→E2 conformational change, which reverses when the phosphorous is eventually released. Na^+ and K^+ binding sites cycle between high affinity (semicircles in the schematic diagram) and low affinity (rectangles) states that are coordinated with the conformational change.

and the net rate of solute transport across the membrane approaches a maximal value. Both facilitated transport and simple diffusion depend on concentration gradients: Net solute transport always occurs from high to low concentration. Unlike diffusion, facilitated transport systems are specific because they depend on binding of the solute to a site on the transport protein. For example, the D-glucose transporter is very inefficient at transporting L-glucose, but it can accommodate D-mannose and D-galactose.

ACTIVE TRANSPORT

Active transport systems are similar to facilitated transport systems; both involve the participation of transmembrane proteins that bind a specific solute. In active transport, however, energy is provided—most often by the hydrolysis of ATP—to drive a conformational change in the transporter that leads to solute transport.

The Na-K pump is the most well-characterized active transport system (Figure 5.9). The pump is an assembly of membrane proteins, with three Na^+ binding sites on the cytoplasmic surface, two K^+ binding sites on the extracellular surface, and a region of ATPase activity on the cytoplasmic surface. In one cycle of the pump, energy from the conversion of one ATP molecule to adenosine diphosphate (ADP) is used to transport three Na^+ ions out of the cell and two K^+ molecules into the cell. Both ions are moving "up" their gradients, from a region of low concentration to a region of high concentration. A cell that has many copies of the Na-K pump can maintain an intracellular space that is high in K^+ and low in Na^+.

SECONDARY ACTIVE TRANSPORT

The Na-K pump is an example of a **primary active transport** system. It converts the energy derived from a chemical reaction into movement of molecules against their concentration gradient. **Secondary active transport** systems also move solutes "up" their concentration gradient, but they gain the energy for this from a different source. In secondary active transport, the movement of a molecule

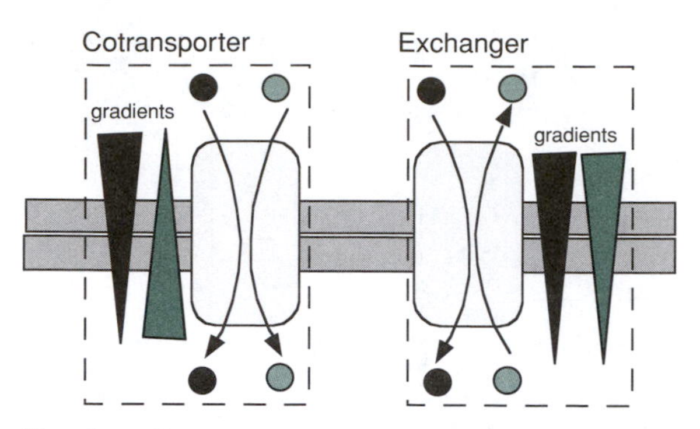

Figure 5.10 **Secondary active transport systems.** Exchangers and cotransporters use the energy stored in the concentration gradient of one solute (indicated by black circles) to move a second solute (colored circles) against its concentration gradient. In cotransporters, the movement of the black solute is coupled to the movement of the colored solute, in the same direction, but against its concentration gradient. In exchangers, the secondary solute (colored) is moved in the opposite direction as the primary solute (black).

"down" its concentration—which can occur spontaneously—is coupled with the movement of another molecule "up" its concentration gradient.

Secondary active transport systems are classified according to the relative direction of movement of the two solutes (Figure 5.10). The primary solute moves down its concentration gradient. Movement of the primary solute is spontaneous: The molecules are driven by the energy stored in the concentration gradient, the same kind of energy that drives all diffusion processes. In **cotransporters**, the spontaneous transport of the primary solute is coupled with the transport of a secondary solute in the same direction. An example is the Na/glucose cotransporter, which is present on many cells but is particularly important in the epithelial cells of the proximal tubule in the kidney, where it functions to help recover glucose that might otherwise be lost in the urine (this will be discussed in Chapter 9, see Figure 9.15). In **exchangers**, the primary and secondary solutes move in opposite directions. An example is the Na-Ca exchanger, which uses the energy stored in the Na^+ concentration gradient (high outside the cell, low inside the cell). Na-Ca exchangers are common in cells throughout the body, where they act to maintain the low intracellular Ca^+ levels of most cells.

Secondary active transport systems often get their energy from the action of a primary active transporter. In both of the examples described here—Na/glucose cotransporter and Na-Ca exchanger—the Na^+ gradient that drives the transport is created by the action of the Na-K pump.

ION TRANSPORT, MEMBRANE POTENTIALS, AND ACTION POTENTIALS

Lipid bilayers are impermeable to ions and other charged species, but ions such as Na^+ and K^+ are found in abundance in both the extracellular and intracellular environment. Active transport proteins can move ions across membranes, even moving molecules against a concentration gradient. Cells and microorganisms are exquisitely sensitive to changes in their ionic environment, making ion movement across membranes profoundly important in physiology.

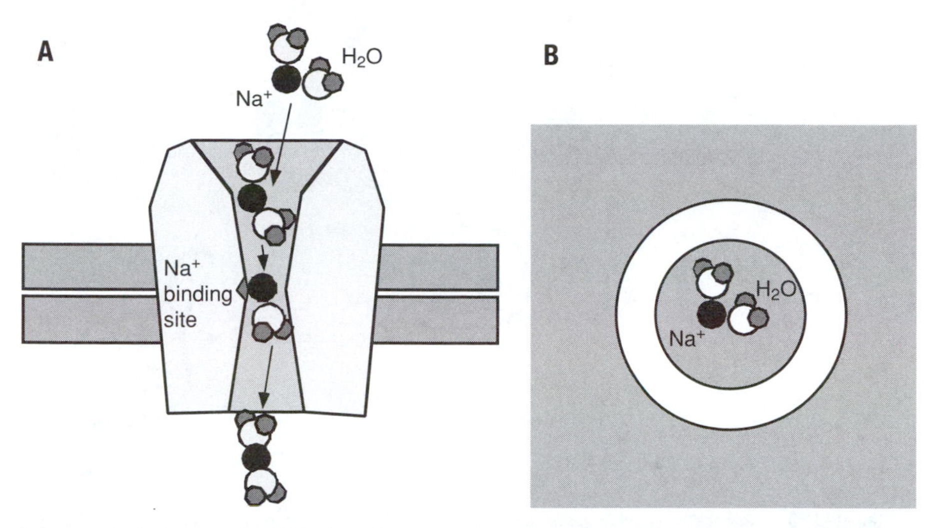

Figure 5.11 **Ion channels. A.** Cross-sectional side view of hypothesized mechanism for selectivity in a sodium channel. A weak binding interaction with the wall of the channel permits temporary release of water for hydration. **B.** Top view of same structure.

Ion movement across membranes is regulated by specialized membrane proteins called ion channels (Figure 5.11). Some ion channels are selective, permitting the permeation of only specific ions, such as Na^+, K^+, or Cl^-. Selectivity is accomplished by a combination of channel characteristics including molecular sieving (which selects for ions of certain size), binding to the protein surface, and stabilization of the nonhydrated ion. Some channels are gated; gated channels exist in multiple states that either permit or exclude ion movement across the membrane. In the open state, conductance (or permeability) to the ion rapidly increases. The closed/open state of the ion channel is regulated by extracellular and intracellular conditions. Some channels are voltage-regulated; others are regulated by binding of a chemical ligand or mechanical stretching of the membrane.

Both facilitated transport systems and active transport mechanisms can be saturated, and the rates of transport are similar. In addition, both carrier-mediated transport systems have less transport capacity than channel-mediated transport or simple diffusion. That is, more molecules per minute can move via an ion channel than by active or facilitated transport.

RECEPTOR-MEDIATED ENDOCYTOSIS AND SIGNALING

Passive diffusion, facilitated and active transport, and diffusion through channels account for most of the molecular transit across membranes. A process called receptor-mediated endocytosis is also important in the life of a cell; biomedical engineers are particularly interested in this mechanism, as it may represent an important target for drug delivery.

Binding of some molecules (call them **ligands**) to membrane receptor proteins can lead to rapid internalization of both the ligand and the membrane protein

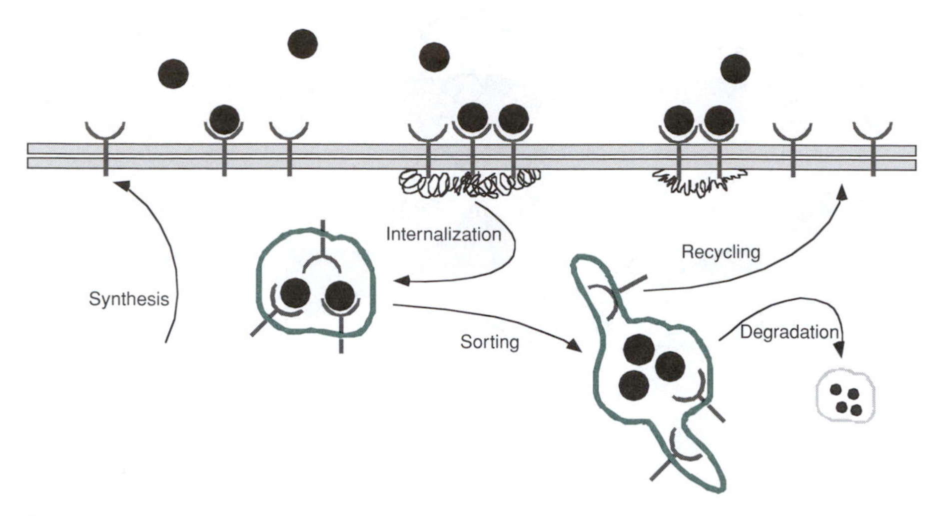

Figure 5.12 **Receptor-mediated endocytosis.** Schematic diagram of the main steps in the endocytotic pathway. Binding at the cell surface induces clustering of receptor–ligand complexes in specialized regions of the membrane. Endocytotic vesicles (colored) form in these regions; separation of receptor and ligand occurs as the endosome matures. Separated receptors can recycle to the surface, whereas remaining material is degraded within lysosomes.

by a process called endocytosis (Figure 5.12). This process has been studied extensively for a variety of systems including low-density lipoprotein (LDL), transferrin, and epidermal growth factor (EGF). Endocytosis frequently leads to accumulation of ligand in intracellular vesicles, including lysosomes, in which degradation of the ligand can occur. Ligand–receptor complexes have different fates after endocytosis. In the case of LDL and LDL receptor, the LDL dissociates from the LDL receptor within the endosome, and the LDL receptor is recycled to the cell surface. LDL receptors are conserved in this process, and an essential nutrient is brought into the cell without sacrificing the carrier protein. Transferrin, an iron-carrying serum protein, releases bound iron in the endosome (becoming apotransferrin, the iron-free form of the protein); the transferrin–receptor complex is recycled to the cell surface, after which transferrin is released into the extracellular environment. Both protein and receptor are conserved in this process. EGF and EGF-receptor are both degraded within the lysosomes. Because the receptor is degraded, EGF binding eventually leads to a decrease in the number of EGF receptors at the surface, a process called receptor down-regulation.

5.4.2 Membrane proteins that regulate cell adhesion

Many different types of proteins are associated with the plasma membrane and positioned so that they are exposed to the extracellular space (see Figure 2.10). Some of these proteins form strong noncovalent associations with either components of the ECM or molecules extending from the surface of other cells. These membrane-associated proteins are called **adhesion molecules**. Table 5.3 lists the families of both classes of cell adhesion molecules.

Table 5.3 Cell adhesion molecules

Cell-to-cell adhesion	Type of binding
Cadherins	Homophilic binding
Ig family members (e.g. N-CAM)	Homophilic and heterophilic binding
Selectins	Heterophilic binding
Cell-to-ECM adhesion	
Integrins	Heterophilic binding

Notes. Homophilic binding means that the receptors bind to other receptors of the same type (i.e., cadherins on one cell bind to cadherins on another cell). Heterophilic binding means that the receptor binds to a different kind of molecule (e.g., some integrins bind to the extracellular matrix [ECM] protein fibronectin). N-CAM, neural cell adhesion molecule.

Integrins are the main adhesion molecules that adhere a cell to ECM components but are also capable of binding to adhesion molecules on the surface of other cells (Figure 5.13). Because they bind to other types of molecules (non-integrin), integrin interactions are heterophilic. Integrins can also act as receptors that allow cells to communicate with their environment. When an integrin binds to its ECM component, it transmits signals within the cell via an intracellular signaling cascade. Chapter 6 provides a detailed description of various types of receptors and their signal transduction pathways. Ligand binding to various integrin receptors triggers downstream activation of genes that can influence the behavior of a cell including shape, movement, and differentiation.

Selectins recognize carbohydrate groups of glycoproteins on the surface of other cells. This interaction is a critical step in the recruitment of leukocytes, which help fight infection, to a local site of inflammation. Selectins on the surface of leukocytes bind to carbohydrate groups on the surface of endothelial cells lining the blood vessel (Figure 5.14). These interactions cause the leukocyte to slow down and roll along the blood vessel wall. Slow rolling along the surface allows integrins on the leukocyte to bind to intercellular adhesion molecules (ICAMs) located on the surface of endothelial cells. Downstream signaling causes the leukocyte to change shape so that it can spread and move in between the endothelial cells and to the site of inflammation. Another type of cell–cell adhesion molecule is **cadherin**. Cadherin interactions are homophilic because they bind to other cadherins on neighboring cells. For example, E-cadherins facilitate selective adhesion of epithelial cells to one another.

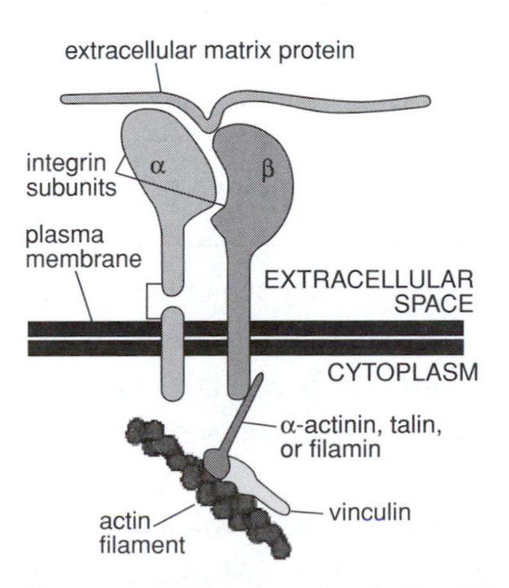

Figure 5.13 **Membrane proteins for cell adhesion.** Some membrane proteins extend into the extracellular space and are able to associate, or bind with, molecules of the extracellular matrix.

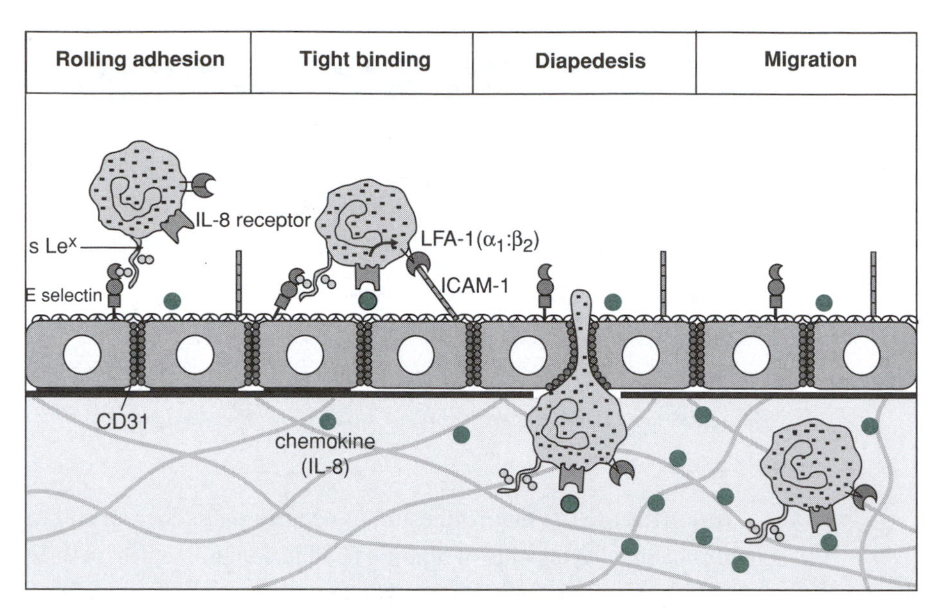

Figure 5.14

Selectins. Cell adhesion proteins called selectins are involved in the adherence of white blood cells to the blood vessel wall. Adherence of blood cells is the first step in their migration into the tissue surrounding the blood vessel, where they play a role in inflammation. ICAM-1, intercellular adhesion molecule 1; LFA-1, lymphocyte function-associated antigen-1; sLe^x, sialyl Lewis x antigen ; IL-8, interleukin 8.

5.5 Cell proliferation

The ability of a cell to divide, and therefore to duplicate itself, is important during embryonic development as well as throughout life. Many cell divisions are necessary as a fertilized egg, a single cell, develops into a newborn human with $\sim 200 \times 10^9$ cells. In adults, cell proliferation is required to replenish cells that die or are destroyed because of injury, to replace cells that are lost as a normal consequence of tissue function, and to increase the number of immune cells in response to a foreign pathogen.

5.5.1 The cell cycle

Cell division in animal cells occurs in an orderly sequence of steps, although the rate of progression through this sequence varies considerably among cell types. The cell cycle (Figure 5.15a) is continuously repeated as each new daughter cell—formed as a consequence of cell division—develops, replicates its DNA, and eventually divides.

Consider the progression of a newly formed cell, which begins in a phase called $\mathbf{G_1}$ because it represents the gap between **mitosis (M phase)** and **DNA synthesis (S phase)**. Cells can exit the cell cycle and remain in an indefinite period of rest called $\mathbf{G_0}$; some cells (such as neurons) are suspended in G_0 for the lifetime of an individual. Thus, they are referred to as nondividing cells.

A

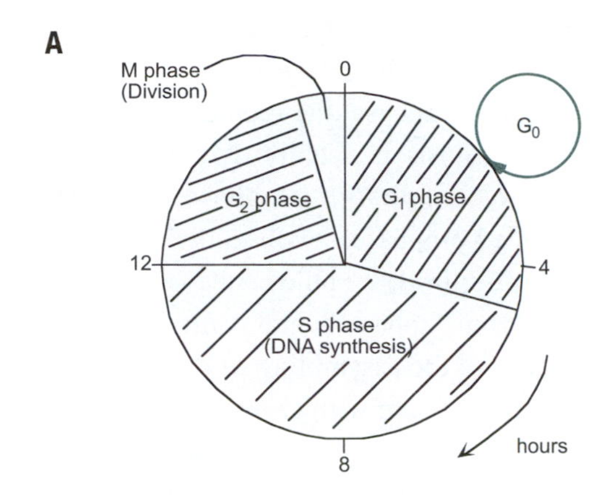

B

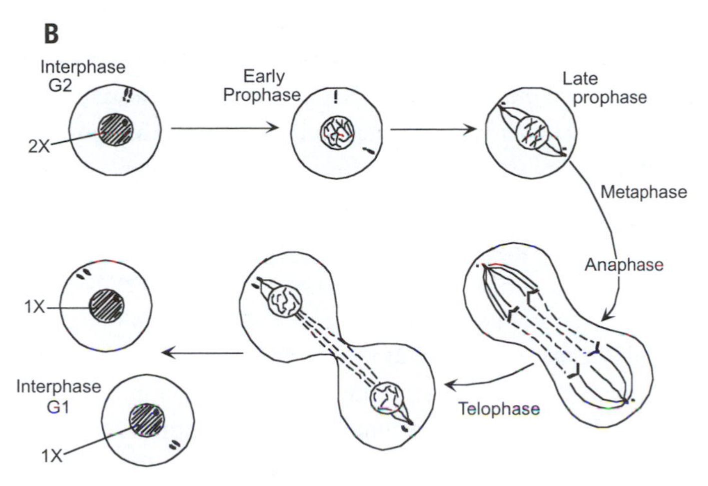

Figure 5.15 **The cell cycle.** Cells divide in an orderly sequence of events. **A**. The cell's DNA is replicated during the S (synthesis) phase of the cell cycle. Before and after the S phase are time lags (G or gap phases). Mitosis, or the event of cell division, occurs during M phase. **B**. M phase occurs in stages.

Re-entry into G_1 from G_0 is usually stimulated by interactions of the cell with its environment or through the action of growth factors, which are discussed later. G_1 ends as the period of DNA replication, called S phase, begins. During S phase, the cell produces an exact copy of its genetic material so that equivalent genetic information is available for transmission to each daughter cell. Because DNA replication precedes the physical division of the cell, there is a period during which the cellular material has twice the normal amount of DNA. In most cells, the G_2 phase is associated with a substantial increase in cell volume; the cell prepares for division by making sufficient cytoplasm for each new cell. The collective period between mitosis events is called **interphase**. The kinetics of interphase—that is, the period of time a cell spends in each of the phases G_1, S, and G_2—is highly variable among different cell types. During cleavage in the early embryo, for example, the G phases are essentially absent; **blastomeres** synthesize DNA and divide without any substantial delay or increase in cell volume.

The most dramatic changes in cell shape and structure occur during mitosis, or M phase, as the cell undergoes the physical transformations that are necessary to divide. At the end of interphase, the mother cell has replicated its DNA and is ready for division. Mitosis is usually divided into stages—**prophase**, **prometaphase**, **metaphase**, **anaphase**, **telophase**, and **cytokinesis** (Figure 5.15b). The first stage, called prophase, involves the movement of cytoskeletal elements to form the **mitotic spindle** and the condensation of nuclear DNA into organized chromosomes. Prometaphase begins with dissolution of the **nuclear membrane** and continues with the assembling of the spindle fibers through the nuclear region. These fibers associate with chromosomes via protein complexes called **kinetochores** at a chromosome region (with a specific DNA sequence) called the **centromere**. During metaphase, the chromosomes become aligned along a plane that is midway between the two poles of the mitotic spindle. At this point, each of the chromosome–kinetochore complexes connects two **chromatids**, which are the replicated DNA complements destined for each daughter cell. Anaphase begins with the separation of each chromosome complex into the two chromatids (which will become the chromosome of the daughter cell). Each member of the chromatid:chromatid pair moves toward its corresponding spindle pole at a speed of ~ 1 μm/min. When the new chromosomes reach the pole, the nuclear envelope begins to re-form in each emerging daughter cell. The period of envelope formation and spindle dissolution is called telophase. Finally, the plasma membrane at the midplane between the cell poles begins to constrict causing dramatic narrowing along the equator. This narrowing, called the **cleavage furrow**, continues until the constriction reaches the middle of the cell at the site of the disappearing mitotic spindle. Eventually, constriction at this location is completed by breakage and resealing of the membrane to form two individual and separate cells. Each daughter cell, now in G_1 phase, contains a complete set of chromosomes and nearly equal volume of cytoplasm. The chromosomes decondense to re-establish the architecture of an interphase nucleus.

The above description of mitosis applies to somatic cells, which are most of the cells in the body. The exceptions are cells of the germ line—sperm cells and ova—which are produced through a process called **meiosis** (Figure 5.16). During mitosis, the cell's genome is replicated and the cell splits once. Meiosis involves one genome replication and two cell splittings, so that four haploid cells are produced from one parent cell. Meiosis occurs in two main stages: **meiosis I** and **meiosis II**. In meiosis I, homologous pairs of chromosomes (one from each parent) are separated into two daughter cells. Each of the first-generation daughter cells then undergo cell division involving segregation of sister chromatids in each second-generation daughter cell. During meiosis I, independent assortment and chromosomal crossover mix genetic information from the paternal and maternal chromosomes to form the chromosomes of the haploid daughter cells (Figure 5.16). One germline cell produces four daughter cells (called gametes).

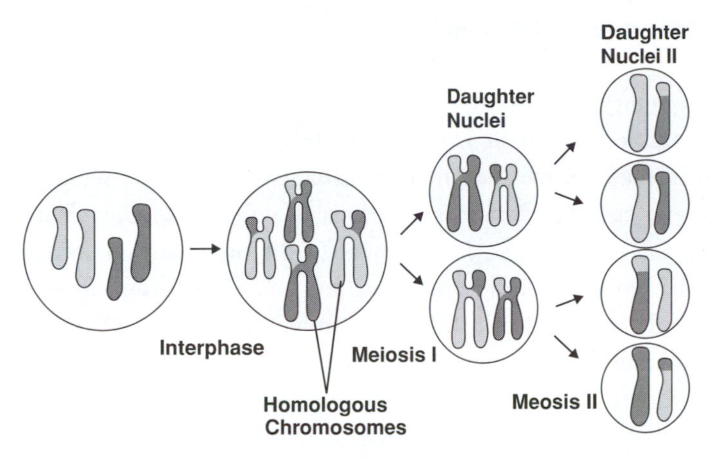

Figure 5.16

Meiosis. Chromosomes are organized into pairs (shown here in different shades of grey to indicate their source from each parent). The chromosomes are replicated before meiosis. Crossing over between the two parental chromosomes occurs at several stages of meiosis. At the end of meiosis, four gametes are produced. Each of the gametes has an unmatched set of chromosomes, with a mixture of genes from both parents.

5.5.2 Control of cell cycle

Many types of cancers are caused by a malfunction in the checkpoints of the cell cycle, resulting in uncontrollable cell proliferation and tumor formation. How do cells know when to initiate DNA synthesis and mitosis? How do they know when to stop the growth process? Some decision-making factors may be present within the cell. For example, certain cell proteins appear to function as biological clocks that encode an internal program for timed cell fate. Interestingly, the number of doublings that can be achieved by fibroblasts in culture correlates with the life span of the organism donating the cells, suggesting a link between cell life span and aging of the animal. Additionally, proteins promote or inhibit the cell cycle; accumulation or disappearance of these proteins influences the ability of the cell to divide. For example, cyclins are intracellular proteins, produced and degraded in a cyclic fashion inside of cells that provide an important element of the control of cell cycle. Signals from the cell's external environment are also known to be critical in the decision-making process. Diffusible protein **growth factors**, such as EGF or vascular endothelial cell growth factor (VEGF), can bind to receptors on the cell membrane and trigger signaling cascades inside the cell that eventually lead to increased expression of growth-promoting genes.

Telomeres are repetitive DNA sequences that appear at the end of chromosomes and function to ensure accurate replication of the DNA sequences near the end. Telomere shortening is frequently observed in cultured cells; the repetitive sequence is reduced by ~50 base pairs in length with each division. Telomere shortening may represent a biological mechanism for programmed cell death or senescence in culture. If this speculation about the role of telomeres in overall life span is correct, then telomerase—an enzyme that preserves telomere length and is found in stem cells—may endow cells with an extended capacity to divide.

5.6 Cell differentiation and stem cells

During embryonic development, as cell division proceeds from the fertilized egg to the multicellular organism, several changes occur in cells. Cell division leads to an increase in the total number of cells. Cells move within the developing organism, causing certain populations of cells to aggregate in regions of the developing embryo. As development of the organism proceeds, cells begin to specialize in function through a process called **differentiation**. Humans begin as a single fertilized cell, which transmits copies of its DNA to all subsequent cell generations. Although all cells in our bodies were derived from this single cell, cells in different organs vary considerably in shape, size, and function. Differentiation is the complex process that leads to diversity in cell properties from a population of cells with the same genetic material.

Stem cells are unspecialized precursor cells that are capable of differentiating or changing into more specialized cell types. Stem cells are capable of both self-renewal and differentiation into more specialized cells. These two defining characteristics—limitless self-renewal and multilineage differentiation—were first identified in precursor cells of the hematopoietic system (i.e., the precursors to blood cells, which occupy the bone marrow). A third defining characteristic of stem cells is their capacity to populate tissues as functional cells after transplantation into a recipient.

Stem cells are classified with respect to the age at which they are isolated. The most potent stem cells are isolated from embryos and are called **embryonic stem cells**. **Adult stem cells** have been isolated from peripheral blood, bone marrow, nervous system, muscle, liver, intestine, skin, and other tissues of adult organisms. The classification of these cells as "stem cells" is achieved by testing their capacity for self-renewal, multilineage differentiation, and repopulation of tissues in recipients. Stem cells play an important role in normal physiology; they exist in small quantities within the tissues of all adult organs, serving as a perennial source of differentiated tissue cells. In addition, scientists are testing their potential for use in repopulation and tissue regeneration.

A stem cell that differentiates into one type of **progenitor** is called **unipotent** (Figure 5.17). Unipotent stem cells display an **asymmetrical division**; the stem cell divides to produce a new stem cell (for renewal) and a progenitor cell that is committed to differentiation along a certain pathway. Although successive rounds of division and differentiation may follow this initial asymmetric division, the progenitor is irreversibly committed to formation of differentiated cells.

Hematopoiesis is an example of **pluripotent** stem cell differentiation (Figure 5.18), in which a single stem cell can produce multiple cell lineages. Progression from a stem cell to a fully differentiated cell occurs in a series of steps of successively greater commitment. At each step, cell proliferation provides for renewal of the stem cell population. Some of these new cells differentiate into progenitor cells that choose one of the fates possible for this stem cell. Further proliferation and differentiation lead some of the cells down alternate pathways

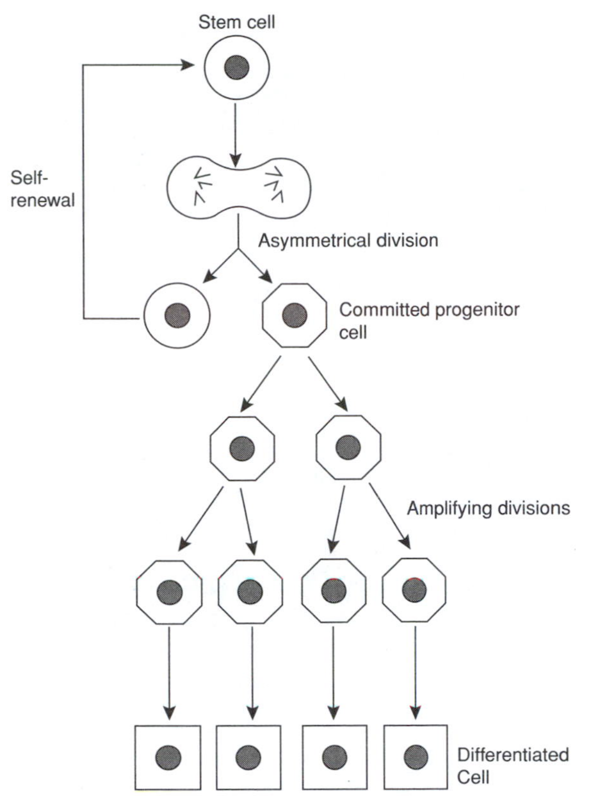

Figure 5.17 **Example of lineage in a unipotent stem cell.** Stem cells are characterized by the ability to undergo asymmetrical division. Stem cells must also be capable of self-renewal.

of differentiation to form separate cell lineages. Each of these steps involves amplification by division and differentiation. These processes are controlled by specific regulatory proteins.

What controls the differentiation of a stem cell down a lineage path? The path for differentiation appears to depend on a variety of factors, including asymmetry in the distribution of intracellular contents to daughter cells, and interactions with soluble and insoluble signaling molecules in the environment. Secreted factors are essential for control of development in hematopoiesis; fate-specific proteins appear to enhance the survival of certain cells. These factors—such as granulocyte colony-stimulating factor, interleukin-2 (IL-2), and erythropoietin—appear to act by selection; presence of the factor prevents the death of certain cells, which are generated randomly from a pool of precursor cells. Secreted factors in other tissues may act by instructive mechanisms, in which the presence of the factor initiates differentiation along a specific lineage.

Stem cells are a potential cell source for cell replacement therapy for diseases as well as seed cells for engineered tissues. Embryonic stem cells are the most controversial but also the most versatile and potent. Cells within the embryo have enormous potential for differentiation because they are the earliest precursors to the hundreds of different cell types in the body. Some scientists are investigating

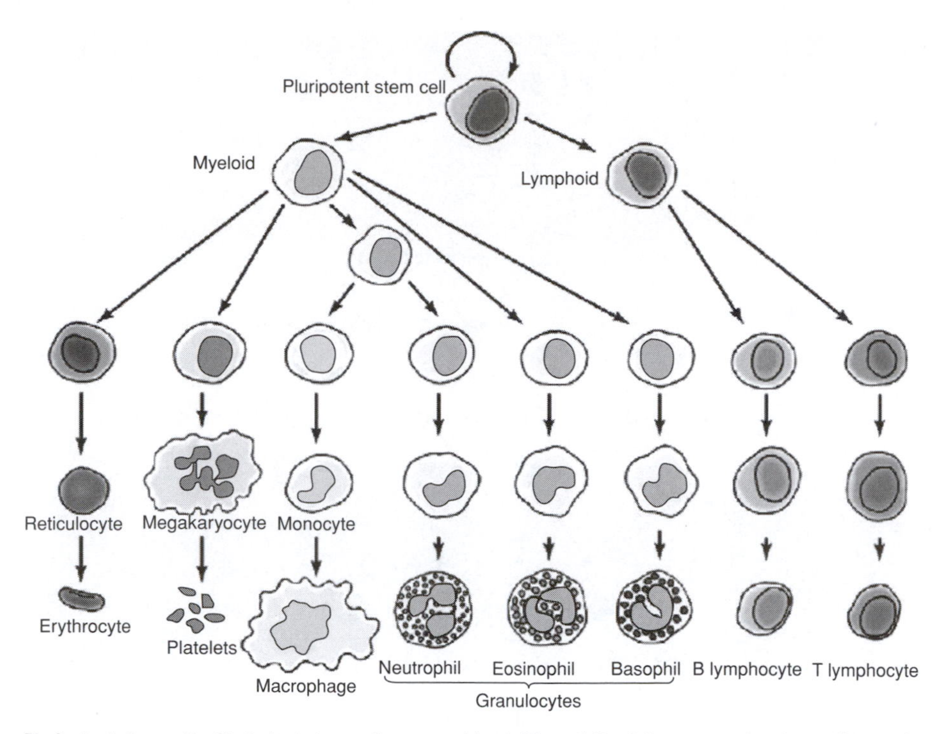

Figure 5.18 **Pluripotent stem cells.** Pluripotent stem cells are capable of differentiating into a range of mature cell types. In this figure, a stem cell in the bone marrow is shown differentiating—through a series of divisions—into many of the cells of adult blood. The curved arrow at the top of the figure indicates self-renewal.

methods to control the differentiation of these early cells into tissue-specific stem cells. Both adult human and mouse hematopoietic stem cells (HSC) can now be isolated from donor tissue using surface antibody markers. When these HSCs are transplanted into patients whose bone marrow has been ablated by radiation and chemotherapy, these cells will repopulate the marrow and initiate production of neutrophils, platelets, and other blood cells. HSCs can also be maintained in culture and expanded to form new HSCs (self-renewal) or differentiated into mature, functional cells. The control of hematopoiesis outside of the body is an important example of tissue engineering, in which the function of a human tissue (in this case, bone marrow) is replicated through cell culture technologies. More recently, scientists have been investigating the differentiation potential of stem cell populations resident in adult tissues. The abundance, capacity for self-renewal, developmental potential, and capacity for engraftment of adult stem cells is not clear at this time. Initial studies indicate that some of these adult stem cells can be coaxed to differentiate into cell types differing from the original lineage of the stem cell—a process called **transdifferentiation**.

5.7 Cell death

Cell death is a critical component of embryo development and a normal, valuable part of the health of many adult tissues. For example, the selective death of cells

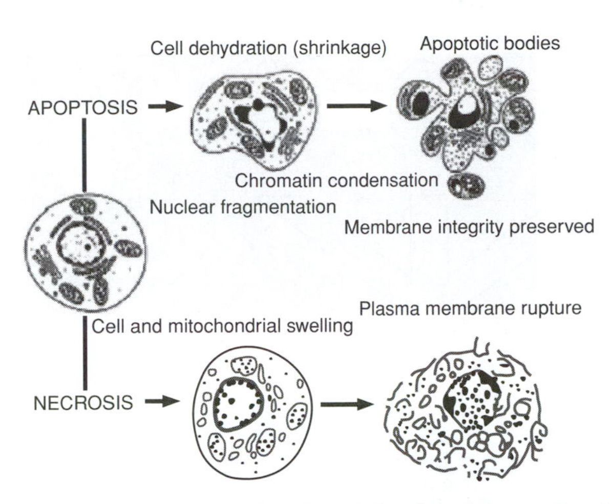

Figure 5.19 **Differences between necrotic and apoptotic cell death.** In necrotic cell death, a period of cell and mitochondrial swelling precedes plasma membrane rupture and eventual cell disintegration. Apoptotic cell death is a more orderly process, with cell shrinkage and characteristic chromatin changes occurring before conversion to apoptotic bodies.

in the primordial Müllerian duct (an embryonic tissue region that develops into the upper portion of the reproductive tract) is the critical step for the embryo in determination between male and female: The cells die in the male embryo as a result of a factor secreted by the developing testes. A balance between cell proliferation and cell death is required for the maintenance of many tissues. To maintain effective barrier properties, the skin is in a constant state of renewal: Dead cells are shed continuously from the top layer of human skin, and new cells are created from stem cell populations in deeper layers. In all, it is estimated that 50 to 70 billion cells die in an average human each day.

There are two types of cell death: **necrosis** and **apoptosis** (Figure 5.19). Cells that die because of tissue damage undergo necrosis; the cells swell and burst, releasing their contents and often triggering an inflammatory response. In contrast, apoptosis, often referred to as **programmed cell death**, follows a sequence of steps in which the chromatin condenses, the cell shrinks, and the plasma membrane pinches off to form apoptotic bodies. These apoptotic bodies are then phagocytosed or engulfed by specialized cells of the immune system called macrophages. Because the cellular contents remain membrane-bound during apoptosis, there are no inflammatory effects on neighboring cells. Thus, apoptosis is also known as cell suicide. Some cells can commit cell murder by inducing cell death in other cells. For example, T cell–mediated killing of infected host cells is an important mechanism of defense by the immune system (see Chapter 6).

5.8 Cell culture technology

The prelude to this chapter introduced the technology of cell culture. Perhaps more than any other technique, cell culture has enabled the rapid scientific and technological advances in human health in the 20th century. Human and animal

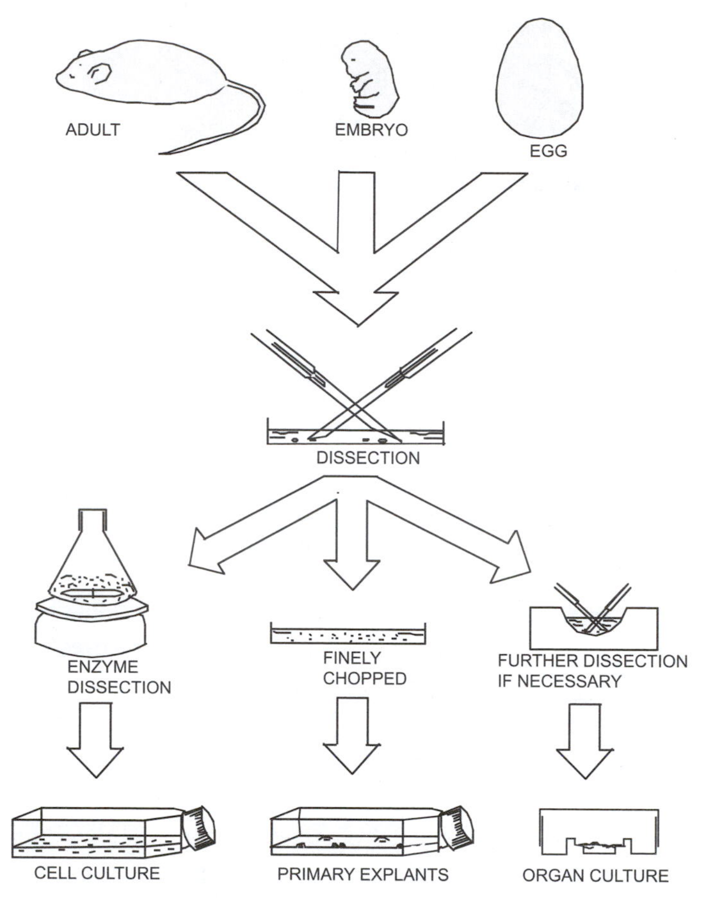

Figure 5.20 **Methods of cell isolation from multicellular organisms and tissues.** Cultured cells are obtained from source material such as adult animals, embryonic animals, or human biopsy specimens. The tissues are dissected into small pieces and are placed directly into culture (organ culture), further dissected into smaller pieces (explant culture), or further digested by enzymes that breakdown extracellular matrix and cell adhesion proteins to yield single cells (cell culture).

cells can be grown outside of an intact organism, in cell culture, to produce biological materials: Culture techniques can be used to produce many more cells or to produce cell-related products such as proteins, DNA, or viruses. The biotechnology industry is based on cell culture.

5.8.1 Sources of cultured cells

Mammalian cell cultures are obtained from the tissues of an animal, usually by one of two general methods (Figure 5.20): 1) explantation or 2) enzymatic degradation. These techniques are usually used to isolate and grow cells from rodents (embryos or adults) or from human biopsy materials. The culture containing cells first isolated from the tissue of origin is called a primary culture. Most mammalian cells have complex **medium** requirements: A variety of different vitamins, essential amino acids, glucose, and salts must be present in the liquid culture medium for cells to survive and proliferate. In most cases, the

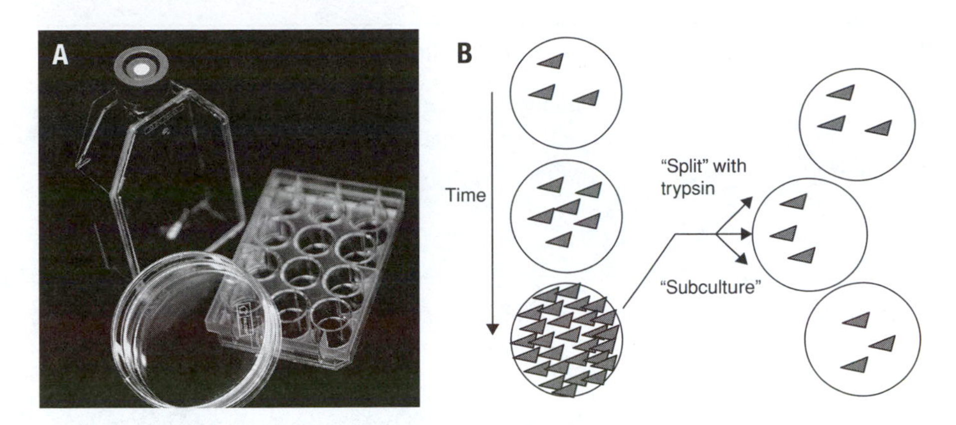

Figure 5.21

Cell culture and propagation by subculturing. A. Specially treated polystyrene dishes and flasks are most often used for mammalian cell culture. The surface treatment is required to provide for cell attachment, which is essential for growth of most cells. **B**. Cells are often maintained in culture until the cell number reaches a certain level, then the cells must be subcultured and diluted into new culture vessels. Subculturing is often accomplished by detachment of the cells from the culture surface—using enzymes such as trypsin—prior to dilution and reseeding.

minimum essential nutrients for the cells are not known; serum from an animal (which contains hormones and growth factors such as insulin, transferrin, IL-2, and others) is frequently added to the medium as a supplement to facilitate cell survival and growth.

In general, these cells grow following attachment to a solid surface, usually glass or specially modified polystyrene (see Box 5.1). In many cases, the cells will grow to form a monolayer on the surface; the cells will stop growing when a critical surface density is achieved. To further propagate these cells, they must be subcultured by detaching the cells, using either enzymatic or chemical treatments, and reinitiating new cultures in a fresh vessel (Figure 5.21). These new subcultures can be maintained under conditions that permit cell growth and then subcultured again. Theoretically, the process can be repeated indefinitely to produce an unlimited supply of daughter cells.

With repeated subculture and growth, a large number of cells can be produced from a single tissue source (Figure 5.22). However, the properties of the cells in culture invariably change as the process of subculturing and propagation is continued. Cells with characteristics that are most well-suited to the particular culture environment will dominate. Among the important factors that contribute to selection are: sensitivity to trypsin (or other detachment agent); nutrient, hormone, or substrate limitation; relative growth rates; and influence of cell density on individual cells. At a certain point in its history, the cell line may undergo a "**crisis**." At crisis, the cells will either completely "transform," which means adapt to the culture environment (becoming a permanent cell line), or die. Transformed cells have a number of distinguishing characteristics (immortality, anchorage independence, loss of contact inhibition, loss of serum dependence, faster growth, tumorigenicity) not found in untransformed cell lines.

Box 5.1 Models of cell proliferation

Cell culture is one of the most important experimental methods used by biological scientists and engineers. If actively dividing cells are maintained in a uniform environment, with no constraints to their growth, the rate of growth is proportional to the number of cells:

$$\frac{dx}{dt} = \mu X, \qquad \text{(Equation 1)}$$

where X is the total number of cells, and μ is the rate constant for cell proliferation. The solution of Equation 1 indicates that the number of cells increases exponentially with time:

$$X = X_0 e^{\mu t}, \qquad \text{(Equation 2)}$$

where X_0 is the initial cell number (i.e., the number of cells at $t = 0$). The rate constant is related to the doubling time for the cell population, t_D:

$$t_D = \frac{\ln 2}{\mu}. \qquad \text{(Equation 3)}$$

Most cells that are derived from animals are anchorage-dependent; cell adhesion and spreading on a solid surface are required for proliferation and function. In addition, as the surface becomes crowded with cells, the rate of cell growth decreases; the rate of cell division is, therefore, dependent on cell density. This phenomenon is called density-dependent cell growth.

In practice, cell proliferation will be constrained by a variety of environmental factors: Cell density, nutrient availability, and waste product accumulation all contribute to the overall rate of growth. For anchorage-dependent cells, the overall rate of cell growth and death is related to cell spreading on the surface. Although it is difficult to separate the influence of cell spreading from the influence of chemical interactions between the cell and the surface, cell spreading alone can account for the variation in apoptosis and DNA synthesis that is observed in certain cells. Microfabrication techniques—adopted from the microelectronics industry and applied to biomedical engineering (BME) systems—have been useful in sorting out some of these issues in cell culture (see Chen C, Mrksich M, Huang S, Whitesides GM, Ingber DE. Geometric control of cell life and death. *Science*, 1997; 276: 1425–1428).

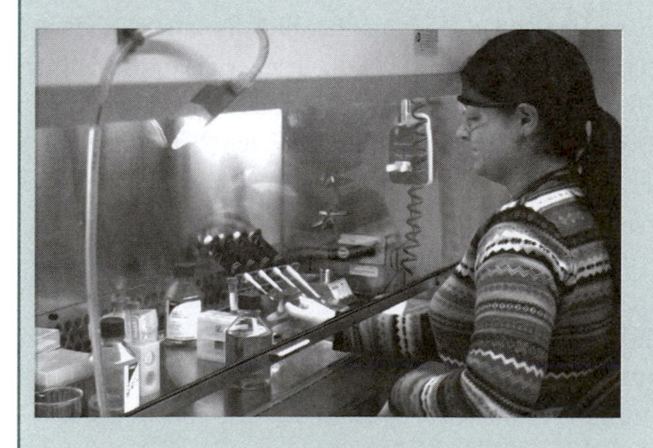

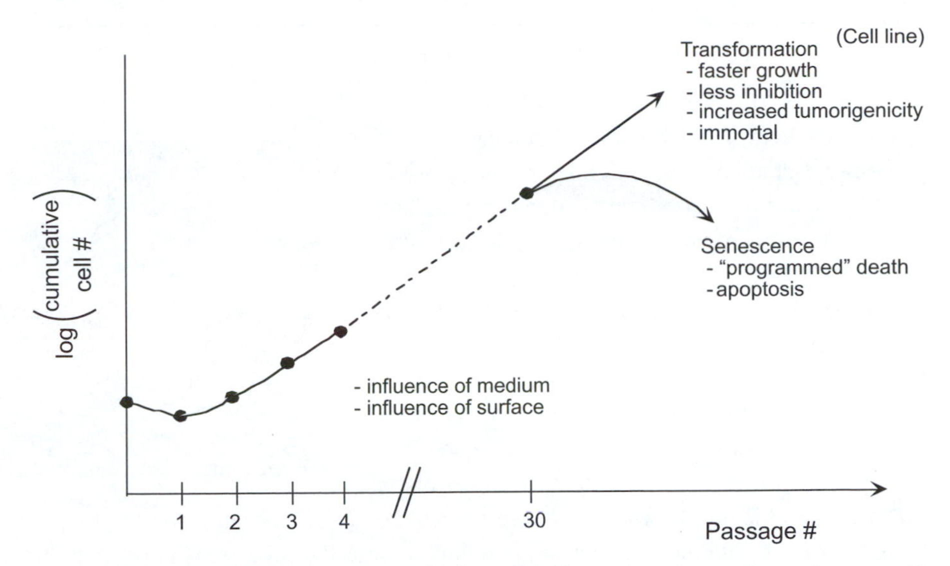

Figure 5.22 **Propagation of cells in culture.** The total number of cells generated from a given primary cell source increases with time. The numbers on the *x*-axis represent subcultures or "passages." The properties of the cells that survive and continue to divide in culture depend on the culture conditions, and can be manipulated by changing the cell culture surface or the composition of the medium. Cells will undergo a crisis after ~30 passages, either transforming into a permanent cell line or undergoing senescence and death.

Summary

- Eukaryotic cells have lipid bilayer plasma membranes that separate their contents—including cytoplasm and organelles—from the extracellular environment.
- All cells use similar elements: The cytoskeleton is essential for determining cell shape; ribosomes, rER, and Golgi are essential for protein synthesis; mitochondria are involved in energy production; lysosomes are important for digestion of unwanted material; genetic information is encoded on chromosomal DNA in the nucleus.
- ECM is important for maintaining tissue structure, holding cells in position within tissues, and providing mechanical/chemical support for cell function.
- Some membrane proteins allow cells to regulate the transport of molecules between the internal and external environment; other membrane proteins allow the cell to adhere to other cells or to ECM; other membrane proteins serve as receptors for ligands and hormones.
- The cell cycle is an orderly sequence of events describing the life of a cell, from birth to division, with checkpoints that control critical cell transformations. The phases of the cell cycle are useful for describing cell activities.
- Stem cells are immature cells that are capable of self-renewal and differentiation into more mature forms.
- Cell culture is one of the most important techniques of modern biology; it has also developed into an important tool for production by biomedical engineers.

Profiles in BME: E.E. "Jack" Richards II

In high school, I liked math and science. Consequently, I studied BME in college. That story is not original for most biomedical engineers. However, becoming a patent attorney . . . that is a little original.

In 1994, I graduated from Texas A&M University with a degree in biomedical engineering. I then worked at Prucka Engineering, Inc., a medical device company that makes electrophysiology monitoring equipment. I traveled throughout the world as a field engineer. I loved it. However, about 3 years into the job, I wondered where I would be 10 years later. I did not see myself continuing to tour the seven seas in my late thirties; I would hopefully be married with kids. I also did not see myself happy moving along the engineering, management, or sales career paths at my company. Great paths—just not for me. I then contacted a friend who studied mechanical engineering in college. For some strange reason, he had also gone to law school.

I visited my friend to see what an engineer/lawyer actually did. He showed me a valve sitting on his desk. "Jack, a company wants me to see if this valve is an invention." He was getting paid to research technology and then write about it. I liked the sound of that. I graduated from law school with a Doctor of Jurisprudence degree in 2001.

I have since passed my state bar exam and the federal patent bar exam. I am now a patent attorney in the Austin, Texas, office of Trop, Pruner & Hu, P.C., a law firm focusing on patents, trademarks, copyrights, and trade secrets. Being a patent attorney can mean many different things. Some patent attorneys focus on trying to obtain patents for inventors. These attorneys draft patent applications and submit them to the United States Patent and Trademark Office (PTO). The PTO has examiners who evaluate the application and often reject it as not being novel. The examiner cites journal articles or prior patents to illustrate that the idea is not novel. The attorney then tries to rebut the PTO's position by distinguishing the proposed idea from the PTO's evidence. This "patent prosecution" can be enjoyable and can be highly technical. However, if you love "hardcore" engineering, note that patent lawyers typically leave the engineering to the engineers. In other words, if detailed engineering is your passion, you may be more satisfied pursuing more traditional engineering career paths.

In contrast to patent prosecution attorneys, other patent attorneys focus more on patent litigation between parties. Unsurprisingly, this practice is more contentious than patent prosecution. An obvious example of patent litigation is when one company sues another company for patent infringement. Although the lawsuit is similar to other lawsuits (e.g., typical commercial disputes), it has parts very specific to patent law. After all, one of the most important facts (hopefully) in such a lawsuit is whether an allegedly infringing product actually infringes the patent. To answer that question, the patent attorney becomes immersed in the relevant technology. The attorney then makes written and oral arguments showing how the product in question is or is not the same as the asserted patent. I enjoy both patent prosecution and patent litigation.

Still other attorneys steer clear of prosecution and litigation and instead focus on "deals." In other words, such attorneys may focus on licensing a patent or technology. When all goes well, all parties to the agreement are happy. This type of practice focuses less on technology and more on business aspects of law. However, such a lawyer may rely on prosecution/litigation patent attorneys to evaluate the technology involved in the agreement.

Specifically regarding patent law and BME, I wish my practice only involved medical devices. However, patent attorneys with electrical, mechanical, and medical backgrounds also prosecute, litigate, and license medical devices. So, in all likelihood, if you practice patent law your practice will likely need to extend into general electrical and mechanical technologies.

One final note about the "hours" . . . maybe you've heard that lawyers work a lot. They do. Law, however, is no different than engineering, medicine, or business. To be good, it takes hard work and long hours.

Regardless of all of the above, the question remains: *Am I happy?* Yes. I enjoy reading and writing, and patent law supplies plenty of both. I enjoy clients and their excitement about their ideas. I also enjoy the critical thinking of patent litigation. Finally, I enjoy that patent law is part of my life but that it is not my life. I work hard but I do have a life, wife, and two children. I even get to watch college football on occasion.

REFERENCE

1. Park YD, Tirelli N, Hubbell JA. Photopolymerized hyaluronic acid-based hydrogels and interpenetrating networks. *Biomaterials*. 2003;24:893–900.

FURTHER READING

An excellent, detailed description of cell structure and function is Chapter 2 of:

Caplan MJ. Functional organization of the cell. In: Boron WF, Boulpaep EL, eds. *Medical Physiology*. Philadelphia, PA:Saunders; 2003: 9–49.

More detailed descriptions are also contained in Chapters 1 and 6 from:

Lodish H, Berk A, Zipursky SL, Matsudaira P, Baltimore D, Darnell JE. *Molecular Cell Biology*. 4th ed. New York: W. H. Freeman; 1999. This book is available free online at the NIH book collection website: http://www.ncbi.nlm.nih.gov:80/books/bv.fcgi?call=bv.View..ShowTOC&rid=mcb.TOC. For more current information, consult the fifth edition of this book, which was published in 2003 or the fourth edition of a similar book:

Alberts B, Johnson A, Lewis J, Raff M, Roberts K, and Walter P. *Molecular Biology of the Cell*. New York: Garland Publishing; 2002.

USEFUL LINKS ON THE WORLD WIDE WEB

http://www.ncbi.nlm.nih.gov/entrez/query.fcgi?db=Books
NCBI Bookshelf

http://www.cellsalive.com/
CELLS alive! Web site

http://users.rcn.com/jkimball.ma.ultranet/BiologyPages/W/Welcome.html
Kimball's Biology Textbook Free Online

http://www.accessexcellence.org/AB/GG
Graphics Gallery

http://bcs.whfreeman.com/immunology5e/content/anm/kb04an01.htm
Cell Death Animation

KEY CONCEPTS AND DEFINITIONS

adhesion molecules a family of extracellular and cell surface glycoproteins involved in cell–cell and cell–matrix adhesion, recognition, activation, and migration

anaphase a stage in meiosis and mitosis in which sister chromatids are separated by the mitotic spindle fibers. The chromatids are pulled away from one another to opposite poles of the dividing cell, the nuclear membrane begins to reform, and the cleavage furrow begins to constrict

apoptosis programmed cell death that enables the body to dispose of damaged, unwanted, or unneeded cells

asymmetrical division a division process of unipotent cells in which one of the two offspring maintains stem cell characteristics and the other differentiates and maintains a specific function

blastomere a cell formed by the division of a fertilized egg in which the G phase is absent, so replication occurs quickly, making up the blastula

centromere the region of a nuclear chromosome to which the spindle fibers attach to the kinetochore during cell division

chromatids a pair of replicated chromosomes produced during mitosis or meiosis; separate during anaphase of meiosis II or mitosis when the centromeres divide and each becomes its own chromosome

cleavage furrow the constriction of the cell membrane during anaphase at the equator of the cell that marks the beginning of cytokinesis in animal cells. As the furrow deepens, the cell divides.

cotransporter a secondary active transport system in which both the primary and secondary solute move in same direction across the membrane

crisis the critical point in culture of tissues in which the cell must adapt to the culture environment or die

cytokinesis division of the cytoplasm of a cell in which two daughter cells result

cytoskeleton a structural support of the cell composed of protein filaments that facilitate cell division, movement, and shape; the protein filaments that compose the cytoskeleton also serve as tracks in which substances are transported within the cell

differentiation changes in cell shape and physiology associated with the production of the mature cell types of a particular organ or tissue

DNA synthesis the process of copying a double-stranded DNA strand prior to cell division resulting in two copies of the original DNA strand

eukaryotic a cell that maintains distinct organelles, a cytoskeleton, and nucleus such as fungi, protozoa, plants, and animals

exchanger a secondary active transport system in which the primary and secondary solute move in opposite directions across the membrane

extracellular matrix (ECM) any material produced by the cell and secreted into the surrounding medium, generally characterized as a three-dimensional scaffold embedded in a gel containing proteins and filament fibers

G_0 phase the period of time in which the cell has completely exited the cell cycle

G_1 phase the period of time representing the gap between mitosis (M phase) and DNA synthesis (S phase)

glycosaminoglycans (GAGs) long, unbranched polysaccharide molecules that are found on the cell membrane and help give various tissue desired structure

Golgi apparatus an organelle composed of stacks of separate intracellular membrane compartments that function to modify and package secreted and integral membrane proteins

growth factors serum proteins that stimulate cell division when they bind to a corresponding cell-surface receptor

hematopoietic stem cells undifferentiated cells in the bone marrow that have the ability to both multiply and differentiate into specific blood cells

in vitro biological or chemical work done in a test tube rather than in the organism itself

integrins the largest family of adhesion molecules that mediate cell–cell, cell–ECM, and cell–pathogen interactions by binding to various non-integrin molecules

interphase stage in the cell cycle between nuclear divisions in which chromosomes are extended and functionally active; the stage in which the cell does not actively do anything for mitosis or meiosis

invariant asymmetrical division a mechanism of stem cell division in which a differentiated progenitor is produced as well as a constant number of stem cells

kinetochore the attachment point of the spindle fibers to the centromere on the sister chromatids during prometaphase in the cell cycle

lysosome a membrane-enclosed vesicle consisting of hydrolytic enzymes used to breakdown cellular components and proteins found in the cytoplasm of eukaryotic cells

medium any material on or in which experimental cell cultures grow; generally, these materials consist of a variety of nutrients or other compounds that yield desired results in the culture

meiosis I first stage in the cell life cycle of gametes in which homologous chromosomes do not replicate but separate into two distinct daughter cells, which further leads to meiosis II

meiosis II second stage in the cell life cycle of gametes in which the daughter cells produced in meiosis I undergo steps the same as mitosis, only with half the number of chromosomes

metaphase the stage of mitosis or meiosis in which the spindle fibers align the chromosomes in an equatorial plane in the cell

mitochondria an eukaryotic organelle that is the site of ATP synthesis via oxidative phosphorylation, the Krebs cycle, and electron transport reactions

mitosis (M phase) see Chapter 3, Key Concepts and Definitions

mitotic spindle microtubule-based structure present during mitosis in which the chromosomes attach and are separated toward opposite poles of the cell

necrosis one of two types of cell death in which the cell swells and ruptures, releasing its contents within the body and often causing an inflammatory response

nuclear membrane also known as the nuclear envelope, a membrane system that surrounds the nucleus of eukaryotic cells. The membrane is perforated with pores that allow transport of genetic material, fluids, and other important compounds

nucleus a membrane-bound organelle containing the entire genetic material, genome, of eukaryotes

organelle an intracellular substructure having a specialized utility essential to proper cellular function

oxidative phosphorylation a process in which ATP is produced in conjunction with the mitochondrion from the breakdown of carbohydrates and fatty acids; see Chapter 4

peroxisome an organelle containing enzymes that catalyze the production and breakdown of hydrogen peroxide throughout the cell

pluripotency property of a stem cell to develop into more than one type of differentiated cell

primary active transport movement of molecules across a biological membrane that is driven by a chemical reaction, usually the hydrolysis of ATP

progenitor one of two types of cellular results from stem cell division which is characterized as the differentiated cell with a specific function

prokaryotic primitive cells, such as bacteria, that lack cytoplasmic organelles, a cytoskeleton, and nuclear membrane, which therefore leaves no distinct nucleus

prometaphase the phase of the cell cycle in which the nuclear membrane dissolves and the spindle fibers attach to the kinetochore of the chromosomes at the centromere

prophase initial stage in cell division in which the chromosomes condense and become visible in the nucleus and the cytoskeleton begins to form the mitotic spindle

ribosome a complex organelle composed of various proteins and ribosomal RNA that catalyzes the translation of messenger RNA into an amino acid sequence

rough endoplasmic reticulum (rER) extensive membranous network, continuous with the outer nuclear membrane and studded with ribosomes that give it a rough, or bumpy appearance

S phase see DNA synthesis

secondary active transport movement of molecules across a biological membrane that is driven by the spontaneous movement of another molecule

selectins a family of adhesion molecules that recognizes and interacts with glycoproteins on the surface of other cells

smooth ER extension of the ER responsible for lipid synthesis

stem cell a cell produced early on after fertilization that is undifferentiated and maintains two defining characteristics: limitless self-renewal and multilineage differentiation

telomere the ends of chromosomes that are necessary for replication and stability; the tip or end of the chromosome

telophase final stage of mitosis or meiosis in which chromosomes uncoil, the mitotic spindle breaks down, and cytokinesis occurs; nuclei also re-form

transdifferentiation conceptual theory that stem cells from one type of tissue may be able to differentiate into another type of tissue when environmental cues are altered

unipotent a property of cells that are only capable of developing into one type of cell or tissue

vector in DNA cloning, the plasmid or phage chromosome used to carry the cloned DNA segment

NOMENCLATURE

μ	Rate constant for cell proliferation	1/s
t	Time	s
t_D	Doubling time for a cell population	s
X	Total number of cells	
X_0	Initial cell number	

QUESTIONS

1. Assume that you have a flask of cells in culture that, when viewed through a microscope, looks like the image in Figure 5.1b. How would you estimate the total number of cells in the flask?
2. Sodium bicarbonate is often used as the primary buffer system in cell culture medium. Why do you think that it is popular? How would you use bicarbonate to buffer the pH near 7.4?
3. Cell cultures are often used to investigate mechanisms of cell function in humans. Do you have any concerns about this practice?
4. The structure of prokaryotic cells is very different from that of animal cells or eukaryotes. Compare these structural differences.
5. Describe the major differences between mitosis and meiosis.
6. The cells in your body are constantly proliferating. How is this important to growth of an organism and the repair of tissues?
7. Facilitated diffusion and active transport are two different mechanisms that the cells use to transport molecules into and out of the cell. Compare and contrast these two types of cellular transport.
8. Bobby has been culturing epithelial cells for experiments that he is performing in the laboratory. He has subcultured the cells many times, and has always fed the cells the proper medium. He recently ran an experiment and found that his cells are not growing the way that he previously observed. What are some reasons why these cells may not be acting as they did previously?

PROBLEMS

1. A spherical cell with the diameter of 10 μm has a protein concentration of 20 mg/mL. Determine the number of protein molecules within the cell if the molecular weight of an average protein is 50,000 daltons (g/mol). Recall that Avogadro's number is $N_A = 6.0221367 \times 10^{23}$ molecules/mol.
2. In Section 5.7, it is stated that 50 to 70 billion cells die in the average human per day. If this is true, what percentage of your mass would you lose each year, if this loss was uncompensated by cell division?
3. Graph the cell concentration as a function of time for the following conditions:
   a. $\mu = 0.10\ h^{-1}$; $X_0 = 1{,}000{,}000$ cells
   b. $t_D = 24$ h; $X_0 = 10$ cells
   c. $\mu = 0.00001\ h^{-1}$; $X_0 = 1{,}000{,}000$ cells
   d. From what you know about rates of cell turnover and division in various human tissues, what kinds of cells are best described by the graph in a? in b? in c?
4. The sphere, cylinder, and rectangular parallelepiped are common shapes that could be used to model different living cells. Assume that you have three cells: a sphere, a cylinder, and a rectangular parallelepiped. Each cell has the same volume (1 μm^3), and the radius of the sphere and the cylinder are equal to the width of the two sides of the rectangular cell.
   a. What are the surface/volume ratios for these shapes?
   b. Which shape is better? Why?
   c. Why might a given weight of small cells be more metabolically active than the same weight of large cells? (Assume the density is constant.)
   d. Does the answer in c change if you compared an equal number of cells (rather than an equal weight)?
5. For a specific type of cell after 3 hours, the concentration of cells per milliliter of solution is about 400/mL. After 10 hours, the concentration has increased to 2000/mL. Determine the initial concentration of cells. Please state all of your assumptions.
6. Equal numbers of fibroblasts and endothelial cells are present initially in a culture.
   a. If endothelial cells double every 40 hours and fibroblasts double every 20 hours, draw a graph showing the percentage of cells in the culture that are fibroblasts as a function of time.
   b. What is the time required for the culture to contain 90% fibroblasts? Write an equation that describes this situation and, when solved for time, will provide the correct answer.
7. A bacterial culture is initially composed of 100 cells. After 1 hour, the number of bacteria is 1.5 times the initial population.
   a. If the rate of growth is proportional to the number of bacteria present, determine the time necessary for the number of bacteria to triple.

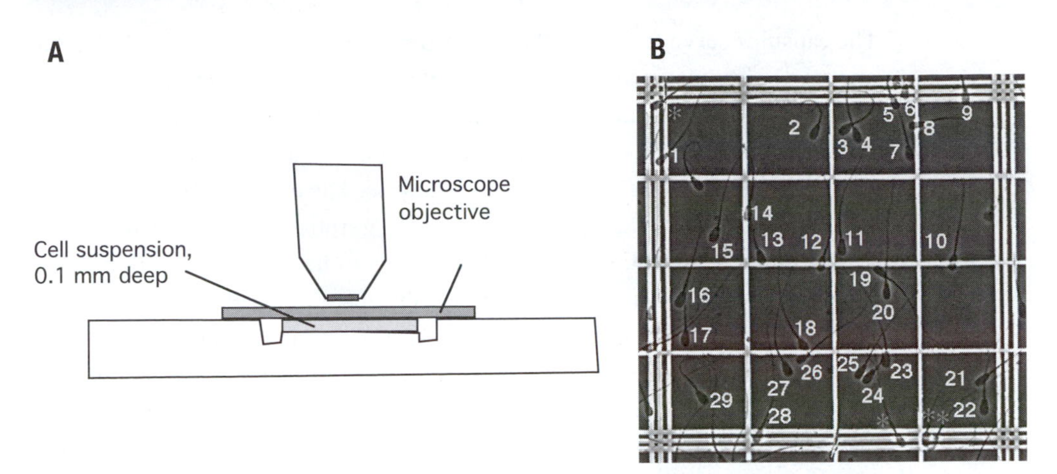

Figure 5.23 **A**. Side view of hemocytometer. **B**. View through microscope objective. Each ruled square is 0.25 mm on a side (i.e., the distance between the three thick lines is 1 mm). Photo courtesy of Dr. John Parrish.

b. What is the time required for a culture with 1×10^6 of the same bacteria to triple? Explain your results.

c. Under what conditions would the answers obtained in b be invalid?

8. PC12 cells produce dopamine and so they may be useful in treating Parkinson's disease (when dopamine is missing). One approach would be to encapsulate PC12 cells inside a spherical capsule (400 μm in outer diameter [OD]; 25-μm wall thickness) and then implant these capsules inside the brain.

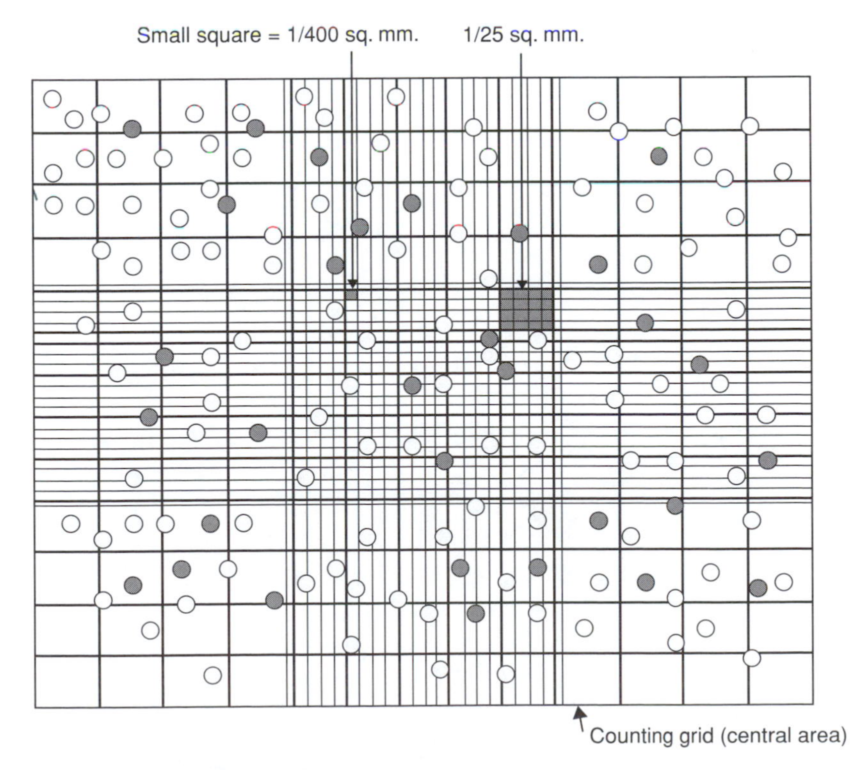

Figure 5.24 **Diagram of a hemocytometer for use in Problem 10.**

The capsules serve to protect the cells from the brain of the recipient. Under one set of circumstances, the cell density in the capsules is initially 1×10^6 cells/mL, and the capsules are incubated in vitro until enough cells grow up to fill up the entire capsule. Calculate the necessary incubation time for 50 capsules to produce enough dopamine for a mouse. The doubling time of a PC12 cell is 8 hours. Data: mass of the mouse: 250 g; ratio of mouse head mass to overall mass: 1/10; ratio of mouse brain mass to head mass: 1/5; dopamine secretion in vitro: 1 g/10^7 cells/day; initial encapsulation cell density: 1×10^6 PC12 cells/mL; dopamine requirement: 0.1 g/kg brain mass/day

9. A device called a hemocytometer is often used to estimate the number of cells in a culture. Cells are suspended in fluid. A small volume of the fluid is placed into a special chamber (see Figure 5.23). For the sperm cells shown in Figure 5.23, what was the concentration of cells in the suspension (cells per milliliter)?
10. You want to fill two culture chambers (surface area = 10 cm^2) with freshly dissected cells, the first at a plating density of 50,000 cells/cm^2 and the second at 750,000 cells/cm^2. A representative sample from the primary cell suspension was first diluted 10:1 (total volume/sample volume) and then counted on a hemocytometer (see Figure 5.24). What volume of the original cell suspension must be added to each culture to achieve the desired plating density?

PART 2

PHYSIOLOGICAL PRINCIPLES

6 Communication Systems in the Body

LEARNING OBJECTIVES

After reading this chapter, you should:

- Understand the concept of affinity of a ligand for its associated receptor.
- Understand the principle of signal transduction, and how signals can be activated by ligand binding to a receptor.
- Understand the role of action potentials in signaling within the nervous system.
- Understand how protein and steroid hormones provide circulating signals in the endocrine system.
- Understand the diverse roles of signaling within the immune system.

6.1 Prelude

Chapter 5 provided background on the structure and function of human cells, which are the main functional units of the body. Most cells are fully independent living entities, capable of consuming nutrients, growing, and functioning autonomously. The human body is a collection of trillions of cells and, amazingly, these units act in a coordinated fashion, so that people can walk (usually without bumping into walls), breathe (without consciously motivating each breath), and kill invading pathogens (without knowing that they are there). How is the operation of all of these cells coordinated? This chapter reviews how cells communicate with each other directly and through signaling molecules to relay signals from outside and inside the cell (Figure 6.1).

Cells communicate with each other directly or indirectly via molecules called **ligands**. In direct cell–cell communication, the ligands are bound to the surface of the cell. Soluble, diffusible ligands are used for communication between cells that are not physically connected or are separated by long distances. The target cell has specialized proteins called **receptors**, which are located on the surface, but anchored in the membrane. Some receptors, such as steroid hormone receptors, are located inside the cell, so the ligand must be able to cross the cell membrane to bind to them. All receptors have two properties: 1) they bind specifically to

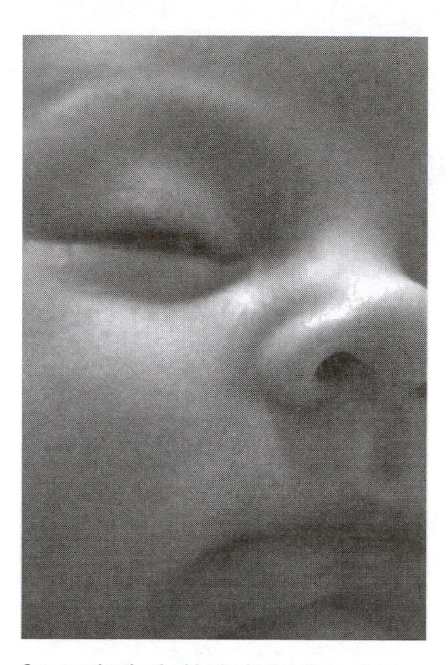

Figure 6.1 **Communication inside the body.** Even when our senses for communication with the external world are apparently silent, the cells within our body are continuously sending and receiving signals. These signals are generated by other cells.

their ligand, and 2) they transmit signals inside the cell. The binding of ligand to its receptor is similar to the "lock-and-key" fit between an enzyme and its substrate described in Chapter 3. The difference is that the binding site does not have catalytic activity. Some receptors do have intrinsic, or built-in, catalytic activity located in other parts of their structure that can be activated upon ligand binding. How tightly a ligand binds to its receptor is called its **affinity** and can be determined as described in Box 6.1.

When a ligand binds to its receptor, structural and (sometimes) chemical changes can occur in the receptor. This original signal, i.e., the ligand binding to the receptor, triggers a series of chemical reactions inside the cell; similar to toppling dominoes, one reaction leads to another (Figure 6.2). The series of reactions that can occur after a ligand binds to its receptor is called a **signal transduction pathway**. Figure 6.3 summarizes the steps in a signal transduction pathway. The pathways that are executed depend on the type of receptor that is involved, as described in Section 6.2. Most pathways use enzymes, which can act as **transducers** to convert the original signal to another form, and **amplifiers** to increase the magnitude of the signal. The response (end result) of these pathways is receptor-dependent and can range from opening an ion pore in the membrane to activating the transcription of specific genes. The general

Figure 6.2 **Pathways.** The binding of a receptor and ligand often leads to activation of a series of chemical reactions called a signal transduction pathway. Physical pathways are common in our experience; people follow a physical pathway to get from one place to another. Signal transduction pathways are analogous to these physical pathways in that they involve a series of choices (Do I go through this doorway?), and each choice influences the outcome (Do I get where I want to go?).

Box 6.1 Basics of receptor ligand binding

The binding of a receptor (R) to its ligand (L) can be modeled by the following reaction:

$$\mathrm{R} + \mathrm{L} \underset{k_r}{\overset{k_f}{\rightleftarrows}} \mathrm{RL}$$

where [R] represents the concentration of unbound receptors, [L] is the free ligand concentration, [RL] represents the concentration of receptors with bound ligand, the rate constant of the forward reaction is k_f, and the rate constant for the reverse reaction is k_r. The **equilibrium dissociation constant** (K_D) is the **affinity** of a receptor for a bound ligand:

$$K_D = \frac{[\mathrm{R}]\,[\mathrm{L}]}{[\mathrm{RL}]} = \frac{k_r}{k_f} \qquad \text{(Equation 1)}$$

Lower values of K_D represent higher affinity (because [RL] increases as K_D decreases). The **equilibrium association constant** (K_A) can also be defined:

$$K_A = \frac{1}{K_D} \qquad \text{(Equation 2)}$$

Values for K_D are often determined by performing an experiment called equilibrium dialysis. In this experiment, a known quantity of receptors is placed in a semi-permeable (i.e., permeable to the small ligands, but not the receptors) bag. The bag is placed in a beaker of solution containing labeled ligands and the two are allowed to reach equilibrium. The labeled ligands can be detected; therefore, the concentration of free ligands outside the bag [L] and the concentration of bound ligands inside the bag [RL] can be measured. Because the total concentration of receptors, $[\mathrm{R}]_T$, is known, the concentration of unbound receptors, [R] can be determined: $[\mathrm{R}]_T = [\mathrm{R}] + [\mathrm{RL}]$. This experiment is repeated for different ligand concentrations and the data be analyzed to determine K_D. The analysis is performed as follows.

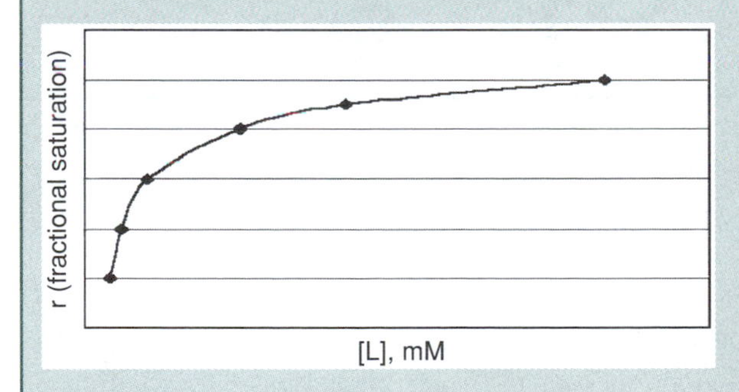

First, define r, the fractional saturation, or the fraction of receptors that are bound to ligands:

$$r = \frac{[\mathrm{RL}]}{[\mathrm{R}]_T} = \frac{[\mathrm{RL}]}{[\mathrm{R}] + [\mathrm{RL}]} = \frac{[\mathrm{RL}]}{[\mathrm{R}]_T} \qquad \text{(Equation 3)}$$

Substituting the definition of the dissociation constant, K_D, gives:

$$r = \frac{[\mathrm{L}]}{K_D + [\mathrm{L}]} \qquad \text{(Equation 4)}$$

(continued)

Box 6.1 *(continued)*

A plot of r versus [L], the saturation curve, is shown here. Often, Equation 4 is expressed in a linearized form; this approach is called Scatchard Analysis. First, Equation 1 is rewritten as:

$$[RL] = \frac{[R][L]}{K_D} \qquad \text{(Equation 5)}$$

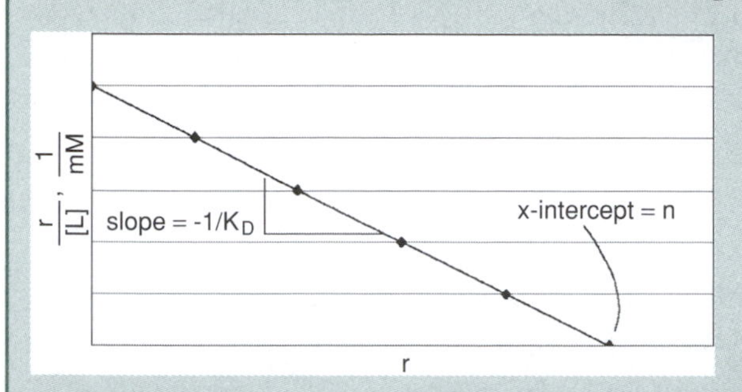

The total number of receptors, $[R]_T$ is constant; the receptors are present in two forms, free and bound:

$$[R]_T = [R] + [RL] \qquad \text{(Equation 6)}$$

Substituting Equation 6 for [R] in Equation 5 yields, after some rearrangement:

$$\frac{[RL]}{[L]} = \frac{[RL]}{K_D} + \frac{[R]_T}{K_D} \qquad \text{(Equation 7)}$$

Dividing by the total receptor number, $[R]_T$, gives:

$$\frac{r}{[L]} = \frac{-r}{K_D} + \frac{1}{K_D} \qquad \text{(Equation 8)}$$

where the slope of the line is $-1/K_D$.

mechanism for signal transduction in a cell is:

$$[\text{ligand} + \text{receptor}] \rightarrow \text{transducer} \rightarrow \text{amplifier} \rightarrow \text{response}.$$

Via this general mechanism, a molecular response (i.e., the transduction after ligand binding to a receptor) can lead to a biological response in the cell, tissue, or body, such as lowering blood glucose. There are many possible variations on this general scheme, however: Sometimes there are multiple transducers before and after the amplification step, or there can be several amplification steps. Some proteins act as both amplifiers and transducers. Also, a ligand can activate multiple transduction pathways in parallel, thereby causing several biological responses.

All functions necessary for growth, differentiation, metabolism, and defense require cells to communicate with other cells. In fact, many diseases can be traced to a defect in cell communication. Cancer, for example, can result from either overexpression of **oncoproteins**, which promote proliferation of cells, or inhibition of **tumor suppressor proteins**, which normally function to prevent tumor development. Moreover, as Table 6.1 illustrates, a majority of drugs act to either enhance or inhibit steps in signal transduction pathways. Because receptors

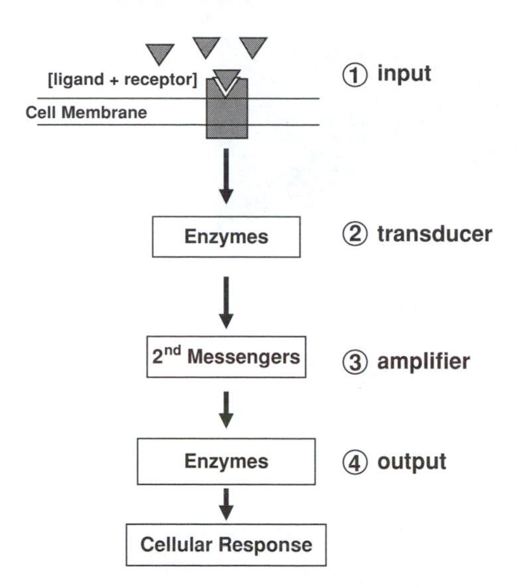

Figure 6.3 **Signal transduction overview.** A signal transduction pathway can be compared to transduction, amplification, and transmission of music from an instrument to a speaker. Signaling molecules, called ligands, bind to receptors located in the cell membrane (or sometimes within the cell nucleus). Similar to how a microphone transduces sound to electrical signals, enzymes inside the cell convert the binding event into other types of signals (such as turning on an enzyme). Often, the signals are amplified using molecules called second messengers. These second messengers in turn relay the signal by activating enzymes or enzyme cascades, which eventually produce a response.

are the gateways to signal transduction, they are the targets of many drugs. **Agonists** act like the natural ligand and enhance signaling; one example is L-Dopa, a dopamine precursor, for treatment of Parkinson's disease. **Antagonists** compete for the receptor binding site, but do not activate signal transduction. Antagonists act as inhibitors because they prevent the natural ligand from binding and activating its normal signal. For example, tamoxifen is an antagonist for the estrogen receptor.

Biomedical engineers can harness the vast knowledge of receptors and their downstream signaling effects in many ways. Many potential drug targets are components of signal transduction pathways. Biomedical engineers have also

Table 6.1 Drugs acting on signal transduction pathways

Disease	Drug	Drug action
Breast Cancer	Tamoxifen	Binds to estrogen receptor and blocks signaling
High Blood Pressure or Heart Failure	Beta Blockers	Binds to beta-adrenergic receptors and blocks signaling
Type I Diabetes	Insulin	Provides the deficient ligand
Parkinson's Disease	L-Dopa	Provides the deficient ligand

Notes: Many drugs act by affecting a component of a signal transduction pathway. Tamoxifen and beta blockers are examples of antagonists whereas insulin and L-Dopa are agonists. Tamoxifen binds to the estrogen receptor and blocks signaling, thereby preventing estrogen-mediated cell growth of breast cancer cells. Beta blockers are common treatments for high blood pressure or heart failure because they decrease heart rate and contractility by binding to beta-adrenergic receptors and inhibiting their normal function. Self-injection of insulin has helped many type 1 diabetics control their blood sugar. Dopamine producing neurons are destroyed in Parkinson's disease, so L-Dopa, a dopamine precursor, can restore some function in patients.

Figure 6.4 **Modes of signaling.** People use different modes of signaling in daily life.

used short fragments of ligands or receptors to achieve a desired function. For example, ligands can be attached to drug-delivery devices to target the drug to a receptor that is found only on the surface of a particular type of cell (Chapter 13). These same ligands can be used to deliver contrast agents to aid in imaging tumor cells (Chapter 12). Receptors themselves can be used in the design of biosensors (Chapter 11). Some receptors that bind to extracellular matrix components have been used to enhance cell adhesion in tissue engineering constructs (Chapter 13). Moreover, ligands that are growth factors can be delivered to a local area to cause differentiation of cells in the construct as well as proliferation of host cells. Receptor–ligand interactions can be altered by mechanical forces in the body, so quantifying and understanding these forces is an area of investigation in cellular biomechanics (Chapter 10).

6.2 Signaling fundamentals

There are several modes of signaling depending on the source of the stimulating signal or ligand (Figure 6.4). Cells can either communicate directly or indirectly with each other (Figure 6.5). Direct cell–cell communication is used in transferring molecules via gap junctions during contraction of heart muscle cells and in signal transmission between cells of the immune system, for example. Secreted signaling molecules are often categorized according to the site of action as autocrine, paracrine, or endocrine. **Autocrine signaling** molecules act on the cells that secrete them. **Paracrine signaling** molecules, such as those used for nerve cell communication, act on target cells in the vicinity of the site of secretion. **Endocrine signaling** molecules, such as the hormone insulin, act on target cells that are distant from the site of secretion. These modes of signaling are similar to the different ways people communicate with each other. For example, Jack wants to get a message to Jill to meet him at the café this evening. Jack writes a reminder note to himself (autocrine). He sends an e-mail message to Jill with the details (endocrine). He writes her a note about a change in time and gives it to her in class (paracrine). After class, they meet and talk to each other to plan how they will get to the café (cell–cell).

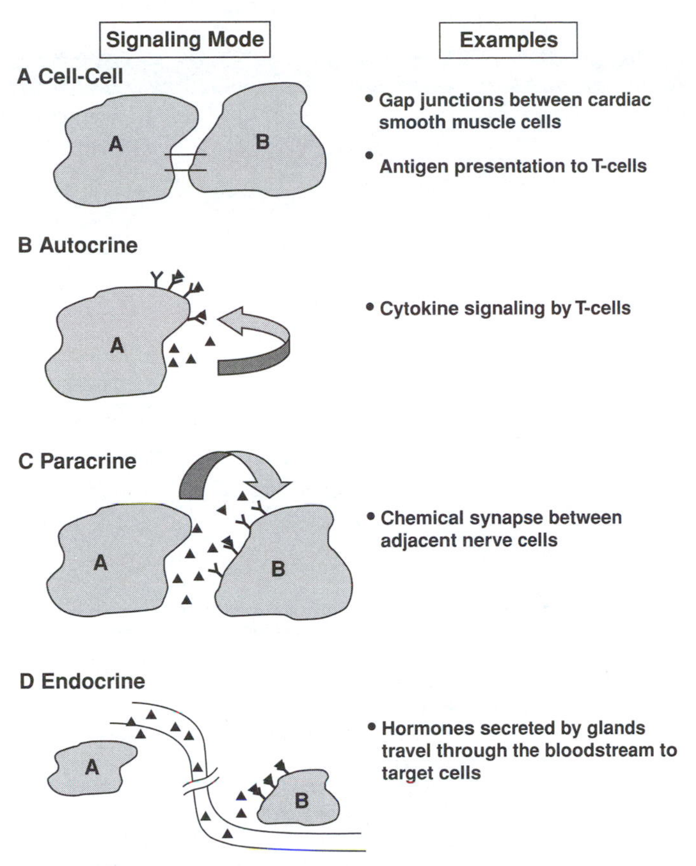

Figure 6.5

Modes of cell signaling. Cells can communicate to each other via direct cell–cell contact through gap junctions or cell surface receptors **A.** Cells also can send signals by secreting signaling molecules (triangles). These molecules or ligands can bind to receptors (Y) located on **B.** the cell's own surface, **C.** a nearby cell, or **D.** a distant cell.

This chapter describes the types of signaling ligands, receptors, and intracellular chemical reactions that are involved in three key communications systems of the body: the nervous system, the endocrine system, and the immune system. These systems function to maintain appropriate conditions in the body by sensing the state of the local environmental conditions (either outside of the body or inside the body) and responding with an appropriate cellular reaction. In the following sections, several signal transduction pathways will be described in detail to illustrate the type of reactions involved in cellular communication. The reader should focus on understanding the general steps that contribute to the transmission of the signal—that is, the function of proteins rather than their exact names. These descriptions are included to illustrate the complexity of the signaling networks and the state of understanding of the molecular details.

6.2.1 Cell surface receptors

Table 6.2 lists the main types of surface receptors: ligand-gated ion channels, G-protein–coupled receptors (GPCRs), receptor tyrosine kinases (RTKs), and enzyme-linked receptors. The ligand-gated ion channels are the simplest type of

Table 6.2 Cell surface receptors

Surface receptor type	When ligand binds. . . .	Examples
(a) Ligand-gated ion channel	Channel opens or closes	– Acetylcholine receptor – Glycine receptor – GABA receptor
(b) G-protein–coupled receptor (GPCR)	G-protein dissociates into Gα and G$\beta\gamma$ subunits; activated subunits can activate or inhibit other enzymes inside cell	– Beta-adrenergic receptors – Indirectly gated ion channels – Glucagon receptor
(c) Receptor tyrosine kinases (RTK)	Receptors dimerize and cytoplasmic tails phosphorylate each other on tyrosine residues	– Insulin receptor – Various growth factor receptors (EGF, VEGF, FGF, etc.)
(d) Receptors linked to enzymes	Receptors dimerize and associate with proteins that have enzymatic activity (kinases or phosphatases)	– Cytokine receptors

Notes: Cell surface receptors can be classified into several categories: (a) ligand-gated ion channels open or close to allow ions to flow in or out of the cell; (b) G-protein–coupled receptors (GPCR) are referred to as seven transmembrane or serpentine receptors because they span the membrane seven times. G-proteins are composed of three subunits (α, β, γ), which normally bind guanosine-5′-diphosphate (GDP) in their inactivated state. Binding of a ligand to a GPCR causes a conformation change in the G-protein and displacement of GDP. The Gα subunit is then free to bind guanosine-5′-triphosphate (GTP) and dissociate from the G$\beta\gamma$ subunits. The Gα–GTP complex then activates an enzyme such as adenylyl cyclase, which catalyzes production of the second messenger cyclic adenosine monophosphate (cAMP). Some G-proteins have inhibitory actions on their downstream effectors. (c) Most receptor tyrosine kinases (RTKs) dimerize (two polypeptide chains come together) when a ligand binds. Dimerization causes the two cytoplasmic tails of the receptor to come in close proximity to each other. Because these receptors have kinase activity in their tails, they are able to cross-phosphorylate tyrosine residues on their partner's tail. These phospho-tyrosine residues are binding sites for various adapter proteins or enzymes. (d) A similar dimerization step happens upon ligand binding to receptors that are linked to an enzyme; however, these receptors do not have any catalytic activity, so they associate with enzymes upon ligand binding and subsequent dimerization. The enzymes that link to these receptors are often kinases, but they can also be phosphatases.

surface receptor because there is no intracellular signaling pathway associated with them. Binding of a ligand causes these ion channels to open or close. Section 6.3.2 discusses ligand-gated channels used in the communication between nerve cells with each other and with muscle cells. The three remaining types of receptors use signal transduction pathways to transmit their signals.

When ligands bind to GPCRs, the G-proteins associated with the receptors are turned on and they, in turn, activate or inhibit an enzyme. An example of a GPCR is the beta-adrenergic receptor, which responds to the hormone epinephrine, commonly known as adrenaline. These are the same beta-adrenergic receptors that are inhibited by beta blockers (Table 6.1). Other GPCRs discussed in this chapter are indirectly gated ion channels (Section 6.3.2) and the glucagon receptor (Section 6.4.2). Another class of receptors, the RTKs, has intrinsic catalytic activity in their cytoplasmic tails. **Kinases** are enzymes that catalyze the addition of phosphate groups (**phosphorylation**) on proteins. As their name suggests, tyrosine kinase receptors catalyze the phosphorylation of tyrosine residues in their substrate. The insulin receptor is an example of an RTK (Section 6.4.2). The RTK family also includes receptors for epidermal growth factor, vascular endothelial growth factor, and fibroblast growth factor. These ligands are important for the

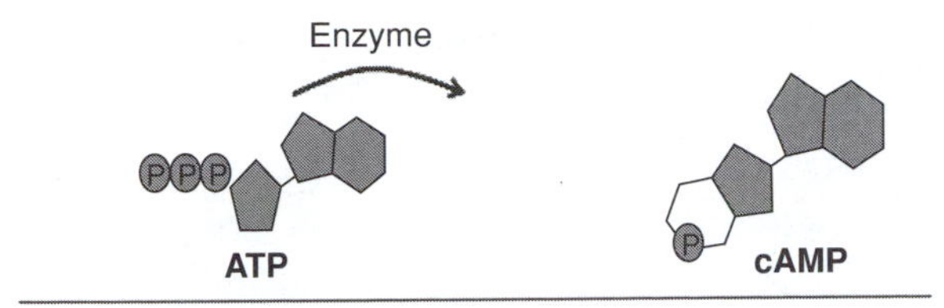

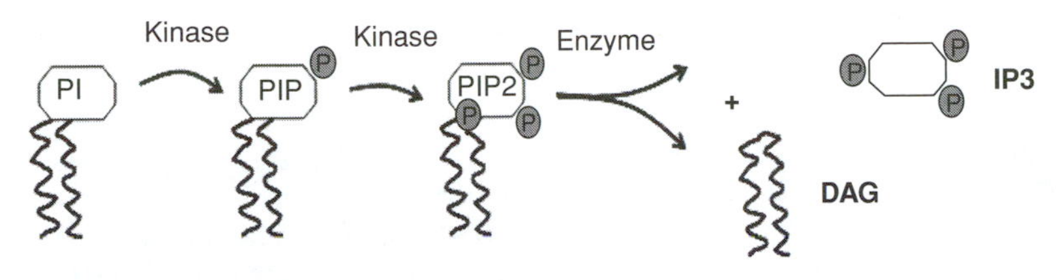

Figure 6.6 **Two common pathways that produce second messengers.** The enzymes that catalyze these reactions are activated by upstream transducers. **A.** In the cyclic adenosine monophosphate (cAMP) pathway, adenosine triphosphate (ATP) is converted to cAMP by adenylyl cyclase. **B.** In the inositol–lipid pathway, a membrane phospholipid, phosphotidylinositol (PI) undergoes two phosphorylation steps to yield phosphatidylinositol (4,5)-bisphosphate (PIP2). An enzyme called phospholipase C catalyzes the hydrolysis of PIP2 to yield (1,2)-diacylglycerol (DAG) and IP3, inositol (3,4,5)-triphosphate. Once synthesized, the second messengers (cAMP, DAG, PIP3) can amplify the signal and participate in downstream signaling events. Ca^{2+} (not shown) sometimes acts as a second messenger.

growth of cells during development and repair. Growth factors are used in promoting cell growth and differentiation in tissue engineering, which will be discussed in Chapter 13. Finally, the fourth class of receptors, enzyme-linked receptors, do not possess catalytic activity themselves but associate with enzymes after their ligands bind. The integrin receptors, which bind to extracellular matrix components, are enzyme-linked receptors (Chapter 3). Cytokine receptors of the immune system are another example of enzyme-linked receptors (Section 6.5.3). Note that steroid hormone receptors are not listed in Table 6.2 because they are not surface bound; they reside in the cytosol or nucleus (Section 6.4.3). Also, an example of cell–cell communication—where the receptor on one cell interacts with a ligand bound to the surface of another cell—will be discussed while reviewing a key event in adaptive immunity, T cell recognition of an **antigen-presenting cell** (APC; Section 6.5.2).

6.2.2 Second messengers

Second messengers serve as molecular transducers; they transfer the original signal to another protein that can help carry out the message. They can also amplify the signal because many second messenger molecules can be produced by a single enzyme. Two common second messenger pathways (cyclic adenosine monophosphate [AMP] and the inositol lipid) are shown in Figure 6.6. The second messengers which are produced in these pathways—cAMP, 1,2-diacylgycerol (DAG), and phosphatidylinositol (1,4,5)-triphosphate (PIP3), and sometimes Ca^{2+}—help to transfer the signal within the cell to the **effector** enzyme, which executes the response.

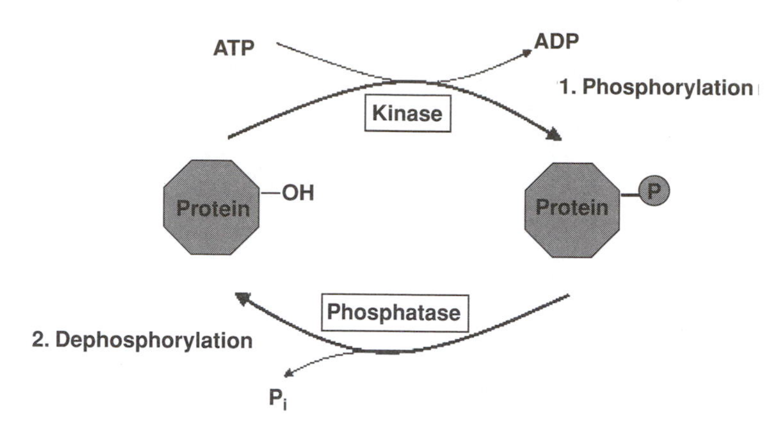

Figure 6.7 **Phosphorylation and dephosphorylation reactions.** Phosphorylation and dephosphorylation are important mechanisms in the transmission of signals. Kinases and phosphatases can either turn on or off enzymes within a signal transduction pathway. Certain amino acid residues (serine, threonine, and tyrosine) with free hydroxyl groups (–OH) can be phosphorylated by kinases. Kinases catalyze the phosphorylation reaction (1) which uses adenosine triphosphate (ATP) as the phosphate source. Adenosine diphosphate (ADP) is a by-product of the phosphorylation reaction. The reverse reaction, dephosphorylation (2) is catalyzed by a phosphatase that cleaves the phospho-protein bond, yielding an inorganic phosphate.

Second messengers provide a powerful physiological tool. Diverse receptor mechanisms can trigger a limited set of second messenger molecules. These molecules, because they are finely regulated under most circumstances, can greatly amplify the initial signal. They can allow for diversity of function among cells because a single ligand might activate different second messengers in different cells.

6.2.3 Phosphorylation and regulation of signal transduction

Phosphorylation (adding phosphate groups) and dephosphorylation (removing phosphate groups) reactions are used frequently in signaling pathways as a way to turn enzymes on and off (Figure 6.7). Phosphorylation of proteins is an important modification that the cell uses to help transfer the signal from the receptor to the final effector enzyme. Kinases catalyze the addition of phosphate groups on certain amino acid residues of proteins. Many signal transduction pathways use a series of kinase reactions or kinase cascades. A "kinase cascade" is a series of phosphorylation reactions in which one protein phosphorylates another protein, which in turn phosphorylates another (e.g., kinase A phosphorylates kinase B, which activates it so that it can phosphorylate kinase C).

6.3 The nervous system

The nervous system functions with the endocrine system to maintain homeostasis. The familiar senses of taste, sight, sound, and smell originate in sensory neurons that send signals directly to the brain. These sensory organs and their neurons

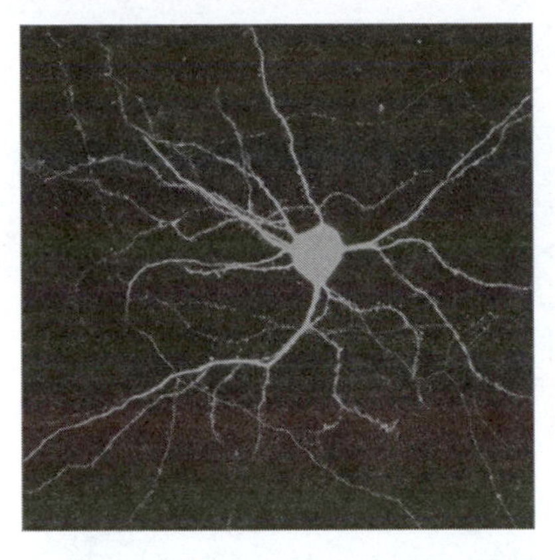

Figure 6.8

Neurons. Neurons are characterized by long processes, which reach out to connect with and collect information from other cells in nearby or distant tissues. The growth of processes called dendrites was stimulated in this cell by the release of brain-derived neurotrophic factor (BDNF) from a biodegradable polymer in the vicinity of the neuron. Photo courtesy of Melissa Mahoney. (See color plate.)

are part of the central nervous system (CNS). The sense of touch uses two types of neurons: those that sense our external environment (e.g., pain sensors on the skin) or those that sense the internal environment (e.g., sensors on the internal organs). Information from all the senses is integrated in the main organs of the CNS, the brain, and the spinal cord. Appropriate responses are transmitted as signals to the muscles and glands via motor neurons. The touch sensory neurons and all the motor neurons make up the peripheral nervous system (PNS). Thus, the nervous system controls the major muscular activities of the body, visceral smooth muscle activity, and secretion of exocrine and endocrine glands.

6.3.1 Neuron structure and action potential

Most neurons have the same basic cell structure and the familiar cellular components that were outlined in Chapter 5, although their shapes can be dramatic and beautiful (Figure 6.8). Neuron function can be depicted by the simple model shown in Figure 6.9. **Dendrites** are extensions that act like antennae to receive input signals from other cells (usually other neurons). The cell body contains organelles, which maintain the cell. The **axon** conducts the integrated signal, and axon terminals carry the output signal. The output signal varies depending on the type of neuron that is transmitting the signal. For example, for a sensory neuron or an interneuron, the output signal could be the release of specialized ligands called **neurotransmitters** at the axon terminal that interact with the dendrite of an adjacent neuron. For motor neurons, the output signal could cause contraction of a muscle cell.

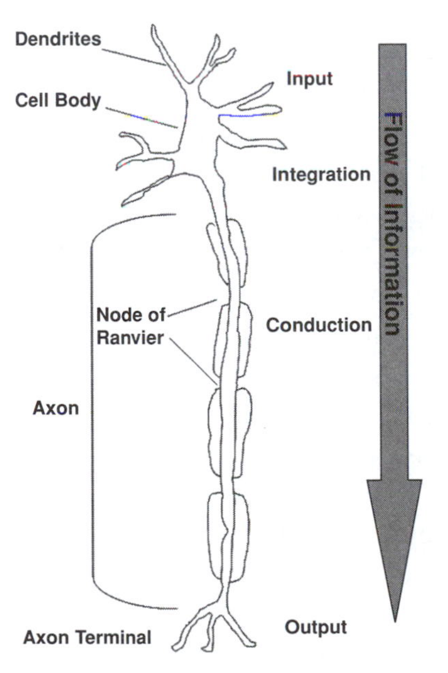

Figure 6.9

Structure and function of a model neuron. Dendrites receive input signals, which are integrated by the nerve cell and conducted by the axon to the output terminals. Nodes of Ranvier facilitate fast conduction of the action potential in some neurons. Flow of information is from the dendrite to the axon.

How is the input signal transmitted from one end of the neuron to the other?

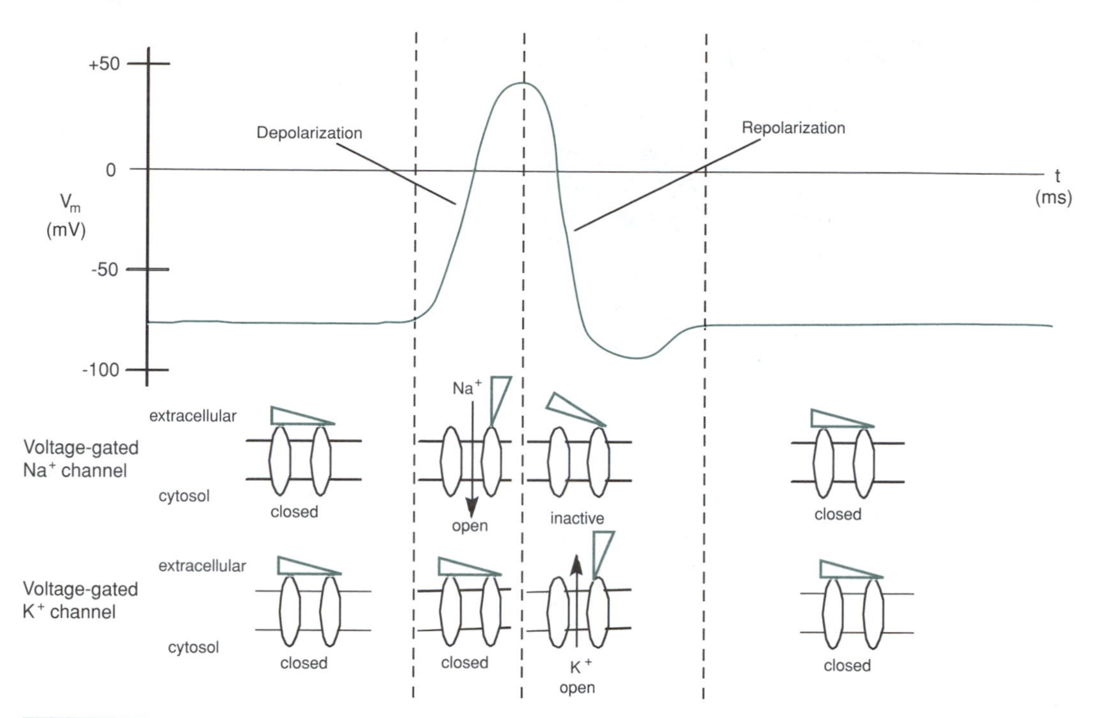

Figure 6.10 **Ion channel gating during an action potential.** An action potential results from depolarization of the plasma membrane when voltage-gated Na^+ channels are open. Repolarization back to the resting membrane potential is caused by the opening of voltage-gated K^+ channels.

Signals are transmitted as an **action potential** or nerve impulse. Cell membranes have an electrical polarity caused by their permeability to some ions, even in the resting state (Box 6.2). The resting membrane potential of a cell is approximately −70 mV caused by separation of charge across the plasma membrane; the inside surface is more negative than the outside of the neuron. The potential difference across the membrane can be changed by the movement of small numbers of ions from inside to outside, or vice versa. Because ions cannot cross the membrane passively, because they are not soluble in the lipid bilayer, they must diffuse through pores called **ion channels**. Some ion channels are always open, allowing the continuous passage or leakage of ions from the side of high concentration to low concentration. In fact, the negative resting membrane potential is mainly caused by leaky potassium (K^+) ion channels that allow passage of K^+ out of the cell, leaving behind excess negative charge. Thus, the resting membrane potential approaches the **Nernst potential** of K^+.

Other channels only open in response to a particular stimulus, such as a local change in membrane potential (voltage-gated channels) or the binding of a specific ligand (ligand-gated channels). An action potential—or the electrical signal that is propagated from one end of an activated neuron to the other—begins with the opening of voltage-gated Na^+ channels at some specific location on the neuron (Figure 6.10).

Box 6.2 The Nernst potential

The movement of ions between the cytosol (the inside of the cell) and the extracellular fluid is driven by two distinct forces: the concentration gradient and the voltage gradient. Consider a simplified model of a cell as shown below in which two compartments of a beaker are separated by a membrane that is permeable to K^+ ions, that is, the cell membrane normally has leaky K^+ channels which are always open. Assume that side A is inside the cell and side B is the extracellular fluid, at time zero, the K^+ ions will begin to flow out of the cell down the concentration gradient by diffusion. Eventually, inside of the cell will become more negative because the positive K^+ ions are leaving the cell. The voltage gradient attracts the positive K^+ ions back into the cell (electrostatic attraction).

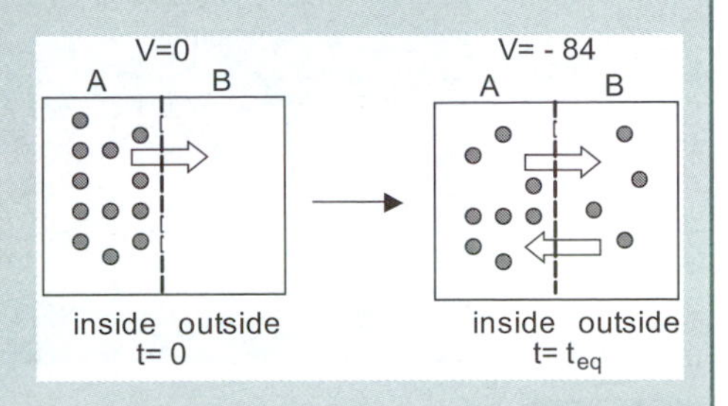

At equilibrium, the ion distribution across the membrane does not change and the concentration gradient and the voltage gradient equal each other. This equilibrium potential is known as the **Nernst Potential** for an ion and can be calculated as shown:

$$V = \frac{RT}{zF} \ln\left(\frac{c_o}{c_i}\right) \qquad \text{(Equation 1)}$$

where V is the Nernst potential, R (1.987 cal/K-mol or 8.314 J/K-mol) is the universal gas constant, T is the absolute temperature in Kelvin, z is the valency or charge of the ion, F (23,062 cal/mol-V or 96,000 C/mol-V) is Faraday's constant and c_o and c_i are the ion concentration outside and inside the cell, respectively.

For experiments performed at room temperature (20°C), the Nernst equation for a monovalent ion reduces to:

$$V = (58\text{ mV}) \log_{10}\left(\frac{c_o}{c_i}\right) \qquad \text{(Equation 2)}$$

Given that the intracellular concentration of K^+ is 140 mM and the extracellular concentration is 5 mM, the Nernst potential for potassium (V_{K+}) equals −84 mV.

During depolarization of an action potential, K^+ channels are closed and Na^+ channels are open. The typical concentration of Na^+ inside the cell is 10 mM and in the extracellular fluid is approximately 150 mM. Calculate the Nernst potential for Na^+. How does this relate to Figure 6.7?

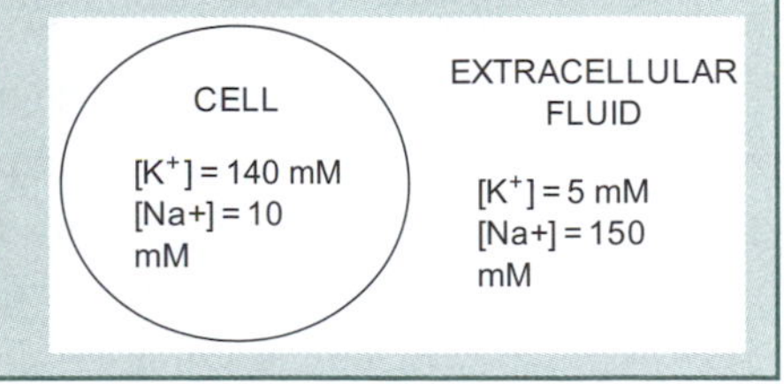

Sequential opening of voltage-gated Na^+ channels results in an inward flow of Na^+ ions, which causes the inside of the cell to become more positively charged compared to the outside. This increase in positive charge results in **depolarization** of a region of the plasma membrane. After opening of the voltage-gated Na^+ channels, another set of voltage-gated K^+ channels opens, allowing the outward flow

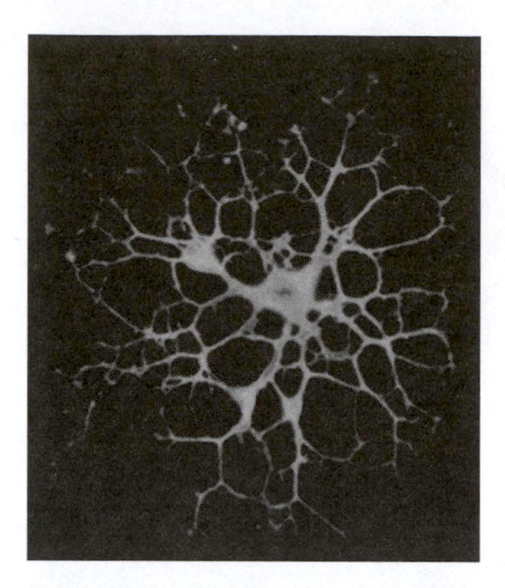

Figure 6.11 **Oligodendrocyte.** Ninety percent of the cells in the brain are glial cells, which support the function of neurons. Cells such as this oligodendrocyte provide myelin coatings for neuron processes, which enable efficient propagation of action potentials. Photo courtesy of Dr. Regina Armstrong, Uniformed Services University of the Health Sciences. (See color plate.)

of K^+ ions and **repolarization** of the membrane in which the charge inside the cell returns to negative. While the K^+ channels are open, the Na^+ channels are closing and inactive—they cannot be reactivated for a brief period called the the **refractory period**. This refractory period ensures that conduction of the action potential is unidirectional from the cell body to the axon terminus. Also, the magnitude of the action potential does not diminish while it is traveling down the axon.

Neurons are not the only cells in the nervous system. Nerve cell bodies and axons are surrounded by supporting cells called **glial cells**. These cells support, protect, insulate, and nourish the neurons. In fact, glial cells outnumber neurons by 10 to 50 times in the body. Specialized glial cells named **oligodendrocytes** (Figure 6.11) in the CNS and **Schwann cells** in the PNS wrap around the axon to form segments of **myelin** sheath. The intervening spaces between the myelin are called **nodes of Ranvier** (Figure 6.9). These node regions are uninsulated and allow for the action potential to travel from node to node. This jumping from node to node or **saltatory conduction** increases the speed of conduction so that action potentials can travel long distances in a short period of time along myelinated neurons. The importance of myelin as an intermittent insulator is seen in multiple sclerosis (MS) where destruction of myelin protein leads to symptoms such as vision impairment and muscle weakness.

6.3.2 Neurotransmitter signaling

How is this electrical signal passed from one neuron to another? When an action potential reaches the axon terminal, it is transmitted to the dendrite of a neighboring neuron. Neurons use a variety of different types of specialized junctions, called **synapses**, to accomplish signal transfer.

Electrical synapses allow direct transmission of action potentials through gap junctions shared by the two cells. **Gap junctions** are composed of channel-forming proteins called **connexins** through which ions can flow freely. Gap junctions are not prevalent in the CNS but play a key role in transmitting signals between visceral smooth muscle fibers or cardiac muscle cells (Chapter 8).

Most of the signal transmission in the CNS occurs via chemical synapses. In a chemical synapse, an action potential arriving at the axon terminal of a presynaptic neuron (the neuron bringing in the signal) triggers the release of

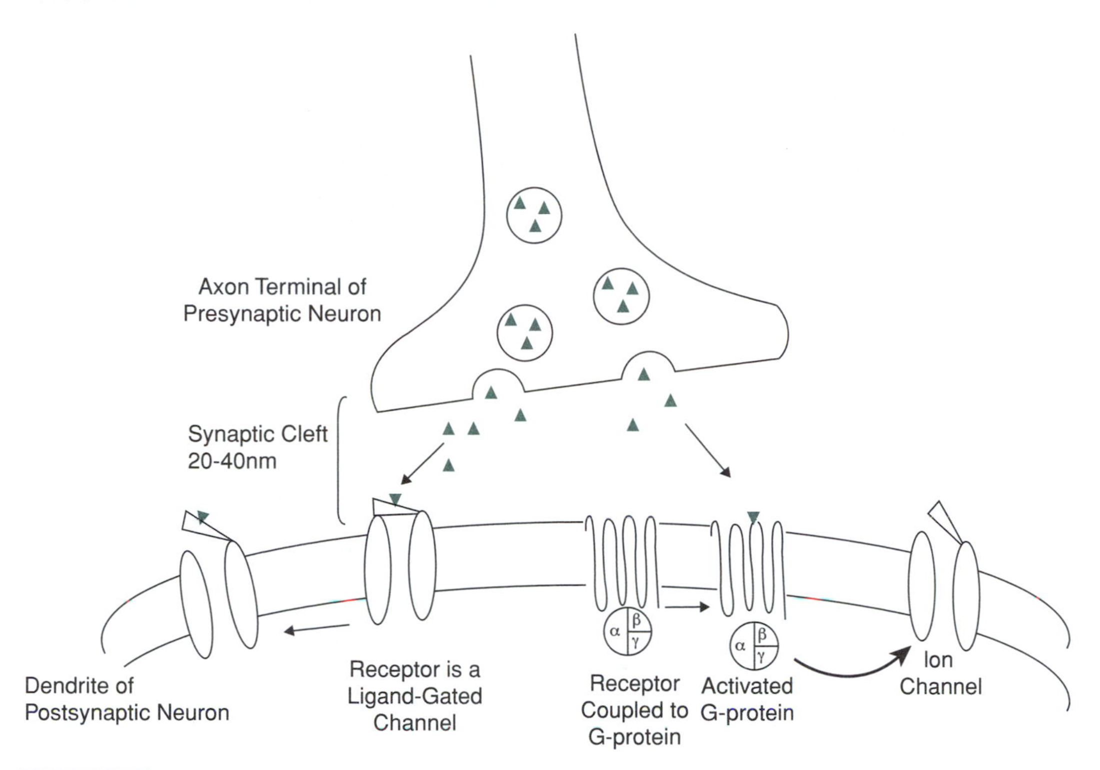

Figure 6.12 **The synapse.** Neurotransmitters are released from vesicles in the presynaptic terminal. The neurotransmitters diffuse across the synaptic cleft and can bind to directly gated or indirectly gated receptors on the postsynaptic terminal resulting in opening or closing of ion channels.

neurotransmitter molecules from vesicles into the extracellular space between the two neurons (Figure 6.12). This space is called the **synaptic cleft** and is approximately 20–40 nm wide. Neurotransmitters are *hydrophilic* chemical messengers that transmit signals across the synaptic cleft from the presynaptic neuron to the postsynaptic neuron (the neuron receiving the signal). The neurotransmitter molecules, acting in this case as **paracrine signaling** molecules, diffuse across the synaptic cleft and bind to receptors on the postsynaptic membrane. Chemical synapses transmit signals in only one direction, from the presynaptic neuron to the postsynaptic neuron.

Binding of the neurotransmitter (ligand) to the receptor causes gating (i.e., opening or closing) of an ion channel (Figure 6.10). There are two types of neurotransmitter receptors: They can either act as the gate themselves (directly gated) or they can act as the gatekeeper and send a signal to another protein that is an ion channel (indirectly gated). Thus, a directly gated receptor is itself a ligand-gated ion channel, which opens or closes upon ligand binding. Indirectly gated receptors are GPCRs that transmit their signal via G-proteins that help to transduce the signal to open or close a nearby ion channel. Directly gated receptors are found on neurons requiring fast synaptic transmission whereas indirectly gated receptors take longer to transmit their signal and are found on the surfaces of neurons involved in slow synaptic transmission. Neurotransmitters

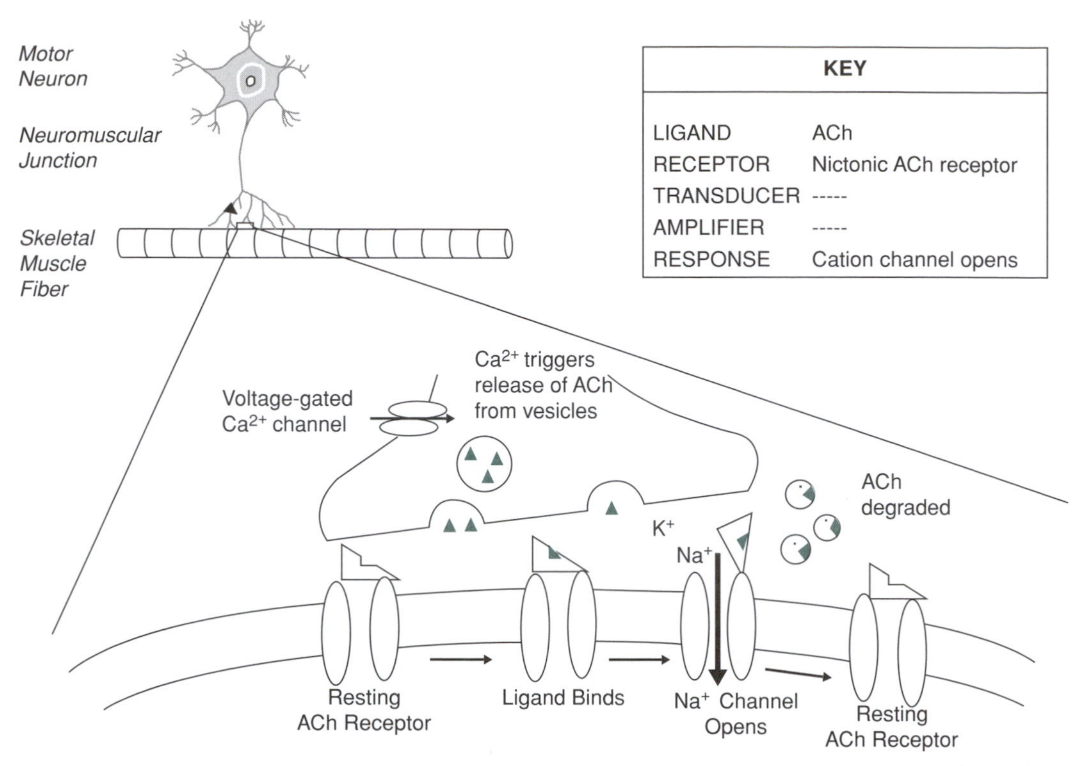

Figure 6.13 **The nicotinic acetylcholine (ACh) receptor is a ligand-gated ion channel.** The nicotinic ACh receptor is expressed on the surface of the muscle cell membrane. In the motor neuron axon terminal, an increase in intracellular Ca^{2+} triggers the release of ACh from storage vesicles. ACh diffuses across the synapse and binds to its receptor. Binding causes opening of the cationic channels allowing Na^+ and K^+ to flow into the cell. ACh is enzymatically degraded in the synaptic cleft, and the receptor returns to its resting (closed) position.

can bind to both directly and indirectly gated channels. A neurotransmitter can inhibit or excite the postsynaptic neuron, depending on which types of channels are opened. Opening Na^+ channels causes depolarization of the neuron and transmission of the action potential. In contrast, opening K^+ or Cl^- channels causes **hyperpolarization** and inhibition of the action potential. A single neuron can receive many synaptic signals (either excitatory or inhibitory), and it is the sum of these depolarizations and hyperpolarizations that determines whether an action potential is generated. If the threshold potential is exceeded, then an action potential is induced. This phenomenon of complete action potential activation upon achievement of a threshold is sometimes referred to as the "all-or-none" firing of response.

SIGNALING THROUGH DIRECTLY GATED RECEPTORS

Some aspects of synapse structure and function can be seen more clearly by examining one particular and abundant type of receptor, called the nicotinic acetylcholine (ACh) receptor. The nicotinic ACh receptor is an example of a ligand-gated ion channel. These directly gated receptors are located on the muscle cell membrane at the neuromuscular junction, which means that the synapse is between a motor neuron and a skeletal muscle cell (Figure 6.13). The ACh

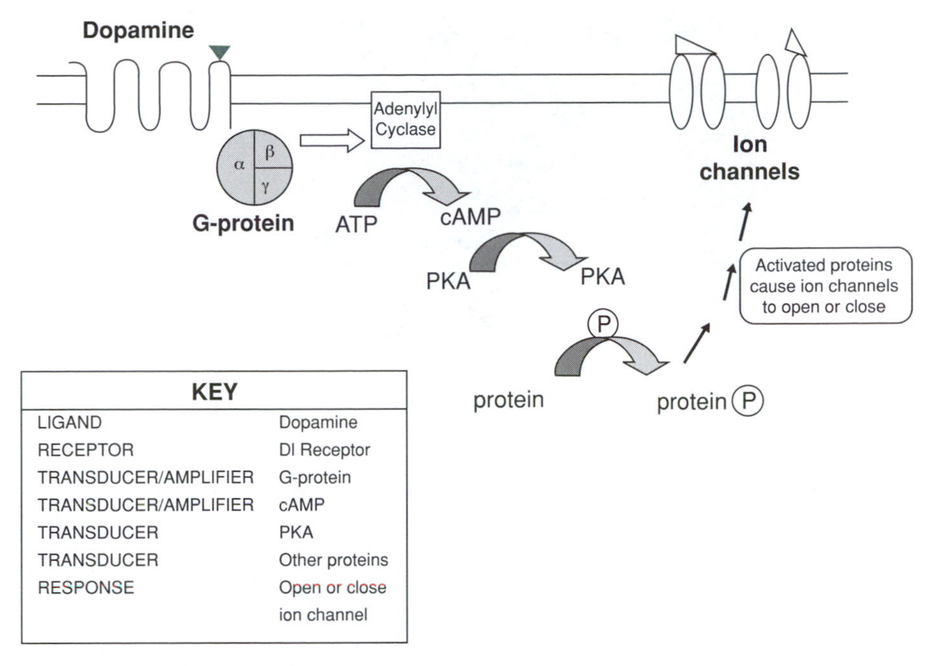

Figure 6.14 **The D1-dopamine receptor is an indirectly-gated ion channel.** Binding of dopamine to the seven-transmembrane receptor activates the associated G protein. The activated Gαs subunit binds to adenylyl cyclase and turns it on. Activated adenylyl cyclase catalyzes the synthesis of the second messenger cyclic adenosine monophosphate (cAMP) from adenosine triphosphate (ATP). The second messenger cAMP activates a kinase called protein kinase A (PKA). The initial signal is amplified because hundreds of G-protein activations can result from a single ligand–receptor complex, and a single activated adenylyl cyclase can produce several cAMP molecules. The activated PKA then starts a kinase cascade that eventually results in the opening or closing of an ion channel.

receptor is composed of five protein subunits that are arranged in a cylinder in the plasma membrane to form a pore or channel. The ACh receptor is a cationic channel, allowing only positively charged ions to flow through. Binding of ACh to the receptor triggers a conformational, or structural, change in the subunits to open the pore, allowing the passage of Na^+ and K^+ ions and transmission of an excitatory signal. In the muscle cell, this influx of positive ions causes membrane depolarization, which triggers the release of Ca^{2+} from intracellular storage depots into the cytosol. Elevated cytoplasmic Ca^{2+} stimulates muscle cell contraction. To restore the postsynaptic (muscle cell) membrane to its resting membrane potential, the neurotransmitter must be removed. This removal is achieved by an enzyme that degrades Ach (acetylcholinesterase), which is localized in the synaptic cleft. In synapses involving other receptors, the neurotransmitter might also be removed by diffusion or taken up by the presynaptic neuron.

SIGNALING THROUGH INDIRECTLY GATED RECEPTORS

For slower-acting synapses, neurotransmitter receptors (indirectly gated) are coupled to G-proteins. They belong to the GPCR family. In the case of the dopamine D1 receptors, binding of dopamine activates a GPCR, resulting in eventual opening or closing of an ion channel (Figure 6.14).

Figure 6.15 **Hormones influence metabolism and behavior.** Some hormones regulate responses that occur over many days, weeks, or months, such as the female menstrual cycle and growth during puberty. Others regulate more rapid events, such as preparation of the body for "fight or flight" responses, when danger is perceived.

6.4 The endocrine system

The endocrine system works in concert with the nervous system to maintain homeostasis in the body. **Hormones** are the chemical messengers of the endocrine system. In response to a specific demand, an endocrine gland secretes hormones into the blood for distribution throughout the body to stimulate a response in a target organ. For example, during "fight or flight" situations, the hormone epinephrine, also called adrenaline, activates beta-adrenergic receptors on muscle cells to stimulate glycogen breakdown so that fuel is readily available. Epinephrine also increases heart rate, dilates the pupils, and constricts blood vessels in certain tissues. All of these actions serve to prepare the body for rapid action. Signaling through the beta-adrenergic receptor is coupled to G-proteins and activation of adenylyl cyclase (as illustrated in Figure 6.14).

The endocrine system primarily controls the metabolic functions of the body, such as the rates of chemical reactions in cells and the transport of substances across membranes. In general, the endocrine system responds slowly and exerts longer-lasting effects than the nervous system does. A common feature of hormone secretion in the endocrine system is control by negative feedback (Box 6.3). In negative feedback, a signal is returned to the organ that secreted the hormone; the signal is negative, therefore it causes the organ to decrease the secretion of hormone. This negative signal is important because the hormone action must not persist indefinitely; negative feedback provides a mechanism for turning off a signal that is no longer needed.

6.4.1 Hormones

Hormones are chemical messengers that control growth, differentiation, and metabolic activities of cells and tissues (Figure 6.15). Secreted hormones bind to receptors located on the surface of cells of their target tissues, which may be located up to 2 meters from the secretion site. Hormones that bind to cell surface

Box 6.3 Negative feedback control

The secretion of hormones by the endocrine system, and many other physiological functions, is controlled by negative feedback. For example, control of blood pressure involves a complex and coordinated network of events; a simplified element of that control system is shown at the right. The block diagram illustrates a control sequence. In this case, a fall in blood pressure stimulates pressure-monitoring cells, which stimulate the heart and blood vessels (through the action of nerve pathways), which act to increase blood pressure. The increase in blood pressure provides a negative, or inhibitory, signal to pressure-monitoring cells, which serves to shut down the response as the blood pressure rises to an appropriate level.

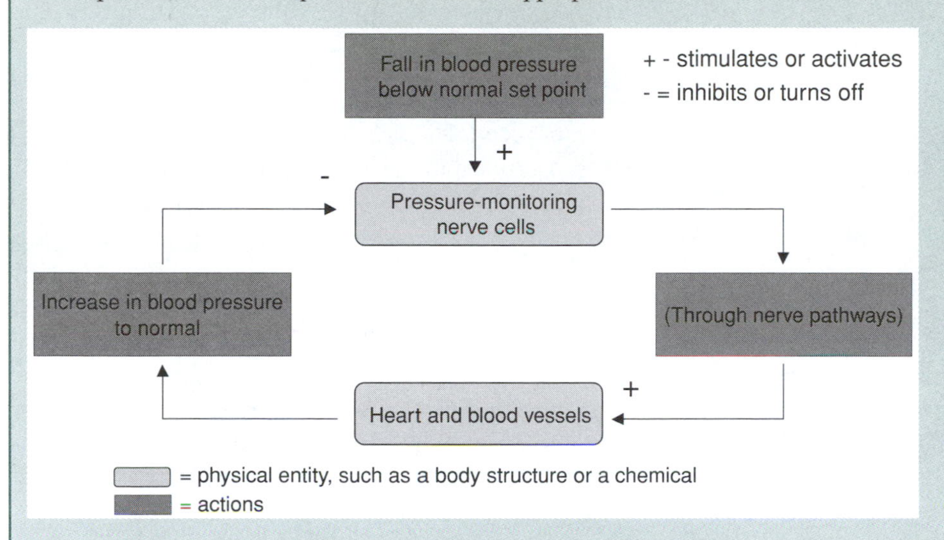

Engineers develop models of control systems and use them as tools to predict the response of the system when subjected to an input perturbation. Let's consider the simpler example shown here, which is attempting to heat up a supply of water to a desired temperature T_D (there are some similarities between this idealized system and the way that body temperature is controlled). The water is mixed completely in a vessel that is equipped with a heating element, through which heat can be added, at a rate q.

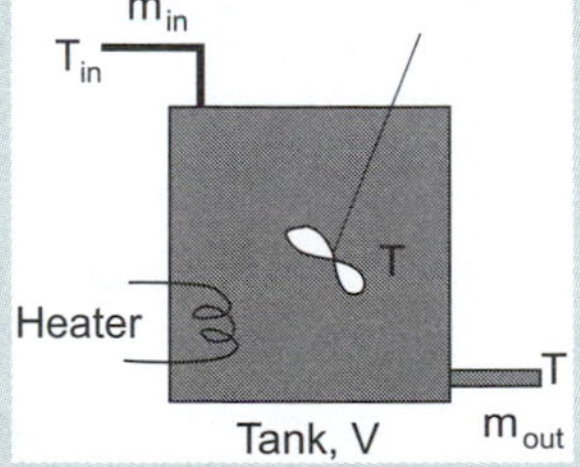

First, think about the flow of water into and out of the vessel. At steady-state, $m_{in} = m_{out}$. (Why is this true?) For simplicity, define $m = m_{in} = m_{out}$.

If everything worked perfectly, and the input stream was always exactly T_{in}, then the amount of heat needed to supply to achieve the desired temperature in the output from the tank would be:

$$q_{ss} = mC(T_{ss} - T_{in,ss}) \quad \text{(Equation 1)}$$

where q is the amount of heat supplied, C is the specific heat of the water, and the subscript ss indicates a steady-state operating condition. This equation provides a complete description of the behavior of the vessel, as long as steady-state is maintained (i.e., as long as none of the variables in the system change with time). But what if the temperature of the input stream changes, even slightly? How can the system maintain a constant output temperature if the input temperature is not perfectly steady?

(continued)

Box 6.3 *(continued)*

The second diagram shows one way to control the output temperature in the face of an unpredictable input temperature. If the temperature at the output is measured, then the deviation from the desired temperature, $T_D - T$, can be calculated. This deviation, which is often called the **error**, is equal to zero if everything is working perfectly and non-zero otherwise. The amount of heat input, q, can then be adjusted based on this measured error:

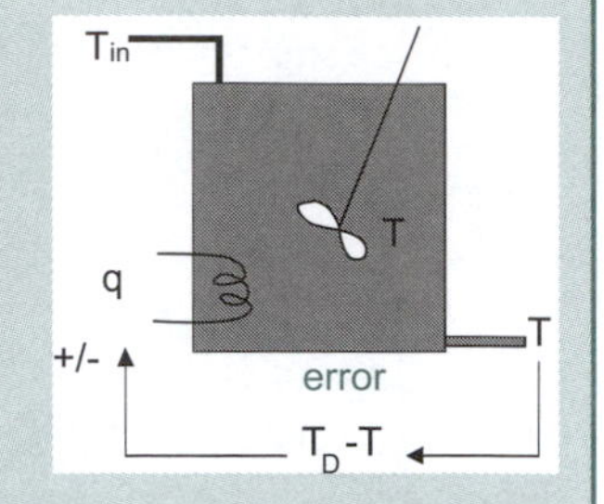

$$q(t) = \underbrace{mC(T_D - T_{in,ss})}_{q_{ss}} + K(T_D - T) \qquad \text{(Equation 2)}$$

where the first term on the right-hand side of the equation is equal to the steady-state heat input and K is a constant that can be adjusted to achieve control of the system. Because the steady-state heat input is adjusted by an amount that is proportional to the error measured, this approach is called proportional control. Equation 2 provides a method for predicting the heat input that must be supplied (as a function of time) to correct for an input temperature, T_{in}, that is an unpredictable function of time.

How is this problem related to BME? Our bodies have control mechanisms for maintaining a contact temperature, even when moving between environments of different temperature (which people do—often and unpredictably throughout the day). As seen in later chapters, models of control processes such as the simple one that is introduced in this Box, are useful for understanding our normal body function and for designing new methods to improve control.

receptors are usually proteins or peptides as shown in Table 6.3. In contrast, steroid hormones are hydrophobic and can cross the cell membrane to bind to their receptors located within the cell.

6.4.2 Protein hormone signaling

Secretion and signaling of the peptide hormones **insulin** and **glucagon** will be described to illustrate how the endocrine system works to maintain carbohydrate homeostasis (Figure 6.16). These hormones are secreted by specialized cells in the pancreas, and work reciprocally to control blood glucose levels within a constant range (80–90 mg/dL). The glucose-sensing beta cells secrete insulin in response to the elevated blood glucose concentration (**hyperglycemia**). In contrast, glucagon secretion from the alpha cells is stimulated by a decrease in blood glucose concentration (**hypoglycemia**). Thus, insulin is secreted in response to the "fed state" whereas glucagon is secreted in response to the "starved state." Glucagon produces its effect by signaling via a GPCR, which activates adenylyl cyclase and production of the second messenger cAMP. Downstream kinases activate an enzyme that stimulates the liver to breakdown glycogen and to produce glucose for release into the circulation. Insulin, in contrast, promotes glucose uptake into muscle and fat cells and increases the synthesis of glycogen, a polymer of glucose. The schematic diagrams of insulin and glucagon action

Table 6.3 Types of immune responses and key players

Immune response	Pathogen type	Main effector cells	Main soluble molecules
Innate	Extracellular (e.g., bacteria)	Macrophages, neutrophils, natural killer cells	Complement cytokines
Adaptive humoral	Extracellular (e.g., bacteria) – Vesicular antigen – pathogen phagocytosed by APCs (e.g., bacterial protein) – Presented with MHC class II	B cells T_h cells $CD4^+$	Antibodies complement Cytokines, chemokines
Adaptive cell-mediated	– Cytosolic antigen – Endogenous protein produced by host cell (e.g., viral protein) – Presented with MHC class I	T_c cells $CD8^+$	

Notes: The innate immune response is active in the early stages of infection and recognizes common features on extracellular pathogens. If the pathogens persist, the adaptive immune response can recognize antigens specifically through either a humoral response or a cell-mediated response. Humoral immunity involves B cells, which secrete antigen-specific antibodies. There are two types of cell-mediated responses (T_h or T_c) depending on the type of pathogen. Vesicular antigens are presented with major histocompatibility complex (MHC) class II molecules and recognized by helper T_h cells that express CD4 on their surface. Cytosolic antigens are presented with MHC class II molecules and recognized by cytotoxic T_c cells that express CD8 on their surface. APC, antigen-presenting cells.

(Figure 6.16) are similar to the block diagram of the control process for blood pressure regulation in Box 6.3; both systems illustrate the action of negative feedback. Negative feedback controls insulin secretion by the pancreas: When glucose is high, insulin release is stimulated; when glucose is low, insulin release is inhibited, through the action of glucagon.

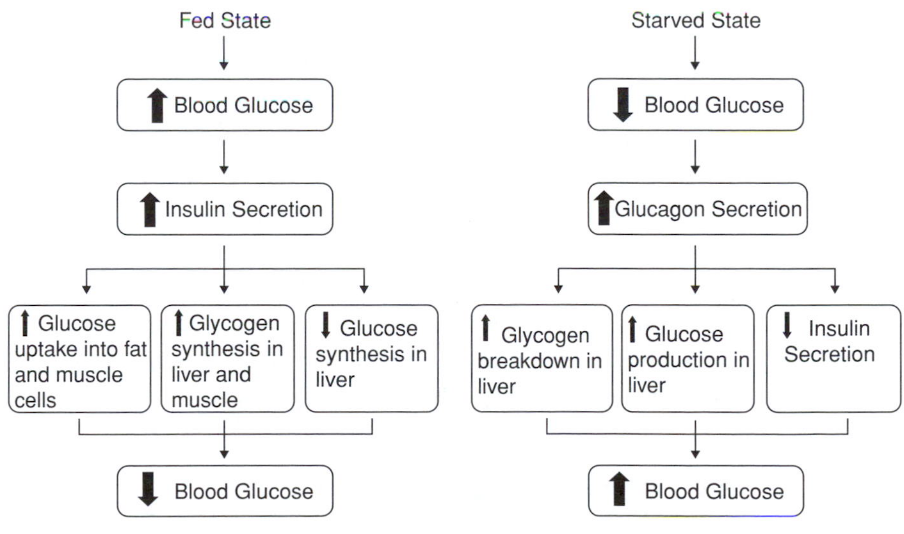

Figure 6.16 **Role of insulin and glucagon in maintaining glucose homeostasis.** Hormone secretion during the fed and starved states maintains homeostasis of blood sugar levels. During the fed state, high blood glucose levels stimulate insulin secretion from the pancreas. Insulin acts on various target cells to lower blood glucose levels into the normal range. In contrast, when blood glucose levels are low (starved state), glucagon secretion is stimulated. Glucagon acts on its target cells to increase blood glucose.

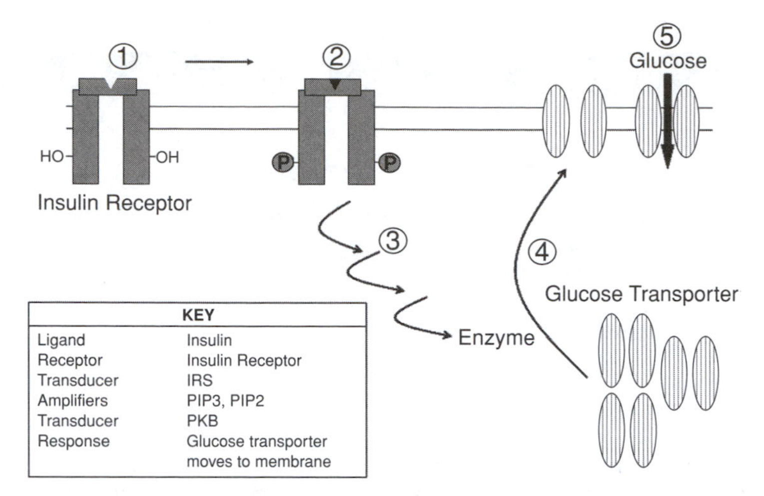

Figure 6.17 **Insulin promotes glucose uptake in fat and muscle cells.** The insulin receptor is a receptor tyrosine kinase (RTK) that contains kinase activity in its cytoplasmic tails. Upon insulin binding to its receptor, a conformation change occurs, bringing the two tails closer together so that they can phosphorylate each other. Insulin receptor substrate (IRS) binds to these sites and becomes phosphorylated. The phosphotyrosine residues on activated IRS are binding sites for the domains of adapter or transducer proteins. These proteins either recruit additional proteins (adapters) to the membrane or propagate downstream signaling pathways (as amplifiers or transducers). Phosphotidylinositol (PI)-3 kinase binds to activated IRS to produce second messenger phosphoinositides (PIP3 and PIP2), which act as amplifiers. These phosphoinositides recruit protein kinase B (PKB), which is activated by phosphorylation by PIP3-dependent kinase (PDK). PKB activates other enzymes that cause translocation of a glucose transporter (GLUT4) from the cytoplasm to the cell membrane. GLUT4 facilitates glucose entry into the cell.

The receptor for insulin is a member of the RTK family (Figure 6.17). Signaling through the insulin receptor is complex. Although multiple parallel pathways can be activated upon insulin binding, only one pathway will be described here. Insulin binding to its receptor leads to a change in its structure. This conformational change brings the two cytoplasmic tails closer together so that they can phosphorylate each other. After this autophosphorylation step, the kinase activity of the receptor is turned on. The activated receptor is now able to phosphorylate tyrosine residues in downstream enzymes. A series of reactions ensues, with the final step being activation of an enzyme that helps to promote translocation of glucose transporters from intracellular pools to the plasma membrane. Thus, glucose uptake into the target cells (fat and muscle cells) is increased by an enhancement in transport across the membrane. Because cells take up more glucose, glucose concentrations in the blood are reduced. Other cellular responses to insulin signaling are shown in Figure 6.16.

6.4.3 Steroid hormone signaling

Steroid hormones have various functions, ranging from stimulating development of reproductive organs and sexual characteristics (estrogen, testosterone) to promoting glucose production and reducing inflammation (cortisol). Steroids taken by some athletes are agonists of the androgen receptor, the signaling of which leads to enhanced development of lean muscle mass. Steroid hormones

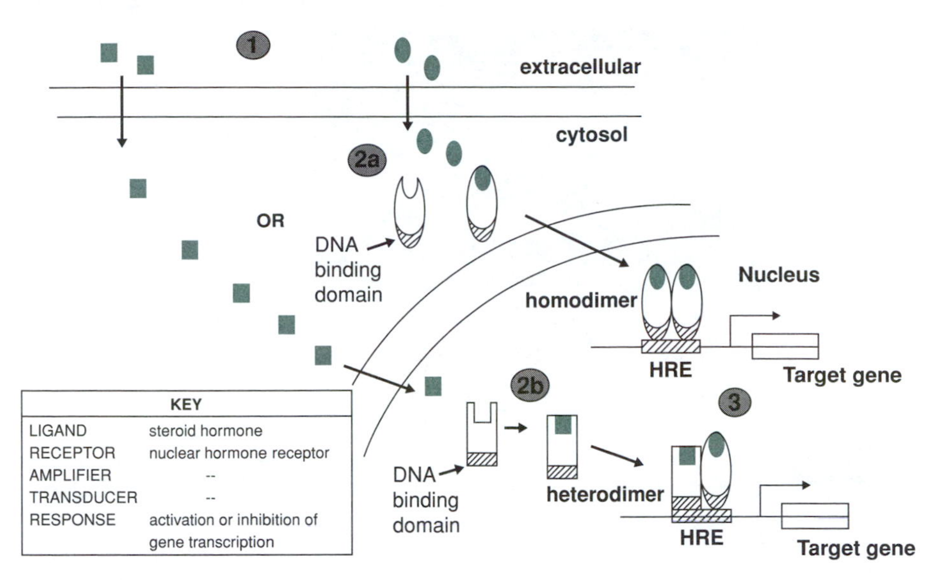

Figure 6.18 **Signal transduction pathways for steroid hormone receptors.** (1) Steroid hormones are lipophilic, so they can cross the membrane to bind to their receptors. Steroid hormone receptors can be located in the cytosol (2a) or the nucleus (2b). (3) After ligands bind, activated steroid hormone receptors can form homodimers or heterodimers and bind to steroid response elements located upstream of their target genes. Binding of a steroid hormone receptor to its deoxyribonucleic acid (DNA) binding site activates transcription of a steroid-responsive gene.

are hydrophobic molecules synthesized from cholesterol; therefore, they can diffuse through the plasma membrane to bind receptors located in the cytosol or nucleus. Nuclear hormone receptors are single polypeptide chains that have a ligand-binding domain separated by a hinge region to a DNA-binding domain. When a ligand activates its nuclear hormone receptor, the receptor–ligand complex binds to hormone response elements (HREs), which are regulatory DNA sequences located upstream of genes (Figure 6.18). Depending on the type of receptor, activated receptors can bind to their HREs as **monomers, homodimers**, or **heterodimers**. Nuclear hormone receptors are also used by thyroid hormones, vitamin D, and active forms of vitamin A (retinoids). They function as transcription factors by recruiting other proteins called coactivators or corepressors, which enhance or inhibit transcription of their target genes, respectively. For example, binding of estrogen to the estrogen responsive element induces transcription of genes that encode proteins involved in promoting cell proliferation.

6.5 The adaptive immune system

6.5.1 Overview

The immune system monitors the internal and external environments for the presence of foreign invaders and implements "attack plans" to eliminate them from the body (Figure 6.19). The cells of the immune system are capable of distinguishing self (cells belonging to the organism) from non-self (cells that do not belong, such as invading microorganisms or cells from a transplanted organ). In the normal immune response, only cells that are non-self or foreign

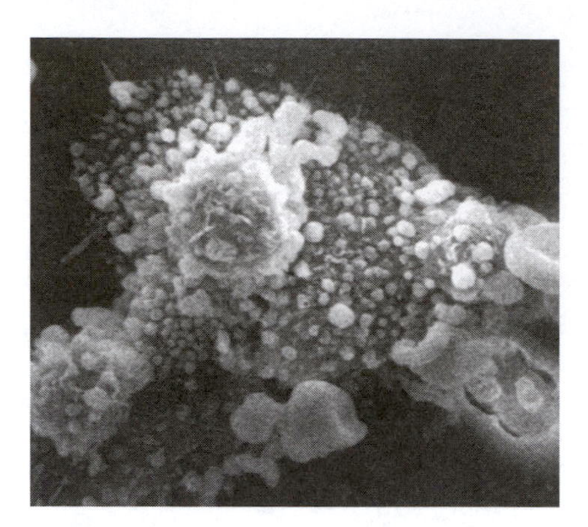

Figure 6.19 **The immune system protects individuals from foreign matter.** This photo shows a macrophage (the smaller, spherical cell) killing an invading cancer cell (the larger, oblong cell under the macrophage). Photo credit: Susan Arnold.

are destroyed, whereas self cells are left intact. Communication between different cell types of the immune system is critical in the recognition of self, surveillance, defense, and clearance of foreign invaders. The signaling mechanisms involve direct cell–cell signaling as well as autocrine and paracrine signaling. The specific mechanisms involved in the immune response are complex, so a brief overview of innate immunity and adaptive immunity will be presented. In the context of this chapter on cell communication, more emphasis will be given to describing some key events in activation of cell-mediated immunity during the adaptive immune response.

Similar to a defense department for a small country (your body), the components include: command headquarters (primary lymphoid organs—bone marrow and thymus), field command posts (secondary lymphoid organs—lymph nodes, spleen), and soldiers (macrophages, natural killer cells, neutrophils, B cells, dendritic cells, and T cells). The first line of defense (i.e., the early stages of an immune response) includes components of the innate immune system, which act like infantry units surveying and recognizing (via receptors) common features in the pathogens. The response of the innate immune system is nonspecific; that is, the response is triggered by recognition of common molecular components of the pathogen. For example, macrophages and neutrophils have receptors that can detect certain surface features, such as lipopolysaccharides, which are characteristic of microbial surfaces but are not found on host cells. In addition to physical barriers such as the skin, the innate immune response uses general mechanisms, which include inflammation and activation of the complement system to help destroy pathogens before they can infect host cells. The **complement system** involves a cascade of **proteolytic**, or cleavage, reactions that activate plasma proteins called complement proteins. In a process called **opsonization**, the complement proteins can tag pathogens for destruction by other cells. However, innate immunity is not always completely effective in keeping pathogens out of the body. Some pathogens have developed ways to avoid detection by the

Table 6.4 Common hormones secreted by glands in the endocrine system

Type	Properties	Examples
Protein hormones	– Hydrophilic – Binds to surface receptors	– Insulin – Glucagon – Growth hormone
Steroid hormones	– Hydrophobic – Binds intracellular receptors	– Testosterone – Estrogen – Progesterone – Corticosteroids

Notes: There are two general types of hormones: Protein (or peptide) hormones can bind to cell surface receptors; steroid hormones are lipophilic and can cross the cell membrane and bind to receptors in the cytosol or in the nucleus.

innate immune system. In addition, the innate immune system might simply fail to work effectively enough to clear the pathogen before an infection is established.

If the pathogen persists in the body, the "special forces" of the adaptive or acquired immune system are recruited. These forces include APCs, B cells, and T cells, among other cells. The adaptive immune system exhibits two responses—the **humoral** response and the **cell-mediated** response—that work together and depend on each other to mount an attack against different pathogens. Both B cells and T cells are capable of uniquely recognizing specific pathogens. Another feature of the adaptive immune response is that it has "memory," so it is capable of protecting against re-infection from pathogens that have entered the body previously. As a result, subsequent encounters with the same pathogen are dealt with swiftly and effectively. The responses implemented by the adaptive immune system depend on the type of pathogen. Table 6.4 summarizes the key players and responses that frequently occur in response to typical pathogens.

An extracellular pathogen can stimulate a humoral immune response in which B cells secrete antibodies that are specific for that antigen. Antigens are molecules that bind to antibodies or receptors on T cells and elicit immune responses; they are often derived from foreign pathogens, although any molecule can potentially act as an antigen, even molecules from the host's own cells. To begin secretion of antibodies, a B cell must be activated. Full B-cell activation requires signals from **helper T cells** (T_h cells), which have themselves been activated by exposure to the same antigen.

Antibodies secreted by B cells aid the body in elimination of the pathogen. Binding of an antigen to an antibody can have several effects. The antibody can neutralize the microbe and prevent it from damaging host cells; it can tag the antigen, making it more attractive for phagocytes to engulf (opsonization); or it can activate the complement system leading to **lysis** of the microbe. Chapter 14 will discuss in more detail the humoral immune response and the different classes of antibodies as well as ways in which vaccines have taken advantage of this machinery to eliminate or mitigate infectious diseases.

Some pathogens elicit a cell-mediated immune response. Pathogens that evade the innate immune system either infect host cells or are **phagocytosed**

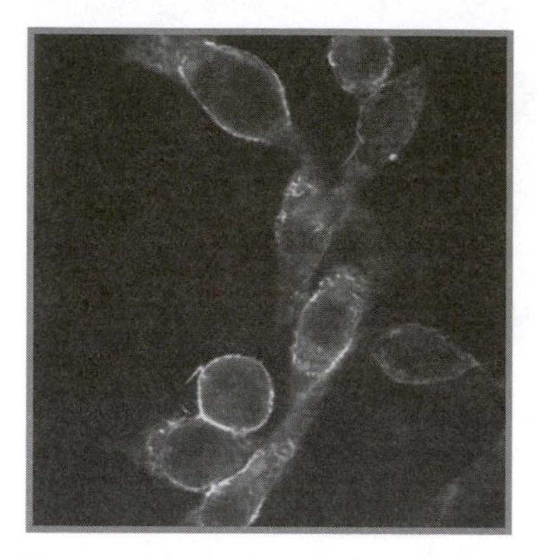

Figure 6.20 **Phagocytosis by antigen-presenting cells.** Many cells of the body are capable of phagocytosis—or ingestion of particulate matter—but antigen presenting cells, such as this dendritic cell, are especially good ingesters. These cells have ingested tiny polymer particles (red), which were placed in the environment surrounding the cells. Photo courtesy of Stacey Demento and Dr. Tarek M. Fahmy, Yale University. (See color plate.)

(Figure 6.20) by professional APCs. In both cases, the foreign proteins are degraded, and the resulting peptide fragments are processed and presented on the surface of the infected cell or APC. Antigenic peptides are loaded onto special receptor molecules called major histocompatibility complexes (MHCs). The MHC molecules, loaded with foreign peptides, are subsequently arranged on the surface of the APC to form a plasma membrane–anchored ligand (Figure 6.14). This movement of the MHC–peptide complex to the surfaces is sometimes called "display" or "presentation" of the peptide.

There are two types of MHC molecules, which are each derived from the host and are characteristic of the host—that is, each individual has different and distinctive versions of the MHC. The type of MHC molecule used for displaying the antigenic peptides depends on the source of the protein (Figure 6.21). Peptides from endogenously synthesized proteins (or cytosolic proteins because they are present in the cytoplasm), such as those from a virus that infects a host cell, are processed and presented in combination with **MHC class I molecules**. MHC class I molecules are expressed by all nucleated cells of the body, but at different levels, depending on the cell type. After extracellular

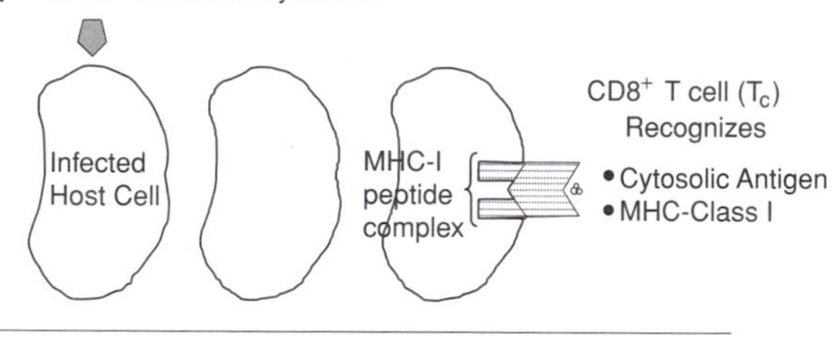

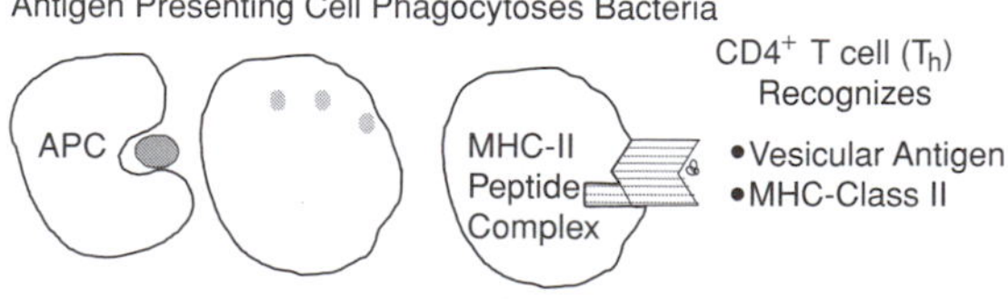

Figure 6.21 **Antigen processing and peptide display. A.** When a host cell is infected by a virus or some bacteria, the foreign proteins are synthesized by the host cell and reside in the cytosol. These proteins are cleaved into peptides and subsequently displayed or presented on the surface of the infected cell by an major histocompatibility complex (MHC) class I molecule. The MHC I peptide complex is recognized by cytotoxic (T_c) T cells that lyse the infected host cell. **B.** Pathogens that are phagocytosed directly by professional antigen-presenting cells (APC). The foreign proteins are cleaved into peptides that are presented on the surface of the APC with an MHC class II molecule. The MHC II peptide complex is recognized by helper (T_h) T cells that subsequently release cytokines that trigger the proliferation of other immune cells, such as B cells and macrophages.

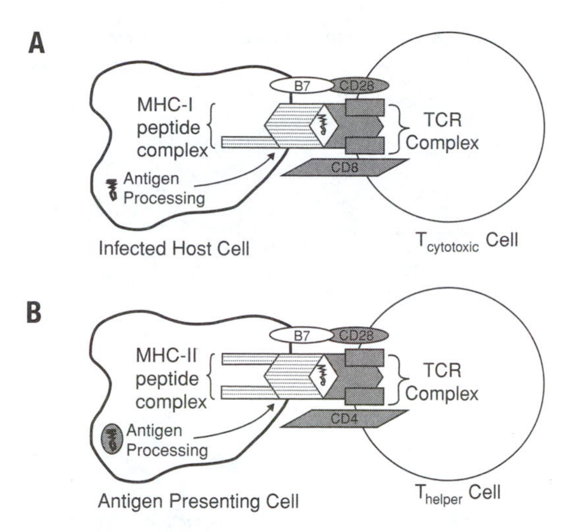

Figure 6.22 **Receptor interactions in T cell recognition of antigenic peptides.** The T cell receptor (TCR) complex recognizes an antigen in the presence of major histocompatibility complex (MHC) molecules. Those antigens bound to MHC class I are recognized by cytotoxic T (T_c) cells. Antigens bound to MHC class II are recognized by helper T (T_h) cells. When a positive identification occurs, the TCR and co-receptor CD4 (on T_h cell) or CD8 (on T_c cell) bind with moderate affinity to the peptide/MHC ligand, other subunits of the TCR complex (namely, CD3 and ξ–chain) activate signal transduction. To achieve full activation, a costimulatory ligand on the APC, called B7, must also bind to its receptor, CD28, on the T cell. This costimulatory signal enhances the TCR interaction with the peptide/MHC ligand.

pathogens are phagocytosed by APCs, their exogenous proteins are degraded in acidic vesicular compartments (such as the lysosome, discussed in Chapter 5), and the resulting peptides are processed and presented in association with **MHC class II molecules**. B cells, macrophages, and dendritic cells express MHC class II molecules and, therefore, can serve as professional APCs.

The peptide–MHC ligand displayed on the surface of the cell serves as a signal to the T cell (Figure 6.22). By recognizing different elements of the peptide–MHC complex, the T cell can determine the source of the intruder (i.e., did it induce MHC class I or MHC class II?) and the chemical nature of the pathogen (i.e., what is the structure of the peptide?). There are two types of T cells, the receptors of which can bind to the peptide–MHC surface ligand and activate downstream signal transduction pathways. Peptides associated with MHC class I are recognized by **cytotoxic T cells** (T_c) whereas those associated with MHC class II are recognized by T_h cells. These two types of T cells are distinguishable by the surface receptors they express as well as their response to binding of the peptide–MHC complex (Figure 6.21). A coreceptor molecule called **CD8** is present on T_c cells; a coreceptor called **CD4** appears on T_h cells. Because of the universal presence of these coreceptors, T_c cells are also called $CD8^+$ T cells and T_h cells are also called $CD4^+$ T cells.

6.5.2 T cell receptor signaling

During antigen recognition, cell–cell communication is mediated by a ligand on the surface of the APC (that ligand is the peptide–MHC complex) binding to a

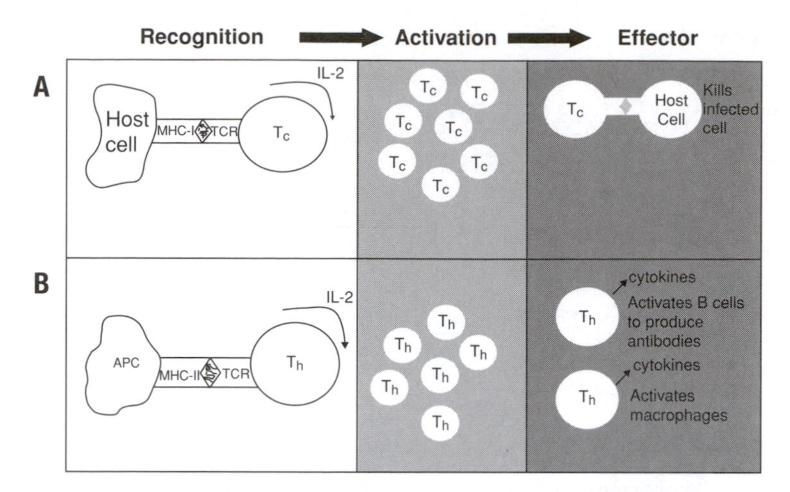

Figure 6.23

Phases of cell-mediated immune response. Once a T cell recognizes its peptide/major histocompatibility complex (MHC) ligand on the surface of an infected host cell (a) or an APC (b), it goes through an activation phase in which autocrine signaling via interleukin-2 (IL-2) stimulates proliferation of the antigen-specific T cell. During the effector phase, the T cells are fully functional and either kill infected cells (T_c cells) or help activate other cells of the immune system (T_h cells).

receptor on the T cell (the T cell receptor complex or TCR complex). The body contains an inventory or repertoire of T cells bearing a large variety of different TCRs; each version of the TCR is capable of recognizing a single antigen, so that the population contains T cells that will recognize virtually any antigen. (How this inventory is acquired in each individual is a fascinating story, which is beyond the scope of this chapter. The interested reader should refer to immunology tests, such as Janeway et al., see Further Reading.) Each T cell expresses only one type of TCR, which is specific for a particular antigen. Millions of T cells, each with an antigen-specific receptor, continually sample the surface of APCs to see whether the presented peptide matches the binding site for the receptor they carry. This scanning process takes place in the draining lymph nodes where circulating APCs and T cells meet. T cells briefly bind to the peptide–MHC complex on the surface of the APC. (It is not a simple matching process, but requires multiple signals; e.g.,, costimulatory molecules B7 and CD28 must also bind to each other for full T-cell activation.) When a match occurs, signal transduction pathways are activated. Thus, multiple receptor–ligand interactions must simultaneously occur before the specialized signal of antigen recognition can be transduced into the T cell (Figure 6.22).

After the recognition phase, in which the T cell recognizes its matching antigen, the activation and effector phases follow as shown in Figure 6.23. During the activation phase, the T cell undergoes proliferation of the antigen-specific clone or **clonal expansion** caused by autocrine signaling of a cytokine called interleukin-2 (IL-2), which is a growth factor. Thus, many copies of the antigen-specific T cell are produced. The effector phase differs for T_c cells and T_h cells because of the different roles they play.

If a T_c cell is activated through binding to its peptide–MHC class I ligand, it kills the target cell. Acting like assassins, activated T_C cells release granules that form pores in the membrane of the target cell. Pore formation facilitates the entry of enzymes that induce apoptosis of the target cell. Recall from Chapter 5 that **apoptosis** is programmed cell death or cell suicide; individual cells die from apoptosis, which leaves neighboring cells unharmed. Thus, the virus-infected cell is destroyed and it is no longer a source of new viral particles, freeing other host cells from further infection.

There are multiple consequences of a T_h cell engaging its matching peptide–MHC class II ligand. A T_h cell interacting with an APC leads to activation of transcription factors that increase transcription of genes that encode for cytokines, cytokine receptors, effector molecules, and proteins involved in cell cycling. The "helper" functions of T_h cells are mainly mediated by cell–cell communication with macrophages and B cells, which is accomplished by paracrine signaling: The T cell secretes cytokine ligands, which bind to receptors on nearby macrophages or B cells and activates them. Activated macrophages kill phagocytosed antigens whereas activated B cells produce antigen-specific antibodies to neutralize and eliminate the antigen.

The TCR is an example of an enzyme-linked receptor. As described in Section 6.2.2, this means that the receptor itself does not possess enzymatic activity but associates with another protein that does. How does binding of the peptide–MHC ligand to the TCR complex stimulate cytokine secretion? Consider the case of IL-2 transcription. When the TCR binds to the matching peptide–MHC ligand on the APC or infected host cell, the co-receptor (CD4 or CD8) also binds to part of the MHC (Figure 6.24). Multiple parallel signaling pathways are activated by binding of the co-receptors, which results in the synthesis of various transcription factors that bind to a promoter region upstream of the IL-2 gene. These transcription factors enhance transcription of IL-2. As illustrated in Figure 6.16, IL-2 is an autocrine signaling molecule that stimulates proliferation and differentiation of naïve T cells into armed effector cells: T_c or T_h cells.

6.5.3 Cytokine signaling

As the previous section illustrated, soluble mediator molecules such as **cytokines** play an important role in adaptive immunity. For example, autocrine signaling through IL-2 is required for T-cell activation and proliferation. Paracrine signaling via local secretion of cytokines is used by T cells in the effector phase. An example of paracrine signaling is that of interferon-γ (IFN-γ).

Cytokines secreted by T cells bind to the receptors located on the surface of APCs. Many cytokine receptors are enzyme-linked receptors. In the case of IFN-γ, the interferon receptor is linked to a kinase (Figure 6.25). The kinase is bound to the cytoplasmic domain of the interferon receptor. When IFN-γ binds to its interferon receptor, two monomers dimerize, bringing the associated

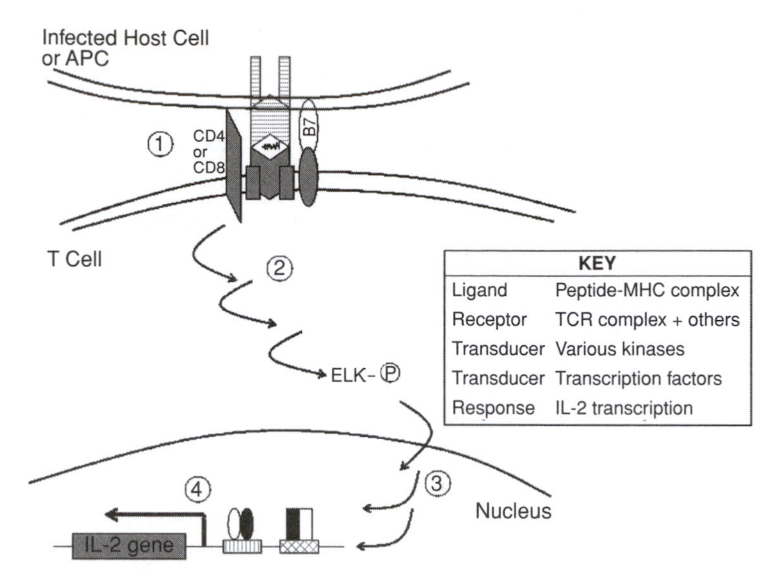

Figure 6.24

Stimulation of interleukin-2 (IL-2) transcription by the activated T cell. (1) The T cell receptor (TCR) complex and co-receptor CD4 or CD8 recognize the peptide/major histocompatibility complex (MHC) on the infected host cell or APC. These interactions cause several TCRs to come together on the surface (clustering), which allows a kinase to phosphorylate tyrosine residues on neighboring proteins. (2) A series of adapter proteins are recruited to the cell membrane, and a guanosine-5′-triphosphate (GTP)-binding protein called Ras is activated, which triggers the mitogen-activated protein (MAP) kinase cascade. (3) A transcription factor (ELK) is phosphorylated, and translocates to the nucleus, and stimulates transcription of yet other transcription factor. (4) These newly synthesized transcription factors activate transcription of the IL-2 gene. Thus, through complex signal transduction pathways, peptide/MHC binding to the TCR complex induces transcription and subsequent secretion of IL-2.

kinases in closer proximity to each other. The activated kinases phosphorylate a transcription factor, STAT (which stands for Signal Transducers and Activators of Transcription). The activated STAT dimerizes and translocates (or moves) into the nucleus where it binds to its sites upstream of IFN-γ–responsive genes to activate their transcription. The products of these genes can contribute to multiple effects in the cell expressing the IFN receptor: increased MHC expression and antigen presentation, macrophage activation, as well as increased antimicrobial activity and B-cell production of opsonizing antibodies.

6.6 Connections to biomedical engineering

This chapter introduced a number of biological concepts related to cell communication and the coordination of body functions. Many of these concepts are probably new to most readers, and some of them are difficult to understand because the vocabulary is also new and specialized. This difficulty is particularly apparent in learning about the immune system, where much of the terminology is dense and filled with abbreviations that appear to be obscure.

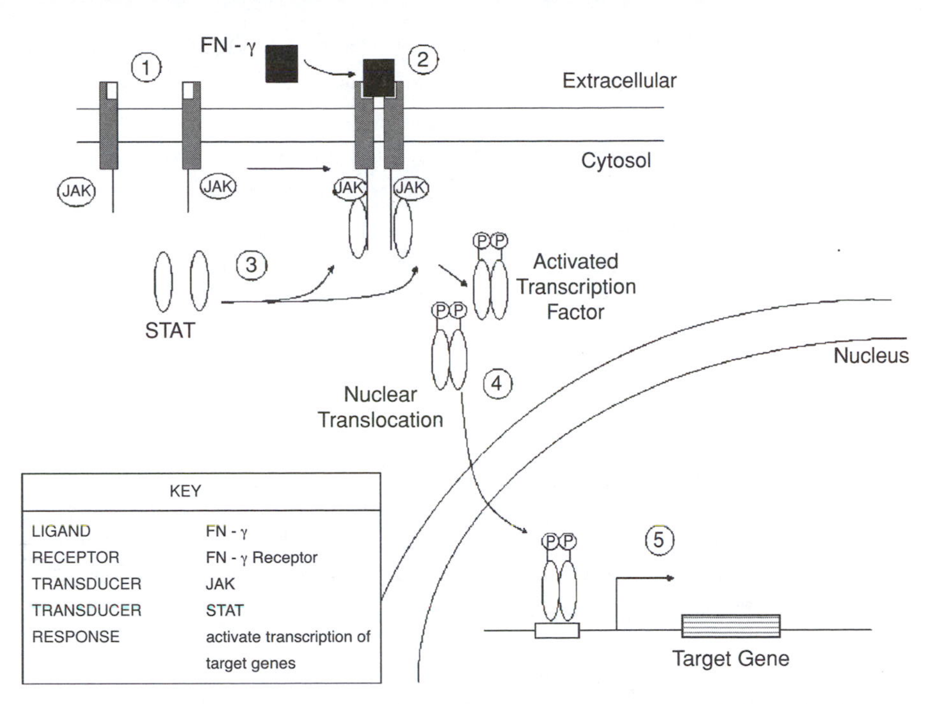

Figure 6.25 **Interferon receptor signaling results in activation of transcription of target genes.** (1) The inactive interferon receptor exists as a monomer that is associated with an enzyme called Janus kinase (JAK). (2) Interferon-γ (IFN-γ) binding causes the receptor to bind to another identical receptor (or *dimerize*), bringing the two JAKs in close proximity so that they can phosphorylate tyrosine residues on the IFN receptor tail. (3) These phosphorylated residues create binding sites for a transcription factor called STAT. STAT becomes phosphorylated by JAK and dimerizes. (4) The STAT dimers then translocate into the nucleus and (5) bind to regulatory deoxyribonucleic acid (DNA) sequences upstream of their target genes and activate their transcription.

Why should biomedical engineers bother with all of this biology, particularly in areas that are not historically important to engineers such as immunology and endocrinology? The science of cell communication represents one of the important frontier areas for biomedical engineering (BME) and, soon, most biomedical engineers will be deeply immersed in cellular and molecular biology. Many examples of this future emerge in the chapters that follow:

- Biomedical engineers are beginning to learn how to build instruments that can interface directly with cells and tissues of the nervous system; these new instruments might lead to artificial retinas that can connect directly with visual pathways in the brain or brain–machine interfaces that enable paralyzed people to control robotic arms and computers through their thought processes. For many decades now, biomedical engineers have been learning how to design artificial organs that secrete or respond to normal hormones (such as insulin and glucagon) for treatment of diabetes.
- The tools of molecular biology have advanced to the point at which engineers can now routinely add or subtract key signaling genes in cells and tissues to improve artificial organ function (Chapter 15) and design cancer diagnostics

and therapy (Chapter 16). These kinds of engineering manipulations cannot be done properly without an understanding of the biology of signal transduction.

- The immune system participates in almost every aspect of physiology, confounding the efforts of biomedical engineers who make implantable electronics, prosthetic joints, imaging contrast agents, artificial organs, and so on. Some aspect of immune system function will emerge in every chapter of this book. Biomedical engineers are making tremendous progress in control and exploitation of immune system function, as described in Chapter 14.

Summary

In this chapter, cell signaling was presented within the context of three physiological systems that use communication extensively: the nervous system, endocrine system, and immune system.

- The nervous and endocrine systems are the main control systems of the body, helping to maintain homeostasis.
- The immune system constantly monitors for the presence of foreign invaders and acts to rid them from our bodies.
- Cells are capable of executing controlled responses to internal and external changes in the environment through intricate signal transduction pathways.
- Cells communicate with each other via signaling ligands that interact with receptors located on the surface or inside the target cell.
- Receptor–ligand binding can activate different signal transduction pathways depending on the type of ligand, receptor, and target cell.
- Although the signal transduction pathways and responses may differ, all cellular communication proceeds in a cascade fashion: A ligand binds to its receptor, enzymes and second messengers transduce and amplify the signal inside the cell, and the cell responds to the signal.
- The consequences of signal transduction pathways (or cellular response) vary, ranging from opening or closing of an ion channel to activating transcription of target genes.

Many diseases are the result of failures or alterations in signal transduction. Any component of a signaling pathway can be affected. Understanding these pathways has also enabled development of therapeutic drugs that can inhibit or enhance the biological responses. Biomedical engineers can use the knowledge of receptor–ligand interactions in the design of new therapeutics, diagnostic methods, and biosensors.

Profiles in BME: Douglas Lauffenburger

Doug Lauffenburger writes of his career: I was born in Des Plaines, IL, in 1953, and remained an Illinois native through my 1971–1975 undergraduate experience. My favorite subject through high school and my early college years was always mathematics, though I enjoyed physics and chemistry reasonably well also. Biology was not so interesting to me early on, mainly because it emphasized a qualitative, descriptive, categorical approach. For the same reason, I was never attracted toward medical school. However, late in my undergraduate experience I found that my chemical engineering major did not excite me, either, in terms of its application areas. Fortunately, during my senior year at Illinois I took a cell physiology course that was an ideal mix of fascinating biological phenomena and quantitative analysis, and I was hooked by what an engineering approach could do in understanding how living systems operate.

Thus, I entered graduate school at University of Minnesota (Twin Cities campus) with the objective of studying biology from an engineering perspective. I did this in the chemical engineering program, under the joint tutelage of Professors Kenneth Keller and Rutherford Aris. My work there fortunately coincided with the beginning of the molecular biology and biotechnology revolutions, and I eagerly took all the mechanism-focused biology courses I could find – biochemistry, molecular biology, cell biology, microbiology, immunology, and virology. My doctoral thesis focused on mathematical analysis of chemotactic cell migration, with applications in both environmental microbiology and the immune and inflammatory host defense systems.

There were no industries looking to hire engineers focused on molecular/cell biology at the time of my PhD graduation (1979), so to pursue a research career attempting to fuse engineering with modern molecular cell biology I had no choice but to find a faculty position in academia. Fortunately, I found an amazingly open-minded and progressive home at University of Pennsylvania in the chemical engineering department with joint appointment in the cell biology program; it has always been crucial to me to have a home in both an engineering department and a biology department for the kind of biology-based engineering research and education program I found exciting and compelling. However, in the early 1980s, neither engineers nor biologists thought there was any future in an engineer trying to develop quantitative models for how cells operate, so it was quite a challenging time for finding funding support, journals to publish in, or meeting sessions to present in. I would tell my own doctoral students in the 1980s (including some superb early pioneers in this new area of bioengineering, including [not exclusively] Bob Tranquillo, Dan Hammer, Jennifer Linderman, Helen Buettner, and Cynthia Stokes) that they would be hard-pressed to find someone willing to hire them to do what they were learning to do in my laboratory.

Happily, in the 1990s this situation changed, slowly but substantially, and now in the post-genomic era it has become clear to engineers and scientists alike that the interface of engineering with modern biology is going to be one of the most important areas for technology and society in the 21st century. A biology-centric bioengineering discipline is rapidly emerging (which we term at MIT "biological engineering" for emphasis), which will likely offer an exceptionally

(continued)

Profiles in BME *(continued)*

powerful new set of approaches to problems in medicine and human health. My own laboratory worked hard to pursue foundational aspects of this new discipline in the 1990s, focusing on how receptor-mediated signaling networks govern all kinds of cell functional behaviors (e.g., death, proliferation, differentiation, and migration) involved in physiology and pathology along with diagnostics and therapeutics. My teaching career moved from Pennsylvania first to the University of Illinois (Urbana-Champaign) during the period 1990–1994 with appointments in chemical engineering, cell biology, and biophysics, and then more recently since 1995 at MIT. The founding of the biological engineering division at MIT in 1998 is one of the high points of my career, for it manifests the vision of a truly biology-based engineering discipline that I had sensed for roughly 20 years. My laboratory members and I are especially gratified by the strong collaborations we are undertaking with a number of the key molecular and cellular therapeutics companies nationally and internationally, as it is these kinds of companies which will represent a large opportunity for bioengineers seeking to have great impact on medicine and health care in the coming decades.

Douglas A. Lauffenburger is the Uncas and Helen Whitaker Professor of Bioengineering and co-director of the Biological Engineering Division at the Massachusetts Institute of Technology. He also holds appointments in the Department of Chemical Engineering and the Department of Biology, is the director of the Biotechnology Process Engineering Center and the Center for Biomedical Engineering, and is an affiliate for the Center for Cancer Research. His research focus has been in quantitative cell biology and he is well-known for his work in chemotaxis and cell motility. Dr. Lauffenburger is a founding fellow of the American Institute of Medical and Biological Engineering. In 2001, he was named to both the National Academy of Engineering and the National Academy of Arts & Sciences. Dr. Lauffenburger has been a prolific author and was recently awarded the William H. Walker Award for Excellence in Contributions to Chemical Engineering Literature by the American Institute of Chemical Engineers. He has advised over 50 doctoral students, many of whom have gone on to become leaders in the field of Bioengineering. He received his Bachelor's Degree from the University of Illinois Urbana-Champaign in 1975 and Ph.D. from the University of Minnesota in 1979, both in Chemical Engineering.

FURTHER READING

CELL SIGNALING

Alberts B, Bray D, Lewis J, Raff M, Roberts K, Watson JD. *Molecular Biology of the Cell*. 3rd ed. New York: Garland Publishing; 1994.

Berg JM, Tymoczko JL, Stryer L. *Biochemistry*. New York: W. H. Freeman; 2002.

Cooper GM. *The Cell—A Molecular Approach*. 2nd ed. Sunderland, MA: Sinauer Associates, Inc.; 2000.

Ganong WF. *Review of Medical Physiology*. 21st ed. New York: McGraw Hill; 2003.

Lodish H, Berk A, Zipursky SL, Matsudaira P, Baltimore D, Darnell JE. *Molecular Cell Biology*. 4th ed. New York: W. H. Freeman & Co.; 2000.

NERVOUS SYSTEM

Kandel ER, Schwartz JH, Jessell TM, eds. *Principles of Neural Science*. 3rd ed. East Norwalk, CT: Appleton & Lange; 1991.

Matthews GG. *Neurobiology: Molecules, Cells, and Systems*. 2nd ed. Malden, MA: Blackwell Science; 2001.

Siegel GJ, Agranoff BW, Albers RW, Fisher SK, Uhler MD, eds. *Basic Neurochemistry, Molecular, Cellular, and Medical Aspects*. 6th ed. Philadelphia, PA: Lippincott, Williams & Wilkins; 1999.

Webster RA, ed. *Neurotransmitters, Drugs and Brain Function*. New York: John Wiley & Sons; 2001.

ENDOCRINE SYSTEM

Aranda A, Pascual A. Nuclear hormone receptors and gene expression. *Physiol Rev*. 2001;81(3):1269–1304.

Bast RC, Kufe DW, Pollock RE, Weichselbaum RR, Holland JF, Frei E, eds. *Cancer Medicine*. 5th ed. Hamilton, Canada: BC Decker Inc.; 2000.

Katzenellenbogen BS, Choi I, Delage-Mourroux R, Ediger TR, Martini PG, Montano M, et al. Molecular mechanisms of estrogen action: Selective ligands and receptor pharmacology. *J Steroid Biochem Mol Biol*. 2000;74(5):279–285.

Nussey SS, Whitehead SA. *Endocrinology: An Integrated Approach*. Oxford, UK: BIOS Scientific Publishers, Ltd.; 2001.

Nystrom FH, Quon MJ. Insulin signaling: Metabolic pathways and mechanisms for specificity. *Cell Signal*. 1999;11(8):563–574.

IMMUNE SYSTEM

Aaronson DS, Horvath CM. A road map for those who don't know JAK-STAT. *Science*. 2002;296(5573):1653–1655.

Abul KA, Lichtman AH, Pober JS. *Cellular and Molecular Immunology*. 4th ed. Philadelphia: W.B. Saunders; 2000.

Janeway CA, Travers P, Walport M, Shlomchik M. *Immunobiology*. 5th ed. New York and London: Garland Publishing; 2001.

Paul WE. *Fundamental Immunology*. 5th ed. Philadelphia: Lippincott Williams & Wilkins; 2003.

USEFUL LINKS ON THE WORLD WIDE WEB

National Center for Biotechnology Information (NCBI) Web sites:
http://www.ncbi.nlm.nih.gov/entrez/query.fcgi?db=PubMed
PUBMED allows you to search the "MEDLINE database" of scientific journal articles.

http://www.ncbi.nlm.nih.gov/entrez/query.fcgi?db=books
BOOKS allows you to search several electronic books on the "NCBI Bookshelf"

Membrane Potential Tutorial:
http://sun.science.wayne.edu/~bio340/Applets/
Dr. Robert Stephenson, Wayne State University

Membrane Potential Simulations:
http://www.unm.edu/~toolson/122gld_home.html
Dr. Eric C. Toolson, University of New Mexico

Various Freeware including Membrane Transport and Membrane Potential:
http://www.physiologyeducation.org/
Physiology Educational Research Consortium

KEY CONCEPTS AND DEFINITIONS

action potential electrical signal or nerve impulse

affinity the strength of binding of one molecule to another at a single site

agonist a molecule that can bind to a receptor and activates a receptor

amplifier a device that increases the magnitude of an input

antagonist a molecule that competes with a ligand and inhibits receptor activation

antigen-presenting cells cells that display peptide fragments from antigens on their surface along with other molecules required for the activation of T cells

apoptosis an active process of programmed cell death, characterized by cleavage of chromosomal DNA, chromatin condensation, and fragmentation of both the nucleus and the cell; also referred to as cell suicide

autocrine signaling a type of cell signaling in which a cell secretes a molecule to which it also responds

axon a long process extending from the cell body of a neuron that conducts an electric impulse (action potential) away from the neuron

CD4 a coreceptor protein on helper T cells that recognizes antigens bound to MHC class II molecules

CD8 a coreceptor protein on cytotoxic T cells that recognizes antigens bound to MHC class I molecules

cell-mediated immunity an adaptive immune response in which antigen-specific T cells have the main role

clonal expansion proliferation of antigen-specific lymphocytes in response to antigenic stimulation; an important step in adaptive immunity

complement system a set of proteins activated by the innate immune system that facilitates uptake and destruction of pathogens

connexins a group of homologous proteins that form the intermembrane channels of gap junctions

cytokine any of numerous secreted, small proteins (e.g., interferons, interleukins) that bind to cell-surface receptors on certain cells to trigger their differentiation or proliferation

cytotoxic T cell a type of T cell that kills other cells when it recognizes a peptide fragment on MHC class I proteins with its CD8 receptor

dendrite a process extending from the cell body of a neuron that is relatively short and typically branched and receives signals from axons of other neurons

depolarization a change in the membrane potential in the positive direction from its normal negative level

dimerization a process in which two polypeptide chains bind to each other

effector proteins located inside the cell that carry out the action of the original signal of a ligand binding to a receptor

endocrine signaling a type of cell signaling in which cells secrete molecules that are carried by the circulation to distant target cells

equilibrium association constant (K_a) the ratio of the concentration of the bound receptor $[RL]$ to the concentrations of unbound receptor $[R]$ and ligand $[L]$; the inverse of the equilibrium dissociation constant; the higher the K_a, the tighter the binding between the receptor and ligand

equilibrium dissociation constant (K_D) the equilibrium constant for the dissociation of a complex of two or more biomolecules into its components; the ratio of the concentrations of the unbound receptor $[R]$ and ligand $[L]$ to the bound receptor $[RL]$ (Chapter 2)

gap junctions sites of electrical connection between the membranes of two cells that use connexins to bridge the gap between the insides of the two cells and allow small molecules such as ions to cross directly from one cell to the other

glial cell a cell that provides support to nerve cells, but does not participate directly in synaptic interactions and electrical signaling

glucagon a hormone produced by the pancreas that causes an increase in blood glucose levels, opposing the action of insulin

helper T cell a type of T cell that stimulates other lymphocytes when it recognizes a peptide fragment on MHC class II proteins with its CD4 receptor

heterodimer a protein consisting of two different polypeptide chains

homeostasis the body's ability to maintain a stable internal environment in response to a changing external environment or internal malfunction

homodimer a protein consisting of two of the same polypeptide chains

hormone a ligand that induces specific responses in target cells, especially in the endocrine system; hormones regulate the growth, differentiation, and metabolic activities of various cells, tissues, and organs

humoral immunity an adaptive immune response in which antibodies produced by B cells cause the destruction of extracellular microorganisms and prevent the spread of intracellular infections

hydrophilic polar or charged molecules that dissolve readily in water (Chapter 2)

hydrophobic nonpolar molecules that are insoluble in water (Chapter 2)

hyperglycemia the presence of an abnormally high concentration of glucose in the blood

hyperpolarization a change in the membrane potential, making the cell more negative than its resting membrane potential

hypoglycemia the presence of an abnormally low concentration of glucose in the blood

insulin a protein hormone secreted by beta cells in the pancreas that has a key role in the regulation of carbohydrate and fat metabolism throughout the body, it is especially important in promoting the storage of glucose, which lowers the blood glucose level

ion channels a membrane protein that forms an aqueous pore through which charged ions can cross the membrane

kinase an enzyme that catalyzes the phosphorylation of certain molecules using ATP as a phosphate source

ligand any molecule, other than an enzyme substrate, that binds tightly and specifically to a macromolecule, usually a protein, forming a macromolecule–ligand complex

lysis the breaking of a cell's membrane, which results in the death of the cell

MAP kinases *m*itogen-*a*ctivated *p*rotein kinases that are activated in response to a variety of growth factors and other signaling molecules. MAP kinases phosphorylate serine or threonine residues in their substrates.

MHC class I molecules receptors that present peptides generated in the cytosol to CD8 T cytotoxic cells

MHC class II molecules receptors that present peptides degraded in intracellular vesicles to CD4 T helper cells

monomer a protein consisting of one polypeptide chain

myelin protein and lipid material produced by glial cells that insulates regions of an axon

Nernst potential the potential across a semipermeable membrane caused by a difference in concentration of an ion on each side of the membrane. The Nernst potential occurs when the concentration gradient is equal to the voltage gradient.

neurotransmitter a small, hydrophilic molecule that carries a signal from a stimulated neuron to a target cell at a synapse

nodes of Ranvier unmyelinated regions of an axon that are exposed to the extracellular space and contain the Na^+ channels for the propagation of action potentials

oligodendrocyte a type of glial cell that produces myelin in the central nervous system

oncoprotein a product of an oncogene that can induce one or more characteristics of cancer cells, such as proliferation

opsonization a process by which specialized proteins of the complement system or antibodies bind to pathogens and facilitate phagocytosis of the pathogen by macrophages or neutrophils

paracrine signaling a type of cell signaling in which a molecule secreted by one cell acts on a neighboring target cell

phagocytosis the uptake of large particles by a cell; particles are taken up in vesicles that fuse to the lysosome

phosphatase an enzyme that catalyzes the removal of a phosphate group from a protein

phosphorylation the covalent addition of phosphate group to a protein

proteolysis the hydrolysis of proteins into peptides and amino acids by cleavage of their peptide bonds

receptor a specialized protein on a cell's surface, but anchored in the cell membrane, which binds to a specific ligand and transmits an appropriate signal

refractory period the period after a neuron fires, during which a stimulus will not evoke a response

repolarization a change in membrane potential in the negative direction, making the cell interior more negative and returning it to the resting membrane potential

saltatory conduction rapid, efficient propagation of action potentials caused by myelinated axons

Schwann cell a type of glial cell that produces myelin in the peripheral nervous system

signal transduction pathway a series of events initiated by the binding of a ligand to its receptor that allows a cell to communicate with and respond to other cells or its extracellular environment

synapse a specialized junction that allows two nerve cells to communicate either through chemical or electrical signals

synaptic cleft extracellular space between two adjacent nerve cells into which neurotransmitters are released by the presynaptic cell

transducer a device that converts a signal from one form to another

transfer function a mathematical expression of the relationship between the input signal and output signal of a system

tumor suppressor proteins products of tumor suppressor genes that normally function to suppress tumors so when they are inactivated in some types of cancers, they promote tumor development

NOMENCLATURE

c	Ion concentration	M
C	Specific heat	J $mol^{-1}K^{-1}$
F	Faraday's constant	cal/mol-V
K_D	Equilibrium dissociation constant	M
K_A	Equilibrium association constant	M^{-1}
k	Rate constant	Units vary
k_f	Forward rate constant	Units vary
k_r	Reverse rate constant	Units vary
[L]	Free ligand concentration	M
m	Flow rate	mol/s
q	Heat	J/s
R	Universal gas constant	cal/K-mol or J/K-mol
[R]	Concentration of unbound receptors	M
r	Fractional saturation	Unitless
[RL]	Concentration of receptors with bound ligand	M
t	time	s
T	Absolute temperature	K
V	Nernst potential	V
z	Valence or charge of an ion	V

SUBSCRIPTS

bound	Bound protein
D	Desired value
eq	At equilibrium
f	Forward reaction

i	Inside
in	Into the system
r	Reverse reaction
o	Outside
out	Out of the system
ss	Steady-state operating conditions
total	Total

QUESTIONS

1. Describe each of the following diseases. What components of the nervous system are affected? How do these abnormalities result in impaired function?
   a. Epilepsy
   b. Schizophrenia
   c. Alzheimer's disease
   d. Parkinson's disease
   e. Multiple Sclerosis (MS)
   f. Amyotrophic Lateral Sclerosis (ALS or Lou Gehrig's Disease)
   g. Huntington's disease
   h. Myasthenia Gravis
   i. Stroke
2. Action potential propagation is unidirectional. What ensures that the action potential is passed along the length of the axon in only one direction from the dendrites to the axon terminal? Also, what structural features of the synapse contribute to the one-way information transfer from the presynaptic neuron to the postsynaptic neuron?
3. Juvenile diabetes (Type I) is an autoimmune disease. What does this mean and how does this lead to high blood glucose levels?
4. Human immunodeficiency virus (HIV) infects T cells expressing CD4. How does this infection lead to acquired immunodeficiency syndrome (AIDS)?
5. One of the transducers in many growth factor signaling cascades is *ras*. How can mutations in *ras* (found in some types of cancers) contribute to transforming a normal cell into a tumor cell?
6. Find a receptor that utilizes the MAP kinase cascade. Sketch a diagram showing the proteins and enzymes involved in the cascade and describe the steps.
7. Why is it important to characterize the MHC type of an individual before bone marrow or organ transplantation?
8. Immunosuppressive drugs such as cyclosporin A and tacrolimus are used in organ transplant recipients. What makes these drugs immunosuppressive (i.e., their mode of action)? What are the side effects related to taking immunosuppressive drugs?

PROBLEMS

1. Your body temperature is maintained within a constant range as shown in the graph.

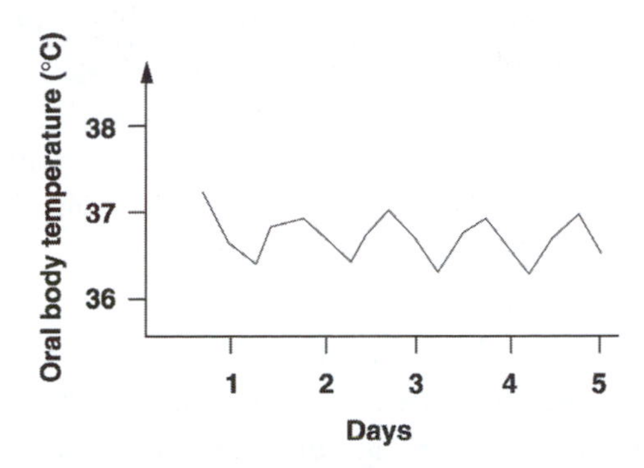

   a. Use a flow chart to illustrate what temperature-increasing mechanisms can raise your temperature when you are too cold. Your flow chart can have multiple branches.
   b. Use a flow chart to illustrate what temperature-decreasing mechanisms can lower your temperature when you are too hot. Your flow chart can have multiple branches.
2. Given that the extracellular concentration of Cl^- is approximately 120 mM, what is the intracellular concentration if the Nernst potential for Cl^- is 39 mV?
3. An equilibrium dialysis experiment was performed to characterize the binding affinity of a mouse IgG or human IgG antibody for an antigen, GAD65. The association constant for the monoclonal mouse IgG is 4.75×10^8 M^{-1} and for a human IgG is 1.3×10^{10} M^{-1}.
   a. Calculate the dissociation constant for each antibody. Which antibody has a higher binding affinity for GAD65? Explain.
   b. For a GAD65 concentration of 0.7 nM, calculate the fraction of free Ab sites for each antibody.

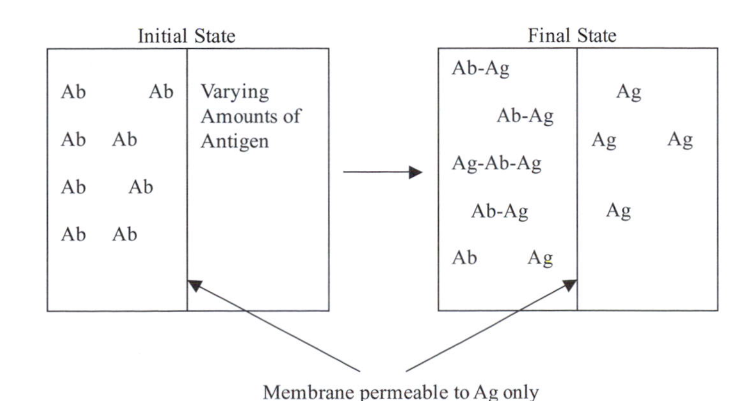

4. An insulin receptor binds its ligand, which is the hormone insulin. The affinity constant is 1.5×10^{-9} M. Construct a Scatchard plot for this receptor-ligand interaction.
5. Figure 6.16 shows the role of glucagon and insulin in the control of blood glucose levels. Convert each of these diagrams to a block control diagram, such as the one shown for blood pressure regulation in Box 6.3. Use the symbols that are defined for stimulation of inhibition (+ and – signs) and the shapes of boxes (square for actions; rounded for physical entities) to differentiate between events in your block diagram. Include as much information from Section 6.4.2 as possible into your block diagram.
6. A proportional control scheme for heating a water stream is illustrated in Box 6.3. Use the information in that box to answer the following questions.
   a. Assume that the input stream is following into the process at 10 L/min at a temperature of 20°C, and that you want to heat the liquid to body temperature (37°C). What is the steady-state heat input to this process?
   b. Assume that, at some time $t = 0$, the temperature of the input stream suddenly changes to 25°C. If the heat input to the unit is not changed, what will the temperature of the output stream be once a new steady-state is reached? How long do you think it will take to reach this new steady-state temperature?
   c. Describe a strategy for controlling the output stream temperature at a desired temperature of 37°C with the possibility that the input stream temperature will be a function of time. Draw a block control diagram for your process. How will you implement it? Draw a diagram (by plotting temperature versus time) for what you would expect to observe if the sudden temperature rise described in (b) were to occur, with your new process in place.

7 Engineering Balances: Respiration and Digestion

LEARNING OBJECTIVES

After reading this chapter, you should:

- Understand the concepts of an engineering system, system boundaries, and the differences between open and closed systems.
- Be familiar with the concepts of homeostasis and steady state and be able to distinguish equilibrium from steady state.
- Understand the concepts of external and internal respiration.
- Be familiar with air volumes and flow rates in the lungs.
- Understand how oxygen is carried by blood and the quantitative relationships describing oxygen concentration.
- Understand the relationship between carbon dioxide, bicarbonate ion, and pH in body fluids.
- Understand the diffusing capacity of the lung and how it relates to the properties of the respiratory membrane.
- Understand how the structure of the digestive organs (stomach, small intestine, large intestine, pancreas, and liver) is related to their functions in digestion.
- Understand the role of enzymes in digestion, and the importance of enzyme activation after secretion (i.e., the value of zymogens).
- Understand the role of reactor models in understanding digestion and absorption of nutrients.

7.1 Prelude

Humans eat, drink, and breathe to bring into their bodies the raw materials for growth, repair, and generation of the energy necessary for life and the actions that bring pleasure to life. This chapter provides an overview of human nutrition and respiration from the perspective of biomedical engineering (BME). The human body is an elegant machine that requires inputs for sustained operation. What are the processes responsible for input of nutrients and raw materials? How are molecular nutrients extracted from ingested materials? How are these processes controlled?

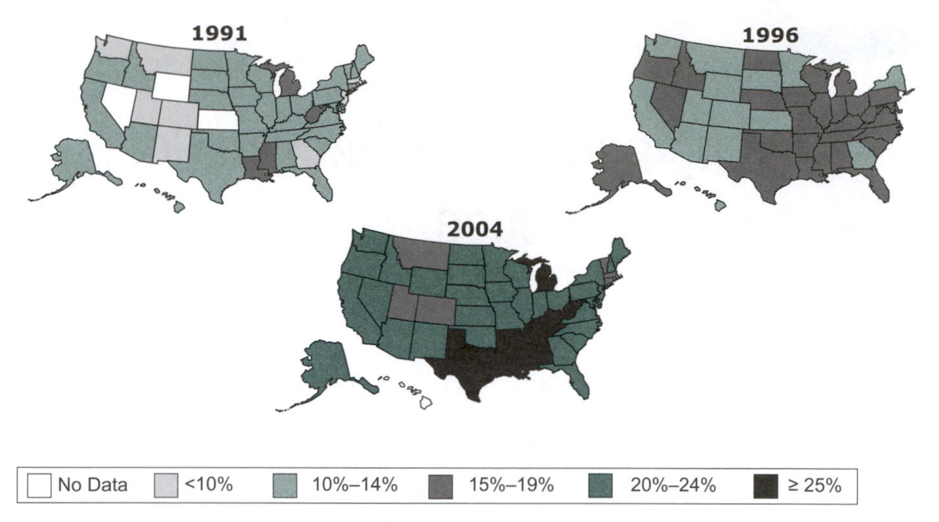

Figure 7.1

Obesity trends among U.S. adults. These maps show the fraction of the adult population that has a body mass index (BMI) $>$ 30 in 1991, 1996, and 2004. Reprinted with permission courtesy of the U.S. Centers for Disease Control and Prevention: http://apps.nccd.cdc.gov/brfss/.

The human machine requires food and water for continued operation, but the relationship between food intake and human health is complex and poorly understood. Some things are clear: Whole foods that we ingest get broken down into components such as amino acids and sugars, which the body uses to synthesize new proteins and to generate or store energy. Some elements of these metabolic processes are presented in Chapter 2. Protein synthesis, energy generation, and metabolic processes occur in cells throughout the body; hence all of these processes are related to the circulation of molecules in the body, which is discussed in Chapter 8. This chapter will look at overall events and establish some general descriptions of the quantity and composition of what enters our bodies.

The issues of food intake, nutrition, and human health are becoming increasingly important in the United States. The Centers for Disease Control and Prevention report a dramatic increase in obesity in the United States during the period from 1985 through 2005 (Figure 7.1). In addition, diseases related to environmental exposure to toxins and pollutants are widespread and the numbers are still increasing. Asthma among children has increased to epidemic proportions, accounting for 1 in 6 of all visits to pediatric emergency rooms in the United States (Figure 7.2).

How can biomedical engineers help address public health problems such as obesity and asthma? One of the central themes of engineering is the analysis and interpretation of mechanisms, often from complex systems in which all of the internal workings of the system are not known. Certainly, the workings of the human body—which convert food, air, and water into energy and body mass—is exceedingly complex. Biomedical engineers can contribute to understanding relationships between intake of nutrients, air, drugs, toxins, and other molecules and human health. As this chapter will illustrate, BME analysis has already provided important insight into the workings of the human lung and kidney.

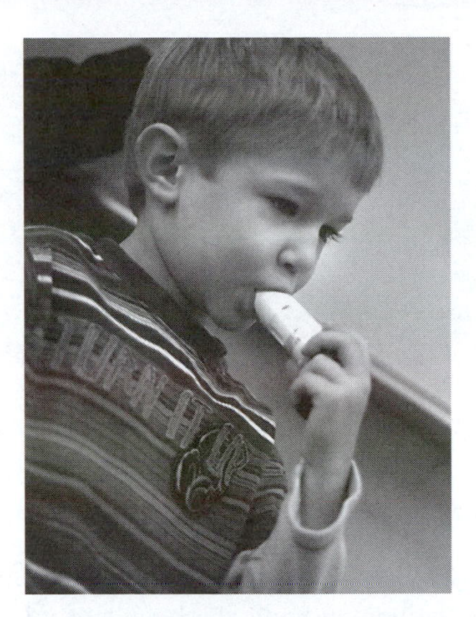

Figure 7.2 **A child uses a nebulizer to treat his asthma.** Approximately 20 million Americans suffer from asthma, including 9 million children. Direct health care costs for asthma are now greater than $11 billion annually.

In this chapter, important concepts are introduced that are essential for the analysis of any engineering: the concept of a system under study and system balances. These concepts are introduced and then illustrated in the context of lung, gastrointestinal, and renal physiology. Because they are general concepts, they are used throughout the remaining chapters of the book.

7.2 Introduction to mass balances

A major goal of this book is to present an approach for solving problems in BME. The use of engineering tools to solve problems in human health is the overall objective of biomedical engineers. The development of problem-solving skills is started early in the book: For example, Box 1.2 of Chapter 1 provided advice about how to set up and solve engineering problems. The important role of problem-solving techniques continued in Chapter 2: Box 2.1 defined some of the essential elements of engineering analysis, including the concept of a system under study.

This section presents another frequent aspect of engineering analysis: the **mass balance**. A mass balance is simply a systematic approach to applying a physical principle: Mass cannot be created or destroyed. Before illustrating a mass balance, the concept of a **system** is presented.

Engineering analysis invariably begins by defining the system under study. The system might be a supporting beam in a bridge, a set of components on an integrated circuit board, the human body, or an individual cell or organ in the body. In describing a system, it is important to pay careful attention to the **system boundaries**, that is, the physical sites of intersection between the system under study and the rest of the world (Figure 7.3). Often there are multiple choices for the

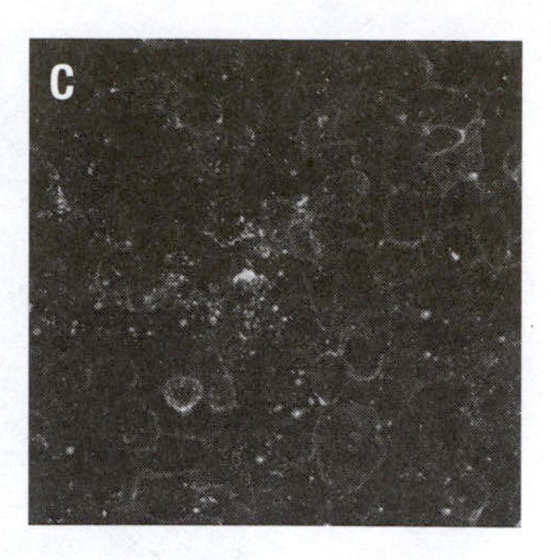

Figure 7.3 **Examples of engineering systems.** Almost anything can be defined as an engineering system and then analyzed. Some examples of engineering systems are buildings **(A)**, people or animals **(B)**, and cells **(C)**. The cells are outlined with a green dye, and they contain red drug-delivery particles. (See color plate.) Photo in panel B courtesy of Karen Haabestad.

system boundaries, which lead to alternate descriptions of the system under study. For example, if the system is the human body, the boundary might be defined as the skin and the physical openings between the body and the environment (such as the mouth, nostrils, and urethra). Alternately, the boundary could be the skin and the mucus epithelial surfaces that line the intestinal, respiratory, and reproductive tracts. In the first case, the contents of the intestines and lungs would be part of the system; in the second case, they would not be in the system, but instead part of the environment in which the system resides (Figure 7.4).

After the system is defined, it becomes possible to examine changes in the system caused by movement of mass, heat, or electrical charge across the system boundaries. Although balances of any conserved variable can be considered, this section considers the analysis of conservation of mass.

In a closed system, there is no movement of mass across the system boundaries during the period of observation. Open systems, in contrast, do allow for

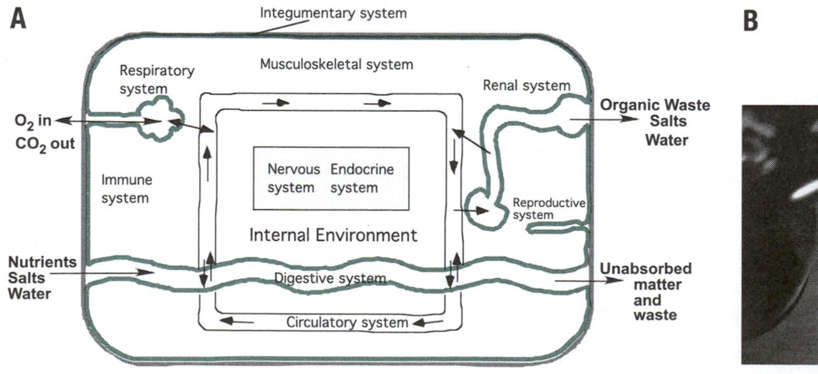

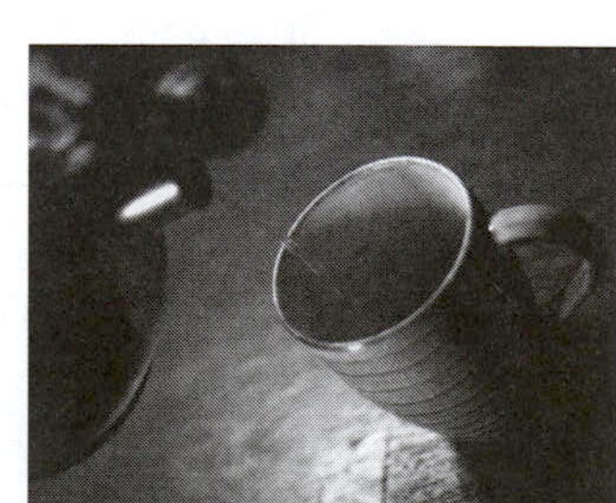

Figure 7.4 **Engineering systems. A.** Schematic of the human body system. Recall the discussion in Box 2.1 and notice again that there are two methods for defining the system under study: The colored boundary does not include the space inside of the intestinal, reproductive, and urinary systems; the grey boundary does include them. Notice also the arrows, which represent the movement of mass across system boundaries. **B.** The volume of water inside of a vessel can be analyzed by defining the volume of the vessel as a system. The vessel could be a mug, the stomach, or a sink; see Box 7.1 for details.

exchange of mass across system boundaries. A mass balance is a useful method for accounting, which allows the analysis of changes in mass within an open or closed system. A mass balance on an open system can be written:

$$ACC = IN - OUT + (GEN - CONS), \tag{7.1}$$

where *ACC* is the rate of accumulation of mass in the system, *IN* is the rate of flow of mass into the system, *OUT* is the rate of flow of mass out of the system, *GEN* is the rate of mass generation within the system, and *CONS* is the rate of mass consumption within the system (i.e., *GEN* – *CONS* is the overall net rate of mass production within the system boundaries).

7.2.1 Assumptions, predictions, and models: water balance in the body

Consider a simple example of this quantitative balance in which the system is the human body, the system boundary is the skin and external openings (i.e., mouth, nostrils), and the conserved quantity is water. Equation 7.1 is used to account for the balance of water over time in the system. The system and boundaries are now defined, the conserved quantity is identified, and now comes the important task of identifying **assumptions.**

Recall Box 1.2, which outlined an approach for solving engineering problems. One of the most important elements to identify as you solve a problem is the set of assumptions that provides the foundation for your analysis. For example, with certain assumptions, the mass balance of Equation 7.1 can be reduced to:

$$\frac{dV}{dt} = \dot{V} \tag{7.2}$$

What assumptions were used to generate this mathematical equation? Several were invoked: 1) there is no output of water, 2) there is no internal generation of water, and 3) there is no internal consumption of water. (These are incorrect assumptions, as experience tells you, but keep that to yourself for now.) One of the most important attributes of the assumptions is that they allow the conversion of a word problem (as in Equation 7.1) into a mathematical problem (as in Equation 7.2). After the problem is posed as a mathematical equation, it can be solved with the tools of mathematics, but the solution that you will obtain (even if it is mathematically correct) is only as good as the assumptions that were used to create it.

In Equation 7.2, the time derivative, dV/dt, indicates the rate of accumulation of water in the system, and $\dot{V}$ represents the volumetric flow rate of water into the system (liters/day). If the flow rate is constant with time, this equation has the solution:

$$V_N = V_{N-1} + \dot{V}\Delta t, \tag{7.3}$$

where V_N is the total volume in the system at the end of the time interval, Δt, V_{N-1} is the total volume at the beginning of the time interval, and Δt is the duration

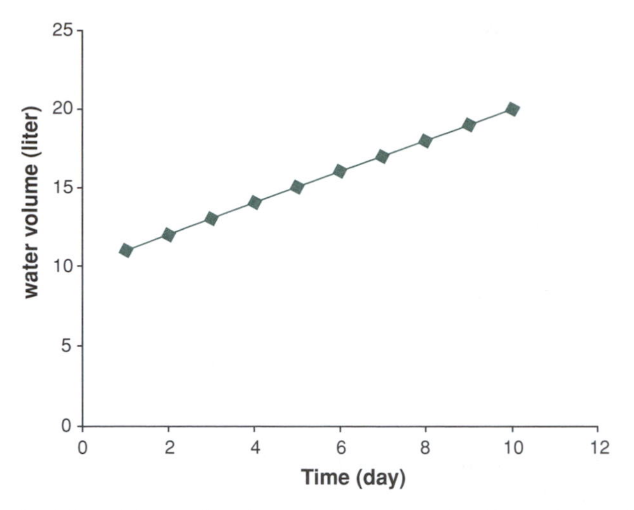

Figure 7.5 **Prediction of the quantity of water in the body as a function of time.** As described in the text, this prediction (from Equation 7.3) assumes that water volume is a function of time with constant input and no output.

of the time interval. This equation indicates that the total amount of water in the system increases linearly with time, as shown in Figure 7.5. In constructing this graph, the initial amount of water in the system was assumed to be 10 liters (i.e., this is the amount of water in the system, V_{N-1}, at the beginning of the observation time).

SELF-TEST 1 Water is flowing into a stoppered 16-L sink at a rate of 1.5 L/min. If the sink contains 5 L of water at 7:34 pm, use Equation 7.3 to calculate the time of overflow.

ANSWER: 7:41 pm

Figure 7.5 represents the result of a simple engineering analysis. The analysis is based on a mass balance; there are examples of analyses based on other methods throughout the rest of the book. Here, nothing is known about the internal operation of the system, but the model is built on assumptions. The assumptions lead to a prediction for how the system should behave: In this example, the volume of water is predicted to increase linearly with time. This **prediction** can now be compared to actual **observations** of the system behavior.

Do the predictions match the observations? If the answer is yes, then the model is a tool that can be used to predict future behavior. The model may also explain something about the internal workings of the system: A correct prediction suggests (but does not prove) that the assumptions of our analysis were true. This type of analysis, in which a description of a system leads to a mathematical equation predicting its behavior, is called a **mathematical model**.

Of course, as is evident from Figure 7.5, the predictions in this case do not match the observations. The human body does not increase in volume linearly as water is consumed throughout the day. The assumptions were incorrect, so the model predictions do not match the behavior. With regard to a model of the human body, many things are wrong with this analysis, including:

1. In the human body, water leaves the system continuously, by crossing system boundaries from the system to the surrounding environment. For example, water leaves the body in urine, sweat, tears, and feces.
2. Water is produced and consumed by chemical reactions with the body. Chapter 2 describes the variety of ways that water is involved in condensation and hydrolysis reactions, for example, in the formation of peptide bonds between amino acids.

How can these other factors be accounted for in applying Equation 7.1 to the problem of water balance in the body? Some of the approaches are illustrated in Box 7.1, which describes water balances by considering the similar (but simpler) analyses of water level in a sink with one input and one output stream. Box 7.1 yields a model that can be used effectively to predict the volume of water within a sink.

Let's extend the analysis by adding more information to the model. The kidney is an important organ for maintaining the normal balance of water and ions such as Na^+, Cl^-, and K^+. (The ways in which the kidney accomplishes this are discussed in Chapter 9.) How much water does the kidney need to excrete to achieve water balance? Water balance implies that the amount of water in the body remains constant with time: This also means that the rate of water accumulation is zero. This general concept will be useful in all analysis: When the value of a variable does not change with time, the rate of accumulation of that variable is zero.

With *ACC* equal to zero, Equation 7.1 becomes:

$$0 = IN - OUT + (GEN - CONS) \tag{7.4}$$

Assuming that water can leave from the kidneys and from other sources, so that *OUT* is equal to $OUT_{\text{kidney}} + OUT_{\text{other}}$:

$$OUT_{\text{kidney}} = IN - OUT_{\text{other}} + (GEN - CONS) \tag{7.5}$$

SELF-TEST 2 A subject is monitored for the following quantities: the amount of water ingested over a 24-hour period, the amount of water lost in urine over 24 hours, and the amount of water lost by sweating, breathing, and other processes. How much water is generated by internal chemical reactions inside this subject?

Our data for this experiment are, for the 24-hour period:

$IN = 2.1$ L of water consumed in food and drinks

$OUT_{\text{kidney}} = 1.4$ L of water lost in the urine (due to the action of the kidneys)

$OUT_{\text{other}} = 0.10$ L of water lost due in sweat; 0.35 L of water lost due to breathing; 0.45 L of water lost due to others

ANSWER: $GEN - CONS = 0.2$ L, therefore *GEN* is greater than *CONS* by 0.2 L per day

Box 7.1 Modeling the water level in a basin with one input and one output stream

Consider, as an example of an engineering system, a simple basin (like the one in your bathroom) that has a faucet, which can be opened to allow a variety of water flow rates into the sink, and a drain, which can allow a variety of water flow rates out of the sink by adjustment of the stopper.

If the basin is selected as the system under study, then there are only two ways for water to get into and out of the system: Water can enter the system through the input stream (the faucet) and leave the system through the output stream (the drain). Water is not produced or generated within the system. Equation 7.1 can be applied to this system to produce:

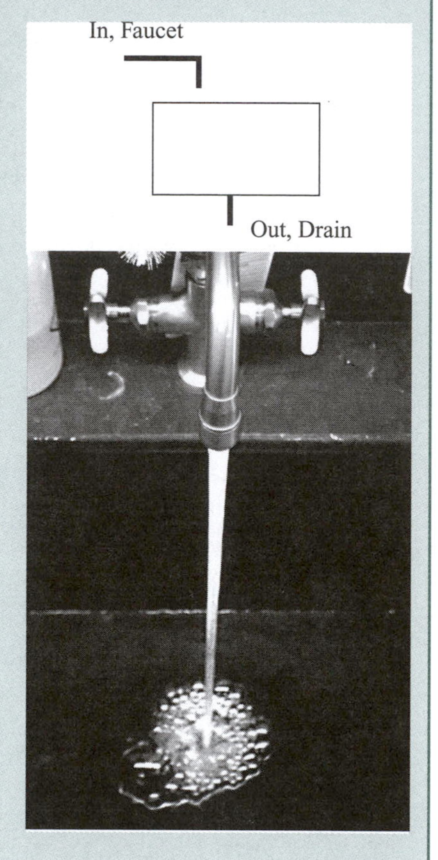

$$\frac{dM}{dt} = \dot{M}_{in} - \dot{M}_{out}, \qquad \text{(Equation 1)}$$

where M is the total mass of water within the system (in g), M_{in} is the flow rate of water into the system (in g/min), and M_{out} is the flow rate of water out of the system (in g/min). If the flow rates are constant with time, then the solution to this differential equation is:

$$M = \left(\dot{M}_{in} - \dot{M}_{out}\right) t + \mathrm{B}, \qquad \text{(Equation 2)}$$

where B is a constant. (Verify that Equation 2 satisfies the differential equation in Equation 1 by plugging 2 into 1.)

To complete the description of water level as a function of time in the basin, the value for B must be determined. If the value of M is known at any specific time t, then B can be calculated by direct substitution into Equation 2. For example, if there is 100 g of water in the sink initially, then $M = 100$ g at time $t = 0$, and therefore B must be equal to 100 g.

The amount of water in the basin would probably be measured in volume instead of mass. In the case of liquid water at a constant temperature, the density of water is constant, so Equation 2 can be converted from mass to volume by multiplying both sides of the equation by the constant density ρ:

$$\rho M = \left(\rho \dot{M}_{in} - \rho \dot{M}_{out}\right) t + \mathrm{B}\rho, \qquad \text{(Equation 3)}$$

which can be expressed in terms of volumes:

$$V = \left(\dot{V}_{in} - \dot{V}_{out}\right) t + B^*, \qquad \text{(Equation 4)}$$

where the total system volume and flow rates are now expressed in volumetric units (mL and mL/min) and B* is a new constant.

Notice that this equation can now be converted into Equation 7.3, in the case where the input flow rate is $\dot{V}$, the output flow rate is zero, and the volume in the system at time $\Delta t = 0$ is V_{N-1}.

7.2.2 Tracer balance in the body

Water is an abundant chemical in the body. Other chemicals or agents that might be of interest are present in much smaller quantities. Regardless of the amount of an agent inside the body, a mass balance can be used to examine it. (Of course, there are practical limits to the methods that can be used to actually *measure* the concentration of a rate agent in the body, but that can be ignored for now.)

Consider another example of a mass balance concerning the human body, which examines a **tracer** that is ingested in the food and passes out of the body in the urine. Furthermore, assume that this tracer is unchanged from its original form during passage through the body—that is, it is neither generated nor consumed by chemical reactions. If M is the total mass of the tracer in the body at any time, then Equation 7.1 can be written:

$$\frac{\mathrm{d}M}{\mathrm{d}t} = 0 - kM + (0 + 0) \tag{7.6}$$

In this equation, the accumulation of the tracer in the body is written in its mathematical form: The rate of accumulation of the tracer is equal to the first derivative of its abundance with time. In addition, a new parameter has been introduced: k is a **rate constant**. In this case, k is a first-order rate constant, which assumes that the disappearance of the tracer obeys first-order kinetics. In first-order kinetics, the rate of output of the tracer agent is proportional to the amount that is currently in the body (remember the definition of reaction rate constants in Box 2.3).

It is important to note that the term *IN* from Equation 7.1 is zero in this example. An additional assumption has been made that there is no introduction of the tracer into the body during the period of observation. Mathematically, the balance equation applies only for times ($t > 0$) after the food or tracer has been consumed. In other words, the material was consumed (or entered the system) at some instant in time $t = 0$, and Equation 7.6 examines the amount in the body at all times after this instant.

Equation 7.6 is an example of a first-order linear differential equation; equations of this type are encountered frequently in engineering analysis. The general solution to Equation 7.6 is:

$$M = Be^{-kt}, \tag{7.7}$$

where B is a constant (this constant arises during the integration step needed to solve Equation 7.6). If some fixed amount of tracer (let's call it D, or dose) was consumed at $t = 0$ and the total volume of the body is V, then the value of the constant B that satisfies the following "initial condition" can be found:

$$at\ t = 0; \quad M = M_{\text{dose}} \tag{7.8}$$

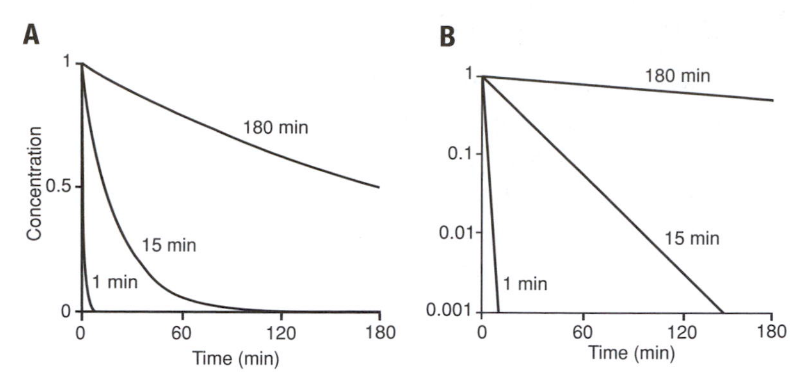

Figure 7.6 **Tracer concentration versus time.** Tracer concentration as a function of time, calculated from Equation 7.12, is shown for half-life values of 1, 15, and 180 min. **A.** Results of the equation on a linear-linear axis. **B.** Results on a semi-log axis (i.e., the *y*-axis has a logarithmic scale).

This value of A is substituted back into Equation 7.7 to yield the complete solution for this particular situation:

$$M = M_{\text{dose}} e^{-kt}. \tag{7.9}$$

SELF-TEST 3 A tea bag contains a dye molecule that seeps out of the bag, with first-order kinetics, after the bag is immersed in water. At some point after immersion in the water, the bag contains 5 units of dye; 10 minutes later, the bag contains 2.5 units of dye. At what time will the bag contain 1 unit of dye?

ANSWER: 23.3 min after the bag had 5 units

The half-life for elimination of a compound from the body can be calculated from Equation 7.9. If the half-life for elimination of a tracer is called $t_{1/2}$, the equation can be rewritten as:

$$\frac{M_{\text{dose}}}{2} = M_{\text{dose}} e^{-kt_{1/2}}, \tag{7.10}$$

which can be solved to yield:

$$t_{1/2} = \frac{\ln(2)}{k}. \tag{7.11}$$

Therefore, the half-life of a compound in the body is related to the rate constant for elimination. This result can be used to write the expression for mass of tracer in the body as a function of time:

$$M = M_{\text{dose}} e^{-\ln(2)\frac{t}{t_{1/2}}}. \tag{7.12}$$

The effect of tracer half-life on the kinetics of disappearance of an agent from the body is shown in Figure 7.6.

7.2.3 Homeostasis, steady state, and equilibrium

Homeostasis refers to the maintenance of the conditions necessary for life within the body's internal environment. In humans, a vast number of chemical concentrations (such as Na^+ concentration) and physical parameters (such as blood pressure and temperature) are maintained within certain ranges that are necessary for cell and organ function. Homeostasis can be observed at different levels of organization within the body. Certain parameters, such as total body water, are relevant for the whole body. Individual cells also have mechanisms to maintain their individual homeostasis. For example, sodium and potassium concentrations in the cytoplasm of neurons are tightly controlled to permit electrical excitability (see Box 6.2).

Negative feedback is one of the most important strategies that the body uses to maintain homeostasis. The principle of negative feedback was introduced in Box 6.3, with an example of temperature control. Each cell in the body, and every organ and tissue, are kept in a vital, functional state by the operation of myriad feedback loops. Although it is possible to study individual feedback mechanisms in isolation, all of these events are operating simultaneously in the body to maintain the body in a **steady-state** condition.

The concept of steady state was introduced in Box 2.3, with respect to diffusion across a membrane, but it applies broadly. A body is in steady state if the parameters that define it (i.e., its mass, volume, temperature, composition) do not change with time. As an example, consider a person's weight or mass. A person who consumes 2,000 calories per day will ingest about 470 gm (~1 lb) of protein, carbohydrate, fat, and fiber and 2.1 L (~4.7 lb) of water (Table 7.1), but a person who consumes this amount of water and food will not gain 5 pounds per day. People use energy, which consumes carbohydrates and fat, and they excrete water and wastes. Body weight stays relatively constant—in a steady-state condition—despite the fact that lots of material is entering and leaving the body each day. Dozens or perhaps hundreds of feedback loops maintain this steady state, with each loop controlling some aspect of metabolism, the state of hunger and thirst, or the urinary output. For most people, body weight remains stable and unchanging, as long as they remain conscious of the signals that the body provides to help them maintain this steady state.

Equilibrium is a different concept from steady state, but they are easily confused. **Equilibrium** represents a state of perfect balance. It can refer to a chemical state, in which all of the differences that would lead to movement of molecules, chemical transformation of molecules, or any other changes in the system are balanced. Molecules are still in motion within the system, but there are no changes in the composition or properties of the system with time. Equilibrium can refer to a physical state, in which a body is at rest because all of the forces acting on it are not changing and add up to zero. It can refer to an energy state, in which all of the components of a system are at the same temperature. If there is not

Table 7.1 Recommended composition of a 2,000-calorie diet

Total Fat	Less than	65 g	585 calories
Saturated Fat	Less than	20 g	
Cholesterol	Less than	300 mg	
Sodium	Less than	2,400 mg	
Total Carbohydrate		300 g	1,200 calories
Dietary Fiber		25 g	
Protein		50 g	200 calories
Vitamin A		5,000 IU	
Vitamin C		60 mg	
Calcium		1,000 mg	
Iron		18 mg	
Vitamin D		400 IU	
Vitamin E		30 IU	
Vitamin K		80 μg	
Thiamin		1.5 mg	
Riboflavin		1.7 mg	
Niacin		20 mg	
Vitamin B-6		2.0 mg	
Vitamin B-12		6 μg	
Folate		400 μg	
Biotin		300 μg	
Pantothenic acid		10 mg	
Phosphorus		1,000 mg	
Iodine		150 μg	
Magnesium		400 mg	
Zinc		15 mg	
Selenium		70 μg	
Copper		2.0 mg	
Manganese		2.0 mg	
Chromium		120 μg	
Molybdenum		75 μg	
Chloride		3,400 mg	
Potassium		3,500 mg	

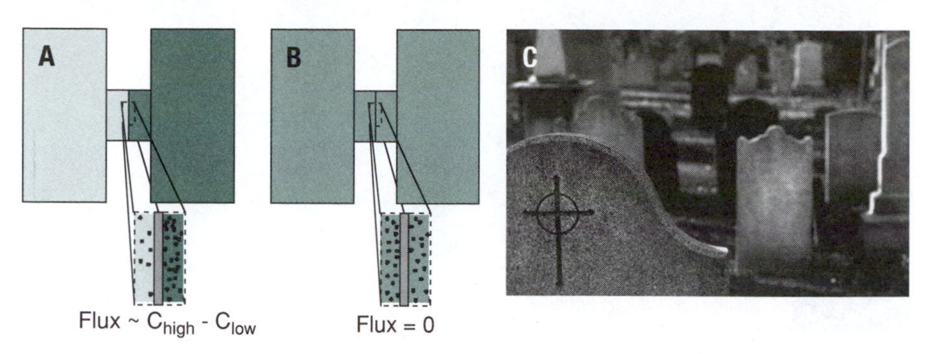

Figure 7.7

Equilibrium. A. A system at steady state, but not at equilibrium. The system is at steady state because the elements under study (the small grey and green boxes that are in contact through a membrane) remain at constant concentrations, because they are connected to large reservoirs of constant concentration. Even as molecules flow from the high concentration region (green) to the low concentration region (grey), the overall concentrations in the green and grey region do not change. Molecules move in and out of the large external reservoirs quickly (at least quickly relative to the rate of movement of molecules across the membrane), which keeps the concentrations in the small reservoirs relatively constant. **B.** The same system at equilibrium. **C.** Biological equilibrium at Grove Street Cemetery, New Haven, CT.

equilibrium (if something is out of balance), then the system will begin to change: Molecules will diffuse or accumulate because of reactions; bodies will move because of gravity; heat will flow from a hot region to a cold one. If a system is at equilibrium, it will stay the same forever—as long as no other forces act on or are introduced into the system.

Biological systems, if they are alive, are not at equilibrium. They ingest food, create energy, and use this energy to create chemical differences. Box 6.2 describes the Nernst potential in living cells, which is the electrical potential created by ion concentration differences across the cell membrane. These chemical differences are created by a continuous input of energy; the energy drives a motor that keeps the concentration of sodium and potassium from reaching equilibrium across the membrane. Equilibrium happens in graveyards, and even then it is only complete after a very long time.

A process can be at steady state, but still not at equilibrium, because steady state always implies that a system is not changing with time *over some interval of time*. One example of this is shown in Figure 7.7; steady state exists within the small green–grey boxes that are separated by a membrane. Although molecules are continuously flowing from the green to the grey box, the concentration in each of these boxes remains constant because the boxes are connected to large reservoirs that provide rapid replacement (or absorption) of molecules that are lost (or gained).

Another example of steady state involves the flow of heat, which naturally moves from regions of high temperature to regions of low temperature. It is logical to consider the steady-state rate of movement of heat into your house on a cold day: If the temperature outside is constant and the temperature inside is constant, the heat is moving from the warm house to the cold environment at a constant rate (Figure 7.8). This steady-state situation cannot be maintained forever, but

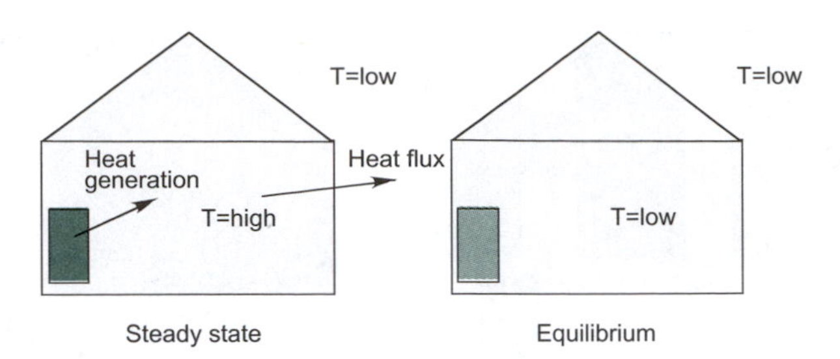

Figure 7.8 **Steady state versus equilibrium.** In the steady-state situation, the house is maintained at a constant temperature; the temperature in the house is higher than the surroundings because of internal heat generation in the house. If the rate of heat generation exactly matches the heat flux from the house (caused by the natural movement of heat down a temperature gradient: heat flux ~ temperature difference), a steady-state condition has been reached. If the furnace no longer supplies heat energy, the house will reach thermal equilibrium with the surroundings.

will eventually end: The temperature outside the house will change; the furnace heating the house will run out of fuel; or the heat flow from the house will occur for long enough to heat up the whole environment.

Almost every organ and tissue within the body contributes in some way to homeostasis at the level of the whole body. For example, the circulatory system (described in Chapter 8) provides oxygen to and removes carbon dioxide from all of the internal body tissues. The gastrointestinal system, described later in this chapter, provides nutrients. The kidneys and the liver are particularly important organs for maintenance of homeostasis. Both the liver and the kidneys are responsible for the elimination of wastes and end products of metabolism from the body fluids (this is discussed more fully in Chapter 9). In addition, the kidneys have an important role in the overall control of fluid volume. The kidneys accomplish this in two ways: 1) by regulation of chemical composition of the body fluids, particularly, concentration of sodium, and 2) by direct elimination of water from the body.

The remainder of this chapter uses the principles of mass balance to examine the physiology of respiration and digestion.

7.3 Respiratory physiology

In many senses, respiration is the basis of human life. The concept of respiration involves two related concepts (Figure 7.9). The first, which is called **internal respiration**, is the metabolic process by which energy is derived from organic materials in the diet. The process of internal respiration occurs in cells within the body. As internal respiration occurs, the cells consume oxygen and produce carbon dioxide as a by-product. Thus, internal respiration—or the normal operation of all the cells in the body—creates a need for the intake of large quantities of oxygen and the expulsion of large quantities of carbon dioxide from the body. The lung, in a process called **external respiration**, accomplishes this intake and expulsion of oxygen and carbon dioxide. External and internal

A

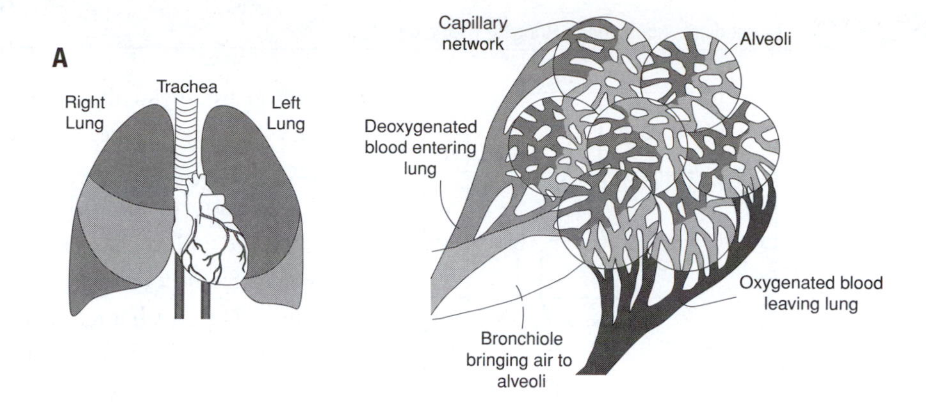

B

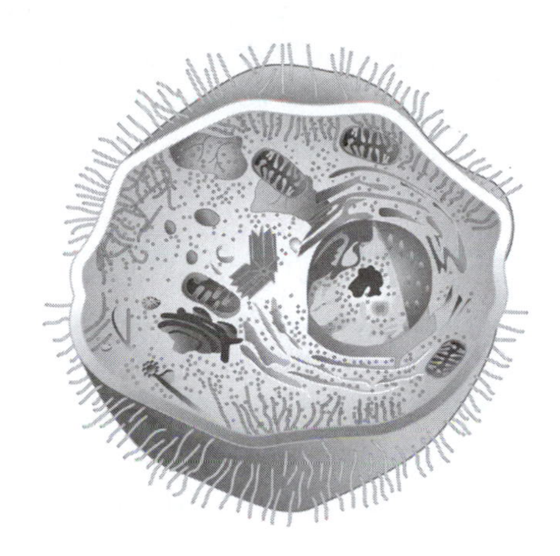

C

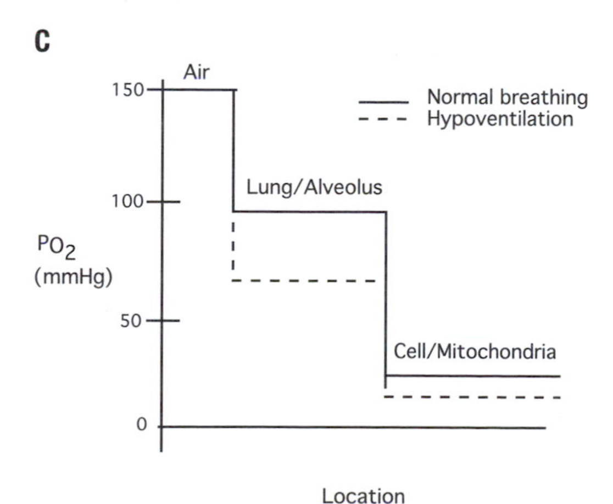

Figure 7.9 **Internal and external respiration are linked by the circulatory system. A.** External respiration brings oxygen into the lung, filling tiny sacs deep in the lung, called alveoli. Oxygen diffuses into capillaries, which surround the alveolus. **B.** The circulatory system carries the oxygen to all of the cells within the body. Mitochondria inside of cells use this oxygen to produce cellular energy. **C.** The concentration of oxygen in the external air is measured in units of partial pressure; it is usually ~150 mmHg. Concentration is lower within the alveoli because of the action of capillary blood flow, which removes oxygen. Oxygen concentration is lower in cells because of the chemical reactions that consume oxygen in generation of energy.

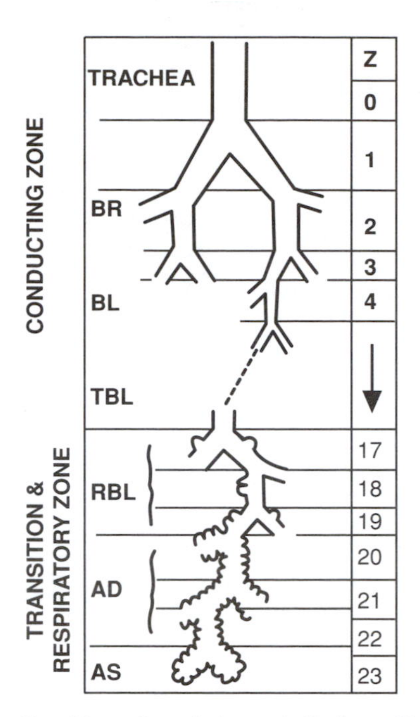

Figure 7.10 Branching pattern of airways in the lung. BR, bronchus; BL, bronchiole; TBL, terminal bronchiole; RBL, respiratory bronchiole; AD, alveolar duct; AS, alveolar sac or alveolus. Adapted from (1).

respiration are physically linked by the circulatory system, the chief function of which is the distribution of oxygen-rich blood from the lungs to all of the cells in the body and the collection of carbon dioxide–rich blood from cells for gas exchange in the lungs. The mechanics of external respiration are reviewed in this section; internal respiration is described briefly in Chapter 2 and is expanded further later in this chapter.

External respiration is accomplished by an elegant engineering system, the human lung. The exchange of oxygen and carbon dioxide occur by the diffusion (recall Box 2.3) across a complex membrane that separates flowing blood from the gases within the **alveolus**, one of 300 million small saclike units that make up the lung. Alveoli are connected to the ends of the last branches of the bronchial network (Figure 7.10), an elaborate treelike structure of ever finer branches that grows from each end of the right and left main **bronchi**, the first two branches of the **trachea**. A dense network of capillaries (Figure 7.9) covers each alveolus; the capillaries carry deoxygenated blood to the alveolus and oxygenated blood away.

Air enters the lung through the trachea, a large-diameter vessel that branches into the right and left main bronchi, which each further branches into bronchi of smaller diameter. The branching pattern continues for many generations, and ends in terminal bronchioles that each connects to an alveolus (Figure 7.10). Each alveolus is a small, gas-filled sac with a diameter ~300 μm. The first 16 branches of the airways are called the conducting airway; its main function is to serve as a passive conduit for bringing air from the atmosphere to the deeper parts of the lung. The conducting airway volume is ~150 mL, although there is substantial variation among individuals. Most of the lung volume is within the alveoli, which contain ~3000 mL of gas.

Lung volumes in individuals can be measured with a device called a **spirometer**. The essential design elements of a spirometer are simple: a bucket filled with water that contains a second inverted bucket. A conducting tube allows the subject to inhale and exhale into the internal bucket; the changes in volume with time during breathing can be measured by monitoring the change in volume of air in the inverted bucket (Figure 7.11).

A number of lung volumes and capacities can be measured from the spirometer. By asking the subject to breathe normally, and also to make maximum

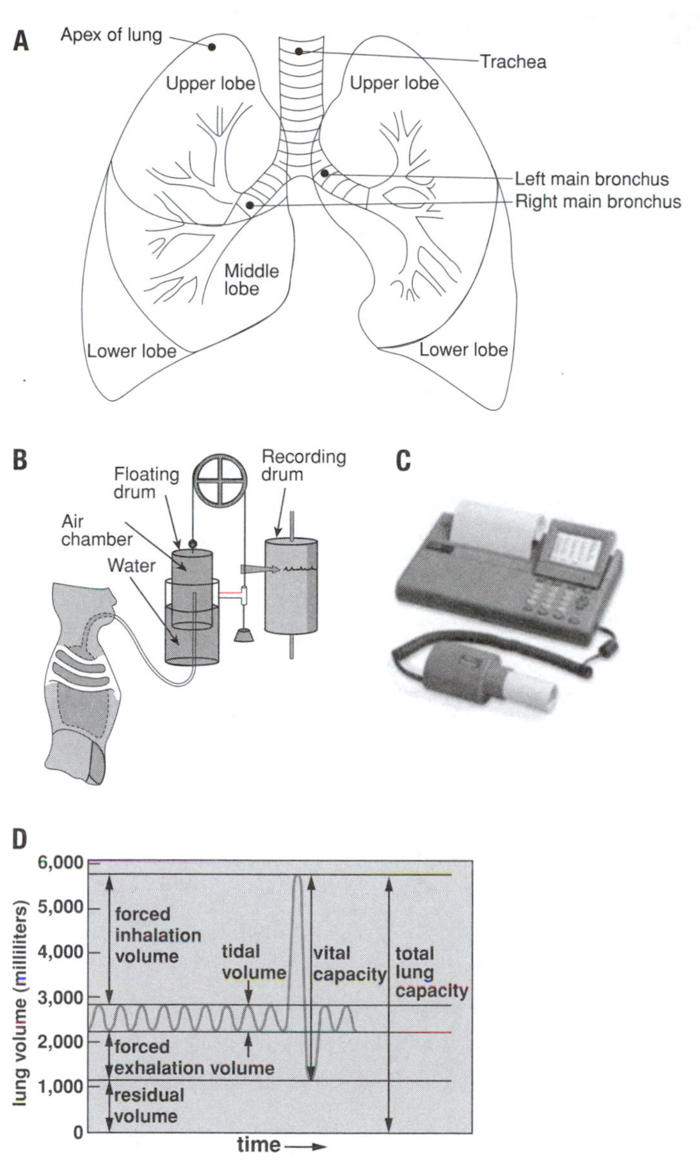

Figure 7.11 **Branching of the first several generations of bronchi.** **A.** The arrangement of bronchi with the lungs. **B.** Schematic diagram showing the operation of a spirometer. **C.** An example of a modern spirometer. **D.** Typical lung volume tracing from a spirometer, illustrating the calculation of lung volumes.

inspiration and expiration efforts, the spirometer can measure **tidal volume**, which is the volume change of the lung during normal breathing, forced inhalation volume, and forced exhalation volume. Other parameters, called **capacities**, can be defined based on sums of these various volumes. For example, **functional residual capacity** is the amount of air that is left in the lung after expiration during normal breathing. Residual volume cannot normally be measured in a spirometer; it is the amount of air that is left in lung after maximum exhalation, ~2 L in most subjects.

SELF-TEST 4 If there are 300 million alveoli in the lungs, and each alveolus is 300 μm in diameter, what is the total surface area of the oxygen exchange surface in the lungs?

ANSWER: Surface area $\sim$80 m^2; volume $\sim$4 L

The exchange of air between the alveolus and the atmosphere is called **ventilation**. The overall process occurs in two phases: Air is moved into the lungs from the atmosphere during **inspiration** and out of the lungs during **expiration**. The lung is bounded by the diaphragm, a muscular layer that separates the thorax from the abdomen on the bottom and from the muscular chest wall on the sides. Inspiration and expiration are aided by the actions of sets of muscles that increase and decrease the volume of the thorax or chest. For the present discussion, assume that flow rate ($\dot{V}$, which is defined as a flow of volume in this example, see Box 7.1) is linearly related to pressure drop, ΔP:

$$\Delta = \dot{V} R, \tag{7.13}$$

where R is the **resistance** to flow and the pressure drop ΔP is measured from the outside air to the inside of the alveoli.[1] The flow rate, $\dot{V}$, is positive when it occurs in the direction of decreasing pressure.

This mathematical relationship between pressure and flow (Equation 7.13) is analogous to the relationship between voltage and current flow: $\Delta E = IR$, where ΔE is a voltage difference, I is the current flow, and R represents the electrical resistance. In the simple electrical circuit, the voltage drop is the driving force that leads to current flow. The amount of current that flows depends on the resistance of the circuit. Analogously, the pressure drop is the driving force for fluid flow, and the rate of fluid depends on the resistance of the vessel that is containing the flow. Chapter 8 includes a more detailed discussion of the relationship between pressure drop and fluid flow in vessels; for now, it is enough to assume that Equation 7.13 is a reasonable description of the pressure and flow in the respiratory system.

SELF-TEST 5 Using a clock to time your own breathing, and using the tidal volume shown in Figure 7.9, estimate the flow rate, $\dot{V}$, into your lungs during a normal inspiration.

ANSWER: $\dot{V} = 12$ L/min (approximately)

In the case of inhalation and expiration, ΔP is the pressure difference between the atmospheric air, P_{atm}, and the gas in the alveolus, P_{alv}:

$$\dot{V} = (P_{\text{atm}} - P_{\text{alv}})/R. \tag{7.14}$$

[1] Chapter 8 provides more detail on the relationship between pressure and flow.

In the period between breaths, when there is no flow of air, P_{alv} is equal to P_{atm}. Note that the pressures in Equation 7.14 are measured relative to P_{atm}; this convention will be used for pressures in the discussion of circulation in Chapter 8, as well. Therefore, P_{atm} is equal to zero. From Equation 7.14, P_{alv} is less than zero during inspiration and greater than zero during expiration. Ventilation is driven by changes in P_{alv}.

SELF-TEST 6 Using the flow rate just determined, and assuming that the pressure drop during inspiration is 5 mmHg (or 7 cm H_2O, when converted into the unit of pressure that respiratory physiologists often use), what is the resistance to inspiration?

ANSWER: R = 0.4 mmHg-min/L (or 0.6 cm H_2O-min/L)

The functional characteristics of the lung—and, therefore, the mechanics of ventilation—are related to its static and dynamic mechanical properties. The lung itself is an elastic object: It is constantly pulling inward, and would collapse if not held open by the chest wall. The chest wall is also elastic, but constantly pulling outward because of the presence of muscles, which are always under a slight tension. Chapter 10 discusses some additional mechanical aspects of lung function.

7.3.1 Ventilation rates

When a tidal volume is inspired, not all of the inhaled air enters the alveoli. Because air movement through the lung involves reciprocal inspiration and expiration events, the first packet of gas that is exhaled on each breath is the same as the last packet of gas that was inhaled (Figure 7.12). This last packet never reaches the alveoli, and therefore is not useful for gas exchange: Because it does not enter the alveoli, it does not participate in gas exchange (which is discussed in the next section). In fact, the entire system of conducting airways (from the trachea to branching level 16 in Figure 7.10) does not participate in gas exchange. Because it is not a useful source of oxygen, the volume of gas within these airways, which is approximately 150 mL, is called the **dead space** or dead volume. With each breath, the lung is working to fill and empty this dead volume.

It is useful to account for the fraction of each inspiration that goes to fill the dead volume and the fraction that goes into alveoli. The total ventilation rate, $\dot{V}_T$, is equal to the number of breaths per minute multiplied by the volume inspired per breath. Assuming a tidal volume of 500 mL and breathing rate of 15 breaths/min, total ventilation, $\dot{V}_T$, is 7.5 L/min. If the tidal volume includes 150 mL of dead volume and therefore 350 mL of alveolar volume, alveolar ventilation, $\dot{V}_A$, is 5.3 L/min.

If a person breathes faster, he or she will increase the delivery of oxygen to the alveoli. The relationship between alveolar ventilation and concentration of

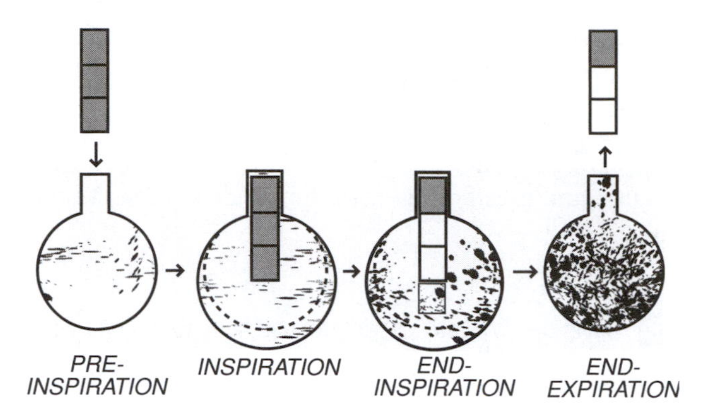

Figure 7.12 **Anatomical dead space.** The volume of the respiratory airflow pathway that is not involved in gas exchange is called the anatomical dead space. The gas within this volume is not exhaled, and it is the first gas to enter the alveoli on inspiration.

oxygen in the alveoli can be determined by a mass balance. If our system is the entire population of alveoli, a mass balance of oxygen can be derived, by starting with Equation 7.1 and adding appropriate substitutions for *ACC*, *IN*, *OUT*, *CONS*, and *GEN*:

$$0 = \dot{V}_A \left(\frac{P_{O_2}}{P_T}\right)_{\text{atm}} - \dot{V}_A \left(\frac{P_{O_2}}{P_T}\right)_{\text{alv}} - \dot{V}_{O_2} + (0 - 0)\,. \tag{7.15}$$

Assuming steady-state conditions, *ACC* is zero. Similarly, it is assumed that there is no generation or consumption of oxygen in the alveolar space, so *GEN* and *CONS* are also zero. Equation 7.15 contains three nonzero terms: one *IN* and two *OUT*s. Oxygen comes into the alveolar volume during inspiration; the volumetric flow rate of oxygen *IN* is equal to the alveolar ventilation rate times the fraction of volume that is oxygen. The volume fraction of oxygen in the inhaled atmospheric gas is equal to the partial pressure of oxygen divided by the total pressure (a more complete description of the concept of partial pressure is in the next section). Similarly, oxygen leaves the alveolus during expiration with a volumetric flow rate equal to the ventilation rate times the fraction of oxygen in alveolar gas. The partial pressure of oxygen in the alveolar gas will be less than the partial pressure in the inhaled gas because oxygen diffuses into the blood and is carried away from the alveolus to be consumed by cells. This is the other *OUT*: Oxygen is lost from the alveolus to the blood, and the rate of loss is equal to the rate of consumption of oxygen by the body, which is designated $\dot{V}_{O_2}$.

Equation 7.15 can be solved to determine alveolar partial pressure of oxygen as a function of the other variables:

$$(P_{O_2})_{\text{alv}} = (P_{O_2})_{\text{atm}} - \left(\frac{P_T}{\dot{V}_A}\right)_{\text{alv}} \dot{V}_{O_2}. \tag{7.16}$$

In Figure 7.13, alveolar partial pressure is plotted as a function of alveolar ventilation rate for two different values of oxygen consumption rate, $\dot{V}_{O_2}$: 250 and 1000 mL/min. The lower value represents a normal oxygen consumption

Table 7.2 Composition of air, alveolar gas, arterial blood, and venous blood (in mmHg)

	Venous blood	Arterial blood	Alveolar gas	Dry air	Atmosphere	Expired air
P_{O_2}	40	100	100	160	159	120
P_{CO_2}	46	40	40	0.3	0.3	27
P_{N_2}	0	0	559	600	597	566
P_{H_2O}	—	—	47	0	3.7	47

rate whereas the higher represents a rate that might be achieved during exercise. The partial pressure of oxygen in the atmosphere is taken from Table 7.2.

Imagine that a person is sitting in a chair, with a $\dot{V}_{O_2}$ of 250 mL/min and a $\dot{V}_A$ of 5 L/min. Figure 7.13 suggests that her alveolar oxygen partial pressure is ~130 mmHg. Suppose that she suddenly stands up and starts to run, raising her $\dot{V}_{O_2}$ to 100 mL/min, and she tries to accomplish this without increasing her ventilation rate, $\dot{V}_A$. What will happen? The partial pressure of oxygen in the alveoli will drop precipitously, falling to almost zero. This situation is not sustainable, as this low partial pressure of oxygen in the alveolus will limit the amount of oxygen that can diffuse into the blood. She will need to increase her ventilation rate to sustain this higher level of oxygen consumption.

SELF-TEST 7 Table 7.2 provides some typical values for oxygen, carbon dioxide, nitrogen, and water partial pressures in gases. Why isn't the partial pressure for oxygen in the alveolar gas the same as the partial pressure in the expired air?

ANSWER: It isn't the same because of the dead volume, in which the partial pressure of oxygen will be equal to that in the atmosphere.

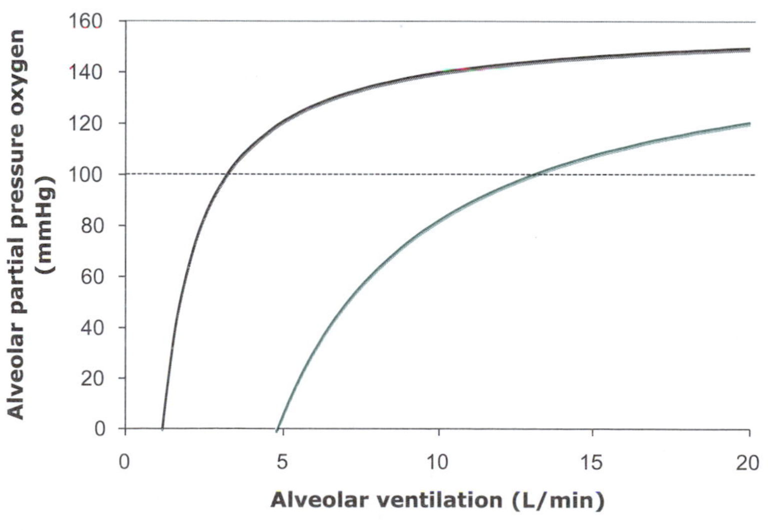

Figure 7.13 **Alveolar oxygen partial pressure as a function of alveolar ventilation.** Both lines were determined using Equation 7.16. The black line represents an oxygen consumption rate of 250 mL/min, and the colored line represents an oxygen consumption rate of 1,000 mL/min. The upper horizontal line indicates maximum alveolar oxygen concentration (160 mmHg), and the lower dashed horizontal line indicates the normal value (100 mmHg).

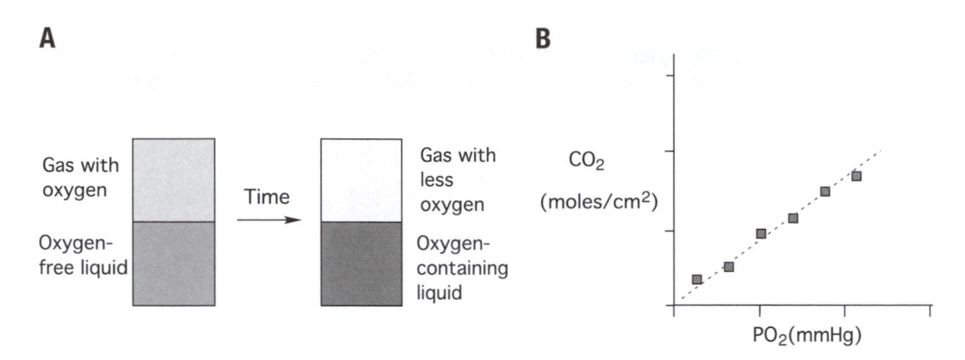

Figure 7.14

Determining the equilibrium relationship between concentration of oxygen in a gas and in a liquid. A. Imagine a simple experiment in which an oxygen-free liquid was placed in a closed vessel with a gas containing oxygen. Over time, oxygen would dissolve in the liquid, decreasing the concentration of oxygen in the gas and increasing the concentration of oxygen in the liquid. After a long time, the concentrations in each phase would remain constant, indicating equilibrium. **B.** The equilibrium relationship for oxygen in a liquid, such as water.

7.3.2 Oxygen carriage in the blood

As described in the previous sections, oxygen and carbon dioxide are components of air, which is moved into and out of the alveolar volume of the lungs by rhythmic breathing. The levels of oxygen and carbon dioxide in the alveolar gases can be altered by changing the rate of breathing, as shown for oxygen in Figure 7.13. These two gases are also present as dissolved species in blood. In fact, the primary function of the lungs is to get oxygen into—and remove carbon dioxide from—the blood. Oxygen is necessary for energy generation in the mitochondria of cells in the body; carbon dioxide is produced as a waste product of metabolism in tissues. The lungs bring oxygen into the body during inspiration and expel carbon dioxide during expiration.

Before exploring the relationship between breathing and levels of oxygen and carbon dioxide in the blood, methods for relating gas concentration in the air and gas concentration in blood are needed. It is common to describe the concentration of oxygen in a gas in terms of **partial pressure**: The partial pressure of oxygen, P_{O_2}, is the fraction of the pressure in gas that is caused by molecules of oxygen (partial pressure is defined mathematically in Box 7.2). Oxygen partial pressure was used in the previous section to calculate the amount of oxygen in the alveolar gases. In contrast, it is common to describe concentration of oxygen in a liquid in terms of a molar concentration (moles oxygen/volume of liquid). How can you relate the composition of a gas and the concentration of gas molecules in a liquid?

It might be helpful to consider how one would measure this relationship by experiments: Assume that 1 L of oxygen-free fluid is sealed in a closed vessel with 1 L of gas containing oxygen (Figure 7.14a). Oxygen will dissolve in the liquid, causing the number of oxygen molecules in the gas to decrease and the number in the liquid to increase. After sufficient time, the system will come to

Box 7.2 Oxygen carrying capacity of water and blood

Oxygen is required for human life, and humans are predominantly water. So it is surprising to learn that oxygen is sparingly soluble in water. It is common to describe concentrations in the gas phase relative to conditions called standard temperature and pressure (STP), which is 0°C and 1.013×10^5 Pa (1 atm). For pure water, oxygen solubility can be expressed as:

$$C_{O_2} = HP_{O_2}, \qquad \text{(Equation 1)}$$

where C_{O_2} is concentration of oxygen dissolved in water and P_{O_2} is the partial pressure of oxygen in the gas. At STP, the coefficient H is equal to 1.005×10^{-11} mole/cm^3-Pa. From this equation, you should be able to demonstrate that water, exposed to pure oxygen gas at STP, will contain 1.02×10^{-6} moles of oxygen/cm^3 of water.

This concentration in water can also be expressed as the volume of oxygen gas that is carried by the water. Because 1 mole of gas at STP occupies 22,400 cm^3 of volume, C_{O_2} of oxygen-saturated water is 2.28×10^{-2} cm^3 O_2/cm^3.

Because the concentration of oxygen in water and the partial pressure of oxygen in a gas phase are related by Equation 1, you can describe an oxygen concentration in solution by referring only to the equivalent partial pressure. Physicians and physiologists often do this, and they make matters more confusing by referring to the partial pressure of oxygen in the gas phase as **oxygen tension**.

Oxygen comprises 21% of our atmosphere. Partial pressure is defined as:

$$P_{O_2} = y_{O_2}P, \qquad \text{(Equation 2)}$$

where y_{O_2} is the mole fraction of oxygen in the air and P is the total pressure. Therefore, the partial pressure of oxygen in a dry atmosphere (with 0% humidity), which has a total pressure of 760 mmHg, is 160 mmHg. The partial pressure of oxygen in the lung gases is lower because lung gas is saturated with water: Here, the oxygen partial pressure is 100 mmHg. Thus, the concentration of oxygen in plasma (which can be approximated as water in equilibrium with the lung gases) is equal to the partial pressure of oxygen in the lung times H, or 1.34×10^{-7} moles of oxygen/cm^3 or 3×10^{-3} cm^3 O_2/cm^3.

The normal cardiac output is 5 L/min. If dissolution of oxygen in water was the only method for blood to carry oxygen, the maximum amount of oxygen delivery would be (3×10^{-3} cm^3 O_2/cm^3) × (5,000 cm^3/min) = 15 cm^3 O_2/min. This is far below the body's demand for oxygen, which is 250 cm^3/min.

The oxygen carrying capacity of blood is augmented by the presence of hemoglobin, a protein that is highly concentrated in the cytoplasm of red blood cells. Each hemoglobin molecule contains four binding sites for oxygen; therefore, hemoglobin can pick up oxygen from the saturated water that surrounds it, and give blood a much higher capacity for oxygen carriage than water. As shown in Figure 7.15, blood can carry up to 0.2 cm^3 O_2/cm^3, or almost 70 times as much oxygen as water at a partial pressure of 100 mmHg. For a normal cardiac output, saturated blood can carry (0.2 cm^3 O_2/cm^3) × (5,000 cm^3/min) = 1,000 cm^3 O_2/min, much more than needed to meet the oxygen demand.

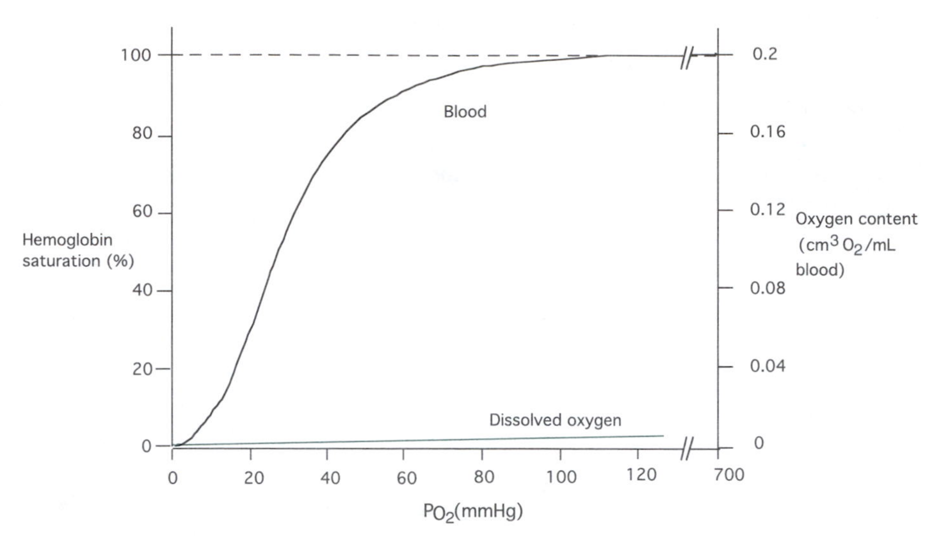

Figure 7.15 **Equilibrium curve for oxygen in blood (black line) and water (colored line).** The curve for red blood cell (RBC)-free blood (also called plasma) would almost be identical to the line for water. The horizontal dashed line indicates 100% hemoglobin saturation.

equilibrium: The number of molecules of oxygen in the gas phase and in the liquid phase will stay constant after that time. Imagine repeating this experiment many times, each time using a different concentration of oxygen in the initial gas phase. Each time, when equilibrium is reached, the concentration of oxygen in each phase is determined. This experimental data can then be plotted, as in Figure 7.14b. Under some circumstances, the relationship is linear, so that—over at least some range of concentrations—the relationship between concentration in the liquid and partial pressure in the gas can be described by a simple expression (Equation 1 in Box 7.2).

SELF-TEST 8 Concentrations can be expressed in units of (moles substance/volume liquid). When the dissolved substance is a gas, the concentration can also be expressed as (equivalent volume of ideal gas at standard temperature and pressure/volume liquid). Assume that a gas is dissolved in water at a concentration of 1 mole/L of liquid. What is the volume of 1 mole of an ideal gas at standard temperature and pressure (0°C and 1 atm)?

ANSWER: 22.4 L; therefore, the concentration is 22.4 L gas/L.

Blood is a much better oxygen carrier than water is because blood contains large amounts of hemoglobin and each hemoglobin molecule can bind and carry four oxygen molecules. Over the range of oxygen partial pressures that are relevant in the body (0–150 mmHg, see Figure 7.9c), the relationship between blood concentration and partial pressure is not linear but sigmoidal, or S-shaped (Figure 7.15). Because binding to hemoglobin is the main source of oxygen carriage, there is a limit to the number of oxygen molecules that can be carried in blood: When the hemoglobin binding sites are filled, the blood becomes **saturated** with oxygen. Arterial blood, which is loaded with oxygen from the alveolar gas at a partial pressure of ∼150 mmHg, is nearly saturated with oxygen: It is ∼97.5% saturated (20 mL/dL or 0.2 mL/mL). With this carrying capacity, the arterial blood delivers a potential of 1,000 mL O_2/min. Box 7.2 describes the oxygen-carrying capacity of blood in more detail.

SELF-TEST 9 If venous blood returned to the heart has an oxygen tension of 20 mmHg, what is the hemoglobin saturation?

ANSWER: ∼20% hemoglobin saturation, from Figure 7.15.

7.3.3 Carbon dioxide carriage and acid–base balance in the blood

Cells throughout the body form carbon dioxide as a by-product of metabolism. Overall, approximately 200 mL/min of carbon dioxide is formed in tissues. This carbon dioxide is collected by blood in the capillaries, and transported back to the lung for removal.

Carbon dioxide is more soluble than oxygen in water, yet only ∼10% of the carbon dioxide load can be carried in this way. Another 30% is carried as complexes of carbon dioxide with proteins, including hemoglobin. The rest of the carbon dioxide—about 60%—is converted into bicarbonate:

$$CO_2 + H_2O \rightleftarrows H_2CO_3 \rightleftarrows H^+ + HCO_3{}^- \tag{7.17}$$

The first part of this reaction should be familiar; the reaction of carbon dioxide and water to produce carbonic acid was introduced in Chapter 2 (see Box 2.2). This first reaction occurs slowly in aqueous solutions, but very rapidly in the presence of the enzyme **carbonic anhydrase**, which is present in the interior of red blood cells (RBCs). (The action of carbonic anhydrase was described in Chapter 4.) As carbon dioxide accumulates in the blood, it diffuses into RBCs, where it is converted into carbonic acid, and therefore bicarbonate. The bicarbonate is then transported back out of the RBCs via a transport protein exchanger with chloride, which serves as the predominant form of carbon dioxide in the blood. These reactions are reversed as the blood enters the lung capillaries. Carbon dioxide diffuses into the alveolar gas, driving the reactions shown in Equation 7.17 to the left: That is, when carbon dioxide concentration decreases, bicarbonate is converted back into carbon dioxide, which can diffuse into the alveolus for elimination.

Removal of carbon dioxide in the lungs is important for another reason: The level of carbon dioxide in tissues is intimately related to the local pH. Before discussing this in detail, recall the discussion in Chapter 2 on the basic concept of acid–base equilibrium. In Box 2.4, the Henderson–Hasselbalch equation was derived; this equation relates the pH of a solution in the presence of a weak acid and its complementary base:

$$\mathrm{pH} = \mathrm{pH}_a + \log\frac{[A^-]}{[HA]}, \tag{7.18}$$

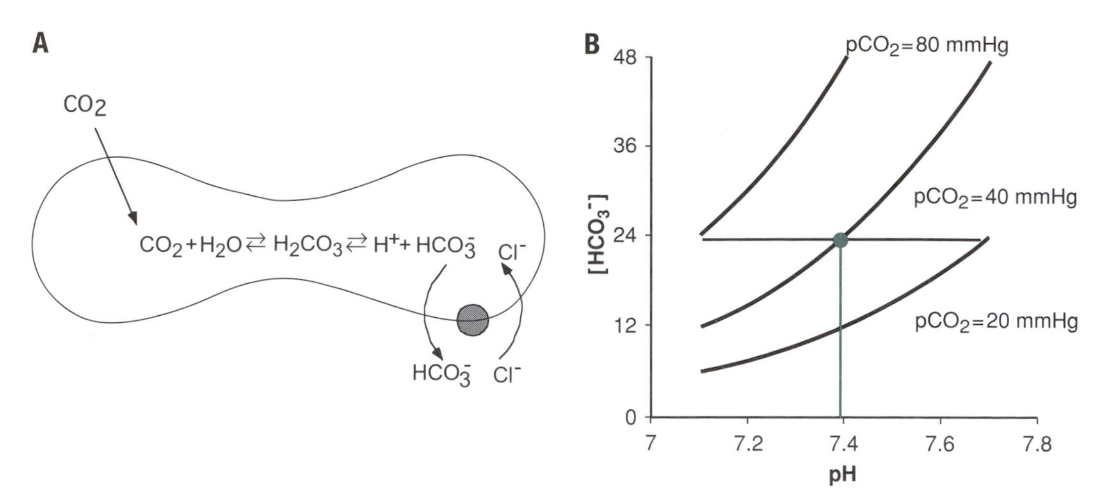

Figure 7.16 **Carbon dioxide in the blood. A.** Carbon dioxide is converted into bicarbonate for transport in the blood. Red blood cells (RBCs) contain carbonic anhydrase, which speeds up the conversion of CO_2 to carbonic acid within the cell. Bicarbonate is transported out of the cell, and into the plasma, by the bicarbonate–chloride exchange protein. Other fates of CO_2, including reactions with hemoglobin, are not shown. **B.** Effect of PCO_2 on pH in blood. Each line shows Equation 7.11 plotted for a fixed value of pCO_2. The colored dot indicates the normal value of the three variables, and the black horizontal line shows the effect of increased or decreased breathing on blood pH (if there were no other buffer systems present in the blood).

where K_a is the dissociation constant for the acid *HA, pK_a* is equal to –log(K_a), [*HA*] is the concentration of the acid, and [A^-] is the concentration of the complementary base (i.e., the acid that has lost its H^+ ion, or proton).

Box 2.2 also introduced the role of carbon dioxide in one of the most important buffering systems in the human body. A **buffer** is a chemical that, when present in a solution, absorbs hydrogen ions, H^+, and therefore resists changes in pH when hydrogen ions are added to the solution (it resists or "buffers" against changes in pH). The bicarbonate ion (HCO_3^-) is present in relatively high concentrations in the blood, where it acts to resist changes in pH of the blood as acidic compounds are created in the body by the processes of metabolism.

For the particular case of bicarbonate, the Henderson–Hasselbalch equation can be written (see Chapter 2):

$$\mathrm{pH} = 6.1 + \log \frac{[\mathrm{HCO_3^-}]}{0.03 P_{\mathrm{CO_2}}}. \tag{7.19}$$

There are three parameters in this equation: pH, [HCO_3^-], and P_{CO_2}. Under normal breathing conditions, the value of P_{CO_2} is 40 mmHg. Figure 7.16 shows the Henderson–Hasselbalch relationship for three different values of P_{CO_2}. At normal values of P_{CO_2} and pH (pH 7.4), the bicarbonate concentration is 24 mmoles/L. If the P_{CO_2} increases to 80 mmHg, by slower breathing, the pH would drop; if the P_{CO_2} decreases to 20 mmHg, by rapid breathing, the pH would increase.

Bicarbonate is not the only buffer present in the blood; many proteins, including hemoglobin, also serve as buffers. Therefore, the change in pH with breathing is not as dramatic as shown in Figure 7.16. The buffering power of a solution is defined as the amount of hydroxide ion that must be added to a solution to change the pH. Quantitatively, it is defined as:

$$\beta = \frac{\Delta[\mathrm{OH}^-]_{add}}{\Delta\,(\mathrm{pH})}. \tag{7.20}$$

For the nonbicarbonate buffering systems in blood, the buffering power is ~25 mM per pH unit.

SELF-TEST 10 What is the total amount of CO_2 generated in the body if the venous blood contains 52 mL CO_2/dL and arterial blood contains 48 mL CO_2/dL?

ANSWER: 200 mL/min

How does the blood carry CO_2 generated throughout the body? CO_2 crosses cell membranes readily and enters the blood. Some of this CO_2 combines with water to form bicarbonate: This bicarbonate accounts for about 11% of the CO_2 in the interstitial fluid. The remainder, or 89% of the CO_2, goes into RBCs. Carbon dioxide in the RBCs has three possible fates: Some remains dissolved (4%), some combines with other molecules (such as hemoglobin, 21%), and some combines with water to form carbonic acid and then bicarbonate (64%).

7.3.4 Diffusion

Previous sections in this chapter illustrated some of the concepts underlying the lung's most critical functions: to bring oxygen into the body and expel unwanted carbon dioxide. Ventilation brings atmospheric gas into the alveoli, where it mixes with gas already present and provides a source for oxygen to diffuse into the blood; the low concentration of carbon dioxide in atmospheric gas provides a sink for carbon dioxide to diffuse out. The previous two sections considered the relationship between gas partial pressures and blood concentrations, but how do these gases get from the alveolar gas into the blood? Figure 7.17 illustrates the intimate relationship of lung capillaries and alveoli: Each alveolus is surrounded by a dense network of lung capillaries. What are the barriers to diffusion in this arrangement? How fast can oxygen and carbon dioxide move across these barriers? How does gas diffusion change in lungs that are diseased? These questions can be addressed by engineering analysis.

The transport of oxygen and carbon dioxide from the alveolar gas into the blood (or vice versa) can be understood by developing a mathematical model. To develop a model, it is important to first understand the geometry. Each alveolus

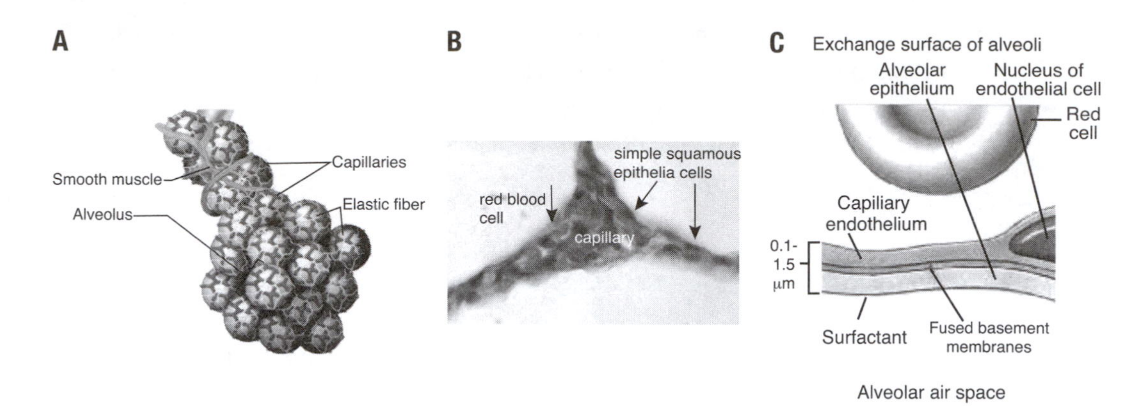

Figure 7.17 **Diffusion across the respiratory membrane. A.** Each alveolus is surrounded by a dense array of capillaries, so that a continuous sheet of blood is flowing over the surface of the alveolar membrane, as shown in **(B)**. This arrangement puts the alveolar gas in close proximity to flowing blood, as shown in **(C)**. Center photo credit: Rose M. Chute. (See color plate.)

is an air-filled sac that is covered by a network of capillaries (Figure 7.17a). The alveolar sac is formed from a continuous monolayer of alveolar epithelial cells. The capillaries that cover this sac are composed of a continuous monolayer of capillary endothelial cells. These capillary tubes are filled with RBCs, the ultimate site for oxygen carriage. Histological thin sections cut from the lung provide a clear view of the intimate relationship between the capillaries and the alveolar air spaces (Figure 7.17b). A simple geometric model that has emerged from these observations is illustrated in Figure 7.17c. In this model, the combined cell barriers and interstitial space are viewed as one continuous, monolithic barrier of thickness 0.1–1.5 μm, which is called the **respiratory membrane**.

Box 7.3 describes the mathematical model for diffusion across the respiratory membrane. The net rate of gas diffusion across the membrane is given by an equation that is sometimes called Fick's law:

$$\dot{V}_{\text{net}} = D_{\text{L}}(P_{\text{alv}} - P_{\text{cap}}), \tag{7.21}$$

where $\dot{V}_{\text{net}}$ is the overall rate of transport of gas across the membrane, P_{alv} is the partial pressure of the gas in the alveolus, P_{cap} is the partial pressure of the gas in the capillary, and D_{L} is the diffusing capacity of the membrane for the gas of interest. The diffusing capacity is related to more fundamental properties of the respiratory membrane:

$$D_{\text{L}} = \frac{AK_{\text{gas}}D_{\text{gas}}}{\delta}, \tag{7.22}$$

where A is the total surface area available for diffusion of the gas, K_{gas} is the relative solubility of the gas in the membrane, D_{gas} is the diffusion coefficient for the gas in the membrane (the diffusion coefficient is defined in Box 2.3), and δ is the thickness of the respiratory membrane. This equation suggests that the rate of transport of a gas from blood to alveolar air increases with the surface area, increases with the molecule's diffusivity or solubility, and decreases with

Box 7.3 Diffusion through the respiratory membrane

Box 2.3 described the basic physics of diffusion, a passive process in which molecules move from regions of high concentration to low concentration. The final equation in Box 2.3 was an expression of Fick's law of diffusion at steady state:

$$J_x = -D\frac{dc}{dx} = \text{constant}. \qquad \text{(Equation 1)}$$

Note that, in this expression, the concentration c refers to the concentration of the gas in the respiratory membrane. The geometric model of the respiratory membrane that was described in the text is shown schematically here. Remember that the assumption of steady-state flux allowed us to demonstrate that the flux through the membrane, J_x, is also a constant.

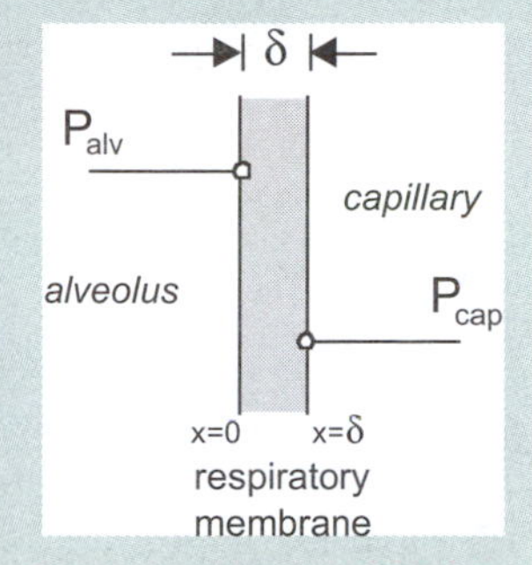

Because the flux is a constant, and the diffusion coefficient D is a constant, the derivative dc/dx must also be a constant. Therefore, the derivative can be approximated as:

$$\frac{dc}{dx} = \frac{(c_{cap} - c_{alv})}{\delta}. \qquad \text{(Equation 2)}$$

In Box 7.2, a coefficient H was used to relate the concentration of a gas in liquid to the partial pressure in the overlying phase at equilibrium:

$$C_{O_2} = HP_{O_2}. \qquad \text{(Equation 3)}$$

This expression is sometimes called **Henry's law**. In our model of the respiratory membrane, a similar expression is used by assuming that the concentration of dissolved gas in the respiratory membrane is related to the partial pressure as shown:

$$C_{O_2} = KP_{O_2}, \qquad \text{(Equation 4)}$$

where K is a solubility coefficient. Substituting Equations 2 and 4 into Equation 1 yields:

$$J_x = -DK\frac{(P_{cap} - P_{alv})}{\delta}. \qquad \text{(Equation 5)}$$

The volumetric rate of flow through the membrane is equal to the flux multiplied by the total surface area, A:

$$\dot{V}_{net} = \frac{DKA}{\delta}(P_{alv} - P_{cap}),$$

	K (mL gas/mL-atm)	$\frac{D_{gas}K_{gas}}{D_{O2}K_{O2}}$
O_2	0.024	1
CO_2	0.570	20
CO	0.018	0.81
N_2	0.012	0.53
He	0.008	0.93

which is equivalent to Equation 7.21, with the diffusing capacity D_L defined as described in the text.

The table shows values for the solubility coefficient for a range of respiratory gases. By using this table, you can calculate the D_L for any of these compounds if the D_L for any one of them is known. How would you do this?

thickness of the membrane. **Fick's law of diffusion** is named after Adolph Fick, a German physiologist, mathematician, and inventor (i.e., biomedical engineer), who—among many accomplishments—published his law of diffusion in 1855 and invented one of the first contact lenses in 1887 (recall Table 1.1).

Equation 7.21 is used to estimate the diffusing capacity D_L in humans. Carbon monoxide is sometimes used for this measurement because it is so sparingly soluble in blood that the partial pressure of carbon monoxide can be neglected ($P_{cap} \sim 0$). The diffusing capacity can therefore be estimated from:

$$D_{\mathrm{L}} = \left(\frac{\dot{V}_{\mathrm{net}}}{P_{\mathrm{alv}}} \right)_{\text{carbon monoxide}} . \tag{7.23}$$

How is this used to estimate D_L? First, it is important to recognize that the net rate of transport across the respiratory membrane is equal to the overall rate of consumption in the body. Therefore, D_L can be estimated by measuring the rate of consumption of CO and the partial pressure in the alveolar gas. Often, this is accomplished with the single-breath method, where a gas containing a small amount of carbon monoxide is inhaled and the partial pressure of carbon monoxide in the alveolus is measured as the breath is held for ~10 seconds.

7.4 Digestion and metabolism

The human diet is complex. This is particularly true in the United States and other developed nations, where large quantities of processed food are consumed. The ingredient list of almost any processed food includes chemical compounds that are unfamiliar to most people (Table 7.3). Of course, also within these ingredient lists are chemical compounds that are essential for health, including water, proteins, carbohydrates, and fats.

Oxygen, which is needed in large quantities for efficient energy generation, is brought into the body by the action of the lungs. Most other materials are absorbed into the body through the digestive system, a set of organs that function collaboratively to absorb ingested food into the body (Figure 7.18). During digestion, solid foods are mechanically pulverized into small pieces, which are further dissolved into molecules. Large molecules, such as proteins and polysaccharides, are broken down into monomer units to allow for absorption into the body.

The digestive system has a number of functions that are familiar to engineers: storage and controlled emptying, mixing, secretion, digestion, and adsorption. In its simplest description, the digestive tract is 8 m of tubing connecting the mouth to the anus. Curiously, the inside of the tube—although deep within the body—is really part of the external environment; a person could swallow a ball bearing and it would pass through his or her body without ever entering it. This is important because some of the contents of the digestive tract would be hazardous if they

Table 7.3 List of ingredients for some foods

Twinkie (snack cake)	Whole wheat bread
Enriched wheat flour—enriched with ferrous sulfate (iron), B vitamins (niacin, thiamine mononitrate [B1], riboflavin [B12], and folic acid)	Whole wheat flour
Sugar	Water
Corn syrup	Wheat gluten
Water	High-fructose corn syrup
High fructose corn syrup	Soybean oil
Vegetable and/or animal shortening—containing one or more of partially hydrogenated soybean, cottonseed or canola oil, and beef fat	Salt
Dextrose	Molasses
Whole eggs	Yeast
Modified corn starch	Mono- and diglycerides
Cellulose gum	Ethoxylated mono and diglycerides
Whey	Sodium stearoyl lactylate
Leavenings (sodium acid pyrophosphate, baking soda, monocalcium phosphate)	Calcium iodate
Salt	Calcium dioxide
Cornstarch	Datem
Corn flour	Calcium sulfate
Corn syrup solids	Vinegar
Mono- and diglycerides	Yeast nutrient (ammonium sulfate)
Soy lecithin	Extracts of malted barley and corn
Polysorbate 60	Dicalcium phosphate
Dextrin	Diammonium phosphate
Calcium caseinate	Calcium propionate
Sodium stearoyl lactylate	
Wheat gluten	
Calcium sulfate	
Natural and artificial flavors	
Caramel color	
Sorbic acid (to retain freshness)	
Color added (yellow 5, red 40)	

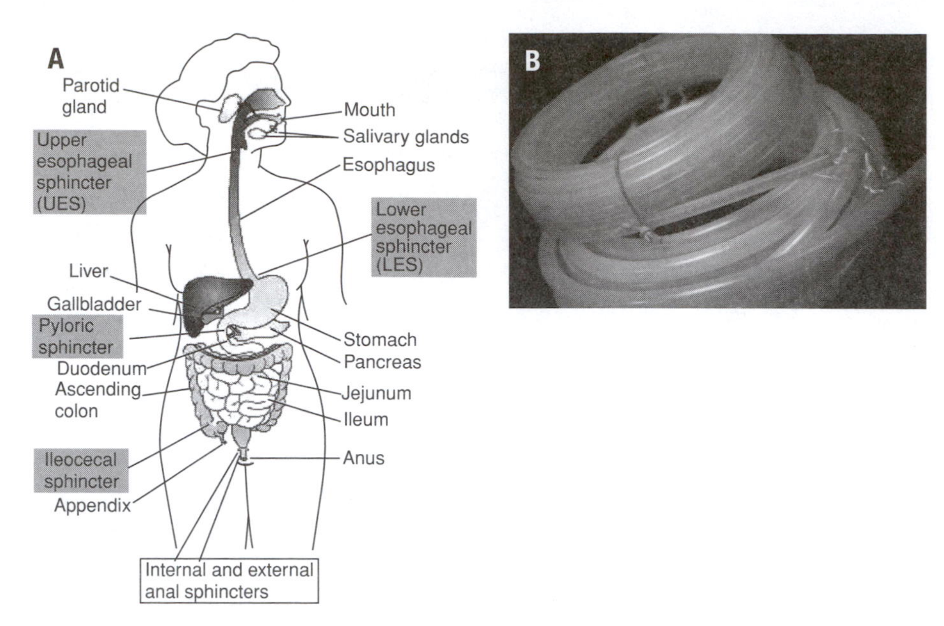

Figure 7.18 **The digestive tract. A.** Schematic illustration of the digestive tract showing the many organs involved in its coordinated function. **B.** Eight meters, about 30 feet, of tubing.

entered the body. One of the most important functions of the digestive tract is to perform digestive actions—to break food into its molecular components—without harming the tissues inside the body. For molecules to enter the body, they must move through the wall of the tube.

The wall of the intestinal tube is a layered structure (Figure 7.19). Two layers of smooth muscle cells—one oriented along the intestinal axis and another oriented circumferentially—surround the intestinal tube and produce movement. Muscular movements provide both mixing of intestinal contents and propulsion along the

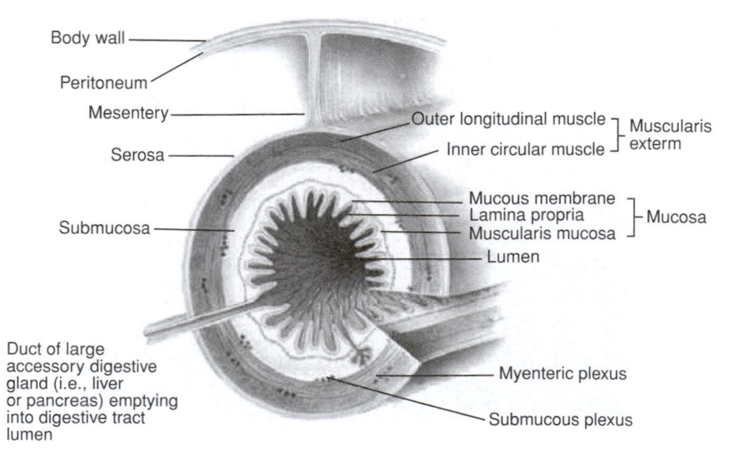

Figure 7.19 **Anatomy of the intestinal wall.** The microscopic anatomy of the wall of the intestine varies in esophagus, stomach, and small and large intestine, but many features are retained. From outside to lumen, the layers are: serosa, longitudinal muscle layer, circular muscle layer, submucosa, and mucosa. Two systems of nerves, one from the central nervous system and another called the enteric nervous system, control movements of the intestinal wall. (See color plate.)

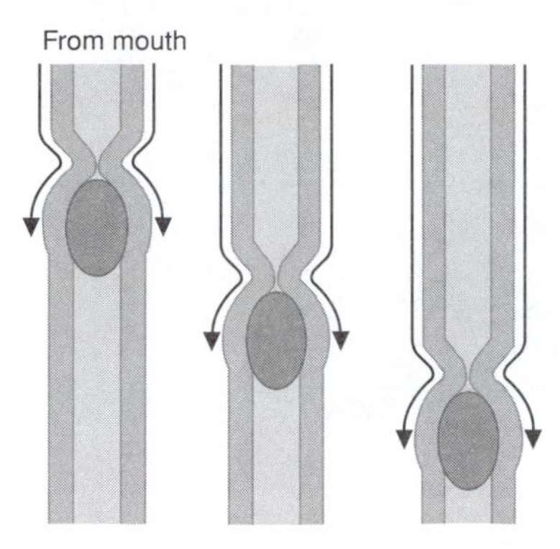

Figure 7.20 **Peristalsis.** The intestinal tract is lined with muscular tissue. The contraction of this tissue is regulated by a special set of nerves (called the enteric nervous systems) that produce coordinated movement such as peristalsis. Peristalsis is the rhythmic wave of contraction that pushes food forward through the tract.

intestine. Different regions of the intestine contract with different underlying rhythms: For example, the stomach can contract up to 3 times/min, whereas the duodenum can contract 8–10 times/min.

Peristalsis is the muscular event that propels food through the intestine: It involves a coordinated pattern of distal relaxation and proximal constriction (Figure 7.20). Peristalsis first occurs with swallowing, which is a largely involuntary process. Valves and sphincters, which are specialized regions of smooth muscle, occur at locations along the intestine; these serve to control the rate at which intestinal contents move from region to region and to control intestinal emptying. For example, the lower esophageal sphincter is 3 cm of thickened smooth muscle that prevents reflux of fluid from the stomach into the esophagus.

7.4.1 Organs of the digestive system and their function

Food that is swallowed passes down a straight muscular tube, the esophagus, and enters the stomach. The stomach is an expandable vessel, which can hold up to 1 L of material. It has an important engineering function. It acts by mechanical mixing, crushing, and chemical digestion to convert solid food into small particles, which increases the surface area of the food to allow extraction of nutrients. Cells of the stomach lining secrete a large volume of fluid into the stomach lumen. These secreted gastric fluids are highly acidic, causing the pH of the stomach contents to drop as low as 1. Low gastric pH protects people from bacteria in the diet because most bacteria cannot survive in low pH. It is also important for the action of stomach enzymes. The enzyme pepsin initiates the process of protein digestion. Pepsin is formed from the precursor pepsinogen at low pH; pepsin also has maximal proteolytic activity at low pH. To survive this low pH, the lining of the stomach is protected by a variety of mechanisms, including a mucus layer.

SELF-TEST 11 If the pH of the stomach contents is 1, what is the magnitude of the H^+ gradient between blood and gastric contents?

ANSWER: The pH in the blood is 7.4. Because pH is $-\log[H^+]$, the ratio of $[H^+]_{stomach}/[H^+]_{blood}$ is $\sim$1,000,000 (10^6).

The stomach is connected to the small intestine, which is divided into three segments: the duodenum, the jejunum, and the ileum. Transfer of material from

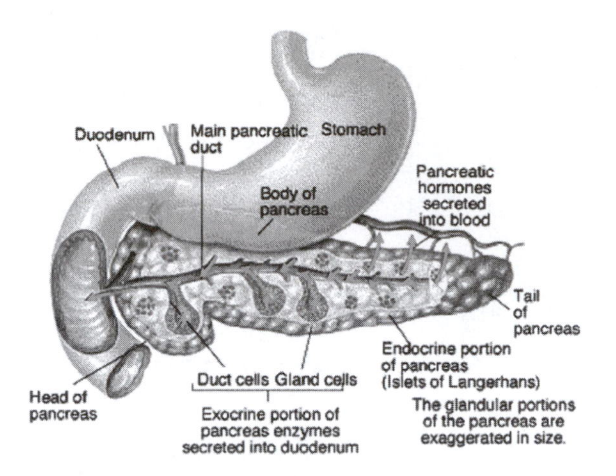

Figure 7.21

Pancreatic secretions. The pancreas has two separate functions. The endocrine function of the pancreas—including, most famously, secretion of the hormone insulin into the blood—is discussed in Chapter 6. The exocrine function of the pancreas is essential in digestion; the pancreas produces digestive enzymes that are secreted through a duct directly into the duodenum.

the stomach to the duodenum is controlled by a sphincter, called the pyloric sphincter. Food is retained in the stomach to allow time for initial digestion by pepsin and gastric acids. Muscular contractions of the stomach also provide mechanical force to aid in breakdown of solids.

A major chemical function of digestion is hydrolysis, to convert large molecules into small subunits. Recall that polypeptides and polysaccharides are produced by condensation reactions and can be broken down into their components by hydrolysis (Section 2.6.1). These reactions are accelerated by the presence of enzymes in the stomach (such as pepsin). The small intestine has a rich array of digestive enzymes, many of which are secreted into the intestine by the pancreas (Figure 7.21). These enzymes include peptidases (such as chymotrypsin and carboxypeptidase) to break down proteins, lipases that break triglycerides into glycerol and fatty acids, amylases to break down carbohydrates, and ribonucleases to break down nucleic acids. The pancreas also secretes a bicarbonate-rich juice, which neutralizes the acidic stomach fluids that enter the duodenum.

Many of the pancreatic enzymes are zymogens (proteins that are secreted in an inactive form and then converted into an active enzyme): Trypsinogen, chymotrypsinogen, and procarboxypeptidase are produced in the pancreas, secreted into the duodenum, and activated by the enzyme enterokinase, which is associated with the inner lining of the duodenum. The highly specialized inner lining of the intestine is sometimes called the brush border.

The liver has many functions in digestion and regulation of metabolism. Despite the wide variety of functions, there is little specialization within cells in the liver; all of the cells perform all of these functions. The liver also has a unique circulatory system: The hepatic artery and hepatic portal vein both flow into the liver. Therefore, cells of the liver receive a mixture of arterial blood from the heart and venous blood from the intestines. To deliver this blood flow

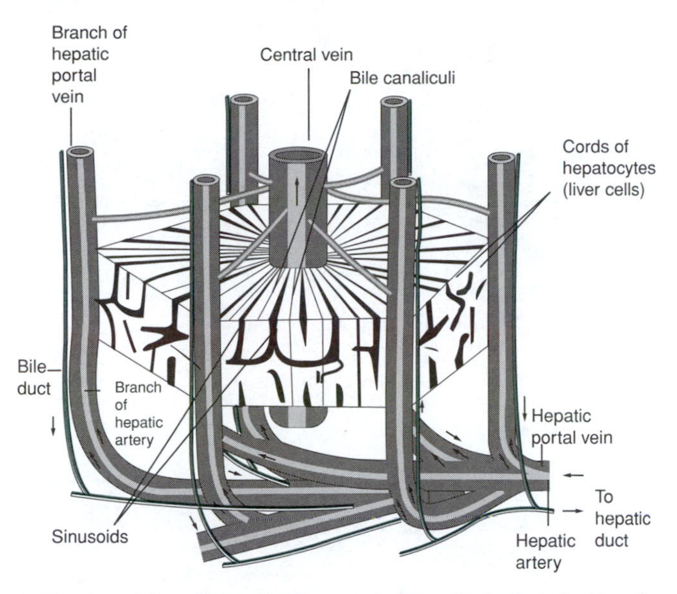

Figure 7.22 **Architecture of liver.** Liver cells are arranged within a "spoke and wheel" configuration. Blood is delivered to the outside of the wheel by branches of the portal vein and hepatic artery, so that blood flows inward—past the cells, to the central vein located at the spoke.

to cells within the liver, the architecture of the liver is orderly. Branches of portal vein and hepatic artery travel into the liver together, arriving at the end of the spokes of the wheel. Cords of hepatocytes lie along the pathway from the hepatic artery/portal vein to the central vein. A characteristic portion of liver is shown in Figure 7.22: In each section, blood flows from the outer into the center region, where it is collected and removed by the central vein. Bile ducts drain in the reverse direction toward the bile-collecting ducts, which parallel the hepatic artery/portal vein. This arrangement provides for: 1) substantial flow of blood—containing both compounds absorbed from the intestine (portal blood) and oxygen (arterial blood) past a large surface area of liver cells and 2) pathways for collection of waste products in bile. Chapter 9 describes the role of the liver in biotransformation and elimination of drugs and toxins in bile.

The small intestine is the longest section of the gastrointestinal tract. The small intestine is responsible for absorption of most nutrients (Figure 7.23); the amount of surface area that is available for absorption is critically important. As described in Self-test 4, the absorptive surface area in the lung is enhanced by its architecture: Many small, spherical alveoli produce a large surface area. The small intestine is about 6 m long, with a diameter of 4 cm. Circular folds on the inner luminal surface increase the surface area by threefold, villi increase the surface area by 10-fold, and microvilli (which constitute the brush border) increase the surface area by 20-fold. Therefore, the total surface area is increased by 600-fold relative to a plain tube. Each villus has its own capillary network. The cells that line the villi are among the most rapidly proliferating cells in the body. Turnover is approximately every 3 days. Because of this high level of proliferation, the

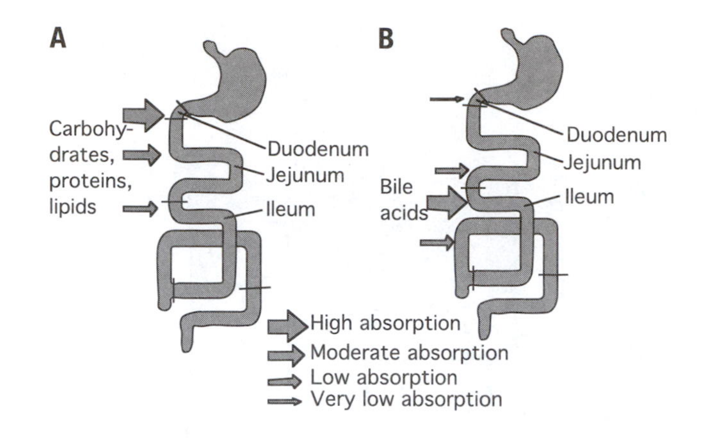

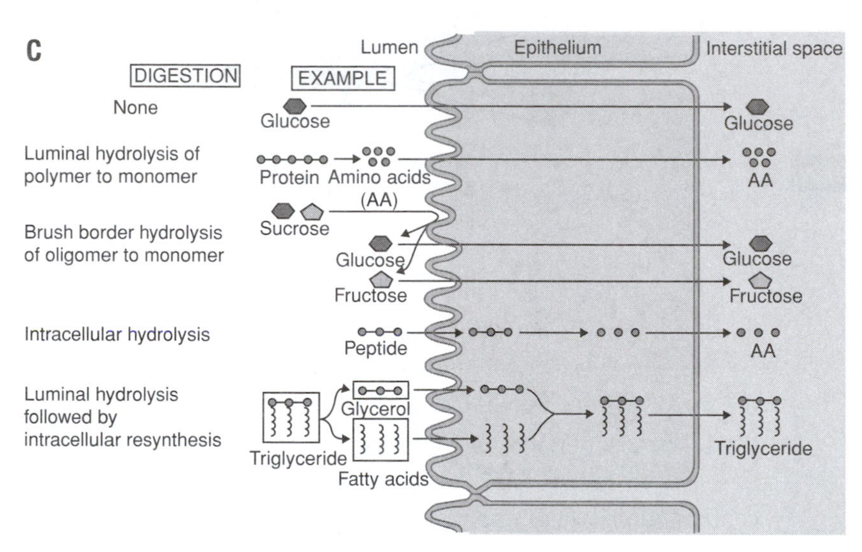

Figure 7.23 **Overview of digestion.** Compounds are absorbed in different segments of the intestinal tract. **A.** Proteins, carbohydrates, and lipids are absorbed predominantly in the duodenum, jejunum, and ileum. **B.** Bile salts are absorbed primarily in the proximal ileum, but also in the jejunum and ascending colon. **C.** Carbohydrates, proteins, and triglycerides are absorbed by different mechanisms.

small intestine is particularly susceptible to insults that influence mitosis, such as chemotherapy drugs and radiation exposure.

SELF-TEST 12 What is the surface area of the small intestine?

ANSWER: If the small intestine were a smooth tube, its surface area would be 0.75 m^2. Folds, villi, and microvilli provide a total surface area of 450 m^2.

The absorptive function of the small intestine is similar to the function of the proximal tubule in the kidney, which is discussed in Chapter 9. In both cases, molecules are selectively transported from one side of an epithelial cell layer to another: In the kidney tubules, molecules are reabsorbed from the tubule fluid (which will become urine) to the blood. In the intestine, molecules are

Table 7.4 Fluid balance in the body

Fluid in food ingested	1,000 mL	
Fluid ingested	1,000 mL	
Saliva produced	1,500 mL	
Gastric juice secreted	2,000 mL	
Pancreatic juice secreted	1,500 mL	
Bile fluid secreted	500 mL	
Intestinal fluid secreted	1,000 mL	
Absorbed by the small intestine		6,500 mL
Absorbed by large intestine		1,900 mL
Water in stool		100 mL
Total	8,500 mL	8,500 mL

absorbed from the luminal fluid into the blood. The function is determined by properties of the epithelial cells. For example, in the small intestine, membrane transport proteins allow the selective uptake of monosaccharides, amino acids, or components of triglycerides (Figure 7.23). This transport occurs through the villus epithelial surface.

The digestive system secretes large volumes of juices to facilitate digestion; the intestines reabsorb most of this fluid (Table 7.4). The small intestine is a high volume absorber, handling between 5 and 7 L of water per day. The large intestine—also called the colon—is primarily a drying organ, reabsorbing the majority of the water remaining, which amounts to between 1 and 1.5 L of water per day.

The large intestine also harbors substantial populations of bacteria, which convert dietary carbohydrates to monosaccharides for absorption. The ileocecal sphincter and ileocecal valve keep the bacterial contents of colon from re-entering the ileum. A major function of the large intestine is to store undigested waste material and allow its release under voluntary control.

7.4.2 Digestion and metabolism

The main organs of digestion—the stomach, small and large intestine, pancreas, and liver—work in coordination to provide nutrients to the body in support of overall metabolism. Metabolism is the broad term for all of the processes involved with energy production, energy use, and body growth or maintenance. Metabolism can be further divided into anabolism, those chemical reactions in which new molecules are formed, and catabolism, processes in which molecules are broken down. A healthy male of average size (70 kg) uses approximately 2,100 kcal/day to support his resting metabolism. Energy to support metabolism comes from the diet. If this average male's diet contains 2,000 kcal/day (as in Table 7.1)—or if he increases his level of physical activity so that he uses more than the resting level of energy—he will lose weight.

A person's overall rate of metabolism changes depending on his or her state of health, level of activity, and environment. To define rates of metabolism, clinicians define the basal metabolic rate (BMR) for a person as the rate under specific, controlled conditions. BMR is measured in a person who has had a full night's sleep, not eaten for 12 hours, rested physically for 1 hour, and is free of physical and mental stimuli. Therefore, BMR is less than the resting or normal metabolic rate defined above. When measured in this standard way, BMR is higher in males than females by about 5%; in both sexes, BMR decreases with age. BMR varies from person to person, and is controlled by hormones released by the thyroid gland.

The energy to support metabolism is derived from the molecules in the diet (Table 7.5). For most individuals, the majority of calories are ingested as carbohydrates. Carbohydrates are consumed in different forms: An average diet might contain 10% monosaccharides, 40% disaccharides, and 50% starch (Table 7.6), but only monosaccharides can be absorbed by the intestinal epithelium (Figure 7.23). Enzymes in the intestines convert disaccharides to monosaccharides. Starch is digested by a two-step process: Intraluminal enzymes (such as α-amylase) catalyze hydrolysis to oligosaccharides, and membrane enzymes at the brush border (such as the disaccharidases lactase, glucoamylase, and sucrase–isomaltase) further digest to monosaccharides. The resulting monosaccharides are absorbed by epithelial cells via membrane transport proteins such as the sodium–glucose cotransporter (SGLT1) or other glucose transporters (e.g., in the case of fructose).

Protein digestion is efficient: Dietary protein consumption is approximately 70–100 g/day, and less than 4% of ingested protein is excreted undigested. Enzymes catalyze the breakdown of proteins and peptides to short oligopeptides and amino acids (Figure 7.23). Some short peptides can be absorbed intact; these are digested by enzymes within the epithelial cytoplasm to amino acids. Protein absorption is more versatile than carbohydrate absorption: Amino acids, short peptides, and even whole proteins can be absorbed in some cases. For example, newborns have special transport systems that allow them to absorb whole, intact proteins, such as antibodies that they ingest in mother's milk. The body can synthesize some amino acids, but not all of them. The nine essential amino acids, which must be absorbed from the diet, are phenylalanine, leucine, methionine, lysine, isoleucine, valine, threonine, tryptophan, and histidine.

Digestion and absorption of fat involve a more complicated series of steps because fat molecules do not dissolve in water (recall Chapter 2). The primary dietary form of fat is triglycerides. Mechanical processes in the mouth and stomach emulsify fat—that is, they create a suspension of tiny, fat-rich globules. Lipases act on these globules to break down triglycerides into fatty acids and diglyceride. Bile salts are derivatives of cholesterol and are secreted by the liver. These salts act as detergents, helping to solubilize monoglycerides and free fatty

Table 7.5 Composition of some common foods

		Hostess® Ding Dongs® (chocolate snack cake, crème filling)	McDonald's® hamburger (regular, single patty, plain)	Skittles® (one 2.17-oz package)	Broccoli, cooked, drained, with salt (1 cup chopped = 156 g)
Water	g	10	33.79	2.37	141.48
Energy	kcal	368	265	251	44
Energy	kJ	1540		1051	183
Protein	g	3.12	13	0.12	4.65
Total lipid (fat)	g	19.36	10	2.71	0.55
Ash	g	1.26		0.6	1.44
Carbohydrate	g	45.36	32	56.2	7.89
Fiber, total dietary	g	1.8	1	0	5.1
Sugars, total	g	32.4	32	47.13	2.17
Minerals					
Calcium	mg	3	63	0	62
Iron	mg	1.84	2.4	0.01	1.05
Sodium	mg	241	387	10	409
Phosphorus	mg		103	1	105
Potassium	mg		145	6	457
Zinc	mg		2	0.01	0.7
Copper	mg		0.09	0.047	0.095
Manganese	mg		0.213	0.017	0.303
Selenium	μg		21.7	1.2	2.5
Magnesium	mg		19	1	33
Vitamins					
Vitamin C	mg		0	41.5	65.5
Thiamin	mg		0.333	0.001	
Riboflavin	mg		0.27	0.014	
Niacin	mg		3.717	0.009	
Pantothenic acid	mg		0.369	0.004	
Vitamin B-6	mg		0.063	0.002	
Folate	μg		53		168
Vitamin B-12	μg		0.89		
Vitamin K	μg			5.8	220.1
Vitamin E	mg			0.27	
Vitamin A	IU				3069
Lipids					
Fatty acids, total saturated	g	11.036	4.141	0.538	0.084
Fatty acids, total monounsaturated	g	3.992	5.456	1.835	0.037
Fatty acids, total polyunsaturated	g	1.198	0.918	0.074	0.261
Cholesterol	mg	14	35	0	0
Other					
Beta carotene	μg				1841
Lutein + zeaxanthin	μg				2367

Table 7.6 Forms of carbohydrate in the diet

Starch	Oligosaccharides	Monosaccharides
Amylose (plant)	Sucrose (table sugar)	Fructose
Amylopectin (plant)	Lactose (milk sugar)	Glucose
Glycogen (animal)		

acids. As might be apparent from this short and incomplete description, the physical and chemical processes that lead to fat absorption are more complicated than those involved in absorption of carbohydrates and proteins.

7.4.3 Modeling the digestive tract

The function of the intestine is to break down (or digest) food into small molecules and to absorb those small molecules into the body. The absorptive function is therefore similar to the function of the lungs, but with important differences. First, the intestine must absorb a much wider range of molecules. Second, the intestinal membrane absorbs molecules from a liquid medium, rather than from the gaseous medium of the alveolus. As is clear from the earlier description of the digestive system, the intestine has evolved an anatomical structure that is different from the lung to accomplish its absorptive function. The digestive function of the intestine has no counterpart in the lung.

When the intestine is digesting food—breaking it down into simpler chemical constituents—it is acting as a chemical reactor, or a vessel in which a chemical reaction is occurring and is under some degree of control. Engineers, particularly chemical engineers, have much experience with the analysis and design of chemical reactors (typically for use in the petroleum, chemical, pharmaceutical, and food industries). Some of these methods can be applied to understanding the action of the intestine on ingested food.

Engineers use different ideal reactor models to describe the progress of chemical reactions (Figure 7.24). The simplest model is a batch reactor, in which the reaction ingredients are loaded into a reaction vessel at some initial time. The vessel is then closed and maintained for some period of time, after which the reaction products are removed from the vessel. During its time of operation,

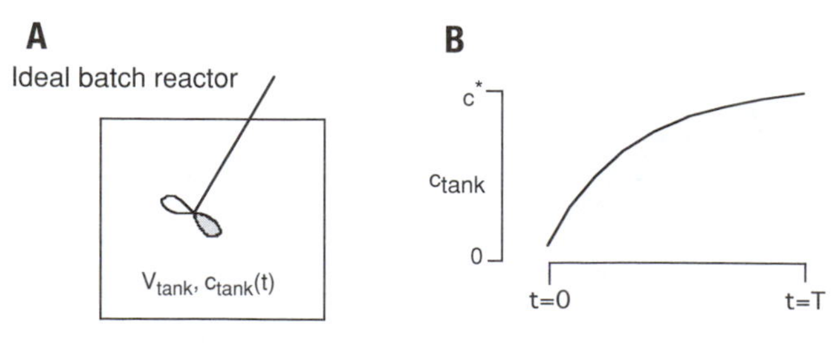

Figure 7.24 **Simple batch reactor. A.** The batch reactor is a closed system, in which an initial reactant mixture is loaded. **B.** The concentration of a product increases with reaction time in the batch reactor.

chemical reactions within the reactor act to change the concentration of key ingredients; for example, the concentration of a product might increase with time because it is generated by the reaction (Figure 7.24b). The batch reactor is an example of a closed system (recall this definition from Section 7.2).

In some situations, the stomach may function as a batch reactor. After a meal, the stomach is loaded with a complex mixture—including the contents of the meal and the secretions of the stomach in response to the meal—which is now held in a fixed volume by the sphincters and mixed by the action of gastric muscle. Applying Equation 7.1 to a mass balance of a key ingredient:

$$V\frac{\mathrm{d}C}{\mathrm{d}t} = GEN, \tag{7.24}$$

where V is the reactor volume, C is the concentration of a product of interest (moles/L), and GEN represents that rate of product generation (moles/min). The change in concentration with time, or accumulation in the reactor, depends on the rate of generation. For an enzyme-catalyzed reaction—such as the digestion of proteins catalyzed by pepsin—the rate of production formation would be determined from Michaelis–Menten kinetics (Box 4.2):

$$V\frac{\mathrm{d}C}{\mathrm{d}t} = V\frac{v_{\mathrm{max}}[S]}{K_{\mathrm{m}} + [S]}, \tag{7.25}$$

where v_{max} and K_{m} are constants and $[S]$ is the concentration of protein. Equation 7.25 can be used to estimate the extent of protein digestion as a function of time in the stomach reactor. Let's assume that pepsin is digesting a polypeptide, which it can cleave at only one site, to produce two equal size peptide fragments:

$$Peptide \xrightarrow{\text{pepin}} Fragment1 + Fragment2 \tag{7.26}$$

If $[S]$ represents the concentration of peptide, and C represents the concentration of Fragment 1, a new variable called X, which represents the extent of reaction, can be defined such that:

$$\begin{aligned} [S] &= [S]_0 - X \\ C &= X \end{aligned}, \tag{7.27}$$

where $[S]_0$ is the initial concentration of peptide. The extent of reaction variable, X, which has units of moles/L, allows easy accounting for the stoichiometry of the reaction: X represents the formation of 1 mole/L of Fragment 1 and consumption of 1 mole/L of peptide. It also allows Equation 7.25 to be written:

$$\frac{\mathrm{d}X}{\mathrm{d}t} = \frac{v_{\mathrm{max}}([S]_0 - X)}{K_{\mathrm{m}} + [S]_0 - X}. \tag{7.28}$$

The concentration of both reactant and product as a function of time can be found by solving the differential equation (Equation 7.28). Solution of this full equation, with Michaelis–Menten kinetics for Fragment 1 formation, is possible, but not easy. The expression is much simpler, however, if the initial concentration

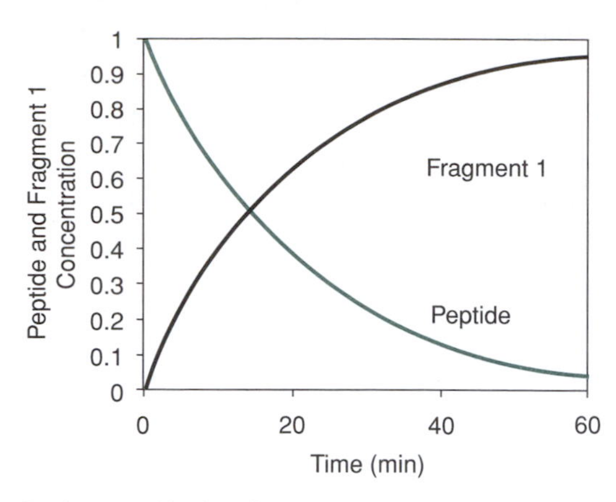

Figure 7.25 **Batch reactor kinetics.** Changes in peptide concentration, $[S]$, and Fragment 1 concentration, C, with time in a batch reactor. Normalized concentrations are plotted ($[S]/[S]_0$ and $C/[S]_0$) for a value of v_{max}/K_m of 0.05 min^{-1}.

of reactant, $[S]_0$, is assumed to be much lower than the K_m. Then, Equation 7.28 reduces to:

$$\frac{dX}{dt} = \frac{v_{max}([S]_0 - X)}{K_m}, \tag{7.29}$$

which has the solution:

$$X = [S]_0\left(1 - e^{-\frac{v_{max}}{K_m}t}\right). \tag{7.30}$$

Figure 7.25 shows the changes in $[S]$ and C with time.

The batch reactor is a useful model for some situations, but it can only describe closed systems. Some biological reactors are open systems, allowing a continuous inflow of reactants and a continuous outflow of products and unused reactants. Engineers have developed a variety of models to describe the performance of continuous reactors; two ideal reactor models are shown in Figure 7.26.

One of the first challenges an engineer might face in trying to apply these reactor models to a real system is this: Which of the ideal continuous reactors best matches the real system? To answer this question, the engineer might examine the behavior of the reactor in a simpler mode, in which a **tracer** species—which is not consumed or produced in any chemical reactions—is allowed to flow through the system. By observing the pattern of outflow of the tracer chemical, it is often possible to determine which reactor model will describe the system most effectively. This tracer method is described in the next few paragraphs and illustrated in Figure 7.26.

Assume that an ideal stirred tank is operating with an inlet stream containing a dye molecule that does not react within the tank. If the tank has been operating for a long enough time, the concentration of the dye within the tank, the concentration of dye in the outlet stream, and the concentration in the inlet stream are all equal (c^*). If the inlet stream is suddenly changed, so that it now contains no dye molecules ($c_{in} = 0$), the concentration of dye in the tank and in the outlet stream will slowly decrease (Figure 7.26a). Our basic assumption regarding the ideal

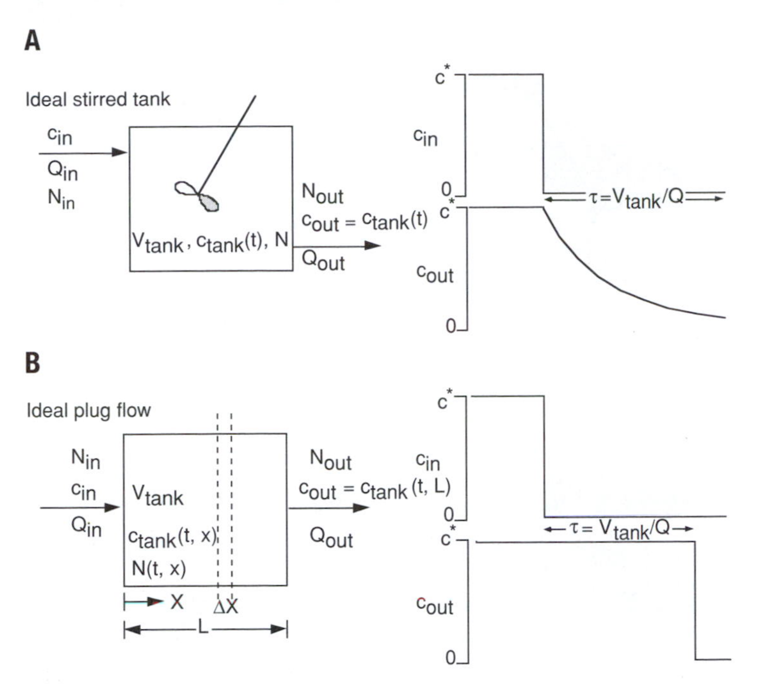

Figure 7.26 **Ideal open reactor models. A.** The ideal stirred tank reactor (in which the contents of the reactor volume are perfectly mixed) shows a characteristic exponential drop in tracer concentration outflow after an abrupt drop in tracer inflow. **B.** The plug flow reactor (in which the flow stream moves predictably through the reactor volume) shows a time delay; a sudden change in inflow concentration produces a sudden change in the outflow emergence after a lag time.

stirred tank leads to several predictions: 1) the dye concentration throughout the tank is always uniform but changing with time, $c_{\text{tank}}(\text{t})$; 2) the outlet concentration is equal to the concentration in the tank ($c_{\text{tank}} = c_{\text{out}}$); and 3) the tank/outlet concentration changes in a predictable way in response to the step change in inlet concentration:

$$c_{\text{out}} = c^* e^{-\frac{t}{\tau}}, \tag{7.31}$$

where t is the time after initiation of the step change in inlet concentration and τ is a characteristic time for the compartment ($= V_{\text{tank}}/Q$). For this example, the characteristic time is equal to the time required for the outlet concentration to decrease from c^* to $0.37c^*$.

In ideal plug flow (Figure 7.26b), molecules moving through the region do not mix, but rather travel in an orderly fashion from inlet to outlet. The simplest physical model of ideal plug flow is slow fluid flow through a long tube: Each "packet" of fluid enters the tube just after some other packet, and remains behind its neighbor from inlet to outlet. Each packet emerges from the tube in the order in which it entered. The time required for a packet to move from inlet to outlet is τ ($= V_{\text{tank}}/Q$); a step change in inlet concentration produces a step change in outlet concentration that is offset by τ.

How do these two continuous flow vessels perform as reactors? Assume that an ideal continuous reactor is used for pepsin-catalyzed degradation of a peptide,

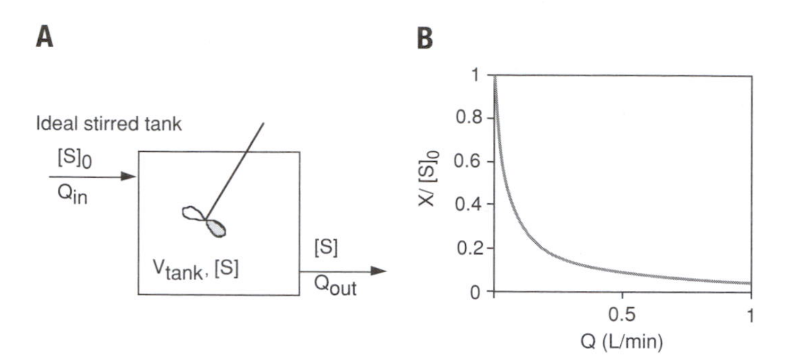

Figure 7.27 **Continuous reaction in ideal stirred tank reactor. A.** Reactor configuration for an ideal stirred tank reactor. **B.** The relationship between extent of reaction and inlet flow rate for a reactor with volume of 1 L and v_{max}/K_m equal to 0.05 min^{-1}.

as described in Equation 7.26. Assume that the peptide concentration is low, so that the rate of reaction is given by:

$$\text{Rate Peptide Digestion} = \frac{v_{max}}{K_m}([S]_0 - X). \quad (7.32)$$

(This is the same simplified Michaelis–Menten kinetics that was used in Equation 7.29.) The reactor will operate at steady state, with continuous inflow of a reactant stream and outflow of products and reactants. To maintain steady state, the inlet and outlet flow rates must be equal ($Q = Q_{in} = Q_{out}$).

An ideal stirred tank reactor can be analyzed by mass balance on one of the reacting species, in which the system is defined as the contents of the reactor (Figure 7.27). A mass balance on our peptide for the ideal stirred tank reactor yields (from Equation 7.4, which applies to all steady-state balances):

$$0 = Q[S]_0 - Q[S] - V\frac{v_{max}}{K_m}[S]. \quad (7.33)$$

Notice that, in an ideal stirred tank reactor, the outlet concentration of peptide is equal to the concentration of peptide in the tank. This must be true because in the ideal stirred tank, the concentration is the same at every point in the reactor. Applying the definition of extent of reaction, X, from Equation 7.27, and solving for X yields:

$$X = [S]_0 \frac{V\left(\frac{v_{max}}{K_m}\right)}{\left(Q + V\frac{v_{max}}{K_m}\right)}. \quad (7.34)$$

The extent of reaction in an ideal reactor increases with reactor volume, V, and reaction rate, $\frac{v_{max}}{K_m}$, and decreases with inlet/outlet flowrate, Q.

SELF-TEST 13 The analysis of an ideal stirred tank reactor focused on changes in substrate concentrations from the inlet to the outlet stream. Because the reaction is enzyme-catalyzed, enzymes must also be supplied to the reactor. How would enzymes be supplied?

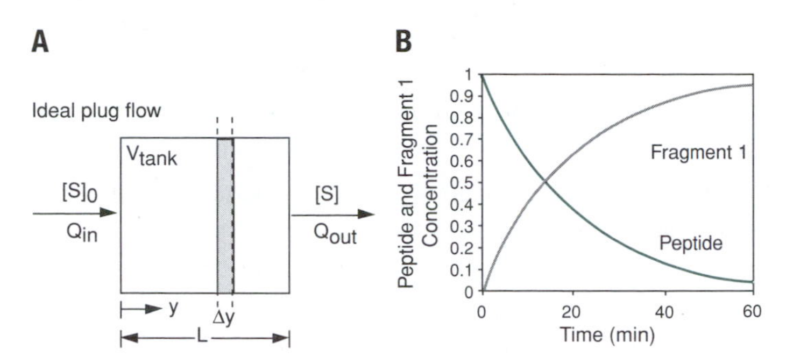

Figure 7.28 **Ideal plug flow reactor. A.** Analysis of this reactor is facilitated by defining the system as an element of fluid that is moving through the reactor volume. **B.** This definition of a system makes the plug flow reactor equivalent to a batch reactor, where the reaction time is now the transit time through the reactor, $T = V/Q$.

ANSWER: Enzymes could be supplied in the inlet stream, but it would also flow out through the outlet stream. Therefore, enzymes would have to be continuously supplied. Alternately, to save the quantity of enzyme that is needed, the enzymes might be immobilized within the reactor. How might this be done?

Ideal plug flow reactors can be analyzed most easily by defining the system differently. Instead of using the entire reactor volume, as in the ideal stirred tank, the reaction volume is defined as an element of fluid that is moving through the reactor contents (the grey area in Figure 7.28). This system moves through the reactor, with time, and eventually emerges from the reactor outlet. The time required for movement of this system volume through the reactor is equal to V/Q, where Q is equal to both the inlet and outlet flow rates, as in the stirred tank.

Because of the unique configuration of the ideal plug flow reactor, in which the flowing stream does not mix and all fluid elements emerge from the reactor in the same order in which they entered, the plug flow reactor is equivalent to a batch reactor. This phenomenon can be visualized in Figure 7.28a; the grey volume represents a closed system because nothing enters or leaves that volume during its transit through the reactor. For this reason, the extent of reaction for an ideal plug flow reactor is identical to the extent of reaction within a batch reactor, except that the reaction time (for the batch reactor) is replaced with the transit time for the control volume (in the plug flow reactor).

Summary

- Engineering analysis requires definition of the system under study, which is accomplished by identifying system boundaries.
- Systems at steady state do not change appreciably with time.
- Internal respiration is the process by which cells obtain energy by oxidation of chemical compounds; external respiration is the process of exchange of oxygen with the external environment, which is accomplished by the lungs.
- Hemoglobin allows blood to carry substantial quantities of oxygen, despite the low solubility of oxygen in water.

- Carbon dioxide dissolved in water forms carbonic acid, which dissociates to bicarbonate and hydrogen ions; the bicarbonate/carbon dioxide system is the primary mechanism for buffering to stabilize against changes in pH in the body.
- The digestive system is responsible for digestion of food into its molecular components and absorption of nutrients into the body.
- The chemical reactions of digestion can be understood in terms of mathematical models of chemical reactors.

REFERENCE

1. West JB. *Respiratory Physiology: The Essentials*. Second ed. Baltimore, MD: Williams & Wilkins; 1979.

FURTHER READING

Boron WF, Boulpaep EL. Medical Physiology. Philadelphia, PA: Sanders; 2004. The respiratory system is described in chapters 25 through 31. The gastrointestinal system is described in chapters 40 through 45.

USEFUL LINKS ON THE WORLD WIDE WEB

http://oac.med.jhmi.edu/res_phys/TutorialMenu.html

The Johns Hopkins School of Medicine has on-line tutorials on respiratory physiology. The tutorial on "The Mechanics of Quiet Breathing" is appropriate for use with this chapter, whereas the tutorial on "Static Elastic Properties of the Lung and Chest Wall" is more appropriate for the material in Chapter 10.

KEY CONCEPTS AND DEFINITIONS

alveolar duct the part of the respiratory passage beyond the bronchioles from which alveolar sacs and alveoli arise

alveolus (*pl.* alveoli) one of the terminal saclike dilations of the alveolar ducts in the lung

assumption a proposition that is treated for the sake of a given discussion as if it were known to be true

bronchiole a minute branch into which the bronchus divides, which lack cartilage

bronchus (*pl.* bronchi) one of the subdivisions of the trachea, serving to carry air to and from the lungs

buffer a chemical that, when present in a solution, absorbs hydrogen ions and thus resists changes in pH

capacities a term used to describe the maximum amount that a system (e.g., lungs) can contain

carbonic anhydrase an enzyme that catalyzes the interconversion of carbon dioxide and water to form carbonic acid, bicarbonate ions, and protons (H^+)

continuously stirred tank reactor a reactor model used to estimate the key unit operation variables when using a continuously agitated-tank reactor or a reactor the contents of which are well mixed, to reach a specified output

dead space the air that is inhaled by the body but does not partake in gaseous exchange

equilibrium a state in which opposing or influencing forces are perfectly balanced

expiration exhalation, the release of gases from the lung

external respiration the bodily process of inhalation and exhalation; the process of taking in oxygen from inhaled air and releasing carbon dioxide by exhalation

Fick's laws a set of laws that describe diffusion and define the diffusion coefficient

homeostasis the ability or tendency of a living organism (or an open system in general) to maintain a stable internal environment by adjusting its physiological processes

ideal reactor model a model reactor that is used to describe a system and its boundaries

inspiration inhalation, intake of gases by the lungs

internal respiration the metabolic processes whereby certain organisms obtain energy from organic molecules; processes that take place in the cells and tissues during which energy is released and carbon dioxide is produced and absorbed by the blood to be transported to the lung

intrapleural a term that refers to a location within the pleura, either the covering of the lungs (visceral pleura) or of the inner surface of the chest wall (parietal pleura)

mass (or material) balance a method of accounting for material entering and leaving a system based on the conservation of mass principle (matter cannot be created or destroyed)

mathematical model an abstract model that uses mathematical language to describe the behavior of a system

observation a specified behavior or event as seen in reality; an actual or real event

partial pressure the pressure that would be exerted by one of the gases in a mixture if that gas alone is occupied the same volume as the mixture

plug flow reactor a model reactor used to estimate the key unit operation variables when using a continuous tubular reactor or a reactor in which contents move in a flow stream to reach a specified output

prediction the forecast of a specified behavior of a system based on underlying assumptions

rate constant a constant of proportionality that relates the rate of a chemical reaction at a given temperature to the concentration of reactants or products

resistance a measure of the degree to which flow of a substance (e.g., fluid or current) is hindered

respiratory bronchiole the smallest portion of the bronchiole connecting the terminal bronchiole to the alveolus

respiratory membrane a thin epithelial layer of squamous cells that contains fluid lining the alveolus

saturated the fraction of total protein binding sites that are occupied at any given time

spirometer a gas meter used for measuring respiratory gases

steady state an unvarying condition (i.e., all state variables are constant) despite ongoing conditions to change that state

system a set of interconnected parts that come together to form a complete whole (e.g., the excretory system)

system boundaries the physical sites of intersection between the system under study and the external environment

terminal bronchiole the last portion of the bronchiole before it subdivides into respiratory bronchioles

tidal volume the volume of air inhaled and exhaled at each normal breath

tracer a substance introduced into a biological organism or other system so that its subsequent distribution can be followed from its color, fluorescence, radioactivity, or other distinctive properties

trachea a large membranous tube supported by rings of cartilage extending from the larynx to the bronchial tubes conveying air from and to the lungs; the windpipe

ventilation replacement of air (or other gases) with air (or other gases)

NOMENCLATURE

A	Surface area	m^2
ACC	Rate of accumulation of mass (or material) within the system	mole/s
B, B*	Constants of integration	Variable
C	Concentration of product of interest	Mol/L
C_{in}, C_{out}	Inflow and outflow concentrations	Mol/L
CONS	Rate of mass (or material) consumption within the system	Mole/s
D_{gas}	Diffusion coefficient of a gas in the respiratory membrane	cm^2/s
D_L	Diffusing capacity	L/s-mmHg
E	Electrical potential	V
GEN	Rate of mass (or material) generation within the system	Mole/s
I	Current	Amperes
IN	Rate of flow of mass (or material) into the system	moles/s
J_x	Flux	L/m^2-s; kg/m^2-s
k	First-order rate constant	1/s
K_a	Equilibrium constant	—
K, K_{gas}	Relative solubility of gas in the respiratory membrane	mL gas/mL-atm
K_m	Michaelis constant	

M	Mass of agent within the system	Mole (or g)
M_{dose}	Initial dose administered	Mole (or g)
$\dot{M}_{in}$	Mass flow rate of material out of a system	g/sec, g/min
$\dot{M}_{out}$	Mass flow rate of material into a system	g/s, g/min
OUT	Rate of flow of mass (or material) out of the system	Moles/s
OUT_{kidney}	Outflow rate of mass via kidney	Mole/s
OUT_{other}	Outflow rate of mass via other means	Mole/s
P	Hydrostatic pressure	Pascal (or mmHg)
P_{alv}	Alveolar pressure	Pascal (or mmHg)
P_{atm}	Atmospheric pressure	Pascal (or mmHg)
P_{cap}	Hydrostatic pressure within the capillary	Pascal (or mmHg)
P_{CO2}	Partial pressure of CO_2	Pascal (or mmHg)
pH	Negative logarithm of $[H^+]$	
pK_a	Negative logarithm of K_a	
P_{O2}	Partial pressure of oxygen	Pascal (or mmHg)
Q	Flow rate	g/s, L/s
Q_{in}	Inflow rate	g/s, L/s
Q_{out}	Outflow rate	g/s, L/s
R	Resistance to flow or current	Pascal-s/L (to volume flow) Ohm (to current)
t	Time	s
V	Volume	L
V_{max}	Maximum reaction velocity	Moles/L-s
V_N	Total volume at a given time interval	L
V_{N-1}	Total volume at the beginning of the time interval	L
$\dot{V}$	Volumetric flow rate	L/s
$\dot{V}_A$	Alveolar ventilation	L/s
$\dot{V}_{in}$	Volumetric flow rate of material in of a system	L/s
$\dot{V}_{net}$	Overall rate of volume flow across the respiratory membrane	L/s
$\dot{V}_{o_2}$	Oxygen consumption rate by the body	L/s
$\dot{V}_{out}$	Volumetric flow rate of material out of a system	L/s
$\dot{V}_T$	Total ventilation rate	L/s
[X]	Concentration of X	Moles/L or g/L
y_{O2}	Mole fraction of oxygen in the air	—

GREEK

δ	Thickness of the respiratory membrane	cm
ρ	Density	g/L, kg/m^3
β	Buffering power of a solution	Moles/pH unit

QUESTIONS

1. Give three examples of biochemical reactions in the body that produce water.
2. Why do you think that physiologists defined a new parameter, the diffusing capacity D_L, to describe diffusion through the respiratory membrane, instead of using the physical parameters listed in Equation 7.22?
3. Cholera is a pathogen that infects the intestine, producing a toxin that blocks one of the G proteins (preventing normal cycling), and locks a Cl channel open (the same channel that is involved in cystic fibrosis: CFTR channel). Water reabsorption is severely impaired. Patients die of dehydration, sometimes losing liters of fluid within a few hours. Why would blockage of an ion channel effect water reabsorption?

PROBLEMS

1. Estimate the fraction of water (as a percentage) that leaves your body per day in urine, feces, sweat, tears, and expired air. Justify the assumptions that you make in this calculation.
2. A kitchen sink has a volume of 15 L. If the drain is closed and the sink is initially empty, plot the volume of water in the sink as a function of time. How long will it take to overflow the sink after the faucet is open, allowing a flow rate of 0.3 L/min?
3. Re-answer the previous problem, but assume that the drain is partially opened 10 minutes after the water flow starts, allowing water to flow out at a constant rate of 0.1 L/min.
4. For the sink in the previous two problems, assume that the drain is partially open, such that the rate of flow out of the sink is proportional to the amount of water in the sink: $\dot{V}_{\text{out}} = \lambda V$. Find an equation for the volume of water in the sink as a function of time.
5. Assume that the sink of the previous problems is operating with the following steady-state conditions: The volume of water in the sink is 5 L, and water is flowing into and out of the sink at a steady rate of 0.3 L/min. A quantity of dye molecule (20 g) is rapidly added to the volume in the sink. Plot the concentration of dye molecule in the sink as a function of time after its rapid introduction.
6. A person consumes 100 μg of a tracer chemical. Assume that the person is able to collect all of the tracer in their urine (and therefore measure the

amount that has come out of the body), as well as the concentration in the blood, as a function of time (see table).

a. Is a first-order rate constant appropriate for describing the process of elimination via the kidneys? Justify your answer.

b. Assuming that the answer to **a** is "yes," find the rate constant k and the total volume V from these data.

Time (h)	Amount collected in urine during the previous hour (μg)	Concentration in the blood (μg/L)
1	9.4	90
2	8.7	82
3	7.9	75
4	6.9	66
5	6.5	61
6	5.7	55
7	5	50
8	4.9	43
9	4.3	39
10	3.9	35

7. In the previous problem, is the value of V that you determined equal to the volume of the person? Does this make sense to you? If not, what does V represent?
8. You consume a soft drink containing 100 mg of artificial sweetener. Assume that this chemical is immediately absorbed into your body and that 30% of it is eliminated from your body in 3 hours. What is the first-order rate constant, k, describing the elimination of the sweetener from your body over this period?
9. Exhaled air is ordinarily at body temperature and 100% humidity (i.e., it is fully saturated with water). Estimate the mass of H_2O that is lost each day because of normal breathing. The saturation vapor pressure (SVP) of water at 37°C is 47 mmHg and contains 44 mg/L of water, whereas at room temperature (20°C) it is 20 mmHg and contains only 18 mg/L.
10. Using the data in Table 7.2, calculate the following values:
    a. The mass of oxygen gain in the body per day
    b. The mass of carbon dioxide loss from the body per day
    c. The mass of water loss from the body per day
    d. The net mass change in a day because of breathing
11. This chapter illustrates the use of a mass balance approach to estimate the relationship between alveolar ventilation rate and partial pressure of oxygen in the alveolar gas (Figure 7.13). Perform a similar analysis on carbon dioxide and plot the alveolar partial pressure of carbon dioxide as a function of alveolar ventilation rate for two carbon dioxide generation rates in the body: 200 mL/min and 500 mL/min.

12. As shown in Figure 7.13, the mass balance model derived and presented in Equation 7.16 predicts that the partial pressure of oxygen is zero when the alveolar ventilation rate is less than 5 L/min (for an oxygen consumption rate of 250 mL/min). In fact, if you use Equation 7.16 to calculate the oxygen partial pressure for alveolar ventilation between 0 and 5 L/min, the equation yields a negative number for partial pressure. Is a negative partial pressure possible? Is the model useful for this range of parameters? Explain.
13. The diffusing capacity D_L is 21 mL O_2/min-mmHg. Using the value for K in Box 7.3, estimate the diffusion coefficient for O_2 in the respiratory membrane. Estimate values for A and δ from information provided in the text, and justify your estimates.
14. Using the Henderson–Hasselbalch equation, if the CO_2 concentration is increased from 40 mmHg to 80 mmHg, how much does the pH change? What is the change in H concentration that accompanies this change?
15. Show that Equation 7.7 is a valid solution to Equation 7.6. For what values of B is the solution valid?
16. A premium ice cream contains 16 g of fat and 20 g of sugar per serving (4 oz). A brand of frozen yogurt contains 4 g of fat and 32 g of sugar per serving. Calculate the fuel value of a serving of each item. Both items contain 5 g of protein.
17. A person of average mass burns about 30 kJ/min playing tennis. Find the time a person would have to spend playing tennis to burn up the energy in a 2-oz serving of cheese.
18. Develop a better model for ingestion of a tracer, which includes two separate systems: one system representing the intestines (call it compartment 1) and another system representing the rest of the body (compartment 2).
    a. Draw a block diagram showing the two systems with arrows connecting the inputs and outputs in appropriate ways.
    b. Write a differential equation for the mass balance of tracer in each open system.
    c. Show that the two equations are satisfied by functions of the form:

$$D = Ce^{-k_1 t}$$
$$M = Ae^{-k_1 t} + Be^{-k_2 t}$$

    where M is the amount of tracer in compartment 2 as a function of time, D is the amount of tracer in compartment 1 as a function of time, and A, B, and C are constants.

For help with Problem 18, or more advanced reading on this subject, see Chapter 7 in Saltzman, W.M., *Drug Delivery: Engineering Principles for Drug Therapy*, New York: Oxford University Press (2001).

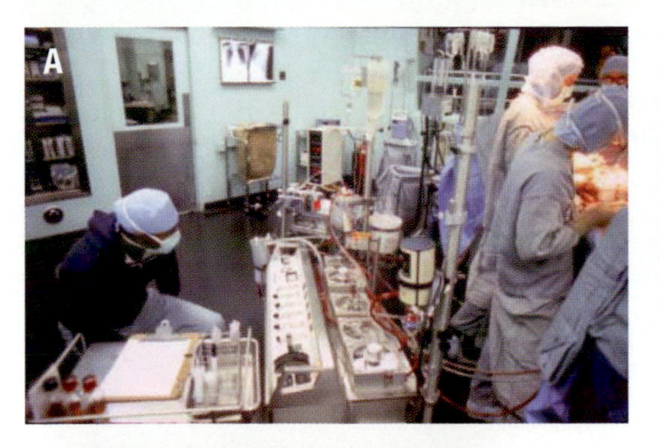

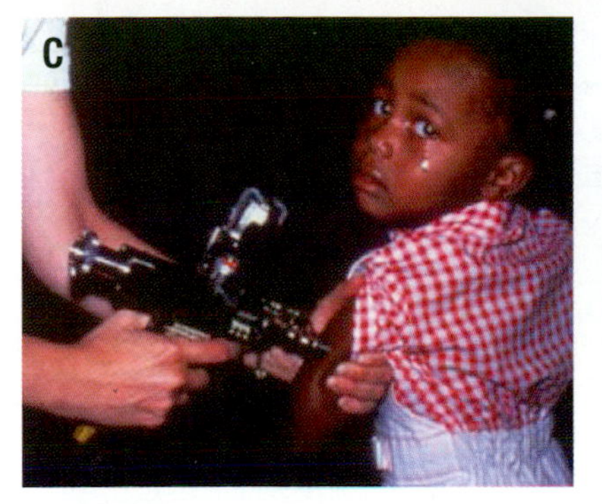

D

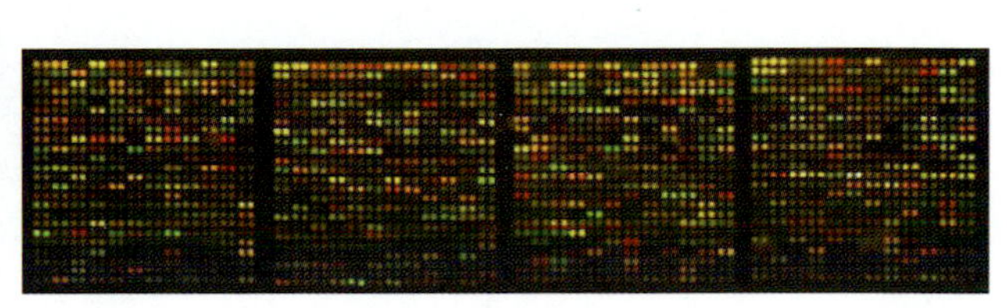

Plate 1.3 **Examples of new technology that permitted medical advances. A.** Heart–lung machine that permits heart transplantation and surgery. Photo courtesy of National Institutes of Health. **B.** Jet airplanes are used for rapid transport of a preserved organ to a distant operating room. **C.** An injector for vaccine delivery. Photo courtesy of The Centers For Disease Control and Prevention. **D.** DNA microarrays can be used to measure the expression of genes in cells and tissues. Photo courtesy of the W. M. Keck Foundation at Yale University.

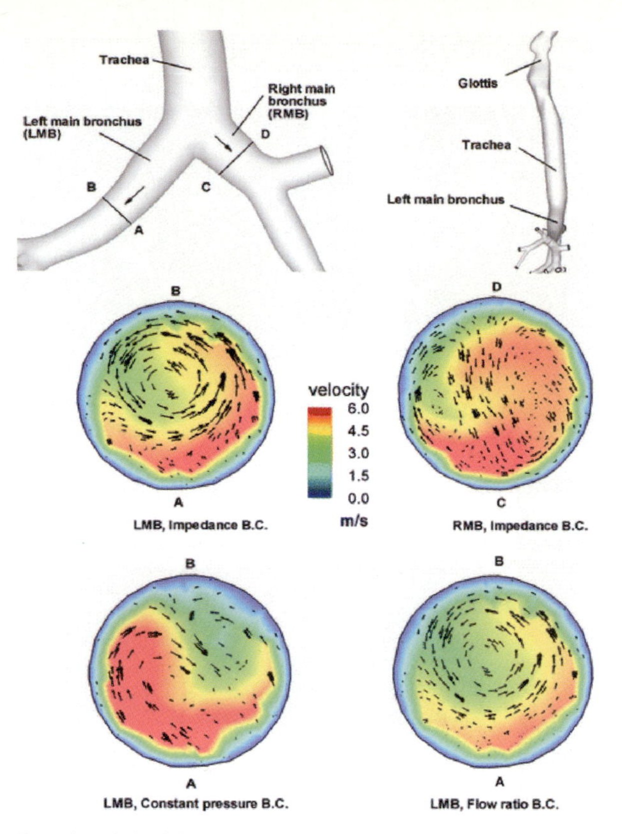

Plate 1.9 **Examples of physiological modeling.** Modeling of air flows in the left main bronchus and right main bronchus under different conditions within the lung. Models such as this one are useful in evaluating breathing patterns for patients with different lung diseases, and quantifying the extent of disease in a particular patient from (7) with permission.

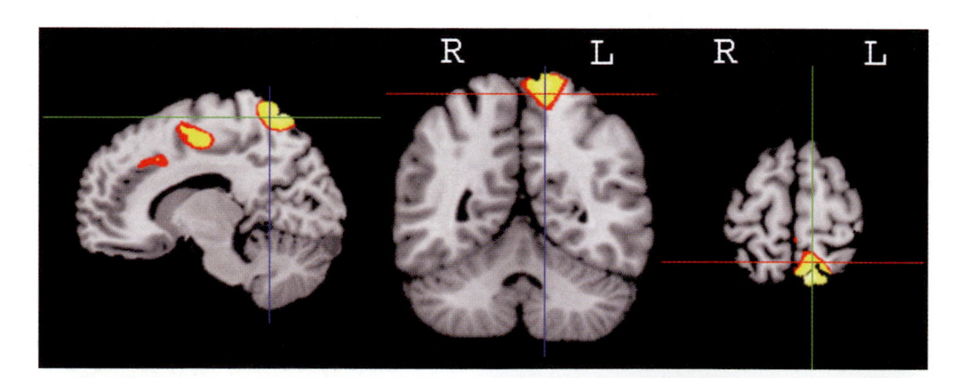

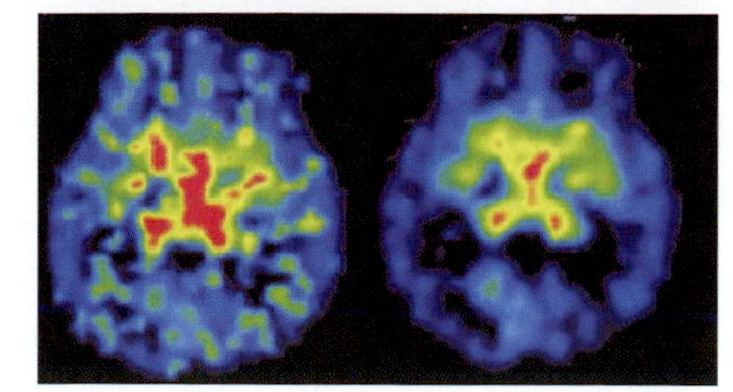

Plate 1.11 **Biomedical imaging.** This field is described in Chapter 12. Functional magnetic resonance (fMRI, top) and positron emission tomography (PET, right) are now providing functional and metabolic information in addition to anatomical information. Top panel courtesy of Todd Constable, Yale University. Bottom panel reprinted by permission from Macmillan Publishers Ltd: *J Cereb Blood Flow Metab.* 2003;23:1096–1112.

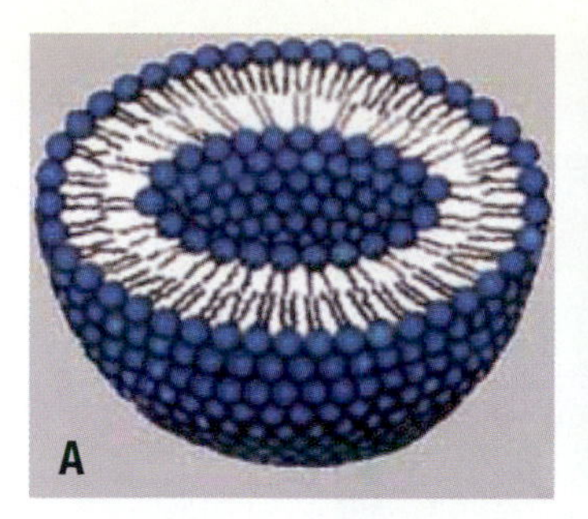

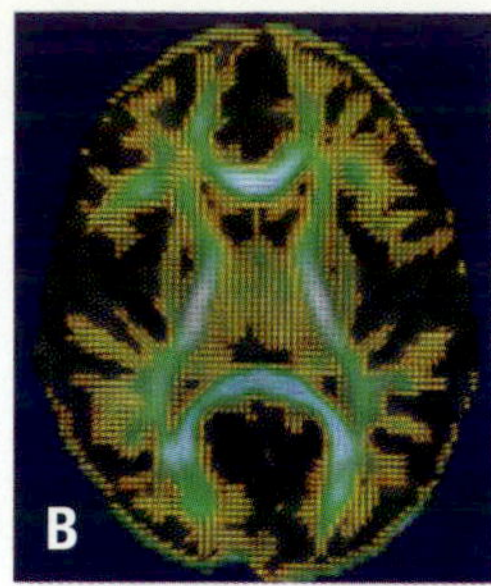

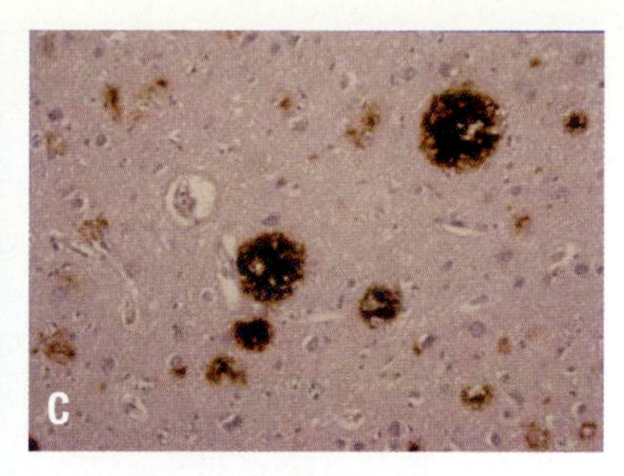

Plate 2.1 **Examples of chemistry in biomedical engineering.** **A**. Liposomes are synthetic structures produced by assembly of lipids into small sacs or vesicles. **B**. Diffusion tensor imaging (DTI), which maps the diffusion of water in the brain using magnetic resonance imaging (MRI). Photo courtesy of A. Brock and L. Staib, Yale University. **C**. Collection of proteins that accumulate in the brains of patients with Alzheimer's disease, called an Alzheimer's plaque. Photo courtesy of BRACE-Alzheimer's Research Registered UK Charity No. 297965.

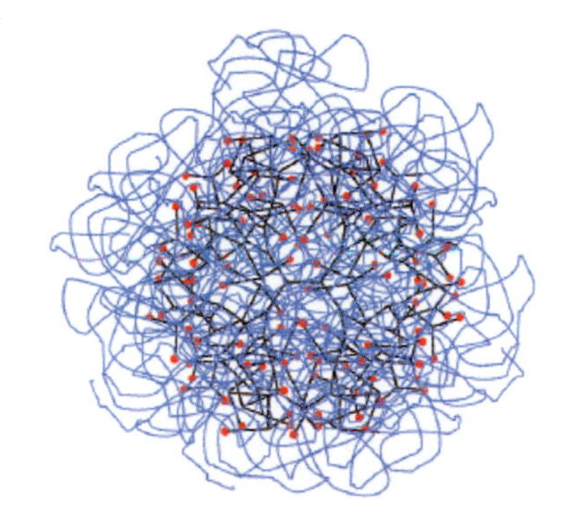

Plate 2.4 **Biomedical engineers design new molecules.** This diagram shows the structure of a polymer built for drug delivery and targeting. A polyamidoamine dendritic polymer (black) with poly(ethylene glycol) arms (blue) attached to the dendrimer via covalent linkages (red). Reproduced from (1) with permission, copyright 2002 American Chemical Society.

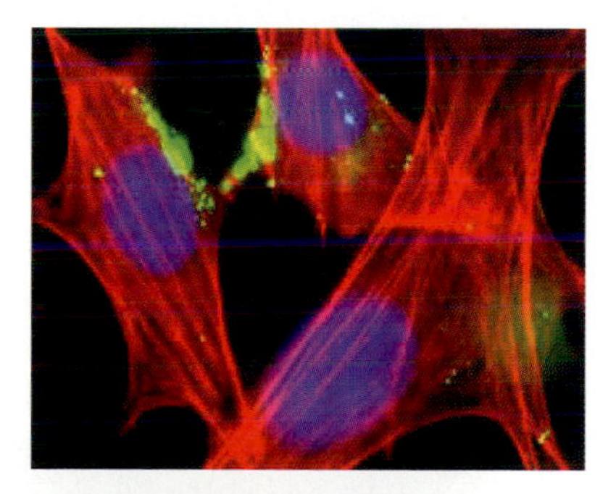

Plate 3.1 **Gene therapy vectors and designer deoxyribonucleic acid (DNA).** Polyoma virus-like particles associating with actin fibers in cultured cells. Reprinted with permission from Macmillan Publishers Ltd: *Gene Therapy*. 2006;7(24):2122–2131.

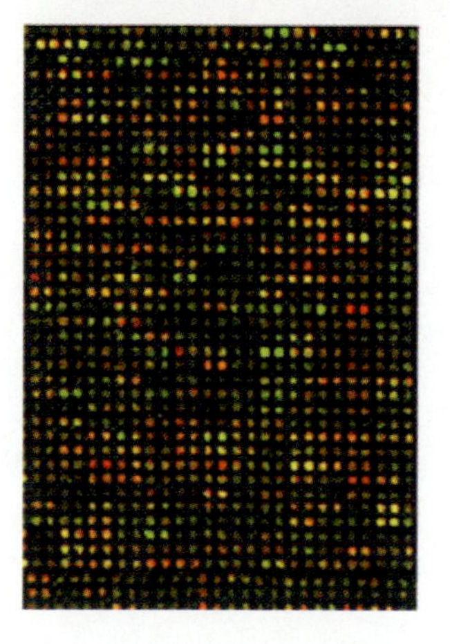

Plate 3.2 **Systems biology is one of the frontier areas for biomedical engineers.** Deoxyribonucleic (DNA) microarray, in which DNA probes are attached to a solid surface, with each dot on the surface containing many copies of one particular known DNA sequence. This microarray allows researchers to screen for the presence of many individual genes in a small sample of tissue or cells. Photo credit: Dr. M. A. Saghai Maroof, Virginia Polytechnic Institute and State University.

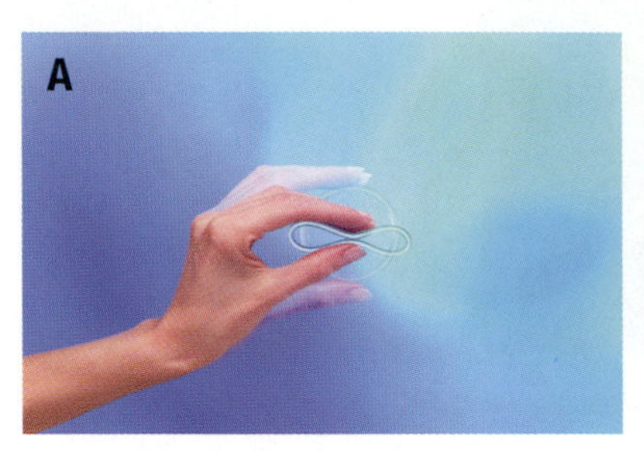

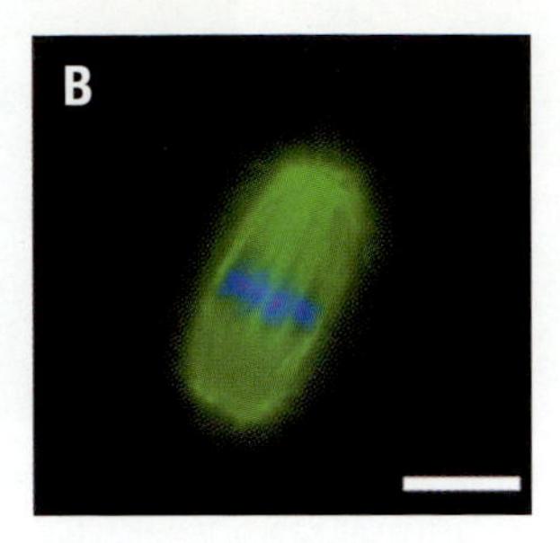

Plate 3.8

Biomedical engineers and reproductive health. A. NuvaRing® contraceptive is a soft polymer vaginal ring that releases hormonal contraceptives over a period of several weeks. Lower doses of hormones can be used, because they are released directly into the reproductive tissues. Photo used with permission. **B.** An oocyte that was maintained in a three-dimensional tissue engineering matrix retains its ability to undergo mitosis. Reprinted from (3) with permission from Elsevier.

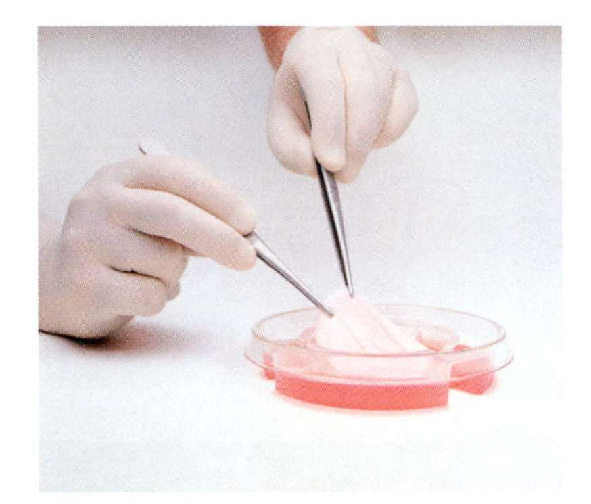

Plate 5.3

Cells can be used in tissue engineering. Apligraf® (Organogenesis Inc) is a skin substitute containing living cells. The skin substitute is constructed by first culturing human dermal cells within a gel of collagen protein. After culturing for 6 days, a layer of epidermal cells is added to the top of the collagen/dermal cell gel. This layered structure can be used to treat skin wounds that are difficult to heal, such as skin ulcers that occur in diabetics. Photo used with permission.

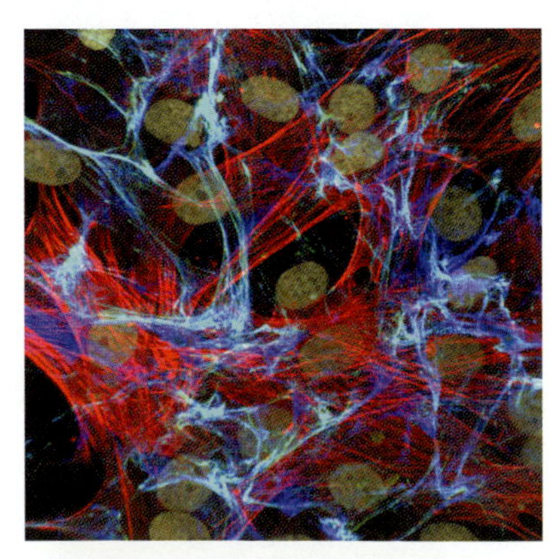

Plate 5.6

Fibroblast cells are surrounded by an extracellular matrix. Image by Dr. Boris Hinz, Laboratory of Cell Biophysics, Ecole Polytechnique Fédérale de Lausanne, Lausanne, Switzerland.

Plate 6.8

Neurons. Neurons are characterized by long processes, which reach out to connect with and collect information from other cells in nearby or distant tissues. The growth of processes called dendrites was stimulated in this cell by the release of brain-derived neurotrophic factor (BDNF) from a biodegradable polymer in the vicinity of the neuron. Photo courtesy of Melissa Mahoney.

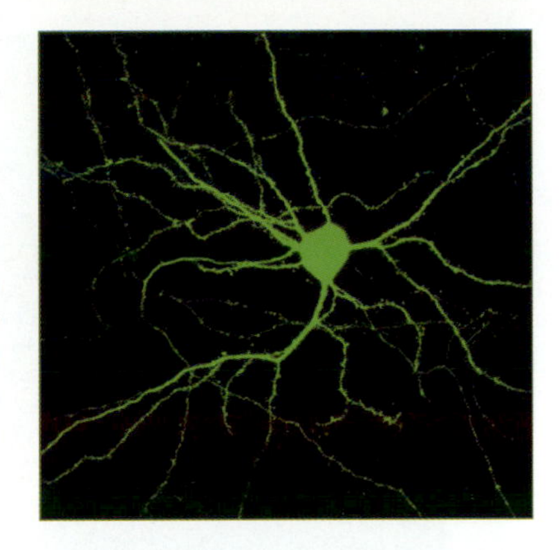

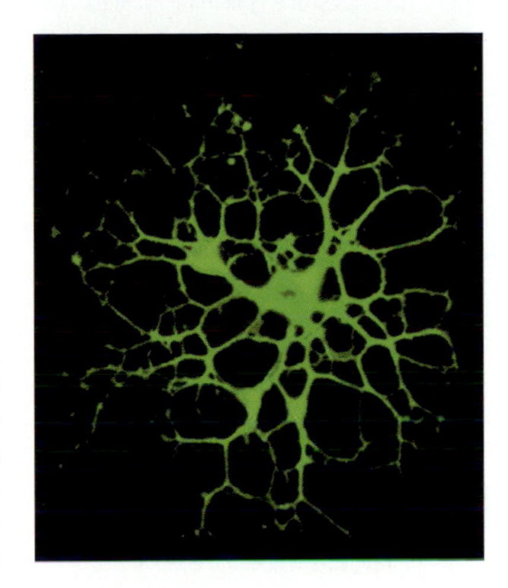

Plate 6.11

Oligodendrocyte. Ninety percent of the cells in the brain are glial cells, which support the function of neurons. Cells such as this oligodendrocyte provide myelin coatings for neuron processes, which enable efficient propagation of action potentials. Photo courtesy of Dr. Regina Armstrong, Uniformed Services University of the Health Sciences.

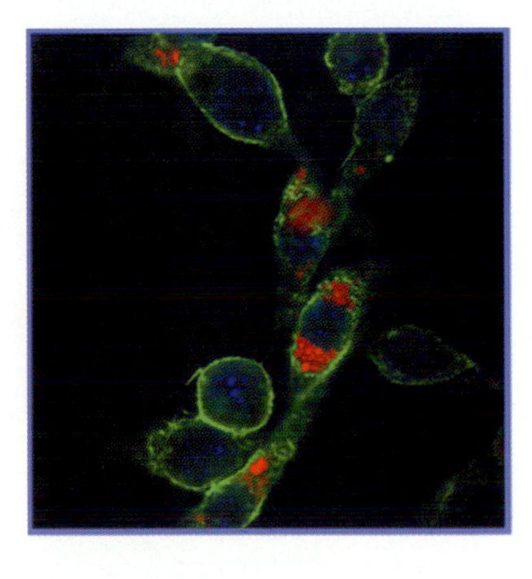

Plate 6.20

Phagocytosis by antigen-presenting cells. Many cells of the body are capable of phagocytosis—or ingestion of particulate matter—but antigen presenting cells, such as this dendritic cell, are especially good ingesters. These cells have ingested tiny polymer particles (red), which were placed in the environment surrounding the cells. Photo courtesy of Stacey Demento and Dr. Tarek M. Fahmy, Yale University.

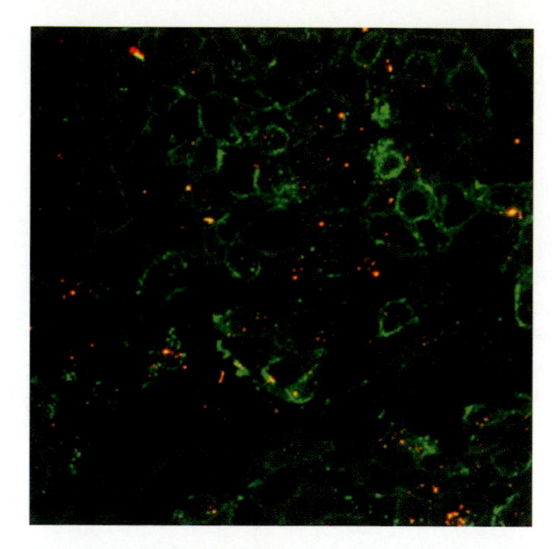

Plate 7.3 **Examples of engineering systems.** Almost anything can be defined as an engineering system and then analyzed. Some examples of engineering systems are buildings, people or animals, and cells (shown here). The cells are outlined with a green dye, and they contain red drug-delivery particles.

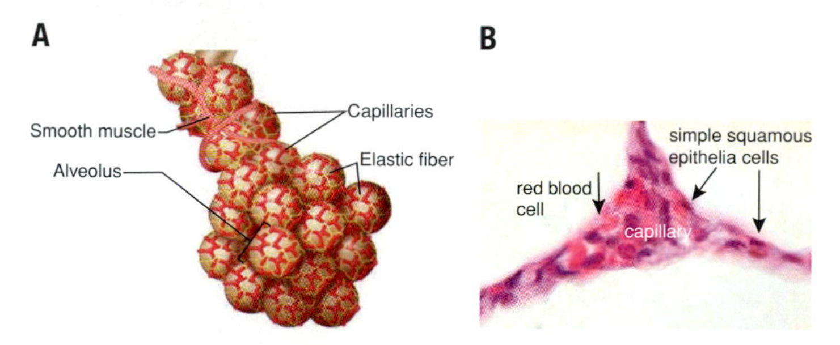

Plate 7.17 **Diffusion across the respiratory membrane. A.** Each alveolus is surrounded by a dense array of capillaries, so that a continuous sheet of blood is flowing over the surface of the alveolar membrane, as shown in **(B)**. This arrangement puts the alveolar gas in close proximity to flowing blood. Photo credit: Rose M. Chute.

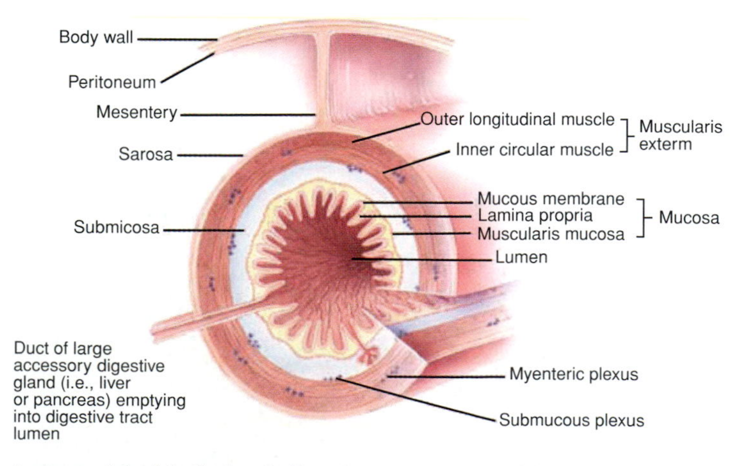

Plate 7.19 **Anatomy of the intestinal wall.** The microscopic anatomy of the wall of the intestine varies in esophagus, stomach, and small and large intestine, but many features are retained. From outside to lumen, the layers are: serosa, longitudinal muscle layer, circular muscle layer, submucosa, and mucosa. Two systems of nerves, one from the central nervous system and another called the enteric nervous system, control movements of the intestinal wall.

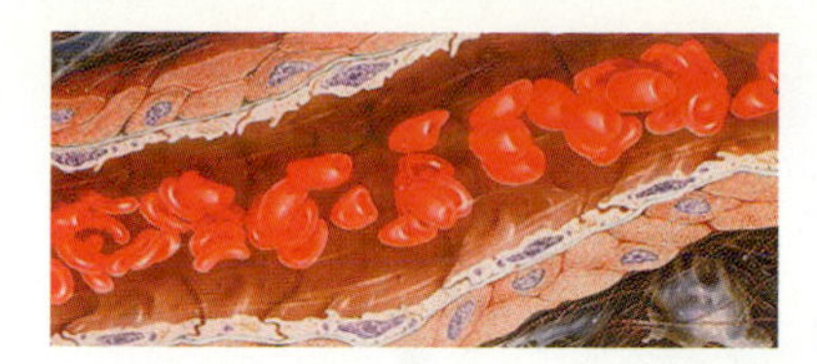

Plate 8.1 **Blood cells flowing through an arteriole.** Cells in some vessels can deform to orient themselves with the flow to reduce resistance. Image from Schmid-Schönbein, H., Grunau, G. and Brauer, H. Exempa hamorheologic 'Das strömende Organ Blut,' Albert-Roussel Pharma GmbH, Wiesbaden, Germany, 1980, and provided by Professor Oguz K. Baskurt.

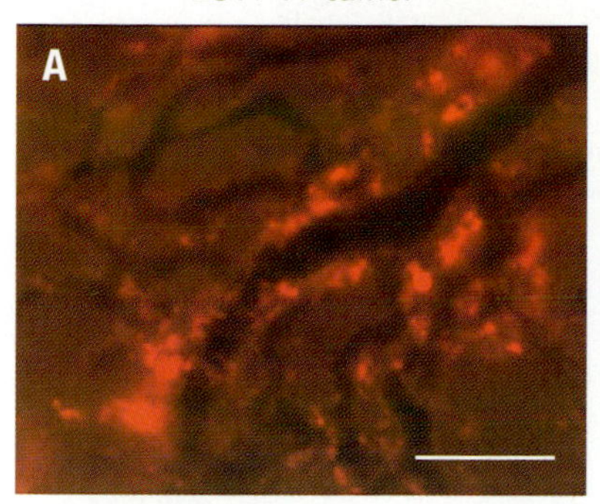

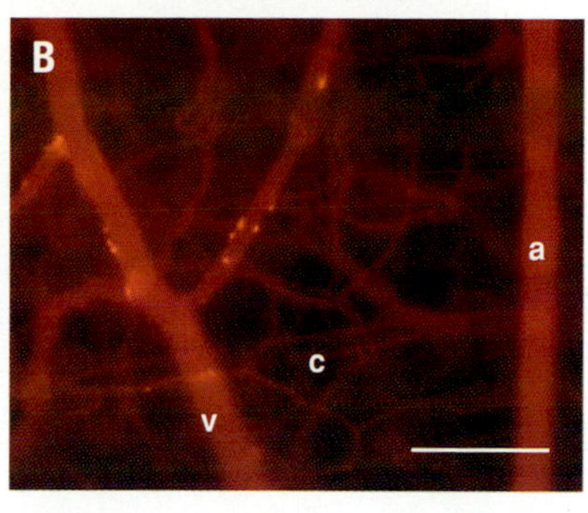

Plate 8.11 **Heterogeneous distribution of liposomes in tumor and normal tissues.** Human colon adenocarcinoma cells (LS174T) were transplanted in mice, and fluorescently labeled liposomes were injected intravenously. The photos were taken 2 days after the injections. **A.** In solid tumor, liposomes accumulated only in perivascular regions. Bar = 100 μm. **B.** In normal subcutaneous tissue, liposomes accumulated only in the wall of small postcapillary venules (6–25 μm in diameter). Neither parallel capillaries (c) nor arterioles (a) and large collecting venules (v, > 25 μm) were labeled by liposomes. Bar = 100 μm. Photo courtesy of Dr. Fan Yuan and modified with permission from: Yuan F, Leunig M, Huang SK, Berk DA, Papahadjopoulos D, Jain RK. Microvascular permeability and interstitial penetration of sterically stabilized (stealth) liposomes in a human tumor xenograft. *Cancer Res.* 1994;54:3352–3356.

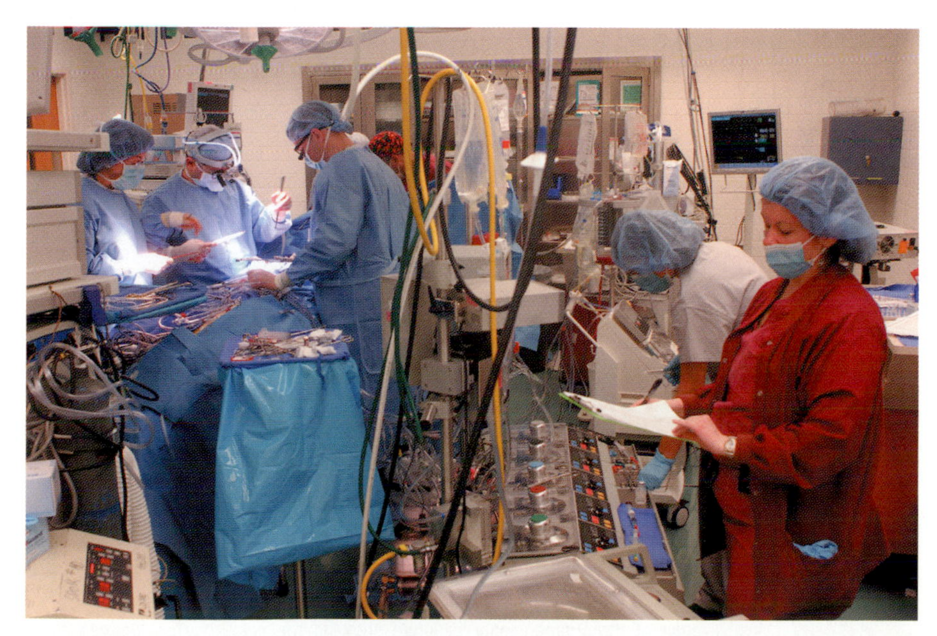

Plate 11.1 **Instruments in an operating room.** Modern surgical suites are filled with instruments that provide continuous monitoring of a patient's vital signs and that aid in the surgical procedures. These instruments make surgery safer for the patient, and enable surgeons to perform operations with speed and precision that would be otherwise unattainable. Photograph courtesy of Yale University Media and Technology Services.

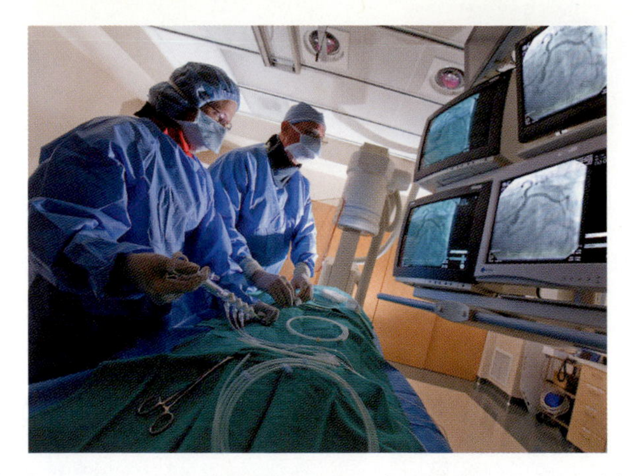

Plate 11.17 **Doctors in a catheterization laboratory, where devices are inserted deep into the body for measurement of physiology.** Photo courtesy of Intermountain Healthcare.

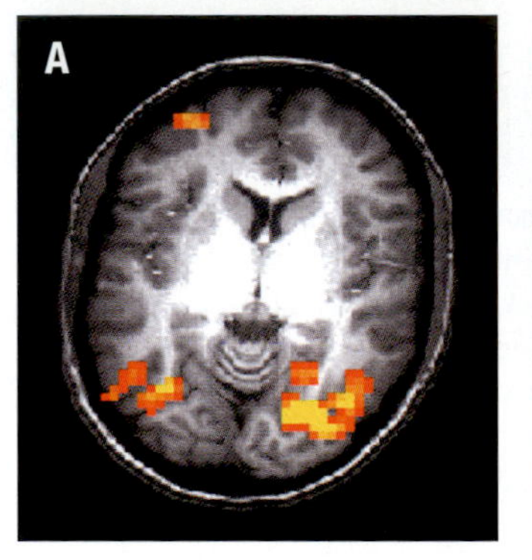

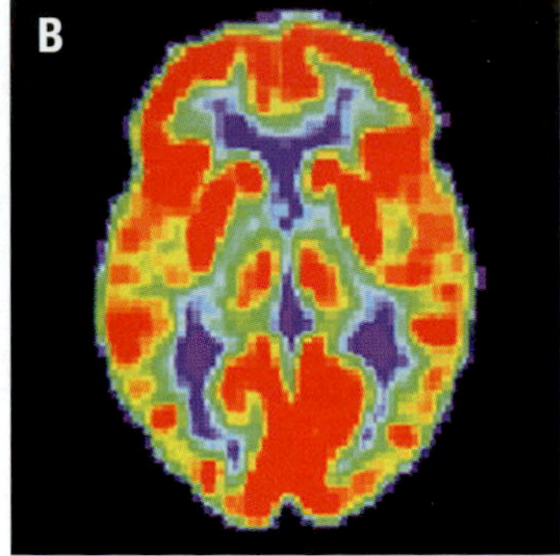

Plate 12.1 **Examples of imaging with new technology. A.** Functional magnetic resonance imaging (MRI; color) superimposed on anatomical MRI. Image courtesy of Dr. Jody Culham. **B.** Positron emission tomography scan obtained from a normal human brain. Image courtesy of the National Institute on Aging.

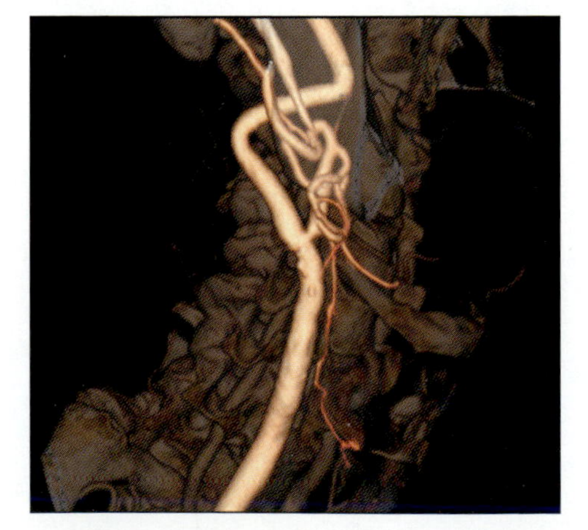

Plate 12.10 **Visualization of three-dimensional structure showing the carotid artery derived from computed tomography.** Image provided courtesy of GE Healthcare.

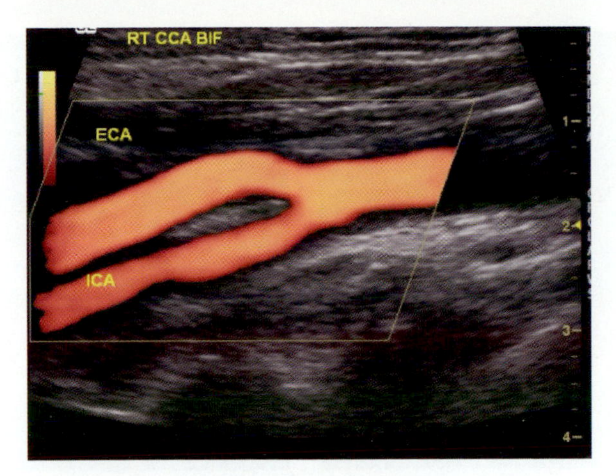

Plate 12.13 **Doppler image of the bifurcation of the carotid artery.** Image courtesy of Dr. James D. Rabinov, Massachusetts General Hospital and used with permission.

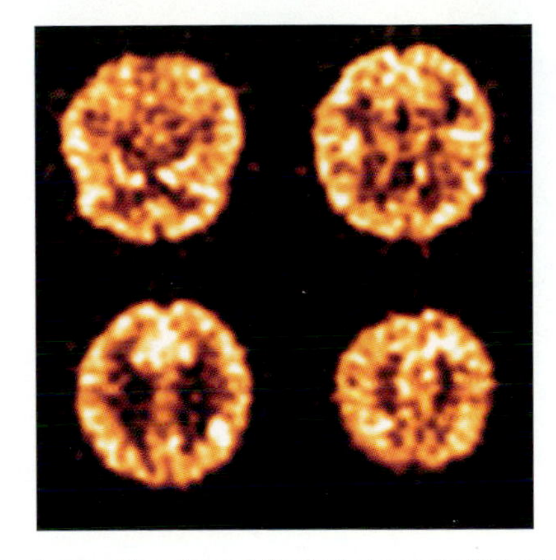

Plate 12.18 **Image slices through the brain imaged with positron emission tomography.**

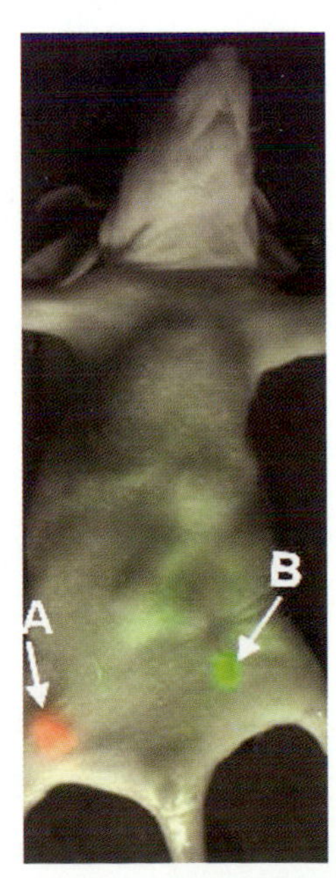

Plate 12.21 **Imaging of cells that express fluorescence in a living animal. A.** This location contains a U87 tumor with GFP fluorescence, emitting at 520 nm. **B.** This location contains a tumor with NB1691 Luciferase fluorescence with peak emission at 480 nm. Image courtesy of M. Waleed Gaber, PhD, Biomedical Engineering and Imaging Department, University of Tennessee Health Science Center, Memphis.

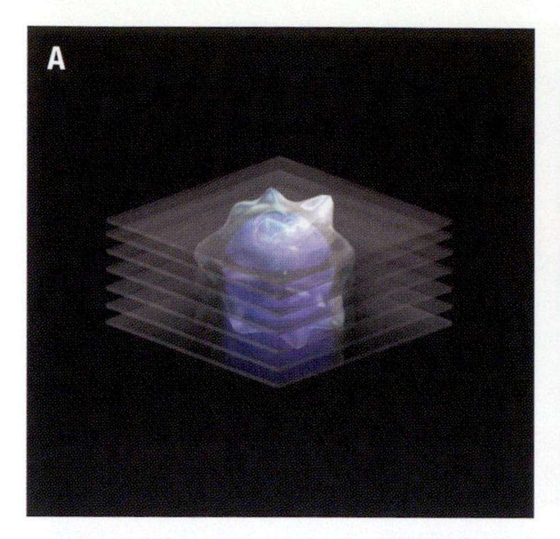

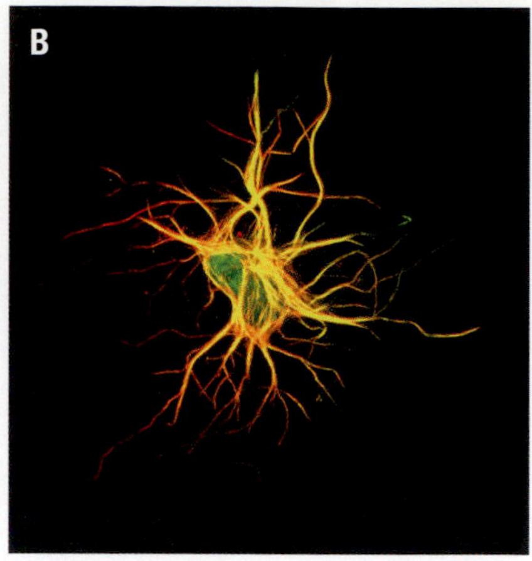

Plate 12.22

Confocal image. A. Series of images is taken on different focal planes. This set of sections—called a serial section—can be assembled into a three-dimensional image, as in this image of a hippocampal neuron **(B)**. Image courtesy of Robert S. McNeil and Dr. Gary Clark, Cain Foundation Laboratories, Baylor College of Medicine.

Plate 12.26

Example of intensity values in a digital image. Zoomed-in image segment showing discrete (quantized) intensity values at discrete (sampled) locations with grey level values displayed at each pixel. Image provided courtesy of USGS/EROS, Sioux Falls, SD.

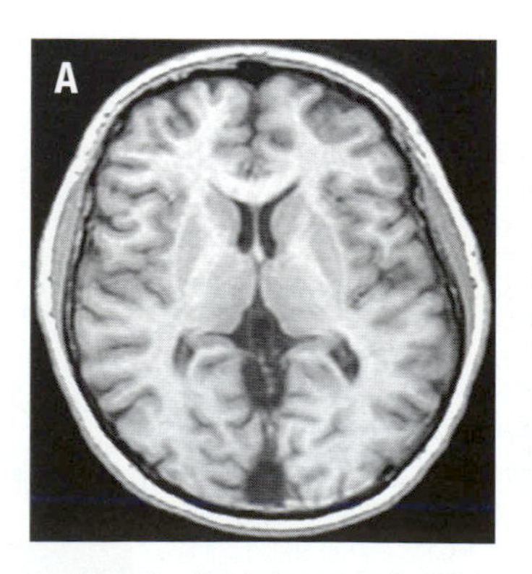

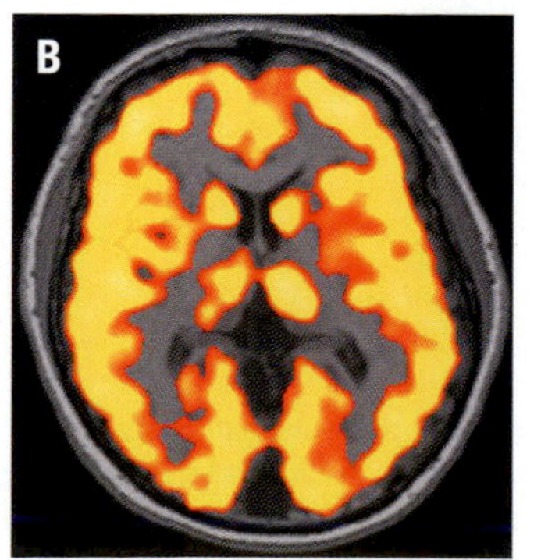

Plate 12.29

An example of brain image registration showing a magnetic resonance image aligned (A) and overlaid **(B)** with a positron emission tomography image from the same subject, showing the corresponding structure and function.

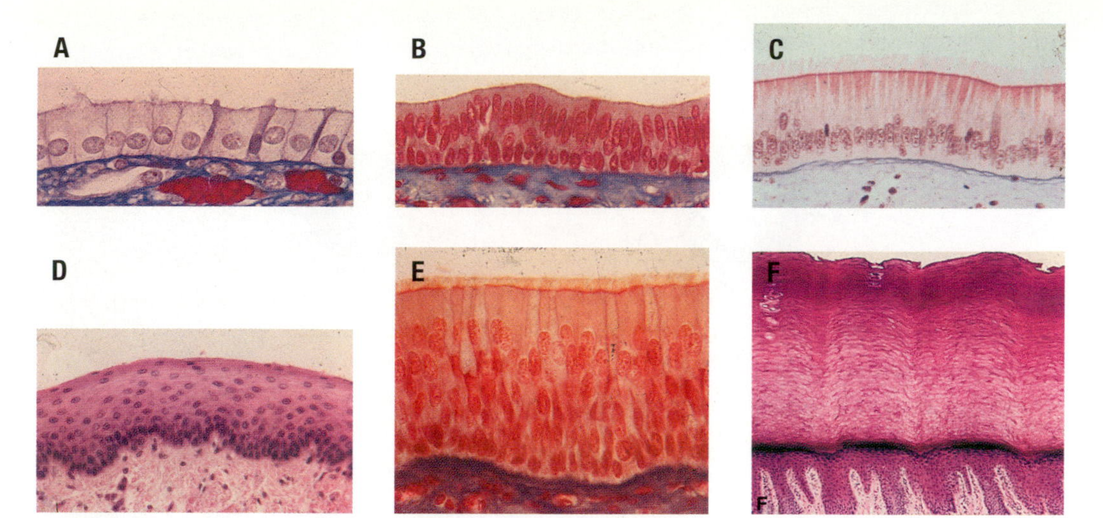

Plate 13.9 **Layered structures in epithelial tissues.** Many tissues within the body are orderly, often layered structures, with different populations of cells arranged in positions that contribute to overall function of the tissue. These microscopic images show simple columnar epithelium (kidney tubule) (A); stratified columnar epithelium (B); ciliated columnar epithelium (C); stratified squamous epithelium (esophagus) (D); ciliated pseudostratified epithelium (trachea) (E); and keratinized stratified squamous epithelium (skin) (F). Reprinted with permission from: *Bloom & Fawcett: Concise Histology* (5).

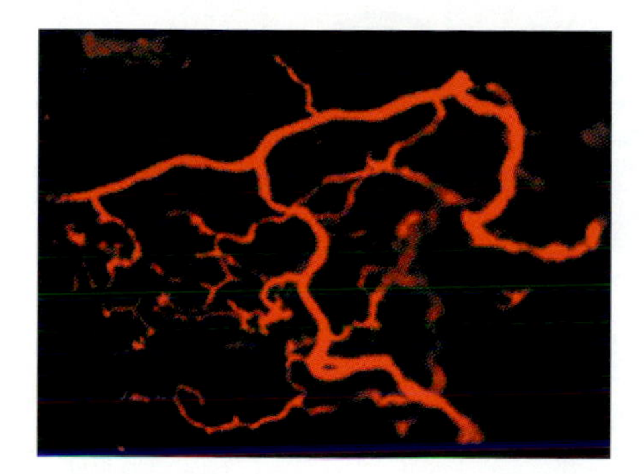

Plate 13.10 **Tissue engineering approach to create new vascular beds.** Human endothelial cells and microspheres that slowly release vascular endothelial growth factor are both suspended in three-dimensional gels. After implantation of the construct into a mouse, new blood vessels are formed in the mouse. These blood vessels are lined with human endothelial cells, but have mouse blood flowing through them. Used with permission. Copyright 2005 National Academy of Sciences, USA. For more details, see (6), (7), and (8).

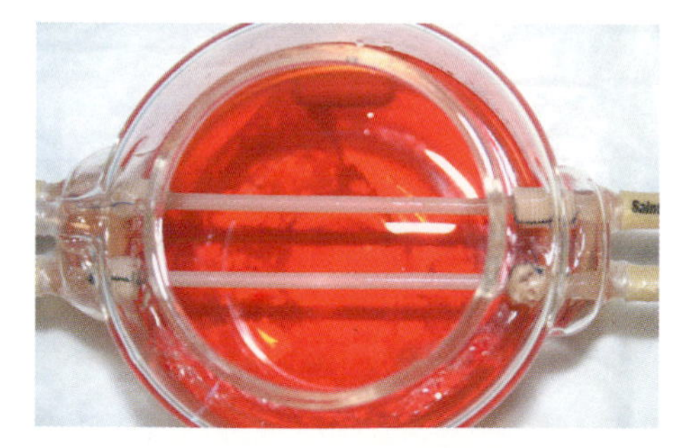

Plate 13.11 **Tissue-engineered porcine arteries created in the laboratory, in a bioreactor, over a period of 8 weeks.** To form these arteries, tubes of a synthetic, biodegradable polymer mesh were seeded with vascular smooth muscle cells and then grown in a special bioreactor under conditions of pulsatile radial strain. The rapidly degrading polymer mesh was replaced by cells and secreted extracellular matrix, including collagens and glycosaminoglycans. The resulting tissue engineered arteries were 8 cm in length and 3 mm in internal diameter. Photo courtesy of L.E. Niklason. For more information, see (9).

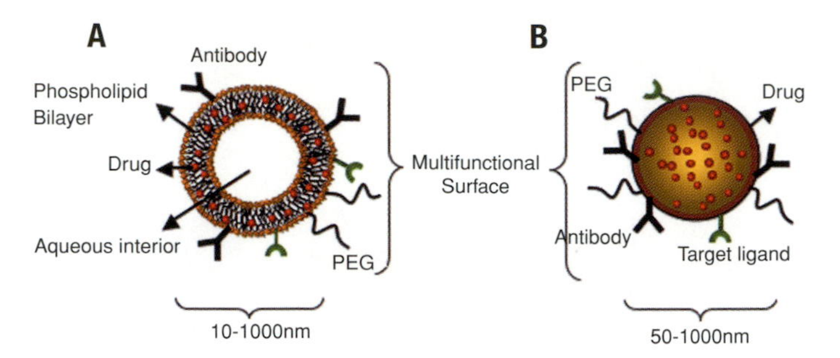

Plate 13.13

Examples of drug-delivery nanoparticles. A. Liposomal systems are vesicular with targeting or stealth groups, such as Poly(ethylene glycol) (PEG), attached to the surface lipids. **B.** Solid biodegradable nanoparticles are formed from a polymer emulsion with drug dispersed in the polymer matrix and targeting or PEG groups attached to the surface.

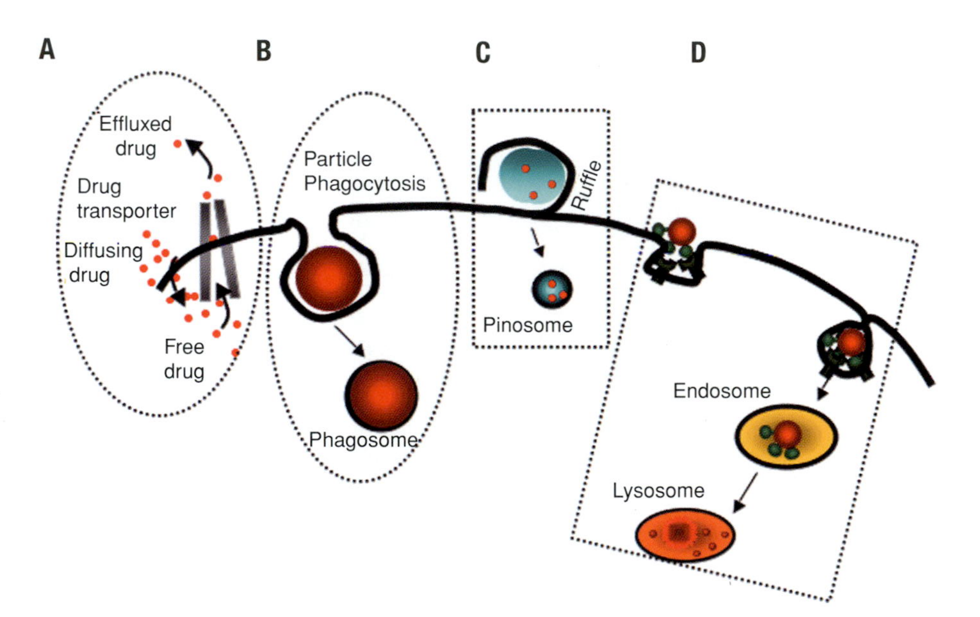

Plate 13.15

Processes leading to cellular delivery of drugs. A. Passive diffusion of free drug. **B.** Nonspecific phagocytosis of a nanoparticle. **C.** Drug entrapped in fluid and uptake by pinocytosis. **D.** Receptor-mediated endocytosis. Adapted from (11).

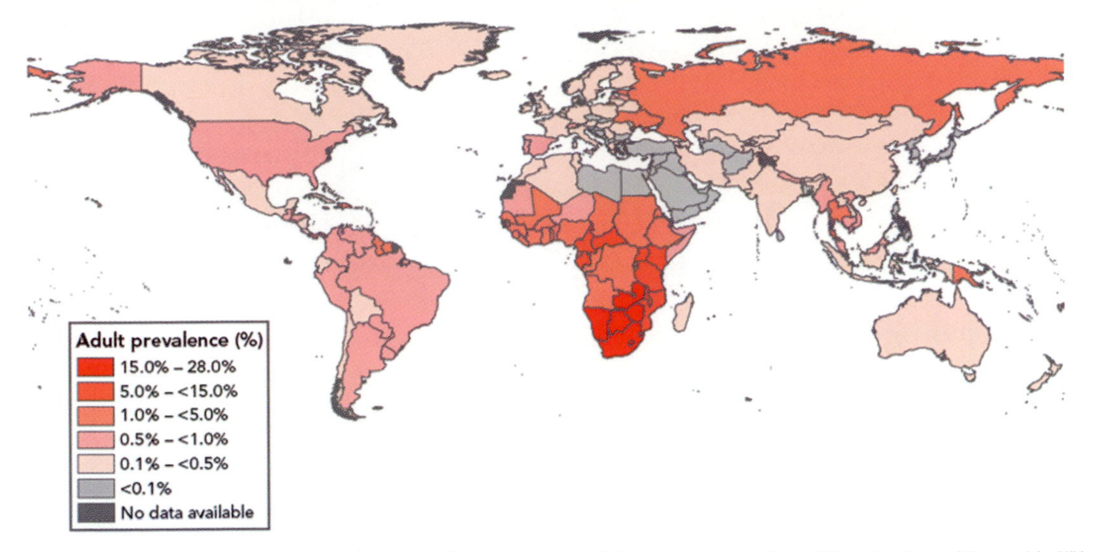

Plate 14.13

Prevalence of human immunodeficiency virus (HIV) infection among adults in different regions of the world. HIV infections are most prevalent in areas of Africa, but there are few regions of the world that are untouched by the virus. In 2007, it was estimated that 33 million people were living with HIV infections Reprinted with permission from the UNAIDS 2008 "Report on the Global AIDS Epidemic."

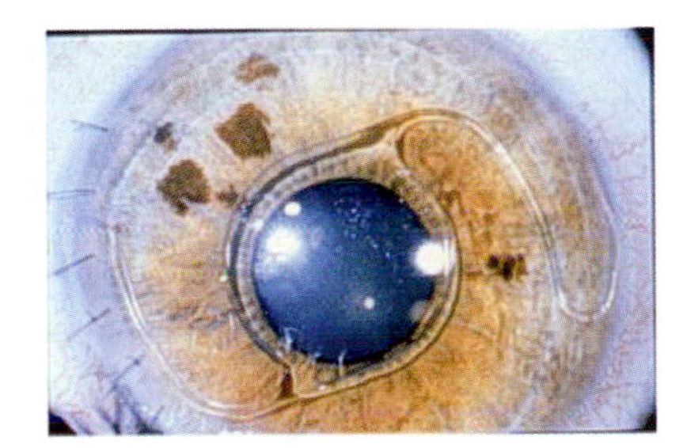

Plate 15.2

Plastic implantable lenses and catheter. An Artisan® phakic intraocular lens, which is implanted in front of the natural lens in extremely nearsighted patients, to allow normal vision without glasses. The artificial lens is made of polymethyl methacrylate. Image courtesy of Dr. James J. Salz and the American Eye Institute.

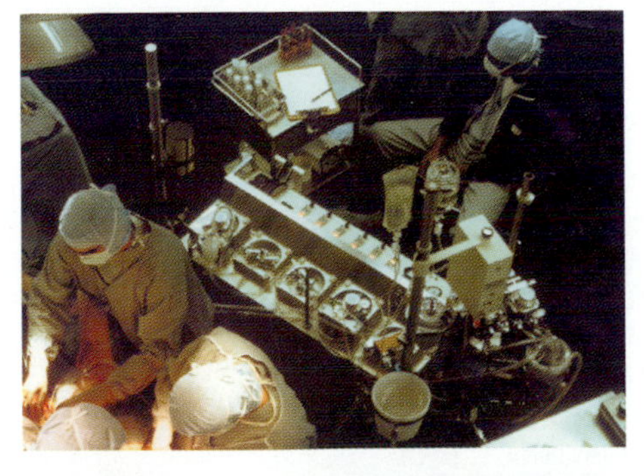

Plate 15.11

Heart–lung machines. Image of a perfusionist operating a heart–lung machine. Photo courtesy of National Institutes of Health.

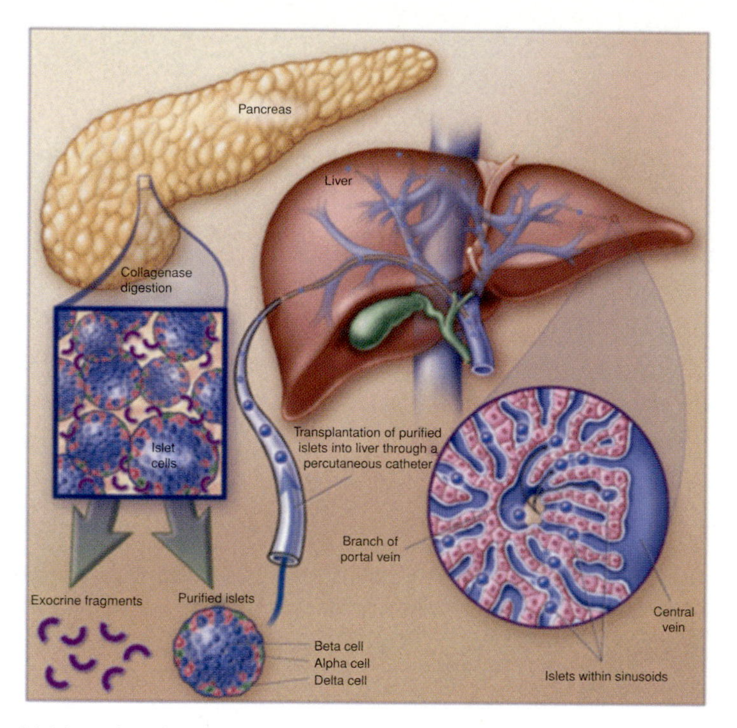

Plate 15.21

Islet transplantation. A pancreas is obtained from a donor. The pancreas is digested with collagenase to free islets from the surrounding exocrine tissue. The freed islets, containing mostly beta and alpha cells, are purified to remove remaining exocrine cellular debris. The purified islets are infused into a catheter that has been placed percutaneously through the liver into the portal vein, whence they travel to the liver sinusoids. Drawing and caption from (17) and used with permission.

Plate 16.2

Histological section showing cervical cancer, specifically, squamous cell carcinoma in the cervix. Tissue is stained with Pap stain and magnified 200×. Photo courtesy of National Cancer Institute.

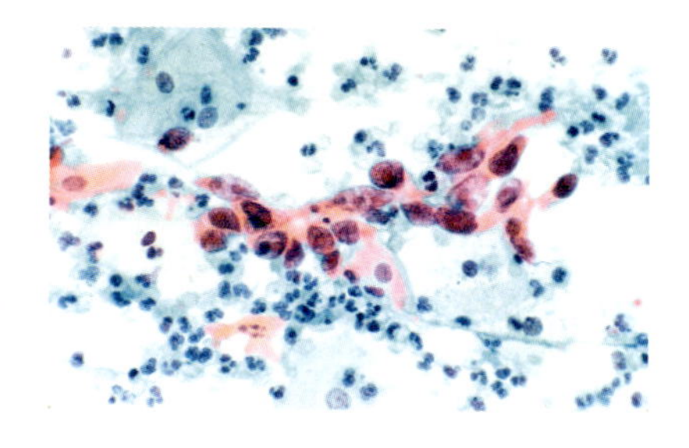

Plate 16.3

Histological slide (hematoxylin and eosin stain at 300×) showing prostate cancer. Right: Somewhat normal Gleason value of 3 (of 5) with moderately differentiated cancer. Left: Less normal tissue with a Gleason value of 4 (of 5) that is highly undifferentiated. The Gleason score is the sum of the two worst areas of the histological slide. Photo credit: Otis Brawley.

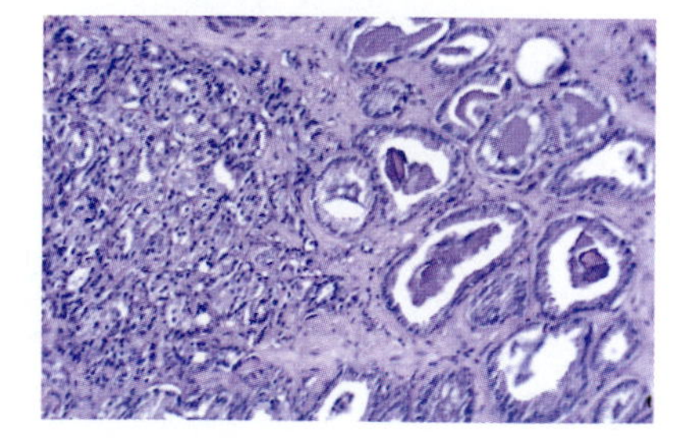

Plate 16.4

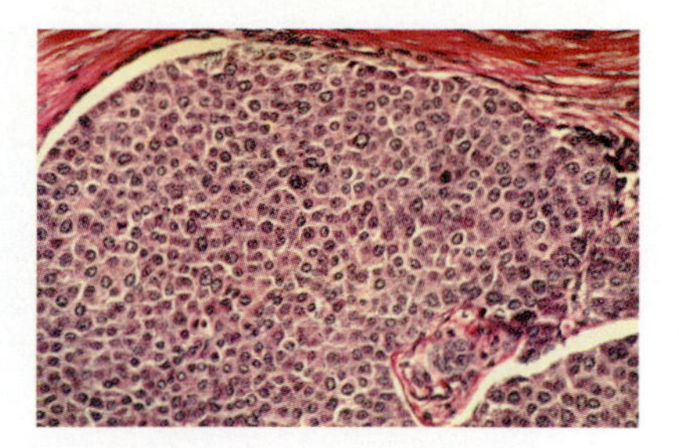

Breast cancer invades normal tissues and establishes new centers of growth. Note that the duct (demarcated by the two long, white areas, which are the walls of the duct) on inside of breast is completely filled with tumor cells. This histological slide was stained with hematoxylin and eosin and magnified to 100×. Photo by Dr. Cecil Fox and provided courtesy of the National Cancer Institute Visuals Online.

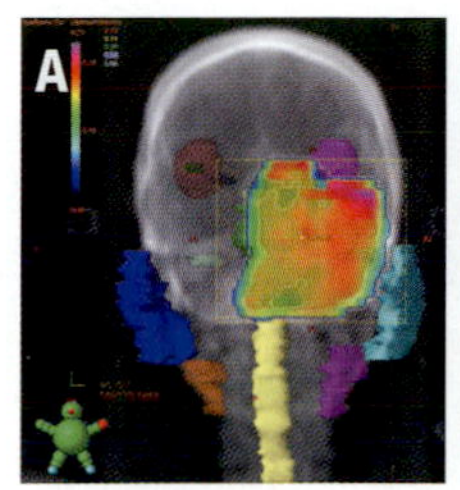

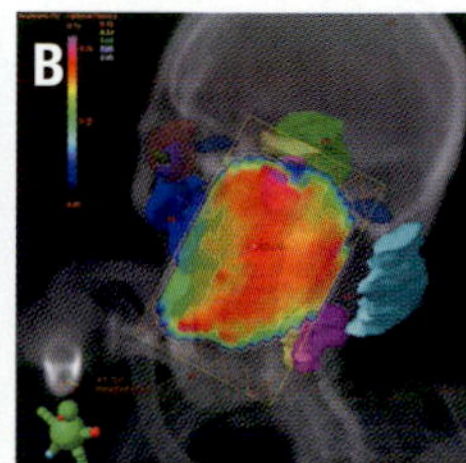

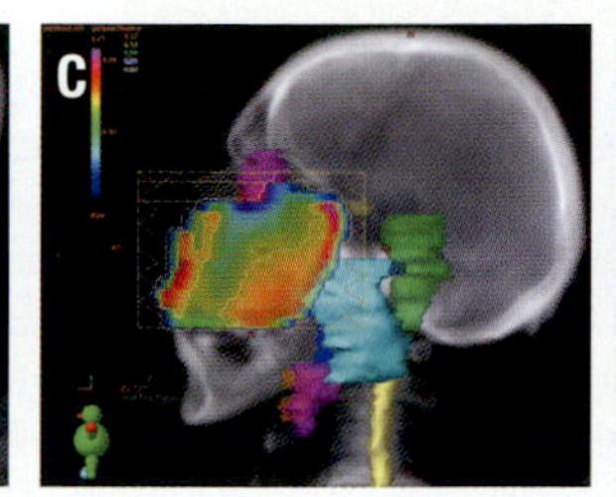

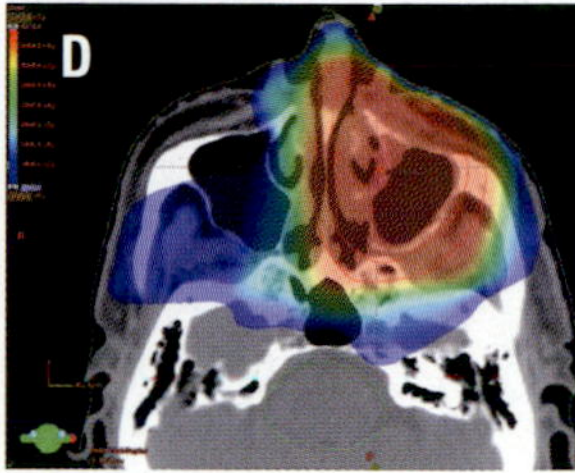

Plate 16.9

Example of intensity-modulated radiation therapy (IMRT). These images show the planning of radiation therapy for a head and neck cancer using IMRT. Anterior **(A)**, left superior oblique **(B)**, and left lateral **(C)** views are shown. Each image shows the pattern of the beam intensities, the surrounding bony anatomy, and manually segmented important radiosensitive anatomical structures such as the eyes, lenses, optic nerves, inner ear, brain stem, and spinal cord. The beam intensity patterns were determined from constraints on dose delivery to the tumor and avoidance of dose delivery to sensitive normal tissues. A single axial CT slice **(D)** shows the radiation dose distribution that was calculated for that slice from these beams. Images courtesy of Dr. Jonathan Knisely, Yale University.

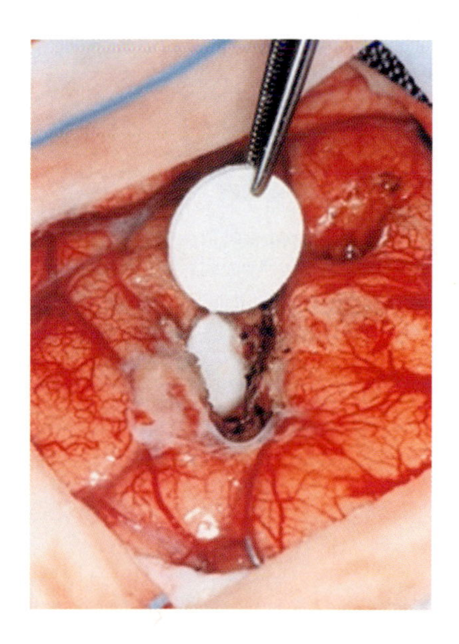

Plate 16.14

Photograph showing GLIADEL® wafer placement into a human brain following surgical resection of a malignant glioma. The walls of the tumor cavity are lined with up to eight such wafers. Each wafer is 14.5 mm in diameter and 1 mm in thickness, and contains 7.7 mg of carmustine. Reproduced with permission from Fleming AB, Saltzman WM. Pharmacokinetics of the carmustine implant. *Clin Pharmacokinet.* 2002;41:403–419.

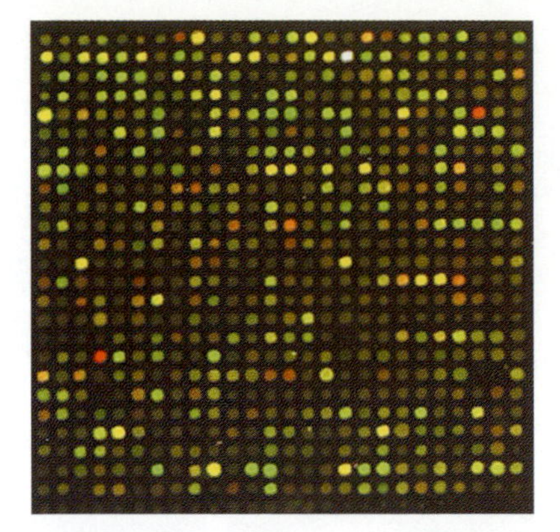

Plate 16.17

Deoxyribonucleic acid (DNA) microarray. In this array, each of the colored spots corresponds to a tiny island filled with single-stranded DNA encoding a particular gene (this 25 $\times$ 25 array contains 625 genes). An experiment is performed by exposing this array simultaneously to samples obtained from two different messenger ribonucleic acids (mRNAs): one mRNA sample (sample 1) is reverse transcribed to complementary DNA (cDNA) with red fluorescence, and the other (sample 2) is reverse transcribed with green fluorescence. The two samples compete for hybridization to the array: On spots that appear red, sample 1 had more mRNA; on spots that appear green, sample 2 had more mRNA; and on spots that appear yellow, both samples had substantial mRNA. Image courtesy of Dr. Valerie Reinke, Yale University.

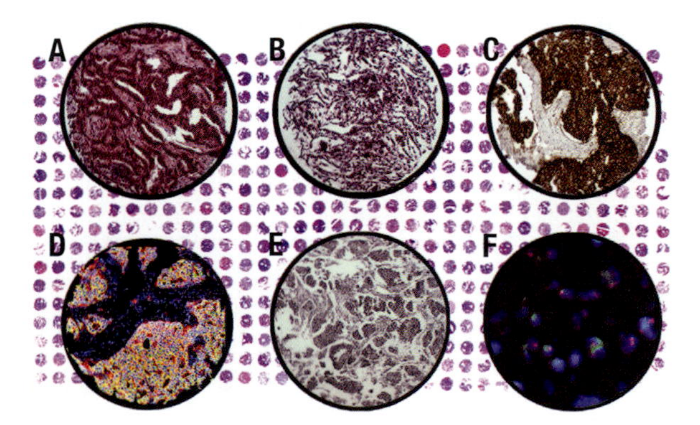

Plate 16.21

Tissues prepared in a variety of formats can be arrayed. A. Breast carcinoma stained with hematoxylin and eosin (H&E). **B.** Cell line A431 stained with H&E. **C.** Breast carcinoma stained for HER-2/*neu.* **D.** Breast carcinoma stained for cytokeratin (green), DNA (blue), and HER-2/*neu* (red). **E.** Breast carcinoma with F-actin detected by in situ hybridization. **F.** Breast carcinoma with *ERBB2* amplification by fluorescence in situ hybridization. Images courtesy of Dr. David Rimm, Yale University and reprinted from (7) with permission from Nature Publishing Group.

8 Circulation

LEARNING OBJECTIVES

After reading this chapter, you should:

- Understand that the circulatory system consists of a circulating fluid, a system of vessels, and a pump.
- Know the composition of blood and the role of cells in determining blood's physical properties.
- Understand the general structure of the vascular system.
- Understand the relationship between vessel radius, resistance to flow, and pressure drop.
- Understand the function of capillaries in the distribution of flow throughout tissues and transport of molecules.
- Understand the anatomy of the heart and the electrical system that generates coordinated contractions.
- Understand the events in the cardiac cycle and how pressure is generated within the chambers and the aorta.

8.1 Prelude

Our bodies appear, from the outside, to be solid masses that are slow to change but, just beneath the surface, the body's fluids are in constant motion. Blood moves at high velocity throughout the body within an interconnected and highly branched network of vessels (Figure 8.1). The human circulatory system is responsible for the movement of fluid (and therefore vital nutrients contained in the fluid) throughout the body.

The purpose of the circulatory system is a familiar one to engineers and bakers; it provides mixing, and good mixing is an essential element of many successful enterprises. Cakes are made from flour, eggs, sugar, and milk (among other things); your birthday will be ruined (or at least a bit tarnished) if the chef does not mix these ingredients well. But why must humans be mixed? Mixing exposes cells throughout the body to oxygen and other nutrients, while simultaneously providing a pathway for removal of waste products. It is the circulatory system that enables us to grow large, so that we have inner regions that are separated from the atmospheric oxygen that all cells need for energy (Figure 8.2).

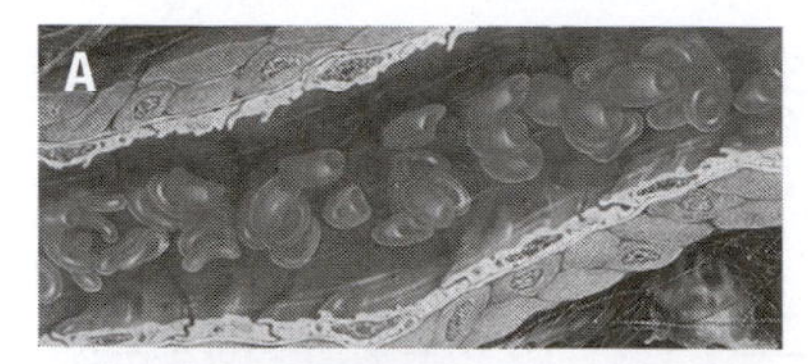

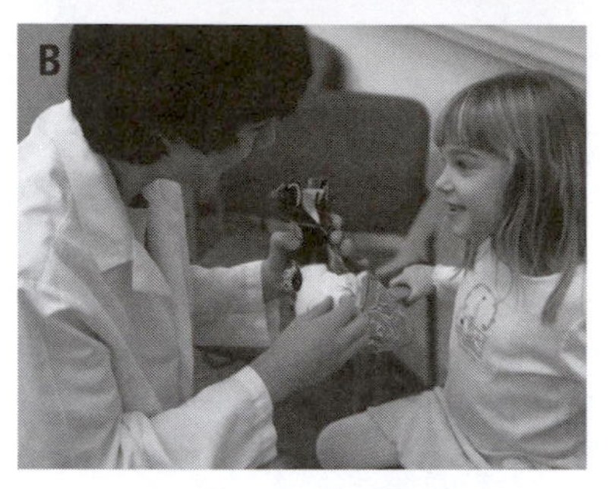

Figure 8.1 **Blood cells flowing through an arteriole. A.** Cells in some vessels can deform to orient themselves with the flow to reduce resistance. Image from Schmid-Schönbein, H., Grunau, G. and Brauer, H. Exempa hamorheologic 'Das strömende Organ Blut,' Albert-Roussel Pharma GmbH, Wiesbaden, Germany, 1980, and provided by Professor Oguz K. Baskurt. (See color plate). **B.** Cardiovascular health is important at all stages of life.

Circulation of fluids is the function of the cardiovascular system. The human cardiovascular system is composed of three separate organs: The heart serves as a pump, the vasculature serves as the plumbing, and the blood is the circulated fluid. Failure in any of these organs can result in failure of the system, so there are many diseases of the cardiovascular system. Biomedical engineers have made contributions to almost every area of cardiovascular health, including construction of artificial hearts and heart valves (which are described in Chapter 15), understanding of the flow properties of blood, and mathematical modeling of the microcirculation. In fact, some of the most important work in understanding the special blood vessels that develop in tumors has been done by biomedical engineers (Figure 8.3).

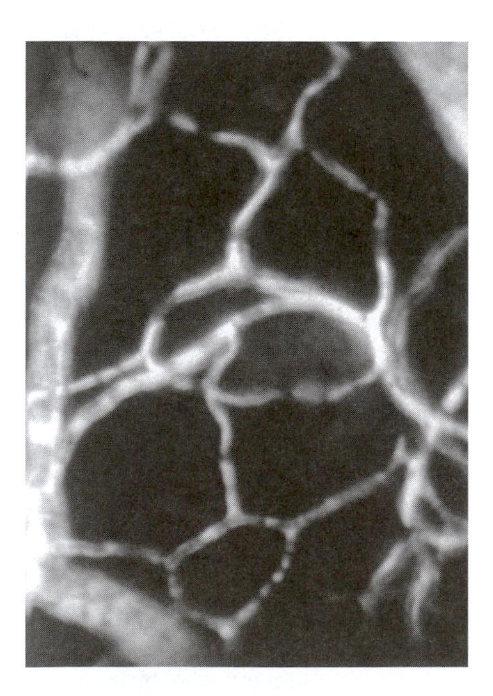

Figure 8.2 **Microcirculation.** The circulatory system carries oxygenated blood to every section of the body. Most cells in the body are less than 100 μm from the nearest small blood vessel. Reproduced, with permission, from: Barker JH, Hammersen F, Bondar I, Uhl E, Galla T, Menger M, et al. The hairless mouse ear for in vivo studies of skin microcirculation. *Plast Reconstr Surg.* 1989;83:948–959.

8.2 The circulating fluid

Blood is a special fluid in many ways. Blood is a suspension of cells, primarily **red blood cells** (RBCs), within a protein-rich fluid called plasma. Chapter 7 described

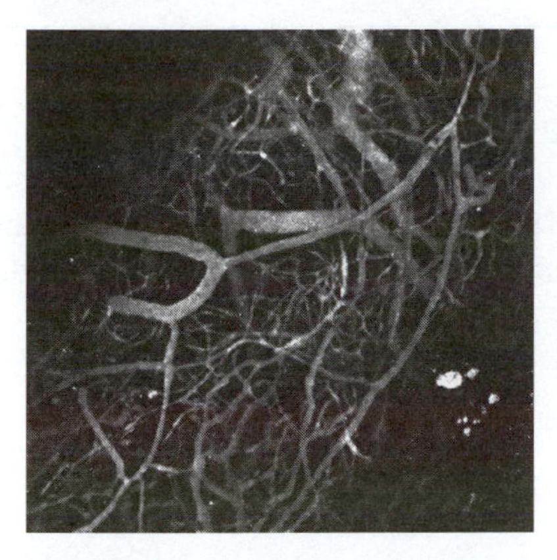

Figure 8.3 **Tumor vessels.** The microcirculation—or network of small vessels—in tumors provides the oxygen that is needed for tumor growth. Perhaps tumor growth can be reduced if the tumor vessels are destroyed. For that reason, exploring the biology of tumor blood vessels—such as the vasculature in this developing P22 sarcoma in a rat—has been a major effort in cancer research. Photo courtesy of Professor Gillian Tozer (University of Sheffield, UK) and Professor Borivoj Vojnovic and Dr. Richard Hodgkiss (Gray Cancer Institute, UK).

the important role of RBCs as mobile packets of hemoglobin, the protein that carries oxygen in the blood; the high RBC content of blood makes it an efficient oxygen carrier. To make the RBCs move throughout the body, the heart creates a pressure drop within the blood vessels, and this pressure drop causes blood to circulate. Because the number of RBCs per volume of blood is so high—a normal female will have more than 5 million RBCs per milliliter of blood—blood is more difficult to pump than pure water is. The presence of RBCs and proteins in blood both contribute to its higher **viscosity**.

Viscosity is a physical property that is related to the resistance to "flow" or deformation of a fluid. Honey, for example, is also more viscous than water. Viscosity is a property that can be quantified, as described in Box 8.1. The viscosity of water at body temperature (37°C) is 0.01 g/cm-s, which is usually translated to the special unit for viscosity of 1 centipoise (cp) (see Appendix C for other units). Blood has a viscosity of ~3 cp. The viscosities of some common fluids are listed in Table 8.1.

Blood contains a number of other cell types in addition to RBCs, which do not contribute substantially to its properties as a viscous fluid, but are of biological importance. The complete cellular composition of blood is listed in Appendix A. RBCs are the greatest in number and are the only cells that contain hemoglobin. Platelets are small cells that undergo dramatic shape and size changes when

Table 8.1 Viscosities of some fluids

Substance	Viscosity (cp)
At room temperature (20°C)	
Air	0.018
Gasoline	0.4–0.5
Water	1.003
Olive oil	81
At body temperature (37°C)	
Water	0.75
Plasma	1.2
Blood	~3

Box 8.1 Blood viscosity

Consider a fluid trapped between two parallel plates (Figure Box 8.1a, left). If the bottom plate is held stationary and the top plate is moved to the right with a velocity v by application of a tangential force F, the fluid within the gap will be subjected to a shearing stress that produces fluid motion. The force applied to the plate is uniformly transmitted over the entire area of plate–fluid contact; therefore, the tangential shear stress τ is equal to F/A, where A is the cross-sectional area of the plate.

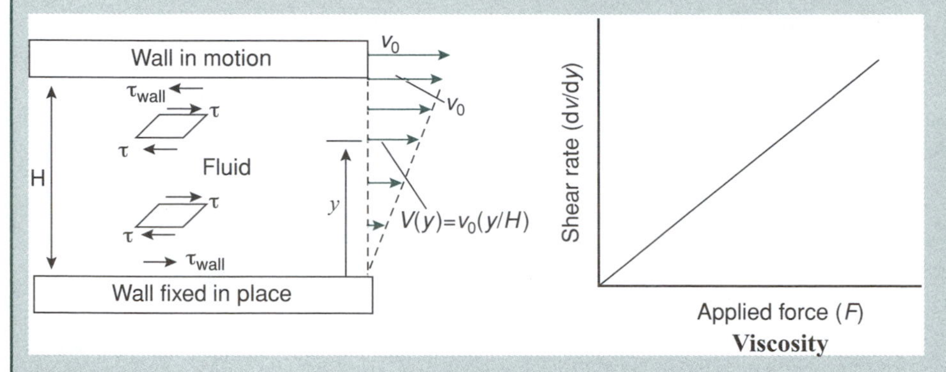

Figure Box 8.1A

Experimentally, the shear stress is directly proportional to the velocity of the upper plate, v and inversely proportional to the gap distance h:

$$\tau \propto \frac{v}{h}. \qquad \text{(Equation 1)}$$

More precisely, the shear stress is equal to a constant multiplied by the first derivative of the velocity with respect to distance normal to the moving plate:

$$\tau_{xy} = -\mu \frac{dv_x}{dy}, \qquad \text{(Equation 2)}$$

where τ_{xy} is the viscous shear stress in the x direction exerted on a fluid surface of constant y. The negative sign must be included because viscosity has a positive value ($\mu > 0$), and the shear stress flux is positive in the direction of decreasing velocity.

The shear stress and the velocity gradient (also called the shear rate) are proportional, with the constant of proportionality $-\mu$, where μ is the viscosity of the fluid. For some fluids, Equation 2 holds over a wide range of shear rates with a constant μ (as in Figure Box 8.1a, right); these are called *Newtonian* fluids. The value of the viscosity μ is a property of the fluid; μ is a function

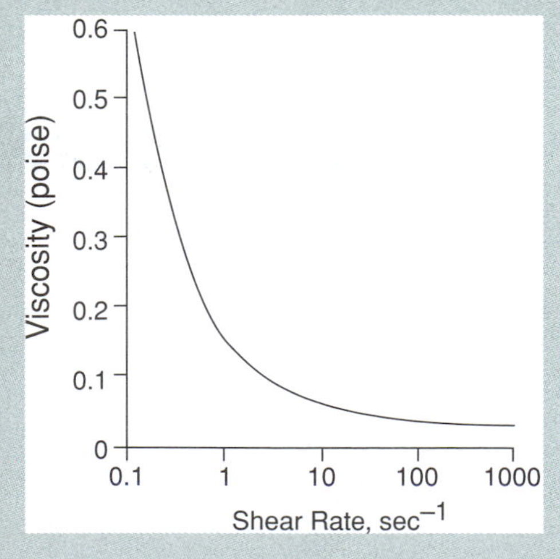

Figure Box 8.1B

of fluid phase composition, temperature, and pressure. Water and blood plasma are Newtonian fluids; at body temperature (37°C) the viscosity of water is 0.75 cp (1 cp = 0.01 g/cm-s) and plasma is 1.2 cp.

Blood is a non-Newtonian fluid because the viscosity changes with shear rate. When the shear rate is low, RBCs tend to form aggregates, which increase the viscosity. Whole blood has a viscosity of 3.0 cp, provided that the shear rate is sufficiently high so that the RBCs do not aggregate. At intermediate shear rates, the viscosity of blood is a function of shear rate, as shown in the Figure Box 8.1b. Viscosity of blood is also a function of composition, depending particularly on the concentration of RBCs and certain key proteins. Casson's equation is often used to describe the viscosity of blood as a function of these key variables:

$$\tau^{1/2} = \tau_y^{1/2} + \left[\frac{\mu_0}{(1-H)^{a\alpha-1}}\right]^{1/2} \dot{\gamma}^{1/2}, \qquad \text{(Equation 3)}$$

where $\dot{\gamma}$ is the shear rate (equal to dv_x/dy), τ_y is the yield stress, H is the hematocrit of blood, μ_0 is the viscosity of plasma, and $a\alpha$ is a constant that depends on the protein composition. Typical values of these parameters for blood are: $H = 0.45$; $\mu_0 = 0.012$ poise; $2 < a\alpha < 4$; and τ_y depends on the concentration of fibrinogen in the blood according to: $\tau_y^{1/2} = (H - 0.1)(C_F + 0.5)$, where C_F is the fibrinogen concentration in grams per 100 mL.

they are activated. They become activated at sites of injury, spreading out and aggregating to seal defects in blood-vessel walls to prevent blood loss. Together with certain blood proteins, such as fibrinogen, platelets are important in clot formation. A variety of white blood cells also circulate within the blood. Many of these cells are components of the immune system, which is discussed in Chapter 6.

8.3 The blood vessels

8.3.1 Overall design of the vascular system

The vascular system is a complex ensemble of connected vessels. Flow of oxygenated blood emerges from the heart into the aorta, on its way to the periphery (Figure 8.4). The aorta branches into large arteries, which further branch into main arterial branches, which further branch into terminal arteries, then **arterioles**, then finally capillaries. The cross-sectional area of these vessels changes significantly as the network progresses through the body (Table 8.2). The network of branching vessels proceeds to vessels of ever decreasing diameter (aorta to artery to arterioles to capillaries). In the smallest vessels, the **capillaries**, vessel diameter is approximately the same as the diameter of cells within the flowing blood.

One essential consequence of this highly branched structure is that every metabolically active cell in the body is within 100 μm of the nearest capillary, and is therefore assured adequate nutrients for metabolism.

Table 8.2 Geometrical properties of the circulatory system*

Structure	Diameter (cm)	Number	Total cross-sectional area (cm^2)
Aorta	1.0	1	0.8
Large arteries	0.3	40	3.0
Main arterial branches	0.1	600	5.0
Terminal branches	0.06	1,800	5.0
Arterioles	0.002	40×10^6	125
Capillaries	0.0008	12×10^8	600
Venules	0.003	80×10^6	570
Terminal veins	0.15	1,800	30
Main venous branches	0.24	600	27
Large veins	0.6	40	11
Venae cavae	1.25	1	1.2

* In dogs, see Appendix A for more details.

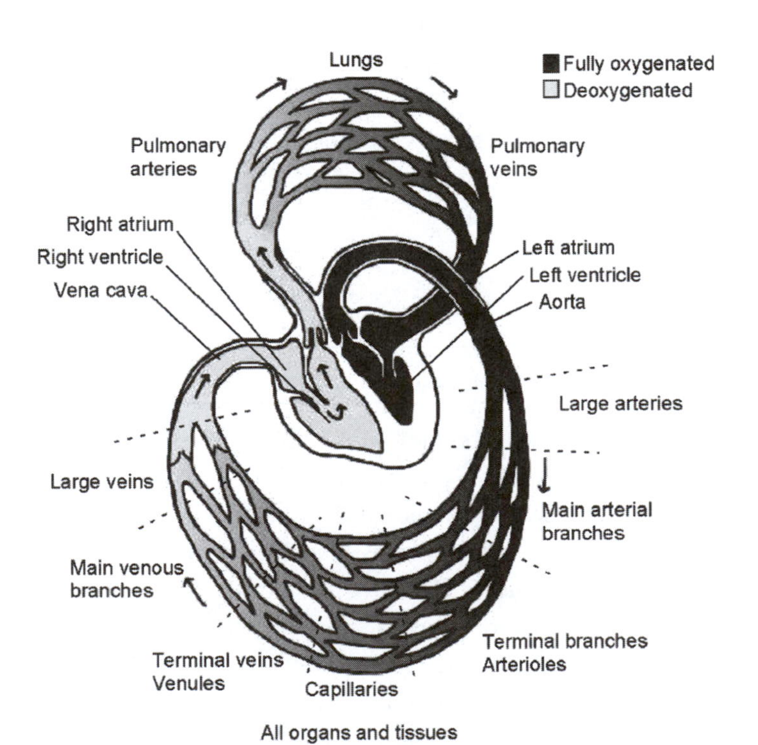

Figure 8.4 Overall structure of the circulatory system.

Although the circulatory network is anatomically complex, many of its important physical properties can be deduced from a simple engineering model: fluid flow through a straight cylindrical tube. As illustrated in Box 8.2, the equations for fluid motion can be used to obtain a relationship between blood flow rate (Q) and pressure drop (Δp) along a simple, cylindrical blood vessel:

$$Q = \frac{\pi r_v^4}{8\mu L}\Delta p, \tag{8.1}$$

where r_v is the radius of the vessel, μ is the viscosity of the fluid within the vessel, and L is the length of the vessel. Recall the discussion in Chapter 7 in which the **resistance** to air flow into the lung was defined by the expression: $\Delta P = \dot{F}R$. (In this book, $\dot{F}$ is used to designate air flows in the respiratory system and Q is used to designate blood flows in the cardiovascular system.) Both of these equations are analogous to Ohm's law, the equation that is used to describe the flow of current (I) through an electrical wire caused by a difference in electrical potential (Δvoltage):

$$IR = \Delta\text{voltage} \quad \text{or} \quad I = \frac{1}{R}\Delta\text{voltage} \tag{8.2}$$

Comparison of Equation 8.1 to 8.2 permits the resistance, R, of the cylindrical vessel to be defined:

$$Q = \frac{1}{R}\Delta p \quad \text{where} \quad R = \frac{8\mu L}{\pi r_v^4}. \tag{8.3}$$

SELF-TEST 1 If a blood vessel changes its diameter, decreasing from 4 mm to 2 mm (or by a factor of 2), what is the relative change in resistance?

ANSWER: R increases by a factor of $16 = 2^4$.

Equation 8.3 indicates that resistance to flow is a strong function of vessel radius: $R \propto r_v^{-4}$. In other words, as blood vessels become smaller, the resistance to flow increases dramatically; this effect is shown graphically in Figure 8.5.

Figure 8.5 illustrates an additional consequence of the branching pattern of blood vessels: The majority of the overall resistance to blood flow resides in the smallest vessels. In fact, ~80% of the pressure drop in the systemic circuit occurs in arterioles and capillaries. This natural consequence of the physics of fluid flow is exploited in the regulation of blood flow to organs of the body. Local blood flow to a tissue is controlled by **constriction** and **dilation** of the arterioles delivering blood to that tissue. Because the greatest overall resistance is provided by arterioles, and because individual arterioles have muscular walls, which permit them to adjust their diameter—and hence resistance—the proportion of blood flow arriving to the tissue served by an arteriole can be regulated with precision (Figure 8.6).

Box 8.2 Blood flow through a cylindrical vessel

The circulatory system is a network of interconnected, branching cylindrical vessels. Although the overall structure of the network is complex, a substantial fraction of the behavior of the system can be predicted by analyzing a simple engineering system: the flow of fluid through a cylindrical vessel (Figure Box 8.2a). Engineers have been studying the dynamics of flow in cylindrical tubes for more than 150 years, and much is known about the characteristics of these systems.

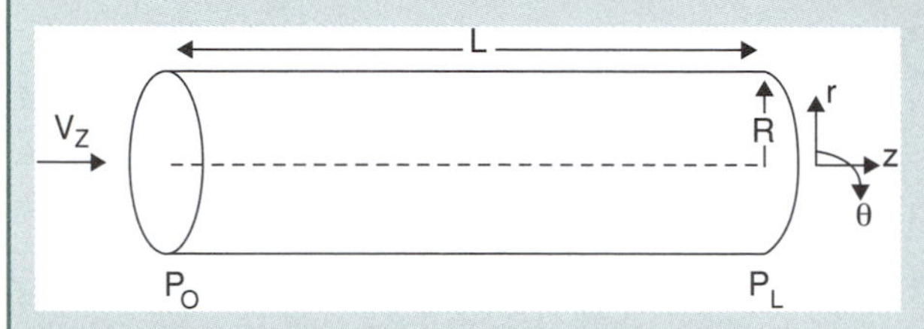

Figure Box 8.2A

Fluid flows can be described quantitatively by the Navier–Stokes equations, a set of equations that can be derived based on conservation of mass and momentum within the flowing fluid. The full set of differential equations is beyond the scope of this text, but it is useful to look at one component of these equations. Consider, for example, a flow that is confined within a cylindrical tube, so that momentum is transferred only in the z direction. The z component of the Navier-Stokes equations, in cylindrical coordinates, can be written:

$$\rho\left(\frac{\partial v_z}{\partial t}+v_r\frac{\partial v_z}{\partial r}+\frac{v_\Theta}{r}\frac{\partial v_z}{\partial \theta}+v_z\frac{\partial v_z}{\partial z}\right)=-\frac{\partial p}{\partial z}+\mu\left[\frac{1}{r}\frac{\partial}{\partial r}\left(r\frac{\partial v_z}{\partial r}\right)+\frac{1}{r^2}\frac{\partial^2 v_z}{\partial^2 r}+\frac{\partial^2 v_z}{\partial^2 z}\right]+\rho g_z,$$

(Equation 1)

where the coordinate system is defined in Figure Box 8.2a, v_z and v_θ are fluid velocities in the z and θ directions, p is the pressure, ρ is the fluid density, and g_z is the acceleration caused by gravity in the z direction. (Do not be concerned about where this equation comes from; you will learn that later, in a class in fluid mechanics or biological transport phenomena.) This equation can be simplified by the following assumptions (which apply reasonably well to blood flows in most vessels in humans):

Steady flow (i.e., $\left(\frac{\partial v_z}{\partial t}=0\right)$); flow only in axial direction (i.e., $v_r = v_\theta = 0$); gravitational forces are negligible (i.e., $g_z = 0$). These assumptions can be used to simplify Equation 1:

$$\frac{\partial p}{\partial z}=\mu\frac{1}{r}\frac{\partial}{\partial r}\left(r\frac{\partial v_z}{\partial r}\right). \qquad \text{(Equation 2)}$$

A few other things are assumed to be true for the flow shown in Figure Box 8.2a: The pressures at each end of the tube are known (i.e., $p = p_0$ at $z = 0$; $p = p_L$ at $z =$ L); the fluid velocity at the vessel wall is zero (i.e., this is the so-called "no-slip" condition," $v_z = 0$ at $r =$ R); and the velocity is a maximum in the center of the tube (i.e., $\frac{\partial v_z}{\partial r} = 0$ at $r_v = 0$). Equation 1 and the three conditions just stated are all satisfied by the following expression for velocity:

$$v_z=\frac{(p_0-p_L)r_v^2}{4\mu L}\left(1-\frac{r^2}{r_v^2}\right). \qquad \text{(Equation 3)}$$

The local velocity is a parabolic function of radial position (Figure Box 8.2b); flow of this type is called Hagen–Poiseuille flow, in honor of Jean Louis Marie Poiseuille and Gotthilf Heinrich Ludwig Hagen. Poiseuille (a physiologist) and Hagen (an engineer) independently published the first systematic measurements of pressure drop within flowing fluids in simple tubes in 1839 and 1840.

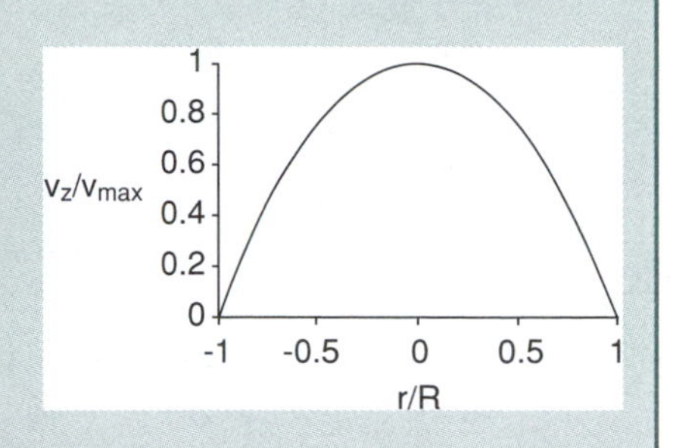

Figure Box 8.2B

To determine the dependence of fluid flow rate, Q, on pressure drop, Equation 8.3 can be integrated over the vessel cross-section:

$$Q = \int_0^{2\pi} \int_0^{r_v} v_z(r) r dr d\theta, \qquad \text{(Equation 4)}$$

which can be solved to yield:

$$Q = \Delta p \frac{\pi r_v^4}{8\mu L}. \qquad \text{(Equation 5)}$$

Note that the flow rate Q increases linearly with pressure drop ($\Delta p = p_0 - p_L$), decreases linearly with length L and fluid viscosity μ, and increases with the fourth power of tube radius R. At a given pressure drop, the rate of flow is a strong function of vessel radius.

The circulatory system is also remarkably efficient at mixing. When drug molecules are injected into a vein, the concentration of that agent at the injection site is suddenly increased. The high velocity of local flow, coupled with the extensive branching and rebranching of the cardiovascular circuit, produces rapid distribution of the injected agent throughout the body. For agents confined to the plasma volume, mixing is completed in several minutes; that is, after several minutes, the blood in every vessel of the body has an equal concentration of the drug.

Engineers and scientists have been studying flow through cylindrical vessels for a long time. One such engineer, Osborne Reynolds (1842–1912), who was

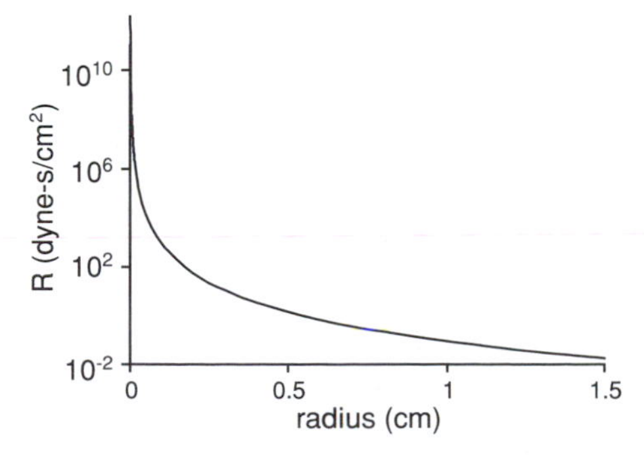

Figure 8.5 **The relationship between resistance to blood flow and blood vessel radius.** The curve represents Equation 8.3.

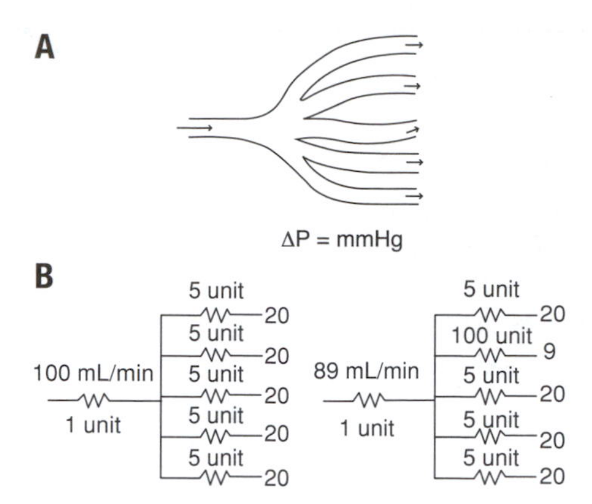

Figure 8.6 **Flow in blood vessel networks. A.** An imaginary blood-vessel network in which one vessel branches into five subvessels. **B.** When the resistance of one of the subvessels increases, the overall resistance of the network changes, and the flow is redistributed.

the first professor of engineering at Owens College[1] in Manchester, England, discovered that flows in pipes undergo a common transition. When the flow rate is low, the flow pattern is well behaved: All of the "particles" of fluid flow in straight lines that are parallel to the axis of the pipe (this is true of the example shown in Box 8.2). This kind of flow is called **laminar flow**. When the flow rate is steadily increased, however, the flow pattern eventually becomes unstable: **Eddies** appear in the fluid, and the flow becomes disordered, no longer following the parallel streamlines of laminar flow. This chaotic flow is called **turbulent flow**. The critical flow rate of the transition point depends on many factors, including the viscosity of the fluid and the diameter and length of the pipe. Reynolds discovered that all pipe flows can be characterized by a dimensionless number, now called the **Reynolds number** (Re):

$$\mathrm{Re} = \frac{\rho v_{\mathrm{av}} (2r_{\mathrm{v}})}{\mu}, \tag{8.4}$$

where ρ is the fluid density, v_{av} is the average velocity, and μ is the viscosity. Reynolds found that, for laminar flow, Re $<$ 2,100. Appendix A includes tables that show the Re for flow in vessels throughout the circulatory system. On the systemic side, the highest Re is in the aorta (1,200–1,500), and the lowest is in the capillaries (0.0007–0.003). With the exception of the ascending aorta (where the Re can be as high as 5,800), flows in the circulatory system are usually laminar. Turbulence does occur, however, particularly in areas of disruption of the normal flow pattern (e.g., when there is a **bifurcation** in a vessel or an obstruction that impinges on the flow within the vessel or a narrowing that creates higher than normal velocities). The eddies of turbulent flow create sounds, which clinicians can sometimes hear (using stethoscopes) and use for diagnosis. These sounds, called **bruits**, can indicate the presence of **atherosclerosis** in, for example, the carotid arteries.

[1] Owens College is now called University of Manchester.

8.3.2 Arteries and veins

Equation 8.1 describes the creation of flow in blood vessels caused by pressure differences. In Chapter 7, pressure was also discussed in relation to the mechanics of breathing. In describing the mechanics of the lung, it was helpful to differentiate between the transpulmonary pressure drop ($P_{alv} - P_{ip}$) and the atmospheric to alveolar pressure drop ($P_{atm} - P_{alv}$). (Recall, also, that all of the pressures are defined relative to the atmospheric pressure, so that $P_{atm} = 0$.) Similarly, in the cardiovascular system, Δp in Equation 8.1 refers to the internal pressure drop measured at different points along the axis of a vessel; this is an **axial pressure drop**. It is also useful to define a **transmural pressure difference**, Δp_{tm}:

$$\Delta p_{tm} = p_v - p_{if}, \tag{8.5}$$

where p_v is the pressure in the vessel, and p_{if} is the pressure in the interstitial fluid outside of the vessel. The transmural pressure drop is important in keeping the vessel inflated (pressure in the vessel must be higher than pressure in the interstitial fluid so that the vessel does not collapse). In addition, as described later in this chapter, the transmural pressure drop determines the rate of water movement out of vessels into the adjacent tissue.

Of course, the pressure in blood vessels varies with time, as the heart contracts and relaxes. Measurement of blood pressure is one of the most frequently performed medical procedures. Biomedical engineers continue to design new instruments for measurement of blood pressure, as discussed in Chapter 11. These changes in pressure can be felt in the pulse (which is most easily palpated in the radial artery in the wrist or the carotid artery in the neck). Cyclic variations in pressure lead to pulsatile flows (i.e., flow rates that vary periodically with the pressure variation). These cyclic variations in pressure and flow are most prominent in the vessels nearest the heart. Typical pressure variations are shown in Figure 8.7. Although Ohm's law does not apply directly in this situation, the variations in the flow rates are similar to the variations in pressure shown in Figure 8.7.

Recall that, earlier, when the resistance to flow was defined, it was stated that about 80% of the resistance of the vascular network occurred in the arterioles and capillaries. This effect can be seen graphically in Figure 8.7b: The pressure drops from ~90 mmHg to 10 mmHg from the beginning of the arterioles to the end of the capillaries. It is also important to remember that many vessels have smooth muscle layers so that they can contract and dilate in response to physiological needs. Hence, vessels can change their resistance to modulate the amount of flow that they receive in response to a given pressure.

The distribution of pressure within the vascular system changes when a person stands up from a recumbent position. Part of this change is caused by physics of hydrostatic pressure. In a simple column of water (Figure 8.8), the pressure at the bottom of the column is higher than the pressure at the top because of the force exerted on the fluid at the bottom by all of the fluid that sits above it.

A

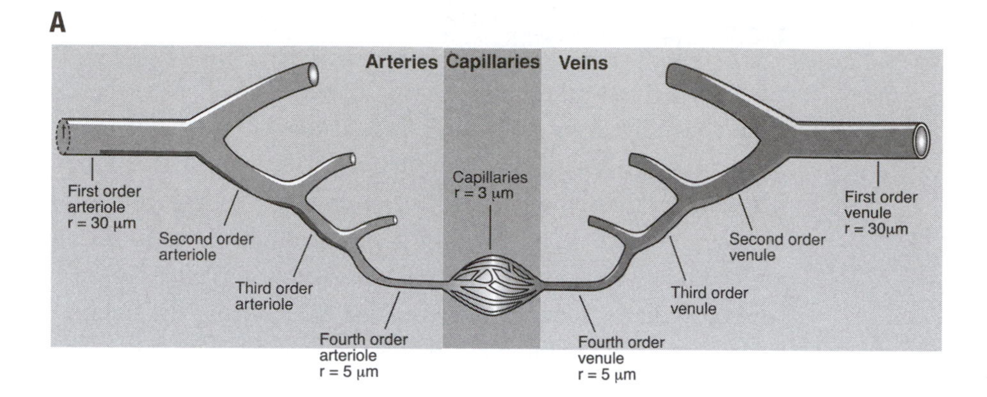

B

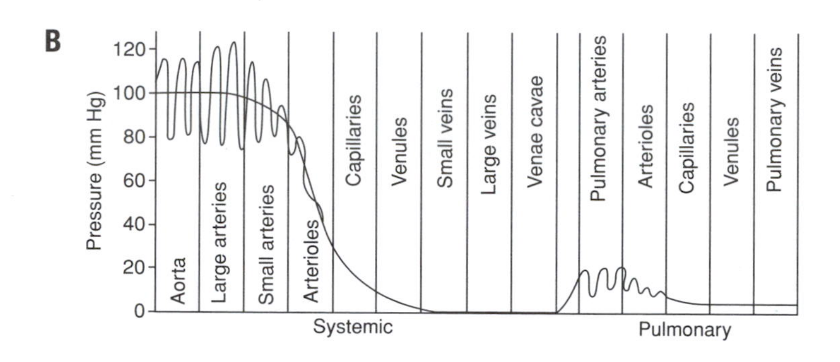

C

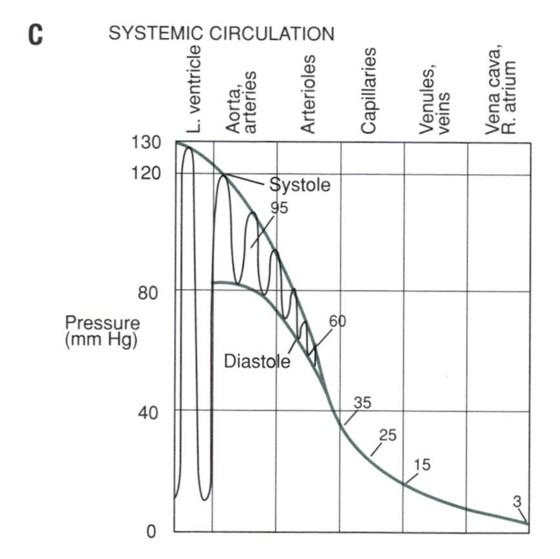

Figure 8.7 **Axial pressure drops in the circulatory system. A.** Schematic diagram of vessel branching with nomenclature. **B.** Pressure measured at various locations along the circulatory path. **C.** Expanded view of ranges of pressure measured during stages of the heartbeat.

The variation in pressure can be described as:

$$p_2 = p_1 + \rho h g, \tag{8.6}$$

where p_1 and p_2 are the pressures at points 1 and 2 within a fluid, h is the vertical difference in height between point 1 and point 2, and g is the acceleration caused by gravity.

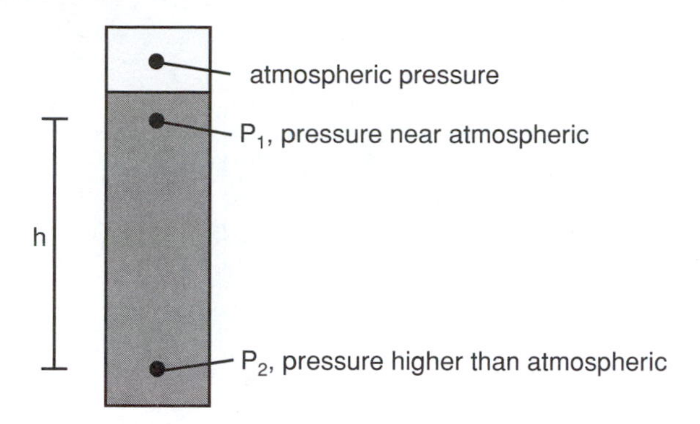

Figure 8.8 Variation in pressure within a column of fluid.

SELF-TEST 2 Find the pulse in your radial artery and feel its force against your fingers. Raise your hand above your head as high as you can. Does the force feel lessened? Explain.

ANSWER: Pressure in the radial artery is ~100 mmHg or 140 cm H_2O. The pressure drops as you raise your hand above your heart, by approximately the length of your arm (in cm H_2O).

Equations 8.1 and 8.3 only apply for pipes with stiff, unchanging walls, but blood vessels exhibit varying degrees of **compliance**. A compliant material is one that deforms with application of a pressure or stress. In blood vessels, which experience changes caused in hydrostatic pressure, the compliance is defined as:

$$C = \frac{\Delta V}{\Delta p}, \tag{8.7}$$

where ΔV is the change in vessel volume that occurs for a given change in hydrostatic pressure, Δp. Figure 8.9 shows the compliance of large vessels: the aorta and vena cava. Compliance is important for the function of these vessels in the circulatory system. A compliant aorta allows this large vessel to serve as a pressure reservoir that maintains a high arterial pressure, even when the heart is relaxing between active contractions. When the ventricle beats, the pressure increases in the aorta (this is described in more detail in Section 8.4.3), and the aorta increases in volume because of its elasticity or compliance. When the ventricle stops contracting, the elasticity of the aorta serves to maintain the pressure within the vessel. Large veins, such as the vena cava, are exceptionally compliant, which allows them to serve as volume reservoirs. In this case, the high compliance translates into an ability to expand greatly in volume (ΔV) with a small increase in pressure (Δp).

Later, in Chapter 10, which discusses biomechanics, we will encounter **Laplace's law**, which describes the relationship between pressure drop and wall tension in inflated vessels such as a capillary or an alveolus. This same concept applies to both objects, except that the geometry of the blood vessel is cylindrical, in contrast to the spherical geometry of the alveolus. The transmural pressure

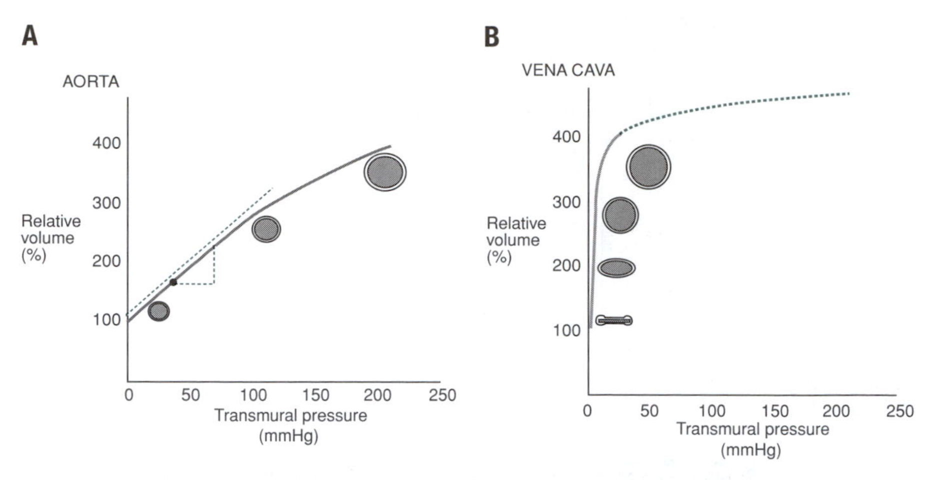

Figure 8.9 **Distention profiles for arteries (aorta; A) and veins (vena cava; B) in response to transmural pressure changes.** The compliance, which changes with transmural pressure, is equal to the slope of a line tangent to the distention curve at a given pressure. In both vessels, the volume of the vessel increases with an increase in transmural pressure. The mechanical properties of the vessels differ, hence the relationship between vessel dilation and pressure differs. For arteries **(A)**, the compliance is relatively unchanged with transmural pressure. For veins **(B)**, the compliance is very high at low transmural pressures but is low at high pressures (which are not encountered under normal physiological conditions).

drop across the blood vessel wall is supported by tension in the wall:

$$\Delta p_{\text{tm}} = \frac{T}{r}. \tag{8.8}$$

When a vessel is subjected to an increase in transmural pressure drop, it will distend; for a given pressure drop, veins distend more than arteries, as shown in Figure 8.9. As it distends, two things happen: The tension in the wall increases according to Equation 8.8, and the wall becomes thinner (because the circumference of the vessel is larger and there is only a fixed amount of total material in the wall). If the transmural pressure becomes too high, the vessel may rupture because the material in the wall can no longer support the wall tensions that are generated. When a vessel wall becomes weakened because of disease, it can no longer distend in its usual fashion and an **aneurysm**, or local bulging in the vessel wall, may occur (Figure 8.10). Aneurysms can lead to rupture of the vessel wall because the bulging portion of the vessel (with a smaller radius than the vessel itself) experiences an increase in tension in the wall, as indicated by Equation 8.8. This increase in tension, coupled with the decrease in wall strength, contributes to the potential for rupture.

The previous discussion refers primarily to the systemic circulation—that is, the blood-vessel network that carries oxygenated blood from the left side of the heart to all of the tissues of the body. The pulmonary circulation obeys similar principles, except that the pressures generated by the right side of the heart, and transmitted to the pulmonary vessels, are much lower than the pressures generated by the left side of the heart. The venous pressures, however, are the same on each

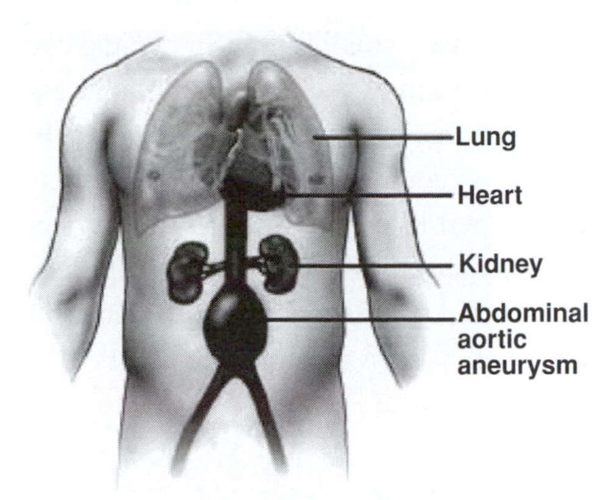

Figure 8.10 **Aneurysm.** When a vessel weakens, it can become abnormally distended, forming an aneurysm.

side. The left heart and right heart deliver the same volume per stroke, but at very different pressures.

8.3.3 Capillary function

The capillaries are the business end of the circulation because it is in the capillaries that the exchange of molecules such as oxygen, carbon dioxide, glucose, and other nutrients occurs (Figure 8.11). Unlike arteries and veins, capillaries have no smooth muscle, no elastic fibers, and very thin walls. The important element of their function is the continuous endothelial cell monolayer, which forms the surface of the inside of the vessel and the junctions that form between cells in this monolayer. It is this cell monolayer that regulates the movement of molecules through the capillary wall. Some capillaries are called **fenestrated capillaries**;

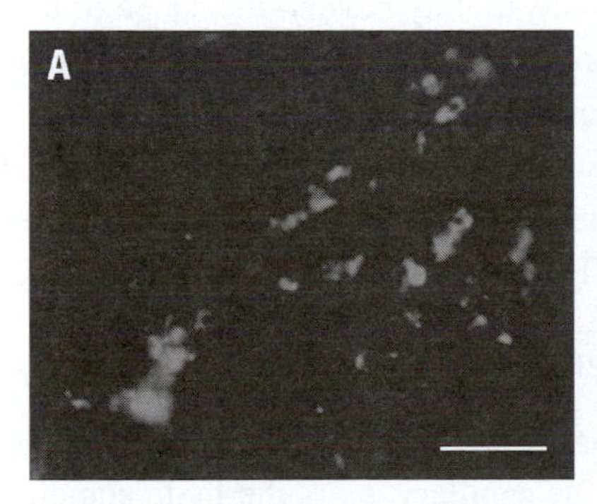

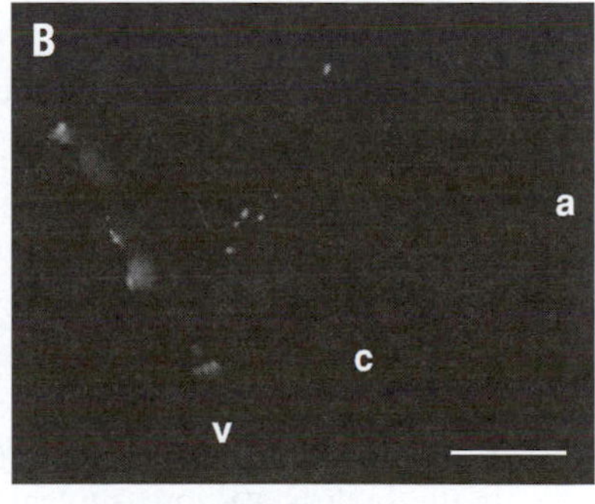

Figure 8.11 **Heterogeneous distribution of liposomes in tumor and normal tissues.** Human colon adenocarcinoma cells (LS174T) were transplanted in mice, and fluorescently labeled liposomes were injected intravenously. The photos were taken 2 days after the injections. **A.** In solid tumor, liposomes accumulated only in perivascular regions. Bar = 100 μm. **B.** In normal subcutaneous tissue, liposomes accumulated only in the wall of small postcapillary venules (6–25 μm in diameter). Neither parallel capillaries (c) nor arterioles (a) and large collecting venules (v, > 25 μm) were labeled by liposomes. Bar = 100 μm. Photo courtesy of Dr. Fan Yuan and modified with permission from: Yuan F, Leunig M, Huang SK, Berk DA, Papahadjopoulos D, Jain RK. Microvascular permeability and interstitial penetration of sterically stabilized (stealth) liposomes in a human tumor xenograft. *Cancer Res.* 1994;54:3352–3356. (See color plate.)

they are leaky because they have many small holes. Sinusoidal capillaries, which are found in some special tissues such as the liver, have large gaps between the cells, which provide big leaks and abundant molecular exchange. It is the structure of the capillary (particularly the structure of the endothelial cell monolayer) that determines the permeability properties of the local vascular system.

Many molecules, such as oxygen, move through the capillary wall and into the surrounding tissue by diffusion. Diffusion was introduced in Box 2.2. The permeability of the capillary wall to a given molecule can be estimated directly from Equation 2 in Box 2.2:

$$\dot{M} = AJ_x = -AD\frac{dc}{dx} = -AD\frac{\Delta c}{t}, \tag{8.9}$$

where $\dot{M}$ is the rate of mass flow through the capillary wall (equal to flux $J_x \times$ area), A is the area, D is the diffusion coefficient of solute in the capillary wall, t is the thickness of the membrane, and Δc is the concentration difference across the capillary wall (out – in). If the concentration is lower outside than inside, the solute will diffuse from the capillary lumen into the interstitial fluid outside of the vessel. The rate of mass flow depends on several variables: It goes up with capillary surface area, down with the capillary wall thickness, and up with increasing concentration drop.

In reality, the delivery of solutes such as oxygen is not adequately described by Equation 8.9. After oxygen permeates out of the capillary, it must diffuse through the tissue space to reach cells that are distant from the capillary. Not all of the oxygen can diffuse throughout the tissue because the cells near the capillary surface use some of it. The role of the capillary in the delivery of molecules such as oxygen to a tissue space can be illustrated by a simple mathematical model, which was developed by August Krogh (1874–1949) and earned him the Nobel Prize in Medicine. The details are slightly beyond the scope of this text (but not much, so the curious are encouraged to look further). An example of the results of this model is shown in Figure 8.12.

Water—and very small solutes that can dissolve in water and move through the fenestrations in capillaries—can move across capillary walls because of the transmural pressure gradient. This effect is often described as follows:

$$J_{\text{fluid}} = K(p_c - p_{if}), \tag{8.10}$$

where J_{fluid} is the flux (mass/area/time) of fluid through the capillary wall, p_c is the hydrostatic pressure in the capillary, p_{if} is the hydrostatic pressure in the interstitial fluid, and K is the permeability of the capillary wall to the flow of fluid. The permeability, K, is the inverse of the resistance. Note the similarity of this expression to Equation 8.1, except that in Equation 8.10 the pressure drop is transmural, rather than axial. The movement of water in the space around capillaries has a number of interesting features. Frank Starling (1866–1927) made a number of important observations about cardiovascular mechanics (and he introduced the concept of hormones). Starling's equation, for example, incorporates

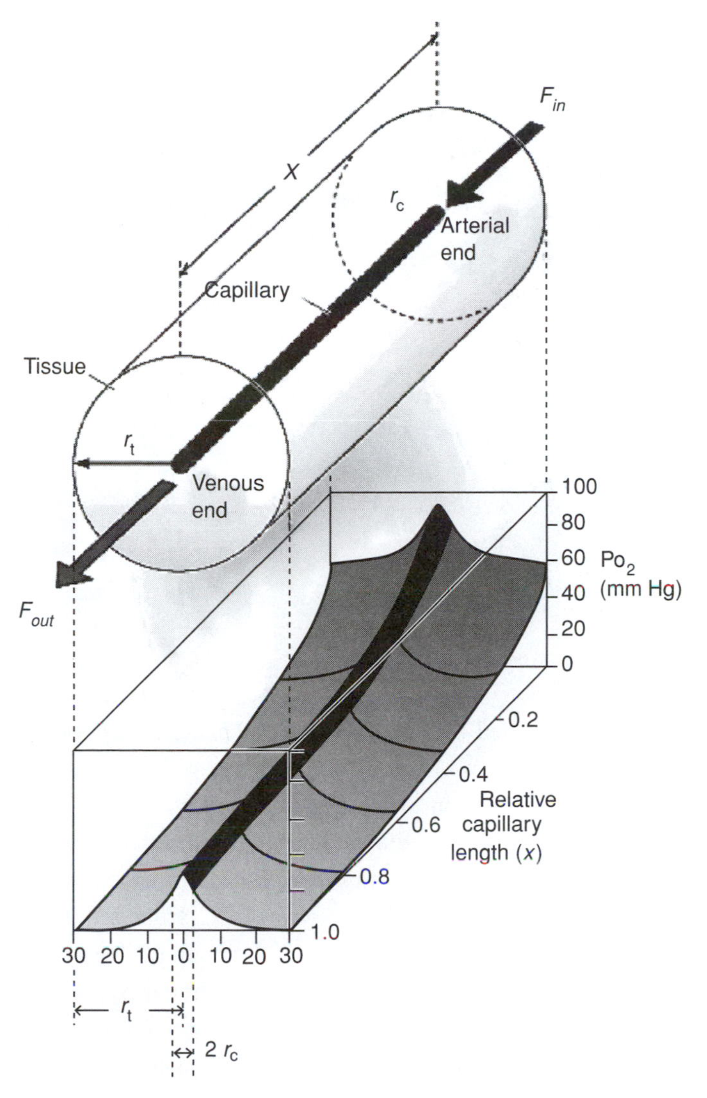

Figure 8.12

Oxygen diffusion from a capillary. Example of the results of a mathematical model developed to describe the role that capillaries play in the diffusion of molecules such as oxygen. These models were first developed by August Krogh.

the fact that the rate of water movement in and out of a capillary is not determined by pressure drop alone (as in Equation 8.10). The effect of osmotic pressure must also be included because the blood plasma contains abundant proteins (usually more than in the interstitial fluid) and the presence of protein in the plasma tends to "pull" water into the vessel because of its osmotic activity (recall Box 2.4). Starling's equation can be written:

$$J_{\text{fluid}} = K[(p_{\text{c}} - p_{\text{if}}) - (\pi_{\text{c}} - \pi_{\text{if}})], \qquad (8.11)$$

where π indicates osmotic pressure in the capillary or interstitial fluid.

The heart continually circulates about 5 L/min of fluid throughout the day. Because of the pressure inside the blood vessels, fluid can leak out of the

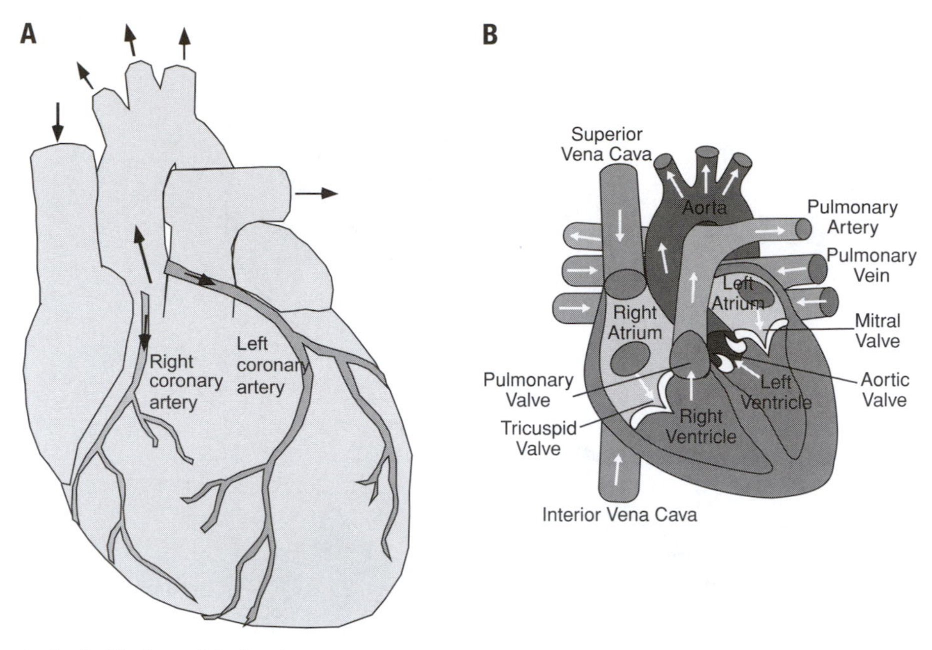

Figure 8.13 Anatomical features of the heart.

capillaries (as described by Equations 8.9): This loss is about 20 L/day. Much of this fluid (16–18 L/day) is recovered in the capillaries because of the effects of osmotic forces (Equation 8.11). The difference between these values is the amount of fluid that is returned to the circulatory system by the lymphatic system. Loss of lymphatic function can lead to accumulation of interstitial fluid, a state called interstitial edema. People who suffer from edema usually have the most significant problems with fluid accumulation in the legs, because the highest pressures are developed in the lower extremities (recall Equation 8.6).

8.4 The heart

The human heart is a marvel of engineering efficiency. It is able to create a near constant driving force for blood flow (i.e., a near constant blood pressure) continuously for many decades. It accomplishes this feat in the face of blood-flow demands that can change dramatically with time, as when a person starts to run from an initial resting position. The healthy heart accomplishes this with only a small change in the pressure drop.

Some of the main features of the heart are illustrated in Figure 8.13. The heart has muscular walls that are formed into four chambers (right atrium, right ventricle, left atrium, and left ventricle). The surface of the heart contains blood vessels, which provide blood flow to the muscular walls. There are also four inflow and outflow tracts, which connect the heart to the vena cava, the pulmonary artery, the pulmonary vein, and the aorta. Many characteristics of the human heart are familiar to most people: The heart has a left side and right side, which serve as

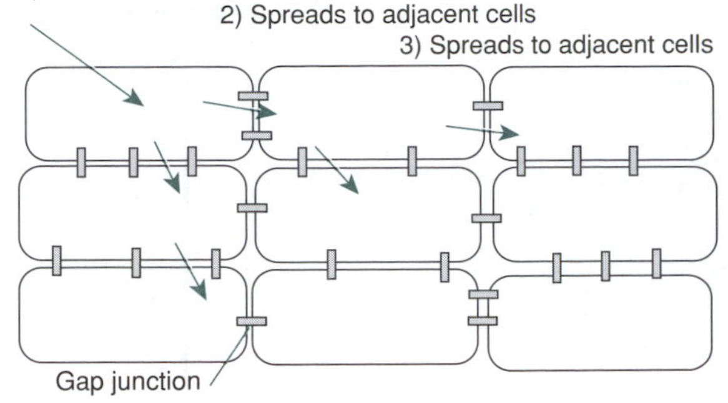

Figure 8.14 **Excitable cells of the heart.** Electrically sensitive cells are connected together to gap junctions to form quiltlike pattern necessary for proper function. Colored arrows illustrate the movement of electrical activity through the tissues.

pumps to drive blood to the systemic circulation and to the lungs (Figure 8.4). Pumping is accomplished by contraction of the muscular walls of each heart chamber, which occurs in a synchronous fashion, described in Section 8.4.3. The left side of the heart, which must send blood to the distant extremities, has to create a larger pressure drop and, therefore, has thicker muscular walls than the right side has. The major outflow tracks into the heart and the openings between the atria and ventricles are guarded by valves: The tricuspid valve sits between the right atrium and ventricle, the pulmonary valve between the right ventricle and the pulmonary artery, the mitral valve between the left atrium and ventricle, and the aortic valve between the left ventricle and the aorta. Contraction occurs rhythmically, with each cycle of contraction (commonly called a beat) lasting about 1 second, so that there are 60–70 beats/min.

8.4.1 Electrical activity of the heart

The contractile cells of the heart wall (called the myocardium) are electrically excitable: Like neurons, they can exhibit action potentials (see Chapter 6). Most neurons receive instructions before they exhibit an action potential. Often, the instructions come to the neuron from another neuron that is close to it, through a specialized junction called a synapse (also described in Chapter 6). The excitable cells of the heart differ in two important ways: First, they are connected together, through gap junctions instead of synapses, to create a quiltlike pattern of electrically sensitive cells (Figure 8.14) and, second, some of the cells can generate action potentials without any external stimulation. The cells with the most rapid self-excitation—or autorhythmic—pattern are found in a region of the heart called the sinoatrial (SA) node. Because cells in this region generate action potentials, which are then propagated to other cells throughout the myocardium, the SA node is the natural pacemaker of the heart; these cells set the rate of rhythmic excitation of the heart. Unlike skeletal muscle, which requires a nerve to initiate contraction, the heart can contract in the absence of nerve input. Nerve input to the heart is present, but it is used to modulate (not initiate) contraction.

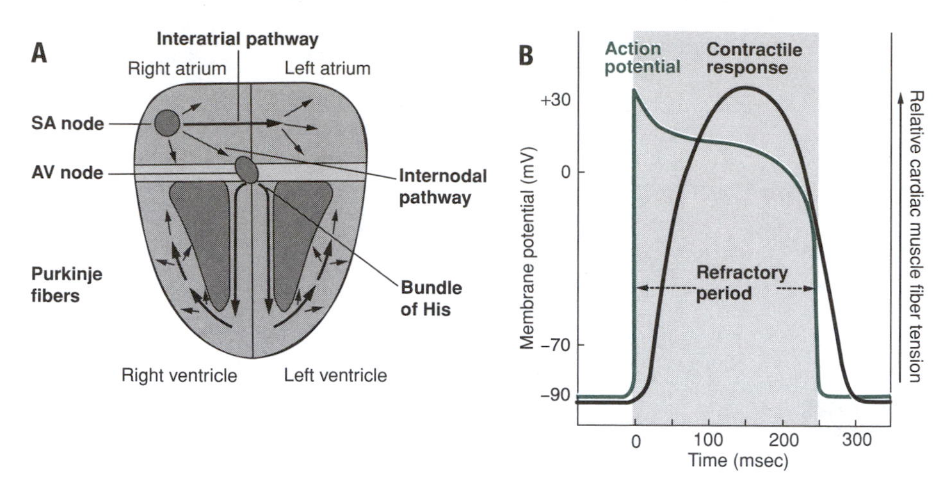

Figure 8.15

Electrical activity of the heart. A. Spread of cardiac excitation. **B.** An action potential (colored line) and the corresponding contractile response (black line). SA, sinoatrial; AV, atrioventricular.

8.4.2 Cardiac conduction and coupling to contraction

After an action potential is generated in the SA node, it is conducted throughout the heart. Specialized conduction pathways allow the signal to travel faster to certain regions of the heart (Figure 8.15a). The signal moves through bundles of specialized fast conducting fibers, in addition to spreading through tissue from contractile cell to contractile cell, illustrated in Figure 8.14. First, an action potential that originates at the SA node spreads through both atria. After a brief delay, it spreads to the atrioventricular (AV) node, which rapidly conducts it to the apex of both ventricles. The AV nodal delay is important for efficient operation of the heart as a pump, as discussed in Section 8.4.3.

There is another important difference between myocardial muscle cells and neurons: Electrical excitement of cells in the myocardium is linked to contraction (Figure 8.15b). When the cell experiences an action potential, it contracts after a small delay. The contraction, or beating, of the whole heart muscle is coordinated by a wave of electrical activity that travels across and through the heart tissue.

The electrical activity of the heart can be measured directly, if specialized cardiac catheters with recording or stimulating electrodes are placed directly within the heart. More conveniently, some aspects of the cardiac electrical activity can be measured from the surface of the body by a technique called **electrocardiography** (ECG). The ECG recording, which is discussed in more detail in Chapter 11, is a reflection of the external currents around the cells, which depend on the number of cells that are synchronously showing electrical activity. Because of the long distance between the source of the signal and the measurement, however, the voltage measured by ECG of ~1 mV is much lower than the voltage changes in individual cells, which are ~100 mV. Biomedical engineers have long been involved with the development of systems for measuring electrical activity,

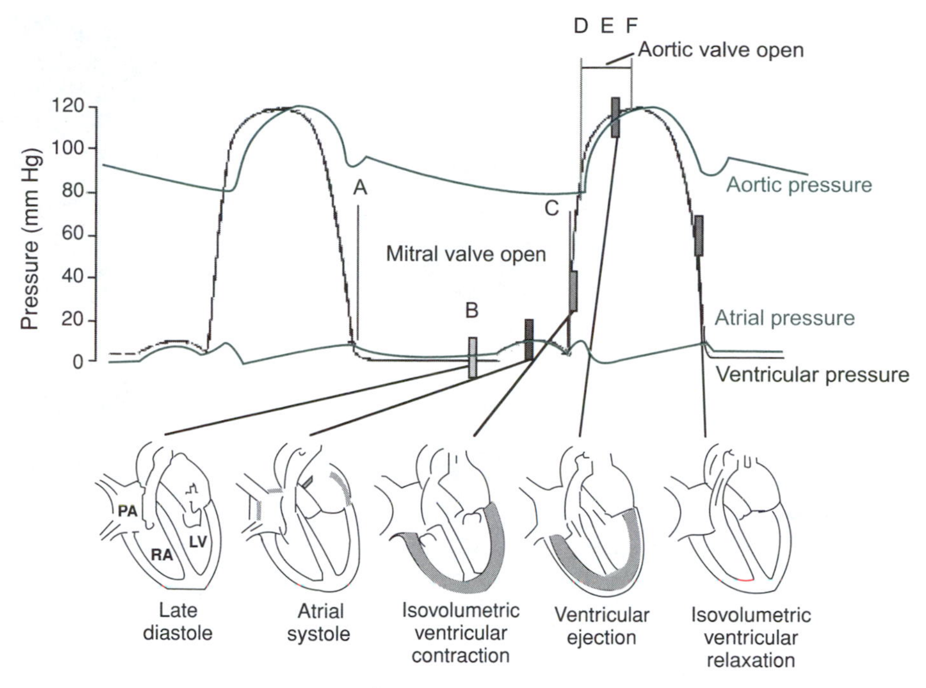

Figure 8.16 **Cardiac cycle.** One heartbeat involves a cycle of electrical activation that elicits contraction and is followed by relaxation.

as well as construction of mathematical models that aid in understanding the dynamics of cardiac conduction.

8.4.3 Cardiac cycle

The highly coordinated sweep of action potentials through the heart muscle and the coordinated contraction of cells that this electrical activity generates produce a highly efficient squeezing of the four chambers of the heart. One beat of the heart, or one round of electrical activation and contraction, is called a **cardiac cycle** (Figure 8.16). The contraction phase of the cardiac cycle is called **systole** and the relaxation phase, or the period in between excitation/contraction events, is called **diastole**.

The cardiac cycle begins with the automatic activation of the action potential. The action potential originates at the SA node, within the right atrium, and spreads through both atria. The two small atria contract as a single unit in an event called **atrial systole**. After a brief delay, it spreads to the AV node and through conduction fibers to both ventricles, which contract as a single unit. As a result of the AV nodal delay, the atria and the ventricles go through separate cycles of systole (contraction and emptying of their contents) and diastole (relaxation and refilling). The atria contract first, sending the blood within the atria into the ventricles, and the ventricles contract shortly after.

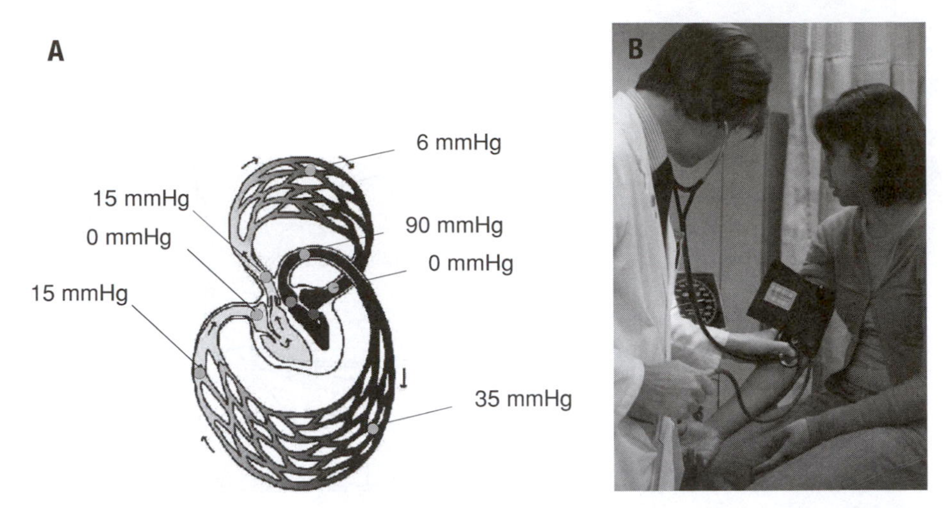

Figure 8.17 **Pressure drops in the circulatory system. A.** The pressure is indicated at several locations within the systemic and the pulmonary circulation. **B.** Blood pressure is most easily measured by compression of an artery in the arm using a device called a sphygmomanometer.

Consider the next step in the cardiac cycle, but focus only on the left side of the heart (Figure 8.16). The goal of the cardiac cycle is for the left ventricle to generate hydrostatic pressure within the aorta. After atrial contraction is complete, contraction of the left ventricle begins. The contraction attempts to decrease left ventricular volume and, therefore, increases blood pressure within the chamber; this phase, called isovolumetric ventricular contraction, occurs while both the aortic and mitral valves are closed. Increasing pressure within the left ventricle eventually causes the aortic valve to open, ejecting blood into the elastic aorta, and initiating blood flow to the rest of the body.

Rhythmic contraction of the heart repeats this ejection sequence; the presence of one-way valves (such as the aortic and mitral valves) ensures blood flow in only one direction (ventricle to aorta). Each cycle has four phases: filling phase (inlet open, outlet closed), isovolumetric contraction phase (valves closed), ejection phase (outlet open, inlet closed), relaxation phase (valves closed). Without valves, the relaxation–contraction cycle of the ventricle would produce no net forward flow, but an endless repetition of blood ejection during the contraction phase and blood return during the relaxation phase. A similar series of events drives blood from the right ventricle through the lung for oxygenation.

Ventricular volume is 120 mL at the end of diastole and 50 mL at the end of systole. The change in volume, which is equal to the volume of blood ejected from the ventricle on each beat of the heart, is called **stroke volume** and is normally 70 mL. Ejection of this volume of blood is accompanied by an increase in aortic pressure from ~80 mmHg (diastolic pressure prior to ejection) to ~120 mmHg (systolic pressure during ejection). The aortic pressure is sufficient to drive flow throughout the circulatory system, with the energy stored in the pressure drop being converted into flow throughout the body (Figure 8.17).

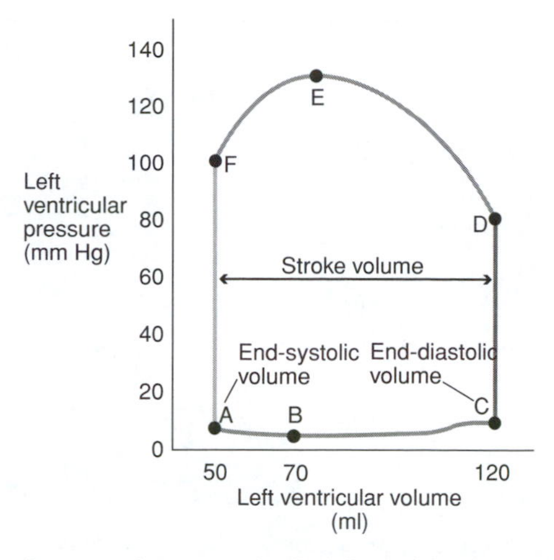

Figure 8.18 **Pressure volume curve for the left ventricle during the cardiac cycle.** Different points in the cardiac cycle (A, B, C, D, E, and F) are indicated in relation to Figure 8.16. The points indicate opening of the mitral valve (A), late diastole (B), closing of the mitral valve (C), opening of the aortic valve (D), ventricular ejection (E), and closing of the aortic valve (F). The line segments indicate ventricular filling (A to C), isovolumetric contraction (C to D), ventricular ejection (D to F), and isovolumetric relaxation (F to A).

Figure 8.18 illustrates another useful approach for examining the cardiac cycle from a mechanical perspective. The heart does work on the blood: It decreases its volume to create the pressure needed to move blood. The work that is done by the heart can be calculated from the pressure-versus-volume diagram:

$$\text{Work} = (PV) + \frac{mv^2}{2}. \tag{8.12}$$

Summary

- The cardiovascular system is composed of a pump, the heart, which circulates a specialized fluid, blood, through an elaborate system of branched vessels.
- The blood is a special fluid composed of cells dispersed in a protein-rich fluid called plasma. White blood cells are involved in the inflammatory response and immune function; RBCs transport oxygen to tissues.
- The blood circulates in the body through a network of vessels including arteries, veins, and capillaries.
- Many of the biophysical properties of the circulation can be deduced using a simple engineering model: fluid flow in a straight cylindrical tube.
- The heart is equipped with a muscular wall that contracts to pump blood from its chambers (atria and ventricles) to other parts of the body.
- The heart's muscular wall is made up of self-excitable cardiac cells that contract in response to electrical stimulation.
- The heart contracts rhythmically to create blood pressure, which drives blood flow.

Profiles in BME: Curtis G. Neason

Born and raised in the suburbs of Houston, TX, near the Johnson Space Center, I was exposed to a variety of careers in science and engineering. One day, during my sophomore year in high school, I was lamenting to a friend that I could not decide between studying medicine and studying computer engineering in college. This friend explained to me that I might not have to choose between the two. His sister had completed a course of study called biomedical engineering (BME), which blended the life sciences with engineering. Sounded perfect!

With this advice I attended Texas A&M University, where I received my BS in Biomedical Engineering. Texas A&M encouraged its students to engage in cooperative education or internships, so I consequently found a "co-op" job assisting a Houston area scientist in designing an epicardial signal acquisition system to record electrical impulse data during open heart surgery. On my first day, I arrived to find the lab door locked and the project closed. Although things seemed pretty bleak, my fortune quickly improved, and one of the cardiologists who was interested in the project took me into his office and gave me a job for my co-op semester. During this short semester, I was exposed to a burgeoning cardiac medical device world. After that summer, I spent three additional semesters working for Intermedics, a pacemaker company located outside of Houston.

Upon graduating from Texas A&M in 1997, I accepted a job as a Clinical Engineer with Prucka Engineering, a small, entrepreneurial company in Houston that designed and manufactured cardiac signal acquisition systems for diagnosing irregular heartbeats. Clinical Engineers did just about everything, including teaching, installation, repair, sales demonstrations, and telephone support. Hospitals all over the world were requesting this type of equipment, so I was able to travel to more than 20 countries on five continents working with some of the world's preeminent cardiologists. I was steadily promoted to higher positions: International Lead Clinical Engineer, International Product Manger, and ultimately Global Product Manager, responsible for the marketing plans for a $30 million business.

Prucka Engineering was acquired by the medical division of General Electric (GE) in 2000. After considerable effort to integrate the Prucka culture into the GE business, we were ultimately successful in reaching out and using resources from all over the GE business, a seemingly endless source of talent and innovation. We were so successful that GE decided to focus more on growing this portion of the cardiology business; I was asked to spend a year exploring new opportunities such as acquisitions, partnerships, and internal innovation. Because I was exposed to multimillion-dollar mergers and acquisitions, I decided I needed to obtain an MBA. By taking classes mostly on the weekends, I obtained my MBA from New York University's Stern College of Business. My wife and I had moved to New York after the GE acquisition, so having access to such a great MBA program at the doorstep of Wall Street was incredibly exciting.

Despite the excitement of expanding the business within GE, I decided I wanted to get back to my original passion, bioengineering. In 2004, I joined St. Jude Medical, a manufacturer of cardiac medical devices. I became a Field Clinical Engineer in Houston for their cardiac pacemaker and implantable defibrillator division. Field Clinical Engineers for St. Jude manage clinical studies for the company in their local area. In other words, they work with the physicians who are interested in conducting the clinical trials for new devices. Field Clinical Engineers also support the local sales staff in conducting highly technical training and clinical troubleshooting.

I live with my wife and twin sons in Houston, where we enjoy the arts, entertainment, and cuisine that the city provides. Besides being fortunate to have such a great family, I feel truly lucky to have discovered bioengineering in high school. It is my hope that others interested in BME will learn early in their lives about the exciting opportunities that exist in the field.

FURTHER READING

Boron WF, Boulpaep EL. *Medical Physiology.* Philadelphia, PA: Sanders; 2004. The cardiovascular system is described in chapters 17 through 24.

USEFUL LINKS ON THE WORLD WIDE WEB

http://www-medlib.med.utah.edu/kw/pharm/hyper_heart1.html
Shockwave animation of the cardiac cycle

http://www.med.ucla.edu/wilkes/intro.html
Tutorial on heart sounds and relation to diseases of the heart

http://students.med.nyu.edu/erclub/ekgexpl0.html
Tutorial on interpretation of the EKG

http://homepage.smc.edu/wissmann_paul/anatomy1/1bloodpressure.html
Description of how to measure blood pressure

http://www.hhmi.org/biointeractive/vlabs/ (select Cardiology lab)
HHMI Virtual Cardiology laboratory

KEY CONCEPTS AND DEFINITIONS

aneurysm an excessive localized enlargement of an artery caused by a weakening of an artery wall

arteriole a small branch of an artery leading to a capillary

atherosclerosis an arterial disease characterized by the deposition of plaques of fatty material on their inner walls

atrial systole the contraction of the heart muscle of the left and right atria

axial pressure drop the internal pressure drop measured at different points along the axis of a vessel

bifurcation the division of a blood vessel into two or more branches

bruits sounds created by the eddies of turbulent flow

capillaries a fine branching network of blood vessels between venules and arterioles

cardiac cycle the sequence of events that occurs in one heartbeat

compliance the property that describes a material's ability to deform with the application of a pressure or stress

constriction the narrowing of blood vessel diameter because of muscular contraction within the vessel wall

diastole a phase of the heartbeat when the heart relaxes to allow the heart chambers to fill with blood from the veins

dilation the widening of blood-vessel diameter caused by relaxation of smooth muscle in vessel wall

eddies the circular flow of fluid in a direction counter to the current generated because of to fluid movement

electrocardiography (ECG) a measurement of electrical activity in the heart using electrodes placed on the skin of the limbs and chest

fenestrated capillaries capillaries that have larger openings to allow larger molecules to diffuse

laminar flow a term used to describe fluids that flow in straight parallel layers without interference between layers

Laplace's law describes the relationship between pressure drop and wall tension in inflated vessels such as a capillary or an alveolus

red blood cell (RBC) one type of blood cell shaped in the form of a biconcave disk that contains hemoglobin to transport oxygen and carbon dioxide to and from the body tissues

resistance a property that characterizes a substance's ability to oppose flow of another substance

Reynolds number a dimensionless number used to characterize whether a fluid will lead to a laminar or turbulent flow

stroke volume the volume of blood ejected when the ventricle contracts

systole a phase of the heartbeat when the heart contracts to pump blood from its chambers to the arteries

transmural pressure difference the difference in pressure within the vessel and pressure in the interstitial fluid outside of the vessel that is necessary to keep the blood vessel inflated

turbulent flow a term used to describe chaotic fluid flow in which the fluid no longer follows the parallel streamline

viscosity the measure of the resistance of a fluid to deform under shear stress

NOMENCLATURE

A	Area	m^2
C	Compliance	$(m^2s)^2/kg$
c, (c_f)	Concentration (of fibrinogen)	Mol/L
D	Diffusion coefficient	m^2/s
E	Electrical potential	V
F	Force	N, $kg\text{-}m/s^2$
$\dot{F}$	Air flow rate	m^3/s
g	Acceleration due to gravity	m/s^2
H	Hematocrit of blood	—
h	Vertical height	M

I	Electric current	Ampere
J_x	Flux	kg/m^2-s
K	Permeability of capillary wall	s/L; s/m^3
L	Length of vessel	m
m	Mass	kg
$\dot{M}$	Mass flow rate	kg/s
$P_1, P_2...$	Pressure at positions 1, 2. . .	Pascal, mmHg, N/m^2
P_{alv}	Alveolar pressure	Pascal, mmHg, N/m^2
P_{atm}	Atmospheric pressure	Pascal, mmHg, N/m^2
P_c	Capillary pressure	Pascal, mmHg, N/m^2
P_{if}	Interstitial fluid pressure	Pascal, mmHg, N/m^2
P_{ip}	Internal pressure	Pascal, mmHg, N/m^2
P_{tm}	Transmural pressure	Pascal, mmHg, N/m^2
P_v	Vessel pressure	Pascal, mmHg, N/m^2
Q	Blood flow rate	m^3/s; L/s
R	Resistance to flow (electric, fluid, etc.)	Electric – ohms; flow – kg/m^4-s
Re	Reynolds number	—
r, r_v	Radius of vessel	m
t	Thickness of membrane	mm
T	Tension in vessel wall	N
V	Volume	m^3
$v, (v_\vartheta v_r v_z)$	Velocity (in directions ϑ, r, z)	m/s^2
v_{av}	Average velocity	m/s^2

GREEK

π	3.14	—
Δ	Change in a quantity or parameter	—
μ, μ_0	Viscosity of fluid, viscosity of blood	centipoises, g/cm-s
ρ	Density	kg/m^3
Π_c, Π_{if}	Osmotic pressure (in capillaries and interstitial fluid)	Pascal, N/m^2
τ	Shear stress	$kg/m\text{-}s^2$
$\dot{\gamma}$	Shear rate	1/s

QUESTIONS

1. Veins and capillaries are both low-pressure vessels. Why do veins typically have thicker, stronger walls than capillaries?
2. How is compliance related to wall tension in the wall of a vessel? Which type of vessel is more compliant: veins or capillaries? What property allows these vessels to be more compliant? What function does this higher compliance serve?

Table 8.3 Fluid flow velocities in various systems

System	v (μ m/min)
Aorta	3.8×10^7 (63 cm/s)
Large artery	$1.2–3.0 \times 10^7$ (20–50 cm/s)
Capillary	$3–6 \times 10^4$ (0.05–0.1 cm/s)
Large vein	$0.9–1.2 \times 10^7$ (15–20 cm/s)
Vena cava	$6.6–9.6 \times 10^6$ (11–16 cm/s)
Mississippi River	1.7×10^7 (28 cm/s)

3. Using library resources, investigate what a pacemaker (the medical device, not the natural pacemaker tissues of the heart) is and how it works. Why is it important to control a patient's heart rate?

PROBLEMS

1. Use data on the dimensions of various vessels, which can be found in Appendix A, to answer the following questions. From the dimensions that are given in ranges, pick a reasonable value.
   a. Calculate the pressure drop per centimeter (for a flow rate of 5 L/min) in the aorta, a terminal artery, an arteriole, and a capillary. Assume that the 5 L/min flow goes through a single vessel.
   b. Repeat the calculation in a, but this time assume that the 5 L/min flow goes through an array of vessels in parallel. The number of vessels in the array at each level (terminal artery, arteriole, and capillary) is the number that is required to have the same total cross-sectional area as the aorta.
   c. Repeat the calculation in a, again assuming that the 5 L/min flow goes through an array of vessels. The number of vessels in the array at each level is the number required to achieve the velocity levels listed in Table 8.3.
2. Intravenous injections are commonly performed with a hypodermic needle and syringe. Assume that the dimensions of the needle are length = L, radius of the needle = 3.3×10^{-4} m, radius of the syringe = 5×10^{-3} m, and that the drug solution has the properties of water.
   a. Assume that 1 mL of a drug solution must be injected over a 3-second period. Will the flow through the needle be laminar or turbulent?
   b. Derive a general expression for the force that must be applied to the plunger in the syringe to accomplish an injection of V milliliters of drug solution in T seconds.

c. What is the force required to achieve the conditions in a?

d. Plot the force applied versus flow rate (V/T) for three values of needle radius: 1×10^{-4} m, 3.3×10^{-4} m, and 10×10^{-4} m. How does the radius of the needle influence the force required to achieve a given flow rate? (Problem adapted from Domach, M.M., *Introduction to Biomedical Engineering*, Prentice Hall; p. 174)

3. What is the overall rate of blood flow, in liters per minute, through the aorta? How many liters of blood move through the heart per day? What fraction of body weight is blood? (density of blood = 1060 kg/m^3)?
4. Estimate the change in hydrostatic pressure within a blood vessel near the ankle when a 6-ft person changes from a lying to a standing position.
5. Examine one of the consequences of branching with a simple model shown in the figure. Assume that there are two blood-flow circuits, which are identical except that one of the circuits has a single vessel, whereas the other has two identical vessels in parallel.

a. If the resistance of a single vessel is R, what is the overall resistance of the two vessels in parallel?

b. If each of the circuits carries the same flow rate, which has the larger pressure drop? How do the two pressure drops compare?

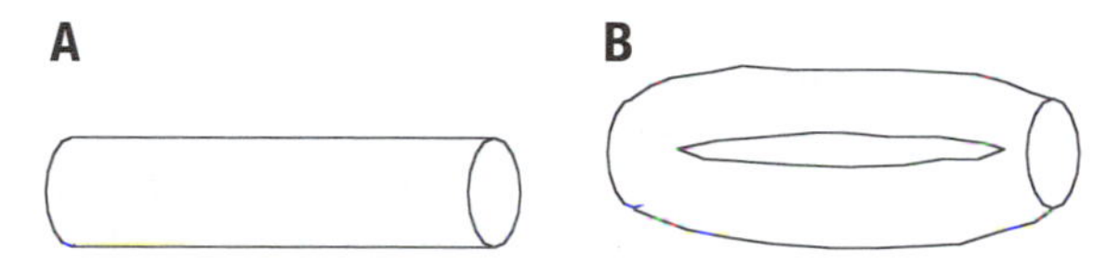

6. What are the physical forces that drive blood flow? Describe the concepts of pressure and resistance. How does pressure vary with volume in an ideal gas? What do you expect is the difference for a liquid? Will pressure still vary with volume?
7. Describe the anatomy of movement of the wave of cell depolarization throughout the heart during a cardiac cycle.
8. Using Equation 8.12 and Figure 8.18, calculate the work done by the heart during one cardiac cycle.
9. A flow loop is set up experimentally, as shown in the figure. Assume that the resistance of the nonbranching segments of tubing is negligible compared to that of segments A and B. The resistances are: $R_A = 1000$ Pa-s/m^3 and $R_B = 3000$ Pa-s/m^3.

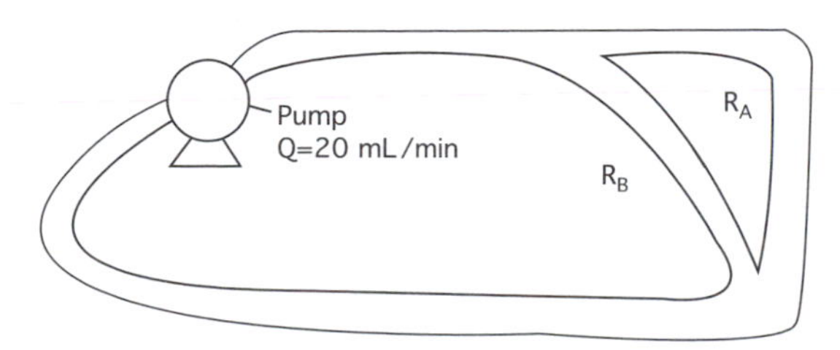

a. Draw an analogous circuit diagram for this flow loop.

b. What is the flow rate through each segment?

c. What is the total pressure drop?
d. What is the pressure drop across each segment?

10. What is the cardiac output for a person whose heart beats 70 beats/min with a stroke volume of 70 mL?
11. As blood flows from the aorta to the arteries, arterioles, and capillaries and through the venous circulation, it encounters various resistances to flow, which results in pressure drops. Using the table, calculate the resistance for an artery and a capillary. (Note: The viscosity of the blood is 4.5×10^4 Pa-s.)

	Aorta	Capillary
Diameter	2.5 cm	8 μm
Length	40 cm	1 mm

12. Using the cardiac output you can calculate in Problem 10, determine the pressure drop (mmHg) in each type of vessel listed in the table in Problem 11. (Watch your units!)
13. Considering normal systemic conditions for blood circulation, match the vessel (from column A) with an appropriate pressure (from column B) and a corresponding radius (from column C).

A. Vessel	B. Pressure	C. Internal radius
Capillary	112 mmHg	3 μm
Vena cava	5 mmHg	15 mm collapsible
Aorta	23 mmHg	12 mm stretchable

14. An unknown liquid is pumped through a cylindrical tube, and both the shear stress and the shear rate are measured in five trials. Plot the shear stress as a function of shear rate. Is the fluid Newtonian or non-Newtonian?

Shear Stress (dyne/cm^2)	Shear Rate (1/s)
2,167	90
2,654	128
5,690	208
13,716	670
20,619	871

9 Removal of Molecules from the Body

LEARNING OBJECTIVES

After reading this chapter, you should:

- Understand the role of the excretory systems in eliminating wastes and toxins and maintaining body balances.
- Understand the concept of biotransformation and the role of the liver in accomplishing the removal of compounds by both direct excretion (through the biliary system) and biotransformation.
- Understand the basic anatomy of the kidney and its functional unit, the nephron.
- Understand the basic processes that underlie kidney function: filtration, reabsorption, and secretion.
- Understand the biophysical processes responsible for filtration and regulation of filtration in the glomerulus.
- Understand the concept of clearance and be able to calculate clearances for typical solutes.
- Understand how proteins in the membrane of tubular epithelial cells—such as channels, active transporters, co-transporters, and exchangers—are responsible for reabsorption and secretion of compounds.
- Understand the role of osmotic pressure as a driving force for water reabsorption in the tubules.

9.1 Prelude

Each person ingests a large number of molecules per day with meals and snacks (Figure 9.1). A similarly large number of molecules enters the body through respiration (Figure 9.2). Body processes—such as building proteins, producing energy, and replenishing lost nutritional stores—use many of these molecules (recall Table 7.1). But a sizeable number of ingested chemicals are either not usable or not needed by the body, and therefore must be eliminated. In addition, metabolic processes generate waste products that are toxic if they accumulate in body tissues. These molecules must also be eliminated.

Figure 9.1 **People eat for the pleasure that it brings, but also to bring vital nutrients into their bodies.** Absorption by the intestines is unregulated—most of what we swallow gets absorbed; body composition is controlled by elimination of unneeded compounds.

The problem of elimination of molecules is staggeringly complex, because the diversity of molecules that can be ingested is enormous (recall Table 7.3). In addition to molecules gained in our diet, each person—in the course of sleeping, walking, or bicycling through his or her normal day—will inhale diverse molecules that float freely in the atmosphere. The systems of elimination in the body must be capable of responding to this onslaught; they must safely manage both the number of molecules and the diversity of molecules to ensure the chemical balance that is necessary for life.

Life processes require a rather stringent set of chemical conditions. For example, Chapter 2 described the importance of pH, which must be maintained within narrow limits for proper function of cells and proteins. The function of communication systems within the body, such as the nervous system and the endocrine system, depends on the release, movement, and recognition of specific chemical messengers. These systems cannot function well if stray molecules compete with normal recognition processes. This is one of the reasons that chemical composition within the brain is tightly regulated by the blood–brain barrier. Most organs and tissues do not have individual protective mechanisms like that of the brain; instead they rely on organs that are specialized for molecular exchange and excretion to maintain an overall homeostasis in the body.

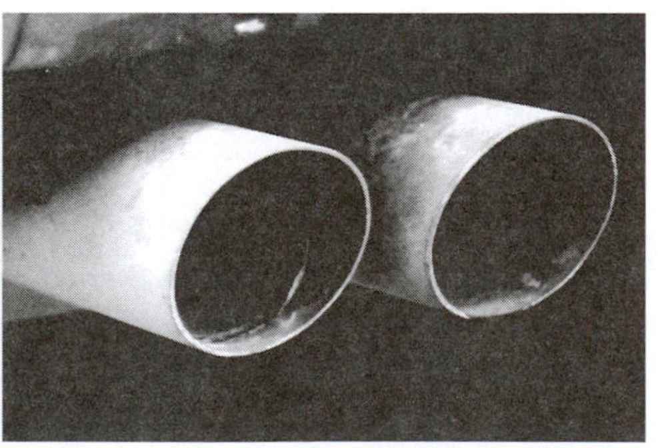

Figure 9.2 **Air contains oxygen, which is necessary for life, but it also contains compounds and particles.** Although these compounds and particles are present at much lower concentrations than oxygen, they can also be toxic. This is particularly true for the agents that are released by gas-powered vehicles and industrial processes.

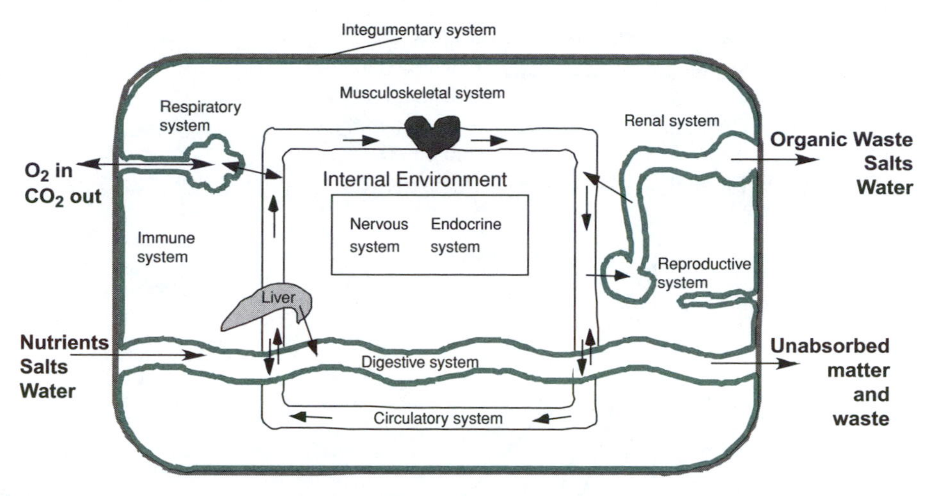

Figure 9.3 **A schematic diagram of body systems illustrating routes of elimination.** See Figure 7.4 for a description of the system boundaries (black and colored lines).

9.2 Examples of elimination of molecules from the body

The human body has a variety of mechanisms for elimination of waste products (Figure 9.3). A few agents, such as carbon dioxide, are eliminated by exhalation from the lung; these agents are small, volatile, and able to permeate through membranes rapidly. Other molecules are chemically converted to other molecules, by chemical reactions that often happen in the liver. This process, called **biotransformation** (Figure 9.4), is critically important for the elimination of many potentially toxic molecules, such as alcohol. Some compounds in food are never absorbed into the body, but are excreted from the body in feces. Other agents leave the body with the feces after being excreted through the biliary system into the intestine. The kidney is the most important organ for maintaining **homeostasis** of the internal environment. It accomplishes this by filtering blood, removing small molecular-weight waste products, and regulating the concentration of the ions that are critical for maintaining the normal function of the heart and brain.

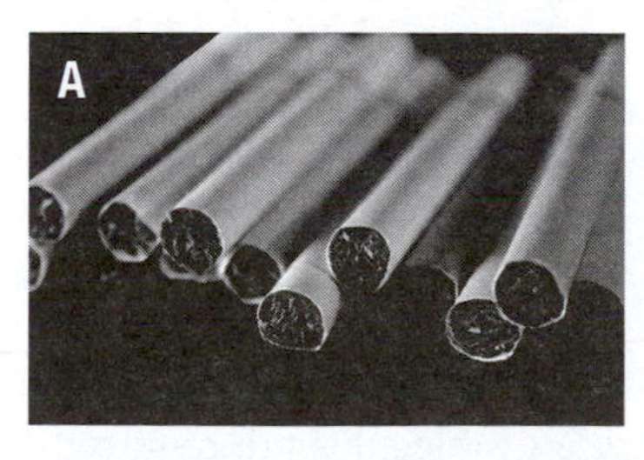

Figure 9.4 **Biotransformation. A.** Most of the biotransformation that is important in elimination of compounds occurs in the liver. The liver has enzymes that convert molecules into related molecules that are easier to eliminate. For example, the liver converts nicotine into other compounds, such as cotinine. **B.** Biotransformation can also convert an inactive compound into a compound with greater biological activity. Image created and provided courtesy of Gino Paull.

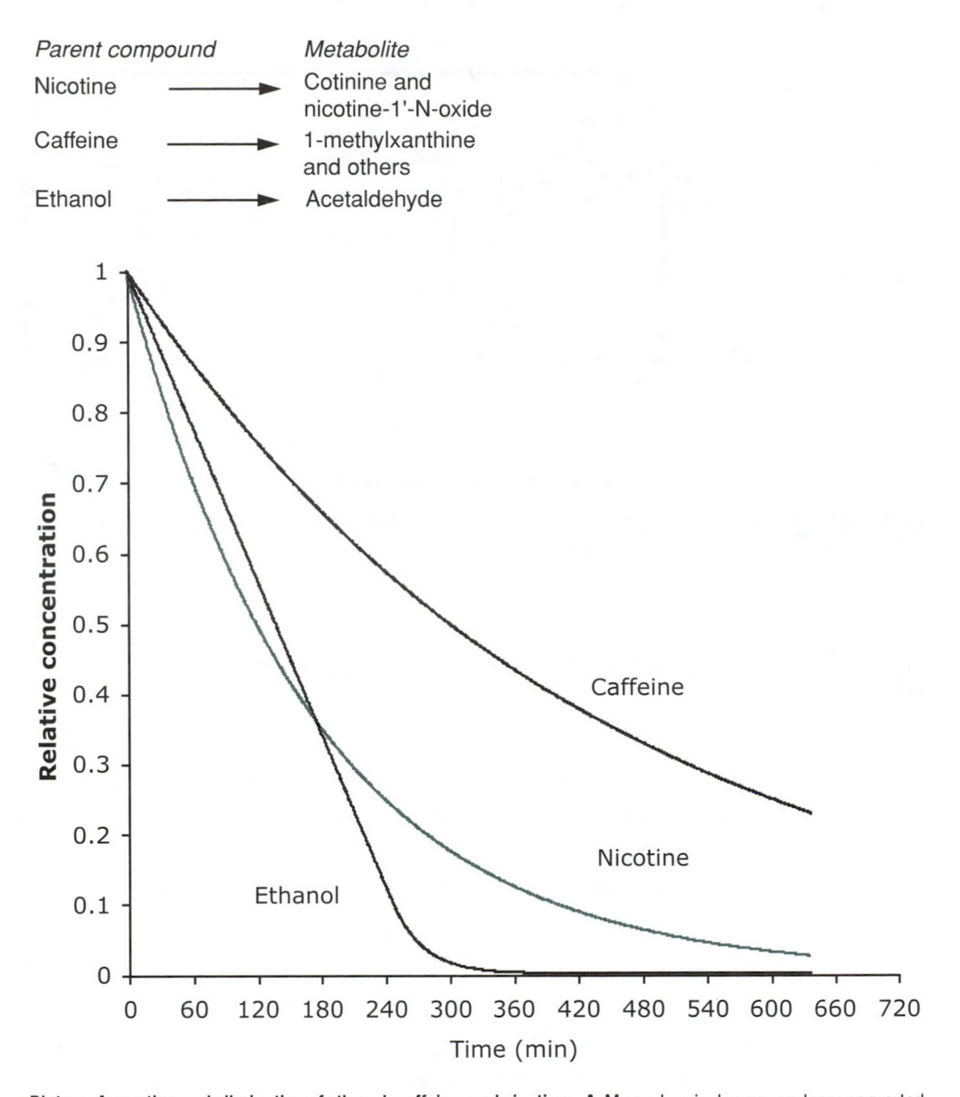

Figure 9.5

Biotransformation and elimination of ethanol, caffeine, and nicotine. A. Many chemical compounds are converted into other compounds prior to elimination. Examples are shown for nicotine, caffeine, and alcohol. **B.** The rate of chemical transformation affects the overall rate of elimination of many compounds from the body. This graph shows the amount of nicotine, caffeine, or alcohol that is left in the body after a single initial dose (at time t = 0).

Recall the tracer balance equation that was introduced in Chapter 7. For a compound that is introduced into the body rapidly and is eliminated by first-order kinetics, the amount of that compound decreases with time according to:

$$M = M_{\text{dose}} e^{-kt}. \tag{9.1}$$

In this chapter, the physiological events that lead to disappearance are examined. Empirically, it is clear that not all molecules are removed from the body at the same rate. If a person inhales a balloon full of helium, and then talks and breathes normally, the helium (and therefore the comical voice it creates) will disappear quickly, as the helium is rapidly removed from the body by breathing. Nicotine, the active ingredient of tobacco, can enter the body in the same way as

Table 9.1 Caffeine content of select common foods and drugs

Product	Serving size	Caffeine per serving (mg)
Caffeine tablet (Vivarin®)	1 tablet	200
Excedrin® tablet	1 tablet	65
Coffee, brewed	240 mL (8 U.S. fl oz)	135*
Coffee, decaffeinated	240 mL (8 U.S. fl oz)	5*
Coffee, espresso	57 mL (2 U.S. fl oz)	100*
Chocolate, dark (Hershey's® Special Dark)	1 bar (43 g; 1.5 oz)	31
Chocolate, milk (Hershey® bar)	1 bar (43 g; 1.5 oz)	10
Red Bull®	240 mL (8.2 U.S. fl oz)	80
Bawls® Guarana	296 mL (10 U.S. fl oz)	67
Soft drink, Coca-Cola® Classic	355 mL (12 U.S. fl oz)	34
Tea, green	240 mL (8 U.S. fl oz)	15
Tea, leaf or bag	240 mL (8 U.S. fl oz)	50

* Estimated average caffeine content per serving. Actual content varies according to preparation.

helium, but its effects last for a longer time. Nicotine, unlike helium, is readily absorbed through the lungs and into the bloodstream. Once in the blood, it stays in the body until it is eliminated, primarily by biotransformation in the liver to compounds such as cotinine that are then rapidly eliminated by the kidney. It is the rate of conversion of nicotine into these other compounds (which are now called **metabolites** of nicotine) that determines its duration of action in the body. The half-life for metabolism of nicotine is about 2 hours (Figure 9.5). Likewise, the physiological effects of smoking a cigarette—such as increased blood pressure and stimulation of the brain—last several hours.

Caffeine, the active ingredient in coffee and many soft drinks (Table 9.1), has a half-life of 5–7 hours (Figure 9.5). Its biological effects also last that long, which explains why many people cannot sleep well after drinking coffee in the afternoon. Caffeine is metabolized by biotransformation into other compounds in the liver. The rate of elimination of some compounds can change, depending on other factors present in the individual. For example, in smokers, the half-life of caffeine is decreased to ~3 hours.

Not all compounds are eliminated by first-order kinetics. Alcohol is converted to a toxic compound called acetaldehyde by the action of enzymes, such as alcohol dehydrogenase, within cells in the liver. Acetaldehyde is further converted to acetic acid by the action of another enzyme. Interestingly, the rate of alcohol oxidation is relatively constant with time (Figure 9.5). Rates of reaction that are constant with time are called zero-order reactions: In this case, the constant rate of reaction indicates that an enzyme is saturated (recall Equation 11 in Box 4.2, Chapter 4, for cases of high substrate concentration). The average rate of metabolism of ethanol in adults is approximately 120 mg/kg/h, or about

30 mL in 3 hours for a 75-kg adult. Whereas metabolism in the liver decreases the concentration of ethanol—and therefore ends drunkenness—it increases the concentration of acetaldehyde, which is toxic. (A long-term consequence of alcohol abuse is liver failure, which appears to be caused by accumulation of acetaldehyde and its direct toxic effect on liver cells.)

When drug companies develop a new potential drug, one of the most important questions that they must answer is: How does the body get rid of it? The safest drugs are those that the body can eliminate easily and predictably. Aspirin is a safe drug that is taken for a variety of conditions, including pain and arthritis. It is frequently used to prevent clotting of the blood. Aspirin is also one of America's favorite drugs. Aspirin is another name for the compound acetyl salicylate. A dose of aspirin within the blood is eliminated with a half-life of 15 minutes; the effects of aspirin can last for several hours. Aspirin is converted into other compounds by the liver; some of these other compounds are biologically active (they are chemically similar to aspirin), and others are not. With time, aspirin and all of the products of biotransformation are excreted in the urine.

SELF-TEST 1 If the half-life of aspirin is 15 minutes, how can the biological effects of aspirin last for several hours?

ANSWER: Recall the relationship between elimination of an agent from the body and the half-life (Figure 7.6). The biological effect will last as long as the concentration of the agent is above its therapeutic concentration (i.e., the concentration in the blood that is necessary for a biological effect). If a large dose is administered—so that the initial concentration is 1,000 times the therapeutic concentration—the effect can last for several hours. Notice that this can be done only if the drug is not toxic at the high initial dose.

9.3 Biotransformation and biliary excretion

The liver is a major site of drug metabolism, but it is also an important excretory organ. Chapter 7 showed the position of the liver with respect to the rest of the digestive system (Figure 7.18a). Most of the venous blood from the intestine flows through the portal vein to the liver. This positioning is important for liver function: The liver has immediate access to all of the nutrients absorbed from the intestine and all of the drugs and toxins that are potentially absorbed with those nutrients.

The liver directly excretes some compounds. These compounds are transported by **hepatocytes** from the blood into the liver biliary system, a branching network of vessels and reservoirs that collects **bile** and empties it into the small intestine (Figure 9.6). Hepatocytes, the functional cells of the liver, extract the compounds from the blood and transport them into the biliary system. In humans, the liver produces 250–1,000 mL of bile per day, which is transported through the biliary vessels, stored and concentrated in the gall bladder, and delivered into the small

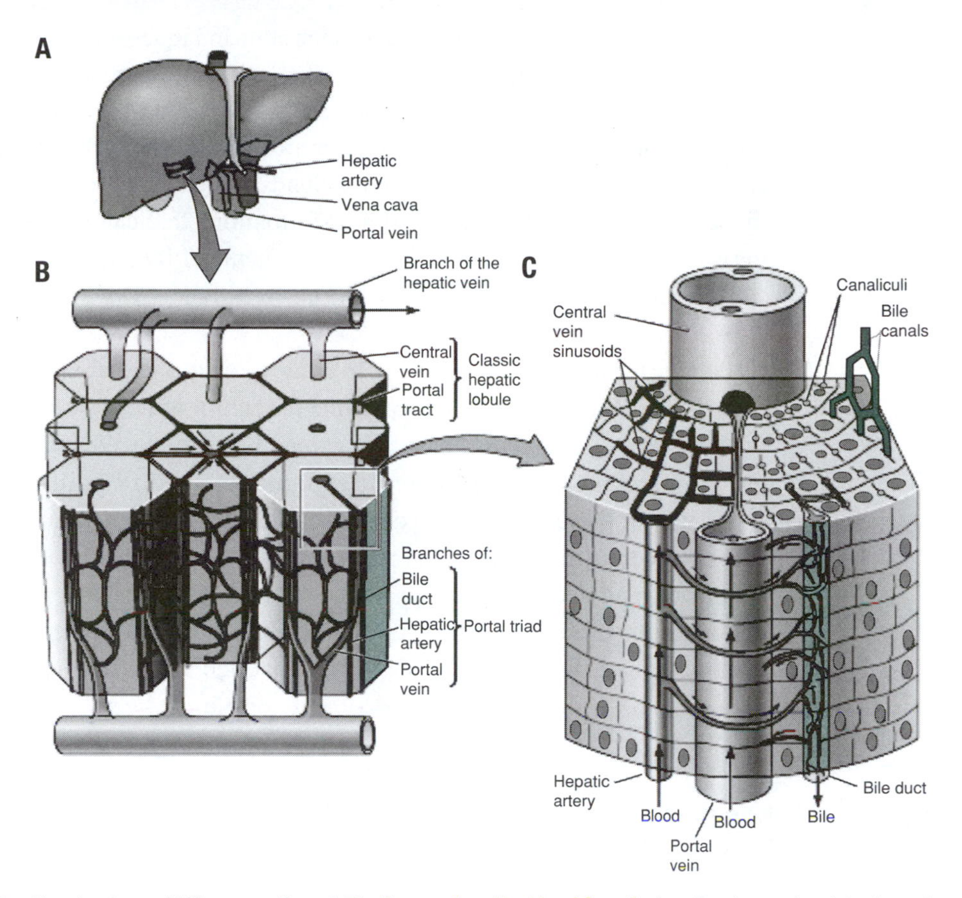

Figure 9.6 **Hepatocytes and biliary secretion. A.** The liver receives blood input from the hepatic artery and portal vein, and all output blood leaves though the hepatic vein but molecules can also leave via the bile duct. **B.** The liver is divided into many lobules, which receive oxygenated blood from a branch of the hepatic artery and nutrient-rich blood from a branch of the portal vein. **C.** Flow of compounds through the liver lobule is highly organized. See also Figure 7.22.

intestine. Once in the intestine, molecules in the bile can be excreted in feces, reabsorbed into the blood, or transformed through the action of intestinal enzymes.

Only certain kinds of compounds are excreted into bile. Excretion into bile is an energy-consuming process in hepatocytes, occurring by some of the same membrane transport processes that occur in tubules within the kidney (more about this in the next section). In humans, the excreted compounds typically have molecular weights (MWs) greater than 400 daltons; smaller compounds are not excreted. Some compounds are excreted unchanged into bile: For example, more than 95% of the antibiotic drug erythromycin is excreted directly into bile. Other compounds are excreted after biotransformation. Estradiol (used for estrogen replacement therapy and birth control) and morphine (used to control severe pain) are excreted as chemical conjugates, as described in the next paragraph.

Biliary excretion is an important mechanism for elimination of molecules from the body, but the liver's role in preparing compounds for excretion by the kidney is quantitatively more important. Many agents, including drugs and toxic compounds encountered in daily life, enter the body because of their lipid solubility;

lipid-soluble compounds permeate through skin and mucosal surfaces more readily than do water-soluble compounds. The kidney is inefficient at excretion of lipid-soluble molecules (lipid-soluble molecules are reabsorbed into the blood after filtration; see Section 9.4). The liver aids in elimination by converting lipid-soluble molecules into more polar compounds, which are more easily excreted by the kidney. The range of chemical transformations that can be accomplished in the liver is impressive; this diversity of biochemical function partially accounts for the diverse symptoms—and high drug sensitivity—experienced by patients with advanced liver disease.

Chemical transformations in the liver are greatly facilitated by the action of enzymes within the smooth endoplasmic reticulum of hepatocytes (often called the microsomal fraction). Phase I reactions (oxidation, reduction, and hydrolysis) convert molecules to more polar forms that usually differ in biological activity from the parent form. For drug molecules, biotransformation reactions usually produce metabolites that are less active than the parent drug. Occasionally, inactive molecules are administered with the understanding that they will be converted into an active form by the action of liver enzymes. The anticancer drug cyclophosphamide is not cytotoxic but is converted to powerful agents, including phosphoramide mustard, by a series of reactions in the liver.

9.4 Elimination of molecules by the kidneys

The kidneys, and the rest of the urinary system, are famous for their ability to rid the blood of toxic molecules. Although this is a heroic—and life-saving—function, it is only one of the things that the kidney does to ensure health. When we eat, the absorption of molecules into the body is unregulated: Most of the molecules that we ingest get absorbed into the body. People do not control the balance of sodium, water, and potassium that we ingest in foods: Instead, we eat more than we need (usually) and count on the kidney to excrete the excess, to keep us in balance. When these balancing functions are lost, as in the diverse forms of renal failure, health deteriorates rapidly (Table 9.2).

The kidneys of a typical person process 950 L of plasma each day to excrete in the urine 103 mmol of sodium, 50 mmol of potassium, 2 mmol of bicarbonate, and 52 g of urea (Table 9.3). These observed values provide us with some insight into kidney function. For example, a large fraction of the total amount of urea that enters the kidneys per day gets excreted in the urine (11%). If the kidneys were simple filters, then one might expect 11% of the water that passes through the kidneys to be excreted as well. Imagine that: Each person would **micturate** more than 100 L of urine per day to allow for elimination of 52 g of urea. The kidney must have a mechanism for recovery of water and, therefore, concentration of the urine.

In addition, these amounts of excreted ions represent different fractions of the total amount of these chemicals that pass through the kidneys each day. If the

Table 9.2 Functions of the kidney

Function	Disorders in renal failure
Excretion of toxic substances (endogenous wastes, exogenous toxins, drugs)	Uremia (e.g., loss of appetite, nausea and vomiting, lethargy, coma)
NaCl balance (regulation of ECF volume, plasma volume, blood pressure)	Hypertension Edema
Water balance (regulation of osmolality and ECF Na concentration)	Hyponatremia
K balance	Hyperkalemia
Acid–base balance	Acidosis
PO_4 and Ca balance Activation of vitamin D	Bone disease
Secretion of erythropoietin	Anemia

kidney were a simple filter, even a filter that included a water recovery process to allow for concentration of urea, it would be difficult to explain how 11% of the total urea is excreted, whereas only 0.008% of the total bicarbonate and 0.16% of the total sodium is excreted. In fact, the kidney accomplishes its work by a combination of filtration (to produce the fluid that eventually becomes urine), reabsorption (to reclaim water and vital chemicals from the urine into the blood), and selective secretion (to pump some chemicals from the blood into the urine) (Figure 9.7).

The urinary system (Figure 9.8) is composed of two **kidneys**, which selectively remove molecules from the blood as described later; one **urinary bladder**, which collects urine from the kidneys and serves as a reservoir; and two **ureters**, which are tubular vessels that carry urine from the kidneys to the urinary bladder. The ureters are simple flow vessels, and the bladder is a reservoir; the urine that is produced by the kidney is not changed after it leaves the kidney. Urine is excreted from the body through the **urethra**.

Each kidney receives blood through one artery and returns it via one vein. The total renal blood flow is approximately 1 L/min; for persons with a normal

Table 9.3 Total amounts of fluid and selected compounds flowing through the kidneys

Substance	Flow through renal arteries (per day)	Flow through ureters (per day)
Blood	1,730 L	
Plasma	950 L	
Water		1.5 L
Na^+	133,000 mmol	103 mmol (0.16%)
K^+	3,800 mmol	50 mmol (0.08%)
HCO_3^-	25,600 mmol	2 mmol (0.008%)
Urea	475 g	52 g (11%)

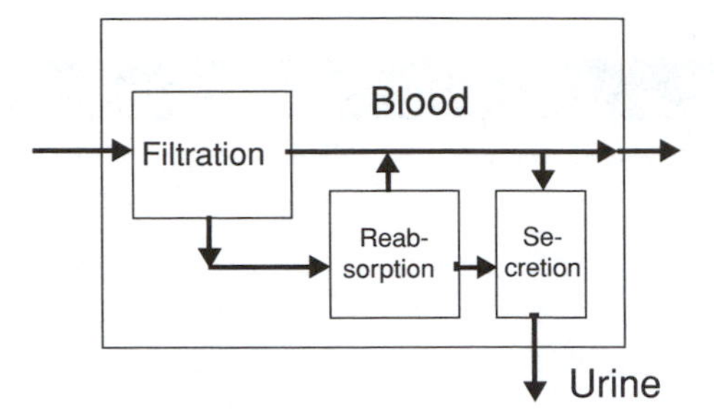

Figure 9.7

Schematic diagram of kidney function. The kidney operates by filtration to produce an ultrafiltrate of blood. Some compounds are reabsorbed from the ultrafiltrate back into the blood; other compounds are secreted directly from the blood into the ultrafiltrate.

concentration of red blood cells, the renal plasma flow (RPF) is, therefore, 0.6 L/min. Because of this high blood flow rate, a tremendous quantity of substances flow through the kidneys each day (Table 9.3). The main functional unit of the kidney is the **nephron**. Each kidney contains about 0.5 million nephrons, which all have similar features, but vary somewhat in their structure. The nephron is a highly organized collection of tubes and vessels (Figure 9.9).

The entry point for fluid into the nephron is the **glomerulus**, which provides the filtration function of the kidney (Figure 9.10). Filtered plasma carries molecules that are sufficiently small through the glomerular membrane into **Bowman's space**. As this **ultrafiltrate** moves through the kidney **tubules**, some molecules—such as water, sodium, and glucose—are reabsorbed into the blood. Other molecules—such as penicillin—are actively secreted into the tubular fluid. Along the tubule system of the nephron, there are 12–13 distinct cell types; each

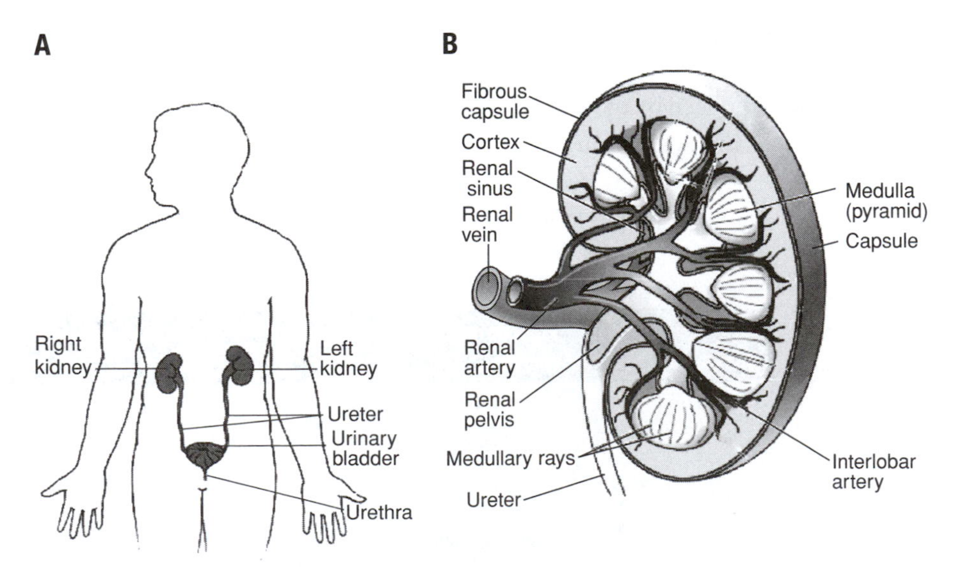

Figure 9.8

Renal/urinary system. A. Humans have a pair of kidneys, which are located on either side of the abdominal aorta. Urine flows from each kidney, through a ureter, to the bladder. **B**. Each kidney receives blood from a renal artery, a direct branch of the aorta. Blood is returned from the kidney to the vena cava by a renal vein. The renal artery, vein, and ureter exit from the kidney in a region called the renal pelvis.

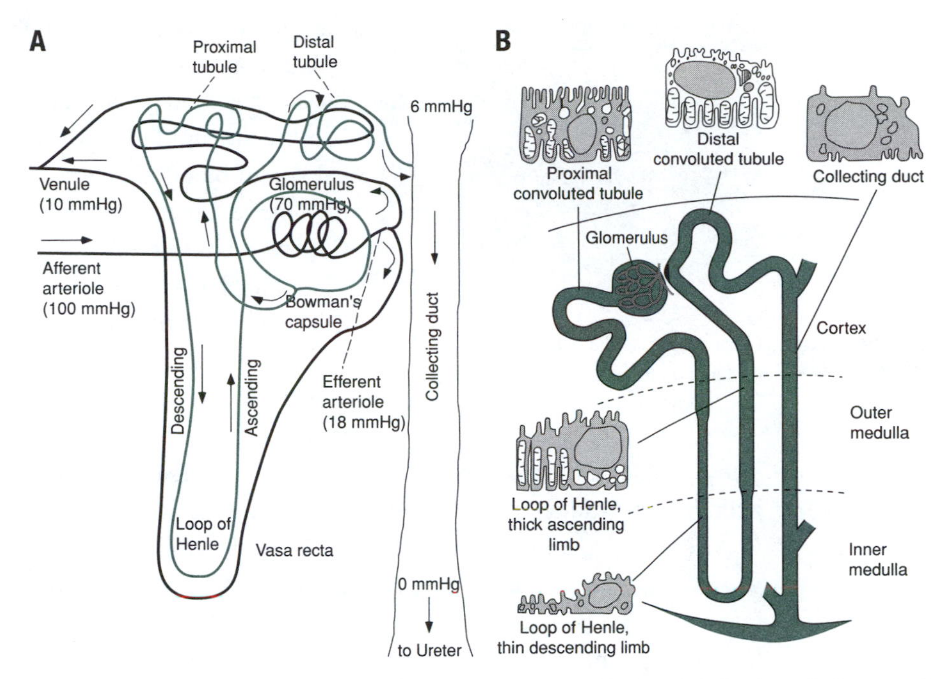

Figure 9.9 **Schematic of a typical nephron. A.** Relationship of the nephron to the blood supply in the kidney. **B.** Structure of the epithelial cells that comprise the kidney tubules change with position.

cell type is important for recognizing and transporting different combinations of solutes, to accomplish the tasks of reabsorption and secretion. These elements act in coordination to produce urine and achieve homeostasis in the blood.

9.4.1 Filtration in the glomerulus

The total RPF is about 0.6 L/min. Within the kidney, each renal artery subdivides into branches, eventually producing more than 1 million arterioles that are needed to feed blood into the individual nephrons. **Afferent arterioles** bring blood into

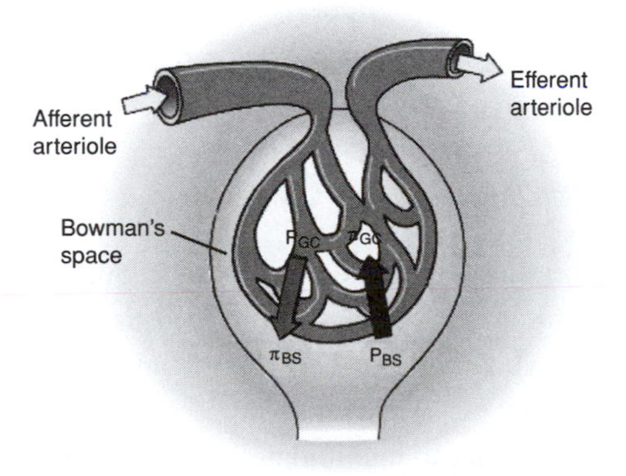

Figure 9.10 **Glomerular filtration.** The rate of flow of filtrate in the glomerulus (called glomerular filtration rate [GFR]) depends on the gradient of hydrostatic and oncotic pressures.

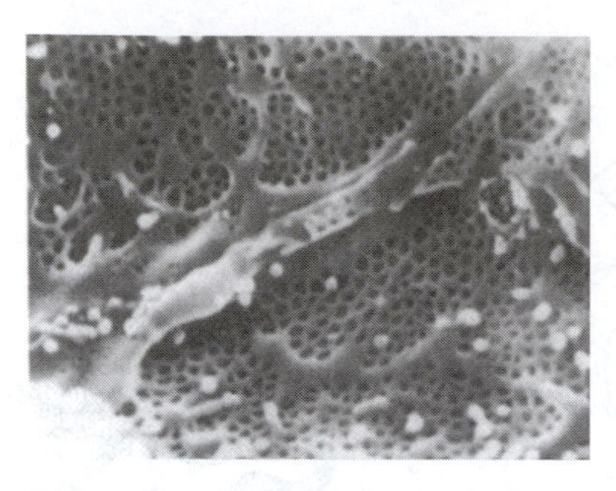

Figure 9.11 **Cross-section of the filtration barrier in the glomerulus.** Reproduced from (1) with permission, copyright Elsevier 2005.

the glomerulus of each nephron, and the **efferent arteriole** moves blood away. Between the afferent and efferent arterioles is a tuft of capillaries, which is the physical site of blood filtration. The glomerular membrane is the filter or sieve in the system (Figure 9.11); the membrane is composed of endothelial cells of the capillary walls, a basement membrane, and other epithelial cells of renal origin, called podocytes, which have branched extensively within the glomerulus and wrap around the capillaries. These podocytes also form part of the filtration barrier of the glomerulus. The endothelium is **fenestrated**, filled with ~90-nm pores in the cell.

Arterioles have muscular walls, which can be used to contract or dilate the vessel. The nephron has an unusual arrangement: Blood flows from an arteriole, into a capillary bed (the glomerulus), and then into another arteriole. (Recall that the usual arrangement is flow from the capillaries and collected into venules; see Chapter 8.) Because an arteriole is present on each side of the glomerular capillary bed, the pressure within the capillary can be regulated by control of the resistance through the arterioles (Figure 9.10). The pressure in Bowman's space, P_{BS}, is always ~14 mmHg; the rate of filtration across the glomerular membrane is determined by the pressure drop across the glomerular membrane, ΔP_G. The hydrostatic pressure drop across the membrane is:

$$\Delta P_G = P_{GC} - P_{BS}. \tag{9.2}$$

where P_{GC} is the pressure within the glomerular capillary. Hydrostatic pressure is not the only driving force for fluid movement across the membrane. Because proteins are generally too large to pass through the small pores in the filtration barrier (as discussed later in this section), protein concentration in the blood is much higher than protein concentration in the filtrate. The protein concentration difference creates an osmotic pressure driving force, which tends to pull water back into the capillary from Bowman's space (Figure 9.10). The overall driving force for filtration is, therefore:

$$\Delta P_G = (P_{GC} - P_{BS}) - (\pi_{GC} - \pi_{BS}), \tag{9.3}$$

where π_{GC} is the oncotic pressure within the glomerular capillary, and π_{BS} is the oncotic pressure in Bowman's space. (This osmotic pressure difference is often called the **oncotic pressure** when it arises because of the presence of proteins, not electrolytes such as Na^+ and Cl^-.)

Box 9.1 illustrates how the ability to manipulate resistances in the afferent and efferent arteriole allows the nephron to adjust the driving force for glomerular filtration.

Box 9.1 Glomerular filtration

Blood flows into and out of the glomerulus through the afferent and efferent arterioles, shown in Figures 9.9 and 9.10. This arrangement, with an arteriole on each side of a capillary bed, allows each nephron to control the rate of glomerular filtration.

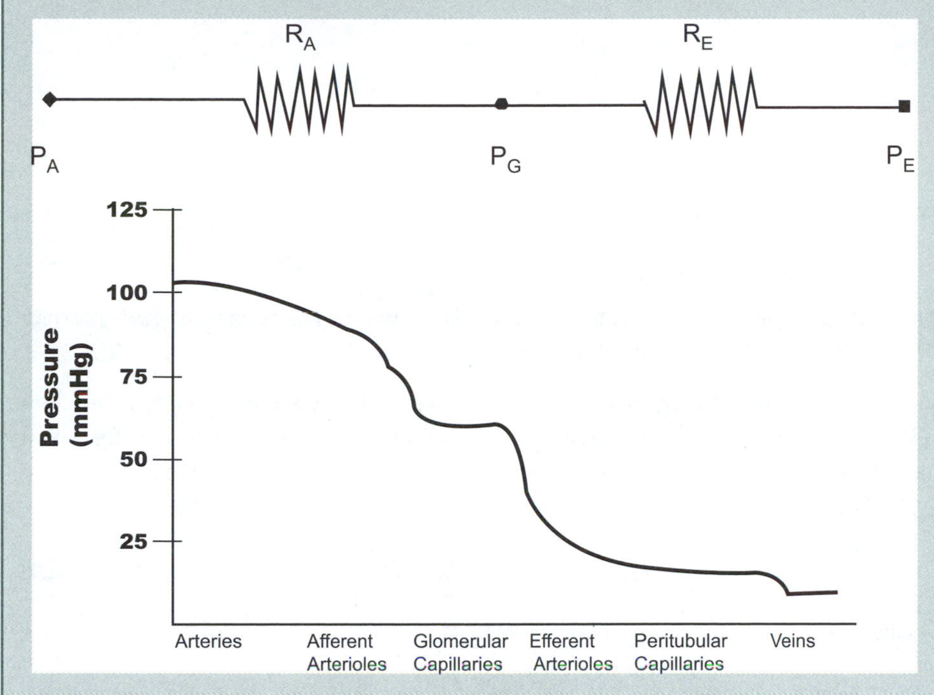

Figure Box 9.1A. The afferent and efferent arterioles behave as two resistors in series.

Combining Equations 9.2 and 9.4 yields an expression for the GFR:

$$GFR = (P_{\mathrm{GC}} - P_{\mathrm{BS}}) \times K. \qquad \text{(Equation 1)}$$

Under most circumstances, the pressure in Bowman's space is constant, ~14 mmHg. The permeability of the filtration barrier, K, is also a constant. Therefore, GFR depends on the pressure in the glomerular capillary, P_{GC}. A nephron that has control over P_{GC} has control over GFR.

Recall from Chapter 8 the relationship between pressure, flow, and resistance for a vessel:

$$Q = \frac{1}{R}\Delta p \quad \text{where} \quad R = \frac{8\mu L}{\pi r_{\mathrm{v}}^4}, \qquad \text{(Equation 2)}$$

which is Equation 8.3. The flow through the afferent and efferent capillaries can be modeled using Equation 2 (Figure Box 9.1a). Assume that the resistance of the afferent arteriole, R_{A}, is equal to the resistance of the efferent arteriole, R_{E}. Equation 2 can be written for each arteriole:

$$Q_{\mathrm{A}} = \frac{1}{R_{\mathrm{A}}}(P_{\mathrm{A}} - P_{\mathrm{G}}) \quad \text{and} \quad Q_{\mathrm{E}} = \frac{1}{R_{\mathrm{E}}}(P_{\mathrm{G}} - P_{\mathrm{E}}). \qquad \text{(Equation 3)}$$

(continued)

Box 9.1 *(continued)*

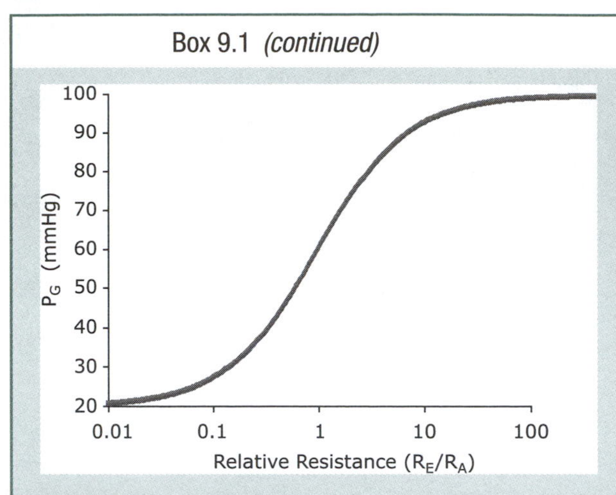

Figure Box 9.1B. Typical pressure drop within the glomerular vessels. Notice that the pressure does not drop over glomerular capillary distance, but drops substantially over the two arterioles.

The flow rate through the efferent arteriole, Q_A, is equal to the flow rate through the afferent arteriole, Q_E, minus the GFR for this single nephron. (Why is this true?). Assume, for a first estimate, that the GFR for the nephron is much less than the flow rate through the arterioles, so that $Q_A \sim Q_E$. Then it follows that:

$$\frac{1}{R_A}(P_A - P_G) = \frac{1}{R_E}(P_G - P_E). \qquad \text{(Equation 4)}$$

Equation 4 can be solved for the glomerular pressure:

$$P_G = \frac{\frac{R_E}{R_A} P_A + P_E}{\frac{R_E}{R_A} + 1} \qquad \text{(Equation 5)}$$

Figure Box 9.1b shows how glomerular pressure, P_G, changes when the ratio of efferent to afferent resistance changes (R_A/R_E). In this calculation, it is assumed that the total pressure drop is constant: $P_A = 100$ mmHg and $P_E = 20$ mmHg, as shown in Figure Box 9.1a. When the efferent resistance is very low, compared to the afferent resistance, P_G is low: The majority of the pressure drop occurs over the afferent arteriole. In contrast, when the efferent resistance is high, P_G is high. By regulating the relative resistance of the arterioles, the nephron can change the glomerular pressure drop over a wide range, producing a correspondingly wide range of GFR.

SELF-TEST 2 The oncotic pressure difference is approximately 25 mmHg, but it rises slightly to about 35 mmHg, as blood flows through the glomerular capillary. Why does it increase?

ANSWER: As blood flows through the capillaries, filtration depletes water and small molecules. Because proteins such as albumin are not filtered, their concentration increases with removal of water.

Filtration in the glomerulus is based on size; water and small molecules pass easily through the pores in the glomerular membrane (Figure 9.11). Larger

Table 9.4 The filterability of molecules within the glomerulus

Substance	Molecular weight	Filterability
Water	18	1.0
Sodium	23	1.0
Glucose	180	1.0
Insulin	5,500	1.0
Myoglobin	17,000	0.75
Albumin	69,000	0.005

molecules cannot pass through these pores. Table 9.4 lists some compounds in order of their "filterability," which is roughly a measure of the fraction of these molecules that would pass through the membrane. As the size of the molecules increases, the filterability decreases. Negatively charged molecules are more hindered to filtration—that is, they are less likely to be filtered—than are uncharged or positively charged molecules of the same molecular weight (Figure 9.12). This effect of charge is particularly important for molecules of the size of albumin (MW = 69,000, which corresponds to a ~3-nm radius). If size were the only determinant of filtration, more than 10% of it would be filtered in the glomerulus; because of its negative charge, less than 1% is filtered.

The glomerular filtration rate (GFR) is the total volume of fluid that is filtered across the glomerular membranes in the kidneys per unit time. GFR increases in direct proportion to ΔP_G:

$$GFR = \Delta P_G \times K, \tag{9.4}$$

where K, which is sometimes called the ultrafiltration coefficient, is the inverse of the resistance of the glomerular membrane to filtration. Note that this expression is

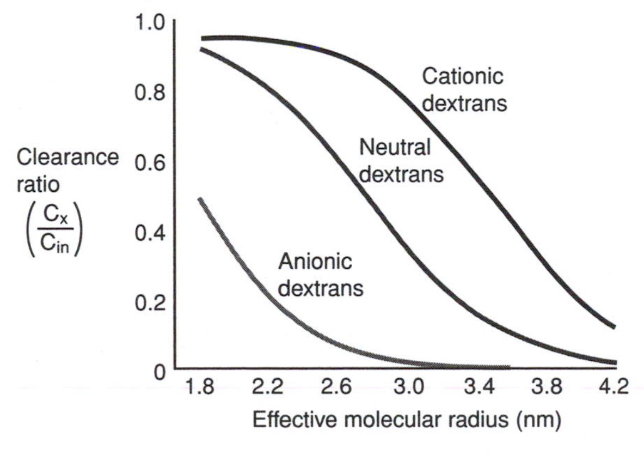

Figure 9.12

Filterability of charged molecules. The clearance ratio is a measure of the ease with which a molecule crosses the glomerular filtration barrier: The value is 1 for a molecule that crosses freely and 0 for a molecule that does not cross at all. Dextrans are polysaccharides of variable molecular weight; they can be rendered with positive charge (cationic dextrans) or negative charge (anionic dextrans). The clearance ratio decreases with increasing molecular weight for all dextrans. At a given molecular weight, the positive dextran is filtered more easily than the neutral or anionic dextrans.

analogous to the equations for pressure drop and flow in blood vessels (Equations 8.2 and 8.3), although in this case the GFR is moving out of the vessels and through the vessel wall.

For a typical 70-kg person, the GFR is approximately 125 mL/min, which is a significant fraction (25%) of the RPF of ~600 mL/min. This ultrafiltrate of plasma, created in the glomerulus, is processed within the nephron, by reabsorption and secretion, to become urine. The workload of the kidney is notable. In an average day, the nephrons of the kidney will create and process ~180 L of ultrafiltrate to create ~1 L of urine.

9.4.2 Clearance and excretion in the urine

The concept of **clearance** is valuable for understanding renal physiology. Clearance is most easily understood using the mass balance approach introduced in Chapter 7. Assume that the system under analysis is the kidneys, as shown schematically in Figure 9.7, and that one chemical species is of interest. A mass balance (Equation 7.1) on the species of interest indicates that, at steady state:

$$IN_{\text{plasma}} - OUT_{\text{plasma}} = OUT_{\text{urine}}. \tag{9.5}$$

This equation states that the mass of our species removed from the plasma per time is equal to the amount excreted in the urine per time. This equation can be expressed in terms of the following symbols: $\dot{V}$ is the volumetric rate of urine production, $[X]$ indicates the concentration of chemical X in the plasma or urine, and rate of plasma flow into the kidney through the renal arteries and vein is RPF_{a} and RPF_{b}, respectively:

$$RPF \times ([X]_{\text{plasma, artery}} - [X]_{\text{plasma, vein}}) = (\dot{V})[X]_{\text{urine}}, \tag{9.6}$$

which assumes that $\dot{V}$ is much lower than RPF (see Self-test 3). Equation 9.6 shows that excretion of compound X with urine is accompanied by a drop in the plasma concentration: Concentration of X is lower in the vein than in the artery.

How much of a drop in concentration is created? Imagine that the incoming plasma stream was split into two streams—one that the kidney cleans perfectly and one that it ignores (Figure 9.13). Clearance (CL) is the imaginary volume of the renal blood flow that is completely cleansed (or cleared) of compound X per unit time. Mathematically, following from Equation 9.6, it is defined as:

$$CL \times ([X]_{\text{plasma, artery}} - 0) = (\dot{V})[X]_{\text{urine}} \tag{9.7}$$

or, simplifying:

$$CL = (\dot{V})\frac{[X]_{\text{urine}}}{[X]_{\text{plasma, artery}}}. \tag{9.8}$$

How is this imaginary clearance useful? For a molecule X that is freely filtered (i.e., filtered without any hindrance by the glomerular membrane) but not

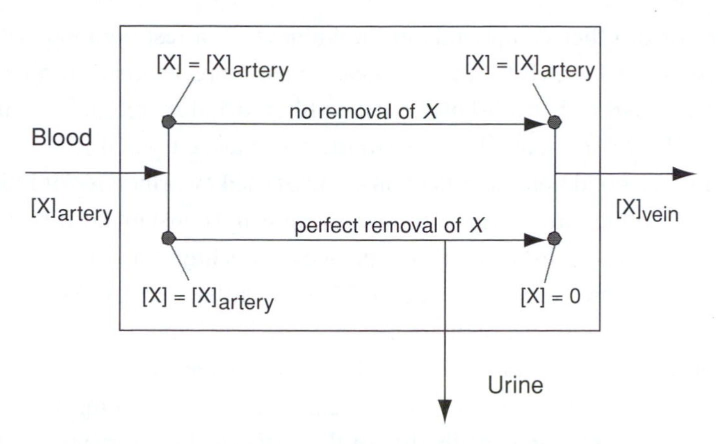

Figure 9.13 **Schematic diagram to illustrate the concept of renal clearance.** Imagine a blood flow that can be separated into two fractions: The compound is removed completely from one stream and not removed from the other.

reabsorbed or secreted, every molecule that is removed from the plasma during filtration appears in the urine. The clearance for this compound is equal to the GFR. A polysaccharide called inulin behaves this way: It is small enough to filter easily with the plasma, but is not further managed by the kidney. Therefore, the GFR can be experimentally estimated by measuring the clearance of inulin:

$$CL_{\text{inulin}} = GFR = \dot{V}\frac{[\text{inulin}]_{\text{urine}}}{[\text{inulin}]_{\text{plasma}}}. \tag{9.9}$$

Another compound, called para-aminohippuric acid (PAH) is so avidly secreted that it is completely removed from the plasma perfusing the kidney. In this case, the clearance is equal to the *RPF*:

$$CL_{\text{PAH}} = RPF = \dot{V}\frac{[\text{PAH}]_{\text{urine}}}{[\text{PAH}]_{\text{plasma}}}. \tag{9.10}$$

Experimental measurements of PAH clearance can be used to estimate *RPF*.

SELF-TEST 3 Equation 9.5 assumes that the volumetric flow rate of urine, $\dot{V}$, is much lower than the RPF. How does this simplify the equation? Is this assumption justified?

ANSWER: A water balance over the kidney, using Equation 9.5, indicates that $RPF_{\text{artery}} - RPF_{\text{vein}} = \dot{V}$. Therefore, a more complete expression for the balance of compound X is:

$$RPF \times [X]_{\text{plasma,artery}} - (RPF - \dot{V}) \times [X]_{\text{plasma,vein}} = \dot{V} \times [X]_{\text{urine}}. \qquad \text{(Equation 9.5}')$$

By assuming that $\dot{V}$ is much lower than *RPF*, the equation is simplified considerably. Our experience suggests that this is a good assumption under most conditions. *RPF* is ~600 mL/min, and urine flow is typically about 1 mL/min.

Clearance is a useful concept because the clearance of certain substances (such as inulin and PAH) can be used to estimate parameters that are important in renal function within patients. It is also useful because it can be used to classify the behavior of other compounds in the kidneys: If a test compound that is freely filtered has CL < GFR, that substance must be reabsorbed; if a test compound has CL > GFR, that substance must be filtered and secreted. Substances that are incompletely or not at all filtered would also have CL < GFR.

This is a good point to reflect on some of the key principles of kidney physiology. The kidneys maintain an incredibly high and constant GFR (150–180 L/day) to excrete toxic substances. Maintenance of this high filtration rate requires high renal blood flow, which amounts to 25% of cardiac output. As the next sections will show, the kidneys regulate excretion of NaCl and water to achieve balance of these substances in the body. Balance is achieved by regulating reabsorption of these substances, not by changing GFR. In regulating water and sodium levels, the kidney protects the life of the individual in emergency situations. For example, the kidneys act to preserve extracellular fluid (ECF) volume—which determines the plasma volume and, therefore, the blood pressure—over regulation of all other substances. This strategy is sensible because a person cannot survive a loss of blood pressure, but can survive periods of loss of pH, sodium, and potassium control.

9.4.3 Reabsorption and secretion in the tubules

The glomeruli of the kidney nephrons produce a large volume flow, which is an ultrafiltrate of plasma. This fluid will eventually become urine, which is excreted, but first it flows through the series of tubules that lead from Bowman's space to the ureters. The renal tubules have a life-saving and demanding function; they must recover from the ultrafiltrate the molecules that are needed in the blood to maintain life processes. Some of these molecules (e.g., water, sodium, and bicarbonate) are recovered in large quantities. Water and sodium recovery are essential for maintaining ECF volume and blood pressure. Bicarbonate is the principal buffer used to maintain body pH. Failure in regulation of any of these chemical systems can lead to death.

The renal tubules maintain balance of solutes and water by transport of solutes through the walls of the renal tubules. Transport of molecules from the fluid inside the tubule to the blood is called **reabsorption**; transport of molecules from the blood to the tubular fluid is called **secretion** (Figure 9.14). Compounds do not move directly from the tubular lumen to the blood; they must pass through the epithelial cell layer that forms the wall of the tubules and through the interstitial fluid (ISF) that bathes the tubules and blood vessels (Figures 9.13 and 9.15). The tubular epithelium is more than a barrier to the movement of molecules: It is these tubular epithelial cells that direct the processes of reabsorption and secretion that determine the overall rates of molecular movement and, therefore, maintain body balances.

This chapter does not attempt to describe all of the processes that are used by tubular epithelial cells to coordinate reabsorption and secretion. (An excellent physiology textbook is listed at the end of the chapter for the interested reader.)

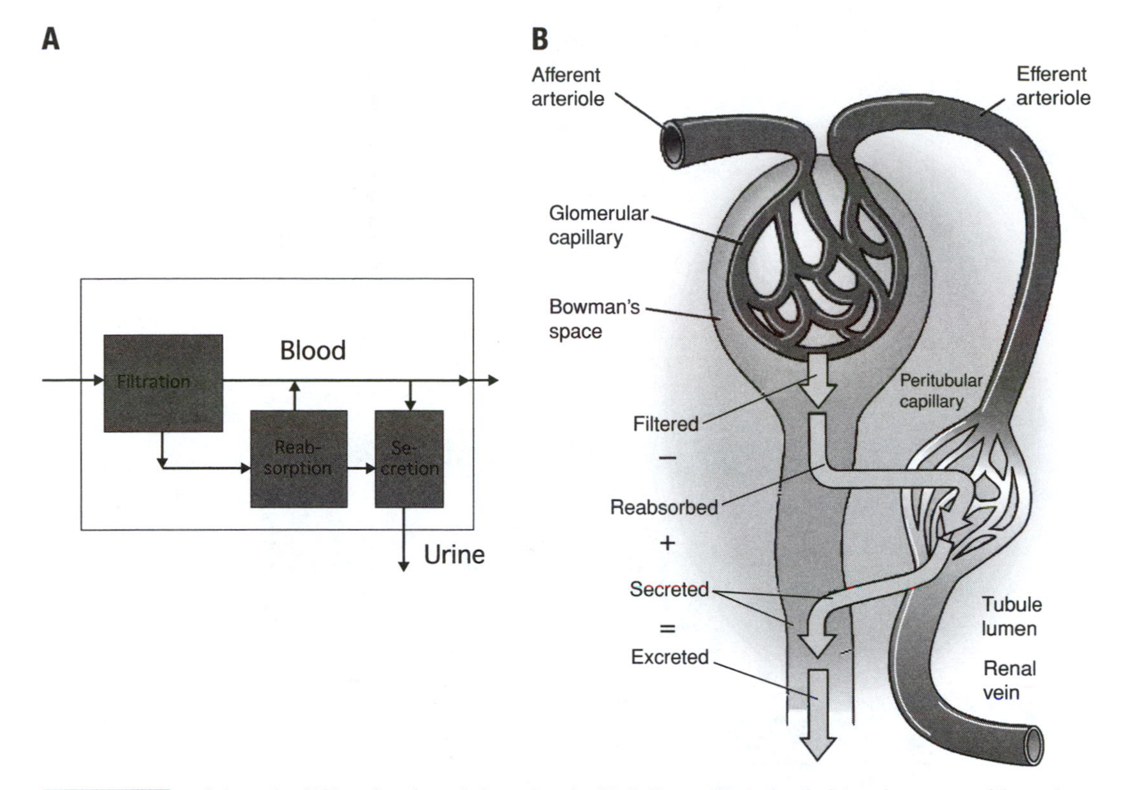

Figure 9.14 **Schematic of kidney function. A.** An engineering block diagram illustrating the internal processes of the nephron. **B.** A schematic diagram showing where these processes occur in the nephron.

Instead, some of the basic principles will be outlined, focusing on the transport of sodium and water.

First, the properties of the tubular epithelium vary with position in the tubule (Figure 9.9). The flow path from Bowman's space is divided into regions called—in order of appearance from Bowman's space to ureter—the proximal tubule, the loop of Henle (which has descending and ascending limbs), the distal tubule, and the collecting duct. Each of these regions has distinctive types of tubular epithelial cells, and each cell type differs in its ability to transport solutes and water.

Second, molecules can move through the epithelium by either a paracellular or a transcellular pathway (Figure 9.16). Note that the transcellular route requires movement through two membranes and crossing of the cell, whereas the paracellular route is more direct and depends solely on passive diffusion. Adjacent epithelial cells are joined by tight junctions, which govern the movement of molecules in the space between cells. Different portions of the nephron have different relative amounts of paracellular compared to transcellular transport, depending on the leakiness of the tight junctions.

Third, each segment of the tubules handles different amounts of solute or water. Consider the process of sodium reabsorption. Large amounts of sodium are removed from the blood by the GFR; most of this sodium is recovered, so that it is not lost in the urine. Under typical conditions, 67% of sodium reabsorption

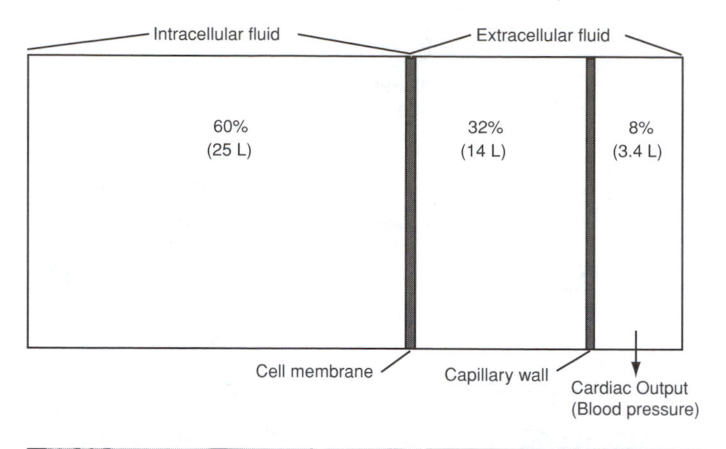

Figure 9.15 **Distribution of fluid volumes in the body of a 70-kg human.** The intracellular fluid is separated from the extracellular fluid by the cell's plasma membranes. The extracellular fluid is separated from the plasma volume by the capillary wall.

occurs in the proximal tubule, 25% in the thick ascending limb of Henle, 5% in the distal tubule, and 3% in the collecting duct.

Fourth, each set of tubular epithelial cells contains molecular transporters—ion channels, active transport systems, co-transporters, and exchangers (see Chapter 6)—that endow the epithelium with the power to reabsorb or secrete. Certain attributes are common for all renal tubule cells. Throughout the nephron, Na–K–ATPase active transporters reside on the basolateral membrane (in other words, the epithelial cell membrane that is most distant from the lumen of the tubule) (Figure 9.16). Cells vary, however, in their mechanism for uptake of Na from the lumen into the apical membrane. Segments of the renal tubule differ with respect to: 1) mechanism of apical membrane Na entry; 2) permeability to water; 3) leakiness of tight junctions and importance of paracellular pathway; 4) pathways for Cl^- reabsorption; and 5) additional pathways for potassium transport resulting in reabsorption or secretion.

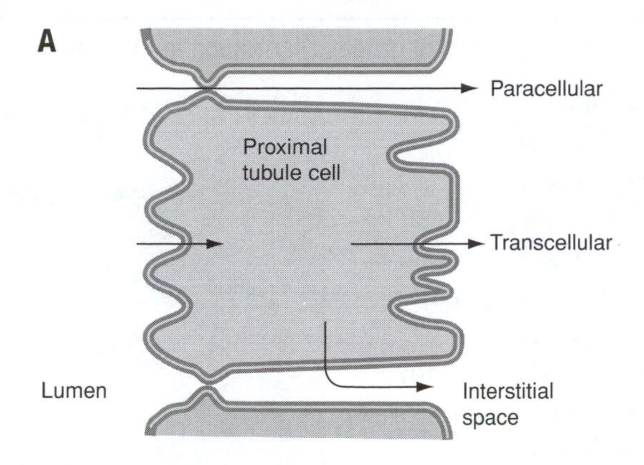

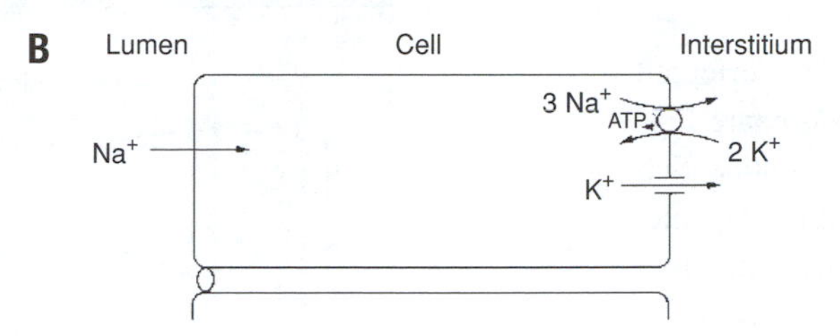

Figure 9.16 **Tubular transport processes. A.** Movement of solutes and water from the tubular fluid to the blood is regulated by tubular epithelial cells. The movement of chemicals can occur via paracellular pathways between cells or by transcellular pathways through the epithelial cells. **B.** All tubular epithelial cells transport sodium ions through the basolateral membrane by a Na-K-ATPase active transport system. Cellular energy is used to move sodium from the cell to the interstitial fluid (ISF) and potassium from the ISF to the cytosol. The movement of sodium out of the cell is balanced by uptake of sodium from the luminal fluid via different pathways. These pathways for Na are particular to each nephron segment. **C.** For example, tubular cells of the proximal tubule have sodium–glucose co-transporters on the apical membrane, so that glucose is reabsorbed with the sodium gradient. They also have Na-H exchangers that facilitate bicarbonate reabsorption.

Osmotic forces drive the reabsorption of water from the tubular fluid. The nephron uses an interesting mechanism to create a high osmotic driving force. A spatial gradient of osmolarity is created by two simultaneous effects: 1) membrane transport systems for moving solutes up a concentration gradient from the tubules to the ISF, and 2) countercurrent flow of tubular fluid in the loop of Henle. Box 9.2 describes this countercurrent-multiplying mechanism in more detail. The net effect of this action is the establishment of a substantial gradient in osmolarity, extending from osmolarity equal to that of the plasma and ISF in the cortex of the kidney to high osmolarity in the ISF of the medulla.

Box 9.2 Countercurrent mechanism of gradient formation in the kidney

Concentration of urine requires a mechanism for reabsorbing water. Osmotic forces provide the driving force of water movement. Removal of water from the fluid within the tubules requires: 1) that some segments of the tubule are permeable to water and 2) that the water-permeable segments of the tubule have an osmotic gradient, arranged so that the osmolarity is lower in the tubular fluid than in the surrounding interstitial fluid.

The tubules of the nephron are oriented with respect to the kidney: Compare Figure 9.2a to Figure 9.8b and notice that the glomerulus is near the kidney capsule, with the long loops of tubules extending down toward the renal pelvis. The most prominent oriented feature of the nephron is the loop of Henle: The descending limb starts at the level of the glomerulus diving deep toward the pelvis, and the descending limb rises back to the glomerulus. One of the important functions of the loop of Henle is to establish a gradient of osmolarity in the interstitial space surrounding the tubules; lower osmolar interstitial fluids are found near the glomerulus, whereas high osmolar fluids are found near the pelvis. The kidney uses this osmolar gradient to adjust the concentration of urine. As the nearly complete urine is making its way down the collecting duct, water can be reabsorbed from the fluid in the duct into an interstitial fluid that is becoming increasingly higher in osmolarity.

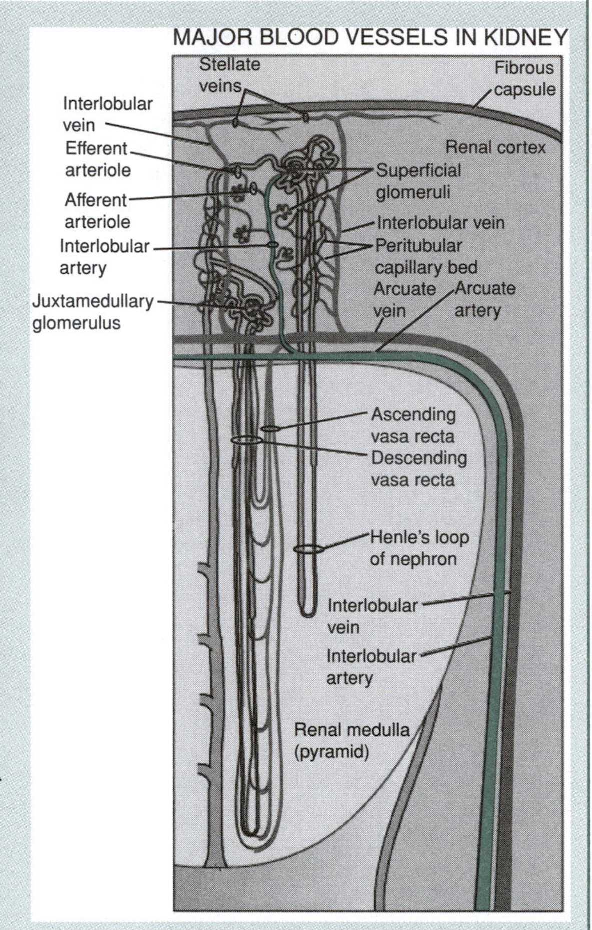

Figure Box 9.2a. The relationship of blood vessels surrounding the nephron to the renal tubules.

The mechanism for creation of this osmolar gradient by the loop of Henle is not completely understood, but it is believed to be accomplished by a fascinating mechanism involving the countercurrent flow in the limbs of the loop. A simple way to envision the mechanism is through a thought experiment, in which you imagine the gradient as it is forming (Figure Box 9.2b). Therefore, assume an initial condition in which all of the fluids are 300 mOsm (the same as plasma). Now, assume that the ascending limb can create—via membrane transport proteins—a 200-mOsm gradient by moving molecules into the interstitial space. Assume further that the fluid within the descending limb will equilibrate with the interstitial fluid (step 1). Because there is fluid flow through the two connected limbs, the more concentrated fluid in the descending limb will flow into the ascending limb (step 2). This more concentrated fluid in the ascending limb will be treated to the same 200-mOsm gradient formation, followed by re-equilibration (step 3), followed by flow (step 4). This cycle can continue (steps 5–7), leading to amplification of the osmolarity within the interstitial space.

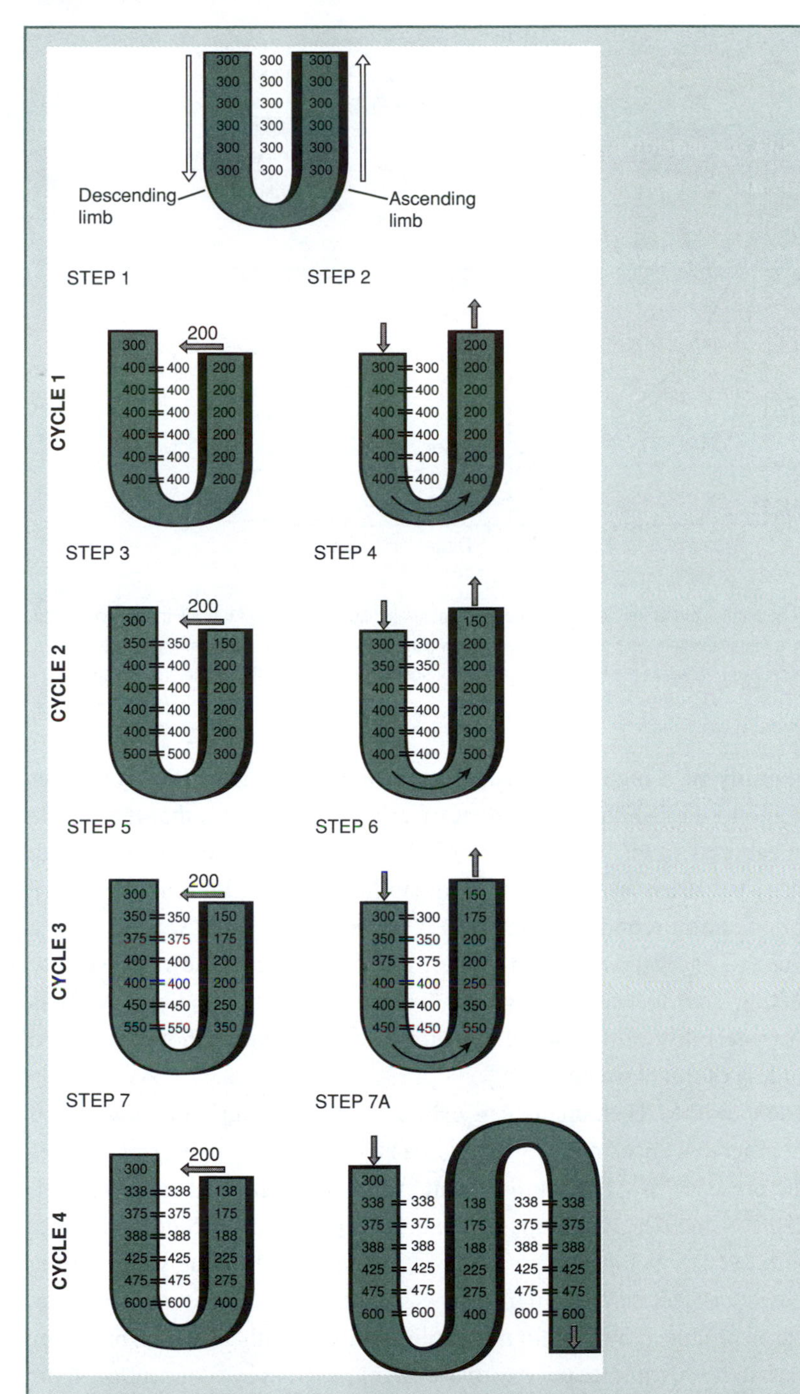

Figure Box 9.2B. The counter-current exchange mechanism for generating large osmotic gradients in the kidney interstitium.

The interstitial gradient that is created by the loop of Henle is now available to the collecting duct, with its sensitivity to ADH (see text). The gradient provides the concentrating potential: The upper limit for osmolarity in the urine is determined by the maximum osmolarity in the interstitial space. The presence or absence of ADH determines the permeability of the collecting duct to water, thereby providing a switch; the collecting duct can either ignore or exploit this gradient.

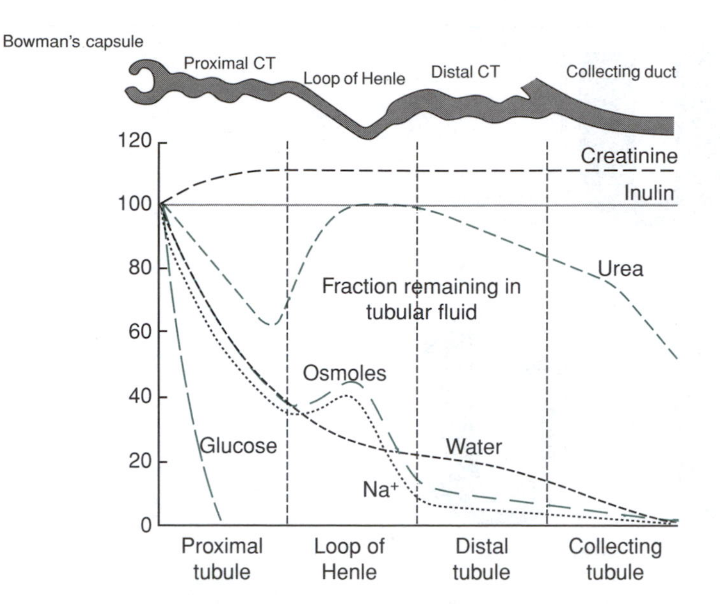

Figure 9.17 **Reabsorption of molecules along the nephron.** The upper diagram shows a series of tubules from the nephron, stretched out in a linear fashion. The graph shows the concentration of glucose, sodium, water, urea, inulin, creatinine, and total osmoles within the tubular fluid, as a function of position along the tubule.

The availability of a high osmolarity creates the potential for flow of water from the tubule to the ISF, but movement of water also requires that the tubule wall is permeable to water. The collecting tubule uses this gradient to regulate the dilution of the urine by having a permeability that can be regulated by a hormone called antidiuretic hormone (ADH). Diuresis is loss of water from the body; antidiuresis is water retention. When ADH is present, the permeability of the collecting duct to water is high, so water is reabsorbed. When ADH is absent, the permeability of the tubule to water is low, so water is not reabsorbed and, therefore, is excreted in the urine.

In the absence of ADH, humans can excrete very dilute urine as low as 40 mOsm (remember that the osmolarity of plasma is ~300 mOsm, or 300 mOsmoles per L). In the presence of ADH, humans can excrete concentrated urine, as high as 1200 mOsm. Some desert animals can excrete even more concentrated urine, allowing them to survive for long periods of time without drinking water. Some excretion of water is necessary, even if water is scarce, to get rid of waste products, such as the urea produced by protein metabolism: A diet with 70 g of protein per day will generate ~300 mmoles per day of urea. Animals that can concentrate their urine more effectively can excrete these wastes in a minimum amount of water.

Our kidneys work to maintain a balance in the body fluids: Water and sodium are excreted to match the amount of water and sodium that we consume in our diet. If a person is eating a diet that contains 150 mmol/day of salt (NaCl) and 70 g/day of protein (which is typical of people in the United States), that person must excrete ~600 mOsmoles per day. Because the body can regulate the

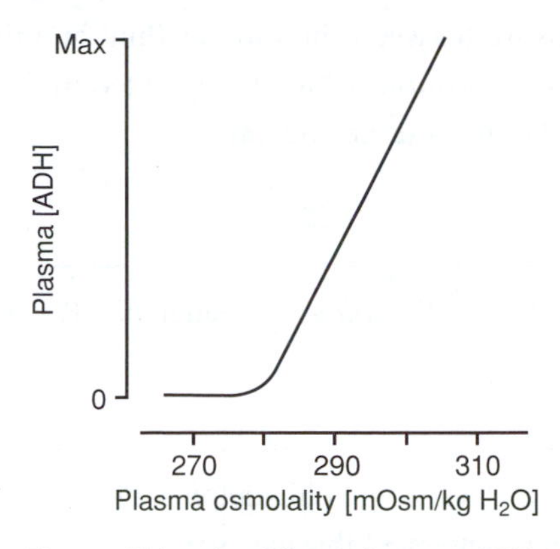

Figure 9.18 **The sensitivity of antidiuretic hormone (ADH) secretion to changes in plasma osmolarity.** When even small amounts of extra water are consumed, urinary excretion of water is increased.

concentration of urine, through ADH, these molecules can be excreted in dilute urine. How much urine would need to be excreted? If urine is dilute (40 mOsm or 40 mOsmoles/L), the 600 mOsm from that diet would require 15 L/day of urine volume. That same diet can be handled with as little as 0.5 L/day of urine volume (1200 mOsm), in the presence of maximal ADH, because of the profound ability of the nephron to concentrate urine. The net result of these principles is shown for a few solutes and water in Figure 9.17.

Figure 9.15 shows the distribution of total body water for a 70-kg individual. Ingested water distributes throughout the body water, slightly decreasing the osmolarity of the fluids. This small change in osmolarity is detected by special cells called osmoregulators, which then create a signal to decrease ADH secretion. Less ADH causes lower permeability in the collecting tubules, which causes the urinary excretion of water to increase. ADH secretion is extremely sensitive to changes in plasma osmolarity. Even drinking a little water (which should create a small change in the osmolarity) creates a noticeable change in ADH concentration (Figure 9.18).

Summary

- Excretion of molecules by the liver and kidney, and biotransformation of compounds in the liver, are responsible for elimination of wastes (such as urea), elimination of toxins (such as drugs), and maintenance of homeostasis.
- Biotransformation occurs primarily within hepatocytes, the primary functional cells of the liver.
- The kidneys receive 25% of the cardiac output: This high flow rate of plasma feeds 1 million nephrons, which are organized collections of tubules, blood vessels, and interstitial spaces.
- The nephrons create, by filtration, an ultrafiltrate of plasma, which is processed during flow through the tubules to reabsorb vital compounds and secrete unneeded compounds.
- Filtration occurs in the glomerulus, which creates an ultrafiltrate of blood by pressure-driven flow.
- Reabsorption and secretion of compounds within the renal tubules is determined by the properties of tubular epithelial cells.

- Differences in osmotic pressure between the tubular fluid and the ISF create the driving force for water reabsorption, which is regulated by hormones such as ADH because of its effects on water permeability.

REFERENCE

1. Boron WF, Boulpaep EL. *Medical Physiology*. Philadelphia, PA: Sanders; 2004.

FURTHER READING

Boron WF, Boulpaep EL. *Medical Physiology*. Philadelphia, PA: Sanders; 2004. The urinary system is described in Chapters 32 through 39.

KEY CONCEPTS AND DEFINITIONS

afferent arterioles the branch of the artery that supplies blood to the glomerulus of each nephron

bile a greenish-brown alkaline fluid that aids digestion of lipid substances and is secreted by the liver and stored in the gall bladder

biotransformation the alteration of a substance, such as a drug, by chemical reactions within the body usually from a toxic state to a less toxic state

Bowman's space a capsule-shaped membranous structure surrounding the glomerulus of each nephron in the kidneys of mammals that extracts wastes, excess salts, and water from the blood

clearance the imaginary volume of the renal blood plasma from which a substance X is completely cleansed (or cleared) per unit time

countercurrent multiplication the process by which mammalian kidney is able to produce hyperosmolar urine

dialysance the number of milliliters of blood completely cleared of any substance by an artificial kidney or by peritoneal dialysis in a unit of time

dialysate fluid used on the other side of membrane during dialysis to remove impurities

distal tubule a portion of the kidney tubule connecting the ascending limb of the loop of Henle to the collecting duct

efferent arteriole a blood vessel that carries blood filtered through the glomerular capillaries away from the nephron

fenestrated having larger openings to allow larger molecules to diffuse

glomerulus a cluster of blood capillaries around the end of the kidney tubule where waste products are filtered from the blood

glomerular filtration the filtration of blood plasma that flows from the afferent arteriole through the glomerulus to the efferent arteriole

glomerular filtration rate (GFR) the volume of fluid filtered from the renal glomerular capillaries into the Bowman's capsule per unit time

hemodialysis a type of renal dialysis whereby the blood of a person who suffers from kidney failure is cleansed outside the body

hepatocyte a liver cell

homeostasis the process whereby the organism is able to maintain a stable internal environment through physiological processes

kidney a bean-shaped organ present in mammals, birds, and reptiles designed to excrete nitrogenous waste from the organism's body

kidney tubule a small tube in the kidney that filters blood and produces urine

loop of Henle part of the kidney tubule that forms a long loop in the medulla region of the kidney, from which water and salts are reabsorbed into the blood

mass balance a method of accounting for material entering and leaving a system that operates on the conservation of mass principle that states that matter can neither disappear nor be created

metabolite a small molecule that is formed by or needed for metabolism

micturate to urinate

nephron a functional unit of the kidney consisting of a glomerulus and its associated tubules from which urine is produced

oncotic pressure a term used to refer to the osmotic pressure difference when it arises because of the presence of proteins instead of ion electrolytes

osmolarity a measure of the total number of solutes per liter of solution

plasma clearance see clearance

plasma clearance rate see clearance

proximal tubule the first portion of the kidney tubule with a striated border that connects and carries fluid from the glomerulus to the loop of Henle

rate constant a proportionality constant in an expression relating concentration of reactants with the reaction rate

renal artery an artery that arises from the abdominal aorta and supplies blood to the kidneys

renal pelvis the broadened top part of the ureters into which kidney tubules drain

renal vein a vein that carries filtered blood away from the kidneys

tubular reabsorption a process whereby water and solutes such as glucose, amino acids, and ions are transported back into the blood from tubular fluid in kidney tubules

tubular secretion a process whereby materials from the peritubular capillaries are transported into the tubular fluid in the renal tubules

ultrafiltrate a term used (in renal physiology) to describe material that is unfiltered and remains in the liquid phase after filtration

ultrafiltration a term used (in renal physiology) to describe glomerular filtration during which very small particles are separated from proteins, cells, and other larger components of the blood

ureters the duct by which urine passes from the kidney to the bladder of cloaca

urethra the duct by which urine passes to the outside of the body from the bladder. It is also the duct through which semen passes to the outside in males.

urinary bladder a hollow muscular distensible sac that serves as a reservoir for urine prior to excretion by urination

NOMENCLATURE

CL	Clearance	L/s; m^3/s
CL_{inulin} CL_{PAH}	Clearance rate of inulin and para-aminohippuric acid (PAH)	L/s; m^3/s
GFR	Glomerular filtration rate	L/s; m^3/s
IN_{plasma}	Mass inflow rate of a species in plasma	kg/s
K	Ultrafiltration coefficient	m^4-s/kg
OUT_{plasma}	Mass outflow rate of a species in plasma	kg/s
OUT_{urine}	Mass outflow rate of a species in urine	kg/s
P_A	Pressure in afferent arteriole	mmHg, Pascal
P_{BS}	Blood pressure in the Bowman's space	mmHg, Pascal
P_E	Pressure in efferent arteriole	mmHg, Pascal
P_{GC}	Blood pressure in the glomerular capillaries	mmHg, Pascal
ΔP_G	Blood hydrostatic pressure drop across the glomerular membrane	mmHg, Pascal
Q	Flow rate	L/s
Q_A	Flow rate in the afferent arteriole	L/s
Q_E	Flow rate in the efferent arteriole	L/s
R_A	Resistance in the afferent arteriole	kg/m^4-s
R_E	Resistance in the efferent arteriole	kg/m^4-s
RPF	Rate of plasma flow	L/s; m^3/s
$\dot{V}$	Volumetric rate of urine production	L/s; m^3/s
[X], e.g., [inulin]	Concentration of species X, e.g., Concentration of inulin	kg/m^3
$[X]_{plasma,artery}$, $[X]_{plasma,vein}$	Concentration of species X in arterial plasma and venous plasma, respectively	kg/m^3
$[X]_{urine}$	Concentration of species X in urine	kg/m^3

GREEK

π_{GC}	Oncotic pressure in the glomerular capillaries	Pascal, mmHg, atm
π_{BS}	Oncotic pressure in the Bowman's space	Pascal, mmHg, atm

QUESTIONS

1. Give three examples of chemical reactions in the body that produce water molecules.
2. Why is a high-pressure capillary system required in the kidney?
3. Describe the characteristics of the glomerular membrane and its permeability to various substances.

4. Write a short essay that distinguishes the process of excretion from secretion.
5. A substance is present in the urine. Does this prove that it is filtered in the glomerulus?
6. The concentration of urea is always higher in urine than in plasma. Does this suggest that urea is secreted?
7. For the following steps in urine formation, state where it occurs and what is the purpose: a) tubular reabsorption and b) tubular secretion.
8. The nephrons of some animals, particularly those adapted to survive in the desert, have loops of Henle that are much longer, relatively, than those of humans. What is the potential benefit of a longer loop of Henle?
9. Explain why some clinicians might describe the HH equation as pH = constant + (kidney/lung).
10. At steady state, how much sodium chloride is excreted per day in the urine of an individual who is consuming 12 g of sodium chloride per day: a) less than 12 g; b) 12 g; or c) more than 12 g?

PROBLEMS

1. Estimate the fraction of water (as a percentage of total body water) that leaves the body per day in urine, feces, sweat, tears, and humidity in expired air. Justify your assumptions in making this approximate calculation.
2. Death occurs if the plasma pH falls outside the range of 6.8–8.0 for an extended time. What is the concentration range of H^+ represented by this pH range?
3. If the GFR is 100 mL/min, use Equation 9.4 and the information in Box 9.1 to determine K.
4. Assume that the plasma concentration of substance X is 2 mg/mL and that it is freely filtered in the glomerulus.
   a. If the GFR is 125 mL/min, what is the rate of substance X flow into the renal tubules?
   b. If the maximum tubular reabsorption rate for substance X is 200 mg/min, how much X will be reabsorbed and how much will be excreted?
   c. Why do many compounds—including substance X in this example—have a maximum tubular reabsorption rate (i.e., what determines the maximum reabsorption rate for a compound)?
5. In Section 9.2, the text discusses the zero-order kinetics of alcohol disappearance from the blood.
   a. Using the approach outlined in section 7.2.2 (Tracer balance in the body), find an equation describing the concentration of alcohol in the plasma versus time, given that the initial concentration in the plasma (at time = 0) is 10 mM.
   b. Graph the results of (a).
   c. Over what time interval is the expression that you derived in (a) valid? Why is it not valid for all time?

6. Using the model for resistors in series provided in Box 9.1—and assuming that the total resistance, P_A, and P_E are the same as in Box 9.1—calculate the relative resistance of arterioles (R_E/R_A) for a glomerular pressure of 80 mmHg.
7. Inulin is a polysaccharide that happens to be neither secreted nor reabsorbed by the tubules of the kidney, and its MW (5200 daltons) is low enough to permit it to pass freely through the glomerulus. It is infused at a steady rate into the blood of a person whose GFR is to be determined. After a while, a steady-state plasma concentration is achieved. Assume that blood samples taken after steady state has been reached show an inulin concentration of 0.1 g/100 mL of plasma. If a total of 180 mL of urine is collected over the next 2 hours, and analysis shows that there is 0.08 g inulin per milliliter in the urine, what is the GFR of the person? It is important to note that inulin is not metabolized by the body and is excreted only in the urine.
8. Calculate the rate of urine production in an individual who has an inulin clearance rate of 125 mL/min, given that the concentration of inulin in the plasma and urine are 100 mg/L and 3 mg/L. (Refer to the previous problem for information on the properties of inulin.)
9. Assume that the concentration of a substance in the urine is 10 mg/mL, its plasma concentration is 0.2 mg/mL plasma, and the rate of urine flow is 3 mL/min. What is the clearance rate of the substance? Do you think this substance is reabsorbed or secreted by the kidney tubules?
10. PAH has the property of being actively secreted from the capillaries surrounding the kidney tubules into these tubules (thus into the urine). At low plasma PAH concentrations (up to 8 mg/100 mL), the fraction of PAH carried by the blood to the kidneys, which is lost to the urine via both glomerular filtration and secretion, totals 91%, as determined by sampling the blood just proximal to the kidney (i.e., from the renal artery) and just distal to the kidneys (i.e., from the renal vein). At higher concentrations, the fraction extracted is lower. If the same procedure as in problem 7 (which described using inulin to measure GFR) is carried out using PAH, the plasma concentration is 1 mg/100 mL, and the urine collected over a 1-hour period contains 0.35 mg of PAH, what is the rate of plasma flow to the kidneys?

PART 3

BIOMEDICAL ENGINEERING

10 Biomechanics

LEARNING OBJECTIVES

After reading this chapter, you should:

- Understand the stress–strain curve and how properties of materials can be evaluated by examining their deformations under applied loads.
- Understand the concept of elasticity in materials, and how it can be described by the Young's modulus.
- Understand the importance of the relationship between structure, function, and material properties in human tissues.
- Understand the intracellular structures that contribute to mechanical properties of cells.

10.1 Prelude

Humans can hold their bodies erect, vertically above the earth, because their bodies are solid objects capable of supporting their own weight. The human skeletal system is a collection of 206 bones, connected by soft tissues—cartilage, ligaments, tendons, and muscles—that together provide a mechanical support system for the human body.

Humans are also capable of movement. Muscles—connected to the solid bone framework—contract to generate forces that result in motion. Dancers, high jumpers, and surgeons learn to control these movements precisely to accomplish tasks and transport their bodies with precision (Figure 10.1). Strength, agility, and stamina can all be enhanced by training and, as a species, our understanding of the effects of training improves each year. As a result, humans continually improve performance on certain tasks, such as Olympic events (Figure 10.2).

This chapter describes some of the elements of human body structure and mechanics. To aid in description, the chapter begins with some basic concepts about the mechanical properties of materials. These concepts will be used throughout the chapter to provide quantitative descriptions of the behavior of the biological materials that make up the human body.

Figure 10.1 **Humans are capable of amazing mechanical feats.** Many athletic events, such as high jumping, require great strength and exquisite body control. Photograph courtesy of Penn State Erie, The Behrend College.

10.2 Mechanical properties of materials

Biological materials, such as bones and muscles and the cells that comprise them, are mechanical objects and, therefore, are subject to forces that occur because of the world around them. The human body and its components can be studied by force and moment analysis, which is among the first concepts learned by students in physics (Box 10.1). Physical forces are present throughout the body. Some forces are large: The head of the femur regularly experiences forces that are 2–3 times the weight of the whole body or 1,800–2,700 N (for a 200-lb man). Much smaller forces have biological effects: Muscles typically generate forces of ~1 N. Nonmuscle cells—when attached to solid substrates as in Figure 5.5b—can each generate forces of 100–800 nN (1).

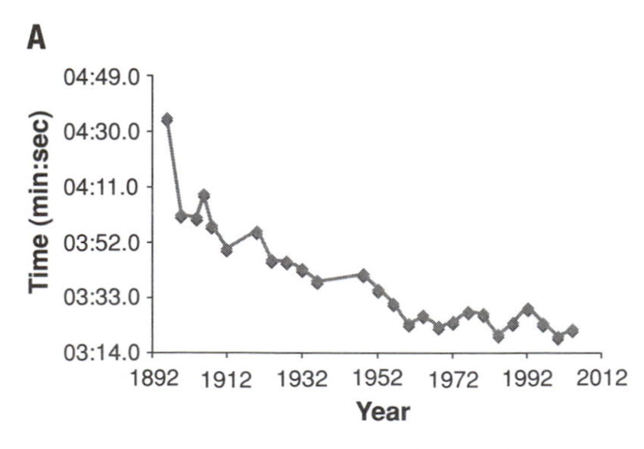

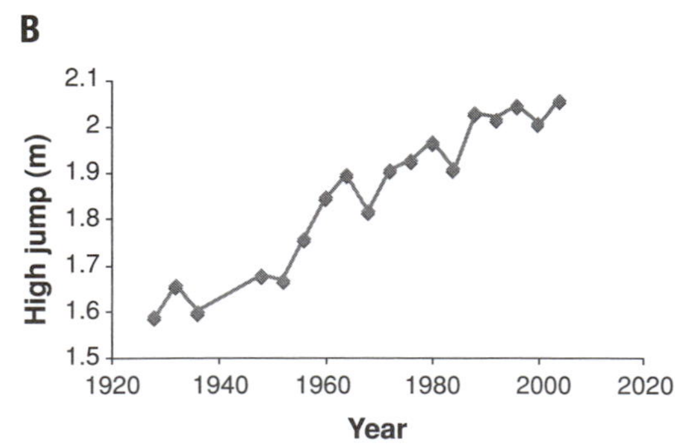

Figure 10.2 **Human performance is improving with time.** As evidence of the improvement in human performance, Olympic gold medal–winning performances for men's 1,500-meter run **(A)** and women's high jump **(B)** are plotted versus time. Surely, there are many factors in the steady improvement in these two events (which were chosen at random; a plot of any Olympic event would yield similar curves): Improvements could be because of better equipment or more able coaching, but at least part of this improvement must be caused by enhanced capability of the human machine.

Box 10.1 Elementary force analysis

One of the most important tools in classical physics is the relationship between force and motion or, more specifically, between force and acceleration. Force analysis, or force balances, emerges from Newton's three primary laws of motion:

Newton's First Law: If no net force acts on a body, the body's velocity cannot change: In other words, the acceleration of the body is zero.

Newton's Second Law: The relationship between an object's mass m, its acceleration $\bar{a}$, and the net force acting on the object $\overline{F}_{\mathrm{T}}$ is Mass × Acceleration = Force.

Newton's Third Law: For every action there is an equal and opposite reaction.

Both force and acceleration are vector quantities: They have a magnitude and a direction.

Newton's Second Law can be expressed mathematically as:

$$\overline{F}_{\mathrm{T}} = m\bar{a} = \sum F_{\mathrm{i}}, \qquad \text{(Equation 1)}$$

where F_{T} is the total force acting on a body, which is equal to the mass m multiplied by the acceleration of the body a, and F_{i} are the individual forces at work on the body. For example, consider an airplane that is moving at a constant speed through the air, at some fixed altitude (Figure Box 10.1). Here the force that is acting to keep the plane aloft and moving forward is separated into two components—F_1 and F_2—which act in the y and x direction, respectively (note that F_2 acts in the negative x direction). Because force is a vector quantity, the total force balance can be broken down into three separate equations, each expressing the balance of forces in the three principal directions (x, y, and z for a rectangular coordinate system):

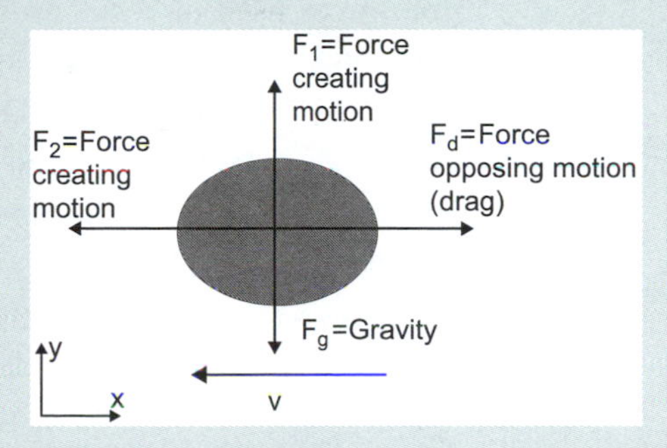

Figure Box 10.1

$$\begin{aligned} F_{\mathrm{T,x}} &= ma_{\mathrm{x}} = \sum F_{\mathrm{i,x}} \\ F_{\mathrm{T,y}} &= ma_{\mathrm{y}} = \sum F_{\mathrm{i,y}} \,. \\ F_{\mathrm{T,z}} &= ma_{\mathrm{z}} = \sum F_{\mathrm{i,z}} \end{aligned} \qquad \text{(Equation 2)}$$

Equation 2 can be applied to the airplane in the figure; because the airplane is experiencing forces with components in only the y and x directions, only two of the equations in Equation 2 are needed:

$$\begin{aligned} ma_{\mathrm{x}} &= F_{\mathrm{d}} - F_2 \\ ma_{\mathrm{y}} &= F_1 - F_g \end{aligned} \qquad \text{(Equation 3)}$$

(continued)

Box 10.1 *(continued)*

If it is assumed that the airplane is maintaining a constant altitude and flying at a constant velocity, both a_x and a_y are zero. Therefore, this analysis reveals that the upward force, F_1, must be equal to the weight of the airplane F_g, and the forward force must be equal to the drag force on the airplane. For a Boeing 747 airliner, the maximum weight is ~900,000 lb (405,000 kg); the drag created by flying through air at a height of 30,000 ft at 600 miles/hour is calculated in Problem 10.

When a material is subjected to a force, it responds in some way. One type of response is motion: A stationary billiard ball that is struck by a moving billiard ball will roll. Billiard balls are hard objects; even when struck with considerable force, the ball will not bend or deform (Figure 10.3). Imagine a billiard ball made of a softer material. When this soft ball is struck, it will probably move, but it will also change shape or **deform**. If billiard balls could deform in response to the forces used on a billiards table, the game would be different. Basketballs, in contrast, are not as hard as billiard balls; basketballs deform when they are slammed into the ground. Deformability helps a basketball to bounce. Good basketball players understand the properties of the ball and, therefore, are able to dribble. A basketball that is too deformable is not good for the game, as anyone who has tried to play with an underinflated ball will appreciate.

Biological materials are constantly exposed to forces. When a person stands, his or her mass—acting under the acceleration of gravity—produces forces on the bones, muscles, and other structures of the legs. Imagine the forces that must be produced within the legs of a high jumper (Figure 10.1) to launch off the ground. These forces can lead to deformation of the materials within the legs, but the amount of deformation depends on the mechanical properties of the materials. Bones are strong materials when compared to muscles, meaning that bones exhibit small deformations when exposed to high forces, even forces that would substantially deform a muscle. In this section, an approach for quantifying the strength and behavior of materials is described.

10.2.1 Elastic deformation and Young's modulus

Consider an elongated object—for example, a segment of a biological tissue or a synthetic biomaterial—that is fixed at one end and suddenly exposed to a

Figure 10.3 **Billiard balls move, but do not deform substantially, when struck.** Photo courtesy of Twm Davies, www.twmdesign.co.uk.

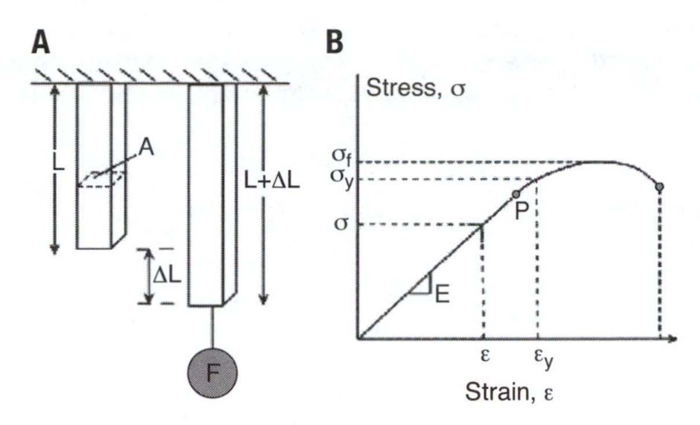

Figure 10.4 **Typical characteristics of an elastic material. A.** Deformation in response to applied load. **B.** Materials can be characterized by a stress–strain curve.

constantly applied **load** (Figure 10.4). The material will change, or deform, in response to the load. The load on the material is defined in terms of the stress, or force per cross-sectional area, that is applied: **Stress**, σ, is equal to the force divided by the area A. The response of the material to this load is measured as **strain**, ε. For a stress applied in one direction (as shown in Figure 10.4), the strain is equal to the fractional increase in length of the material, $\Delta L/L$.

For some materials, deformation in response to an applied load is instantaneous and, under conditions of low loading, deformation varies linearly with the magnitude of the applied force:

$$\sigma = E\varepsilon, \tag{10.1}$$

where σ is the applied stress—equal to the force per unit cross-sectional area, F/A—and ε is the resulting strain. This relationship is called **Hooke's law**, after the British physicist Robert Hooke; this equation describes the behavior of many elastic materials, such as springs, which deform linearly upon loading and recover their original shape upon removal of the load. **Young's modulus** or tensile elastic modulus, E, is a property of the material. Typical values are provided in Table 10.1. The **proportionality limit** of a material (point P in Figure 10.4) is defined as the maximum stress–strain values at which Equation 10.1 applies.

Not all elastic materials obey Hooke's law (e.g., rubber does not); some materials will recover their original shape, even though strain is not linearly related to stress. Fortunately, many interesting materials do obey Equation 10.1, particularly if the deformations are small.

SELF-TEST 1 If a force of 1 N is applied to an elastic specimen that has a cross-sectional area of 1 cm^2 and a Young's modulus of 3 GPa, what strain would be observed?

ANSWER: 3.3×10^{-6} or 3.3×10^{-4}%.

10.2.2 Plastic deformation

Materials "yield," or become irreversibly altered, if they are deformed beyond a critical **yield strain**, which is also called the **elastic limit**, ε_y. This state of

Table 10.1 Mechanical properties of commonly encountered biological materials

	E (MPa)	Density (g/cm^3)
Biological materials		
Bone, long	15,000–30,000	
Human compact bone (longitudinal)	17,000	1.8
Bone, cancellous	90–500	
Bone, vertebrae	100–300	
Dentin	13,000–18,000	
Enamel	50,000–84,000	
Articular cartilage	1–10	
Human knee menisci	70–150	
Brain tissue, grey matter	0.005	
Brain tissue, white matter	0.014	
Spinal cord	0.020–0.6	
Tendon	1,000–2,000	
Tendon, *Tendo achillis*	375	
Human small artery	0.1–4	
Elastin	0.6	
Isolated collagen fibers	1,000	
Formalin-fixed myocardium	101	
Skin (phase I)	0.1–2	
Collagen sponge	0.017–0.028	
Polymers		
Poly(ethylene) (high density)	500–1,000	0.95
Poly(methyl methacrylate)	2,000–3,000	1.18
Polyimides	3,000–5,000	
Polyester	1,000–5,000	
Polystyrene	2,300–3,300	1.05
Poly(tetrafluoroethylene)	400–600	
Poly(lactic acid)	1,000–3,000	
Rubber (average)	2.8	Variable
Metals		
Steel (structural)	200,000	7.86
Aluminum	70,000	2.71
Titanium	107,000	4.51
Others		
Concrete	25,000	2.32
Wood (pine)	11,000	0.61

Note: Data compiled from Ratner BD, Hoffman AS, Schoen FJ, Lemons JE (eds). *Biomaterials Science: An Introduction to Materials in Medicine.* Second ed. San Diego, CA: Academic Press; 2004; Athanasiou KA, Zhu C, Lanctot DR, Agrawal CM, Wang X. Fundamentals of biomechanics in tissue engineering of bone. *Tissue Eng.* 2000;6(4):361–381; Wakatsuki T, Kolodney MS, Zahalak GI, Elson EL. Cell mechanics studied by a reconstituted model tissue. *Biophys J.* 2000;79:2353–2368; and http://silver.neep.wisc.edu/~lakes/BME315N3.pdf. 1 Pa = 1N/m^2.

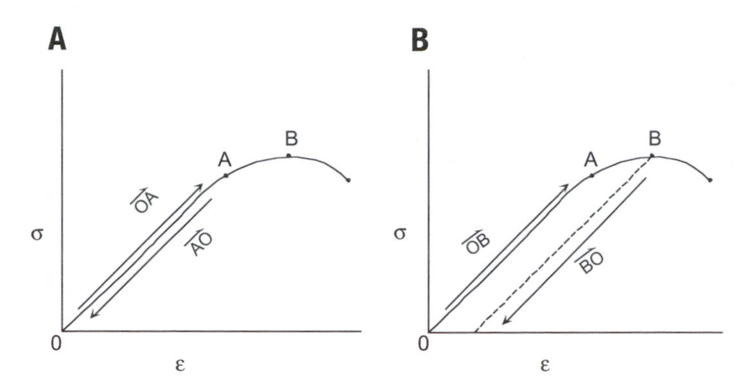

Figure 10.5 **Difference between elastic and plastic deformation. A.** When a material is deformed elastically—stretched from state 0 to state A in this example—the original state (state 0) is recovered when the stress is removed. **B.** When a material is stretched beyond its elastic limit, to state B in this example, it undergoes plastic deformation. The original state 0 is not recovered; the material has been irreversibly altered by the application of stress.

irreversible change occurs at a characteristic **yield stress**, σ_y. Further strain of the material results in **plastic** (rather than elastic) deformation; an irreversible change in the material prevents it from recovering its original state after removal of the applied load (Figure 10.4). The largest stress that a material can endure without failing (i.e., breaking or fracturing) is called the **failure stress** or **maximum stress**, σ_f.

Many materials obey Hooke's law for all strains less than the elastic limit; other materials obey Hooke's law over a more limited range—called the proportional limit—and continue to deform elastically, but not linearly, up to the yield stress.

It is convenient to analyze deformations in a given material by plotting its behavior on the stress–strain plane (Figure 10.5). If an elastic material is subjected to a load producing strain ε_A, which is less than the yield strain ε_y, it will return to its original shape (after removal of the load) by following the same locus of stress–strain coordinates that characterized its deformation. If, however, the material is deformed beyond the elastic limit, to strain ε_B for example, the material will not recover completely. Generally, the relaxation of the material occurs along a line that is parallel to the initial deformation (i.e., with slope equal to E, if it is a linear elastic solid).

10.2.3 Energy storage with deformation

Energy is added to a material when it is stressed; this mechanical energy, called strain energy, is stored in the material. For an elastic material, the **strain energy**, U_o, is calculated from:

$$U_o = \frac{1}{2}\sigma\varepsilon, \tag{10.2}$$

where U_o is the potential energy stored in the deformed material per unit volume. This energy can be determined graphically from the area under the stress–strain curve (Figure 10.6).

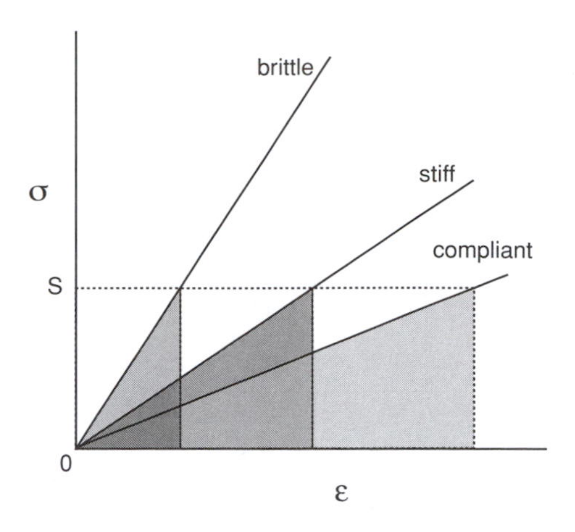

Figure 10.6

Energy storage in elastic materials. The amount of energy stored in each material can be determined by calculating the area under the stress–strain curve, which is shown for the brittle material, stiff material, and compliant material. A compliant material will store more energy than a brittle material for a given value of applied stress, *S* (that is, the area under the curve for the compliant material is greater than the area under the curve for the brittle or stiff materials).

Elastic elongation and relaxation have no net energy cost; all of the energy stored in the material during elongation is returned during relaxation. Energy is lost, however, when deformation goes beyond the elastic limit. The net loss of energy can be calculated as the difference between strain energy required to accomplish the elongation and the energy recovered after removal of the load (net energy loss can be determined graphically, as well).

The ability to store energy can be an important property of biological materials. For example, the aorta is elastic; it stretches when the heart contracts and ejects blood. Because the vessel is elastic, the energy that is stored by stretching the aorta is recovered during diastole, when the vessel returns to its original size. The aorta, because of its ability to store energy during expansion, has a role in the mechanical pumping of blood.

SELF-TEST 2 Assume the three materials described in Figure 10.6 are a long bone, dentin, and knee meniscus, which have Young's moduli of 30,000, 10,000, and 200 MPa, respectively. What strain energy is required to deform each to a strain of 0.1%? How much strain energy is stored in each, if they are each exposed to a stress of 30 MPa? Which is the most brittle tissue, and which the most compliant?

ANSWER: Strain energy = 15, 5, and 0.0001 kPa, respectively, to deform to 0.1%. Strain energy = 15, 45, and 2,250 kJ/m^3, respectively, when exposed to 30 MPa stress. Long bone is the most brittle and knee meniscus the most compliant.

The limit of energy storage for a material can be calculated by the strain energy at failure. Brittle materials have a low U_0 at fracture, whereas compliant materials, which deform readily, can store substantial amounts of strain energy (Figure 10.6).

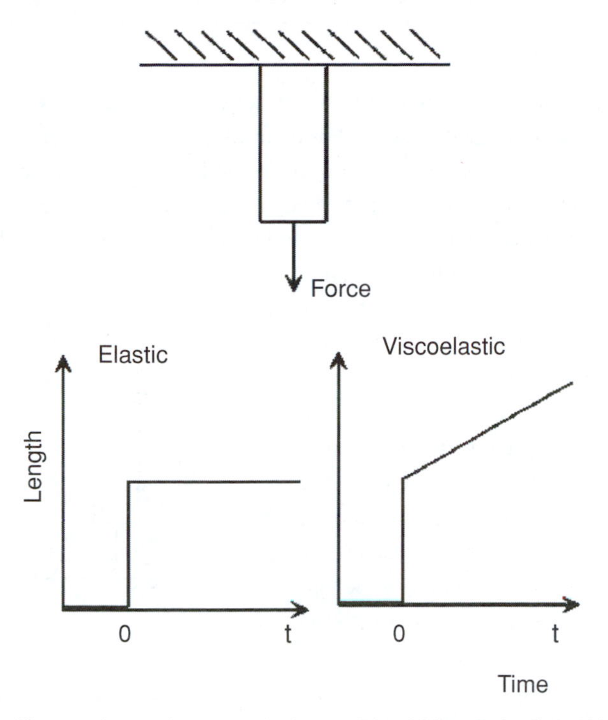

Figure 10.7 **Viscoelastic materials.** Viscoelastic materials exhibit properties that are intermediate to purely viscous or purely elastic material. For example, a purely elastic material will deform immediately to a fixed strain with an applied load (bottom left). A viscoelastic material will deform immediately, similar to an elastic material, but will also continue to deform in response to the load, similar to a viscous material (bottom right). There are many different types of viscoelastic materials: This figure illustrates only one type of response.

10.2.4 Elasticity, viscosity, and viscoelasticity

Equations 10.1 and 10.2 describe the mechanical behavior of idealized elastic materials. Chapter 8 described the behavior of incompressible fluids, another idealized kind of material. Recall Box 8.1, which introduces the concept of fluid viscosity, including the relationship between strain rate (dv_x/dy) and shear stress (τ_{xy}):

$$\tau_{xy} = -\mu \frac{dv_x}{dy}. \tag{10.3}$$

Real materials sometimes behave like one of these idealized models: For example, under certain conditions, cartilage behaves like an elastic material and blood behaves like a Newtonian fluid. In contrast, biological materials, which are frequently complex in composition, can exhibit behaviors that resemble aspects of both the elastic material and an incompressible fluid, but are also unlike either of these idealized models. Materials that exhibit both viscous and elastic natures are called **viscoelastic** (Figure 10.7 and Box 10.2).

10.3 Mechanical properties of tissues and organs

The musculoskeletal system consists of bones, which connect into a skeleton that forms the overall shape of the human body, and muscles, ligaments, tendons, and

Box 10.2 Viscoelasticity

Biological tissues such as skin and muscle are rich in water, which is a fluid at body temperature, but they also have characteristics of an elastic solid. For example, they retain their shape without a vessel. Ideal elastic materials—i.e., materials that are deformed to less than the elastic limit—deform instantaneously (as shown in Figure 10.4). Although they may deform very rapidly after loading, biological materials often continue to deform slowly after the initial period, exhibiting a behavior called creep (Figure 10.7). When this same material is rapidly deformed (Figure Box 10.2a), the force required to maintain this deformation decreases gradually; this process is called stress relaxation (Figure Box 10.2b).

Ideal elastic materials are modeled as springs; their behavior can be predicted by Hooke's law (Equation 10.1). Another element—one that slowly elongates upon application of a force—must be added to model the changes that occur during creep and stress relaxation. Continuous deformation after loading is a characteristic of fluids; the rate of deformation is determined by the viscosity of the fluid (Equation 10.3). A dashpot, or piston within a cylinder, is the mechanical analog for viscosity; the piston slowly moves through the cylinder, at a rate that is determined by friction between the surfaces, in response to an applied load. By combining elastic (i.e., spring) and viscous (i.e., dashpot) elements, models that predict aspects of the behavior of real viscoelastic materials can be developed.

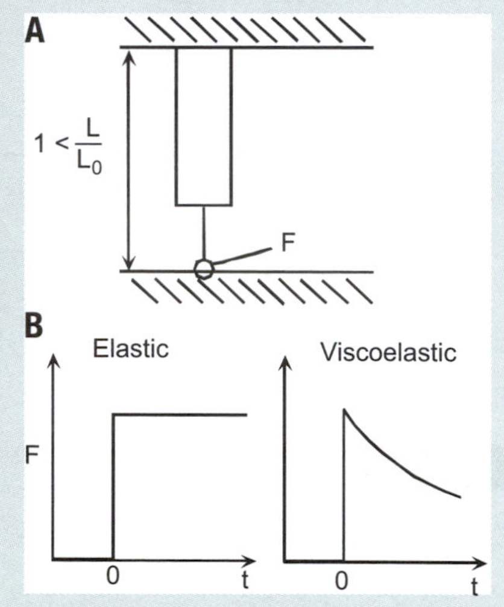

Figure Box 10.2. Elastic and viscoelastic materials respond differently to an instantaneous deformation.

It is often useful to compare the behavior of real materials with that of idealized models. A simple elastic material, for example, can be compared to a perfectly elastic spring. The behavior of the perfectly elastic material is provided by Equation 10.1, which can also be rewritten in the form:

$$F = kU, \qquad \text{(Equation 1)}$$

where F is the instantaneous total force applied to the material, k is the spring constant, and U is the instantaneous displacement of the material. When a force is applied to a spring, the spring instantaneously deforms to the length prescribed by Equation 1 (Figure Box 10.2a). Similarly, a viscous liquid can be compared to another idealized mechanical object, the dashpot (Figure Box 10.2b). The dashpot behaves like a simple viscous liquid:

$$F = \eta \frac{dU}{dt}, \qquad \text{(Equation 2)}$$

which is similar to Equation 10.3, with η equal to the damping coefficient (analogous to viscosity). When a force is applied to a material that behaves like a viscous liquid, it deforms continuously; the deformation of the material will continue for as long as the force is applied (Figure Box 10.3b).

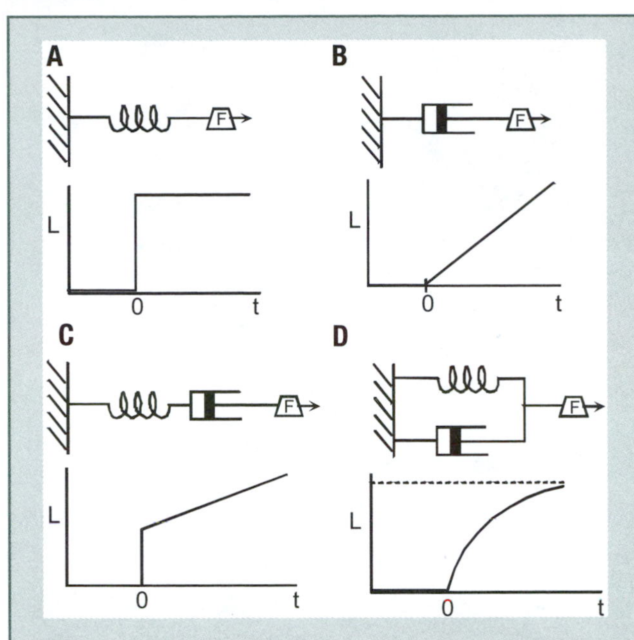

Figure Box 10.3. Spring–dashpot models of an elastic solid (A), a viscous liquid (B), a Maxwell model of viscoelastic materials (C), and a Voigt model of viscoelastic materials (D).

Many materials do not behave like perfect springs or dashpots. When a fixed load is applied, the material might deform with time (like a dashpot), reaching some ultimate deformation that is not exceeded (like a spring). Models of materials can be constructed by combining springs and dashpots in different combinations; the Maxwell model and the Voigt model are illustrated in Figure Box 10.3, c and d, respectively. A Maxwell solid behaves like a spring and a dashpot in series; it will deform instantaneously, like a spring, but the deformation will continue with some steady rate, like a dashpot. The Voigt solid, in contrast, behaves like a spring and dashpot in parallel: There is an initial, slow deformation caused by the dashpot, but the ultimate extent of deformation is limited by the spring.

cartilage that enable its movement. This section briefly reviews the mechanical behavior of these tissues. Other tissues also have important mechanical functions, particularly the lungs and heart, which engage cyclic mechanical deformations to move air and blood. Some aspects of lung mechanics are also described in the paragraphs that follow.

10.3.1 Bone structure and function

Bone is a hard, strong, dense tissue, which is composed of a mineral phase (60%), an organic collagen-rich matrix (30%), and water (10%). Because bone is a composite material, composed of both a soft protein matrix and a hard mineral phase, it has some elasticity and it is also strong. Bone has two typical architectures, compact (or cortical) bone and spongy (or cancellous) bone, which differ in their microscopic structure as well as their mechanical properties (Figure 10.8).

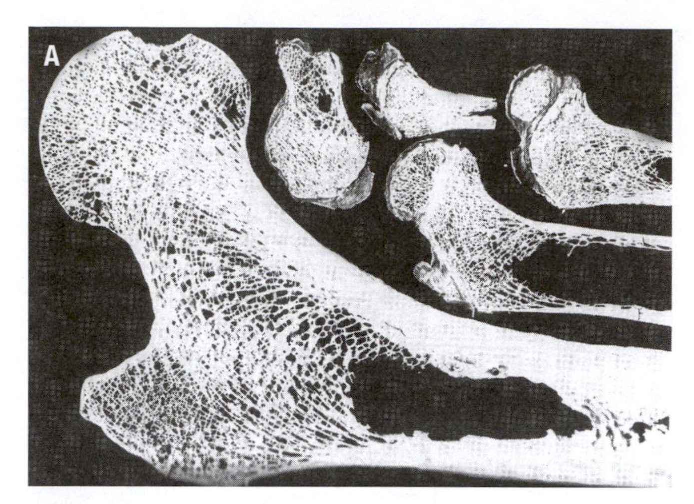
A

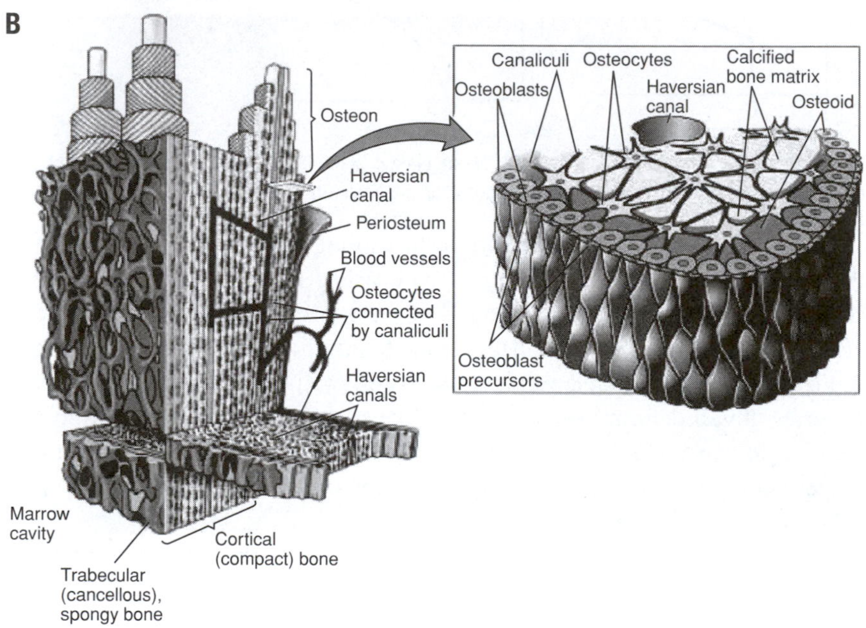
B
Osteon
Haversian canal
Periosteum
Blood vessels
Osteocytes connected by canaliculi
Haversian canals
Marrow cavity
Trabecular (cancellous), spongy bone
Cortical (compact) bone
Canaliculi
Osteocytes
Calcified bone matrix
Osteoblasts
Haversian canal
Osteoid
Osteoblast precursors

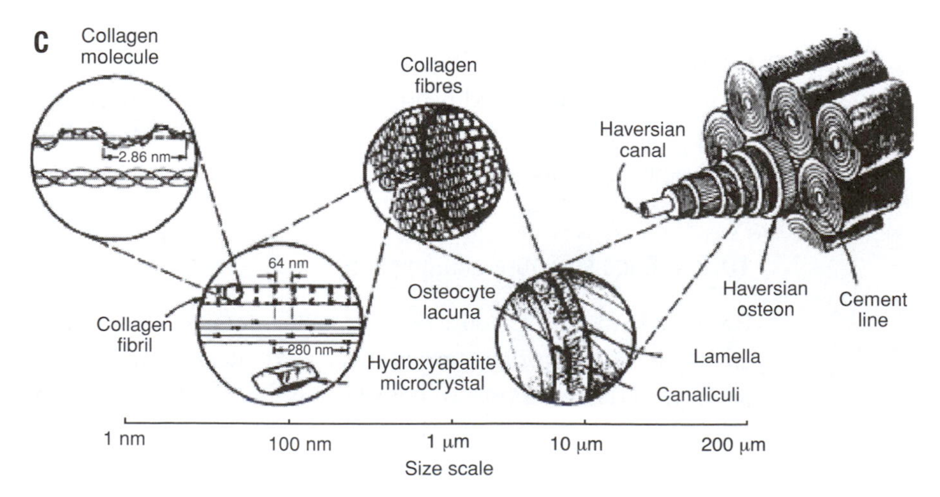
C
Collagen molecule
2.86 nm
Collagen fibres
Haversian canal
64 nm
Collagen fibril
280 nm
Osteocyte lacuna
Hydroxyapatite microcrystal
Haversian osteon
Cement line
Lamella
Canaliculi
1 nm
100 nm
1 μm
10 μm
200 μm
Size scale

Most bones contain both architectures, with the compact bone near the surface and the spongy bone in the interior.

Although most bones contain some regions of compact and spongy bone, the dense, rigid cortical architecture is most prominent in long bones such as the femur and the humerus. The spongy architecture, which is more porous and oriented (note the spongy regions of Figure 10.8a), is most prominent in the bones of the rib cage and spine.

The mechanical properties of bone are the result of its structure, which is built from microscopic components (Figure 10.8c). Because of the highly directional arrangement of microscopic structures, bone has different properties when tested in different directions. This dependence of mechanical properties on the direction of applied loading is called **anisotropy**. Bone and other biological materials differ from common engineering materials, such as metals and many plastics, which display **isotropy**, or similar mechanical properties in all directions.

10.3.2 Structure and function in soft connective tissues

Soft connective tissues surround our organs, provide structural integrity, and protect them from damage. In most soft connective tissues—such as articular cartilage, tendons, ligaments, skin, and blood vessels—cells are sparsely distributed within an extracellular matrix that provides the mechanical property of the tissue. The molecular constituents of extracellular matrix are reviewed in Chapter 5 (see Section 5.3). In contrast to bone, which is a rigid material, soft connective tissues are typically flexible and deformable (notice the difference in Young's modulus between bone and soft tissues such as cartilage, skin, and arteries; Table 10.1).

Soft tissues are often viscoelastic, owing to their heterogeneous structure in which extracellular matrix protein fibers are embedded in a fluid phase. The structure and orientation of the fiber phase (often collagen and elastin fibers) determine the bulk mechanical behavior (2). Figure 10.9 shows schematically the behavior of skin—a typical soft tissue that consists predominantly of connective tissue and shows an orientation parallel to the skin surface. With small tensile deformations (Phase I in Figure 10.9), the tissue behaves as an elastic material; microscopically, collagen fibers within the tissue are deforming without stretching

Figure 10.8

Structure of bone. A. Images showing the mineral phase in typical long bones. From Ethier CR, Simmons CA. *Introductory Biomechanics.* Cambridge, UK: Cambridge University Press; 2007, reprinted with the permission of Cambridge University Press. **B.** There are two types of bone, compact and spongy, both shown here. Most bones contain regions of each bone type. Compact bone has a dense, solid structure with isolated osteocytes (mature bone cells) connected by a network of small channels called canaliculi and larger central Haversian canals. Spongy bone is less dense, with mineralized material forming a porous scaffold that is mechanically strong but lightweight. **C.** All bones have a hierarchical structure. Collagen molecules assemble into fibrils, and then further into fibers, which form the extracellular matrix of bone. Calcium and phosphate ions form hydroxyapatite crystals, the predominant mineral phase of bone that provides mechanical strength. Reproduced from Lakes R. Materials with structural hierarchy. *Nature.* 1993;361:511–515, with permission.

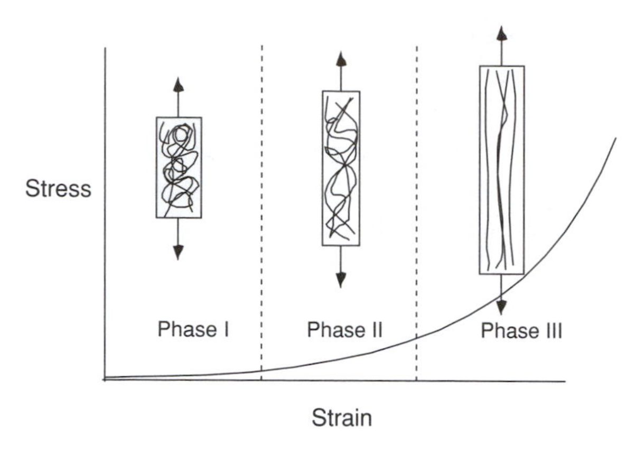

Figure 10.9

Stress–strain curve for a typical soft connective tissue. The mechanical behavior of soft tissues is illustrated schematically for skin, in which fibers that are oriented parallel to the skin surface provide a nonlinear relationship between stress in the material and deformation. Redrawn from Holzapfel GA. Biomechanics of soft tissue. In: Lemaitre J, ed. *Handbook of Material Behavior.* London: Academic Press; 2000.

or large changes in structure. As the strain increases (Phase II), collagen fibers become deformed, straightening in the direction of the strain and increasing the stiffness of the skin. With increased loads, in deformations that are just less than the ultimate tensile strength (Phase III), the collagen fibers are individually aligned in the direction of the applied load, and stretched. Ligaments also exhibit these phases of behavior: In ligaments, Phase I is referred as the toe (or primary) region in which deformation occurs easily with small amounts of stress, Phase II is the linear or secondary region (it is in this region that ligament stiffness is estimated), and Phase III often exhibits kinks or saw-tooth patterns that indicate snapping or fracture of individual fibers, as the applied load approaches the load necessary for failure.

10.3.3 Mechanical aspects of lung function

The lung operates as part of an overall mechanical environment within the **thorax**. An important part of the mechanical environment is the fluid-filled space between the chest wall and the lungs, which is called the **intrapleural space**. The fluid within the intrapleural space serves as a mechanical bridge connecting the lungs and the chest wall.

To illustrate the interplay between the lung, the intrapleural fluid, and the muscles of the thorax, this section will examine the **static** mechanical properties of the lung. Static mechanical properties of a tissue are measured under controlled conditions, when the tissue is not changing with time. Most tissues are actually in a **dynamic** state: For the lung, the volume of the lung is changing, air is moving in or out through the lung bronchial tubes, and the pressures within the fluid and air spaces are changing with time. Although the dynamic conditions of a tissue are ultimately the most important, much can be learned by careful observation of static properties. In any case, a description of tissue dynamics usually depends on a thorough understanding of tissue static properties.

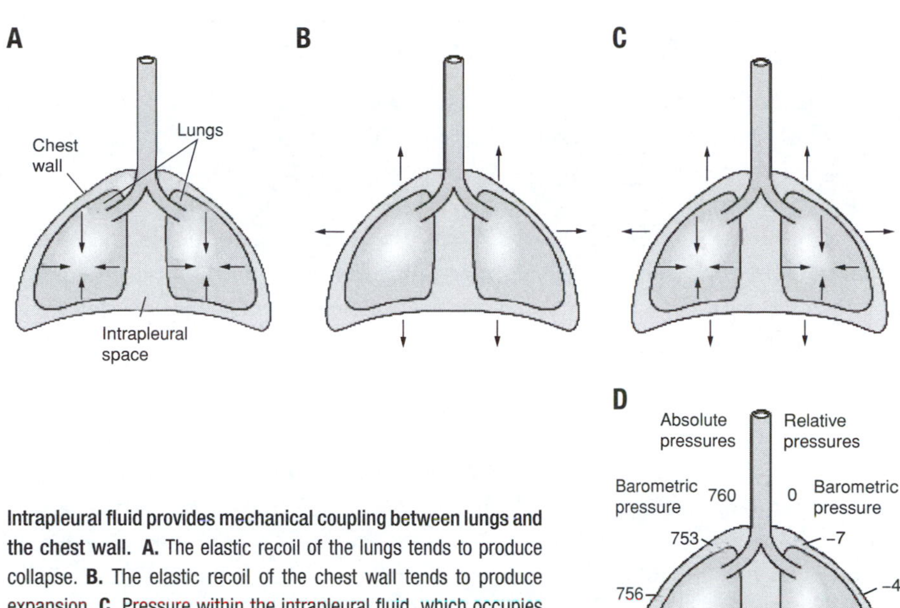

Figure 10.10 **Intrapleural fluid provides mechanical coupling between lungs and the chest wall. A.** The elastic recoil of the lungs tends to produce collapse. **B.** The elastic recoil of the chest wall tends to produce expansion. **C.** Pressure within the intrapleural fluid, which occupies the space between the lung and chest wall, is influenced by the elastic recoil of both. **D.** Typical pressures (in mmHg) within the intrapleural fluid are shown.

If the lung were removed from the chest, it would collapse because the lung has a property called **elastic recoil**. The lung behaves like a balloon; the elasticity of the balloon wall, when unopposed, will force air out through any available opening (see Box 10.3). To remain inflated, a balloon must be either completely closed—in which case the balloon wall exerts force on the entrapped air, increasing the internal pressure—or subjected to a force that opposes deflation. The lung is not completely closed; instead, the chest wall provides elastic recoil that is equal and opposite to the recoil of the lung. While the balloon-like property of the lung is pulling inward (tending to collapse the lung), the chest wall recoil is pulling out, tending to increase the volume of the thorax (Figure 10.10).

The chest wall and the lung are not directly connected; the muscles of the chest wall and diaphragm do not directly increase the volume of the lungs. Instead, the intrapleural fluid between these two elastic objects, the chest wall and the lung, one pulling out and the other pulling in, provides a mechanical connection. Mechanical competition—chest wall pulling out and lung pulling in—expands the volume in the intrapleural space. Because the intrapleural space is closed, a negative intrapleural pressure P_{ip} is created (Figure 10.10). In fact, P_{ip} is about -3 mmHg (-4 cmH_2O) at the end of a normal expiration.

SELF-TEST 3 Figure 10.10 shows that the intrapleural pressure, P_{ip}, is lower at the top of the lung (-10 cmH_2O) than at the bottom (-2.5 cmH_2O). Why is this so?

ANSWER: The intrapleural space is filled with fluid; the hydrostatic pressure is higher at the bottom of a column of fluid.

Box 10.3 Relationship of wall tension and pressure

For a balloon to remain inflated, the pressure inside the balloon must be greater than the pressure outside. The difference in pressure between inside and outside ($\Delta P = P_{in} - P_{out}$) is greater for a "stiffer" balloon. A "stiff" balloon is one in which the wall of the balloon is made of a material that is more difficult to deform (i.e., that has a higher Young's modulus, E).

Pierre-Simon Laplace (1749–1827) was a pioneer of mathematics and physics. Among his many contributions, he determined the pressure drop across the thin, elastic wall of enclosures of different shapes. For a spherical enclosure—like the balloon above—the pressure drop required to maintain a radius r_{sphere} is given by:

$$\Delta P = \frac{2T}{r_{sphere}}, \qquad \text{(Equation 1)}$$

where T is the tension in the thin wall of the balloon. Biomedical engineers and physiologists often refer to this as Laplace's law. For a given tension in the vessel wall, the amount of pressure that must be applied to maintain a given radius increases as the radius decreases. This equation shows why it is difficult to start blowing up a balloon, but gets easier as the balloon inflates: The required pressure goes down as the radius goes up.

A similar expression can be derived for the pressure–tension relationship in a cylindrical vessel of radius $r_{cylinder}$:

$$\Delta P = \frac{T}{r_{cylinder}}. \qquad \text{(Equation 2)}$$

In these equations, T represents a wall tension, or surface tension, or force per unit length of the material. If one is considering a thin material, such as the wall of a balloon, the tension can be converted into stress in the material:

$$\sigma = \frac{T}{t}, \qquad \text{(Equation 3)}$$

where σ is the hoop or circumferential stress (force/area), T is the tension (force/length), and t is the thickness of the material.

These same concepts apply whenever there is an interface between two immiscible fluids, such as air and water. In this case, the tension is called surface tension and is a property of the two fluids. For air–water interfaces, the surface tension is 72 dynes/cm. Surface tension is a measure of the energy that is needed to maintain an interface between two fluids that do not want to mix. To decrease this unwanted energy cost, the interface tends to shrink, but shrinking of the interface can create other forces, such as the increase in pressure that occurs as an air bubble shrinks in an effort to decrease the overall surface energy.

Laplace's equation can be used to determine the pressure inside a bubble of air in water. If the bubble has a radius of 150 microns, then the pressure drop between the air and water is equal to 2 (72 dyne/cm)/0.015 cm, which is 9,600 dyne/cm^2 or 7.2 mmHg. The air pressure within the bubble must be 7.2 mmHg higher than the pressure in the water.

Because an alveolus contains air, and the alveolar cells lining the alveolus are surrounded by a thin film of water, this same calculation also applies to the typical 150-μm-diameter alveolus. If the pressure within the alveolus is atmospheric, then the pressure in the liquid film of the alveolar wall must be −7.2 mmHg. For the alveolus to remain inflated, this pressure must be greater than the intrapleural pressure, P_{ip}. But P_{ip} is greater than −7.2 mmHg, it is closer to −3 mmHg. Why don't the alveoli collapse?

The surface tension in the alveolar fluid is actually much less than the surface tension of water. Type 2 pneumocytes in the alveolus secrete a substance called surfactant, which is 10% protein, 90% phospholipids; the presence of surfactant in the fluid decreases the surface tension by a factor of 10, and also decreases the pressure in the liquid film, allowing the alveoli to remain inflated with less effort from the chest wall. The lung begins to produce surfactant late in gestation, so infants that are born prematurely often lack surfactant and have difficulty breathing. Fortunately, synthetic surfactants are now available that can be used to treat infants with this condition (called respiratory distress syndrome).

The driving force for expansion of the alveolus is the transpulmonary pressure, P_{tp}, which is:

$$P_{tp} = P_{alv} - P_{ip}, \tag{10.4}$$

where P_{alv} is the pressure inside the alveolus. It is this pressure difference—between the air inside the alveolus and the fluid in the intrapleural space—that keeps the lung inflated.

Imagine that a person forces all of the air from his or her lungs and holds his or her breath at this position. (Recall from Chapter 7 that the volume remaining in the lungs is called the residual volume.) The lungs are still inflated, even though the trachea is open and the air in the alveolus has equilibrated with atmospheric pressure. The muscles of the chest wall are working to keep the lungs inflated at this volume. The P_{ip} under these conditions is about -3 cm H_2O. The transpulmonary pressure is, therefore, $+3$ cm H_2O from Equation 10.4.

When there is a puncture wound in the wall of the chest, so that the intrapleural space is now exposed to atmospheric pressure, intrapleural pressure will be lost, and the lungs will collapse like a balloon. In medicine, this event is called **pneumothorax**. The elastic recoil of the lung, which is now unopposed, causes collapse. The lung volume does not reach zero, because total collapse of some airways traps air in some alveoli and small airways.

Assume that this collapsed lung could be slowly inflated by sealing the wound and applying small amounts of suction (to create negative P_{ip}). Assume further that the trachea is open to the atmosphere during this process, so that P_{alv} is 0 (i.e., it is equal to atmospheric pressure). As P_{ip} decreases, P_{tp} will increase according to Equation 10.4. As P_{tp} is increased in small increments, the volume of the lung will also increase; this increase is shown by the bottom curve of Figure 10.11. For very small pressures (phase I), little change in lung volume would be observed. It is difficult to inflate the collapsed airways because of surface tension forces (Box 10.3), which must be overcome before volume can increase. When sufficient pressure is applied, however, the volume of the lung increases linearly with pressure (phase II) until lung volume is near total lung capacity (phase III).

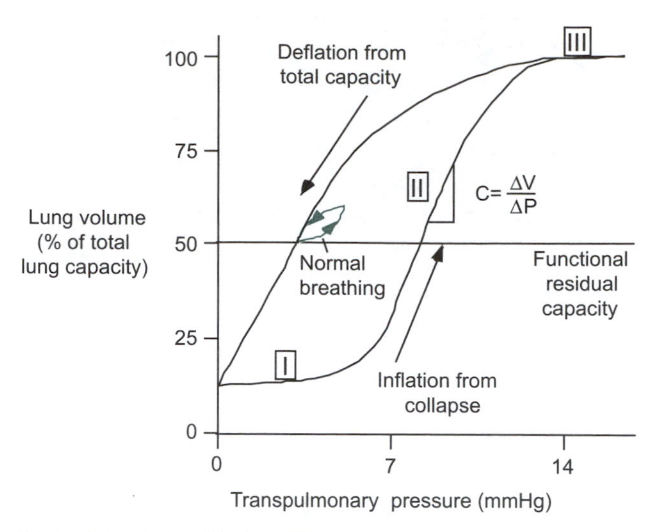

Figure 10.11 **Mechanical behavior of the lung.** The bottom curve shows the inflation of lung from complete collapse. In phase I, P_{tp} must be increased above a threshold to overcome surface tension of the air–water interface on airway surface. In phase II, volume increases linearly with increases in P_{tp}. The mechanical behavior of the lung in this phase is described by the compliance, **C.** In phase III, the volume approaches total lung capacity as the lung stiffens (i.e., the compliance decreases dramatically). The top curve represents the deflation of the lung from total capacity. The difference between the two curves is hysteresis, which occurs because a greater pressure is required to open a closed airway than to keep an open airway inflated. The colored curve represents inflation and deflation with normal breathing: Hysteresis still occurs, but it is less pronounced.

The relationship between lung volume and P_{tp} is different during deflation of the lung (Figure 10.11). The difference is an example of **hysteresis**. Hysteresis, in general terms, occurs when the relationship between two variables—such as pressure and volume in this example—depends on the history of the system under study. As in Figure 10.11, lung volume at P_{tp} of 10 cm H_2O is either 25% or 75%, depending on whether the lung was previously collapsed or fully inflated.

Hysteresis occurs in many biological systems. Neurons that are stimulated by neurotransmitters often do not return immediately to their basal, or prestimulation, state. Cell responses to growth factors often depend on the state of the cell, including its previous exposure to growth factors. Gene expression networks also exhibit hysteresis.

10.4 Cellular mechanics

Cells are complex, deformable objects. Like other objects (Figure 10.12), cells have an internal structure that determines their ability to deform (recall Section 5.2). Red blood cells are slightly larger than the smallest capillaries and, therefore, must deform to move through the circulation (Table 10.2). White cells are substantially larger and less deformable than red cells; therefore, they can have a larger impact on blood flow properties than one would predict from their abundance. The mechanical properties of white cells have a profound influence on their fate in the circulation after infusion. A reduction in the deformability of

Figure 10.12 **Scaffolding and cellular mechanics.** Like the infrastructure that provides mechanical support for buildings, the cytoskeleton contributes mechanical strength to cells.

white cells may also play a role in certain diseases, such as leukemia and diabetes, in which microcirculation can be impaired.

Likewise, the mechanical properties of cancer cells are an important determinant of cancer progression (see Chapter 16). Local invasion of cancer into normal tissue depends on the ability of cancer cells to create the physical forces that allow them to crawl. Metastasis—one of the deadliest elements of cancer—occurs when cells crawl into blood vessels and move through the circulation; the site of deposition of these cells depends on their mechanical properties.

10.4.1 Mechanical properties of cells

The mechanical properties of many cells—including blood cells—have been directly measured by aspiration into a micropipette (Figure 10.13). The properties of the cell are deduced from the deformations observed as the cell is pulled into the pipette using gentle pressure (3, 4). For example, in one method, the suction pressure is gradually increased to a critical pressure, $\Delta p_{critical}$, at which a small, hemispherical section of the cell is pulled into the pipette. Laplace's law permits

Table 10.2 Mechanical properties of blood cells

Cell type	Shape	Characteristic dimensions (μm)	Volume (μm^3)	Cell fraction in blood	Cortical tension (mN/m)
Red cell (erythrocyte)	Biconcave disc	7.7 × (1.4 to 2.8)	96	0.997	
Platelet	Biconcave disc	2 × 0.2	5–10	0.007	
Neutrophil	Spherical	8.2–8.4	300–310	<0.002	0.024–0.035
Lymphocyte	Spherical	7.5	220	<0.001	0.06
Monocyte	Spherical	9.1	400	<0.0005	0.035

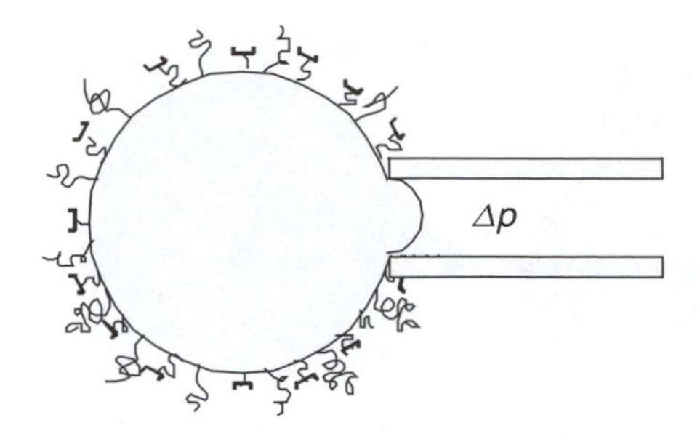

Figure 10.13 Micropipette aspiration to determine cellular mechanical properties.

calculation of cell cortical tension, T, from the critical pressure:

$$T = \frac{\Delta p_{\text{critical}} R_{\text{p}} R_{\text{c}}}{2(R_{\text{c}} - R_{\text{p}})}, \tag{10.5}$$

where R_{p} is the radius of the micropipette and R_{c} is the radius of the cell (5). Similarly, the overall viscosity of a cell can be measured by aspirating the whole cell into a larger micropipette (~4-μm radius for neutrophils) and comparing the time course for cell movement through the pipette lumen to numerical models of individual cell deformation (5). Table 10.2 lists the cortical tension measured for white blood cells.

Other techniques have also been used to measure the mechanical properties of individual cells, by deforming them between parallel plates, for example. The mechanical properties of cells—or the molecules of the cytoskeleton or extracellular matrix—can now be performed using atomic force microscopy (which measures forces of 0.01–100 nN) or optical tweezers (which measure forces less than 200 pN) (6). Table 10.3 lists some values for the Young's modulus of cells measured under different conditions.

10.4.2 Mechanical properties of the cytoskeleton

The intracellular fluid, or cytoplasm, is a much more complicated fluid than plasma. Plasma, or the acellular fraction of blood, is a concentrated protein solution that behaves as a Newtonian fluid with a viscosity of 1.2 cP (or 1.2 mPa-s) at body temperature (37°C). The simplest cytoplasm is probably found in red

Table 10.3 Young's moduli for cells measured by different techniques

Cell type	Measurement technique	E (dyne/cm^2)
Fibroblast (spread on a surface)	Atomic force microscopy (AFM)	16,000
Endothelial cell (spread on a surface and sheared)	AFM	1,575
Endothelial cell (spread on a surface)	Magnetic bead rheometry	45
Endothelial cell (round)	Micropipette aspiration	750
Endothelial cell (round)	Magnetic bead rheometry	22

Note: Adapted from Ethier CR, Simmons CA. *Introductory Biomechanics.* Cambridge, UK: Cambridge University Press; 2007.

blood cells, which have no nucleus or internal organelles. Red cell cytoplasm is a concentrated hemoglobin solution; individual cells contain between 290 and 390 g/L of hemoglobin, which has a viscosity of between 4.2 and 17.2 mPa-s (7).

The basic elements of the cytoskeleton were introduced in Chapter 5, in which the shape of a cell was associated with protein filaments in the cytoplasm. The three most important protein filaments are illustrated in Figure 5.5: actin microfilaments, microtubules, and intermediate filaments. These filaments have intrinsic mechanical properties. In some cases, the mechanical properties of individual filaments have been measured directly, although the measurements are difficult and a clear picture of filament mechanics is not yet available. It is clear, however, that the three main filaments differ greatly in mechanical properties: Actin microfilaments are flexible but unable to withstand tensile forces; microtubules are stiffer than actin filaments (but still have poor tensile strength); and intermediate filaments are more flexible than actin and are stronger under tension (8).

Within the cytoplasm, individual filaments aggregate into bundles; bundling greatly enhances the mechanical strength of the system. In addition, when the ability to bundle and unbundle filaments is regulated, the local mechanical properties of the cytoplasm can be controlled via assembly and disassembly of filaments. Many cytoplasmic constituents participate in the filament-assembly process; actin filaments, for example, are cross-linked by at least four different agents (α-actinin, spectrin, fimbrin, and villin) providing the cell with a variety of tools for dynamically regulating filament assembly. Complex mechanical behavior is observed in simple systems of filaments and filament-binding proteins; in the case of actin filaments and α-actinin, complex rheological behavior is observed in reconstituted samples of purified cytoskeletal elements (9). In general, solutions of cytoskeletal filaments are viscoelastic.

10.4.3 The effects of mechanical forces on cells

It has long been known that cells can produce forces: The actin–myosin machinery within muscle cells produce the force of muscle contraction that allows for human movement (recall Box 4.1), but external mechanical forces can also influence cell behavior. Endothelial cells, which line the lumen of blood vessels, are well-known for their ability to align in the direction of an external fluid flow; these cells are responding to shear forces arising from the fluid motion. In addition, the internal biochemistry of the cell can be influenced by shear forces: Enzymes are activated, mediators of cell responses such as nitric oxide and prostacyclins are produced, and, sometimes, silent genes are activated. The ability to respond to changes in mechanical conditions is an important skill for endothelial cells; these cells live at the interface of the flowing blood with the solid tissue. As flow conditions within the vessel change, the endothelial cell can respond by preparing biochemical messages that lead to vessel constriction, dilatation, or long-term remodeling.

Other cells respond to mechanical forces in other ways. Chondrocytes—the cells within cartilage—reduce their production of extracellular matrix proteins and proteoglycans when exposed to static compressive forces. These same cells, however, increase matrix production when exposed to dynamic (time-changing) forces. Osteoblasts in bone also respond to fluid shear stresses by increasing their production of certain proteins, such as osteopontin. Although there is still much to learn about the role of mechanical forces in the maintenance and function of tissues such as bone, cartilage, and blood vessels, it is clear that cells within these tissues can sense their mechanical environment and respond in significant ways.

Summary

Biomechanics is a central topic of biomedical engineering. This chapter reviewed some of the most important fundamental topics in biomechanics, including the use of force balances to analyze the mechanical function of biological systems, mechanical properties of biological materials, and the diverse role of mechanical forces in cell and tissue function.

- Materials—including biological materials such as organs, tissues, and cells—deform when exposed to physical forces.
- When exposed to small deformations, or strains, many materials deform elastically; these small deformations can often be characterized by an elastic modulus (also called the Young's modulus), E.
- When exposed to deformations beyond the elastic limit, materials undergo plastic deformations, which cause permanent changes in their structure.
- Some biological materials—particularly soft tissues and cells—are viscoelastic; that is, they exhibit both viscous and elastic characteristics when undergoing deformation.
- Bones are strong, dense tissues that provide mechanical support to the body; the two typical architectures of bone—compact and spongy—produce a variety of different mechanical behaviors. In all cases, the mechanical properties of bone are intimately related to its underlying physical structure.
- Soft tissues are usually viscoelastic, with stress–strain behavior that varies depending on the extent of deformation.
- The lung is a biological device for bringing air into the body; the mechanics of breathing can be understood by considering the mechanical coupling between the elastic lung tissue and the elastic chest wall.
- Cells are small (~10 μm), deformable objects. The behavior of cells depends on their mechanical properties; in addition, many cells respond to physical forces in the environment that surrounds them.

REFERENCES

1. Lemmon CA, Sniadecki NJ, Ruiz SA, Tan JL, Romer LH, Chen CS. Shear force at the cell-matrix interface: enhanced analysis for microfabricated post array detectors. *Mech Chem Biosyst.* 2005;2(1):1–16.

2. Holzapfel GA. Biomechanics of soft tissue. In: Lemaitre J, ed. *Handbook of Material Behavior*. London: Academic Press; 2000.
3. Evans E, La Celle P. Intrinsic material properties of the erythrocyte membrane indicated by mechanical analysis of deformation. *Blood*. 1975;45:29–43.
4. Evans E, Hochmuth R. A solid-liquid composite model of the red cell membrane. *J Membr Biol.* 1977;30:351–362.
5. Tsai MA, Waugh RE, Keng PC. Passive mechanical behavior of human neutrophils: Effects of colchicine and paclitaxel. *Biophys J.* 1998;74:3282–3291.
6. Leckband D. Measuring the forces that control protein interactions. *Annu Rev Biophys Biomol Struct.* 2000;29:1–26.
7. Waugh RE, Hochmuth RM. Mechanics and deformability of hematocytes. In: Bronzino JD, ed. *The Biomedical Engineering Handbook*. Boca Raton, FL: CRC Press; 2000:32–1–32–13.
8. Bray D. *Cell Movement*. Second ed. New York: Garland Publishing; 2001: 372 pp.
9. Sato M, Schwarz W, Pollard T. Dependence of the mechanical properties of actin(a-actinin gels on deformation rate. *Nature*. 1987;325:828–830.

FURTHER READING

Ethier CR, Simmons CA. *Introductory Biomechanics*. Cambridge, UK: Cambridge University Press; 2007.

Bray D. *Cell Movement*. Second ed. New York: Garland Publishing; 2001.

USEFUL LINK ON THE WORLD WIDE WEB

http://oac.med.jhmi.edu/res_phys/TutorialMenu.html
Johns Hopkins Tutorials on Respiratory Physiology

KEY CONCEPTS AND DEFINITIONS

anisotropy a property of a material where its mechanical properties depend on the direction of applied loading

deform to change the shape or form of an object

dynamic undergoing change; in motion

elastic deformation a stress-induced change in the shape or volume of an object after which the object returns to its original shape or volume

elastic limit yield strain

elastic recoil the property of a stretched object or organ (e.g., lung to collapse or return to its resting state)

failure stress the stress that a material can endure without failing (breaking or fracturing); also called maximum stress

Hooke's law a law that states that the strain in a solid is directly proportional to the applied stress within the elastic limit of that solid

hysteresis the phenomenon whereby the value of a physical property lags behind changes in the effect causing it (e.g., the inflation of a completely deflated lung lags behind the applied pressure to cause that inflation)

intrapleural space the fluid-filled space between the lungs and chest wall

isotropy a property of a material where its mechanical properties are similar in all directions and do not depend on the direction of applied loading

load a weight or force that is applied to a component of a structure or the structure as a unit

maximum stress the largest stress that a material can endure without failing (breaking or fracturing); also called failure stress

plastic deformation a permanent distortion of the shape or volume of an object caused by an applied stress that strains the object beyond its elastic limit

pneumothorax the presence of air or gas in the intrapleural space that causes the lungs to collapse

proportionality limit the maximum stress–strain values at which the proportionality between stress and strain as defined by Hooke's law applies. It is the point on the stress–strain diagram where the curve becomes nonlinear.

static not changing with time; at rest

strain the fractional increase in material size

strain energy the potential energy stored in deformed material per unit volume equal to the area under a stress–strain curve

stress the force applied to an object per unit area

surface tension the attractive property of the surface of a liquid

thorax the part of a mammal between the neck and the abdomen, including the cavity enclosed by the ribs, breastbone, chest, and dorsal vertebrae

viscoelasticity the property of a substance of exhibiting both viscous and elastic behavior when undergoing plastic deformation. The application of stress to a viscoelastic material causes temporary deformation if the stress is quickly removed, but permanent deformation if it is maintained, thus exhibiting time-dependent strain.

yield strain the strain at which the material becomes irreversibly deformed, also called elastic limit

yield stress the critical stress necessary to cause an irreversible change or deformation in a material

Young's modulus a measure of elasticity equal to the ratio of the stress acting on a substance to the strain produced; also called elastic modulus

NOMENCLATURE

ΔP	Change in pressure	N/m^2, Pa
P_{ip}	Intrapleural pressure	N/m^2, Pa
P_{tp}	Transpulmonary pressure	N/m^2, Pa
P_{alv}	Alveolar pressure	N/m^2, Pa
P_{in}	Inside pressure	N/m^2, Pa

P_{out}	Outside pressure	N/m^2, Pa
$\Delta P_{critical}$	Suction pressure	N/m^2, Pa
r	Radius of sphere	mm, cm, m
R_c	Radius of cell	μm, mm
R_p	Radius of micropipette	μm, mm
T	Tension, cortical tension	N/m
t	Thickness	mm, cm
U_o	Strain energy	kJ/m^3
dv_x/dy	Strain rate	s^{-1}

GREEK

σ	Stress	N/m^2, Pa
σ_f	Failure stress	N/m^2, Pa
σ_y	Yield stress	N/m^2, Pa
τ_{xy}	Shear stress	kg/m-s
ε	Strain	—
ε_y	Yield strain/elastic strain	—
μ	Viscosity	Centipoises (cP)

QUESTIONS

1. Define the following:
   a. Yield stress
   b. Elastic limit
   c. Failure stress
   d. Strain energy
2. Differentiate between elastic and plastic deformation.
3. Define Young's modulus. What property of materials does it measure? How would a material with a high modulus differ from a material with a low Young's modulus?
4. Explain Hooke's law. Do all elastic materials obey Hooke's law? Why/Why not? Give an example of a material that does not obey Hooke's law.
5. What does it mean to say that a material is viscoelastic? Give an example of a viscoelastic material.
6. A student performed an experiment to examine the stress–strain properties for two elastic objects and discovered that Object 1 had a Young's modulus of 600 MPa and Object 2's modulus was 2,300 MPa. Which of these objects is stiffer? Explain. Which of these object stores more energy when deformed by the same amount of stress? Explain your answer.
7. Two plastic rulers (R1 and R2) were subjected to the same stress. Whereas R1 extended by only 10% of its original length, R2 extended by x% of its original length. Calculate the stress applied given the following Young's moduli: R1 = 1,000 MPa and R2 = 600 MPa. Find x%. Determine the strain energy

for each ruler given the applied strain. Which of these rulers is made out of a more compliant material?

8. A college student conducted experiments to examine the properties of mammalian alveoli. He discovered that larger alveoli tend to collapse more easily and are more difficult to inflate than smaller ones. What is the most likely explanation for this phenomenon?

PROBLEMS

1. The forces that act on a body that is swimming through water are similar to the forces acting on an object flying through the air, as shown in Box 10.1, except that swimming bodies are buoyant, which exactly balances the force of gravity. Assume that we are studying the differences between swimmers of different size: a bacterium, a fish, and a human swimmer. Assume further that the drag force, F_d, acting on any swimmer is proportional to its average "size" a:

$$F_d = 6\pi v_x \mu a,$$

where μ is the viscosity of the fluid surrounding the swimmer and v_x is the speed of swimming. Assume that the viscosity of the fluid through which they are swimming is 0.01 g/cm-s and the following are the characteristics of each swimmer:

	Speed of swimming (cm/s)	Size (cm)
Bacterium	0.001	0.0001
Fish	50 (1.1 mph)	10 (3.9 in)
Human	170 (3.8 mph)	170 (5 ft 7 in)

   a. What is the drag force acting on each swimmer? What is the propulsive force that is produced by each swimmer to swim at the stated speed?
   b. Assume that the swimmer suddenly stops exerting energy to swim. At the very instant that it stops, before it has had time to slow down appreciably, what is the drag force on each swimmer? What is the propulsive force?
   c. How long will each swimmer "coast"—that is, how long will it keep moving forward before stopping?
   d. How far did it travel during the "coasting" phase?

2. An elephant and a mouse fall from a 15-foot height. Explain (using as quantitative terms as possible) why the elephant will likely be injured, but the mouse will not.
3. The elastic modulus for three different materials is given as:
   a. Bone = 15,000 MPa
   b. Brain tissue = 0.005 MPa
   c. Artery = 3 MPa

   Draw stress versus strain diagrams for each of these materials.

4. A student conducted an experiment to determine the behavior of poly(ethylene) and poly(tetrafluoroethylene) to stress. The tables here depict the results of his experiments.

Poly(ethylene)		Poly(tetrafluoroethylene)	
Stress (MPa)	Strain	Stress (MPa)	Strain
20	0.022	20	0.033
50	0.056	50	0.083
100	0.111	100	0.167
250	0.278	250	0.417
400	0.444	400	0.667
550	0.611	550	0.917
800	0.941	800	1.510
950	1.167	950	1.950
1,100	1.667	1,100	2.433
1,150	2.111	1,150	3.011
1,170	2.412	1,190	3.502
1,185	2.712	1,200	4.000

a. Plot the stress–strain curves for both types of material. Calculate their respective Young's moduli. How do they compare?
b. From their respective plots, determine their approximate strain energy before plastic deformation begins. How do they compare?
c. What do these calculations of Young's modulus and strain energy tell us about the properties of both materials? Compare the properties of both materials using these calculations.

5. By measuring the pressure inside a balloon that is slowly inflated until it bursts, it should be possible to estimate the maximum tensile strength of the balloon material. Three balloons are tested, and they burst under the following conditions:

Balloon	Diameter prior to burst	Pressure prior to burst
1	20 cm	2 atm
2	10 cm	1.5 atm
3	100 cm	2 atm

6. A Boeing 747 is flying at a constant speed of 600 mph (270 m/s) at an altitude of 30,000 ft (9,100 m). As the plane moves through the atmosphere, air flowing over the surface of the plane creates a drag force, acting in the opposite direction of motion. The drag on the aircraft can be calculated from the following formula:

$$F_{\mathrm{d}} = \frac{1}{2}\rho v_{\mathrm{x}}^2 C_{\mathrm{d}} A.$$

Assuming that the drag coefficient for a 747 is 0.031, calculate the total drag on the moving airliner. Remember that the density of air at 30,000 ft is much different than the density of air at sea level.

7. Spider webs are made of silk. The biomechanics of the sticky capture spiral thread from an orb-weaving spider *Araneus diadematus* were investigated by Köhler and Vollrath. The table here shows the stress and strain values for the sticky silk thread.
   a. Plot a stress versus strain curve for the data shown (note: 1 Pascal = 1 N/m^2).
   b. Using the linear part of the curve, calculate the Young's modulus for the spider silk.
   c. How does this value compare to that of bone (E = 3 MPa) or structural steel (E = 200 MPa)?
   d. Based on your answer in (c), describe how the Young's modulus is a measure of the stiffness of a material.

Stress (MPa)	Strain
0	0
20	0.5
55	1.0
130	1.5
249	2.0
430	2.5
631	3.0
831	3.5
1,031	4.0
1,338	4.76

11 Bioinstrumentation

LEARNING OBJECTIVES

After reading this chapter, you should:

- Describe the common components of a measurement system.
- Understand the different types of sensors and the mechanism by which each converts its detected signals into electrical signals.
- Describe the principle of operation of instruments used to monitor patient body temperature, blood pressure, oxygen saturation, cardiac function, and blood glucose levels.
- Describe the principle of operation of instruments used in the laboratory such as a pH meter and spectrophotometer.
- Understand the importance of the emerging areas of biosensors and microelectromechanical systems (MEMS).

11.1 Prelude

Modern health care has benefited enormously from the work of biomedical engineers to create instruments that are used in clinical monitoring and laboratory analysis. Hospital operating rooms, emergency rooms, and doctors' offices each contain an array of instruments used to measure and record a patient's vital signs such as temperature, blood pressure, pulse, and oxygen saturation (Figure 11.1). Many of the most popular instruments enable non-invasive monitoring of vital signs of patient health: The stethoscope allows doctors to listen reliably to the beating heart, the sphygmomanometer allows them to estimate pressure within vessels deep in the body (Figure 11.2), and the ophthalmoscope allows them to see structures on the retina. It is impossible to estimate the number of lives that have been lengthened or improved by these devices.

The medical device industry—the constellation of large and small companies that design, manufacture, and sell medical devices and instruments—is one of the largest and most rapidly growing sectors of the U.S. economy. Medical device companies employ biomedical engineers to invent, design, build, and test devices for use in medicine. Biomedical engineers also provide technical training to physicians on the use of such devices. Some biomedical engineers work in

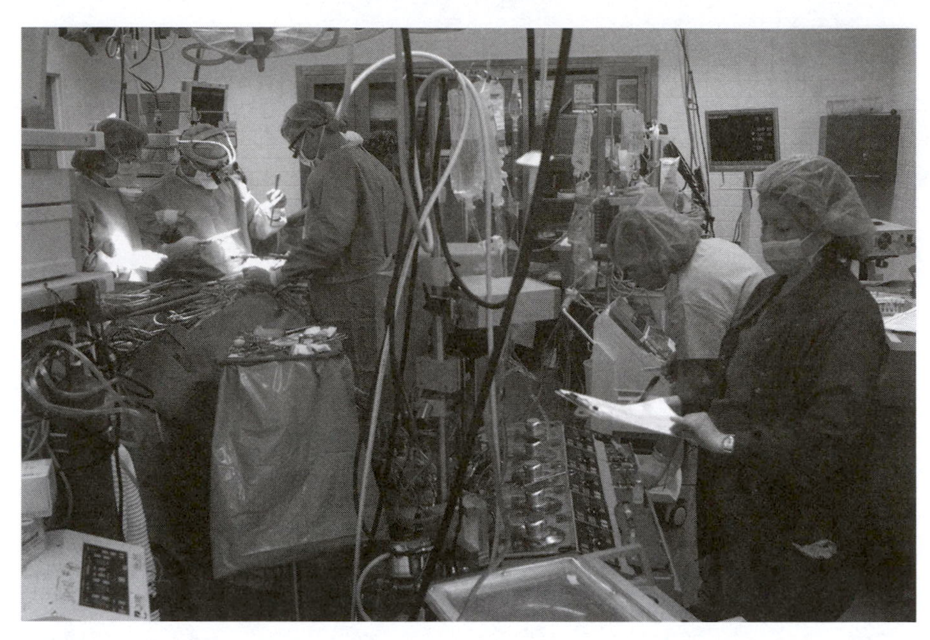

Figure 11.1 **Instruments in an operating room.** Modern surgical suites are filled with instruments that provide continuous monitoring of a patient's vital signs and that aid in the surgical procedures. These instruments make surgery safer for the patient, and enable surgeons to perform operations with speed and precision that would be otherwise unattainable. Photograph courtesy of Yale University Media and Technology Services. (See color plate.)

hospitals overseeing maintenance of medical instruments and adapting instruments to serve patients and doctors: These individuals are often called clinical engineers. In research laboratories, biomedical engineers and other scientists routinely use instruments to take measurements (Figure 11.3). For example, researchers use instruments called spectrophotometers to measure light absorption within liquid samples to quantify the rates of enzymatic reactions or to measure the composition or viability of cells. Microscopes are used to view cells in preparations such as blood smears, which are routinely used to diagnose blood diseases.

The next generation of medical devices and laboratory instruments will provide patients, physicians, and researchers with more information, more rapidly than the instruments of today. Many clinical tests now take many days to complete

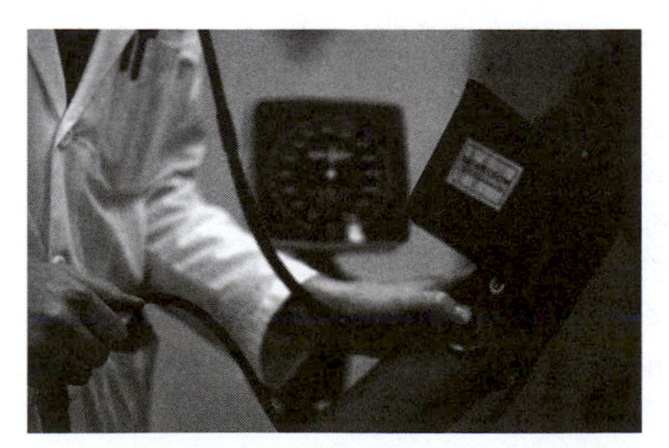

Figure 11.2 **Medical instruments are a major part of modern medicine.** The sphygmomanometer and the stethoscope were created by biomedical engineering tools and are used by almost every health care professional.

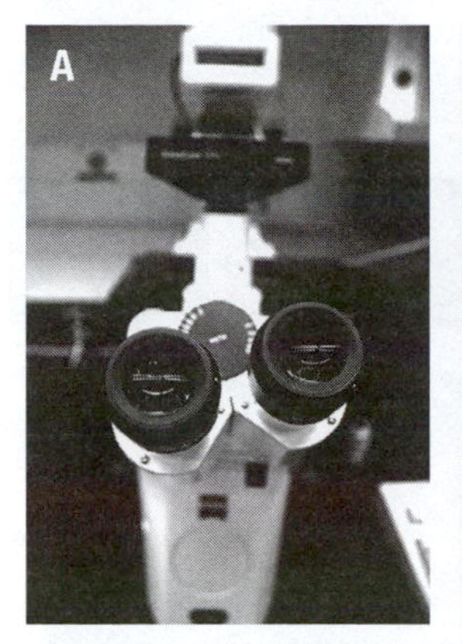

Figure 11.3

Many biomedical instruments are used by both doctors and researchers. A. The microscope has long been used for medical diagnosis, such as identification of bacteria. **B.** The technology for measuring precise volumes of fluid—which is accomplished by instruments like these automated pipetters—allows scientists and health care workers to dispense tiny amounts of liquid reproducibly and safely.

because samples are sent from the doctor's office or clinic to a separate testing laboratory. In the near future, most of these measurements will be performed in minutes by instruments within the doctor's office or clinic (or at the "point-of-care," as this is sometimes called). Some of these instruments might be similar to home pregnancy test kits, in which a sample of fluid is used to produce a visual signal that appears within a few minutes (recall Figure 2.30). In the laboratory, integrated lab-on-a-chip devices—in which small volumes of blood or fluid are simultaneously subjected to multiple measurements—will eventually replace current laborious methods (Figure 11.4). Also in the not-so-distant future, hospitals and clinical laboratories will be equipped with instruments capable of new analyses, such as microarray analysis chips; these new techniques will provide rapid information on the genes and proteins present in patient samples. Information from these microarrays may someday predict a patient's susceptibility to disease before the symptoms manifest or an individual's response to drugs before they are taken. These instruments will increasingly "personalize" medical

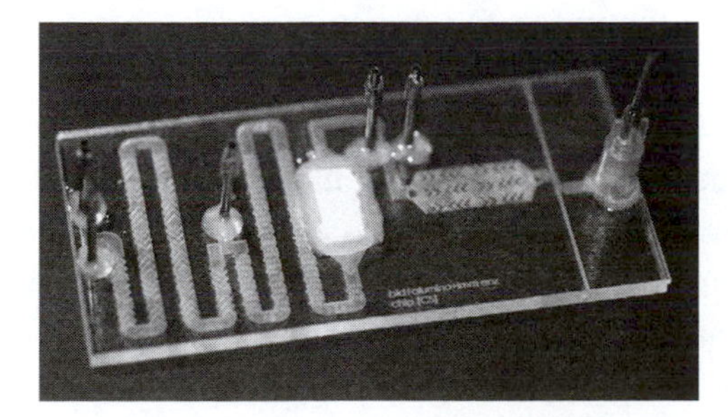

Figure 11.4

Bioinstrumentation of the future. A lab-on-a-chip device for early detection of sarin, an agent found in nerve gas. This device was created to develop a field-deployed handheld method for identifying trace amounts of sarin in the blood of potential victims of chemical warfare. The entire device is 80 mm × 38 mm in size; each device contains a reactor for gas regeneration, a compartment for cell lysis and filtering, a chamber for removal of fluoride ions, and a port for optical detection. From: Tan HY, Loke WK, Tan YT, Nguyen N-T. A lab-on-a-chip for detection of nerve agent sarin in blood. *Lab Chip.* 2008;8:885. Reproduced with permission of The Royal Society of Chemistry.

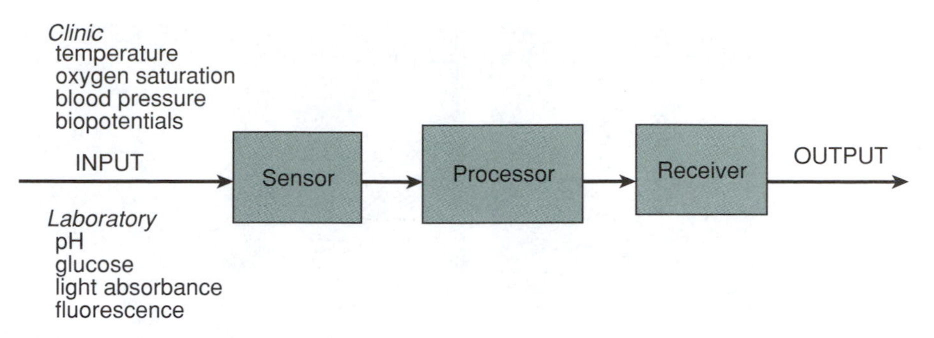

Figure 11.5 Schematic of a typical measurement.

care by allowing physicians to ask and answer questions such as: Will this drug be effective for this person? Will this treatment cause dangerous side effects? What is the optimal dose for this individual?

This chapter describes some of the instruments common to biomedical engineers including new technologies, such as biosensors and lab-on-a-chip devices, which are not yet routinely available. One of the most important roles of the biomedical engineer is to exploit basic physical and chemical principles to create instruments that can be used reliably and safely in patients. This chapter does not include instruments—such as magnetic resonance imaging (MRI) and computed tomography (CT) scanners—that are used for biomedical imaging; these methods are discussed in Chapter 12. Also not included in this chapter are the principles of operation of artificial organs, both implanted (such as the artificial heart) and extracorporeal (or outside the body, such as the cardiopulmonary bypass machine), which are covered in Chapter 15.

11.2 Overview of measurement systems

Most instrumentation systems contain common elements, which are present regardless of the parameter being measured (Figure 11.5). A typical system measures a parameter of interest (labeled as INPUT in Figure 11.5) and creates a reading (or OUTPUT) that the user can comprehend. There are many types of inputs that might be measured in patient care, such as temperature, oxygen saturation, blood pressure, electrical potentials, pH, or glucose concentration. Sometimes the instrument is used to detect an input, such as light absorbance or fluorescence, which is coupled to, or related to, the actual parameter of interest.

The part of the instrument that detects the input is called a **sensor**. The sensor converts the input parameter into a **signal**, usually an electrical voltage, which varies in a predictable and reliable way with changes in the input parameter. There are many types of sensors, which use different mechanisms to convert an input variable into a measurable quantity. After an input parameter is sensed—by conversion of the parameter of interest into a signal—it is further modified by a **processor**. Processing may include, for example, amplification of the signal,

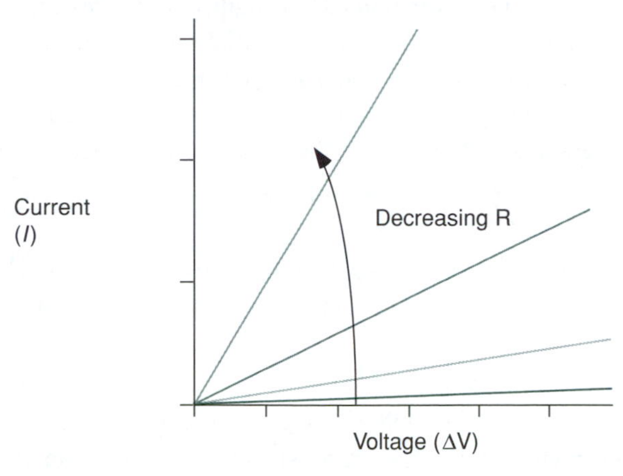

Figure 11.6 **Ohm's law for an electrical resistor.** In a simple resistance element, the current through the element (called a resistor) is proportional to the voltage drop across the element. The constant of proportionality is resistance *R*. For a given resistor, the current increases predictably with voltage drop. For two resistors, one with a higher resistance *R*, the higher resistance allows a smaller current if both are exposed to the same voltage or potential.

filtering to remove unwanted information, or comparison to signals from previous measurements or control signals. The resulting output can be displayed, stored, or communicated via a **receiver**, or a device that presents an interpretable message to humans. The receiver is often a digital readout or a computer display.

According to Ohm's law, the **potential difference** or **voltage drop** (ΔV) across an ideal conductor is proportional to the **current** that flows through it (i), and the **resistance** of the conductor (R) is the constant of proportionality:

$$i = \frac{\Delta V}{R} \quad \text{or} \quad \Delta V = iR. \tag{11.1}$$

Recall that Ohm's law was used in Chapter 8 to evaluate the resistance of flow pathways: An equation similar to 11.1 was used to establish the relationship of pressure drop to flow rate. Here, Ohm's law is helpful for understanding basics of instrument design and operation.

Equation 11.1 (or the analogous Equation 8.3) can be interpreted in terms of **driving forces** and **flows**. In a simple electrical circuit, the driving force is the voltage or **electrical potential**, ΔV, which might be provided by a battery. The electrical potential creates a current, or flow of charge i. (In Equation 8.3, the driving force was pressure drop, ΔV, which created a fluid flow, Q.) Electrical potential is measured in units of volt (V), and current is measured in units of amperes (amp); resistance is therefore measured in units of ohms (where 1 ohm is equal to 1 V/amp).

Ohm's law indicates that the larger the difference in electrical potential, or the greater the voltage driving force, the higher the current flow through the wire. Alternately, an element with high resistance—such as a very thin wire—allows only a low current flow (Figure 11.6). Ohm's law and additional current and voltage rules are used in the analysis of electrical circuits as shown in Box 11.1.

Box 11.1 Circuit analysis

Electrical circuit analysis is an essential skill for biomedical engineers who design or use medical devices and instruments. Input signals, which arise from a variety of sources, are converted to electrical signals that can be processed, amplified, and displayed (Figure 11.5).

A familiar mechanical system is analogous to a simple electrical resistance circuit. Blood flow rate (Q) through a cylindrical vessel is proportional to the pressure drop (Chapter 8):

$$Q = \frac{\Delta p}{R}, \qquad \text{(Equation 1)}$$

where Δp is the pressure drop across the vessel, and R is the resistance to flow. The analogous expression for an electrical system is known as Ohm's law, which defines the current (i) as

$$i = \frac{\Delta V}{R} \quad \text{or} \quad \Delta V = iR. \qquad \text{(Equation 2)}$$

For a circuit consisting of a single resistor and a voltage source, Ohm's law can be applied directly to calculate the unknown current, voltage, or resistance. Ohm's law can also be applied to each individual resistor in a circuit, if the voltage drop or current across the resistor is known.

One method for analyzing circuits containing multiple resistors is to determine an equivalent circuit using the rules for series and parallel resistors. For resistor circuits, no matter how complex the configuration, the circuit can be simplified to an equivalent circuit containing a single voltage source (V_T) and equivalent or total resistor (R_T). For a group of resistors in **series**, all components must have the same current because there is only one flow path for electrons. For a group of resistors in **parallel**, each resistor experiences the same voltage drop across it. The equations for calculating the equivalent resistance for resistors in series or parallel is provided in the figure: R_n, i_n, and V_n represent the resistance, current, and voltage across resistor n in the circuit.

Two additional rules may also be used in the analysis of circuits: Kirchhoff's current law (KCL) and Kirchhoff's voltage law (KVL). KCL states that the sum of the currents at any node in a circuit must equal zero; KVL states that the net voltage around any closed circuit (or loop) is zero. A node is a junction of two or more branches (e.g., nodes are points a, b, c, and d in the bottom panel of Figure Box 11.1). A loop is any closed connection of branches. An example of a loop is the closed loop in the bottom panel of Figure Box 11.1 from point a to b to c to d to a. Any given circuit may contain multiple loops. Thus, as an alternative to calculating an equivalent circuit, KCL and KVL can be applied directly by writing equations for each node and loop. These equations are solved to determine unknown variables.

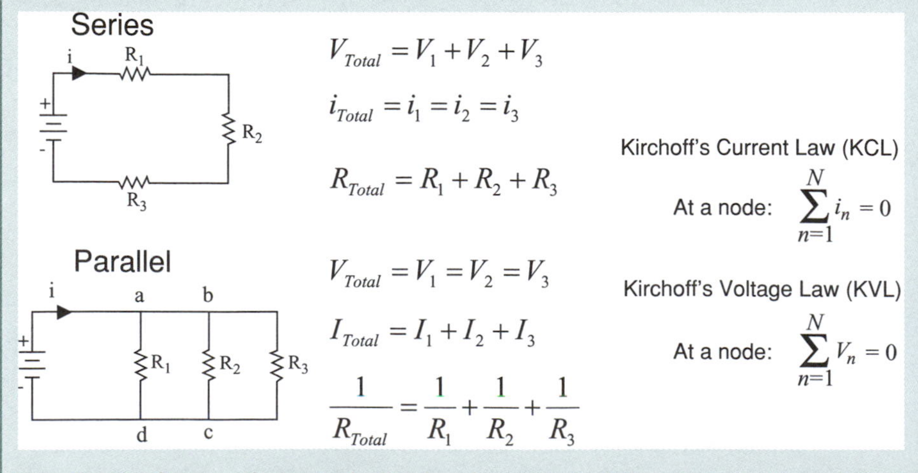

Figure Box 11.1 Summary of current and voltage laws for resistor circuits in series (top) and in parallel (bottom).

Table 11.1 Examples of sensors used in biomedical instruments

Sensor type	Sensing element	Example
Thermal	Thermocouple, thermistor	Electronic thermometer
Mechanical	Strain gauge, piezoelectric sensor	Pressure transducer
Electrical	Electrode	Electrocardiograph (ECG), electroencephalograph (EEG)
Chemical	Electrode	pH meter
Optical	Photodiode, photomultiplier	Pulse oximeter

SELF-TEST 1 A current of 1 microamp flows through a device after it is connected to the poles of a 9 V battery. What is the resistance of the device?

ANSWER: 9×10^6 ohm or 9 Mohm (megaohm)

11.3 Types of sensors

Sensors are now available to measure many parameters of clinical and laboratory interest. Some types of sensors are summarized in Table 11.1. The principles underlying the operation of each of these sensors are described in this section.

11.3.1 Thermal sensors

Control of body temperature is critical to life processes; therefore, the body has multiple mechanisms for maintaining an appropriate internal temperature that, for most people, is near 37°C (98.6°F). Deviation of body temperature from this normal value is frequently used as an indicator of disease. Elevation of body temperature, called **fever**, is a sign of illness; it can be one of the first signs of infectious disease, for example. The development of instruments for measuring temperature has been extremely successful: Many devices are both accurate and inexpensive. As a result, instruments for measuring temperature are now available in every doctor's office and in most homes.

One of the most commonly used temperature-sensing elements is a **thermocouple**. A thermocouple is formed by fusing two dissimilar metals to produce two junctions, as shown in Figure 11.7. If a complete circuit is formed from

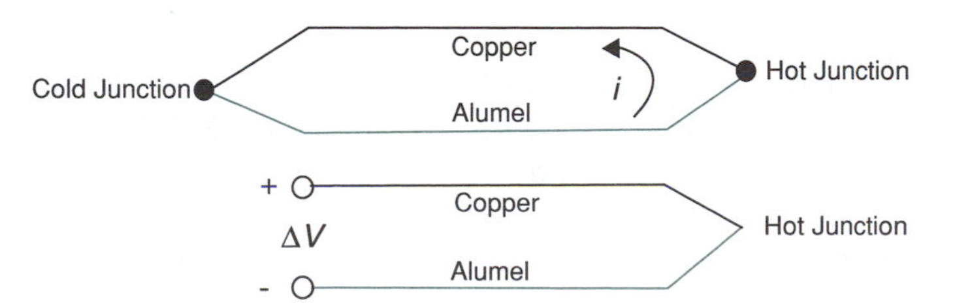

Figure 11.7 **Example of a thermocouple formed by fusion of two metals:** copper and alumel (an alloy containing 5% nickel, 2% manganese, 2% aluminium, and 1% silicon).

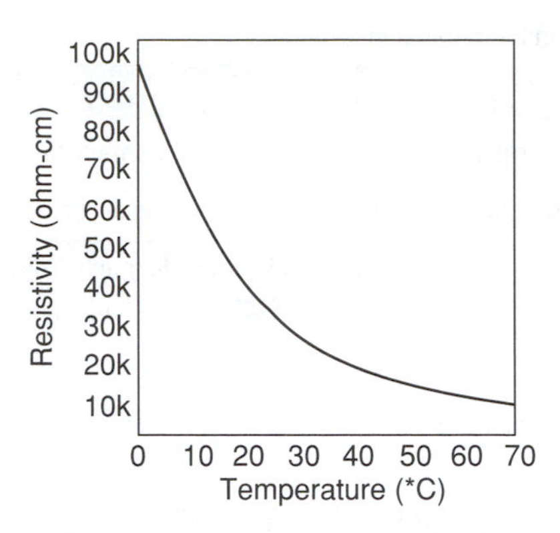

Figure 11.8 **Resistivity-Temperature characteristics for a common thermistor.** The resistance varies exponentially with temperature, as in Equation 11.2.

the two metals, and one of the junctions is maintained at a lower temperature, a current will flow through the metals. This flow of current, which is produced by a temperature difference, is known as the Seebeck effect, and was named after T.J. Seebeck, who first observed the phenomena in the 1820s. The magnitude and direction of the current are determined by both the temperature difference between the junctions and the properties of the metals that make up the junction. When the circuit is broken (Figure 11.7), a voltage difference (called the **Seebeck voltage**) can be measured; this voltage depends in a predictable way on the temperature of the remaining junction. The voltage output is commonly between 15 and 40 microvolts per degree Celsius, but depends on the properties of the metals used. Thermocouples are used widely in manufacturing and other engineering applications because they are inexpensive and reliable and can be used over a wide temperature range, with some capable of measuring temperatures as low as $-200°C$ and others as high as $1{,}250°C$.

Thermistors are homogeneous composites of dissimilar metals that form thermally sensitive resistors. They are distinct from thermocouples, which are heterogeneous combinations of different metals. For any particular thermistor, the relationship between the resistance and temperature is exponential:

$$R_T = R_0 e^{\beta\left(\frac{1}{T} - \frac{1}{T_0}\right)}, \tag{11.2}$$

where T is temperature in degree Kelvin (K), R_T is the measured electrical resistance, R_0 is the resistance at reference temperature ($T_0 = 298$ K), and β is a coefficient that is characteristic of the material. Although the resistance–temperature relationship is exponential (not linear over the entire range of temperatures; Figure 11.8), it is often roughly linear over the much smaller range of temperatures that is of interest for biomedical engineering (i.e., 35–39°C).

SELF-TEST 2 Some thermistors have positive temperature coefficients (i.e., β is greater than zero), and some have negative temperature coefficients (i.e., β is less than zero). In which category is the thermistor illustrated in Figure 11.8?

ANSWER: It has a negative coefficient, because resistance goes down as temperature goes up.

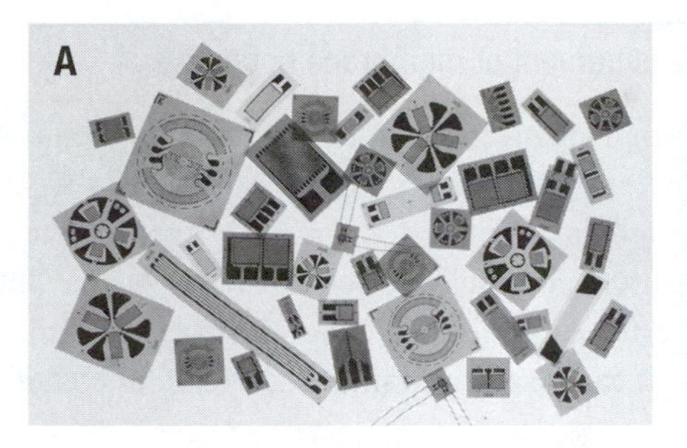

Figure 11.9

Mechanical sensors. A. A wide variety of materials are available for measuring strain over different ranges with different levels of sensitivity. Photo courtesy of Vishay Micro-Measurements, Raleigh, NC, USA, a division of Vishay Intertechnology. **B.** Sensors that convert sound waves into electrical voltages form the basis of many common devices, such as electric guitars.

11.3.2 Mechanical sensors

Mechanical sensors measure force and, therefore, can be used to measure pressure. These sensors often consist of materials, such as metal wires or films, which exhibit a change in resistance in response to a change in shape. Recall from Chapter 10 that certain kinds of shape deformations can be reported as changes in **strain**, ε. For example, a material that becomes elongated in the presence of an applied force exhibits strain that can be measured as the fractional change in length ($\varepsilon = \Delta L/L$). An instrument that measures strain is called a **strain gauge** (Figure 11.9). In biomedical applications, a strain gauges typically measure strains in the range 10^{-6} to 10^{-3}.

Mechanical strain can be measured using sensing elements composed of **piezo-electric** materials. *Piezo* literally means pressure and derives from the Greek word *piezein*, which means "to squeeze." When a piezoelectric element, such as quartz, is placed under pressure, polarized ions within the crystal are deformed, leading to generation of an electric charge. The converse is also true: Application of an electric field to a piezoelectric material will create strain (or deformation) in the material. Piezoelectric materials are used in various applications including pressure transducers and acoustic transducers. Acoustic transducers convert strains that occur because of sound waves into electric charge, or vice versa, and are commonly found in telephones, microphones, electric guitars, and automated blood pressure machines at the local drugstore.

11.3.3 Electrical sensors

Electrical sensors or electrodes are used in both research and clinical settings. In biological applications, electrodes are often used to detect the electric potential generated by cellular ionic currents. The electrodes that accomplish this can range in size from micro-sized probes to larger adhesive pads.

All cells have a resting membrane potential caused by the difference in ion concentrations across a cell membrane that is permeable to these ions (see Chapter 2); electrodes are useful for measuring these potentials (Box 11.2). In excitable

Box 11.2 Electrodes for measurement of membrane potentials

Electrophysiology is the study of the behavior of electrically excitable cells and membranes. Cells and tissues are studied by inserting probes or electrodes into the cell or tissue, in such a way that the biological function of the system is not altered, and measuring its electrical behavior. The probes that are used for these measurements can be electrical wires or needles, thin glass cylinders (called micropipettes) that are filled with conductive electrolyte solutions, or other devices. Often, the probe is so small that it can be inserted into a cell without changing the cell's behavior significantly.

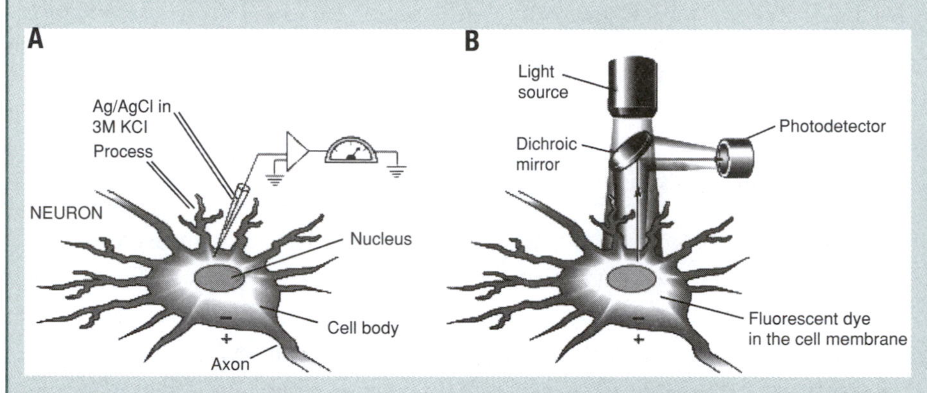

Figure Box 11.2A Measurement of membrane potential. The membrane potential of a cell can be measured directly by inserting an electrode through the membrane (left) or indirectly, by first loading the cell with fluorescent dyes that are sensitive to local electrical potential and then observing the cell using fluorescence microscopy (right).

Electrophysiologists use a variety of techniques to explore different cell behaviors. In a **voltage clamp** technique, the potential (or voltage) drop across a membrane is held constant so that the resulting current can be measured. In a **current clamp**, a known current is introduced through the electrode, and the resulting change in potential is measured.

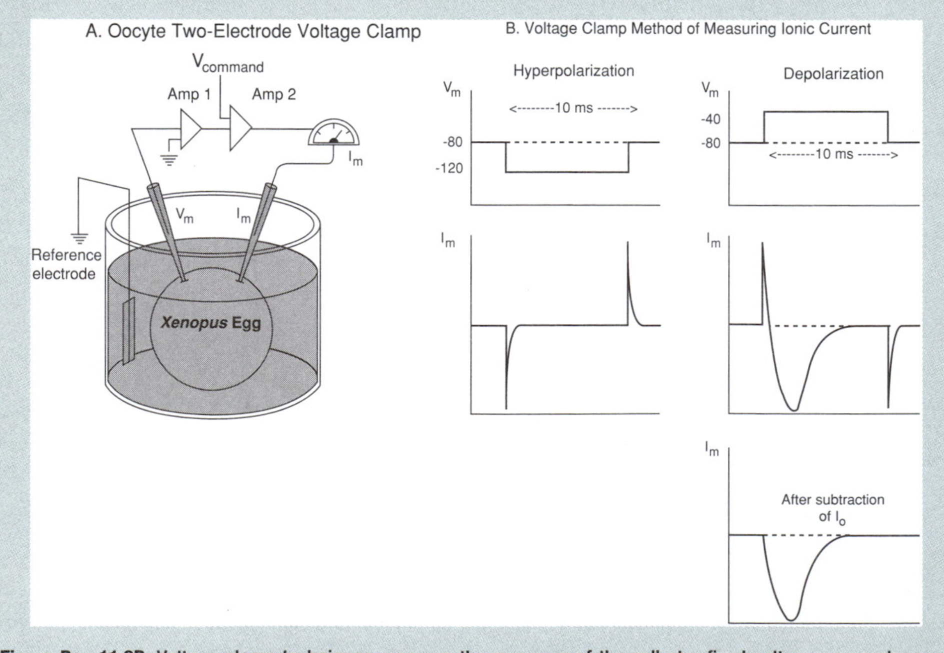

Figure Box 11.2B Voltage clamp techniques measure the response of the cell at a fixed voltage or membrane potential.

Fluorescent dye molecules are now used for indirect measurement of electrical activity. Here, chemicals with certain properties are introduced into the cell: The chemicals are **fluorescent** (i.e., they emit light at a certain wavelength after they are excited by light of another wavelength), and the fluorescence of the chemicals is altered by the electrical state of the surrounding fluid. With these chemicals inside a cell, microscopes that detect fluorescence are used to monitor the changes that occur in the cell as it experiences an event, such as an action potential. Fluorescent dye methods do not require inserting a device into the cell, and they can be used to observe electrical activity in any part of the cell that can be viewed with a microscope. Therefore, very small or confined cell appendages, such as the narrow processes found in many nerve cells, can be studied.

Patch clamp techniques were developed by Erwin Neher and Bert Sakmann, who won the 1991 Nobel Prize for this work. In a patch clamp, a micropipette—typically with a larger diameter than the pipettes used to impale cells—is used to remove or isolate a small section of a cell membrane. Importantly, the edge of the pipette is sealed to the membrane, so that the region within the patch (defined by the circumference of the pipette) is electrically isolated from the surroundings. This method allows the study of the properties of the membrane within the patch.

cells such as neurons and muscle cells, certain movements of ions into or out of the cell can trigger action potentials, which can also be detected using electrodes. The electrocardiograph, or ECG, indirectly measures the electrical activity of cardiac muscle cells using surface electrodes attached to the skin. Similarly, the electrical activity of the brain is sometimes measured using electrodes placed on the surface of the skull, often by using specially designed, electrode-loaded caps.

In the laboratory, specialized electrodes measure the electrical currents through individual ion channels using the patch-clamp method (Box 11.2). These electrophysiological measurements are performed on a microscope stage using specialized glass microelectrodes. A "patch" of the membrane is suctioned and held in place while measurements are taken on single ion channels. By creating membrane patches in different configurations, including patches that are removed from the cell and patches that are left attached to the cell, the patch-clamp technique can measure a variety of properties of ion channel proteins (Figure 11.10).

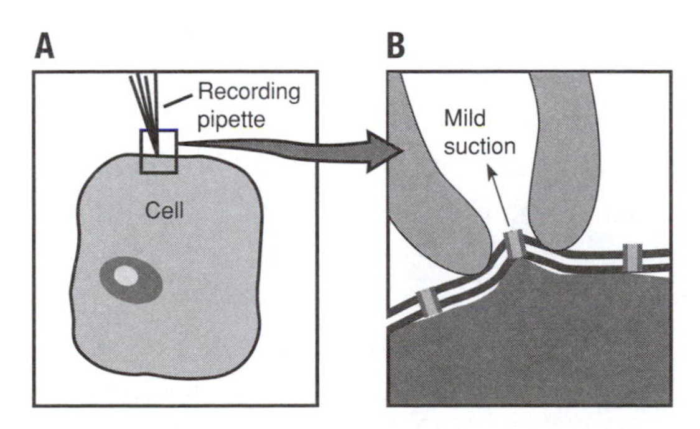

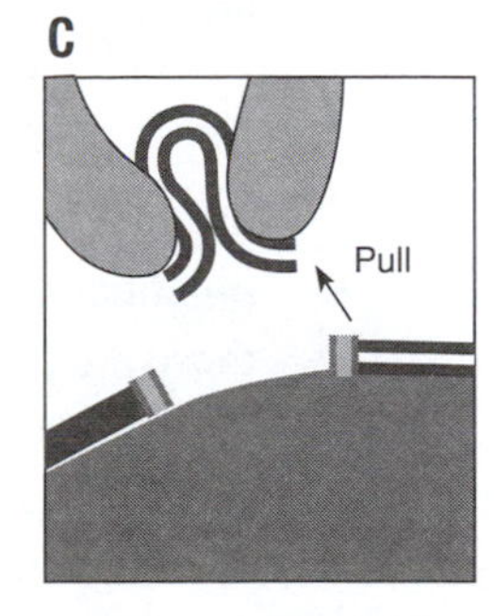

Figure 11.10 **Patch-clamp techniques permit the measurement of currents through individual ion channels.** Researchers have learned how to collect patches of membranes using different techniques, allowing the measurement of a variety of ion channel properties. **A.** Overview of pipette and cell. **B.** Tight contact is created between the pipette and the plasma membrane. **C.** The pulled membrane breaks away from the cell and allows access to the cytoplasmic domain of the ion channel.

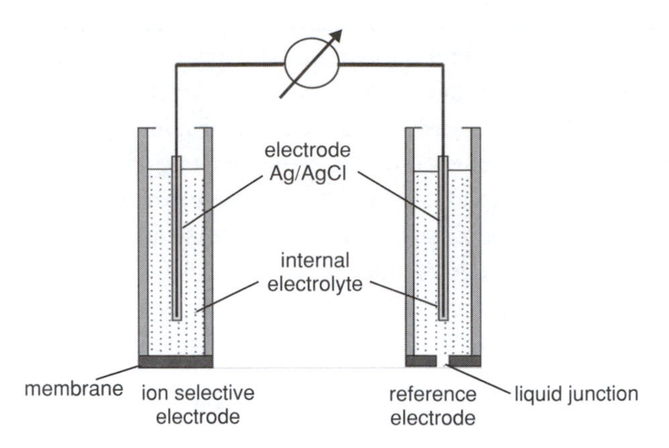

Figure 11.11 **Ion-selective electrodes measure the concentration of particular ions.** The selectivity of the electrode comes from an ion-selective membrane, which allows only ions of interest to enter the electrolyte chamber at the measurement electrode (left).

11.3.4 Chemical sensors

Chemical sensors are designed to measure the presence—and sometimes the concentration—of specific chemicals. Because of the great diversity of chemical species (i.e., ions, gases, chemical or biochemical agents)—and the need to create sensors that detect only one or a small number of related species—there are many types of chemical sensors.

Ion analyzers or detectors are used broadly for monitoring the quality of air, water, food, and consumer and pharmaceutical products; they are also used in both clinical and biomedical research laboratories. This section focuses on one type of chemical sensor, which uses an ion selective electrode (ISE) to detect ions. Sensors that use a biological molecule as the sensing element are classified as biosensors and are discussed in Section 11.6.

ISEs acquire their specificity from membranes that are permeable to a particular ion species. For example, glass ISEs, which are selectively permeable to H^+ ions, are commonly used in pH meters (this use is discussed in more detail in Section 11.5.1). Other ISEs contain membranes that are selectively permeable to ions such as Ca^{2+}, K^+, Na^+, Cl^-, F^-, and NO_3^-. A working cell consists of an ISE, a reference electrode, and a voltmeter (Figure 11.11). These sensors produce a potential, or voltage, that is proportional to the ion concentration. For that reason, these are called potentiometric sensors.

Another type of chemical sensor is an amperometric sensor, in which the current is proportional to the concentration of the species generating the current. An example of an amperometric sensor is the Clark electrode, which is used for oxygen measurement. Operation of an amperometric sensor generates a current flow, which requires the use of a polarizing voltage source. Development of an electrode to measure oxygen—which was accomplished in 1953 by Leland Clark, a biochemist who became known as the "Father of Biosensors" for this invention—revolutionized the study of oxygen metabolism in tissues.

The operation of a Clark oxygen sensor is shown in Figure 11.12. If oxygen is present in the environment around the end of the electrode, it diffuses through the membrane (D) and accumulates in the electrolyte solution (C). Oxygen in the solution reacts at the cathode; it is reduced to hydroxide ions. This reaction can only occur if there is a flow of current (or electrons) to the cathode; these electrons are generated at the anode by the oxidation of silver (Ag). The silver ions that are generated react with chloride (Cl^-) to form AgCl salt, which accumulates on the anode surface. In this overall sequence of reactions, a current of electrons from anode to cathode is produced: The magnitude of that current depends on the number of oxygen molecules that are reduced at the cathode. The rate of reaction depends on the concentration of oxygen in the electrolyte solution. Therefore, the concentration of oxygen can be determined by measuring the current that is produced.

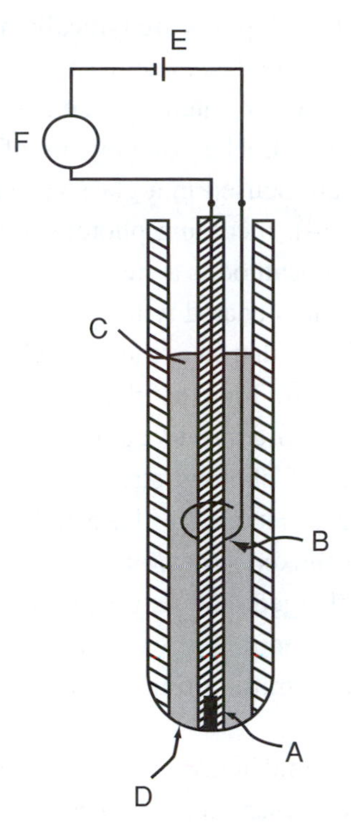

Cathode: $O_2 + 2H_2O + 4e^- \longrightarrow 4OH^-$

Anode: $Ag \longrightarrow Ag^+ + e^-$

Figure 11.12 **The Clark oxygen sensor, schematic diagram.** This amperometric sensor contains a platinum cathode (A), an Ag/AgCl anode (B), an electrolyte solution (KCl; C), an oxygen-permeable membrane (D), a voltage supply (E), and a device for current measurement (F). Oxygen is reduced at the platinum cathode, and silver is oxidized at the anode, as shown in the chemical reaction equations.

11.3.5 Optical sensors

Optical sensors are able to detect light in the visible, infrared (IR), or ultraviolet (UV) regions of the electromagnetic spectrum. This range of the electromagnetic spectrum spans a wide range of wavelengths, from 10 nm up to 1 mm (for a review of the electromagnetic spectrum, see Box 12.1 in Chapter 12).

Light can be detected by either **photodiode** arrays or **photomultiplier** tubes; both of these devices convert light energy into an electrical signal. Photodiodes are made of semiconductor materials such as silicon (Si) or gallium arsenide (GaAs) (Figure 11.13). When a photon of light strikes this material, a current is generated. The resulting voltage is proportional to the intensity of the incoming light. Photodiodes are used in many biomedical devices, including the finger pulse oximeter, which is described in Section 11.4.3.

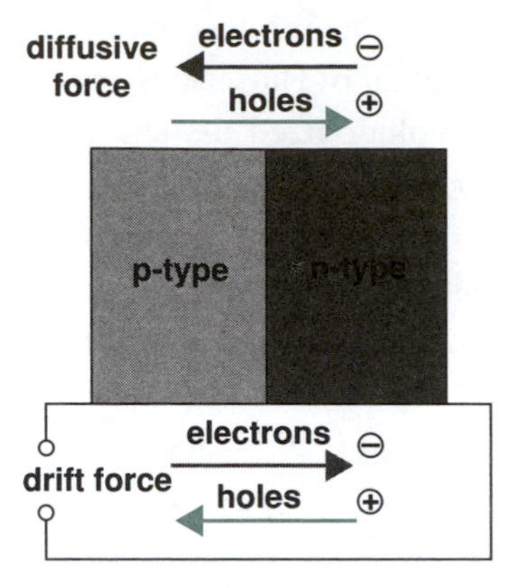

Figure 11.13 Structure of a photodiode.

Photomultipliers are typically more sensitive than photodiodes, meaning that they can detect light at lower intensity (i.e., a smaller photon density). The components of a photomultiplier are housed in a glass vacuum tube (Figure 11.14). Incoming photons are absorbed by the photocathode, a negatively charged electrode that is coated with a photosensitive compound, so that it generates electrons in the presence of light by the photoelectric effect. Electrons are produced in proportion to the number of photons striking the cathode. The current produced by these light-induced electrons is amplified through a cascade, or linked series, of electrodes called **dynodes**. Each dynode is more positively charged than its predecessor, resulting in an increasing number of electrons being emitted at each stage. Therefore, the current reaching the anode can be up to 1 million times greater than the initial photon energy incident on the photocathode.

Photodiodes and photomultipliers are found in UV/visible spectrophotometers, which are commonly found in biomedical research laboratories.

11.4 Instruments in medical practice

Previous chapters have described physiological parameters that can be used to track the health or progress of disease in patients. For example, Chapter 7 introduced the concept of oxygen saturation, Chapter 8 discussed the importance of blood pressure, and Chapter 9 presented analysis of concentrations of compounds in the urine as a reflection of kidney function. Physiological parameters such as temperature, pulse rate, oxygen saturation in the blood, blood pressure, and electrical potentials are routinely measured in doctors' offices, emergency

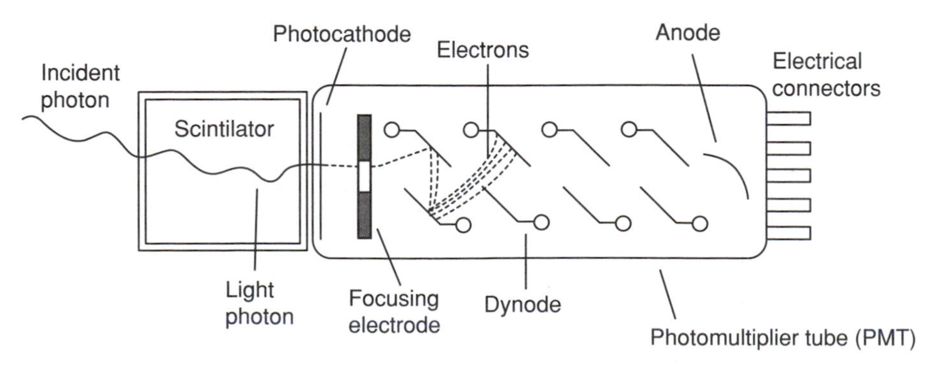

Figure 11.14 **Components of a photomultiplier.** All of the components are housed in a vacuum tube. A series of dynodes amplify the current created by the photocathode; the overall sensitivity of the photomultiplier depends on the efficiency and number of dynodes.

Table 11.2 Some typical medical instruments

Input	Instrument	Sensor	Output	Range*
Temperature	Oral digital thermometer	Thermistor	Temperature display	32–40°C
Blood pressure	Digital sphygmomanometer	Stethoscope or strain gauge	Pressure	0–400 mmHg
Blood oxygen	Pulse oximeter	Photodiode	Percent oxygen saturation	0–100% SpO_2
Biopotentials				
Cardiac biopotentials	Electrocardiograph (ECG)	Skin electrodes	Electrocardiogram	0.5–5 mV
Neural biopotentials	Electroencephalograph (EEG)	Scalp electrodes	Electroencephalogram	5–300 mV
Retinal biopotentials	Electroretinograph (ERG)	Contact lens electrodes	Electroretinogram	0–900 mV
Muscle biopotential	Electromyograph (EMG)	Needle electrodes	Electromyograrn	0.1–5 mV

* Information on the range of measured values from Webster JG. *Bioinstrumentation.* Hoboken, NJ: John Wiley & Sons; 2003.

rooms, outpatient clinics, and hospitals. During surgery, for example, many of these vital signs are monitored continuously as a measure of the health of the patient under anesthesia. Table 11.2 lists some common physiological parameters and the instruments that are used to measure them. The following sections briefly describe the principles that underlie the operation of these instruments.

11.4.1 Measurement of body temperature

Measurement of an elevated body temperature is often the first indication of an abnormal condition; for example, parents measure the temperature of their children as an early sign of illness. The normal body temperature is 98.6°F (or 37°C), although normal temperature varies somewhat from person to person. Usually, a temperature above 99.5°F (or 37.5°C) is a sign of an underlying illness or infection (or a child who needs to stay home from school). Hyperthermia, or elevated body temperature, can also occur as a result of prolonged exercise or exposure to excessive heat.

Why do infections or inflammation cause a state of hyperthermia, or what is usually called a fever? Macrophages at a site of infection or inflammation release molecules called **pyrogens** into the blood. Pyrogens act on cells in the hypothalamus, which then release prostaglandins, local messengers that raise the body's temperature set point, sometimes to 102°F or higher. The body uses the temperature set point to define the desired temperature level for homeostasis (Chapter 6). To reach the newly elevated set point, the hypothalamus also initiates a series of responses for the body to raise the temperature, such as shivering to increase heat production and vasoconstriction in the skin to reduce heat loss. For these reasons, a patient may experience sudden chills at the onset of fever. At the other extreme, hypothermia, or lower-than-normal body temperature, can

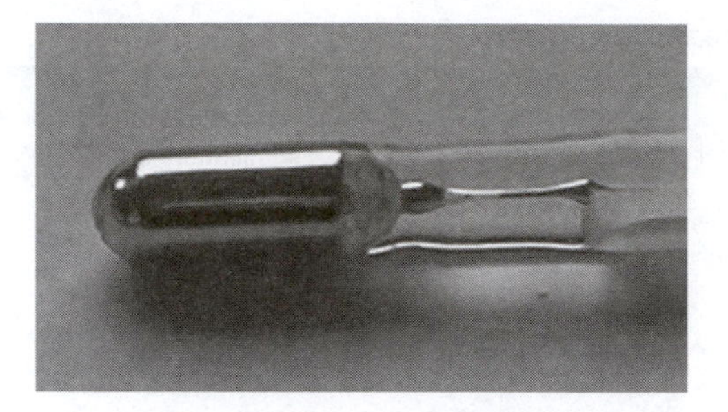

Figure 11.15 **A glass thermometer is a familiar method for measurement of body temperature.** Thermometers can be placed under the tongue, under the arm, or in the rectum to estimate internal body temperature. This photo shows the mercury reservoir and capillary of a classic glass thermometer.

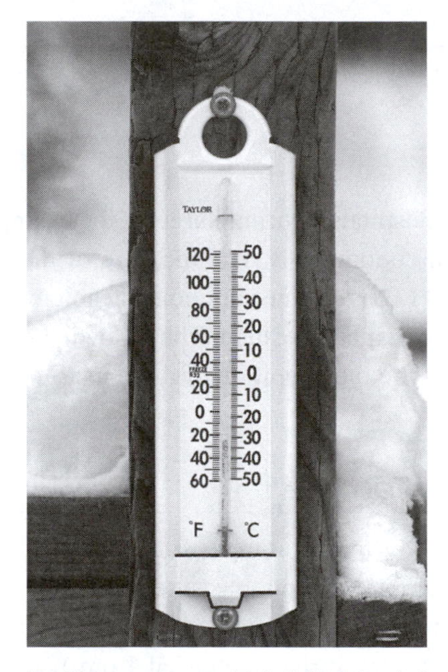

Figure 11.16 **Most thermometers provide both the Fahrenheit and Celsius temperature scales.** Photo courtesy of Chris Kimball.

also occur as a result of underlying illness or exposure to an excessively cold environment.

Several different types of instruments are used to measure temperature, including glass thermometers, electronic thermometers, and IR thermometers. All of these instruments rely on a temperature-sensitive sensor. In 1714, the German physicist Daniel Gabriel Fahrenheit invented the mercury-in-glass thermometer (see Figure 11.15). In glass thermometers, a working fluid, such as mercury or alcohol, is confined to a reservoir connected to a narrow glass tube, or capillary. As the reservoir and its fluid are warmed, the fluid expands and travels up the capillary. These thermometers are precalibrated by the manufacturer—marks on the tube are scaled in degrees Fahrenheit and Celsius. The level at which the fluid stops gives an indication of the temperature. The Fahrenheit scale was introduced in 1724 by Daniel Gabriel Fahrenheit, who set 32°F as the temperature at which ice freezes and 212°F as the temperature at which water boils. The Swedish astronomer Anders Celsius (1701–1744) invented the Celsius scale, which uses 0°C as the freezing point and 100°C as the boiling point of pure water. Although the United States continues to rely on the Fahrenheit scale, most of the world and all of the science community use the Celsius scale (Figure 11.16). Converting a temperature from degrees Fahrenheit to degrees Celsius is simple and can be calculated using the following equation:

$$T_{\text{deg C}} = \tfrac{5}{9}(T_{\text{deg F}} - 32). \quad (11.3)$$

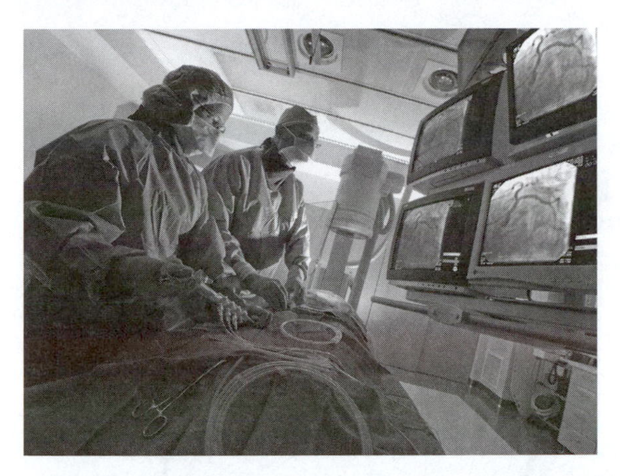

Figure 11.17 **Doctors in a catheterization laboratory, where devices are inserted deep into the body for measurement of physiology.** Photo courtesy of Intermountain Healthcare. (See color plate.)

Conversion of a temperature in degrees Celsius to degrees Farenheit can easily be calculated by rearranging Equation 11.3.

SELF-TEST 3 Room temperature is usually about 75°F. Convert this number to °C.

ANSWER: 24°C

Measurement of body temperature via a glass thermometer provides only an estimate because the thermometer's sensing mechanism cannot be placed deep into the body. Actual core body temperature can be measured directly by using **catheters** to place temperature sensors into deep body compartments, such as the pulmonary artery. Placement of a catheter—a long, thin tube made of a biocompatible, flexible material, usually some kind of plastic—is accomplished by a procedure called a **catheterization** (Figure 11.17). The catheter is inserted into an easily accessible artery or vein, usually in the arm or leg, and guided to the heart. By this procedure, the tip of the catheter can be precisely placed in the heart, or into the lumen of vessels near the heart (Figure 11.18). The tip of the catheter can be loaded with sensors for temperature, blood pressure, or other variables (Figure 11.19). Other specialized features can also be added to the catheter. For example, patients diagnosed with atherosclerosis (a buildup of plaque that narrows the lumen of an artery) may receive a balloon **angioplasty**: a catheterization procedure in which a balloon at the tip of the catheter is inflated to push the plaque against the vessel wall and, therefore, increase the diameter of the flow path. This procedure is now frequently followed by a procedure called stenting, in which a wire mesh cylindrical support, or **stent**, is placed at the lesion site to maintain the open lumen (see Chapter 15 for a description of drug-eluting stents, which are stents that also provide a slow release of drugs for helping to maintain the opening of the vessel after angioplasty).

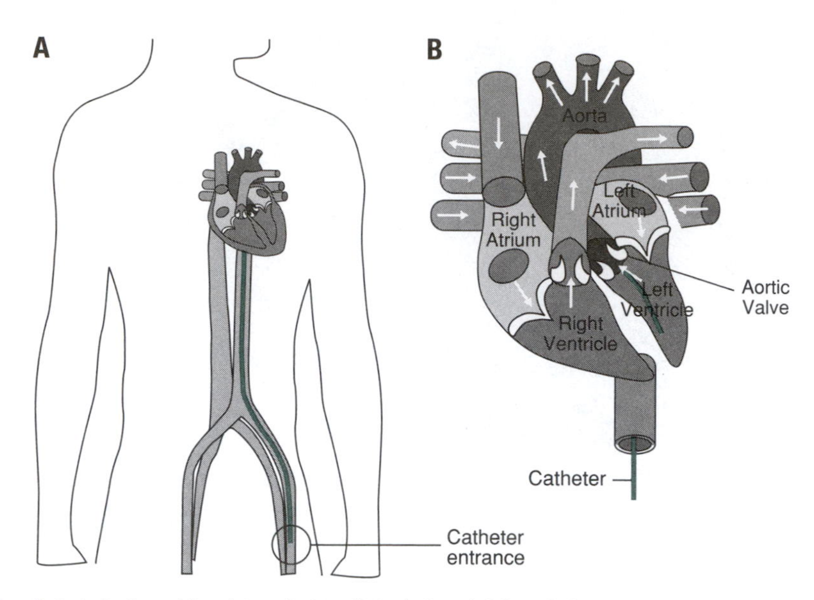

Figure 11.18 **Catheterization.** A long tube called a catheter is threaded through the circulatory system to reach locations deep inside the body.

For catheter-based measurement of temperature, a thermistor wire within the lumen of the catheter transmits blood temperature: This measurement is the most accurate reflection of the core body temperature. However, catheterization is invasive, expensive, and risky. Therefore, other methods, such as the glass thermometer, are used for routine temperature measurements.

Glass thermometers, because they are inexpensive and reliable, are still used in the laboratory and in meteorology for measuring the outdoor ambient temperature. However, their use in medicine is becoming increasingly rare. Mercury is highly toxic to the nervous system, and many countries ban their use in medical applications; manufacturers have replaced mercury with liquid alloys of gallium, indium, and galinstan. Even with the availability of non–mercury-based glass thermometers, other technologies for temperature measurement are becoming more commonplace (Figure 11.20).

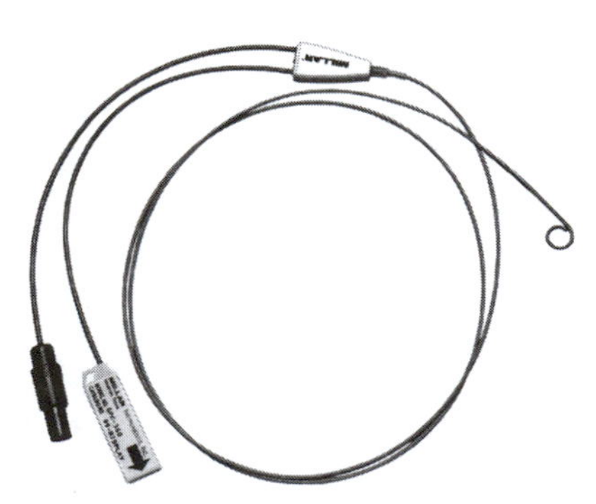

Figure 11.19 **Catheters are made of biocompatible polymers and can be equipped with sensors and other devices. A.** Image of a Multi-Segment Pressure-Volume Mikro-Tip® Catheter. Photo provided by, and used with permission from, Millar Instruments, Inc.

Electronic thermometers use thermocouples or thermistors as the sensing element; the sensing element is placed in direct contact with the body surface. Ear thermometers, now often used at home, measure body temperature through a noncontact probe that senses IR radiation emitted from the tympanic membrane. The emitted IR radiation is measured with a pyroelectric crystal. **Pyroelectricity** is a property of certain materials; these materials respond to heat by the generation of an electrical potential

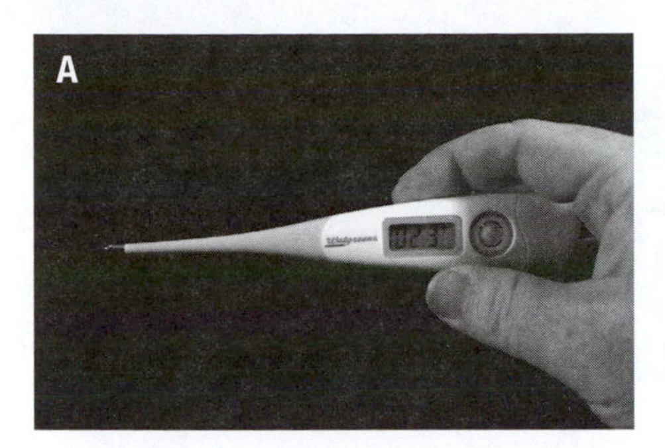

Figure 11.20 **Thermometers for safe and rapid measurement of body temperature. A.** Digital thermometers use thermocouples or thermistors as their sensing element. Photo courtesy of Mike Kroemer. **B.** Ear thermometers make rapid readings of temperature easy and safe. Using infrared technology first developed by NASA to measure the temperature of stars, the device measures body temperature in 2 seconds with the push of a button. Photo used with permission.

(this generation of potential is similar to piezoelectric materials, which generate an electrical potential in response to deformation). Some natural materials such as quartz and bone exhibit pyroelectric behavior. Measurement of body temperature by pyroelectric materials is rapid and safe, takes only a few seconds, but is prone to underestimation of the core body temperature.

11.4.2 Measurement of blood pressure

Blood pressure is critically important in the operation of the cardiovascular system and human health. Measurement of blood pressure provides rapid information about the health status of the cardiovascular system. A patient with chronic high blood pressure (called **hypertension**) increases his or her risk of cardiovascular disease: Hypertension is a risk factor for atherosclerosis, heart attack, and congestive heart failure. High blood pressure may also cause small blood vessels in the brain or the retina of the eye to weaken and burst or bleed, increasing the risk for stroke and vision damage or blindness. The small blood vessels in the kidney are also affected by high blood pressure; narrowing and thickening of the vessels can lead to inefficient glomerular filtration and eventual kidney failure. Alternately, blood pressure can become too low (a condition called **hypotension**) because of rapid blood loss after trauma or hypothermia. Hypotension can threaten all of the body's organs because of a lack of blood and oxygen delivery.

Blood pressure is often reported as systolic pressure over diastolic pressure. As described in Chapter 8, systolic pressure is the maximum pressure exerted on the arterial walls during contraction of the ventricles of the heart, whereas diastolic pressure occurs during the relaxation of the ventricles. In an average adult, the systolic/diastolic blood pressure is 120/80 mmHg. Individuals who are diagnosed with hypertension typically have blood pressure of 140/90 mmHg or higher.

The most direct method for measuring blood pressure is through catheterization. To enable pressure measurement within a blood vessel, the catheter is filled

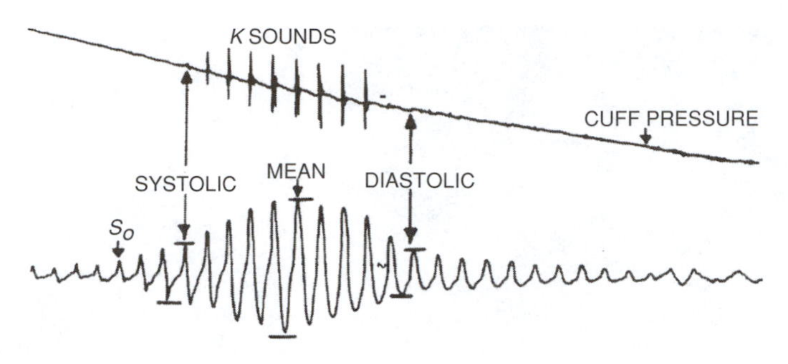

Figure 11.21

Blood pressure measurement. As the cuff pressure is deflated, Korotkoff (K) sounds appear between systolic and diastolic pressures and oscillations in the pressure waveform can be detected. Adapted from: Webster JG. *Bioinstrumentation.* Hoboken, NJ: John Wiley & Sons; 2003.

with a fluid, which transduces the arterial pressure to an external pressure transducer. This method is invasive and, therefore, is only used when it is necessary to measure pressures within specific vessels, such as the pulmonary vessels for detection of pulmonary hypertension.

The most common method of measuring blood pressure is based on the detection of **Korotkoff sounds**, also known as the **auscultatory method**. Auscultation refers to listening to sounds within the body. During a blood pressure measurement, an air-filled sphygmomanometer cuff is wrapped around the patient's upper arm. The nurse or physician places a stethoscope under the cuff and inflates the cuff to a pressure above the systolic pressure; a fully inflated cuff completely stops the blood flow by compression of superficial arteries in the arm. As the cuff pressure is slowly decreased, the operator listens for the appearance (the start of systole) and disappearance (the start of diastole) of Korotkoff sounds (see Figure 11.21). These sounds correspond to the turbulent flow of blood as it spurts through the occluded artery during cuff deflation. A disadvantage of the auscultatory method is clinician bias: Different operators have different levels of skill in hearing and interpreting the Korotkoff sounds. Some automated machines use microphones to detect the sounds, but even those machines cannot overcome the difficulty of measuring blood pressure in cases where the late Korotkoff sounds are barely audible, such as in patients who have hypotension.

Another method to measure blood pressure non-invasively involves analyzing variations in arterial blood pressure as the cuff is deflated. This technique is called the oscillometric method (Figure 11.21). The oscillometric method uses pressure sensors—strain gauges or piezoelectric sensors placed on the patient's arm—to detect fluctuations. The oscillometric method can be easily automated; patients with hypertension can use automated devices at home to monitor their blood pressure in response to drug treatment.

11.4.3 Measurement of oxygen saturation in the blood

Proper blood O_2 and CO_2 levels are critical for maintenance of pH and healthy function of cells. Lung diseases including asthma, emphysema, severe

Figure 11.22 **Pulse oximeter.** Image provided by, and used with permission from, Nonin Medical, Inc.

pneumonia, and pulmonary edema (fluid in the lungs) can cause low oxygen levels in the blood, a condition known as **hypoxemia**. Blood oxygen is monitored for all patients during surgery and often for patients during recovery in the hospital. One way to measure oxygen content in blood (which is usually expressed as the percentage of O_2 saturation; $\%SpO_2$) is by taking an arterial blood sample: This procedure is more difficult than is collection of a venous blood sample because the arteries in the arm are deeper, and the muscular arterial wall is more difficult to puncture with a needle. As an alternative, O_2 saturation can be measured by a pulse **oximeter**. The pulse oximeter also monitors the heart rate, or pulse; in fact, measurement of the pulsations of blood in the tissue is essential for its operation. The oximeter device, which consists of a light emitter probe coupled with a photodiode, is attached to a fingertip or earlobe (or leg of a neonate) (Figure 11.22).

The pulse oximeter measures the difference in light absorbance at two wavelengths: one in the red portion of the visible region and another in the IR region. These measurements provide information on the ratio of oxyhemoglobin (hemoglobin [Hb] bound to oxygen) to deoxyhemoglobin (Hb without oxygen) in the tissue (Figure 11.23). Red light is not absorbed well by oxygenated blood, but IR light is absorbed. By using photodiodes to produce small light beams at these two wavelengths, and detectors to measure the fraction of each light

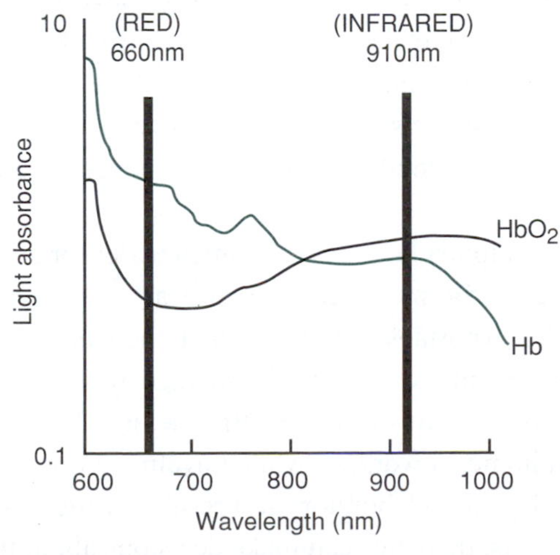

Figure 11.23 **Measurement of oxygen saturation in tissue.** The pulse oximeter measurement uses light absorbance at two different wavelengths: 660 nm (in the red portion of the visible region) and 910 nm (in the infrared region). Oxyhemoglobin (HbO_2) and deoxyhemoglobin (Hb) have different absorbance characteristics at these two wavelengths.

Figure 11.24 Glucose monitors allow point-of-care testing of blood sugar levels.

beam that passes through the tissue, the device calculates the ratio of red to IR absorbance. To isolate the absorbance caused only by flowing arterial blood, and not absorbance caused by other components of the tissue or venous blood, the electronics within the device track the pulsations in absorbance with arterial blood flow; the increase in absorbance that accompanies arterial pulsation is caused by the addition of arterial blood to the region being sampled. The ratio of red/IR absorbance (R/IR ratio) during pulsation is correlated to the $\%SpO_2$ of the arterial blood. The oximeter device displays the $\%SpO_2$ value, together with the pulse rate. Typically, an R/IR ratio of 0.5 corresponds approximately to $100\%SpO_2$, although the relationship between R/IR ratio and $\%SpO_2$ is nonlinear and different for devices from different manufacturers.

11.4.4 Measurement of blood glucose

The incidence of diabetes is increasing rapidly in the United States and other countries. People can now live long lives with diabetes because drugs such as insulin are available to allow them to control blood sugar levels. But insulin—and other drugs to control glucose metabolism—are often difficult to use, in that the dose that is needed varies with things that are difficult to estimate accurately, such as the effect of diet and exercise on blood glucose concentration. Therefore, many diabetics measure their blood glucose levels frequently, and adjust their medication accordingly.

Frequent self-testing of blood glucose is now much easier for diabetics because portable, accurate systems for measuring glucose are now available (Figure 11.24). In fact, several dozen portable U.S. Food and Drug Administration (FDA)-approved blood glucose monitors were on the market by 2007. These glucose meters are an example of a **point-of-care testing device**. Patients can use the glucose meters while at home, at work, or while traveling without the need to go to the clinic or wait for days for a laboratory test result. By measuring and tracking their own glucose levels, diabetics can make decisions about the appropriate timing and dosage of self-injected insulin to maintain normal blood glucose levels between 70 and 125 mg/dL.

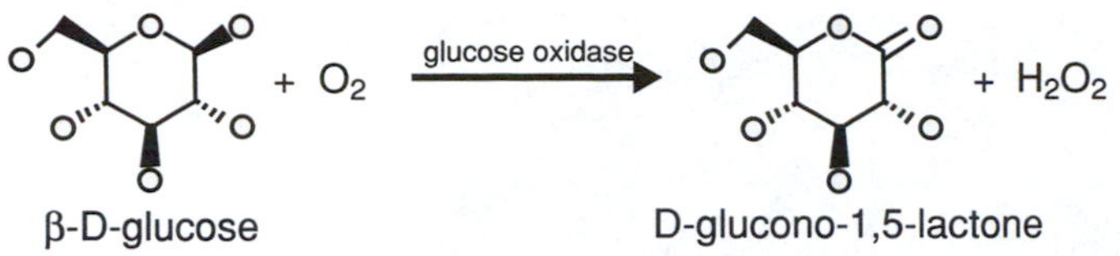

β-D-glucose + O_2 —glucose oxidase→ D-glucono-1,5-lactone + H_2O_2

Figure 11.25 Glucose oxidase catalyzes the conversion of glucose to a lactone with the production of hydrogen peroxide.

Use of the current home glucose monitors is relatively simple. First, the individual obtains a small sample of blood by pricking a finger; this drop of blood is placed on a test strip. The test strips are precoated with a glucose-sensitive enzyme (such as glucose oxidase, glucose dehydrogenase, or hexokinase, depending on the manufacturer). The enzymes on the test strip catalyze a reaction: For example, glucose oxidase catalyzes the oxidation of glucose to D-glucono-1,5-lactone and hydrogen peroxide (Figure 11.25). The resulting hydrogen peroxide reacts with a dye catalyzed by peroxidase to produce a blue color. Because of the action of the enzyme, the intensity of the color is directly proportional to the concentration of glucose that was present in the blood sample. The strip is placed into a small device, which uses light transmission through the colored sample to determine glucose concentration: This calculation of glucose concentration is usually based on a manufacturer's precalibration of the device with known samples of glucose. Some glucose meters use alternate methods of detection; for example, some meters measure the electrical resistance of the sample, which also correlates with glucose concentration.

Newer model glucose monitors are in development. Some of these models enable continuous monitoring via a subcutaneously implanted sensor, which eliminates finger sticks and allows diabetics to predict the need for food or insulin. It is possible that non-invasive systems for continuous glucose monitoring will be available someday. In the future, these continuous monitors will be coupled with drug-delivery systems (discussed in Chapter 13, in particular, in the discussion about Figure 13.2) to simulate the feedback control that is normally provided by beta cells in the pancreas (discussed in Chapter 6, about Figure 6.16).

11.4.5 Measurement of cardiac electrical potential by ECG

Some tissues in the body produce electrical potentials, such as the action potential of neurons and cardiac muscle cells, that are intimately related to their normal function. Measurement of electrical potentials that arise from these tissues can, therefore, be important indicators of tissue function. Electrical potentials can be measured invasively, by insertion of recording electrodes mounted on a catheter into the heart (Figure 11.17) or implantation of electrode arrays into the brain (Figure 11.26). These are risky and expensive procedures, which must be performed by highly specialized physicians in operating rooms or catheterization laboratories. Alternate methods are useful for screening for and diagnosing conditions that are not urgent enough to require brain surgery or invasive catheterization.

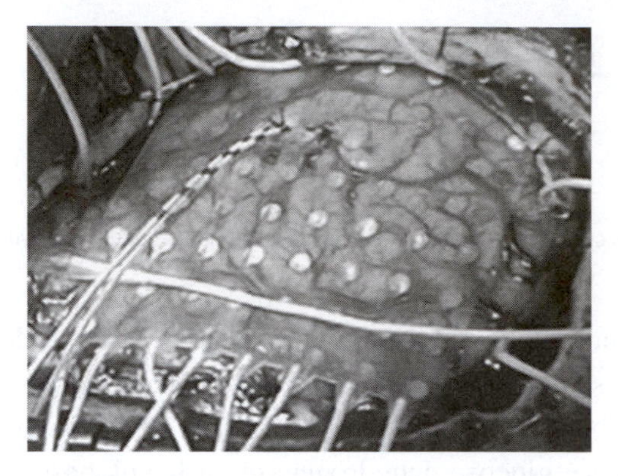

Figure 11.26 Surgical insertion of an electrode array into the brain of a patient with epilepsy.

The ECG (or EKG) provides a non-invasive, reliable method for screening and preliminary diagnosis of heart disease. The ECG does not measure the electrical activity of the heart directly, but measures the cumulative cardiac electrical activity by skin electrodes, or leads, that are attached to the surface of the body.

By analyzing the ECG tracings, or electrocardiogram, the physician can determine if there are abnormalities in the timing of events in the cardiac cycle or the amplitude of the electrical signal. A Dutch physiologist, Willem Einthoven, first adapted the string galvanometer—a thin, low mass, conducting fiber that carried the small current from the heart through a magnetic field for measurement—for the recording of ECGs in 1903; he was awarded the Nobel Prize in Physiology or Medicine in 1924 for that work. Modern direct ECG instruments use the same principles as the Einthoven string galvanometer, but now amplifiers and other electronics are used to enhance and clarify the signal.

A standardized 12-lead system is typically used for ECG recordings: Standardization of lead placement allows cardiologists using different instruments at different sites to compare their results. Because the ECG is an indirect measurement made at a distance from the source of the electrical activity, multiple leads are compared to produce multiple views of a complicated electrical event, which is occurring deep within a three-dimensional volume. The assembly of an ECG record from multiple leads is similar to the assembly of a three-dimensional image of an object from multiple two-dimensional photographs, each taken from a different angle. The configuration of the standard limb leads surrounds the heart, with the primary leads arranged in a pattern called Einthoven's Triangle (Box 11.3).

The ECG recording is displayed as voltage (usually in mV) as a function of time (usually in seconds) (Figure 11.27). The typical ECG waveform is analyzed in segments that comprise a waveform called a PQRST wave: a small P wave, a short delay, a large QRS wave, a second delay, and a small T wave. The phases of the cardiac cycle can be correlated with the structure of this wave (see Figure 8.16 and surrounding text for a description of the cardiac cycle). The P wave corresponds to the depolarization of the right and left atria. The PR segment, which includes the time delay between the P wave and the QRS complex, measures

Box 11.3 Non-invasive measurement of the electrical activity of the heart

Cardiac electrical activity can be monitored from the surface of the body by a technique called **electrocardiography** (ECG, or EKG from the German word *elektrokardiogramm*). The ECG recording is a reflection of the external currents around cardiac muscle cells, which depend on the number of cells that are synchronously showing electrical activity. Because of the long distance between the source of the signal and the measurement, the voltage measured by ECG of ~1 mV is much less than the voltage changes in individual cells, which are ~100 mV (recall Figure 8.15).

The ECG requires an instrument for measuring the small currents that are associated with heart activity. Einthoven achieved the breakthrough in measurement of these small currents with his string galvanometer. Today, skin electrodes are attached to the skin with the help of an adhesive and conductive gel layer, which ensures good electrical contact with the skin. These electrodes are connected to electronics for processing and recording of the ECG signals.

Twelve leads—or signals—are collected for a complete ECG recording. Six of these leads are produced from three electrodes placed on the arms and left leg: These are called the standard limb leads, which are placed in a pattern called Einthoven's triangle (see Figure Box 11.3a). The other six leads, called the chest leads, are placed on the chest, in different positions over the heart.

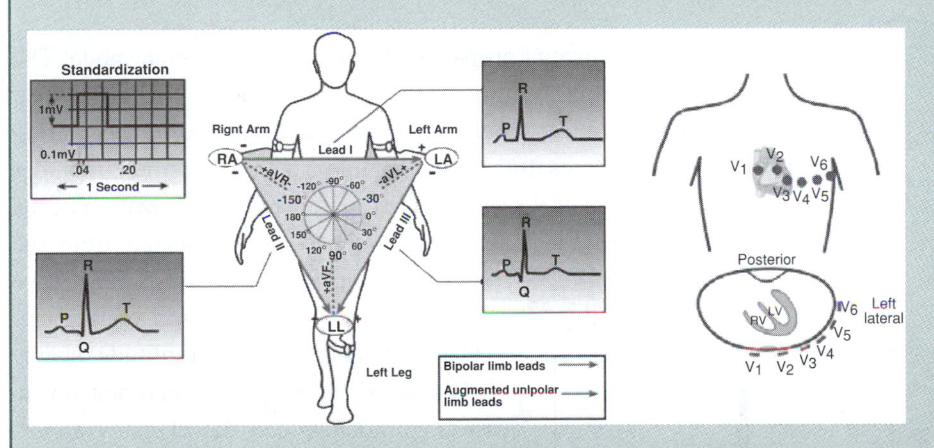

Figure Box 11.3A Surface placement of the standard electrodes in electrocardiography. Left: The standard limb leads are placed in a pattern surrounding the heart that allows measurement of different aspects of heart electrical activity. With electrodes attached to the right arm (RA), left arm (LA), and left leg (LL), several different potentials can be examined. Lead I provides the potential difference between RA (negative) and LA (positive); lead II provides the potential difference between RA (negative) and LL (positive); and lead III provides the potential difference between LA (negative) and LL (positive). These leads are all *bipolar*, in that a potential difference between two of the limb electrodes is measured. In addition, augmented leads are produced with each of electrode (RA, LA, and LL) serving as a single positive. These leads, called augmented leads (aVR, aVL, and aVF), are *unipolar* and use a combination of the other two electrodes as a composite negative. Right: The six chest (or precordial) leads are placed in a pattern on the chest overlying the heart. Each of these leads (V_1 through V_6) is unipolar.

(continued)

Box 11.3 (*continued*)

Cardiologists can use the information from the ECG to diagnosis arrhythmias (disorders in the cardiac rhythm), heart block (disorders in conduction of signals through the cardiac conduction system), and other disorders of the heart (see Figure Box 11.3b).

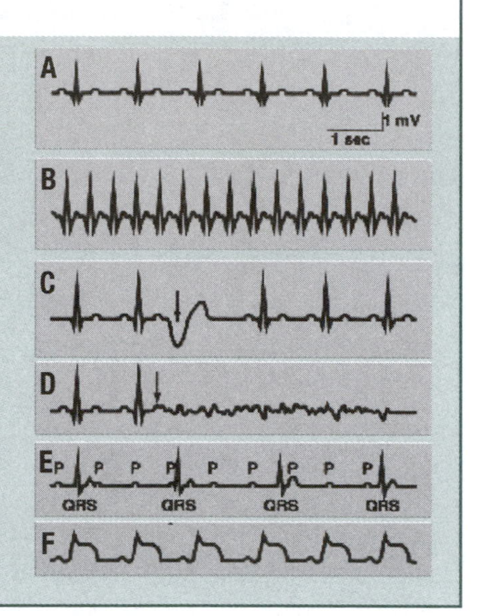

Figure Box 11.3B Examples of heart conditions that can be detected with an electrocardiograph (ECG). A. Normal rhythm. **B.** Tachycardia, an abnormally rapid heart rate. **C.** Extrasystole, premature initiation of a ventricular contraction. **D.** Ventricular fibrillation, uncoordinated ventricular contraction. **E.** Heart block, impulses from the atria not reaching the ventricle. **F.** Myocardial infarction, death of cardiac tissue causing irregular activity.

the time delay for action-potential propagation at the atrioventricular node. The QRS complex, the most striking feature of the ECG, corresponds to the electrical depolarization of the ventricles; repolarization of the atria also occurs during this time period and contributes to the QRS complex. The ST segment includes the time between completion of ventricular contraction and repolarization of the ventricles; the blood is ejected from the ventricles during this period. The T wave corresponds to repolarization of the ventricles. By analyzing the amplitude, shape, and timing of the events recorded on an ECG, physicians can detect many abnormal heart conditions.

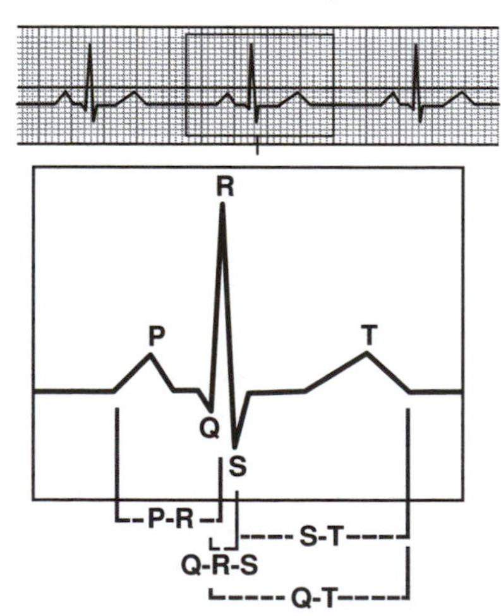

Figure 11.27 **Electrocardiograph (ECG) for measurement of cardiac electrical activity.** When printed on standard ECG paper, each block indicates a time interval of 0.04 seconds, and a voltage interval of 0.1 mV. Therefore, the PR interval in this diagram is ~0.22 seconds and the amplitude of the R wave is ~1 mV above the baseline.

Electrical potentials can also be measured in other tissues. Electrical potentials at the surface of the brain or outer surface of the head are recorded using an electroencephalograph (EEG). A normal EEG reveals continuous oscillation of electrical activity because of activity of neurons in the brain. Abnormal patterns in frequency and amplitude of the EEG waves are found in patients with epilepsy and other diseases. An electroretinograph uses special electrodes, mounted on contact lenses, to

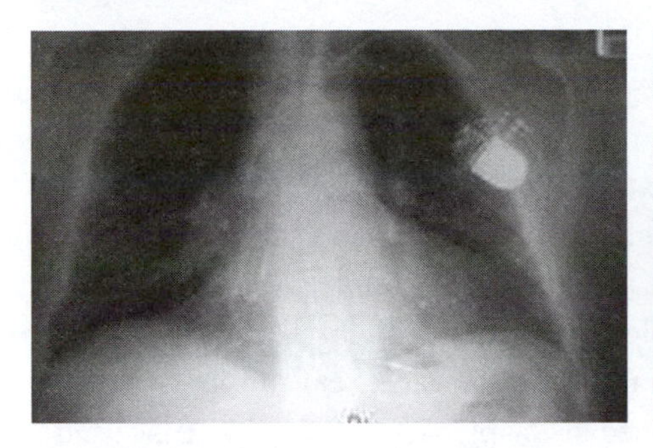

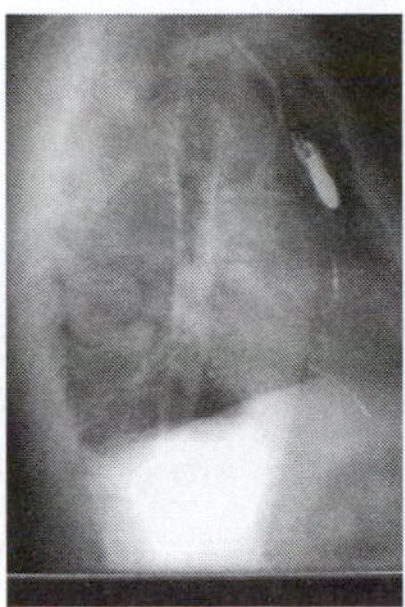

Figure 11.28 **Placement of a cardiac pacemaker.** These x-ray images show the placement of a cardiac pacemaker in a patient. The device is implanted in the chest, usually just below the collarbone. Insulated wires, which carry the current produced by the pacemaker, connect the device to the heart. Photos courtesy of www.helenacardiology.com.

measure the electrical response of photoreceptors to light. An electromyograph measures the electrical activity of skeletal muscle fibers. Often, needle electrodes are used to reach the muscle by penetrating skin; electrical activity is measured as the subject is asked to contract and relax the muscle. Abnormal muscle electrical activity is found in diseases such as muscular dystrophy, inflammation of muscles, pinched nerves, peripheral nerve damage, and amyotrophic lateral sclerosis.

11.4.6 Devices for electrical stimulation of tissues

Electrically excitable tissues—heart, brain, and muscle—can be stimulated by the controlled introduction of current. For example, in a patient whose heart rate is too slow (a condition called **bradycardia**), a pacemaker can restore normal cardiac rhythm. A pacemaker is a small, battery-powered device that is implanted in the chest; insulated electrical leads connect the pacemaker to the lining of the heart (Figure 11.28; a photograph of a pacemaker outside the body is shown in Figure 15.14). The pacemaker can both detect activity in the heart and induce activity. The device monitors the heart rate of the patient and, when it detects an abnormally long interval between beats, it delivers electrical pulses to generate extra heartbeats. The pacemaker initiates cardiac conduction in place of the sino-atrial node, the heart's natural pacemaker. The first pacemakers were implanted in patients in 1958. Now, more than 250,000 are implanted each year in the United States.

In the procedure called **cardiac defibrillation**, a shock of electricity is used to convert an abnormal cardiac rhythm—usually, ventricular fibrillation, in which the ventricles are contracting chaotically (Box 11.3)—to a normal rhythm. The first, and still the most common, defibrillators are external: Paddles, which are connected to a box of electronics, are placed on either side of the patient's chest, and the shock is supplied (see Figure 1.2). A number of different types of defibrillators are now available, including automated external defibrillators (AED), which first record the patient's ECG and determine, automatically, if a shock is appropriate to treat the rhythm of the patient's heart. AEDs can be placed in public locations because they do not require medical professionals for operation. AEDs—located in train stations, airports, or shopping malls—provide an opportunity for immediate resuscitation of victims of sudden-onset ventricular fibrillation: Before these devices were available, portable defibrillators in ambulances were the most readily available source. In addition, totally implantable defibrillators are available for patients at high risk for defibrillation.

Other devices are already in use for electrical stimulation of tissue. The deep-brain stimulator (mentioned in Chapter 1, see Figure 1.10) is used to treat patients with Parkinson's disease: Direct electrical stimulation of tissue in a particular region of the brain reduces the severity of some of the symptoms of Parkinson's disease, including tremors and other movement disorders. Similar approaches are useful for treating some kinds of chronic pain and may be useful in other diseases in the future.

11.5 Instruments in the research laboratory

Many biomedical engineers work in research laboratories, designing and testing new medical technologies, or applying engineering methods to understand human physiology. Some of the instruments that engineers—and all other scientists—use for their research work are described in this section.

11.5.1 Measurement of pH

Measurements of pH are performed almost daily in a biomedical research laboratory. For example, pH control is essential for the solutions used for cell culture and for many other fluids prepared and used in the laboratory; often these solutions must be adjusted to a pH of 7.4. General concepts of acid–base balance and pH are introduced in Chapter 2. Recall that pH is another way of expressing the hydrogen ion concentration in a solution; it is defined as the negative log of the hydrogen ion concentration:

$$pH = -\log[H^+]. \tag{11.4}$$

Most current pH meters consist of a single electrode, which contains both a reference and sensing electrode, and electronics that permit an easy-to-read digital display (Figure 11.29). The pH sensor is a potentiometric sensor: A voltage is generated without a current flow (contrast this type of sensor with the Clark oxygen sensor, introduced earlier in this chapter). The sensing electrode is an ISE with a glass membrane tip that is permeable to H^+ ions. When exposed to a solution of unknown pH, the ISE produces a voltage.

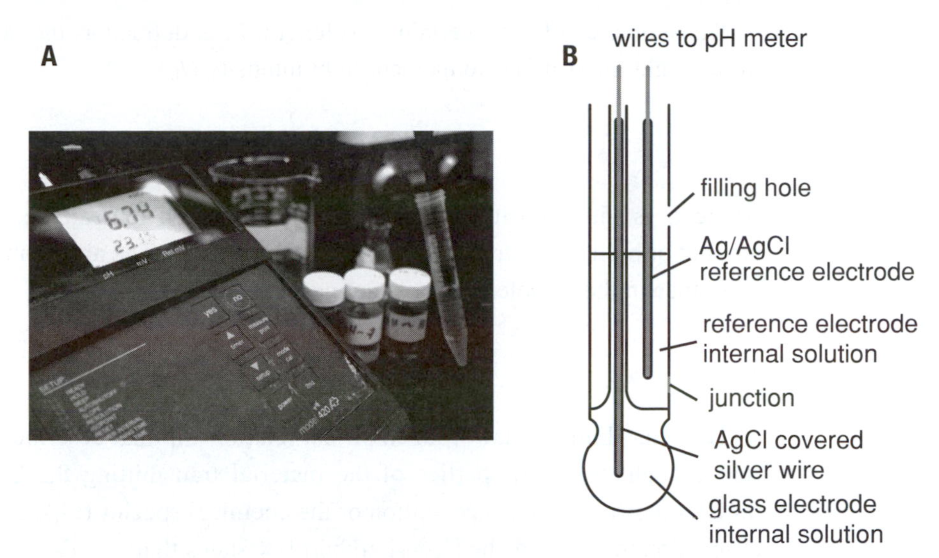

Figure 11.29 **pH meters are among the most common instruments in biomedical research laboratories. A.** The electronics provide for digital display of the pH of a solution. **B.** The sensing element of the pH meter is typically a combination electrode, in which the sensing and reference elements are housed in the same glass cylinder.

11.5.2 Spectrophotometry

A spectrophotometer is an instrument for measuring the intensity of light. Most instruments are designed to allow for the measurement of the extent of light transmission through a biological sample, often fluid that is contained in a transparent cuvette (Figure 11.30). Many biological molecules absorb light, but they absorb light better at some wavelengths than others: For example, Hb absorbs light throughout the range of wavelengths from 600 to 1000 nm, but it absorbs best near 600 nm (Figure 11.23). Therefore, spectrophotometry is a useful technique for measuring the concentration of biological molecules in solution.

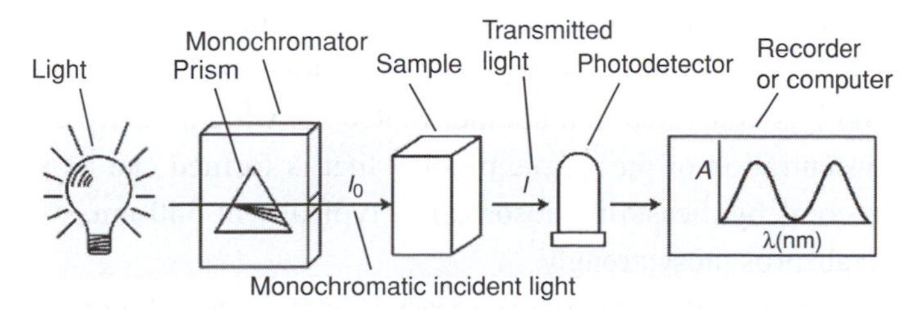

Figure 11.30 **A typical design for a spectrophotometer.** Most spectrophotometers contain a light source that produces high-intensity light at a range of wavelengths that span the visible and ultraviolet spectrum. The light shines through a monochromator, which allows the passage of only specific wavelengths. In this way, light of a selected wavelength passes through the sample, which is typically held in a transparent quartz cuvette. The fraction of light that passes through the sample is measured by a photodetector on the other side of the sample.

Transmittance (T) at a certain wavelength, λ, is defined as the ratio of transmitted light intensity (I) to incident light intensity (I_0):

$$T_\lambda = \frac{I}{I_0}, \tag{11.5}$$

where I_0 is the intensity of light that is incident on the sample, and I is the amount of light that is transmitted through the sample. The absorbance (A) of the substance in the sample cuvette is defined as:

$$A_\lambda = \log\left(\frac{I_0}{I}\right) = -\log T_\lambda. \tag{11.6}$$

The Beer–Lambert law is an empirical relationship that describes the absorption of light to the properties of the material transmitting the light. If c (in units of mol/L) is the concentration of the chemical species responsible for light absorption in the fluid, the Beer–Lambert law states that:

$$A_\lambda = \varepsilon c l, \tag{11.7}$$

where l is the path length (or distance that the transmitted light travels through the sample, typically 1 cm in most spectrophotometers), and ε is a proportionality constant known as the molar extinction coefficient (L/mol-cm). The absorbance A is also called the optical density, or *OD*. The extinction coefficient is dependent on the chemical species and the wavelength of the incident light. Many common biochemicals absorb light at UV and visible wavelengths: A higher extinction coefficient indicates more light absorbance per quantity of chemical present.

Spectrophotometric readings in the visible wavelength range (380–780 nm) are used to detect **colorimetric reagents**. These reagents are molecules that contain regions called **chromophores**, which absorb light of certain wavelengths. Chromophores, therefore, are the parts of molecules that are responsible for their color. Many molecules are colored. Here, chromophores are used to measure the results, or end points, of biochemical reactions. For example, in a common laboratory experiment for measuring cellular growth in culture, the molecule 3-(4,5-dimethylthiazol-2-yl)-2,5-diphenyltetrazolium bromide (MTT) is added into a cell culture. In its initial state, MTT is yellow. In the mitochondria of living cells, MTT is converted into another molecule, which is purple (Figure 11.31). The concentration of the purple product that is formed can be measured using spectroscopy, by measuring absorbance, typically at 540 nm, where the purple product absorbs most strongly.

A number of routine laboratory measurements—including protein assays, cellular assays, enzyme-linked immunoassays, and enzymatic assays—use measurement of absorbance to quantify concentration. For measurements in the visible range, the typical assay involves measurement of the *OD* of known samples to produce a standard curve in the linear range. For example, most proteins absorb

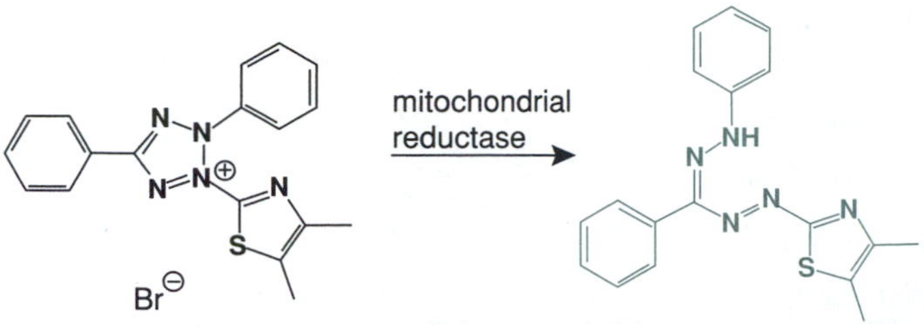

Figure 11.31 **The basis of a common colorimetric assay.** MTT (3-(4,5-dimethylthiazol-2-yl)-2,5-diphenyltetrazolium bromide; left), a tetrazole structure, is converted into the purple formazan (right) in the presence of living cells. By adding a known amount of MTT to a cell culture, and using spectroscopy to measure the amount of the purple product that is formed after a fixed amount of time, the number of living cells in the culture can be calculated.

UV light at a wavelength of 280 nm; this UV absorbance is caused by aromatic rings on amino acids such as tyrosine, so the amount of absorbance varies from protein to protein depending on the amino acid composition of the protein. To measure the concentration of any particular protein in solution, the first step is to construct a standard curve by preparing a series of solutions of known protein concentration—over a range of 0.05–1 mg/mL, for example. The A_{280} of each of these solutions is measured and plotted as a function of concentration (Figure 11.32).

SELF-TEST 4 If a solution of unknown concentration of bovine serum albumin (BSA) gives an A_{280} reading of 0.4, estimate the concentration of BSA, using Figure 11.32.

ANSWER: ~0.6 mg/mL

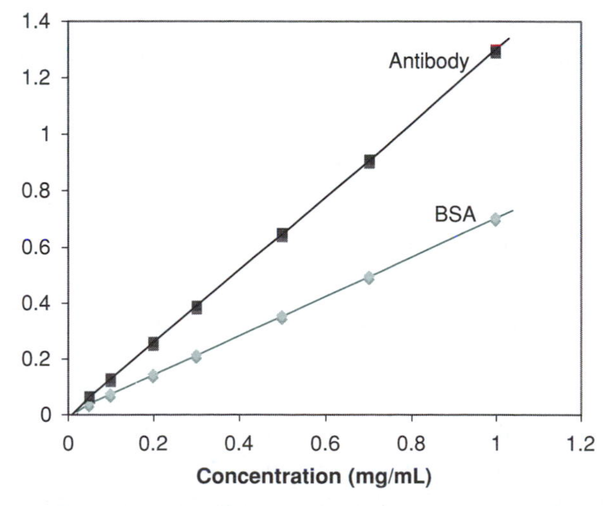

Figure 11.32 **Assays for protein concentration in solution.** The absorbance at 280 nm, A_{280}, is shown for known concentrations of pure solutions of bovine serum albumin (BSA) and human immunoglobulin G (IgG) (Antibody) in saline solution. Lines indicate linear fits to the measured values.

Table 11.3 Extinction coefficients for nucleic acids at 260 nm, in units of mL/(ug-cm)

Double-stranded DNA	0.02
Single-stranded DNA	0.029
Double-stranded RNA	0.025
Oligonucleotides	0.04

Notes: Calculated from http://www.piercenet.com/Proteomics/browse.cfm?fldID=17859183-EC46-4FA1-A5EE-1E257534AD13#table1. DNA: deoxyribonucleic acid; RNA: ribonucleic acid.

UV absorbance readings can also be used to determine the concentration of nucleic acids in solution: Extinction coefficients for nucleic acids are high enough to allow determination of concentrations as low as 1 μg/mL using A_{260} (Table 11.3).

11.6 Biosensors

A **biosensor** is an analytical device that uses a biological sensing element—such as a protein, cell, or section of tissue—that is coupled with a physical or chemical transducer. Biosensors are used to detect the presence of specific agents, which might be chemical groups, DNA, or other biochemical compounds. They typically gain their specificity—or their ability to detect only the desired agent—from the biological sensing element. In this way, biosensors harness the specific interactions of biology to make a technological measurement.

In medicine, it is easy to understand the value of sensors that can specifically measure the presence of viruses, proteins, or drugs from within a complex biological sample (such as a blood or urine sample or a tissue sample obtained by biopsy). The challenge is to find reliable techniques for coupling the biological element to the physical sensor and to find methods for doing this to achieve measurements of high sensitivity (i.e., measurements that can detect small amounts of the agent or substance of interest).

Biosensors are often built from sensing elements that are already well established, such as pH sensors or optical sensors (Figure 11.33). Biological molecules—such as antibodies or enzymes—are often used as the sensing element because the specificity of these molecules is well established and because the characteristics of many enzymes and antibodies are already well understood. The glucose sensor that uses glucose oxidase for detection (discussed earlier in this chapter) is an example of an enzyme-based biosensor. The home pregnancy test kit described earlier in the book is an example of an antibody-based sensor because it uses an antibody that binds specifically to human chorionic

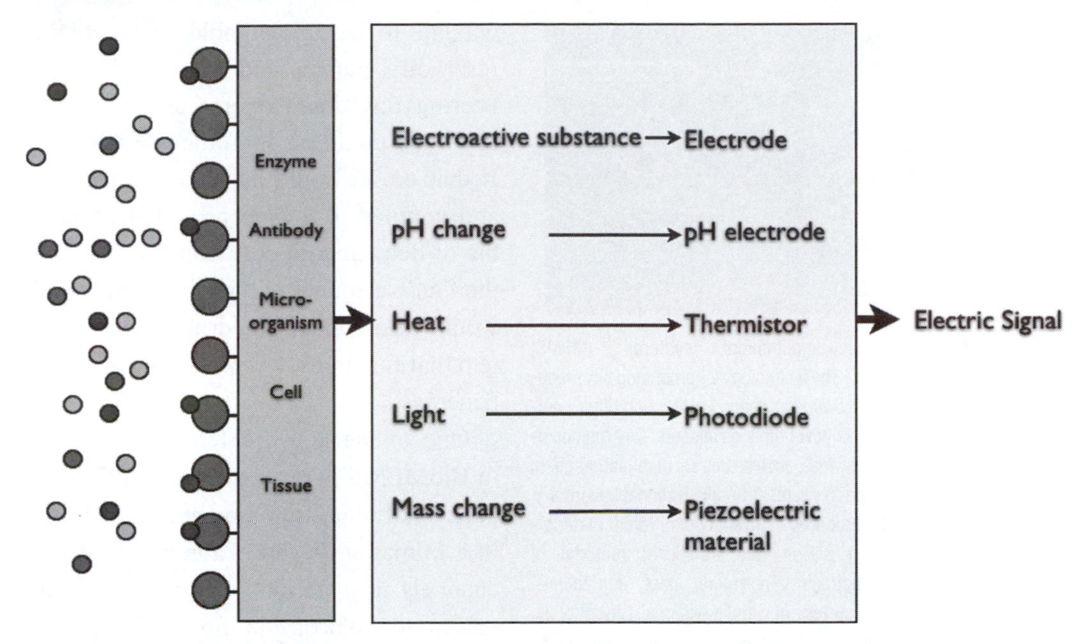

Figure 11.33 **Biosensor design.** A biosensor uses a biological sensing element (such as an enzyme or tissue) to create a change that can be converted into an electrical signal. Often, the biological sensing element is connected to one of the other sensors (such as a thermistor or photodiode) discussed earlier in the chapter.

gonadotropin (hCG), a protein hormone abundant in pregnancy. In the case of the home pregnancy test kits, binding of hCG to the antibody produces a color change that can be detected by the human eye.

11.7 Biomicroelectromechanical systems and lab-on-a-chip devices

It is now possible to build sophisticated miniature machines. Miniaturized systems that combine mechanical and electrical components are sometimes called *m*icro*e*lectro*m*echanical *s*ystems, or MEMS (Figure 11.34). Most MEMS devices are made by variations on techniques that were first developed by the microelectronics industry to build components for computers and other electronic devices. Several key fabrication technologies—such as molding and plating, photolithography, wet etching (the selective removal of material using chemicals), and dry etching (the selective removal of material using ion beams)—have become routine and are now available in laboratories and manufacturing sites throughout the world. Some techniques—particularly molding methods involving polymers such as poly(dimethylsiloxane) (PDMS, or silicone)—are simple enough that they can be performed in most laboratories.

MEMS devices can also be coupled to biosensors, creating a new class of devices called BioMEMS. For example, imagine a MEMS device that is small

Figure 11.34 **Microelectromechanical systems (MEMS) devices.** The technology for producing tiny machines, with working gears, valves, and other components is now highly developed. Tiny multicomponent devices, such as this set of miniature gears produced in silicon at Sandia National Laboratory, can be produced in a variety of different materials including silicon (the traditional material of microelectronics), polymers, glass, and metals. Image courtesy of Sandia National Laboratories, SUMMiT™ Technologies, www.mems.sandia.gov.

enough to be implantable and contains reservoirs, pumps, and valves for administering tiny doses of a drug solution at rates that could be continuously adjusted. If that device could be filled with insulin and coupled to a biosensor that is capable of detecting glucose concentrations in the fluid surrounding the device, the result would be a BioMEMS drug-delivery system that functions to replace the endocrine pancreas.

One of the key problems encountered in BioMEMS is the movement of fluid, or packets of fluid, from one place to another in a miniature device. Fabrication of tiny channels in materials (Figure 11.35) and valves to control the flow of fluids in devices (Figure 11.36) are becoming routine. The need for fluid handling in BioMEMS devices has motivated new discoveries in **microfluidics**, the science and technology of movement of small volumes of fluid through channels of small diameter.

These tools—**microfabrication**, BioMEMS, and microfluidics—have led to the concept of **lab-on-a-chip**. The idea of a lab-on-a-chip is to miniaturize the diverse procedures that are needed to perform a laboratory analysis, from the beginning to the end. Many clinical laboratory methods are time-consuming and labor-intensive, and require significant amounts of sample, reagents, and expensive instrumentation systems. For example, Figure 11.4 shows a device that is capable of accepting a drop of blood and detecting the presence of the chemical warfare agent sarin. To accomplish this, the device has separate, very small chambers for preparation of the blood, chemical reaction, and detection of the final product. There are many examples of common procedures in medicine that could be reduced to a lab-on-a-chip device, such as the measurement of blood chemistry values from a drop of blood (rather than the tube of blood that is now needed).

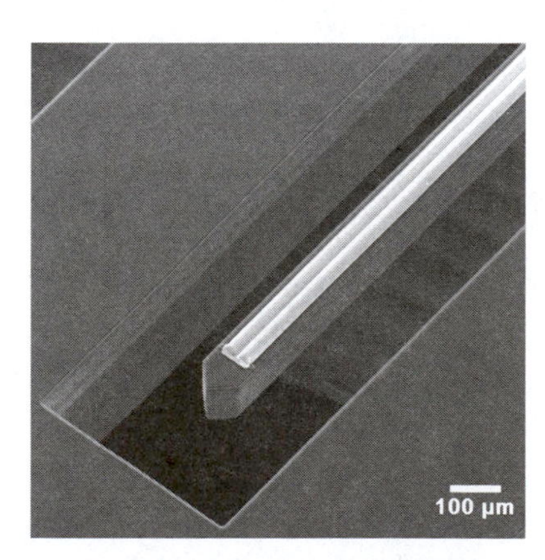

Figure 11.35 **Microfluidic channels enable the movement of small volumes of fluid through biomicroelectromechanical systems (BioMEMS) devices.** This image shows two parallel microfluidic channels on a probe that was designed for infusion of drugs into the brain. Reprinted from Neeves KB, Sawyer AJ, Foley CP, Saltzman WM, and Olbricht WL. *Brain Research.* 2007;1180:121–132, with permission from Elsevier.

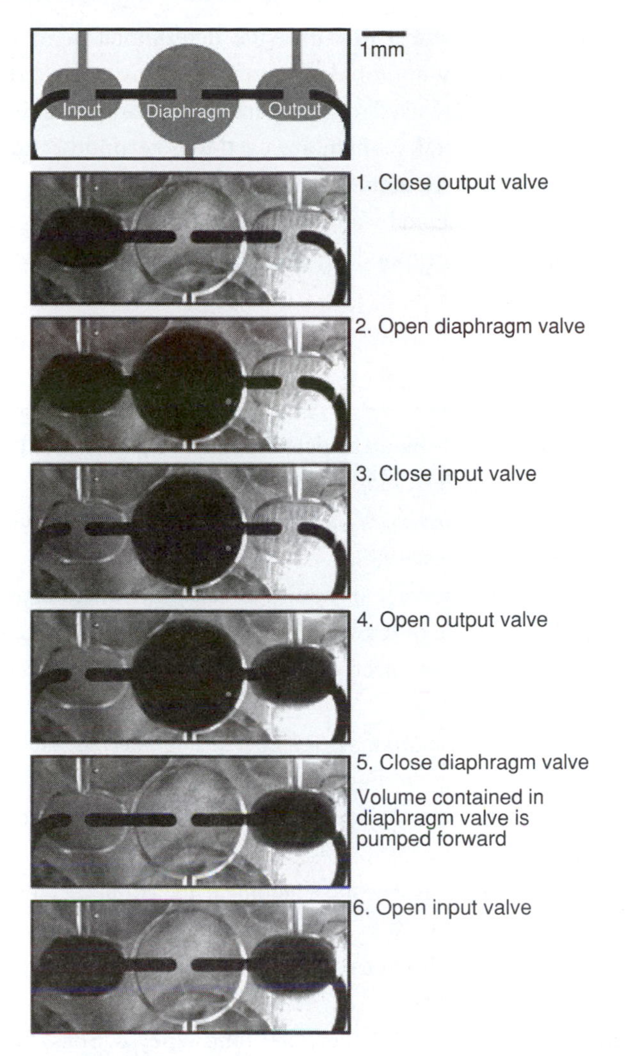

Figure 11.36 **Operation of valves in a microfluidic device.** This sequence of images shows the movement of dye in a 340-nL fluorinated ethylene propylene (FEP) Teflon pump at each of the six steps in a single pump cycle. At the optimal actuation rate for this pump, this cycle repeats every 90 ms and pumps 1.1 μL s-1. From: Grover WH, von Muhlen MG, Manalis SR. *Lab Chip*. 2008;8:913–918. Reproduced by permission of The Royal Society of Chemistry.

Summary

- Bioinstrumentation has changed medical practice by improving care of patients in operating rooms, hospitals, doctors' offices, and—increasingly—at home.
- Measurement systems require a sensor, to convert the measured input to an electrical signal, as well as processors and electronics.
- The technology for sensing temperature, mechanical strain, electrical potential, chemicals (such as oxygen), and light are well developed and dependable.
- Sensors can be deployed deep in the body by the use of catheters and other medical devices.

- ECG measurements provide a non-invasive determination of the electrical behavior of the heart, allowing rapid diagnosis of many cardiac disorders.
- A variety of instruments are used routinely in biomedical research laboratories, including spectrophotometers, which are used to determine concentrations of molecules in solution via their ability to absorb light.
- Biosensors can be constructed by adding a biological sensing system—usually a molecule or a cell—that responds in a predictable and measurable way to the parameter of interest.

FURTHER READING

Bashir R. BioMEMS: State-of-the-art in detection, opportunities and prospects. *Adv Drug Deliv Rev*. 2004;56(11):1565–1586.

Garson A Jr. *The Electrocardiogram in Infants and Children: A Systematic Approach*. Philadelphia: Lea & Febinger; 1983.

Hooper VD, Andrews JO. Accuracy of noninvasive core temperature measurement in acutely ill adults: the state of the science. *Biol Res Nurs*. 2006;8(1):24–34.

Kissinger PT. Biosensors—a perspective. *Biosens Bioelectron*. 2005;20(12):2512–2516.

Rizzoni G. *Principles and Applications of Electrical Engineering*. 4th ed. New York: McGraw-Hill; 2003.

Situma C, Masahiko H, Soper SA. Merging microfluidics with microarray-based bioassays. *Biomol Eng*. 2006;23(5)213–231.

Smith JP. Medical and biological sensors: a technical and commercial review. *Sensor Review*. 2005;25(4):241–245.

Walker HK, Hall WD, Hurst JW, eds. *Clinical Methods*. 3rd ed. Stoneham, MA: Butterworth Publishers; 1990.

Webster JG. *Bioinstrumentation*. Hoboken, NJ: John Wiley & Sons, Inc.; 2003.

Webster JG. *Medical Instrumentation: Application and Design*. 3rd ed. Hoboken, NJ: John Wiley & Sons, Inc.; 1998.

USEFUL LINKS ON THE WORLD WIDE WEB

http://www.allaboutcircuits.com/
All About Circuits

http://www.chm.davidson.edu/ChemistryApplets/spectrophotometry/EffectOfConcentration.html
Beer–Lambert law simulator

http://medlib.med.utah.edu/kw/ecg/
ECG Learning Center

http://www.ibiblio.org/obp/electricCircuits/
Lessons in Electric Circuits

Profiles in BME: Bill Hawkins

Bill Hawkins was elected President and Chief Operating Officer of Medtronic, Inc., in May 2004. Previously, he served as Senior Vice President and President of Medtronic's Vascular business.

Hawkins joined Medtronic in January 2002 from Novoste Corporation, where he had been President and Chief Executive Officer since 1998. Earlier positions included Corporate Vice President and President of the Sherwood–Davis & Geck organization of American Home Products; President of the Ethicon Endo-Surgery organization of Johnson & Johnson; President of the Devices for Vascular Intervention group and U.S. Operations, Guidant Corporation; and several increasingly responsible executive positions culminating in the presidency of the IVAC organization for Eli Lilly & Company. He began his medical technology career with Carolina Medical Electronics in 1977.

Hawkins received his Bachelor of Science degree in electrical and biomedical engineering from Duke University in 1976 where he also conducted medical research in pathology. He received a Master of Business Administration degree from the Darden School of Business, University of Virginia, in 1982. Hawkins is a member of the Board of Visitors of the Engineering School of Duke University and the Guthrie Theatre Board; and a Trustee of the University of Virginia Darden School Foundation.

Bill Hawkins writes:

I grew up in Durham, North Carolina, where my father was mayor. I knew early on that the one thing I did not want to do was to go into politics. Although I didn't follow in his footsteps, my father has been an inspiration to me as a mentor, a friend, and an individual who demonstrates the highest integrity. Living near Duke University, with its formidable medical center, I developed an interest in medicine and ultimately attended Duke.

I graduated from Duke with a degree in electrical and biomedical engineering and started my first job with Carolina Medical Electronics (CME). One reason I was interested in working at CME stemmed from my biomedical studies at Duke under Professor Olaf von Ram, who was working on ultrasound. CME had acquired the rights to the first continuous wave Doppler ultrasound right around the time I graduated. I was fascinated by this new technology that had the potential to diagnose stroke in the carotid arteries. I learned a lot about myself and business at CME and, after three and a half years, decided to go back to business school. I wanted to learn more about how companies become successful. I earned my MBA from the Darden School of Business at the University of Virginia.

Equipped with a business degree, I decided that I wanted to go to a company that had a proven track record for success and leadership development. After a summer internship with IVAC (owned by Eli Lilly & Co.) in San Diego, I started work for Lilly in Indianapolis and stayed there

(*continued*)

Profiles in BME (*continued*)

for 14 years. Now, fast-forward to today, after having broad medical device experiences at Lilly, Johnson & Johnson, American Home Products, Guidant, and Novoste, to my role as President and Chief Operating Officer (COO) at Medtronic. What I've learned during my career is that finding a company that matches your personal values is extremely important. It was appealing to me to work at Medtronic, where I could impact patients around the world.

Medtronic has many great businesses, and my job as COO is to make the whole of the company greater than the sum of its parts. A key part of that is getting the right people in the right jobs. In the end, success depends on having the right people in place to leverage the capabilities of the company and enable the businesses to be successful. This leverage is especially important today because we are taking a broader view toward managing different disease states through the convergence of devices, biologics, and information management. I feel fortunate that my broad range of experiences—at large, small, and start-up companies in medical technology—are all relevant to my role in moving Medtronic into the future.

http://www.devicelink.com/mddi/
Medical Device Link

http://www.ncbe.reading.ac.uk/NCBE/PROTOCOLS/menu.html
National Centre for Biotechnology Education—Practical Protocols

http://www.pacemakerproject.com/
Pacemaker Project

http://www.sensorland.com/
Sensorland (The link "How they work" contains useful descriptions of sensors.)

KEY CONCEPTS AND DEFINITIONS

angioplasty a catheterization procedure in which a balloon at the tip of a catheter is inflated to push arterial plaque against the vessel wall to increase the diameter of the flow path

auscultatory method a method of measuring blood pressure by listening for the appearance and disappearance of Korotkoff sounds within the body

biosensor an analytical device that uses a biological sensing element—such as a protein, cell, or section of tissue—that is coupled with a physical or chemical transducer

bradycardia an abnormally slow heart rate

cardiac defibrillation a procedure in which a shock of electricity is used to convert an abnormal cardiac rhythm to a normal rhythm

catheter a long, thin tube made of a biocompatible, flexible material, usually some kind of plastic. It can be equipped with sensors and other devices to accomplish procedures deep within the body.

catheterization a procedure in which a catheter is inserted into an easily accessible artery or vein, usually in the arm or leg, and is guided to the heart

chromophore a part of a molecule that absorbs light of a certain wavelength, thereby giving the molecule its characteristic color

colorimetric reagents molecules that contain chromophores

current a flow of positive electrical charge

current clamp a technique in electrophysiology in which the changes in voltage produced by a cell are measured by keeping the current flowing through the recording electrode constant

driving force when used in reference to an electrical circuit, this term denotes the voltage provided by the power source

dynodes one unit within a cascade of electrodes of increasing positive charge, used in a photomultiplier tube to amplify the current produced by the photomultiplier

electrical potential a measure of the amount of potential energy per unit of charge; voltage

fever elevation of body temperature

flows movements of air or liquid

fluorescent emitting light of a certain wavelength upon excitation by light of another wavelength

hypertension chronic high blood pressure

hypotension chronic low blood pressure

hypoxemia low levels of oxygen in the blood

Korotkoff sounds sounds that can be detected during the measurement of blood pressure (between systolic and diastolic pressures)

lab-on-a-chip a microfabricated device, integrating several laboratory processes on a single chip

microfabrication manufacturing microstructures by methods basically applied in microelectronics, such as micromachining, microlithography, injection molding, and embossing

microfluidics transporting and manipulating minute amount of fluids through microchannel manifolds (microscopic fluid flow)

oximeter a device that measures the oxygen saturation of the blood, as well as the pulse

parallel circuit a circuit in which the current has more than one path to follow

patch clamp an electrophysiology technique in which a micropipette is used to remove or isolate a small section of a cell membrane. The technique allows the investigator to study the properties of the membrane within the patch, such as the activity of individual ion channels.

photodiode a device made of a semiconductor material that generates a current when it is struck by a photon of light

photomultiplier a device that converts light energy to electrical energy with the help of a photocathode and amplifies the signal using a series of dynodes

piezoelectric having the capacity to generate an electric current when placed under pressure

point-of-care testing device a device that takes measurements on a patient at or near the site where he or she is receiving care, such as a doctor's office or medical clinic

potential difference the difference in the amount of work necessary to move a unit of electric charge from one point to another; also called the voltage drop

processor the part of an instrument that modifies the initial signal picked up by the sensor

pyroelectricity the ability of certain materials to generate electrical potential in response to heat

pyrogens molecules released by macrophages that induce a fever; their release into the blood causes the hypothalamus to release prostaglandins, thereby raising the body's temperature set point

resistance the degree to which a material opposes the flow of an electric current, usually measured in ohms

resistor a component in a circuit that modifies the current flow by increasing the resistance

Seebeck voltage the voltage difference produced when a circuit exhibiting the Seebeck effect, such as a thermocouple, is broken

sensor the part of an instrument that detects the input

series circuit a circuit in which the components are connected sequentially such that the current must flow through every part

signal an indicator, usually an electrical signal, that varies in a predictable and reliable way, with changes in the input parameter measured by an instrument

stent a wire mesh cylindrical support used to maintain the width of a bodily cavity, such as a blood vessel

strain the fractional change in the size of a material in response to an applied force

strain gauge an instrument used to measure strain

thermistor homogeneous composites of dissimilar metals that form thermally sensitive resistors

thermocouple a temperature-sensing device formed by fusing two dissimilar metals at two junctions, thereby creating a current if the junctions are at different temperatures

voltage clamp a technique in electrophysiology that measures the current produced by a cell when the voltage drop across its membrane is held constant

NOMENCLATURE

ΔV	Voltage drop	
I	Current	Amperes
R	Resistance	Ohms
R_T	Measured electrical resistance	Ohms
T	Temperature	Kelvin
R_0	Resistance at reference temperature	Ohms
E	2.71828	—
T_{degC}	Temperature in Celsius	°C
T_{degF}	Temperature in Fahrenheit	°F

T	Transmittance	
T_λ	Transmittance at a certain wavelength (λ) of light	
I	Transmitted light intensity	
I_0	Incident light intensity	
A	Absorbance	
A_λ	Absorbance at a certain wavelength (λ) of light	
l	Path length	cm
c	Speed of light	m/s

GREEK

β	Coefficient characteristic of the material	
ε	Strain (mechanical) or	—
	Molar extinction coefficient (in measuring absorbance)	L/mol-cm

QUESTIONS

1. Describe the difference between an electrode and a lead in ECG. How can three electrodes be used to generate the six standard limb leads?
2. Complete the table for Ohm's law and its analogous expression for describing fluid flow, so that each expression satisfies the word equation: Potential = Flow × Resistance. Provide the units for each quantity.

	Potential	Flow	Resistance
Ohm's law			
Pressure-driven flow in a channel			

PROBLEMS

1. Consider a thermocouple that gives the voltage versus temperature in the table. Find a mathematical relationship that could be used to calculate temperatures for any measured voltage.

Temp (K)	Voltage (mV)
3.5	−4.66760
8	−4.60670
13.5	−4.52590
18	−4.45710
24	−4.37030
30	−4.28690

Temp (K)	Voltage (mV)
52	−3.99280
60	−3.88300
65	−3.81260
70	−3.74110
80	−3.59480
90	−3.44360
105	−3.20260
115	−3.03740
125	−2.86890
135	−2.69570
145	−2.51840
160	−2.24680
170	−2.06150
180	−1.87250
195	−1.58390
210	−1.29050
225	−0.99120
240	−0.68470
265	−0.16700
275	0.0378
285	0.2387
305	0.635
325	1.0387

2. Figure 11.8 shows the relationship between resistance and temperature for the thermistor.
   a. From the curve provided in the figure, find the coefficient β.
   b. Show that the variation in resistance in the temperature range 35–39°C can be approximated by a linear expression.
3. A 9-volt battery supplies power to a toy car with a resistance of 18 ohms. How much current is flowing through the car?
4. What voltage battery is needed to produce of 500 mAmps current in a circuit with 24Ω of resistance?
5. Four resistors ($R_1 = 100\Omega$; $R_2 = 250\Omega$; $R_3 = 350\Omega$; $R_4 = 200\Omega$) are connected as shown in Figure 11.8. The voltage V is 24 V.
   a. What is the total resistance of the circuit?
   b. What is the total current?
   c. What is the current through each resistor?
   d. What is the voltage drop across each resistor?
6. A circuit has three identical resistors connected in parallel across a 30 V battery. Each resistor is made of a copper wire that is 2 cm long with a diameter of 5.4 μm and a resistivity of 1.724×10^{-8} Ωm.
   a. Draw a diagram of the circuit.

b. What is the resistance of the resistor?
c. What is the total resistance of the circuit?
d. What is the total current?
e. What is the current through each resistor?
f. What is the voltage drop across each resistor?

7. An unknown resistor (R_3) is connected in series with the two other resistors ($R_1 = 20\ \Omega$ and $R_2 = 25\ \Omega$) across a 15 V battery in the order $R_1 \rightarrow R_2 \rightarrow R_3$. An ammeter, which is placed in between the battery and R_1, reads a current of 200 mA.
   a. What is the total resistance of the circuit?
   b. What is the resistance of the unknown resistor, R_3?
   c. What is the total current?
   d. What is the current through each resistor?
   e. What is the voltage drop across each resistor?
8. The standard curves shown in Figure 11.32 have the following equations:

$$\text{BSA: } A_{280} = 0.69\,[\text{BSA}] \text{ and Antibody: } A_{280} = 1.19\,[\text{Ab}]$$

   a. Assume that you measure the *OD* of five unknown solutions that contain BSA and get the following readings: $A_{280} = 0.35$, 0.02, 0.58, 1.23, and 0.44. What are the concentrations of BSA in each of these solutions?
   b. What would the concentration of antibody be for the same set of measurements?
   c. Do you have confidence in all of the estimated concentrations that you have calculated? If not, which values are suspicious to you, and why?
   d. How would you apply this method to a mixture that contained both BSA and Ab?

12 Bioimaging

LEARNING OBJECTIVES

After reading this chapter, you should:

- Be familiar with current biomedical imaging technology.
- Understand the principles behind x-ray, ultrasound, nuclear medicine, optical, and magnetic resonance imaging (MRI) techniques.
- Be familiar with some of the scientific and medical applications of these imaging modalities.
- Understand the basics of digital image processing and analysis.

12.1 Prelude

Biomedical imaging has revolutionized medicine and biology by allowing us to see inside the body and to visualize biological structure and function at microscopic levels. Images are representations of measurable properties that vary with spatial position (and often time). Images can provide exquisitely detailed information about biological structures; the most powerful imaging modalities provide functional information as well, allowing the recording of molecular or cellular processes, or physical properties (such as elasticity or temperature). Methods to visualize and quantify these properties are now available at the macroscopic (i.e., of a size visible to the human eye) and microscopic level. This information can be used clinically for diagnosis and monitoring of treatment as well as scientifically for understanding normal and abnormal structure and physiology.

Technology has brought about remarkable changes in imaging (Figure 12.1). Gene expression can now be imaged using **positron emission tomography (PET) imaging**—an image creation method that depends on injection of special radioisotopes—coupled with methods from genetics. The brain can be imaged at work on cognitive tasks with **functional MRI (fMRI)**, and that information can be used to guide neurosurgery. The mechanical action of the heart can be mapped using high-frequency sound waves (**ultrasound imaging**); these maps identify areas of injury after a heart attack. Images are essential tools in medicine

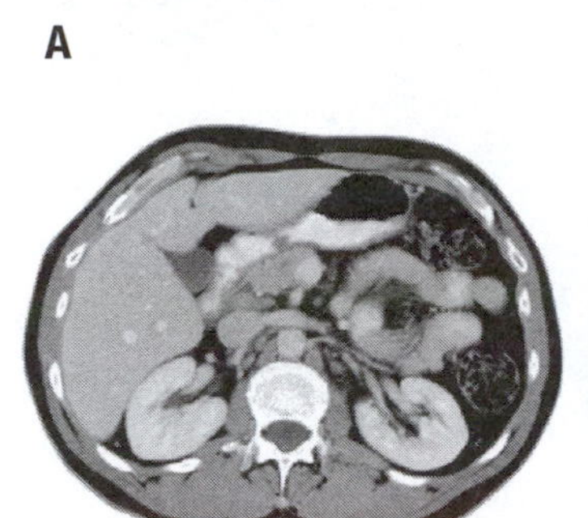

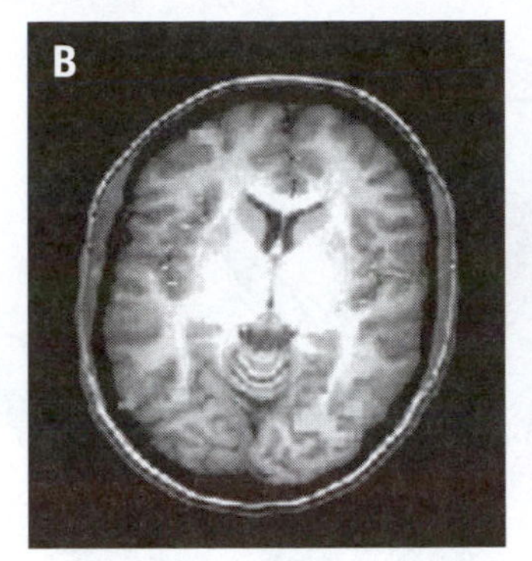

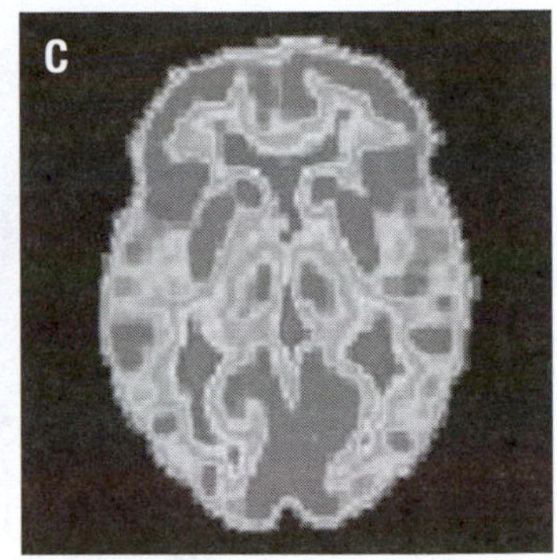

Figure 12.1 **Examples of imaging with new technology. A.** Cross-sectional image of abdomen obtained with x-rays using computed tomography. **B.** Functional magnetic resonance imaging (MRI; color) superimposed on anatomical MRI. Image courtesy of Dr. Jody Culham. **C.** Positron emission tomography scan obtained from a normal human brain. Image courtesy of the National Institute on Aging. (See color plate.)

because they provide a spatial map, enabling physicians to localize the biological phenomena being examined in space as well as time.

The many different imaging **modalities**—or types of imaging methods—fill different scientific and/or clinical niches. Every modality has limitations: A particular method may be low in quality, slow to acquire images, expensive, or not suitable for all patients. The set of advantages for a particular technology (e.g., high quality, faster, cheaper, or dynamic) will make it suitable in the right situations. Different modalities operate over different time or length scales and—because of the physics that underlie their operation—can measure different structures or functions (Table 12.1).

Table 12.1 Characteristics of biomedical imaging modalities

	X-ray	sMRI	fMRI	Ultrasound	SPECT	PET	Endoscopy	Microscopy	Confocal microscopy	MRS
Millimeter resolution	×	×	×	×	×	×	×			
Micron resolution								×	×	
Millisecond resolution			×	×			×	×		
Projection		×								
Surface							×	×		
3D		×	×	×	×	×			×	×
Structure	×	×		×			×	×	×	
Blood flow			×	×	×	×				
Molecular					×	×		×	×	×

Notes: This is a summary of some of the key features of the imaging modalities discussed in this chapter related to spatial and temporal resolution, type of spatial imaging, and the type of measurements made. sMRI: structural magnetic resonance imaging; fMRI: functional magnetic resonance imaging; SPECT: single photon emission computed tomography; PET: positron emission tomography; MRS: magnetic resonance spectroscopy; 3D: three dimensional.

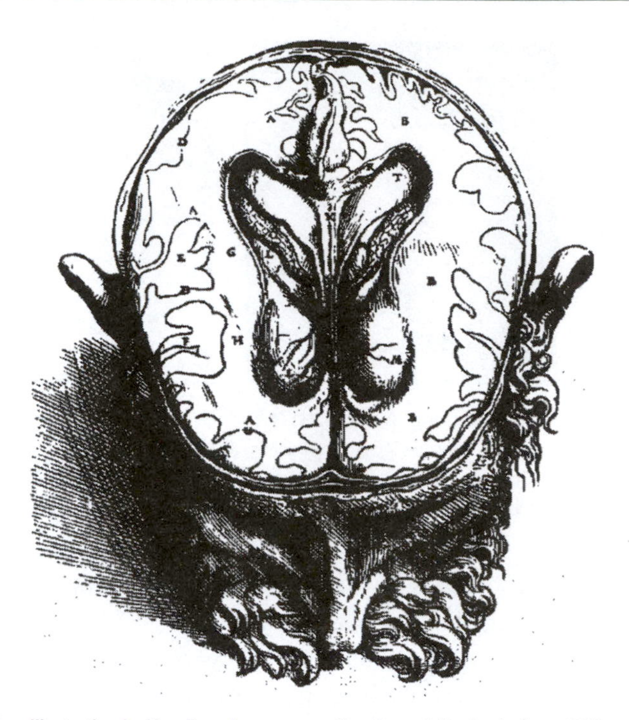

Figure 12.2 Illustration by Vesalius of a cross-section through the brain from 1543.

Virtually all imaging modalities are now **digital**; the images are acquired by a computer and are made up of individual picture elements, or **pixels**. Digital images can be readily processed to improve their quality, make measurements, and extract features of interest. Despite the similarity in the final digital form of the image, there are many different ways to create images. In general, an energy source interacts with the target (such as the human body) and produces a signal. The energy can come from electrons, ultrasound, light, x-rays, or even radio frequency (RF). Transducers are needed to convert the signal into a measurable form, such as a voltage. Often, it is necessary to introduce contrast material into the subject to provide a signal difference highlighting a particular structure or function.

Before the advent of imaging technologies, illustration was used to record biological structure. Andreas Vesalius, the 16th-century Belgian physician and anatomist, brought the art to a new level with his detailed and accurate renderings based on dissection (Figure 12.2). Drawings remained the state of the art until the invention of photography in the 19th century (Figure 12.3). Because it allowed for recording of images, photography was a first step in the revolution in bioimaging. Photographic technology was soon coupled with microscopes to record pictures of cells (Figure 12.4).

The discovery of x-rays late in the 19th century led directly to the first way of seeing inside the human body. An x-ray is capable of producing an image on film, allowing the x-ray image to be recorded and saved. A revolution in bioimaging followed this discovery with the development of many new imaging methods. Advanced microscopic techniques were developed, including

Figure 12.3

A. First photograph by Joseph Niépce taken in 1827. **B.** Example of modern digital photography.

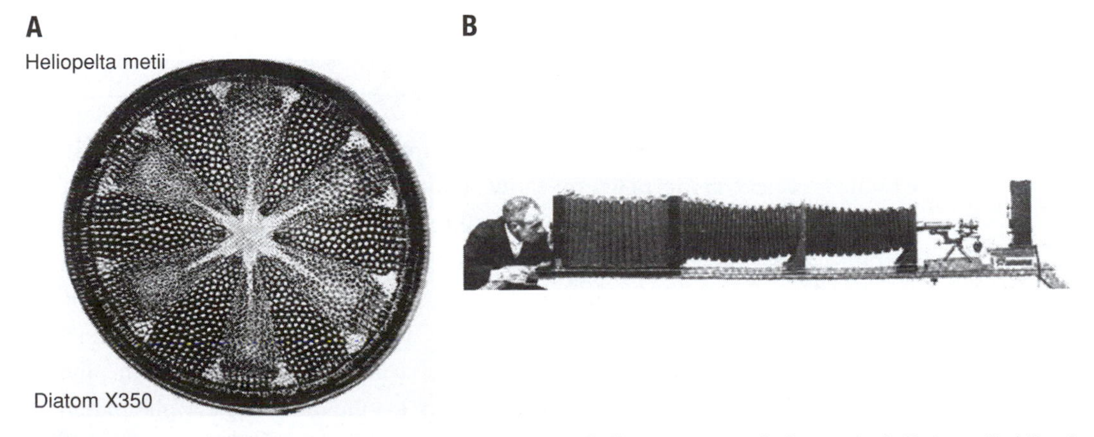

Figure 12.4

Biological photography. A. An early micrograph. **B.** The apparatus required to acquire it. Photo credit: Arthur E. Smith, reprinted with permission of Lutterworth Press.

Table 12.2 Biomedical imaging inventions and some of the key people involved in their development

1600s	Compound microscope	2D microstructure	Zacharias Janssen, Galileo Galilei, Robert Hooke, and Antonie van Leeuwenhoek
1827	Photography	Surface structure	Joseph Niépce
1840	Photomicrograph	2D microstructure	Albert Donné
1896	X-ray	2D (projected) structure	Wilhelm Röntgen
1946	MRS	Function	Felix Bloch and Edward Purcell
1957	Endoscope	2D internal structure	Basil Hirschowitz, Wilbur Peters, and Larry Curtiss
1957	Ultrasound	2D/3D dynamic structure	Tom Brown and Ian Donald
1957	Confocal microscope	3D microstructure	Marvin Minsky
1958	Gamma camera	2D (projected) function	Hal Anger
1963	SPECT	3D function	David Kuhl and Roy Edwards
1972	Computed tomography	3D structure	Godfrey Hounsfield and Allan Cormack
1973	PET	3D function	Michael Phelps
1973	MRI	3D structure, function	Paul Lauterbur and Peter Mansfield

Notes: 2D: Two dimensional; MRS: magnetic resonance spectroscopy; SPECT: single photon emission computed tomography; PET: positron emission tomography; MRI: magnetic resonance imaging.

electron microscopy, which uses electrons instead of visible light to create images of objects much smaller than the wavelength of light (~500 μm). New diagnostic modalities include **gamma camera** imaging (which can create images from radioactive substances injected into the body), ultrasound imaging, PET imaging, **computed tomography** (**CT**) to create three-dimensional images from x-rays, and **MRI**, which now provides high-resolution images of the inside of the body without the hazards of x-rays or other forms of ionizing radiation. These new forms of imaging technology were produced by the work of scientists and engineers (Table 12.2).

12.2 X-rays and CT

In 1895, Wilhelm Röntgen presented a paper to the Würzburg Physical and Medical Society in Germany on x-rays entitled, "On a New Kind of Ray." The amazing ability of these rays to pass through solid objects was immediately recognized as more than a curiosity: X-rays can be used to create images of solid objects, in which the image is a spatial map of the object's susceptibility to penetration by the rays (Figure 12.5). The first Nobel Prize in Physics in 1901 was awarded to Röntgen for this work. The prize was given in physics because of the importance of these rays as a fundamental new phenomenon, but x-rays would prove to be of enduring practical significance in medical imaging.

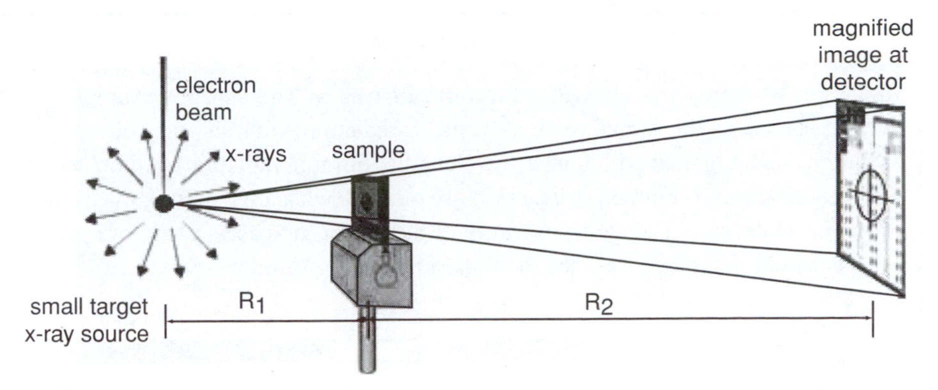

Figure 12.5 Image creation using x-rays.

X-rays are electromagnetic radiation with wavelengths of about 0.1 nanometer (Box 12.1). Because the rays have sufficient energy to knock electrons out of atoms within material as they penetrate, x-rays are a form of **ionizing radiation**. Loss of electrons turns atoms into ions or charged particles; if the material exposed to x-rays is human tissue, and ions are generated within the tissue, these ions can subsequently damage cells. Because of the potential for damage, x-ray exposure times are minimized to that necessary for clinically meaningful results. In addition, the extent of exposure is limited to regions of the body that are necessary to minimize potential side effects.[1]

X-rays have two properties that make them useful for imaging. First, the human body is translucent to x-rays of the correct wavelength (and corresponding energy): The rays pass through the body, but, as they penetrate, they are partially absorbed. The more tissue there is in the path of the x-rays—and the denser that tissue is—the greater the fraction of radiation that is absorbed. X-rays have the "Goldilocks" energy for imaging: Electromagnetic radiation with higher energy will pass through the body without significant absorption, and radiation with lower energy will be completely absorbed, but x-rays are just right. Second, x-rays have the fortuitous ability to expose photographic film, just as visible light does. This property played a role in their discovery: Röntgen recorded their presence on film. X-ray images were originally acquired on film before the development of the digital x-ray camera. In practice, fluorescent screens are used in front of the detector (either film or digital) to amplify the effect of the x-rays (Figure 12.6).

12.2.1 Conventional x-ray imaging

A beam of x-rays directed through a body will expose film (Figure 12.6). An x-ray image is a "negative" image: The film is darker where the body is less dense and lighter where the body is denser. Dark regions occur where the body has lighter elements (such as the flesh of your leg), allowing more x-rays to penetrate through

[1] The biological hazards of ionizing radiation, as well as the use of these rays for treatment of cancer, are described in Chapter 16.

Box 12.1 The electromagnetic spectrum

Radiation is energy transmitted in the form of waves. The sun, for example, emits radiation that is visible to the human eye, as well as radiation with shorter wavelengths (ultraviolet radiation) and radiation with longer wavelengths (infrared radiation). All objects emit radiation. Thermal radiation is emitted strictly because of an object's temperature; the primary wavelength of radiation decreases as the temperature of the object increases. Therefore, the temperature of objects can be estimated from the thermal radiation they emit.

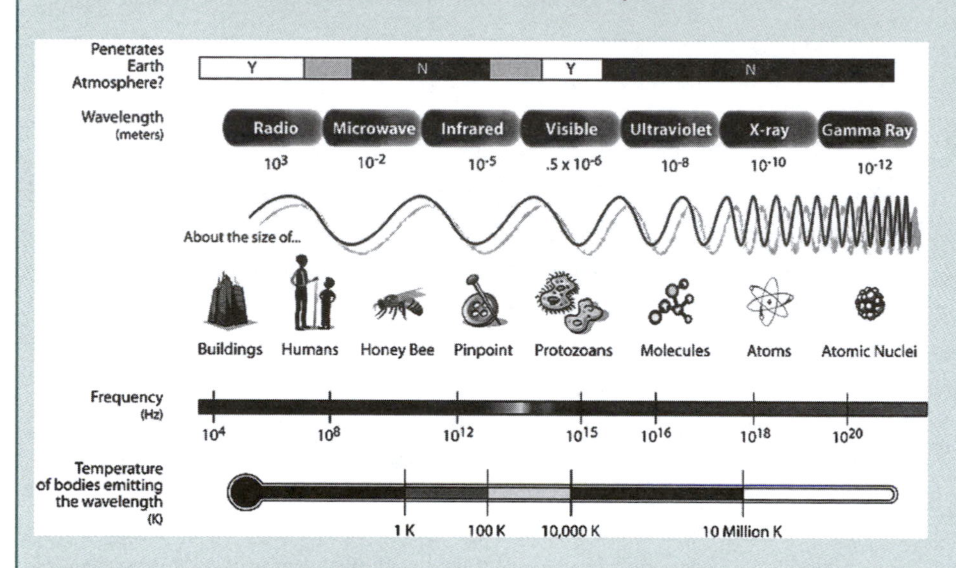

Figure Box 12.1 Properties of the electromagnetic spectrum. Image courtesy of NASA.

The electromagnetic spectrum is the continuous set of all forms of electromagnetic radiation, usually presented as a line divided into regions of decreasing wavelength (Figure Box 12.1). The wavelength of radiation is related to its frequency:

$$\lambda = \frac{c}{f}, \qquad \text{(Equation 1)}$$

and the amount of energy carried by electromagnetic radiation depends on its frequency:

$$E = hf, \qquad \text{(Equation 2)}$$

where c is the speed of light (299,792,458 m/s) and h is Planck's constant (6.63×10^{-34} J-s). Although characterized by wavelength, electromagnetic radiation also behaves like a particle: Each photon of radiation carries a quantum of energy:

$$E = h\frac{c}{\lambda}. \qquad \text{(Equation 3)}$$

Radio waves have the longest wavelengths, and are used for broadcasting radio and television (FM radio, e.g., uses frequencies between 88 and 108 MHz, where 1 MHz is equal to 10^6 Hz). Microwaves are shorter in wavelength; they range from 1 mm to 1 m, with frequencies ranging between 300 MHz and 300 GHz. They are also used for communications, and for power transmission, as in microwave ovens. Infrared radiation has wavelengths in the range of 1 mm to 750 nm; the peak radiation emitted by heated bodies between 1 and 1,000 K (−272 to 728°C) is infrared.

The human eye can detect a narrow range of the electromagnetic spectrum: The wavelengths between 400 and 700 nm produce chemical changes in the cells of the retina. Specialized cells in the retina detect subranges of the visible wavelengths, so that humans can see colors ranging from violet (400–450 nm), blue (450–490 nm), green (490–560 nm), yellow (560–590 nm), orange (590–630 nm), and red (630–700 nm)

Ultraviolet (UV) radiation has wavelengths shorter than visible light and, therefore, carries more energy. Cells in human skin are affected by UV radiation, which can cause burns with prolonged exposure. Gamma rays and x-rays are even shorter in wavelength, and more energetic, than is UV radiation. Because of their short wavelength, x-rays and gamma rays will pass through most substances, hence their usefulness for medical imaging. But they also can cause chemical changes, such as ionization, which is caused by the ejection of electrons from molecules struck by this high-energy radiation. In human tissues, the effects of short term, low doses of ionizing radiation are minor. Higher doses of ionizing radiation can cause immediate biological effects, called radiation poisoning, with irritation and local destruction to tissues, similar to burning. Continued exposure to low doses can cause damage to DNA, creating mutations that lead to local disease in the tissue, particularly cancer.

it to expose the film. Light or white regions occur where the body is dense with heavier elements (like bone), allowing fewer x-rays to expose the film. The x-ray intensity is decreased (attenuated) by the structure through which it passes: In this way, an x-ray image is like a shadow. The conventional x-ray process collapses the three-dimensional structure of the body into a two-dimensional image. Each position on the two-dimensional image represents the penetration of x-rays through the body—which has different thicknesses and different tissues of different densities—onto a flat plane. This kind of image, in which a shadow of a

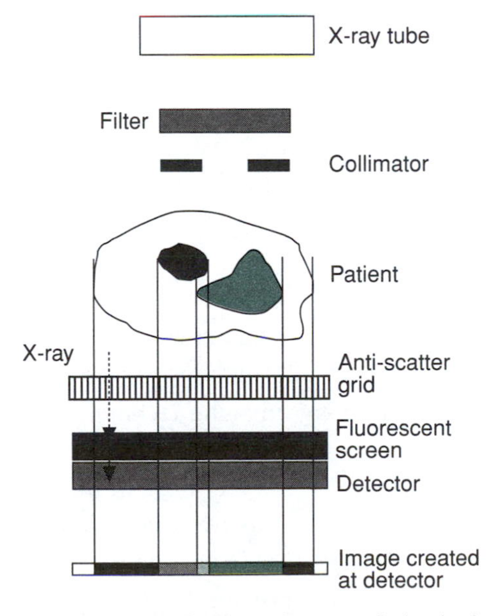

Figure 12.6 **Formation of a medical image by x-rays.** During development of the film, silver halide particles in the film (but only particles that have absorbed energy) are converted to metallic silver, which appear as black grains. The image created on the film (or other detector) is a projection of the density of the three-dimensional object onto a two-dimensional planar image.

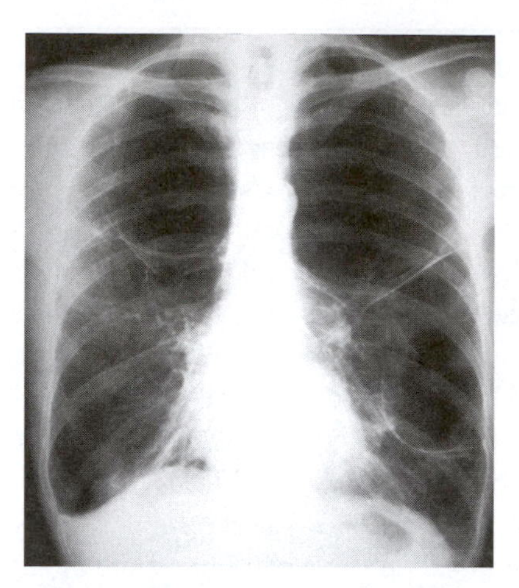

Figure 12.7 **Chest x-ray showing emphysema bubbles in the left lung.** Photo courtesy of the National Institutes of Health/Department of Health and Human Services.

three-dimensional object is created on a plane, is called a **projection**.

X-ray images can show fractures and breaks in bones, cavities in teeth, fluid in lungs, and cancer in breasts. X-rays provide good **contrast** when there is a difference in the density of the tissue, such as with soft tissue and air, bone and soft tissue, or water and soft tissue. Soft tissue structures (e.g., in the brain or abdomen) do not vary much in density and, therefore, do not have as much contrast. One of the most common applications of x-ray imaging is the chest x-ray (Figure 12.7), which is used by physicians to look for infection within the lungs (pneumonia), fractures in the bones of the rib cage, and certain kinds of heart disease.

Although it is composed of structures that do not vary greatly in x-ray absorption, images of the digestive system can be created in another way. A **contrast agent** is often added to enhance the differences in density within the body. For example, the digestive tract can be imaged after asking the patient to swallow a dense element, such as a solution containing barium, which makes the inside of the digestive tract more dense, or brighter on the x-ray. Alternatively, air can be introduced to make the tract dark. Using these methods, small abnormalities in the intestinal wall, such as diverticuli and polyps, can be visualized. Contrast agents are now widely used in biomedical imaging. Dense elements can provide contrast in x-ray images: Other kinds of agents enhance contrast in images created by other imaging modalities.

X-rays can be taken over time, to watch for dynamic changes within the body. These dynamic x-rays are acquired using a fluorescent screen instead of film.

12.2.2 CT

X-ray imaging experienced a profound reinvention with the development of CT. The key invention that led to CT was algorithmic; it arose from a mathematical method that was applied to the x-ray imaging technique. The central idea was to take x-rays at many different angles and use the many images that were thus acquired to reconstruct a three-dimensional image of the body. This three-dimensional image could then be presented as two-dimensional cross-sections, which revealed the relationship of tissues within the body. Instead of projections or shadows of density, CT images could provide the first detailed view inside the body. Godfrey Hounsfield was an English engineer who developed CT in 1972 while working at EMI. (EMI was The Beatles's record company; the huge profits of EMI helped fund the development of CT.) Independently, Allan Cormack, at Tufts University, had developed the underlying mathematics for image reconstruction in the early 1960s. The two inventors won the Nobel Prize in Medicine in 1979 for this work.

In a CT scanner, the patient lies down on a table, which moves through a circular opening in the scanner (Figure 12.8). The x-ray generator and the detectors move in a circle around the patient and create an image of one cross-section through the body at a time. The table moves the patient to image each slice. The mathematical techniques of reconstruction from projections are used to create a three-dimensional image from all of the data collected (Box 12.2).

Whereas the contrast of CT is limited in soft tissue, CT is quite versatile and valuable for imaging the head, lungs, abdomen, pelvis, and extremities (Figures 12.9 and 12.10). More recently, spiral CT scanners have been developed where the x-ray continuously travels along a spiral path. In this way, three-dimensional images are generated rapidly with excellent resolution. The speed also enables dynamic imaging of moving structures, such as the heart. MicroCT imaging has also been developed that allows imaging of very small structures and is particularly useful for medical research in animals, such as rodents.

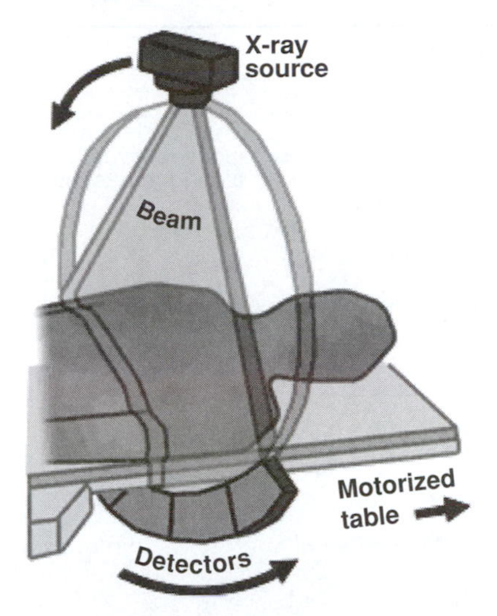

Figure 12.8 Schematic of the acquisition of a computed tomography image.

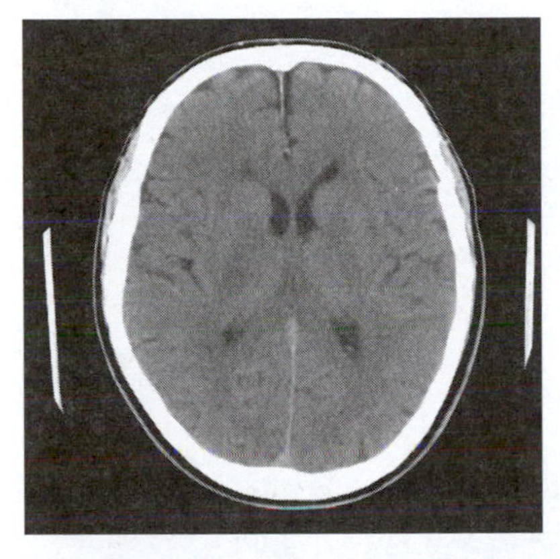

Figure 12.9 X-ray computed tomography slice through the human brain.

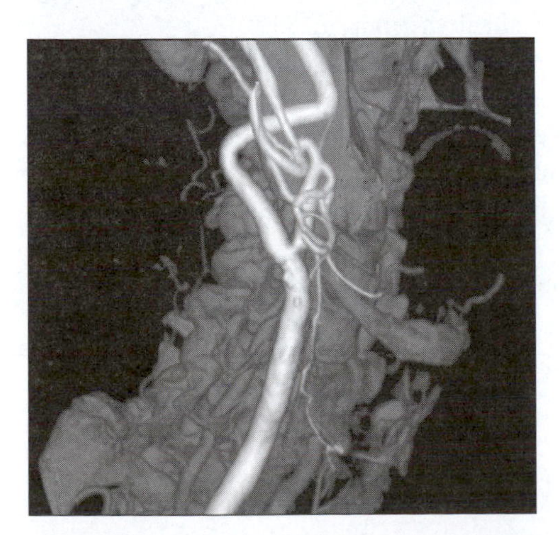

Figure 12.10 **Visualization of three-dimensional structure showing the carotid artery derived from computed tomography.** Image provided courtesy of GE Healthcare. (See color plate.)

Box 12.2 Reconstruction from projections

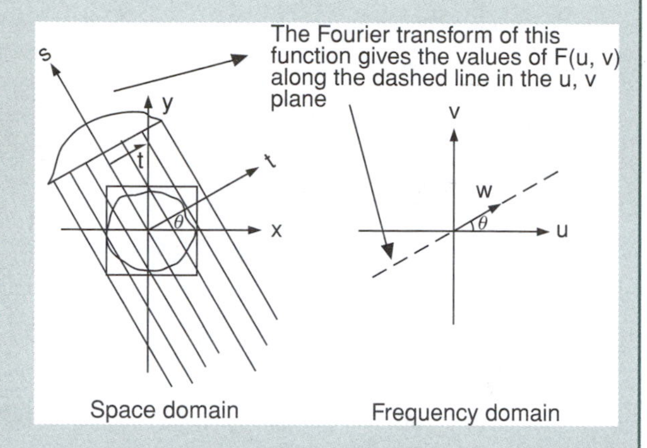

As x-rays pass through the body, they are absorbed, or attenuated, by the tissue. The degree of x-ray attenuation—also called the **radiodensity**—depends on the local density and composition of the tissue. Bones attenuate strongly and absorb most of the incident x-rays; water attenuates much less. Therefore, each tissue type (muscle, fat, bone) has a characteristic radiodensity.

To calculate the fraction of x-ray energy that penetrates completely through the patient, the amount of attenuation produced by each object in the x-ray path must be added together. In mathematical terms, the strength of the transmitted x-ray that is observed is related to the strength of the incident beam by:

$$N_o = N_i e^{-\sum_{\text{ray}} \mu(x,y)ds}. \qquad \text{(Equation 1)}$$

The effect of the attenuation is summed over the path of the ray. This equation can be rewritten as:

$$\ln \frac{N_i}{N_o} = \sum_{\text{ray}} \mu(x, y)ds. \qquad \text{(Equation 2)}$$

By imaging at many different angles, the sum of the values of μ in Equation 2 is obtained from a variety of orientations around the object.

The goal of reconstruction from projections is to determine the values of μ at each location in the actual object being imaged. There are many ways to solve for these values. It is similar to a complicated version of the number game Sudoku, in which the sums must add up as specified by the data. One way to solve the problem is to make an initial guess at each attenuation value and then carefully adjust the guesses until all the sums match.

It also turns out that, if a set of projections of an image is taken at the same angle with varying position, t, the **Fourier transform** of this is equal to the Fourier transform of the image along a line at the same angle. In principle, with projections along many angles, we can reconstruct the Fourier transform (and therefore the image itself) of the slice. In practice, other algorithms, using this relationship, can be developed.

Knowing the value of μ at many slices through the body, we can form a three-dimensional image. Computer visualization techniques can then be used to show the complex relationships between structures in the body.

12.3 Ultrasound imaging

Ultrasound imaging is based on the propagation of high-frequency sound waves through tissue. Ultrasound imaging is fast: Structure and dynamic function can be recorded from rapidly moving tissues, such as a beating heart, and visualized at the same speed in which they happen (this is called imaging in **real time**). Ultrasound imaging is also safe: It can be used on almost any subject, including pregnant women and critically ill persons. With high-powered focused beams, however, ultrasound can also be used for therapy, such as for breaking up kidney stones.

The machines that perform ultrasound imaging are portable and can be brought quickly and directly to the patient, making them extremely versatile. Ultrasound imaging is excellent for emergency medicine because of its ability to image internal bleeding and localize internal injuries. The imaging systems are also relatively inexpensive. One important limitation, however, is that ultrasound imaging is difficult through bone and air (such as in the lungs).

12.3.1 Image generation

Special materials are used to generate a pulse of ultrasound, typically materials called **piezoelectric crystals**. Piezoelectric crystals have a remarkable property in that they will vibrate in response to an electrical signal, or driving voltage. They will also work in reverse: A vibration received by the material will induce a voltage. For ultrasound imaging, the crystals are focused using a concave shape so that the wave is concentrated in one direction. Furthermore, the materials are tuned to have a **resonant frequency**—or the frequency at which the material naturally vibrates—in the megahertz range, from 2×10^6 to 13×10^6 sec^{-1} (or Hz), or from 2 to 13 MHz. Sound waves in this frequency range are known to interact safely with tissue. Megahertz frequency sound waves are not audible to humans; human hearing is typically limited to the range 20–20,000 Hz.

Structure is detected by echoes of the sound waves that bounce off tissue interfaces, such as the head of a fetus (Figure 12.11) or the wall of the heart. At each interface, only part of the wave is reflected; the rest continues on through the body. These echoes, or reflections, return to the transducer, which measures them using the piezoelectric crystal in reverse. The voltage that each echo induces in the crystal is recorded. Because the transducer is producing a series of waves, and the waves are penetrating and only partially reflected, a train of echoes is received back at the transducer and converted into a series of voltages.

The magnitude of the transduced voltage indicates the strength of the echo. The spatial location of each echo is determined by the time-of-flight, $t =$ the time between the generation of the pulse and the reception of the echo. In ultrasound, the speed of sound is assumed to be constant in tissue—at about the same speed as sound in water (1540 m/s). The pulse travels at this speed, c, for a distance of

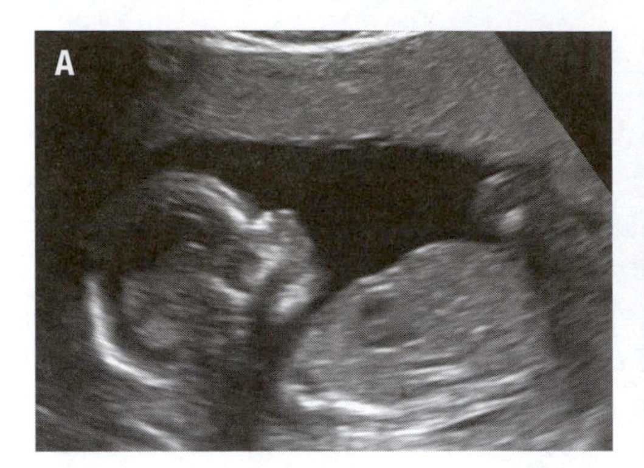

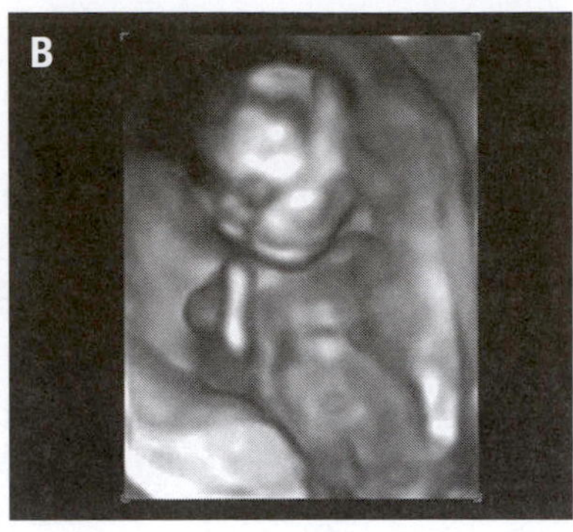

Figure 12.11 **Fetal imaging is one of the most common ultrasound imaging procedures. A.** Conventional fetal ultrasound image. **B.** Three-dimensional image, also of a fetus, generated by ultrasound imaging techniques.

$2d$, twice the distance from the transducer to the structure (it makes a round-trip, there and back). The time is therefore related to the distance by:

$$c = \frac{2d}{t} \quad \text{or} \quad d = \frac{ct}{2}. \tag{12.1}$$

SELF-TEST 1 What is the time required to receive an echo from the aorta, assuming that the transducer is abutted to the abdominal wall, and the aorta is 6 cm deep?

ANSWER: $t = 0.00008$ seconds

With the transducer pointed in one direction, the positions of objects that the pulse encounters in that direction can be determined by Equation 12.1. To form an image, the transducer is pointed in many different directions, by rotating with a motor in back-and-forth sweeping motions. The image lines are combined into a pie-wedge–shaped image, built up from each line of reflections. This process occurs quickly; images can be formed continuously, in real time, creating a movie as either the probe or the subject moves. A snapshot from a movie of a human fetus, moving within the uterus, is shown in Figure 12.11.

Ultrasound images tend to appear "noisy": there are many seemingly random bright spots in the image. Echoes occur not just at organ boundaries but also at small tissue elements (called "speckles") that may not correspond to relevant structures. The signals produced by these speckles can obscure important features. However, because ultrasound imaging provides a real-time video stream, and the speckles are not consistent over time, the perceived image quality is aided by motion.

12.3.2 Doppler imaging

It is often valuable to measure the velocity of a tissue, such as the rate of blood flow or the rate of wall motion in the heart. Velocity can be measured with ultrasound imaging by exploiting the **Doppler effect**. The Doppler effect occurs

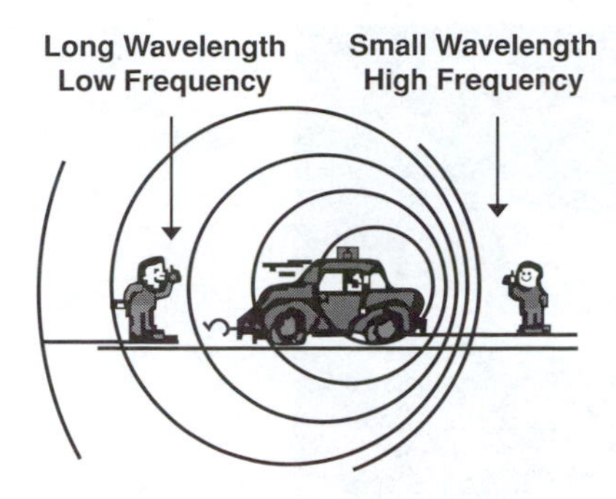

Figure 12.12 **The Doppler effect.** A horn sounding from a moving car will sound different to observers in front of the car than to those behind the car. The observer in front will hear a high-frequency sound; the observer in back will hear a lower frequency sound.

commonly with sounds waves. The sound of a honking horn from a moving car will be higher in frequency for a pedestrian in front of the car (Figure 12.12). For a person on a train platform, the whistle of an approaching train will appear high in frequency (or pitch), but when it rushes past, the pitch of the sound will drop (i.e., it will get lower in frequency).

To create Doppler images with ultrasound, the probe produces an ultrasound signal of a fixed wavelength; however, the wavelength of the echo can change depending on the direction and angle of the movement of the objects producing the echo. In the case of blood flow, the echo comes from the moving particles (blood cells) in the blood.

If the object is moving toward the transducer, the wavelength will be shorter and, therefore, the frequency will be higher; if moving away, the opposite will be true. Faster motion produces a larger change in frequency. If the direction of motion is the same as the direction of the ultrasound pulse, the change will be larger than if the direction is at an angle. The change in frequency can be measured to compute the velocity.

Remember that the wavelength and frequency of wave propagation in a medium are related:

$$\lambda = \frac{c}{f} \quad \text{and} \quad f = \frac{1}{T}, \tag{12.2}$$

where c is the velocity of ultrasound, λ is the wavelength, f is frequency, and T is the period. The observed frequency shifts are often small, but can be optimized by positioning of the transducer because the effect is greatest when the motion and transducer axes are parallel. The wavelength at the moving object is given by:

$$\lambda_1 = \lambda_0 + (v \cos\theta) T_0, \tag{12.3}$$

where θ is the angle between the direction of flow and the ultrasound pulse, v is the velocity of the object, and T_0 is the period (one divided by the frequency). Because the objects that reflect the ultrasound are moving relative to the ultrasound probe, the frequency of the wave propagated back to the ultrasound probe is given by:

$$f_2 = \frac{c - v\cos\theta}{\lambda_1} = f_0\left(1 - \frac{2v\cos\theta}{c}\right), \tag{12.4}$$

assuming $v \ll c$. Figure 12.13 shows an ultrasound Doppler image of a carotid artery.

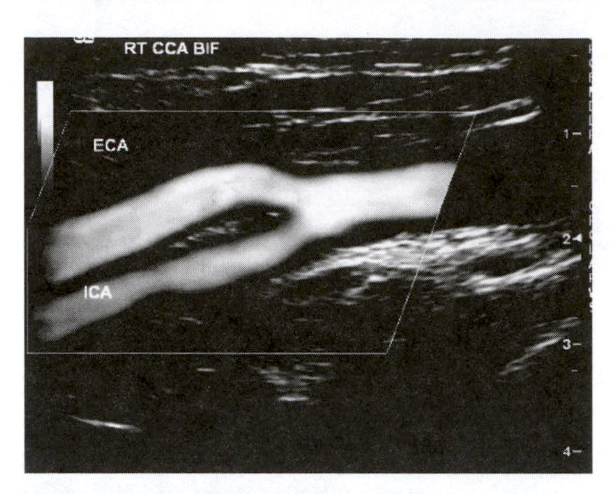

Figure 12.13 **Doppler image of the bifurcation of the carotid artery.** Image courtesy of Dr. James D. Rabinov, Massachusetts General Hospital and used with permission. (See color plate.)

12.3.3 Further ultrasound techniques

Whereas ultrasound is primarily used to create images of body structure, functional images can also be formed. Elastic properties can be determined by measuring the change in structure (strain) when tissue is compressed (stress). Microscopic bubbles can be used to increase the contrast (relative brightness) in ultrasound images. The gas in the bubbles causes strong reflections and thus can act as a contrast agent, creating bright areas in the image. Particular areas can also be targeted by chemically tagging the bubbles to mimic physiologically significant molecules and thus highlight areas of importance. In addition, three-dimensional ultrasound imaging has become possible using either an additional rotation of the transducers or arrays of transducers (Figure 12.11b).

12.4 Nuclear medicine

Nuclear medicine is the first imaging modality that was designed to measure function within the body rather than structure. It is based on the detection of **radioactive molecules**, which are molecules that are unstable and spontaneously decay to release radiation energy, such as gamma rays (Box 12.1). The molecules are selected to engage physiologic mechanisms in the body. The field began in the 1950s and has evolved into a major medical specialty for diagnosis—and now therapy—of disease.

To create a nuclear medicine image, a radioactive compound is ingested, inhaled, or injected. Like x-rays, the radioactivity results in ionizing radiation. At the levels that are typically used, the radioactivity does not damage the body. Moreover, the nuclear medicine technique is always optimized so that the amount of radiation that the patient gets is minimized.

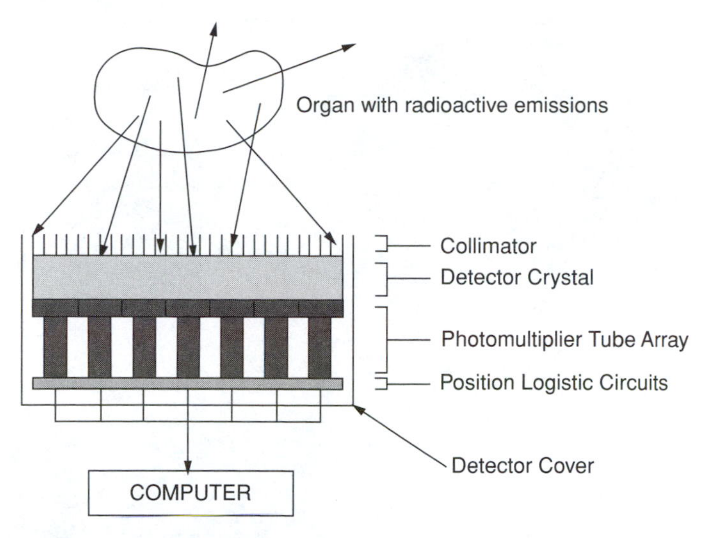

Figure 12.14 **Schematic of gamma camera.** An image of the intensity of radioactivity, which is emitted from the subject, is captured by a gamma camera.

12.4.1 Nuclear medicine imaging methods

The radioactive compounds used in nuclear medicine (which are often called **tracers**) are designed to reflect a particular function, such as blood perfusion or metabolism; typically, this is accomplished by attaching a radioactive isotope to a molecule that participates in that function. For example, **iodine-131** (**^{131}I**) is used because the thyroid has physiological mechanisms for concentrating iodine. The imaging process then measures the distribution of radioactivity in the subject. Areas of the body that reveal high radioactivity indicate the presence of the molecule. The principal types of nuclear medicine imaging methods are planar imaging, **single photon emission CT (SPECT)**, and PET.

Planar imaging and SPECT use molecules that are chemically linked with gamma-ray–emitting elements such as **technetium-99m**. The **half-life** of this technetium isotope is short (6 hours), so the radioactivity is cut in half every six hours and therefore does not last long after injection into the body. A **gamma camera** is a specialized instrument that detects gamma rays and produces an image of the radioactivity (Figure 12.14); the operation of the gamma camera is described in Section 12.4.3.

PET uses very short-lived radioisotopes that emit particles called positrons: For example, **oxygen-15** emits positrons and has a half-life of 2 minutes. Because of the very short half-lives, the compounds must be generated close to their site of use. They are produced using a **cyclotron**, which is a type of particle accelerator, and special facilities for manipulating the radioactive material safely. Positron emitters are typically of low atomic number and isotopes of common physiological elements, such as ^{11}C, ^{13}N, and ^{18}F.

When these isotopes decay, and a positron is emitted in the body, it will almost immediately encounter an electron. These two particles annihilate each other;

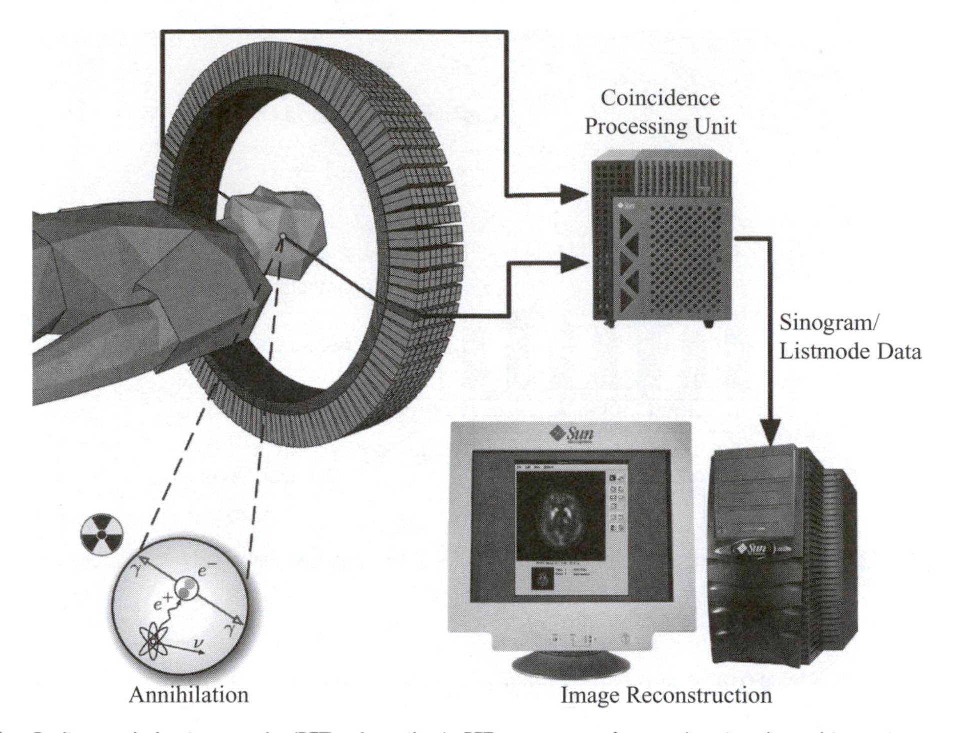

Figure 12.15

Positron emission tomography (PET) schematic. In PET, a sequence of processing steps is used to create an image. During the annihilation process, two photons are emitted in diametrically opposing directions. These photons are registered by the PET scanner as soon as they arrive at the detector ring. After the registration, the data are forwarded to a processing unit, which decides if two registered events are selected as a so-called coincidence event. All coincidences are forwarded to the image-processing unit, where the final image data are produced via image reconstruction procedures.

this process generates two gamma rays moving in exactly opposite directions. To find where the annihilation occurred, detectors are placed in a ring around the imaging field (Figure 12.15); the two gamma rays will encounter the detectors almost at the same time unless one or both are scattered. Scattering occurs when a gamma ray interacts with matter in its path and typically loses energy and changes direction. The detectors, however, expect the events to come in pairs and will ignore events that are likely caused by scattered gamma rays.

12.4.2 Nuclear medicine applications

The tracers used in nuclear medicine are designed to reflect bodily function, such as blood circulation, and are often directed to a particular organ. The first routine clinical application was for thyroid imaging using radioactive iodine (^{131}I) to detect cancer, and to evaluate over- or underactive thyroids. The patient swallows a sodium iodide solution made using ^{131}I, which is absorbed through the digestive system, circulates with the blood, and collects in the thyroid.

With the development of other radioactive tracers, the applications of nuclear medicine have expanded. In the heart or brain, these images can indicate the location of damage caused by a heart attack or stroke. Bone growth, fractures, tumors, and infection can all be visualized using bone scans with special tracers. The urinary tract, lungs, and liver can also be imaged.

Abnormalities in nuclear medicine images may appear as either increased activity (which appears bright in the image) or decreased activity (which appears dark). For tracers that indicate blood flow, increased activity can indicate tumor or fracture, both of which show an increase in blood flow. Decreased activity can indicate the location of damage or restriction of blood flow (stroke, infarction, etc.).

In the heart, blood flow to the heart wall can be measured. Exercise before imaging can be used to measure changes in blood flow associated with stress. The imaging process can be synchronized with the patient's heart rhythm by using an ECG ("cardiac gating") to allow for a movie of the radioactivity pattern to be acquired to assess, for example, wall motion and cardiac function. Without gating, the motion will blur the image.

Exams of the liver and spleen account for about three-fourths of all nuclear medicine scans and are used to show size, shape, position, and irregularities. These scans can be used to detect abnormalities of the liver such as abscesses or lesions that indicate hepatitis, cirrhosis, and other disorders. Enlargement of the spleen may be seen indicating, for example, an infection. Imaging of the kidneys and bladder can show dynamic function and abnormalities in the urinary tract.

In the brain, SPECT and PET can be used to measure blood flow, metabolism, and neurotransmitter binding, using appropriately designed radiotracers. These techniques are useful for diagnosing infection, stroke, and tumors and are finding increasingly important use in the study of neuropsychiatric disorders such as depression and schizophrenia.

12.4.3 Operation of a gamma camera

Planar and SPECT imaging are based on the gamma camera, which is also called an Anger camera, after the American engineer, Hal Anger (Figure 12.14). The gamma camera is assembled from scintillation detectors (which can detect the presence of gamma rays), photomultipliers (which convert light energy into electrical energy), and a collimator (which filters gamma rays). A single camera in a single position produces a planar image that is a projection (i.e., it is a summation) of the radioactivity flowing in the direction the camera is facing. Using mathematical methods similar to those in CT, cross-sectional images can be produced by imaging at many angles, possibly with multiple cameras, and reconstructing cross-sections from the projections.

As gamma rays (high-energy photons) are emitted from within the body, they will be concentrated in certain regions of the body, depending on the design of the tracer chemical. As the rays emerge from the body, some will be scattered or absorbed before they make it to the camera. Gamma rays enter the camera through

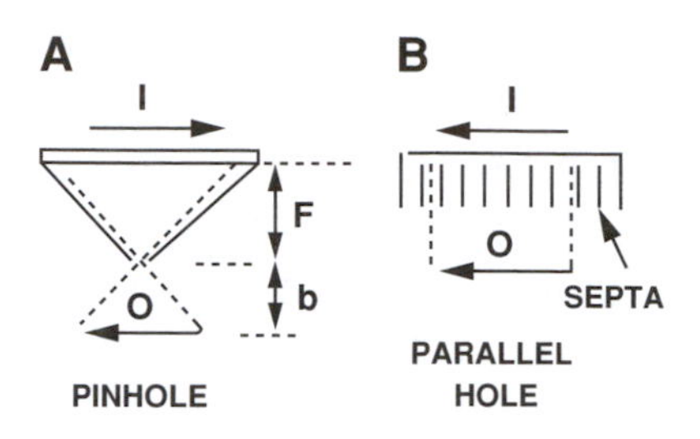

Figure 12.16 Types of collimators.

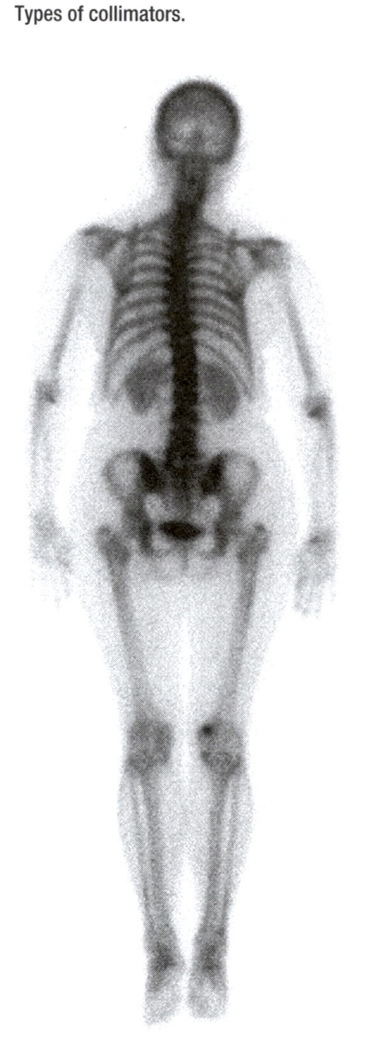

Figure 12.17 **Image from a gamma camera.** Reprinted from Groch, M.W., and Erwin, W.D., Single-Photon Emission Computed Tomography in the Year 2001: Instrumentation and Quality Control. *J Nucl Med Technol.* 2001;29(1):12–18 with permission of the Society of Nuclear Medicine.

the collimator, which is typically made up of channels of lead that only allow penetration of gamma rays that originate directly in line with the channel. In a parallel-hole collimator (Figure 12.16b), the lead strips are parallel to each other and perpendicular to the camera head. They ensure that only photons emerging from directly below the camera are observed. Photons coming in at an angle are absorbed virtually completely by the lead.

Another kind of collimator is the pinhole collimator (Figure 12.16a). Only gamma rays that pass through a pinhole are imaged, resulting in a reversed and potentially magnified image. The magnification factor can be set by adjusting the distance between object and pinhole. The magnification factor is simply the ratio of the detector–pinhole distance to the object–pinhole distance. Pinhole collimators are used for the imaging of small structures that require magnification, such as the thyroid.

After passing through the collimator, gamma-ray photons hit the scintillation crystal, a special material that emits many visible light photons for each gamma photon. The visible light is then detected by the photomultiplier tube and converted to an electric signal. Electronics within the camera keep track of the number and location of these signals and pass that information to the computer to form an image. The image shows the spatial distribution of radioactivity, as well as changes in level of radioactivity as a function of time (Figure 12.17).

To ensure that the image is caused by gamma rays of the proper energy, photons are counted only if they fall within a defined energy range (called the **energy window**) that is based on the characteristic energy of the radioactive tracer. Broadening the energy window will increase the signal (more photons will be counted for the image), but will also allow more

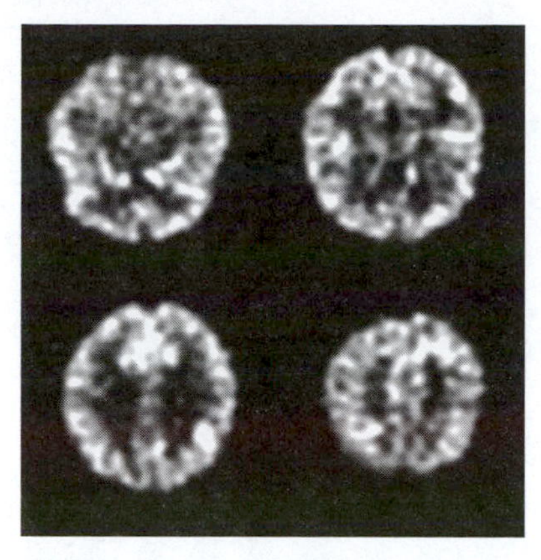

Figure 12.18 Image slices through the brain imaged with positron emission tomography. (See color plate.)

lower-energy, scattered photons—the sites of origin of which are uncertain—to be counted. Because some gamma photons can be absorbed in tissue, thicker body parts will have a decreased signal and more scatter. Larger distances also cause decreased signal.

Radioactive events, such as these, are governed by a **Poisson probability distribution** (Box 12.3). One of the consequences of this distribution is that the standard deviation of the measurements is equal to the square root of the mean activity. If the number of events, or activity, is equal to A, then the standard deviation, or error, is equal to the square root of A. The ratio of signal (activity) to error in the signal (noise) can be calculated:

$$\frac{S}{N} = \frac{A}{\sqrt{A}} = \sqrt{A}, \tag{12.5}$$

where S is the signal level and N is the noise level. For low activity, the signal is not much larger than the noise, and the result is a speckled (noisy) image. As the activity gets higher, the **signal-to-noise ratio (S/N)** increases, and the image quality improves. (See Section 12.7 for more on image quality.) Equation 12.5 predicts that the higher the amount of radioactivity, the better the quality of the image. However, the amount of radioactivity that the subject receives must be as low as possible, to be safe. Therefore, nuclear medicine imaging requires a trade-off between image quality and dose to the patient.

A PET camera works in much the same way, except that **coincidence detection** is needed to identify the position of the positron from the simultaneous (or coincident) detection of gamma rays on opposite sides of the body (Figure 12.15). In addition, physical collimation, as described previously, is replaced with **electronic collimation**. The two photons that are generated by the positron/electron annihilation will reach the detector simultaneously. Thus, the camera is designed so that events are recorded only if they occur simultaneously. The events are localized because they must have occurred on the line connecting the two detectors that recorded the event. Electronic collimation is more sensitive than physical collimation because physical collimation uses lead strips to absorb photons, decreasing the number of photons that reach the detector. PET images, formed in this way, are extremely useful in imaging molecular events in the brain (Figure 12.18).

12.5 Optical bioimaging

Light is the basis for vision in animals. Optical bioimaging mimics these systems and extends our natural ability by allowing vision inside the body and at a

Box 12.3 Poisson probability statistics

Probability is a branch of mathematics that allows calculation of the likelihood (or probability) that events will occur in the future. Probability is different from **statistics**, which is the branch of science that deals with the collection, classification, and analysis of numerical facts.

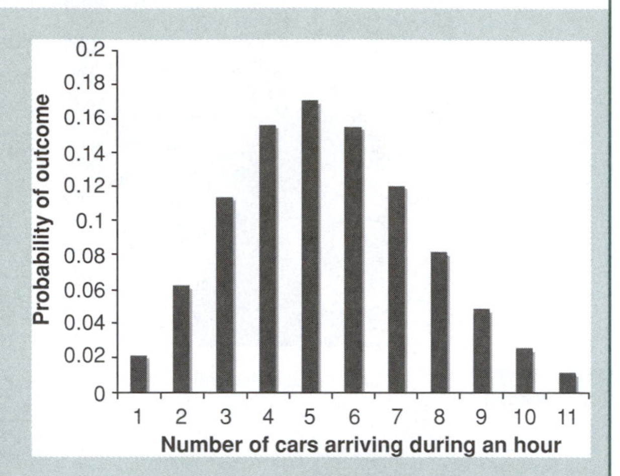

The **Poisson distribution** is an example of a discrete probability distribution. Discrete probability distributions have a finite number of outcomes, such as the flip of a coin (which has only two possible outcomes, heads or tails). The distribution was discovered by Siméon-Denis Poisson (1781–1840); Poisson was studying the likelihood that a certain number of discrete events occur, over a given time interval, if on average the events arrive with a certain frequency. For example, how many cars are likely to enter the freeway via a particular ramp, between 4 pm and 5 pm on June 23, 2025, which is a Monday, if the average arrival rate of cars at that time on a typical Monday is known?

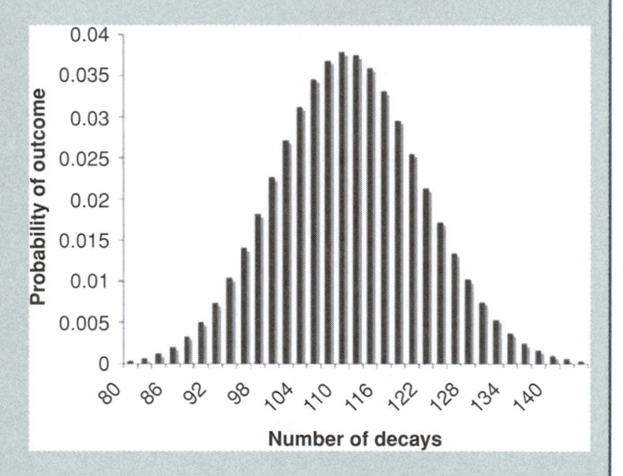

The Poisson distribution is expressed as:

$$f(k;x) = \frac{x^k e^{-x}}{k!}, \qquad \text{(Equation 1)}$$

where $f(k, x)$ is the probability that there are exactly k occurrences and x is the "expected" number of occurrences during that interval. For example, if cars arrive on the ramp on average every 11 minutes, then one would expect 60/11 or 5.4545 . . . cars to arrive during an hour. The Poisson distribution of this situation is shown in the figure.

Radioactive decay is a Poisson process (i.e., a process that follows a Poisson distribution). For a certain radioactive element, the expected number of radioactive decays that occur in any time interval is known. Radioactive chemicals are characterized by the number of radioactive decays that occur per minute: 1 Curie (Ci) of a radioactive substance exhibits (on average) 2.22×10^{12} radioactive disintegrations per minute. Therefore, a sample of 1 pCi (10^{-9} Ci) exhibits 2,220 disintegrations per minute. In a 3-second period, the expected number of decays is 111. The second figure shows the Poisson distribution (calculated from Equation 1) for the probability of observed decays during a 3-second counting interval.

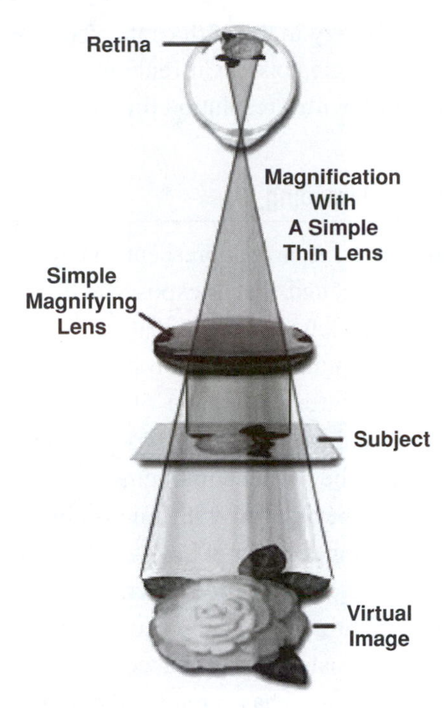

Figure 12.19 Magnification with a single lens.

small scale. Optical bioimaging methods have a long history, beginning with microscopy. The first microscopes were developed during the Renaissance (the 14th to 16th centuries) using convex glass lenses to produce magnified images of objects (Figure 12.19). These devices allowed the first views of cells, bacteria, and other microscopic structures.

Compound microscopes use lenses in combination to increase magnification. The magnification provided by convex lenses is based on the **refraction** of glass. When light travels through a uniform medium, it travels in a straight line. When light passes from one medium to another (see Figure 12.20), such as from air to glass, it may change direction; this is called refraction. Refraction is caused by the change in the speed of light and is governed by Snell's law:

$$\frac{\sin\theta_1}{\sin\theta_2} = \frac{v_1}{v_2}, \tag{12.6}$$

where θ_1 is the incident angle of the light to the interface, θ_2 is the refracted angle, v_1 is incoming speed of light, and v_2 is the outgoing speed of light. The speed of light is inversely proportional to the **refractive index** of the medium. Magnifying lenses refract the rays so that structure appears larger. The first microscopes could magnify only about nine times. Antonie van Leeuwenhoek, who is known

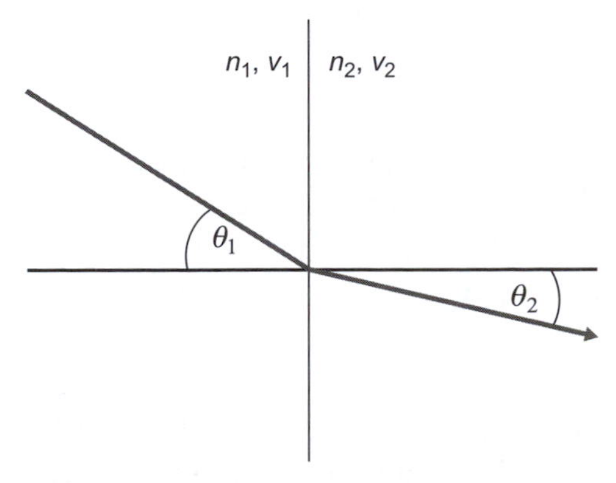

Figure 12.20 Refraction occurs at the interface between media and is described by Snell's law.

for his pioneering work in microbiology in the 17th century, was able to achieve magnifications of 200 times using single lenses. Current optical microscopes can magnify up to factors of 2,000 times with a resolution limit of about 0.2 microns.

12.5.1 Fluorescence and confocal imaging

Certain molecules called **fluorophores** are fluorescent: They emit light at a particular wavelength after they are excited, that is, exposed to light of a particular, shorter wavelength. Fluorescence is viewed by illuminating a specimen with the specific wavelength of light that stimulates the fluorophores. The microscope filters out the illuminating light, creating an image of the emitted light only. Fluorescent dyes can be used to stain specific cell types, which fluoresce (or "light up") when excited by light of the proper stimulating wavelength.

Functional information can also be imaged with microscopy using fluorescence. This technique has powerful implications in biological research. For example, gene expression can be imaged by altering deoxyribonucleic acid (DNA) to produce fluorescent versions of proteins. Green fluorescent protein (GFP), naturally produced by a kind of jellyfish, has greatly facilitated these studies because it is less harmful to the cell than are other synthetic fluorophores that have been developed. Cells that are modified to express GFP or GFP linked to other cellular proteins can be imaged in living animals (Figure 12.21). Variations of GFP that produce other colors have also been developed, enabling multiple objects to be imaged in a single experiment.

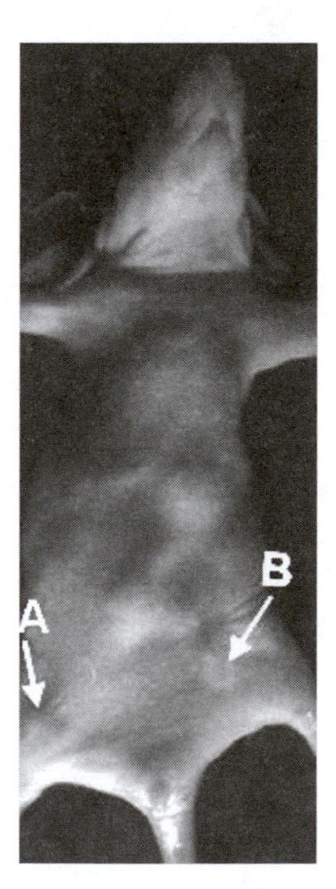

Figure 12.21 **Imaging of cells that express fluorescence in a living animal. A.** This location contains a U87 tumor with GFP fluorescence, emitting at 520 nm. **B.** This location contains a tumor with NB1691 Luciferase fluorescence with peak emission at 480 nm. Image courtesy of M. Waleed Gaber, PhD, Biomedical Engineering and Imaging Department, University of Tennessee Health Science Center, Memphis. (See color plate.)

Two-dimensional structure and dynamics can be seen through traditional microscopes. But fluorescence, when coupled with special microscopes, can yield three-dimensional images. Confocal microscopes, for example, allow imaging of multiple levels through a specimen. To achieve confocal microscopy, light emerging from points above and below a selected focal plane are filtered out by the use of a pinhole. By adjusting the focal plane, different levels can be imaged separately; these images can then be assembled into a three-dimensional image (Figure 12.22).

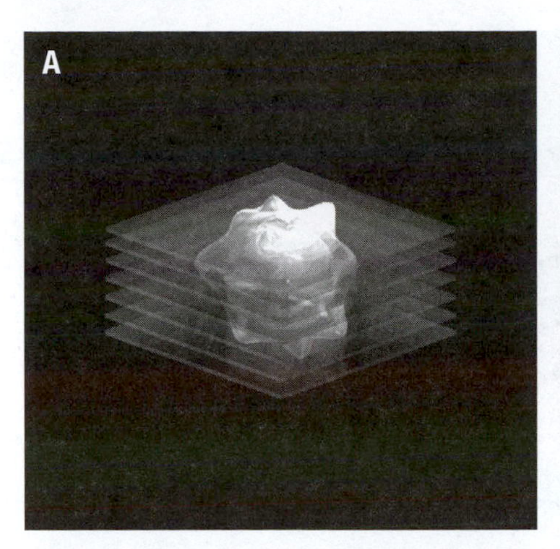

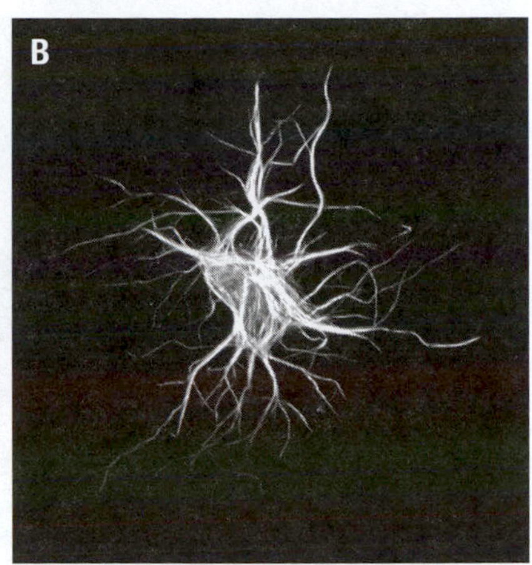

Figure 12.22 Confocal image. **A.** Series of images is taken on different focal planes. This set of sections—called a serial section—can be assembled into a three-dimensional image, as in this image of a hippocampal neuron **(B)**. Image courtesy of Robert S. McNeil and Dr. Gary Clark, Cain Foundation Laboratories, Baylor College of Medicine. (See color plate.)

12.5.2 Endoscopy and fiber optics

Although cameras can easily take pictures of external structure, and microscopes can image tiny specimens, the inside of the body cannot usually be imaged with light because the skin is opaque. However, there are a number of ways to use light to image internal organs. **Endoscopy** uses **fiber optics** to bring light into and out of the body through passageways, allowing views of internal structures. Organs and cavities within the body, such as the digestive system, can be examined with an **endoscope**, which can be flexible or rigid depending on the procedure. Basil Hirschowitz, along with Wilbur Peters and Larry Curtiss at the University of Michigan, developed the first modern endoscope, which was inserted through the mouth and down the esophagus to view the internal lining of the stomach.

Endoscopes—long snakelike devices with internal fiber optics and, often, channels for other instruments—are now available for imaging through every orifice of the body, including the lungs (bronchoscopy) and colon (colonoscopy). In addition, surgeons use tiny scopes to look inside tissues for which there are no convenient orifices. These tiny scopes are often used in the abdomen (laparoscopy) and joints (arthroscopy).

Fiber optics is based on the principle of **total internal reflection**, an optical phenomenon that can occur at boundaries between materials. When light crosses between materials with different refractive indices, light will normally be partially refracted and partially reflected. When light is coming from a high-refractive-index material to a low-refractive-index material (e.g., from water to air), if the angle of incidence is shallow enough (i.e., below a critical angle, so that light is traveling nearly parallel to the surface), total internal reflection occurs and no light is refracted. The critical angle for total internal reflection is determined by

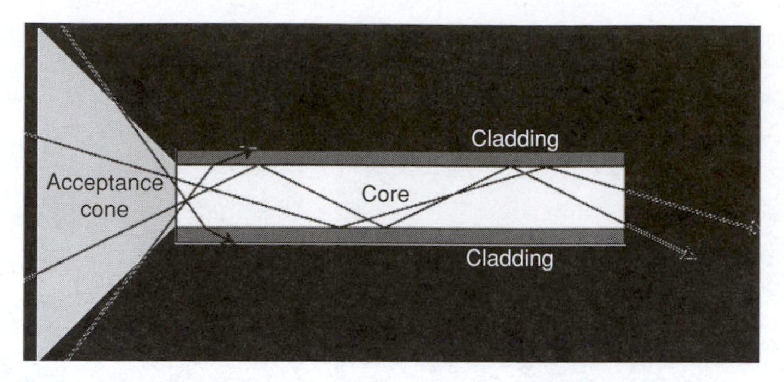

Figure 12.23 Optical fiber with total internal reflection.

solving Snell's law (Equation 12.6) for a refracted angle of 90°. This gives:

$$\theta_1 = \arcsin\left(\frac{v_1}{v_2}\right). \tag{12.7}$$

If this criterion is satisfied, the light does not cross the boundary and no light is lost. The light reflects at the boundary and bounces along, guided by the path of the fiber (Figure 12.23). A stream of water in the air can act as an optical fiber. In practice, optical fibers are typically made of glass with an internal cylindrical core of high refractive index and an outer cladding cylinder of lower refractive index. The fiber is designed so that when light is in the fiber, at each bounce, the critical angle will be satisfied, even if the rod is bent. Without total internal reflection, the light would be absorbed soon after it entered the fiber. Although the same effect could be accomplished by using a flexible tube coated on the inside with a perfect mirror, it could not be as small and it would be extremely difficult to coat perfectly. Total internal reflection fibers can be fabricated out of very thin glass, down to an eighth of a millimeter. Glass so thin will easily bend.

The endoscope uses an optical fiber to bring light into the body and many optical fibers to bring the image information out of the body. Each of the imaging fibers measures one pixel of the image. An array of fibers is used to produce an image. Additional access for drug administration or biopsy may also be provided.

12.6 MRI

Magnetic resonance (MR) provides exquisitely detailed three-dimensional images, especially of soft tissue that cannot easily be imaged in other modalities, such as CT. MRI provides excellent structural images of joints, the brain, and the abdomen with resolution of less than a millimeter. MR is also extremely versatile: It can now be used to measure biochemical composition, tissue function, and molecular diffusion, as well as structure.

MR uses properties of the magnetic dipole (**spin**) of atomic nuclei at high magnetic fields. The magnetic resonance phenomenon was discovered in 1946. MRI began in 1973 with the work of Paul Lauterbur at the University of Illinois

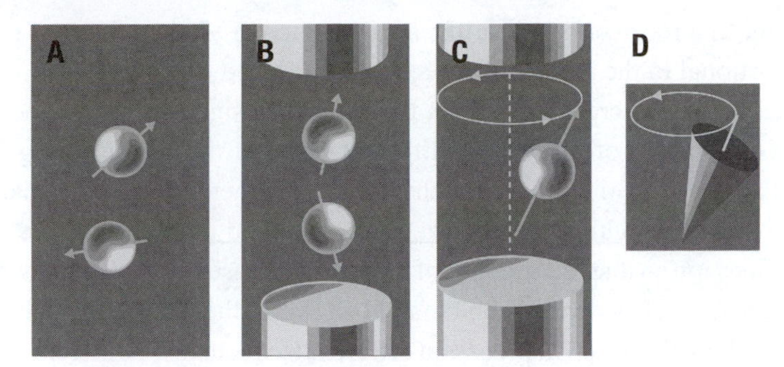

Figure 12.24

Physical basis of magnetic resonance imaging. Schematic of protons randomly oriented **(A)**, aligned after placement in the field **(B)**, and precessing like a top after radio frequency pulse (**C** and **D**).

and Peter Mansfield at the University of Nottingham. It began to be used clinically in the 1980s. Because of the significance of MRI in medical imaging, Lauterbur and Mansfield later won a Nobel Prize in Medicine for their work.

An MR scanner is a large cylinder with a hole running down the center, just large enough to fit a subject. (The tight fit makes some subjects claustrophobic.) The cylinder contains coils of metal wire that are used to generate a strong magnetic field (B_0) and other coils to pick up signals from the subject. The interior of the cylinder can also be quite noisy because of the fast switching of magnetic fields. To achieve the high current levels needed to produce the B_0 field, these coils are wound from wire several miles long, made of special materials, and are cooled to allow for **superconducting**. A standard clinical scanner has a magnetic field of 1.5 **Tesla**, which is about 3,000 times stronger than a refrigerator magnet. Research MR systems have even higher field strengths.

Imaging is usually done with hydrogen nuclei (protons) because they are so abundant (most tissues are predominantly made up of water), although other nuclei are possible. When a subject is placed into the magnetic field of an MR scanner, the magnetic moment vectors of their hydrogen nuclei tend to align parallel to B_0, creating their own net magnetization vector, M. At equilibrium, M is always aligned with B_0 (Figure 12.24).

Short pulses of electromagnetic radiation at a special frequency (called the **Larmor frequency**, which is dependent on the type of nucleus and the field strength) are applied in a plane perpendicular to B_0, creating a transverse magnetic field (B_1) so as to cause some of the hydrogen nuclei to change their alignment, or "tip." The pulses are stopped, and the nuclei spontaneously begin to return to their original equilibrium configuration. The angle of the initial "tip" is called the excitation angle. The change in alignment causes the M vector to wobble (precess) around the B_0 field at the characteristic Larmor frequency. The Larmor frequency is in the RF range (Box 12.1). As the nuclei precess, they release RF energy—which can be detected by the receiver coils surrounding the subject—as the molecules undergo their normal, microscopic tumbling ("relaxation"). The return to equilibrium is characterized by the longitudinal (*T1*) relaxation and the transverse (*T2*) relaxation. The precessing M vector induces an alternating RF current in a receiver coil just like a radio antenna. The strength of the current is proportional to the angle of the precession (Figure 12.24).

To use MR to create an image, the generated signals must be localized in the three dimensions of space using different forms of "spatial encoding." Additional coils are needed to temporarily change the magnetic field by a small amount at different points within the magnet. Magnetic field gradients can be generated in one direction so that the strength of the field increases along that axis. Because the Larmor frequency depends on the field strength, the hydrogen nuclei at different points along the axis will precess at different frequencies. This technique is known as "frequency encoding" because the frequency of the signal that is generated encodes the spatial location along that axis. The M vector has both a frequency and a phase, and the second dimension is encoded using the phase. Another field gradient is used, but only for a short period, just before the frequency-encoding gradient. This gradient changes the phase of the M vector, but not the frequency, and provides information on the second dimension. A different phase acquisition is needed for each location. The third dimension can be localized using an RF pulse designed to select a range of frequencies within a single slice during the corresponding field gradient. By systematic application of these spatial encoding procedures, an image of the whole, three-dimensional volume can be acquired slice by slice.

The signals produced are recorded, and the resulting data are processed to generate an image (Figure 12.25). Relaxation rates differ for different tissues (blood, bone, muscle, etc.), and the image reflects these differences. The brightness in MR is primarily determined by proton density, T1 relaxation, and T2 relaxation within the particular tissue. Proton density is primarily indicative of the presence of water. T1 relaxation shows good contrast for different types of soft tissue, whereas T2 relaxation is sensitive to pathology (edema, infarction, etc.). The versatility of MRI comes from the use of different sequences of gradients and RF pulses, which can be used to weigh these parameters differently and to measure different effects. In this sense, MR is a "programmable" modality. The characteristics of the imaging method can be tuned for good contrast between different tissues in the body or for different measurements. Unlike CT, MR can provide excellent soft tissue contrast. However, MR is not able to image bone well because bone has very little water content and thus produces almost no signal. Because of the greater complexity of the device, MRI is, however, more expensive than CT.

MR can be programmed for study of the chemical composition of tissues. **MR spectroscopy** exploits the fact that the chemical environment of the nuclei modifies the local magnetic field. This change in field will shift the Larmor frequency by a small amount, depending on the compound, and this change can be measured. Initially, spectroscopic information could not be localized and was acquired on an entire sample. Modern techniques allow spectroscopic imaging, although the resolution is currently on the order of 1 centimeter.

A way to measure brain function with MR based on blood flow is to use **fMRI**. Blood flow increases in areas of the brain that are actively performing a cognitive

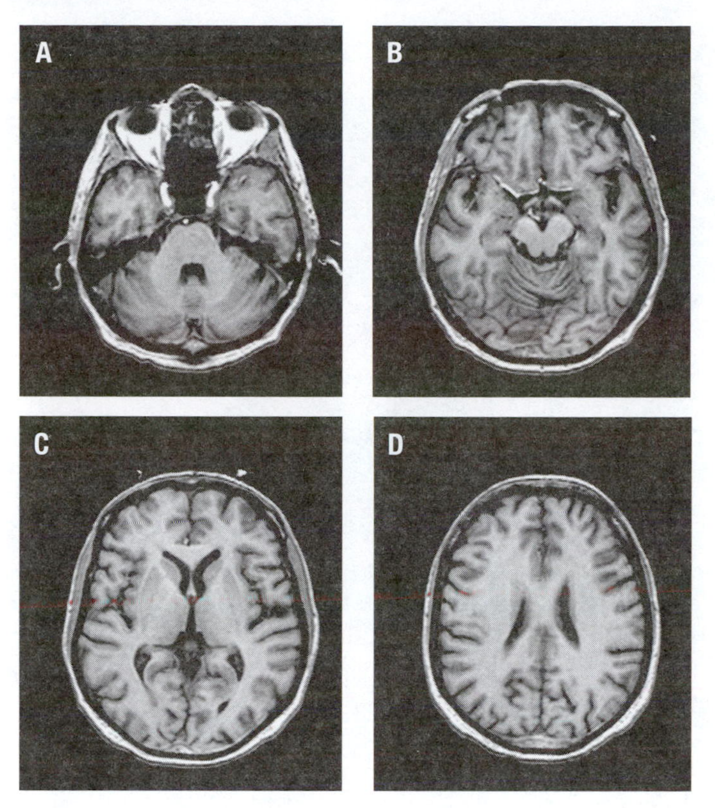

Figure 12.25 **Magnetic resonance (MR) images of the brain.** MR images of axial slices through the brain at four different levels from the level of the cerebellum **(A)** up through the top of the lateral ventricles **(D)**.

or motor task or reacting to a sensation. Blood is magnetic because it contains hemoglobin, an iron compound. MRI of iron causes a reduction in signal caused by the change in magnetic field near the iron. In an active region of the brain, blood flow increases, bringing in oxygenated blood. The oxygenation of hemoglobin, however, changes its magnetic properties and can lead to a small ($\sim$3%) signal change. Because the blood flow changes are very brief, fast imaging sequences must be used. Therefore, these images are lower in quality than typical MR images; the images must be acquired over many repetitions to see the change. The areas with increased blood flow can then be computed from the images. The result is a completely non-invasive method for measuring brain function that has wide applications in neuroscience. It is also being used clinically, for example, to localize areas of functional significance in conjunction with brain surgery.

12.7 Image processing and analysis

Computers and digital imaging play a central role in bioimaging. Much of bioimaging would not be possible without digital image formation, processing, and

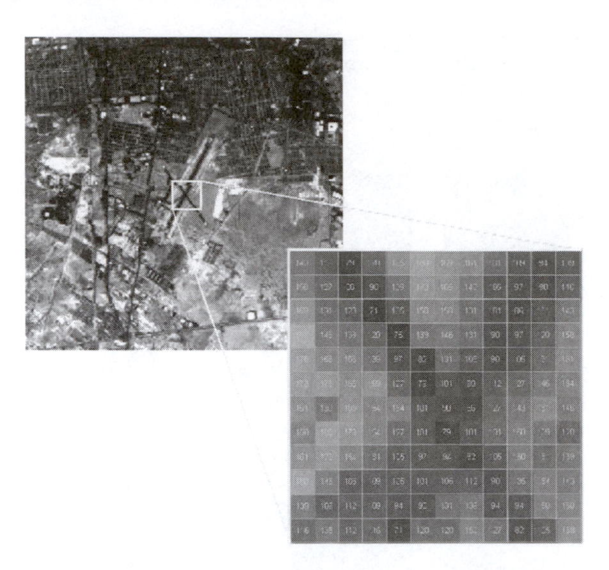

Figure 12.26

Example of intensity values in a digital image. Zoomed-in image segment showing discrete (quantized) intensity values at discrete (sampled) locations with grey level values displayed at each pixel. Image provided courtesy of USGS/EROS, Sioux Falls, SD. (See color plate.)

visualization. Image processing and analysis methods provide ways to improve the quality of images and to quantify what is in them.

12.7.1 Digitization

Digital images are essentially arrays of numbers (Figure 12.26). They are inherently discrete—that is, broken up into elements with a finite number of states—with respect to light intensity (or grey level) and space. For light intensity, images are typically quantized to 256 levels (corresponding to 8 bits of information, because $256 = 2^8$). That is, the intensity of the image at any position can be 0 (black), 255 (white), or any of the 254 shades of grey (1, 2, 3, . . . 254) in between. Although the human eye can distinguish fewer than 100 grey levels in an image, digital image processing on a computer can recognize any level of quantization. Therefore, some medical images have up to 65,536 grey levels ($65{,}536 = 2^{16}$, so this information can be stored in 16 bits). When there are too few quantization levels in a smooth, gradually shaded region, you will see "false contours" at the arbitrary curve where there is a transition from one brightness level to the next.

For two-dimensional images, the individual, usually square, spatial elements of the image are called pixels (short for picture elements); in three dimensions, the spatial elements are usually cubes, called voxels.

The **field of view** of an image is the extent of space that has been imaged. When the field of view is **sampled**, the space within it is discretized and divided into a matrix of pixels. For three-dimensional images, an additional dimension, a slice thickness, is also specified. The size of each pixel is the field of view divided by the matrix size.

SELF-TEST 2 Consider an image of the head, with a field of view of 24 centimeters square. If the image matrix size is 256 × 256, and slices are 1.5-mm thick, what is the size of each voxel?

ANSWER: 0.9375 mm × 0.9375 mm × 1.5 mm

Voxel size is sometimes confused with **resolution**. Resolution is a measure of how much detail can be seen in an image; resolution is defined, therefore, based on the ability to see two objects as distinct when they are a small distance apart. The minimum distance that two tiny objects (or points) can be seen as separate is denoted the resolution. If the objects are closer than the resolution, they blur into one blob.

Resolution is one way of characterizing image quality. Image quality can also be measured by **contrast** and **noisiness**. Contrast is the relative change in intensity (from dark to bright or vice versa) between two neighboring regions. The higher the contrast, the easier it is to perceive the boundary of the object. Noise in images is like noise in sound recordings: It distorts the signal. Noise, in any measurement, can be defined as the difference between the true value and the measured value. All measurements are subject to noise. Typically, image noise appears as random speckles in the image, which are introduced because the image intensities at some points are different than the true values. One way to characterize these unwanted variations in image intensities is to measure the variance of the grey levels within a region of the image that is known to be uniform.

12.7.2 Image enhancement

Computer processing can be used to improve the quality of images. One way to enhance images is to adjust the intensities to improve contrast. The best way to understand contrast in images is in terms of the intensity histogram, which is simply a count of the number of pixels at each intensity or grey level (see Figure 12.27). If the histogram is concentrated in one portion of the grey level range, the image contrast will not be good. If concentrated at the high end, the image will appear overexposed (Figure 12.27b); at the low end, underexposed (Figure 12.27a). The best overall contrast occurs when the histogram is flat or equalized (Figure 12.27d), so that the full contrast range is best used. Computer algorithms have been devised to stretch and shrink the grey levels to transform the contrast of images.

Another way to enhance images is to use **linear filtering** (Figure 12.28). Linear filtering can be defined in terms of small arrays of numbers (kernels) that operate on the image. The filtered image is computed by a simple mathematical operation: The filter is centered at each pixel or voxel in the image, and a new value for that voxel is determined from the sum of the products of the corresponding elements. This process is also called discrete convolution.

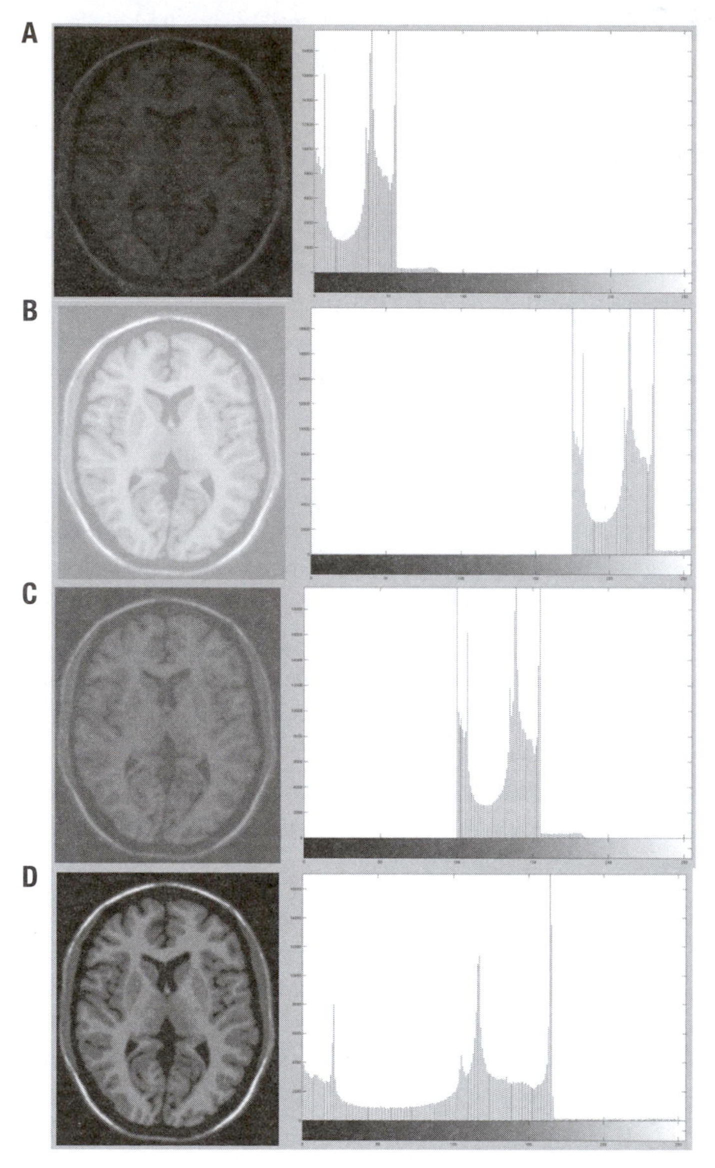

Figure 12.27 **Histograms and contrast enhancement.** The intensity histogram gives an indication of the appearance of an image in terms of brightness and contrast.

For example, a simple smoothing filter can be used to reduce noise in an image. The filter kernel is:

$$\begin{pmatrix} 1/9 & 1/9 & 1/9 \\ 1/9 & 1/9 & 1/9 \\ 1/9 & 1/9 & 1/9 \end{pmatrix}.$$

When this kernel is applied, in a filtering operation, the value of each pixel in the image is replaced by the average of 9 pixel values at and surrounding it. The resulting image will look smoother or blurrier, but the noise will be decreased.

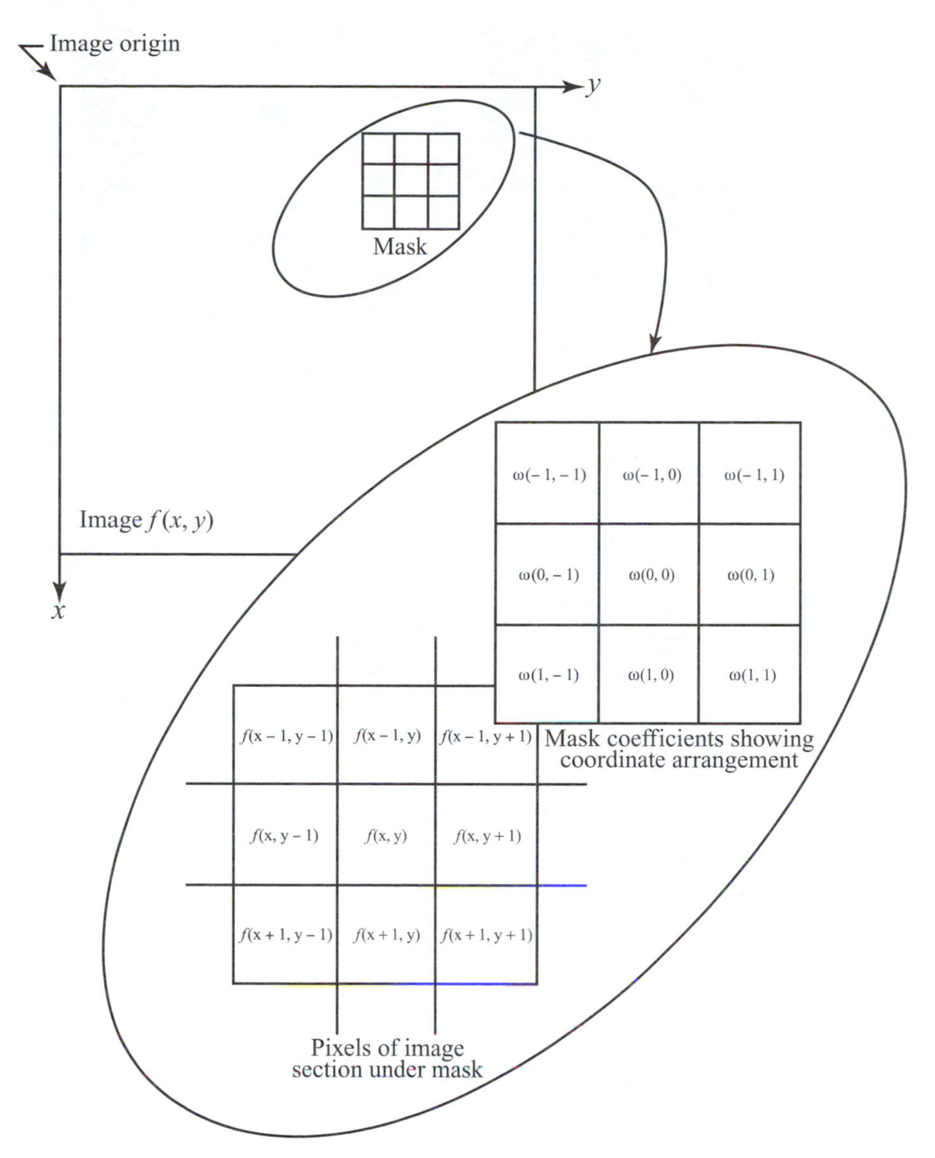

Figure 12.28 Linear filtering with discrete convolution (1).

You can also make an image look sharper using linear filtering. One way is to use a technique called **unsharp masking**. This technique has its origins in darkroom processing of photographs. The idea is to take an image and subtract a blurred version from it. The kernel here is:

$$\begin{pmatrix} -1 & -1 & -1 \\ -1 & 9 & -1 \\ -1 & -1 & -1 \end{pmatrix}.$$

This kernel sharpens because it emphasizes the difference between the pixel and its neighbors.

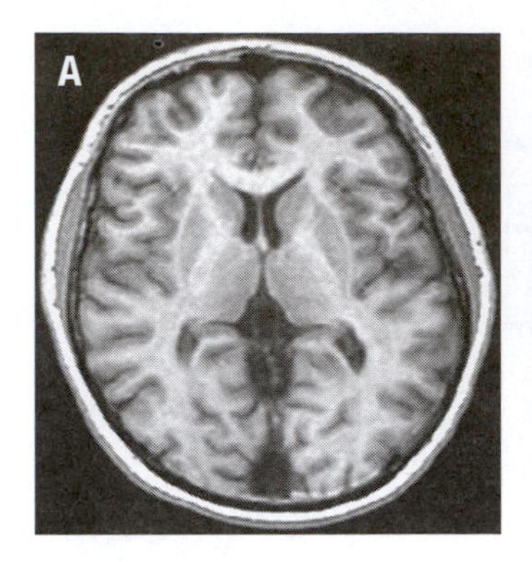

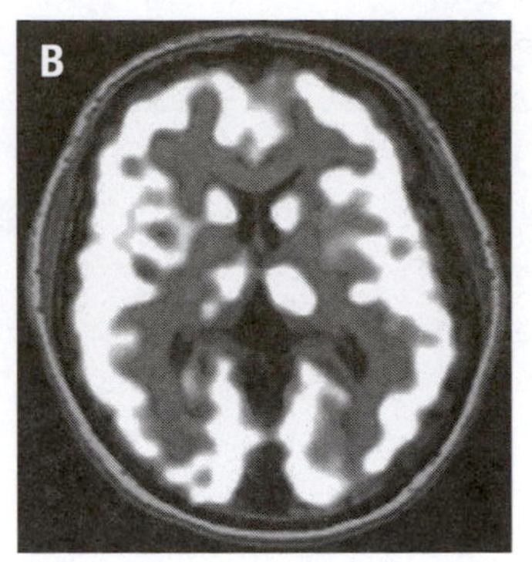

Figure 12.29 **An example of brain image registration showing, a magnetic resonance image aligned (A) and overlaid (B) with a positron emission tomography image from the same subject, showing the corresponding structure and function.** (See color plate.)

12.7.3 Registration and segmentation

Two key tasks for quantifying images are **registration** and **segmentation**. Image registration is the process of aligning images. The goal is to determine the transformation (i.e., relative alignment) between two images that best brings homologous structure/anatomy into correspondence. The registration process involves three elements: a transformation method, a method for quantifying "matching," and an optimization method. The optimization process starts with an initial guess as to how the images align, and then refines the guess in an effort to improve the "matching." For example, the matching might be quantified by correlation of intensity between the images.

The simplest transformation method is rigid, in which the images differ only by a rotation and a shift. In this case, the optimization process searches for the best rotation and shift parameters to make the best match. Nonrigid transformations also allow for warping, and require many more parameters to optimize.

The first practical use of image registration was in image-guided surgery and was accomplished, using x-rays, only two weeks after their discovery was published. In Birmingham, England, in 1895, a woman had a broken needle in her hand. An x-ray was taken and was aligned with her hand; the x-ray image was used to guide the scalpel to remove the needle.

Image registration has many other applications. Images acquired from more than one modality often need to be aligned for interpretation; for example, a structural image from MR and a functional image from PET provide complementary information (Figure 12.29). Studies involving many images over time, such as happens in fMRI, may require alignment because the subject has moved gradually during the imaging process. When the images represent the same subject, and there has been no change in shape, the registration only requires a rigid transformation. When images represent different subjects or changes in shape—for example, for a fetus during development—nonrigid transformations are needed.

Segmentation is the process of identifying smaller structures (organs, cells, etc.) within images. Pixels are then labeled according to the structure in which they belong. Structures may be defined based on their homogeneity of appearance.

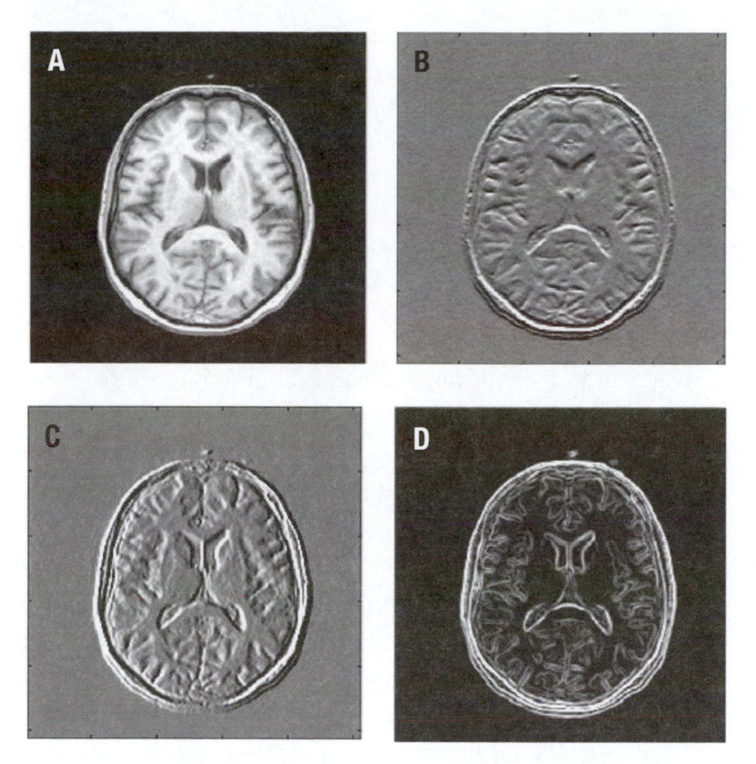

Figure 12.30 Magnetic resonance images of a brain showing the original intensities **(A)**, horizontal **(B)** and vertical **(C)** edges (dark is negative, bright is positive), and edge magnitude **(D)**.

A structure may be, for example, brighter than the surrounding area: The labeling criterion can then be based on that brightness.

One of the most straightforward methods of segmentation is thresholding: All pixels brighter than a threshold value can be labeled as belonging to the object. The threshold brightness may be determined manually or automatically; automatic methods are often based on a criterion that best divides the histogram.

Alternatively, edges between homogeneous regions can be computed. One way to compute edges is by using linear filtering, which can be used to extract certain features, such as edges, from the image. This kernel:

$$\begin{pmatrix} -1 & 0 & 1 \\ -1 & 0 & 1 \\ -1 & 0 & 1 \end{pmatrix}$$

will give a high magnitude at vertical edges. The value may be positive or negative depending on whether the edge goes from dark to light or light to dark. The corresponding kernel:

$$\begin{pmatrix} -1 & -1 & -1 \\ 0 & 0 & 0 \\ 1 & 1 & 1 \end{pmatrix}$$

will respond to horizontal edges. These responses can be combined to give edges at all angles. These edges can then be linked together to form boundaries between regions enabling segmentation (Figure 12.30).

Registration can also be used for segmentation in a process called atlas-based segmentation. One image, the atlas, is fully segmented, perhaps manually. When this image is registered to the subject image, the labeling from the atlas can be applied to the subject, giving a fully segmented image in one step.

Summary

- The technology to image inside the body and at a microscopic scale has greatly evolved and has significantly enhanced our ability to diagnose and treat disease, providing an array of imaging modalities, each with strengths and weaknesses.
- X-ray and CT imaging provided the first true windows into the body, enabling the visualization of internal structure.
- Ultrasound imaging is fast, portable, and inexpensive, providing dynamic images of structure and velocity.
- Nuclear medicine allows the imaging of function within the body, using radioactive compounds to quantify physiologic changes in disease.
- Microscopy has matured in the past hundred years with fluorescence and confocal imaging, enabling the visualization of structure and function at the microscopic scale and in three dimensions.
- Endoscopy literally brings a camera into the body to image internal structure for diagnostic and surgical applications.
- MRI is the most versatile modality, providing structure and function in a programmable way.
- Computer processing of images is essential for the formation and quantification of images, improving their appearance and enabling the measurement of anatomy and function.

REFERENCES

1. Gonzalez RC, Woods RE. *Digital Image Processing.* Boston, MA: Addison-Wesley; 2004.

FURTHER READING

Bankman IN. *Handbook of Medical Imaging, Processing and Analysis*. St. Louis, MO: Academic Press; 2000.

Hendee WR, Ritenour R. *Medical Imaging Physics*. St. Louis, MO: Mosby; 2003.

Kremkau FW. *Diagnostic Ultrasound: Principles and Instruments*. Philadelphia, PA: W.B. Saunders; 2002.

Lacey AJ. *Light Microscopy in Biology: A Practical Approach*. Oxford, UK: IRL Press; 1989.

Russ JC. *The Image Processing Handbook*. Boca Raton, FL: CRC Press; 2002.

Saha GB. *Physics and Radiobiology of Nuclear Medicine*. New York: Springer; 2001.
Smith RC, Lange RC. *Understanding Magnetic Resonance Imaging*. Boca Raton, FL: CRC Press; 1998.

KEY CONCEPTS AND DEFINITIONS

coincidence detection a method for simultaneously detecting two signals at the same time

computed tomography (CT) a medial imaging method that uses x-rays to image internal structures of an object by compiling a series of two-dimensional cross-sections to create three-dimensional images through computer technology

contrast the relative change in intensity (from dark to bright or vice versa) between two neighboring regions in an image

contrast agent a material that is injected during imaging, such as barium sulfate or iodine, to make a particular tissue more visible during radiological imaging tests

cyclotron a type of particle accelerator used to produce radioactive substances by alternating electric fields to add energy to subatomic particles

digital a method of storing information electronically through the use of the binary language, or an array of numbers, typically "1" and "0"

Doppler effect a change in the apparent frequency or wavelength of a wave caused by a change in the distance between a source and an observer

electron microscopy the use of a microscope with the ability to magnify very small objects by deflecting a beam of electrons with electromagnets

electronic collimation an optical device used in a PET camera consisting of channels of lead that only allow penetration of gamma rays that originated directly in line with the channel

endoscope a tubular optical instrument used during surgery to illuminate the organs of the body through the use of fiber optic cables

endoscopy the use of fiber optic cables in a flexible tube (an endoscope) to light the internal organs of the body for examination or to treat a condition

energy window a defined range of energies in which photons are counted to obtain a gamma image

fiber optics thin, flexible fibers made of glass that are used to transmit images or signals in the form of light at high speeds with little signal loss

field of view the area that has been or can be imaged through an optics device

fluorophores certain molecules that have the ability to emit light at a particular wavelength after they are exposed to light of a specific, shorter wavelength

fourier transform a mathematical operation that converts a function into a set of oscillatory functions that can be categorized with respect to their wavelength (or frequency)

functional magnetic resonance imaging (fMRI) a non-invasive imaging procedure that uses radio waves and a magnetic field to image the changes that occur in the brain because of blood flow and cognitive thinking

gamma camera a device used in the medical field to scan patients who have been injected with radioactive materials to produce images of the radioactivity

half-life the time that it takes for half of the atoms of a radioactive element to decay

iodine-131 (^{131}I) a radioisotope of iodine, also called radioiodine, used in nuclear medicine for both imaging and therapy

ionizing radiation electromagnetic radiation with sufficient energy to knock electrons out of atoms within the material they penetrate, thus turning the atoms into ions

Larmor frequency the frequency at which a magnetic nucleus can be excited proportional to the strength of a magnetic field

linear filtering a process to enhance an image by determining a new value for a pixel from the sum of the products of the corresponding elements

magnetic resonance imaging (MRI) a computer-generated method of medical imaging for producing cross-sectional images of internal organs and tissues through the use of radio waves and magnetic fields

magnetic resonance spectroscopy (MRS) a method of imaging human tissues and organs into three-dimensional computer-generated images to study their chemical composition

modalities the different types of methods used in imaging

noisiness the difference between a true value and a measured value that can distort a signal

oxygen-15 (^{15}O) a radioisotope of oxygen with a half-life of 2.04 minutes, commonly used in the studies of respiratory function and in positron emission tomography (PET)

piezoelectric crystal a special type of crystal used to generate ultrasound that will vibrate in response to an electrical signal, or driving voltage, and can also induce a voltage after receiving a vibration

pixel the smallest unit of a digital image that can be displayed and manipulated

Poisson distribution a probability distribution that occurs when counting occurrences of a particular event in a long series of observations

positron emission tomography (PET) a nuclear medical imaging technique used to produce three-dimensional images of the body through the decay of radioisotopes

projection a shadow of a three-dimensional object created on a plane

radioactive molecules molecules that are unstable and spontaneously decay to release radiation energy

radiodensity a property of materials, the relative ability of electromagnetic radiation to penetrate the material

real time recording structure and dynamic function from rapidly moving tissues, such as a beating heart, and visualizing them at the same speed at which they happen

refraction the change in direction of a path of light as it passes from one medium to another

refractive index the ratio of the speed of light traveling through a vacuum to the speed of light traveling through a particular substance

registration the process of aligning images into one coordinate system to better define or compare them

resolution a measure of how much detail can be seen in an image

resonant frequency the frequency at which a material naturally vibrates

sample the process of drawing a group from a larger population and using it to estimate the characteristics of the population as a whole

segmentation the process of identifying smaller structures, such as organs and cells, in images

signal-to-noise ratio (S/N) in imaging, the ratio of signal (activity) to error in the signal (noise), which is usually expressed in decibels

single photon emission computed tomography (SPECT) an imaging technique in nuclear medicine in which gamma rays are used to produce two-dimensional image slices in the body that are converted to three dimensions through computer reconstruction

spin the movement of an electron around its axis in a clockwise or counterclockwise direction

superconducting the process of conducting electric current at very low temperatures in materials that will show no electrical resistance at those temperatures

technetium-99m a man-made gamma-ray–emitting element with a half-life of 6 hours used in medical diagnostic studies

Tesla SI unit used to describe the density of a magnetic field

total internal reflection complete reflection of light at the boundary between two different media at an angle greater than the critical angle

tracers radiopharmaceuticals used in nuclear medicine that are ingested, inhaled, or injected to track the movement of a substance through the body

ultrasound imaging the propagation of high-frequency sound waves (2–13 MHz) through tissue to record the structure and dynamic function of rapidly moving tissues, such as a beating heart or a developing fetus, at the same speed at which they happen

unsharp masking the process of increasing the sharpness of an image through computer manipulation by subtracting out a blurred version of the image to enhance detail

NOMENCLATURE

A	Activity, number of events	
B_0	Magnetic field	Tesla (T)
c	Speed of the wave or photon in a given medium (e.g., sound, light)	m/s
d	Distance	m
f	Frequency	Hz, MHz, GHz
f_2	(In measuring Doppler effect frequency shifts) Frequency of the wave reflected back to the source	Hz, MHz, GHz
f_0	(In measuring Doppler effect frequency shifts) Frequency of the wave created by the source	Hz, MHz, GHz
N	Noise level	
S	Signal level	

S/N	Signal-to-noise ratio	dB
t	Time-of-flight of a wave	s
T	Period	s
v	Velocity	m/s
v_1	(In refraction) Incoming speed of light	m/s
v_2	(In refraction) Outgoing speed of light	m/s

GREEK

λ	Wavelength	m, nm
θ	(In measuring Doppler effect frequency shifts) Angle between the direction of flow and the ultrasound pulse	Degrees
θ_1	(In refraction) Incident angle of light to the interface	Degrees
θ_2	(In refraction) Refracted angle	Degrees

QUESTIONS

1. A patient comes into the emergency room having been shot by a shotgun. What kind of imaging would you do to find the metal pellets so that they can be removed?
2. Imagine a SPECT image of a patient with no radioactivity injected but with a number of capsules filled with radioactivity attached to his skin. These are called "fiducial markers" and can be used for image registration if their locations can be accurately determined. What would the histogram of this image look like? Describe a method of localizing the markers.

PROBLEMS

1. In Doppler ultrasound, given a flow of X m/s, using a Y MHz transducer, what is the largest angle to the flow that would give a change in frequency of 5%?
2. If your detector measures 250 gamma photons every second from a radioactive sample governed by a Poisson distribution, how long do you need to measure to get a signal-to-noise ratio of 10?
3. What are the wavelengths for the radiation used in FM radio transmission, which spans the frequency range of 88–108 MHz? How much energy does this form of radiation carry?
4. You must design an optical fiber for endoscopy. If you use glass with a refractive index of 1.5 for the cladding, what refractive index do you need for the core to achieve a critical angle of 60°?
5. Design an image-processing filter that will give high values for thin, bright horizontal lines in the image.
6. A student performing a nuclear scan notices that he cannot resolve a small vessel that he is trying to image. He determines that this is because of the

small distance between the apertures on the lead collimator. He decides that more counts and thus a better image will be obtained if he removes the collimator and images using only the NaI scintillation crystal. Is this idea sound? Explain (using pictures) why or why not.

7. Because of a malfunction in the hardware of an MRI scanner, the RF pulse amplitude is twice what it should be. The operator of the scanner observes no signal when applying a 90° pulse.
   a. Why, assuming you were unaware of the malfunction, would you find this unusual?
   b. Why is there no signal?
   c. What, assuming you were aware of the malfunction, could you do to get maximum signal? [Hint: $\theta_{\text{flip}} = \gamma B_1 T$, where γ is the gyromagnetic ratio, B_1 is the magnitude of the RF pulse, and T is time.]
8. A radiologist has asked for your help determining whether a tumor in his patient has increased in size. He has had difficulty because the relative intensity of the tumor tissue and surrounding tissue are nearly the same; however, he does know the T1 and T2 relaxation rates of both tissue types. From this information, you plot the relaxation curves shown.

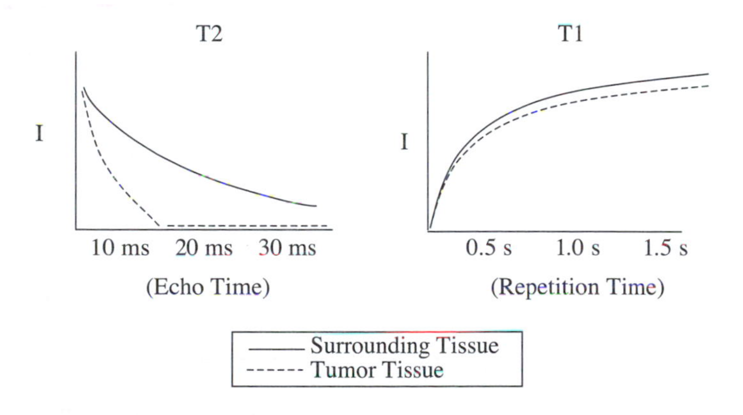

   a. Which (if any or both) of the two plots is useful and why?
   b. To what would you set the echo or repetition time to maximize contrast (approximately)?

13 Biomolecular Engineering I: Biotechnology

LEARNING OBJECTIVES

After reading this chapter, you should:

- Understand the relationship between biomolecular engineering and chemical engineering.
- Understand the concepts of drug effectiveness and toxicity, the limitations of some of the simplest methods of drug administration, and the need for controlled delivery systems that optimize therapy for a particular drug.
- Understand how polymeric materials with different physical properties can be fashioned into matrix, microsphere, transdermal, and other delivery systems.
- Understand how tissue engineering has emerged as a possible solution for organ replacement or healing.
- Understand the biological significance of materials with nanoscale dimensions and how material scientists are learning to assemble materials that use or mimic biological principles of self-assembly.

13.1 Prelude

The early chapters of this book introduce some of the chemicals that are important in human biology. In fact, it is possible to think of the human body as an elaborate bag of chemicals. In the subspecialty called **biomolecular engineering** (or biotechnology), biomedical engineers examine the changes in chemical components within a biological system and develop methods for modifying these chemicals or their interactions. The concept of introducing chemicals to induce a change in a biological system is familiar; for example, we all have some experience with taking purified chemicals such as acetaminophen or ibuprofen as drugs to relieve pain. But new biological tools now make it possible to consider more complex chemical interventions such as gene therapy (in which a new deoxyribonucleic acid [DNA] sequence is introduced to allow expression of a new genetic activity).

The concept of the human body as a bag of chemicals is convenient and comforting. Chemicals can be produced; chemicals can be added or replaced; chemicals are (sometimes) inexpensive. In fact, several decades ago, it was widely

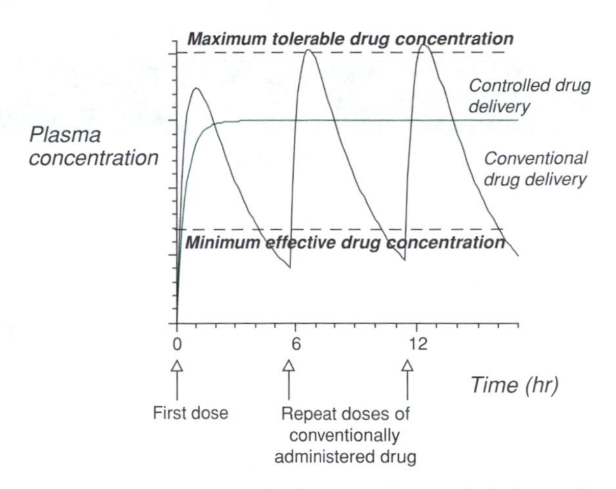

Figure 13.1 **Patterns of drug concentration after administration.** After administration by conventional methods (e.g., subcutaneous injection of a drug solution or suspension or ingestion of a pill or capsule), the drug is absorbed into the bloodstream and available to tissues within the body. As the drug is eliminated from the body, the plasma concentration eventually falls out of the therapeutic range, necessitating repeat dosages of the drug. For many drug therapies, this approach is unsatisfactory because doses must be repeated frequently. By engineering a controlled drug-delivery system, drug levels can be maintained within the therapeutic range for an extended period.

reported that the chemicals in the human body were worth about $1. Other investigators estimate the value at closer to $6 million (particularly if the chemicals are purchased from scientific suppliers). Regardless of the dollar value, we often imagine that our bodies can be supplemented, mended, and improved through the addition of chemicals, but the body contains large numbers of many different kinds of chemicals and, as in many complex systems, unexpected patterns can emerge from the sum of the parts. Although the body is composed of chemicals, we don't understand the identity of all of the chemicals in the system, the forces that hold these chemicals together, or the network of interactions that mold the chemicals into an organism. Yet we often attempt to heal with chemicals. This is the challenge of **drug delivery** (Figure 13.1).

Biomolecular engineering includes drug delivery, but also many similar areas of inquiry in which chemical engineering principles are applied to biomedical engineering (BME) problems. **Chemical engineering** involves mathematical analysis of chemical reactions and transport phenomena; chemical engineering analysis is particularly well suited to problems involving multiple chemical species participating in multiple reactions. This method of analysis has long been used for the development of large-scale chemical-processing methods, such as those involved with the manufacturing of antibiotics and vaccines. But now, given the new tools of molecular biology and the new abundance of genomic information, one can begin to consider new applications of biomolecular engineering, including the adoption of engineering principles to the microscopic scale. We can begin to make predictions on the influence of changes in the chemistry of specific cells within the body. Table 13.1 lists some of the important subdivisions of biomolecular engineering that are either established or emerging.

Table 13.1 Some of the subdivisions of biomolecular engineering

Aspect of biomolecular engineering	Brief definition	Example
Bioprocess engineering	Use of engineering analysis to design large-scale processing of biological materials	Bioreactor design; bioseparations such as chromatography and electrophoresis
Metabolic engineering	Analysis and redirection of the metabolic activity of a cell	Gene therapy to correct metabolic defects
Biomaterials engineering	Design and synthesis of new materials that can be used in biological systems	Materials for prosthetic heart valves
Pharmacological engineering	Analysis of modes of drug administration and the fate of drugs in the human body	Controlled delivery systems for contraceptives; design of dose regimens for chemotherapy drugs
Tissue engineering	Quantifying structure–function relationships in normal and pathological tissues, developing new approaches to repair tissues, and developing replacements for tissues	Tissue-engineered skin; polymer scaffolds for tissue regeneration
Cellular engineering	Use of functional genomic concepts to predict changes in cellular function with gene expression, regulation of cellular function by ligand binding to receptors and intracellular signaling networks	Analysis and manipulation of cell adhesion and motility
Nanobiotechnology	Use of engineering tools to create devices or materials with nanometer scale precision, and the application of these devices in biological systems	Nanoparticles for drug targeting

13.2 Drug delivery

Most people have experienced multiple forms of drug administration: Pills, injections, lotions, and suppositories have been in common use for centuries. People eat to survive and usually enjoy (or at least tolerate) the act of ingestion. For this reason, oral dosage forms are usually preferred: They are painless, uncomplicated, and self-administered. Unfortunately, many drugs are degraded within the gastrointestinal tract or not absorbed in sufficient quantity to be effective. Therefore, these drugs must be administered by intravenous, intramuscular, or subcutaneous injection (injection into a vein or a muscle or under the skin).

A variety of other modes of administration, less common than oral or injection, have evolved because of specific advantages for particular agents or certain diseases (see Table 13.2). The sophistication of drug administration has generally followed discoveries in anatomy and physiology. Egyptian physicians used pills, ointments, salves, and other forms of treatment more than 4,000 years ago. Intravenous injections were first performed in humans in 1665, only a few decades after Harvey's description of the circulatory system. Subcutaneous injections were

Table 13.2 Common routes of drug administration

Route of administration	Example	Advantages	Disadvantages
Intravenous injection	Antibiotics for sepsis	• 100% bioavailability	• Discomfort to patient • Requires health care provider • Risk of overdose or toxicity • Risk of infection
Intravenous infusion	Heparin for anticoagulation	• 100% bioavailability • Continuous control over plasma levels	• Requires hospitalization • Risk of infection
Subcutaneous injection		• Usually high bioavailability	• Discomfort to patient
Intramuscular injection	Insulin for diabetes	• Usually high bioavailability	• Discomfort to patient
Oral	Aspirin, acetaminophen, ibuprofen	• Convenient • Self-administered	• Drug degradation prior to absorption • Limited absorption of many drugs
Sublingual or buccal	Nitroglycerin for angina	• Avoids first-pass metabolism* in liver • Self-administered	• Limited to lipophilic, highly potent agents
Ophthalmic	Pilocarpine for glaucoma	• Local delivery • Self-administered	• Discomfort to some patients • Frequent administration
Topical	Antibiotic ointments	• Local delivery • Self-administered	• Limited to agents that are locally active
Intra-arterial injection	Chemotherapy, in some cases	• Control of vascular delivery to specific regions	• High risk
Intrathecal injection	Pain medication, in some cases	• Direct delivery to brain	• Limited drug penetration into brain tissue • High risk
Rectal		• Avoids first-pass metabolism in liver • Self-administered	• Discomfort leads to poor compliance in some patients
Transdermal	Nitroglycerin patches for angina	• Continuous, constant delivery • Self-administered	• Skin irritation • Limited to lipophilic, highly potent agents
Vaginal	Spermicides	• Self-administered	• Discomfort leads to poor compliance in some patients
Controlled release implants	Norplant® for contraception	• Long-term release	• Requires surgical procedure

Note: *First-pass metabolism occurs when drug molecules enter the circulation through a mucosal surface and then circulate through the liver, where they can be metabolized before distributing throughout the rest of the body.

introduced in 1853, and the modern hypodermic syringe was developed in 1884. Although we now have a good understanding of many aspects of human anatomy and physiology, most current methods for drug administration are direct descendants of these practices, and have changed little over the last few centuries. Despite this long experience with drugs and drug administration, people still die of cancer, infectious disease, and many other conditions for which active drugs are available. Are there better ways to deliver drugs? How can better methods be created?

Controlled drug delivery is the engineering of physical, chemical, and biological components into a system for delivering controlled amounts of a therapeutic agent to a patient over a prolonged period. Controlled drug-delivery systems can take a variety of forms like miniature mechanical pumps, polymer **matrices** or **microparticulates**, externally applied transdermal patches, and transplanted, genetically engineered cells. In most cases, the goal of controlled drug delivery is maintenance of **plasma** or tissue drug levels at a constant level for a prolonged period (Figure 13.1).

Physicians have long used the concept of controlled drug delivery to treat certain diseases. Patients with serious blood-borne infections are treated by continuous intravenous infusion of antibiotics. By monitoring the antibiotic concentration in the patient's blood and adjusting the rate of infusion, the physician precisely controls the delivery of the agent. Because of the high cost and need for hospitalization, this approach is limited to acute, life-threatening illnesses. To apply this same concept more widely, a variety of novel drug-delivery strategies have been developed.

The most obvious approach for controlled drug delivery involves miniaturization of the familiar infusion system. Mechanical pumps, either totally implantable or requiring **catheters**, have been used to deliver **insulin**, **anticoagulants**, **analgesics**, and cancer chemotherapy (Figure 13.2). Although the technology is mature and accepted by many patients, this approach has certain disadvantages. Drugs are maintained in a liquid reservoir prior to delivery, so only agents that are stable in solution at body temperature can be used. The devices are bulky, expensive, and only suitable for highly motivated patients who can visit their physicians regularly. Because of the opportunity for precisely controlled delivery, however, pumps may find additional important applications, particularly when coupled with implanted biosensors for **feedback control**.

The disadvantages of pumps can be avoided by using controlled release polymers as drug-delivery vehicles. In the early 1960s, the first methods for releasing drugs from biocompatible polymers were developed. Since that time, many novel approaches to polymeric controlled release have been tested; as a result, techniques for the production of polymer implants that slowly release drug molecules are now available (Figure 13.3). A variety of polymeric materials have been used as the basis of controlled drug-delivery systems (see Table 13.3). For a specific therapeutic agent, selection of the appropriate biocompatible polymer and method of device fabrication are used to modify the rate, pattern, and duration of drug release.

Table 13.3 Polymers used in controlled drug-delivery systems and tissue engineering

Class of material	Examples
Nondegradable polymers	Poly(siloxane) (e.g., Norplant®) Poly(ethylene-co-vinyl acetate) (e.g., Progestasert® and Ocusert®) Polyurethanes
Biodegradable polymers	Poly(lactide-co-glycolide) (Lupron Depot®) Polyanhydrides Polyorthoesters Polyphosphazenes
Swellable polymers	Poly(2-hydroxyethylmethacrylate) Poly(acrylic acid) Poly(vinyl alcohol)
Biopolymers	Proteins (e.g., collagen, gelatin, albumin) Polysaccharides (e.g., alginate, chitin, starch, dextran)

13.2.1 Nondegradable polymeric reservoirs and matrices

The first controlled release **polymer** systems were based on silicone **elastomers**. In 1964, researchers recognized that certain dye molecules could diffuse through the walls of silicone tubing. This observation led to the development of reservoir

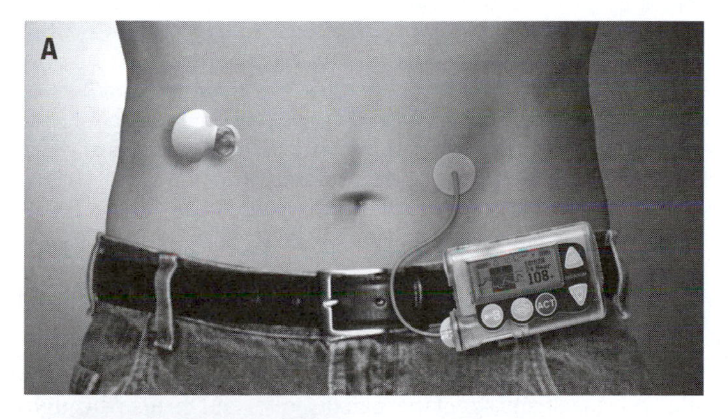

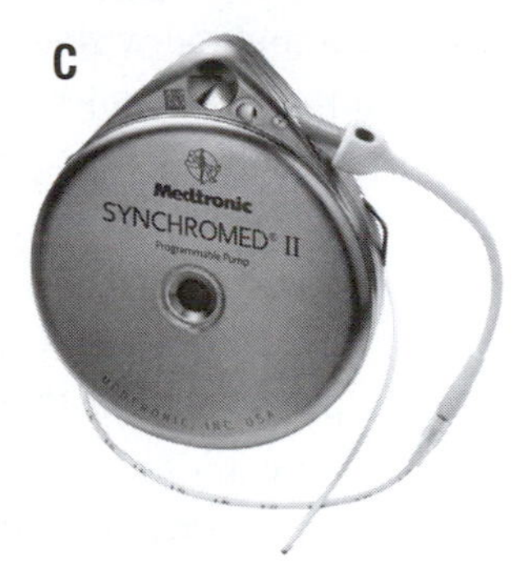

Figure 13.2 **Drug infusion pumps. A.** Medtronics MiniMed Paradigm® REAL-Time insulin pump. **B.** Medtronics MiniMed 2007 implantable insulin pump. **C.** Implantable infusion pump for continuous intravascular infusion of chemotherapy. Reprinted with the permission of Medtronic, Inc. © 2007.

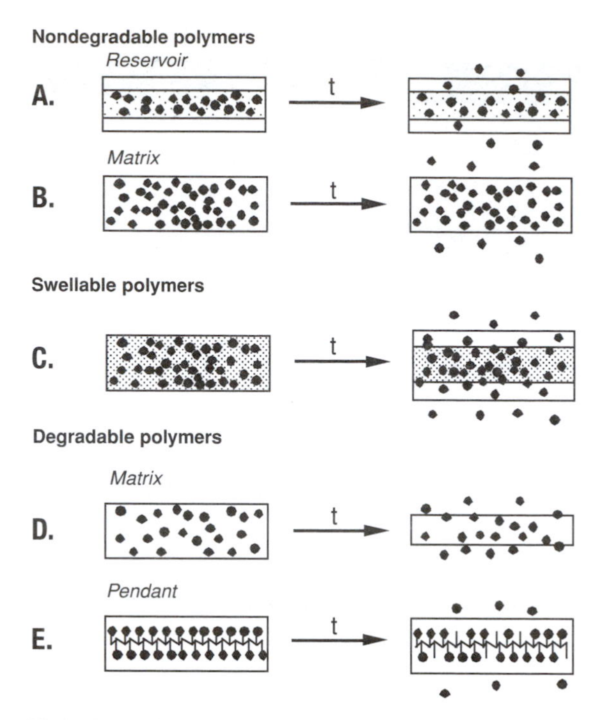

Figure 13.3

Mechanisms of controlled drug delivery by polymers. Polymeric delivery devices can be formulated from non-degradable polymers by enclosing a drug suspension within polymer membranes that control drug release **(A)** or dispersing drug molecules uniformly throughout a polymer matrix **(B)**. In both **(A)** and **(B)**, drug release is controlled by drug diffusion through the polymer. **C.** Water-soluble polymers, like poly(2-hydroxyethylmethacrylate), can be cross-linked and dried to produce matrices that swell when exposed to water. Drug release is activated by the swelling action of the water, which determines the rate of drug release. Similar devices can be formed from biodegradable polymers (polymers that break down in a controlled fashion upon exposure to water) so that drug is released by degradation and dissolution of the polymer matrix **(D)** or cleavage of a covalent bond that links the drug within the polymer matrix **(E)**. In **(D)**, the rate of release is determined by the rate of polymer erosion; in **(E)**, the rate of release is determined by the kinetics of bond degradation.

drug-delivery systems: hollow silicone tubes filled with a liquid suspension of the drug of interest. When immersed in water, the drug is released from the system by diffusion through the silicone wall (Figure 13.3a). This technology permitted the controlled release of any agent that can dissolve and diffuse through either silicone or poly(ethylene-co-vinyl acetate) (EVAc), the two most commonly used non-degradable, biocompatible polymers. One commercial example of this technology is the Norplant® contraceptive system: These cylindrical, silicone-based devices release contraceptive steroids at a constant rate for 5 years following implantation. Norplant provides economical, reliable, long-term protection against unwanted pregnancy; it has been used by women around the world and was approved for use in the United States in 1990. Progestasert®, an **intrauterine** device with a controlled release EVAc coating, delivers contraceptive steroids directly to the reproductive tract, demonstrating another important advantage of polymeric delivery systems: Local release of certain agents can reduce the overall drug dose and, therefore, reduce or eliminate side effects produced by drug toxicity in other tissues. This technology was also used in the Ocusert® (ALZA) system for delivery of pilocarpine to the **cul-de-sac** of the eye. Ocusert® was approved for

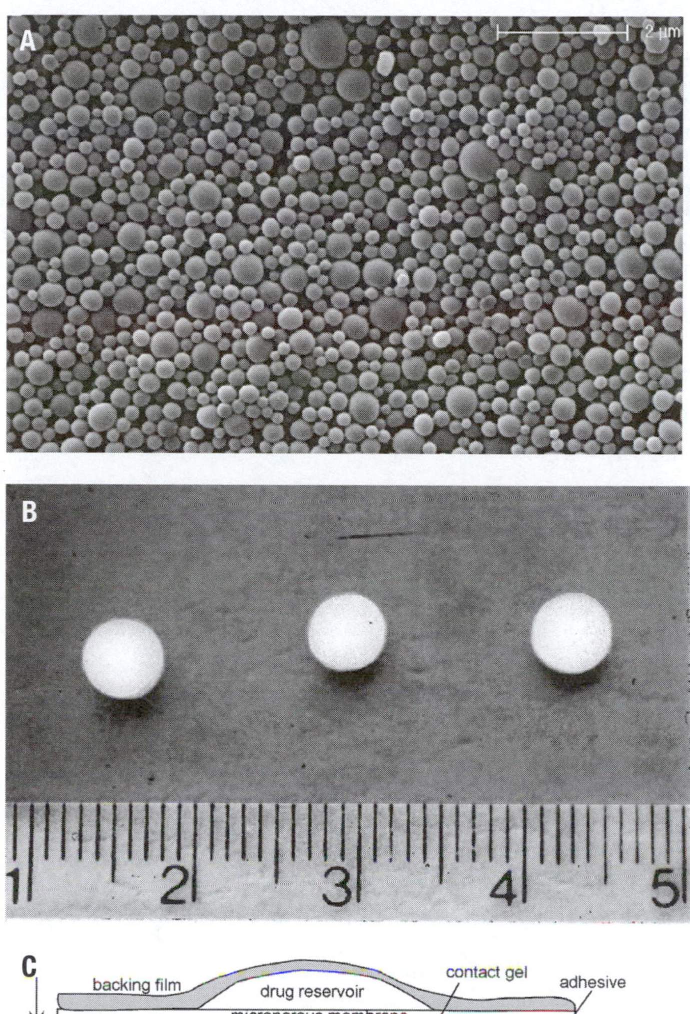

Figure 13.4

Polymer drug-delivery systems. A. Tiny biodegradable particles composed of poly(lactic-co-glycolic acid). Scale bar represents 2 microns. Photo courtesy of Aaron Sin and Kim Woodrow. **B.** Three small drug-delivery discs produced from ethylene vinyl acetate matrix, ruler in centimeters. From (1). **C.** In transdermal delivery systems, a microporous membrane controls the rate of drug diffusion from the reservoir into the skin. A thin layer of contact gel provides a liquid interface between the membrane and the skin surface. From (2) and used with permission of Oxford University Press.

use in the United States but was never widely accepted by glaucoma patients. A recent innovation is the NuvaRing® delivery system for contraceptives (Chapter 3, Figure 3.8b).

Transdermal controlled delivery devices are similar in design to implantable reservoir systems (Figure 13.4c). In this case, a single flat polymer membrane is placed between a drug reservoir and the patient's skin. The polymer membrane is designed to control the rate of drug permeation into the skin over a period of several hours to 1 day. Transdermal delivery systems have been produced for agents that can penetrate the skin easily such as nitroglycerin, scopolamine, clonidine, estradiol, and nicotine; these systems are now used by millions of people. Ongoing research programs are attempting to expand this technology to permit transdermal delivery of agents (such as polypeptide drugs, which do not easily penetrate the skin) and to use switchable components, such as electrical fields, to regulate the rate of drug delivery to the skin.

SELF-TEST 1 A transdermal delivery system is composed of a 1 cm^2 polymer membrane, of thickness 100 μm, that separates a drug reservoir containing 1 mg/mL of a drug from the skin. If the diffusion coefficient for the drug in the polymer membrane is 1×10^{-7} cm^2/s, what is the maximum rate of drug delivery by this device?

ANSWER: Rate of delivery = Area × Flux; Flux = $D(C_2 - C_1)/\Delta x$ (Box 2.5)
Rate of delivery = $(1\ cm^2)(1 \times 10^{-7}\ cm^2/s)(1\ mg/cm^3)/(0.01\ cm) = 10^{-9}$ mg/s (or 9×10^{-5} mg/day)

Solid matrices of **nondegradable polymers** can also be used for long-term drug release (Figures 13.3b and 13.4). In comparison to reservoir systems, these devices are simpler (because they are homogeneous and, hence, easier to produce) and potentially safer (because a mechanical defect in a reservoir device, but not a matrix, can lead to dose dumping). In contrast, it is more difficult to achieve constant rates of drug release with nondegradable matrix systems. Constant release can sometimes be achieved by adding rate-limiting membranes to homogeneous matrices (see Figure 13.3a), yielding devices in which a core of polymer/drug matrix serves as the reservoir. In other cases, water-soluble, **cross-linked** polymers can be used as matrices; release is then activated by swelling of the polymer matrix in water (Figure 13.3c).

13.2.2 Biodegradable polymeric devices

When nondegradable polymers are used as the basis of implanted delivery systems, the polymer must be surgically removed from the patient at the end of therapy. This procedure can be eliminated by the use of **biodegradable polymers**, which dissolve following implantation (Figure 13.3c). Biodegradable polymers, which were first developed as absorbable sutures in the 1960s, are usually used as matrix devices. Many chemically distinct biodegradable polymers have been developed (see Table 13.3); for use as drug-delivery devices these polymers must be biocompatible (provoking no undesirable or harmful tissue response), and the degradation products must be nontoxic. To provide reproducible drug delivery, these polymers must also degrade in a controlled manner. Of the polymers tested, poly(lactide-co-glycolide) (PLGA) has been used most frequently because of its long history of use as a suture material. Other polymers are also important; for example, polyanhydride matrices containing chemotherapy drugs have been approved for local therapy of brain tumors in humans, in a system called GLIADEL® (MGI Pharma, Inc.). Small microspheres of biodegradable polymers can be produced as injectable or ingestible delivery systems (Figure 13.4); Lupron Depot® (Abbott Laboratories) injectable PLGA microspheres,

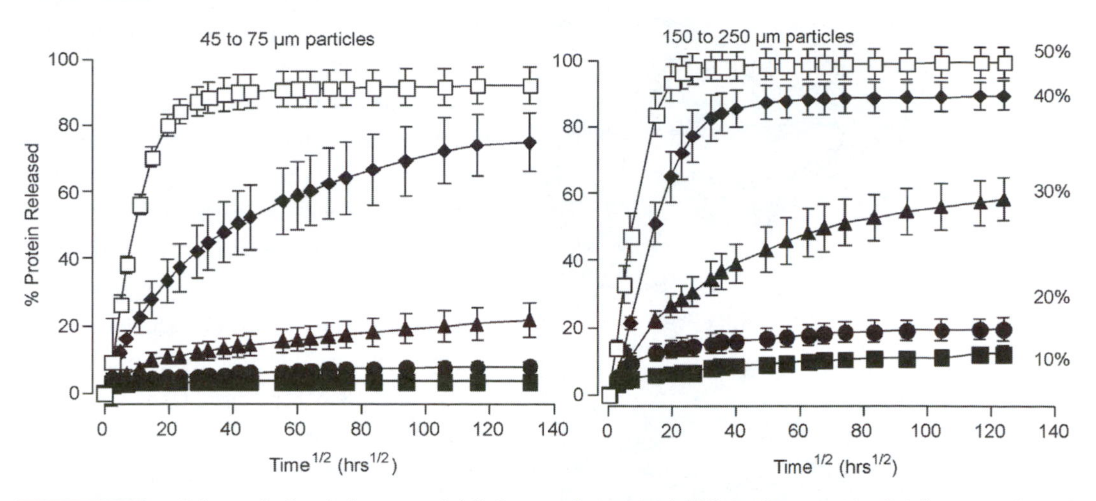

Figure 13.5 **Release of albumin from an poly(ethylene-co-vinyl acetate) (EVAc) polymer matrix.** Details on experimental methods are provided in (3). Solid particles of albumin (either 45–75 μm or 150–250 μm in size) were dispersed in a 1-mm thick EVAc matrix at a total protein loading of 10, 20, 30, 40, or 50%. The cumulative mass of protein released is plotted versus the square root of time (40 $h^{1/2}$ is equal to 67 days; 80 $h^{1/2}$ is 270 days; 120 $h^{1/2}$ is 600 days).

which release a peptide, are approved for the treatment of prostate cancer in humans. The chemistry of the biodegradable polymers can be exploited in other ways to achieve controlled delivery, for example, by linking drugs to the polymer matrix through degradable covalent bonds (Figure 13.3e).

13.2.3 Controlled delivery of proteins and other macromolecules

Recombinant proteins and **polypeptides** are now produced in large quantities by the biotechnology industry. Many of these novel molecules—like **recombinant tissue plasminogen activator**, human insulin, and **human growth hormone**—already have important applications in human health, although they are generally less stable than conventional drugs and more difficult to deliver. Protein drugs are difficult to use in humans because they are either 1) eliminated very quickly when introduced into the body or 2) toxic when delivered **systemically** at the doses required to achieve a local effect. Controlled release polymers may overcome these problems by 1) slowly releasing the protein into the blood over a long period or 2) releasing protein into a local tissue site, thus sparing systemic exposure. The first matrices for protein release were described in 1976. Since that time, many systems for controlled protein delivery have been developed; several are now being evaluated for disease treatment in animals and humans. Proteins can be released for long periods of time from these systems, even many years in some cases (Figure 13.5).

13.2.4 Genetically engineered cells for controlled drug delivery

Future drug-delivery systems will probably use sophisticated biological components that allow self-regulation of drug release. For many proteins, constant rates

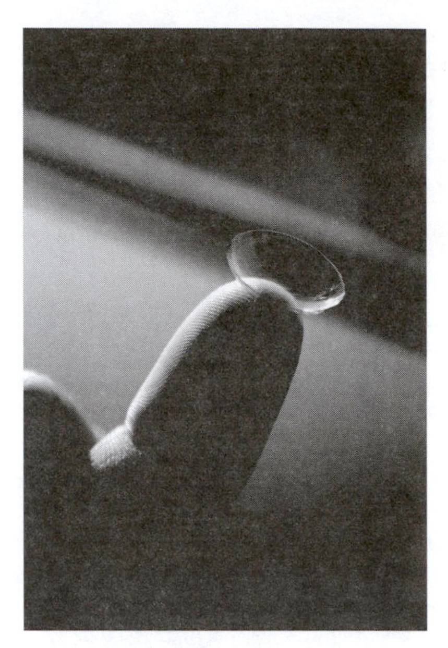

Figure 13.6 **Picture of a hydrogel.** Hydrogels are cross-linked, water-soluble polymer networks that swell in water to form a soft, but stable, material. Hydrogel materials have long been used in medicine. Soft contact lenses are a popular example, but new methods allow for the suspension of living cells within these materials.

of delivery may be insufficient for producing the optimal biological effect. Therefore, advanced polymer systems that provide temporal control over protein release or self-regulation are being studied. Many of these advanced systems use biological components, like **enzymes** or **antibodies**, as the signaling element for regulation of release. Alternatively, it is conceivable that human cells, which already contain complex machinery for regulation of protein production, could be engineered to deliver a protein of interest. For example, transplanted glucose-sensing beta cells from the pancreas could be used for the regulated secretion of insulin.

Drug-delivery strategies may someday be tailored to individuals. Cells harvested from a patient could be genetically modified to increase production of a protein of interest; genes enhancing the cell's ability to control protein expression could also be added. Transplantation of these cells to the patient may restore normal protein delivery. Alternately, methods now being studied may permit the direct introduction of genetic material to cells in the body. Such therapies depend on the development of safe methods for producing stable, high levels of protein expression in human cells, as well as safe and reproducible methods for transplanting human cells. Certain therapeutic cells can be expanded (grown to large numbers) outside of the body for transplantation, to treat a disease (such as cancer), or to form new tissues (as described in the next section). Biocompatible polymers may be important in cell transplantation by serving as a scaffold for attachment of the genetically modified cells. For example, **hydrogel** materials have been used to encapsulate engineered cells as a means of isolating them from direct contact with cells of the immune system (Figure 13.6).

13.3 Tissue engineering

Tissue engineering is a new field of inquiry, defined in the late 1980s, but it is emerging as an important part of biomolecular engineering. Since its period of definition, the field has grown rapidly to the production of clinical products. Tissue engineering combines knowledge from the biological sciences with materials science and engineering to accomplish several goals: to quantify structure–function relationships in normal and pathological tissues, to develop new approaches to repair tissues, and to develop replacements for tissues. Tissue engineering thus involves a combination of disciplines to achieve new therapies and, in some cases, entirely new approaches to therapy.

Tissue or whole-organ transplantation is one of the few options currently available for patients with many common ailments including excessive skin loss and **artery occlusion**. During this century, many of the obstacles to transplantation were cleared: **Immunosuppressive drugs** and advanced surgical techniques make liver, heart, kidney, blood vessel, and other major organ transplantations a daily reality. Transplantation technology has encountered a severe limitation, however. The number of patients requiring transplant far exceeds the available supply of donor tissues. New technology is needed to reduce this deficit.

Tissue engineers are working to develop new approaches for encouraging tissue growth and repair; these approaches are founded on the basic science of organ development and wound healing. A few pioneering efforts are already being tested in patients, including engineered skin equivalents for wound repair and **chondrocyte** implantation for repair of articular cartilage defects. It is clear from these studies that novel tissue replacement strategies will work in certain cases, but hybrid approaches involving cultured cells, manufactured matrices, and advanced materials are required.

13.3.1 Strategies of tissue engineering

Tissue exchange between living humans is an ancient art, of which tissue engineering is the latest genre (for more historical information, see [7]). Ancient Egyptian and Old Testament writers described **blood transfusion**. The folklore of many cultures refers to transplantation of the nose, which was frequently lost by sword or syphilis. Skin and cartilage replacements were performed by Sushruta as early as 1000 BC and by Tagliacozzi during the Renaissance. **Skin grafts** to other sites were common by the 1800s; Churchill donated skin to a comrade in 1898. Although human heart transplantation was not accomplished until 1967, the concept existed with Hua T'o in 3rd-century China. The "modern" age of tissue exchange began in the 1940s, but transplantation did not become widespread until immunosuppressive drugs were discovered. As mentioned previously, success of tissue transplantation today has stringent limits, which are now determined primarily by the availability of donor tissue. Many people with end-stage tissue failure, but with good prospects for survival after transplantation, die waiting for donor tissue. Tissue engineering developed from the urge to create larger supplies of transplantable tissue, perhaps produced from a patient's healthy cells or engineered cell lines.

How can tissue development in culture be manipulated? Some of the mechanisms of cell organization during tissue development (i.e., during the period of

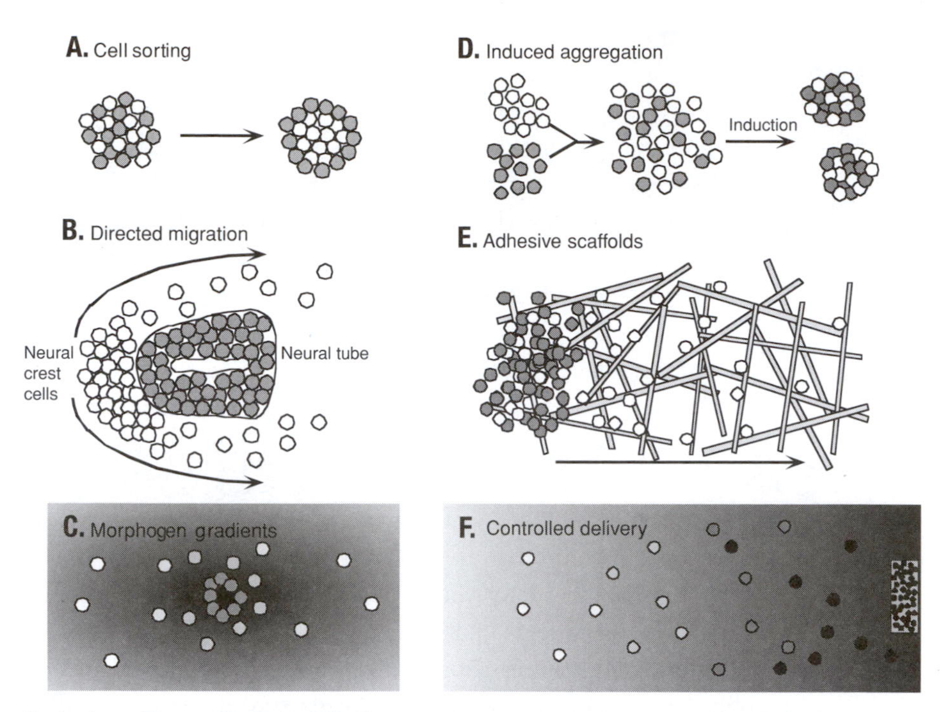

Figure 13.7 **Mechanisms of human development (A–C) compared to strategies for tissue engineering (D–F).** During development and tissue repair, cells (indicated by grey and white spheres) assemble into a functional tissue. Cell assembly and differentiation occur by a number of mechanisms during embryogenesis including cell sorting into layers, which is guided by cell–cell adhesion molecules **(A)**; directed migration along adhesive pathways, which requires interactions with extracellular matrix, cell surfaces, and diffusible substances **(B)**; and differentiation in gradients of morphogens, which provide positional information in the developing tissue **(C)**. In **(B)**, cells from the neural crest migrate around the neural tube and aggregate to form ganglia at distal tissue sites. In **(C)**, the white cells acquire positional information by the local concentration of a factor released by the grey cells. Many approaches to tissue engineering attempt to mimic these natural processes using cells (indicated by white and grey spheres) and synthetic approaches such as induction of aggregation, which can occur by addition of polymers that stimulate cell adhesion **(D)**; provision of scaffolds, which control microarchitecture and adhesion for invasion by specific cells **(E)**; and sustained release of bioactive agents from localized sources **(F)**. In **(D)**, assembly of a suspension of cultured cells, mixed to form the correct proportions, is induced by addition of a soluble cell–cell adhesion promoter. In **(E)**, the matrix of polymer fibers supports the adhesion and migration of the white cells, but not the grey. In **(F)**, the gradient of morphogen, which gives positional information only to the grey cells, is provided by a controlled release implant.

embryonic development in which tissues are formed) are known (Figure 13.7). Laboratory techniques for inducing tissue formation, frequently inspired by natural mechanisms, are available in certain cases (Figure 13.7). Central among these new strategies is the idea, first proposed less than 20 years ago (4), that synthetic materials can serve as degradable templates for tissue regeneration (Figure 13.7e). Blood cells circulate, so they can be transplanted by injection and still find their proper place by native homing mechanisms. Cells of other tissues (such as liver, skin, and brain) are localized at a particular site; transplanted cells need a mechanical foundation and chances for cell–cell communication. In tissue engineering, synthetic materials—particularly biocompatible, degradable polymers—are often used to provide this mechanical support.

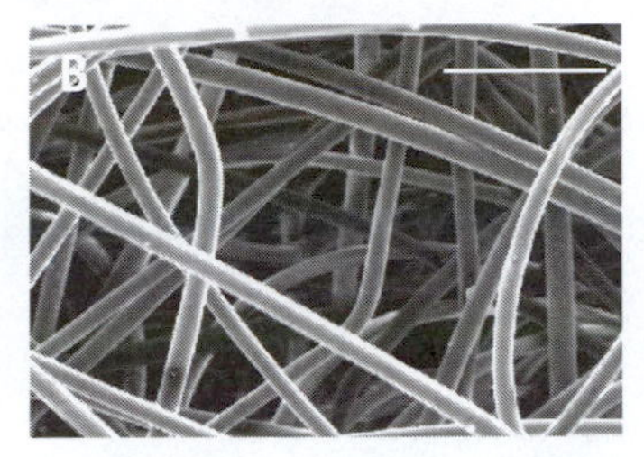

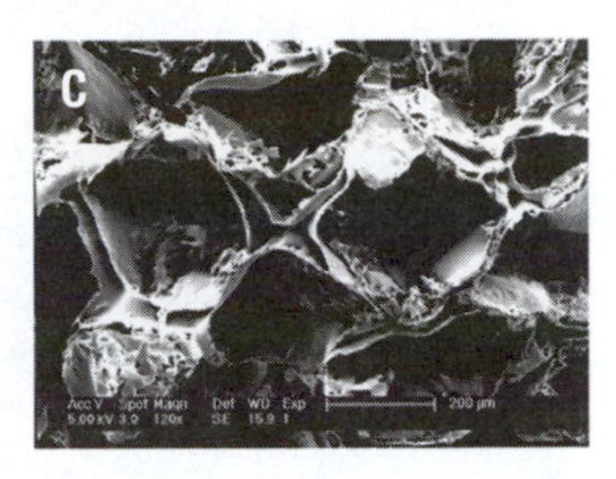

Figure 13.8 **Degradable scaffolds constructed from polymer fibers.** Scanning electron micrographs show low **(A)** and high **(B)** magnification views of a degradable polymer fiber mesh that is used in tissue engineering. The white scale bar in both images is 100 μm. **C**, Scanning electron micrograph of a scaffold produced by salt-leaching technique. Photographs courtesy of Peter M. Fong (**A** and **B**) and Lichuan Qian **(C)**.

13.3.2 Degradable scaffolds for tissue engineering

Degradable scaffolds formed from **synthetic** polymers are attractive materials for tissue engineering (Figure 13.8). By using a biodegradable polymer scaffold, the overall tissue size and shape can be molded; the polymer surfaces provide anchorage points for the cells. The scaffold microgeometry can be optimized for cell recruitment, and the synthetic polymer can be programmed to dissolve as the tissue form emerges. It is now well known that viability and function of surface-attached cells depend on properties of the surface. Therefore, another advantage of synthetic surfaces is that they can be chemically modified to replicate the chemical and physical features of tissues, rendering materials active for certain cell populations.

13.3.3 Example of tissue engineering: Artificial skin

Skin, like many natural tissues, is a layered structure, in which conformation of the different cellular layers is essential for proper tissue function and health (Figure 13.9). Many materials have been proposed and tested as skin grafts: These systems are designed to serve as substitutes for the main functions of skin: to prevent dehydration and infection of large open wound areas and to aid in healing (Table 13.4). Some systems are **acellular** (without cells); one of the earliest systems, for example, uses silicone membranes (to retain fluid) together with a collagen/chondroitin sulfate layer to induce blood vessel and connective tissue integration. These first systems evolved into the clinical product Integra®, which is now the most widely used synthetic skin substitute for burn patients. These materials involve only nonliving components: silicone upper layer and extracellular matrix lower layer. In clinical use, the silicone layer is replaced after three weeks with a thin **epidermal** graft. Subsequent developments have included **autografted cells** (from a small graft) in the lower layer prior to use in the patient.

Another approach is to take a small sample of tissue from the patient, by a surgical procedure called a **biopsy**, and expand the **autologous** epidermal cells in vitro. Epidermal cells are often difficult to grow, however, and producing

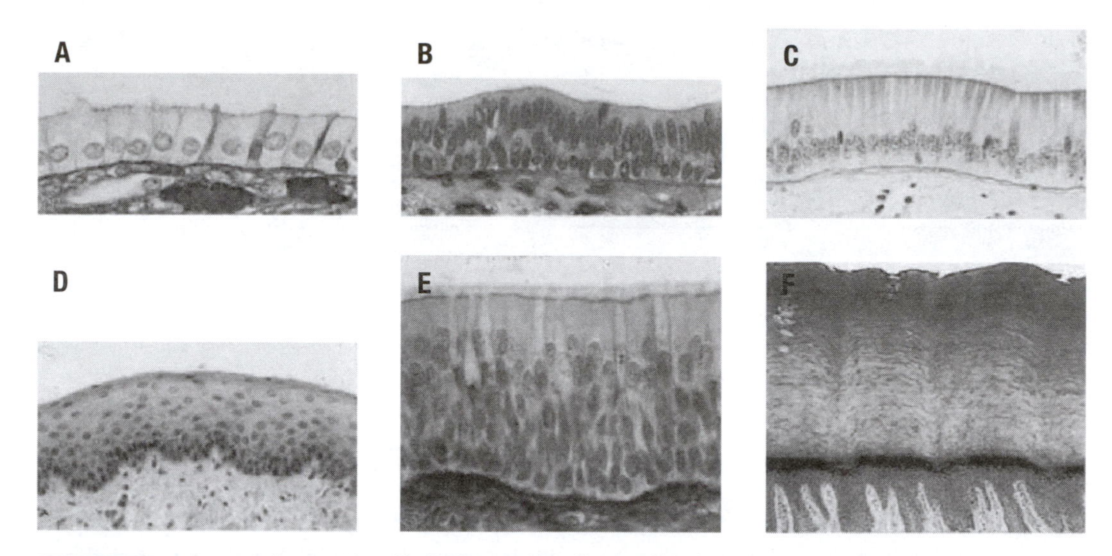

Figure 13.9 **Layered structures in epithelial tissues.** Many tissues within the body are orderly, often layered structures, with different populations of cells arranged in positions that contribute to overall function of the tissue. These microscopic images show simple columnar epithelium (kidney tubule) **(A)**; stratified columnar epithelium **(B)**; ciliated columnar epithelium **(C)**; stratified squamous epithelium (esophagus) **(D)**; ciliated pseudostratified epithelium (trachea) **(E)**; and keratinized stratified squamous epithelium (skin) **(F)**. Reprinted with permission from: *Bloom & Fawcett: Concise Histology* (5). (See color plate.)

large numbers of cells takes time (see Box 13.2). Using optimized cell culture techniques (e.g., by culture with a feeder layer of **irradiated** 3T3 cells—immortal rodent **fibroblast cells**—and medium with **growth factors**), it is possible to achieve a 10,000-fold amplification in cell number by the end of 3–4 weeks, which is still too slow for many clinical situations.

This time delay can be avoided by using **allogeneic cells**, which can be prepared in advance and stored frozen. This approach is used by two biological skin substitutes: Apligraf® (Organogenesis), a collagen-based system containing both allogeneic **endothelial cells** and fibroblasts, and Dermagraft® (Advanced BioHealing, Inc.), a system based on PLGA fiber meshes containing cultured fibroblasts. The fate of the allogeneic cells used in these products after transplantation is not clear and, although these tissue-engineered materials represent important milestones in tissue engineering with potential benefits to patients, their role in clinical practice is still uncertain.

A common problem that is encountered to varying degrees with all of the current tissue-engineered skin products is the lack of formation of a stable **vasculature** that is connected to—and **perfused** by—the blood system of the host. Lack of a robust system of perfused vessels greatly reduces the likelihood of successful engraftment. Indeed, lack of development of an adequate vascular system is a major problem in most tissue engineering products. For skin, this may be an even more important problem to solve because many patients who would benefit from tissue-engineered skin have damaged vessels in the adjacent

Table 13.4 Approaches for tissue engineering of skin

Trade name	Diagram	Composition	Approx. cost per cm^2
Biobrane™ (Dow Hickam/Bertek Pharmaceuticals, Sugar Land, TX)		1. Silicone 2. Nylon mesh 3. Collagen	$0.82
Transcyte® (Advanced Tissue Sciences Inc., La Jolla, CA)		1. Silicone 2. Nylon mesh 3. Collagen seeded with neonatal fibroblasts	$9.16
Apligraf® (Organogenesis Inc., Canton, MA and Novartis Pharmaceuticals, East Hanover, NJ)		1. Neonatal keratinocytes 2. Collagen seeded with neonatal fibroblasts	$16.52
Dermagraft® (Advanced BioHealing, Westport, CT)		1. Polyglycolic acid or polyglactin-910 seeded with neonatal fibroblasts	$8.31
Integra® (Integra Life Sciences Corporation, Plainsboro, NJ)		1. Silicone 2. Collagen and glycosaminoglycan	$3.86
AlloDerm® (LifeCell, The Woodlands, TX)		1. Acellular de-epithelialized cadaver dermis	$6.86
Epicel™ (Genzyme, Cambridge, MA)		1. Cultured autologous keratinocytes	$16.28
Laserskin™ (Fidia Advanced Biopolymers, Italy also marketed as Vivoderm™ by E. R. Squibb & Sons Ltd., United Kingdom)		1. Cultured autologous keratinocytes 2. Hyaluronic acid with laser perforations	n/a
Cadaveric allograft (from nonprofit skin banks)		– Cryopreserved to retain viability – Lyophilized – Glycerolized	$.70

Note: Adapted from: Jones I, Currie L, Martin R. A guide to biological skin substitutes. *Br J Plast Surg.* 2002;55:185–193.

Box 13.1 Pharmacokinetic models

Pharmacology, the study of agents and their actions, can be divided into two branches. Pharmaco-dynamics is concerned with the effects of a drug on the body and, therefore, encompasses dose–response relationships as well as the molecular mechanisms of drug activity. Pharmacokinetics, in contrast, is concerned with the effect of the body on the drug. Drug metabolism, transport, absorption, and elimination are components of pharmacokinetic analysis.

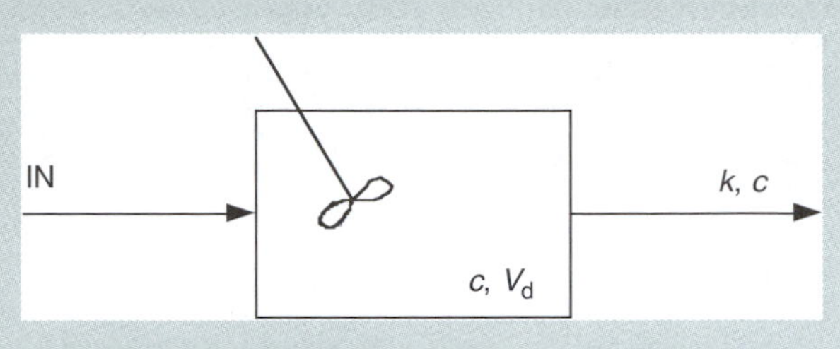

Figure Box 13.1A One compartment model of the body.

The simplest model for distribution and elimination of an intravenously injected drug contains a single compartment representing the volume of distribution, V_d, of the compound (see Figure Box 13.1a). Usually, the elimination of drug from this compartment is assumed to follow a simple kinetic expression; most commonly, first-order kinetics are assumed. A mass balance on the single drug-containing compartment can be expressed as:

$$In - Out + Generation - Consumption = Accumulation, \quad \text{(Equation 1)}$$

which, when applied to this example, yields:

$$-kM = \frac{dM}{dt}, \quad \text{(Equation 2)}$$

where M is the mass of drug within the compartment, and k is a first-order elimination constant. Assuming that an initial mass of drug M_0 is rapidly introduced at time $t = 0$, Equation 2 can be solved to produce:

$$M = M_0 e^{-kt}. \quad \text{(Equation 3)}$$

(If you do not yet know how to solve differential equations like Equation 2, verify that Equation 3 satisfies the differential equation by plugging Equation 3 into Equation 2 and showing that it works.) The concentration of drug within the compartment, c, is related to the total mass within the compartment, $M = cV_d$, so that Equation 3 can also be written:

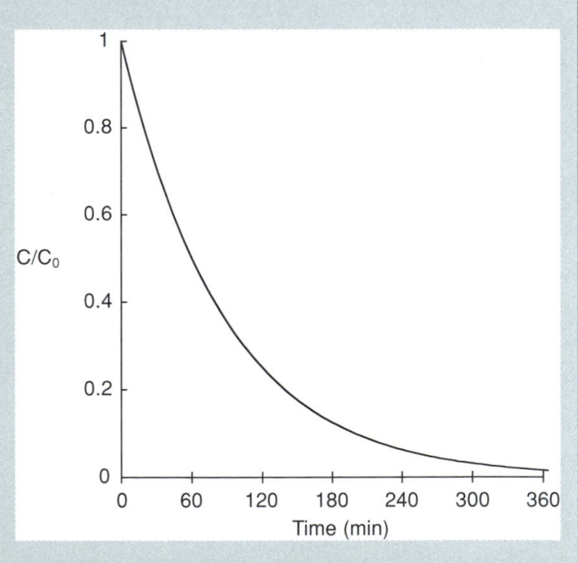

Figure Box 13.1B Change in concentration with time predicted by Equation 4.

$$c = \frac{M_0}{V_d} e^{-kt}. \quad \text{(Equation 4)}$$

Box 13.1 *(continued)*

The concentration of drug within the compartment decreases exponentially with time (see Figure Box 13.1b). We can define the half-life of a drug as the time that it takes for the concentration to be reduced to half of the initial value; if M is M_0 at $t = 0$, then $t_{1/2}$ is the time at which $M = M_0/2$. Using this definition, the half-life for drug residence within the compartment is related to the first-order rate constant:

$$t_{1/2} = \frac{\ln(2)}{k}. \quad \text{(Equation 5)}$$

Simple pharmacokinetic models can be used to tailor therapies to the patient. After measuring the half-life of clearance in a patient, subsequent doses can be provided at intervals calculated to maintain the plasma concentration above a desired level.

host tissue (because of the effect of burns or diabetes, two of the most common indications for skin substitutes) or have limited capacity for growth of new vessels or angiogenesis (because of advanced age).

One approach for enhancing vascularization, which may be applicable to many tissue engineering products including skin, involves the culture of human endothelial cells within a transplantable matrix; matrices of **type I collagen** and **fibronectin** have been used successfully (6). Under the appropriate culture conditions, human umbilical vein endothelial cells (**HUVECs**) will form tubular structures after a brief period of culture (Figure 13.10). When these constructs are transplanted under the skin of a mouse, a HUVEC-lined blood

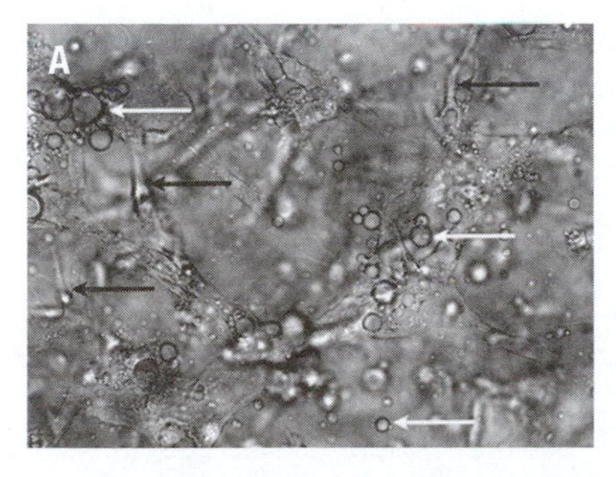

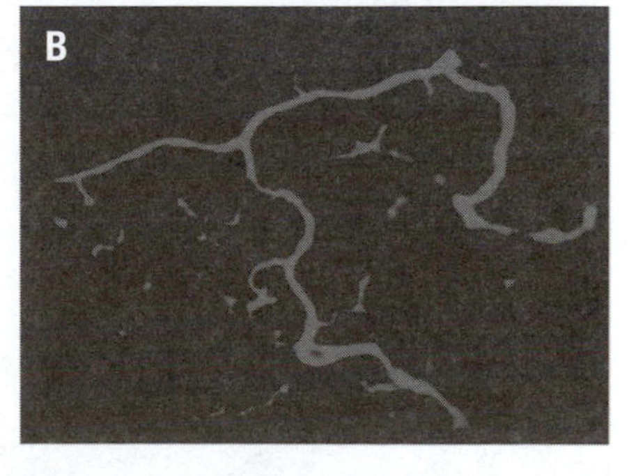

Figure 13.10 **Tissue engineering approach to create new vascular beds. A.** In this approach, human endothelial cells and microspheres that slowly release vascular endothelial growth factor (white arrows) are both suspended with collagen/fibronectin in three-dimensional gels. After a brief period of culture, the cells begin to form cords, which eventually become new blood vessels (black arrows). **B.** After implantation of the construct into a mouse, new blood vessels are formed in the mouse. These blood vessels are lined with human endothelial cells, but have mouse blood flowing through them. (See color plate.) Used with permission. Copyright 2005 National Academy of Sciences, USA. For more details, see (6), (7), and (8).

Box 13.2 Kinetics of cell growth in bioreactors

If actively dividing cells are maintained in a uniform environment, with no constraints to their growth, the rate of growth is proportional to the number of cells:

$$\frac{\mathrm{d}N}{\mathrm{d}t} = k_P N, \qquad \text{(Equation 1)}$$

where N is the total number of cells, and k_P is the rate constant for cell proliferation. Therefore, the number of cells increases exponentially with time:

$$N = N_0 e^{k_P t}, \qquad \text{(Equation 2)}$$

where N_0 is the initial cell number (i.e., the number of cells at $t = 0$). The rate constant is related to the doubling time for the cell population, t_D:

$$t_D = \frac{\ln 2}{k_P}. \qquad \text{(Equation 3)}$$

The graph (Figure Box 13.2) shows the typical increase in cell number that occurs during culture of fibroblasts on a conventional tissue culture plastic surface and some other polymeric surfaces; the doubling time is approximately 1 day.

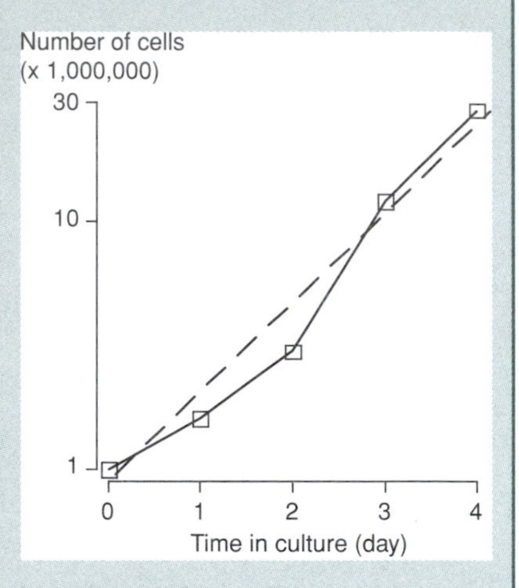

Figure Box 13.2

Most cells that are derived from animals are anchorage-dependent; cell adhesion and spreading on a solid surface are required for proliferation and function. In addition, as the surface becomes crowded with cells, the rate of cell growth decreases; the rate of cell division is, therefore, dependent on cell density; this phenomenon is called density-dependent cell growth.

In practice, cell proliferation will be constrained by a variety of environmental factors: Cell density, nutrient availability, and waste product accumulation all contribute to the overall rate of growth. For anchorage-dependent cells, the overall rate of cell growth and death is related to cell spreading on the surface.

A common phenomenological model, called the Monod model, is used to describe the influence of a substrate or nutrient of limited availability that limits growth:

$$k_P = \frac{1}{N}\frac{\mathrm{d}N}{\mathrm{d}t} = \frac{k_{P,\max} S}{K_M + S} \qquad \text{(Equation 4)}$$

where S is the substrate concentration, and the constants $k_{P,\max}$ and K_M characterize the influence of the substrate on the growth rate. The rate of substrate consumption is related to the rate of cell growth by a yield coefficient, $Y_{N/S}$, which is equal to the ratio of substrate consumption rate, r_S, to cell growth rate, r_N (or $k_P N$): In other words, $Y_{N/S} = -r_S/r_N$. Other models of growth are also available; see the book *Bioprocess Engineering* by Shuler and Kargi for more information.

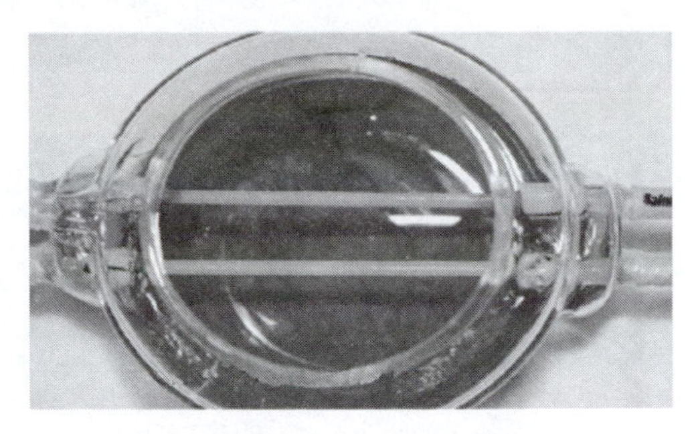

Figure 13.11 **Tissue-engineered porcine arteries created in the laboratory, in a bioreactor, over a period of 8 weeks.** To form these arteries, tubes of a synthetic, biodegradable polymer mesh were seeded with vascular smooth muscle cells and then grown in a special bioreactor under conditions of pulsatile radial strain. The rapidly degrading polymer mesh was replaced by cells and secreted extracellular matrix, including collagens and glycosaminoglycans. The resulting tissue-engineered arteries were 8 cm in length and 3 mm in internal diameter. Photo courtesy of L.E. Niklason. For more information, see (9). (See color plate.)

vessel system—which is perfused by mouse blood—forms within the construct. Genetic engineering of HUVECs, by **overexpression** of the survival gene Bcl-2, led to an increased vessel density in the construct as well as evidence of vascular remodeling to form arteries and veins. Protein delivery systems can be added to the gel phase at the same time that cells are added; these controlled release systems provide a continuous supply of factors that aid in mature vessel formation, such as vascular endothelial growth factor (VEGF). There are many attractive features to this approach, which involves commonly available cells and materials such as collagen that are common in tissue engineering practice.

An alternate approach for making new blood vessels (which is appropriate for making vessels the size of the coronary arteries, which provide blood flow to the muscular tissue of the heart) is to make vessels individually, so that they can be sewn into place by surgeons. In this case, blood vessels must be assembled and grown totally outside the body. There are several promising approaches for making arteries outside the body, and many of them have properties that are nearly equivalent to natural arteries. One of the most promising approaches is to grow smooth muscle cells (which form an inner layer of the functional artery wall) on a degradable mesh in special bioreactors: The bioreactors keep the cells immersed in cell culture medium but also provide a pulsatile flow, which mimics the normal mechanical conditions that arteries experience in the body (Figure 13.11). When these vessels are removed from the bioreactor and sutured into place, they function as arteries, withstanding the stresses of the cardiovascular system and responding appropriately to body signals.

Although the previous sections focused on tissue engineering of cartilage, skin, and blood vessels, research laboratories are working on approaches to produce tissue-engineered muscle, bone, liver, heart valves, and many other tissues [see (2) for more detail].

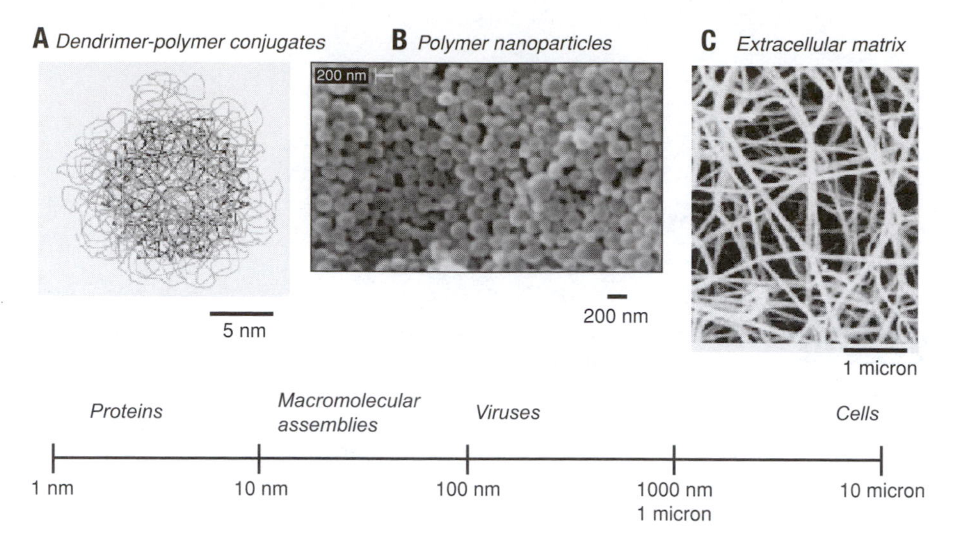

Figure 13.12 **Nanometer objects of biological interest. A.** Dendrimer-polymer conjugates, which are smaller than most viral particles, can be used to deliver drugs and nucleic acids. Reprinted with permission from (10). Copyright 2002, American Chemical Society. See Figure 2.4 and color plate 2.4. **B.** Polymer nanoparticles can be loaded with high concentrations of drugs and targeted to enter specific cells. **C.** The extracellular matrix that surrounds cells and tissues in the body is composed of fibers that are typically 10 nm to 100 nm in diameter.

13.4 Nanobiotechnology

The past few decades have hosted a revolution in materials science. In many cases, it is now possible to manipulate atoms and molecules within materials one at a time, and therefore to construct materials with nanoscale (i.e., the size of individual molecules) precision. This new capability in materials science is called **nanotechnology**.

The potential intersection between nanotechnology and the biological sciences is vast. Biological function depends heavily on units that have nanoscale dimensions, such as **viruses**, **ribosomes**, **molecular motors**, and components of the **extracellular matrix** (Figure 13.12). In addition, engineered devices at the nanometer scale are small enough to interact directly with **subcellular** compartments and to probe intracellular events. The ability to assemble and study materials with nanoscale precision leads to opportunities in both the basic biology (e.g., testing of biological hypotheses that require nanoscale manipulations) and development of new biological technologies (e.g., drug-delivery systems, imaging probes, or nanodevices).

Millimeter-scale and micrometer-scale controlled release systems have been well studied, and some systems have been approved for clinical use, as described earlier in this chapter. One of the major advances in recent years has been further reduction in the size of these systems: It is now possible to make polymer delivery systems that are nanometer in scale, can be easily injected or inhaled, and are much smaller than—and capable of being internalized by—many types of

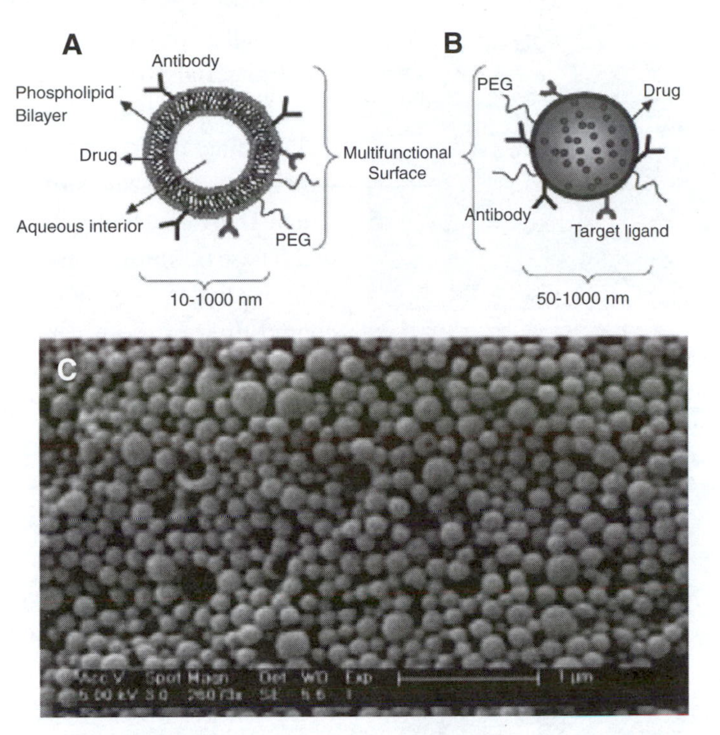

Figure 13.13

Examples of drug-delivery nanoparticles. A. Liposomal systems are vesicular with targeting or stealth groups, such as Poly(ethylene glycol) (PEG), attached to the surface lipids. **B.** Solid biodegradable nanoparticles are formed from a polymer emulsion with drug dispersed in the polymer matrix and targeting or PEG groups attached to the surface. (See color plate.) **C.** Scanning electron micrograph of surface-modified PLGA nanoparticles. From (11) and (7).

human cells. Although there are a variety of ways of achieving nanoscale delivery systems, including self-assembling systems based on **liposomes** or **micelles**, the most stable and versatile systems are miniaturized versions of the synthetic materials that have already been used in drug-delivery applications (Figure 13.13). Construction of such systems is usually accomplished with degradable polymers such as PLGA. These particles can be injected for circulation or used to release drugs locally. The encapsulated drugs can be complex if appropriate methods of fabrication are used to assemble the nanoparticle. For example, it is now possible to make 300-nm particles that have functional DNA within the solid matrix.

Biocompatible and degradable polymers have been around for some time, and much is known about assembling them with different classes of drug molecules. One usually obtains a complex mixture of particles of different sizes and shapes, however; the methods of fabrication are imperfect. Matching methods of particle formation with drugs has been one of the major challenges in this area. Many different ways to make small particles, especially with nanotechnology, have now been described. Unfortunately, few of these methods are compatible with most drugs. Finding better ways to make controlled particles that are compatible with drug incorporation is a challenge for the future.

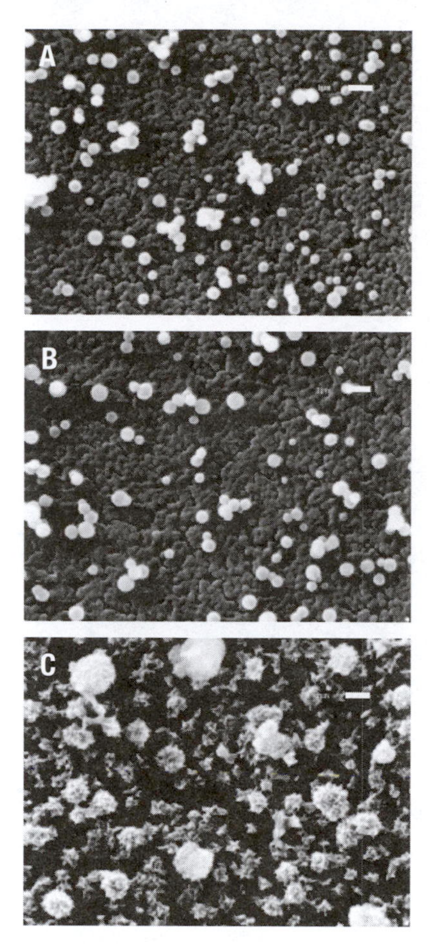

Figure 13.14 **Mineral particles containing deoxyribonucleic acid (DNA).** The particles are formed under mild conditions and, when placed in contact with cultured cells, facilitate uptake and expression of DNA. All pictures are the same magnification (white bar $=1$ μm). Particles range from ~100 nm in diameter **(A)** to ~500 nm in diameter **(C)**. Size of the particles depends on the conditions of the mineralization process, which was different in preparation of **A**, **B**, and **C**. Panel **C** reprinted from (12) with permission from Nature Publishing Group.

It is also possible to make nanostructured materials from minerals and ceramics. Figure 13.14 shows the results of controlled mineral deposition to create surfaces with nanometer-scale structural features in which **DNA molecules** have been embedded. These **nanominerals** are formed under mild biological conditions. The mineral composition of these particles, as well as the loading of DNA, can be controlled. When cells are cultured on these composite surfaces, DNA is taken up into the cell. The overall composition of the surface (and hence its nanostructure) influences the amount of DNA that is taken up and expressed in the cell (12). Other systems have also been used to show that nanoparticles of controlled size and density can be used to facilitate uptake and expression by concentrating DNA at the cell surface.

Some cells will internalize nanoscale particles (Figure 13.15). If these particles are loaded with drugs, such as **chemotherapy drugs**, then the nanoparticles can be used to deliver high drug doses into the cell interior. Polymeric nanoparticles can also be conjugated with cell **ligands** for targeting specific cell populations. In this way, it may be possible to make drug carriers that are much smaller than a cell, but capable of delivering large doses of drug directly to the cell's internal machinery.

New research suggests that these particles can be made from materials that respond to mild external signals (such as light, ultrasound, or magnetic fields) so that the movement of the particles can be directed from outside the body, or the particles could be activated at particular sites. In this way, nanotechnology is providing new methods for using materials in the body; the very small size of the materials makes them suitable for many biological functions. Because of their combination of properties—including subcellular size and controlled release capability and susceptibility to external activation—devices produced by nanotechnology will enable new applications in biological and medical science.

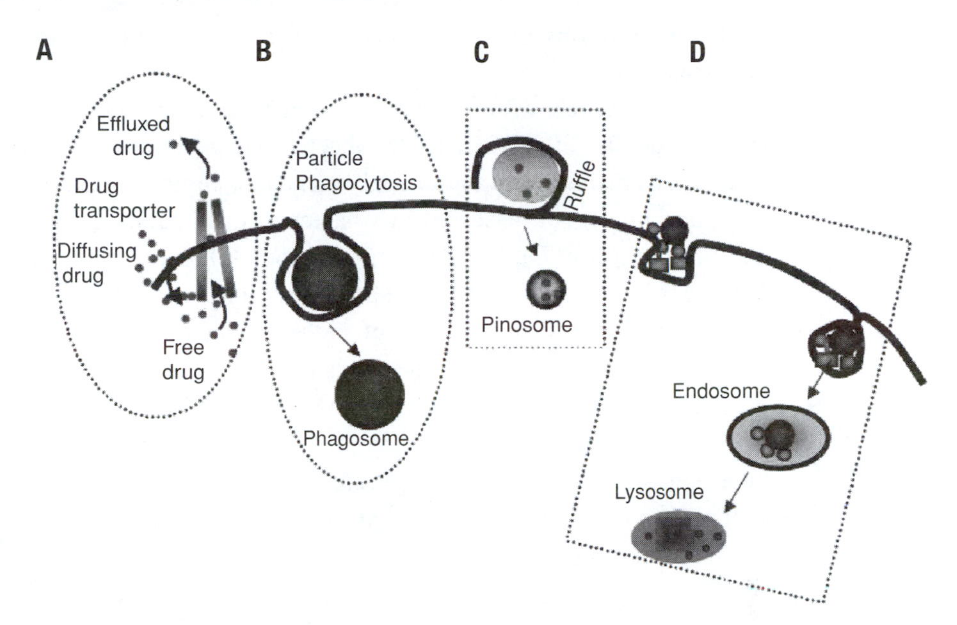

Figure 13.15 **Processes leading to cellular delivery of drugs. A.** Passive diffusion of free drug. **B.** Nonspecific phagocytosis of a nanoparticle. **C.** Drug entrapped in fluid and uptake by pinocytosis. **D.** Receptor-mediated endocytosis. Adapted from (11). (See color plate.)

13.5 Other areas of biomolecular engineering

Table 13.1 lists several aspects of biomolecular engineering beyond the areas of pharmacological engineering, tissue engineering, and nanobiotechnology. Engineers have long been involved in the development of processes and devices that aid in the large-scale processing of biological materials. One of the early advances in this field, now called **bioprocess engineering**, was the development of **bioreactors** for the production of penicillin. Engineers and biologists working together made sufficient quantities of penicillin to treat 100,000 patients per year by the end of World War II (see Further Reading, Shuler and Kargi). More recent developments include bioreactors for the large-scale production of recombinant human proteins, such as insulin and erythropoietin, and the creation of new methods for separation of valuable products from biological mixtures.

In developing these methods, bioprocess engineers began to study the metabolism of cultured cells and learned that it was sometimes possible to control cellular function by manipulating conditions of the culture (e.g., by changing the nutrients that were provided in the medium). The tools of molecular biology created new opportunities for these engineers because now the metabolism of cells could be actively redirected by the addition or deletion of genes in the cell. This field of study, called **metabolic engineering**, has grown substantially in the past few decades. Today, metabolic engineers are concerned with the modification of biochemical pathways inside cells (usually by genetic means) as well as the rigorous evaluation of the resulting cellular phenotypes (i.e., analysis of the state of the cell in terms of the proteins that it expresses and the functions that it performs).

Most of the early work in bioprocess engineering and metabolic engineering involved simple cells: Often, bacterial cells were examined. It is now possible to culture mammalian cells and to apply metabolic engineering principles to these cell populations. As discussed in Chapter 5, mammalian cells exhibit certain behaviors that are not seen in simpler systems. For example, cell adhesion is important to the life of most mammalian cells, and cell motility occurs by unique mechanisms. A new subdivision of biomolecular engineering, called cellular engineering, has emerged from the study of these unique systems. **Cellular engineers** are interested in the basic phenomena such as the influence of physical factors on cellular function, mechanisms of cell adhesion, and characteristics of cellular signaling pathways. In addition, some cellular engineers study the use of cells as therapeutic agents and, in this way, some of the goals of cellular engineering overlap with tissue engineering.

Summary

- Biomolecular engineering uses the principles of chemical engineering and the tools of modern biology to find new solutions for human health care.
- All drugs have side effects that limit their use, but controlled drug-delivery systems can extend and optimize drug use.
- Polymer materials have many potential uses in drug delivery, serving as vehicles for drug distribution and release.
- Tissue engineering is a rapidly evolving area, in which cells and synthetic materials are assembled into tissue-like units for organ replacement and repair.
- Nanobiotechnology involves the design of materials with nanoscale dimensions, which provides them with unique abilities to interact with biological systems.

REFERENCES

1. Mahoney MF, Saltzman WM. Transplantation of brain cells assembled around a programmable synthetic microenvironment. *Nat Biotechnol*. 2001;19:934–939.
2. Saltzman WM. *Tissue Engineering: Engineering Principles for the Design of Replacement Organs and Tissues*. New York: Oxford University Press; 2004.
3. Saltzman WM, Langer R. Transport rates of proteins in porous polymers with known microgeometry. *Biophys J*. 1989;55:163–171.
4. Vacanti JP, Morse MA, Saltzman WM, Domb AJ, Perez-Atayde A, Langer R. Selective cell transplantation using bioabsorbable artificial polymers as matrices. *J Pediatr Surg*. 1988;23:3–9.
5. Fawcett DW, Jensh RP. *Bloom & Fawcett: Concise Histology*. London: Hodder Arnold; 1997.

Profiles in BME: Robert Langer

Robert S. Langer is the Kenneth J. Germeshausen Professor of Chemical and Biomedical Engineering at the Massachusetts Institute of Technology (MIT). Dr. Langer has written more than 725 scientific articles; he also has nearly 500 issued or pending patents. Dr. Langer's patents have been licensed or sublicensed to more than 100 pharmaceutical, chemical, biotechnology, and medical device companies; a number of these companies were launched on the basis of these patent licenses. He served as a member of the United States Food and Drug Administration's (FDA's) Science Board, the FDA's highest advisory board, from 1995 through 2002 and as its Chairman from 1999 through 2002.

Dr. Langer has received more than 100 major awards for his BME research work. In 2002, he received the Charles Stark Draper Prize, considered the equivalent of the Nobel Prize for engineers and the world's most prestigious engineering prize, from the National Academy of Engineering. He is also the only engineer to receive the Gairdner Foundation International Award; 60 recipients of this award have subsequently received a Nobel Prize. In 1998, he received the Lemelson-MIT prize, for being "one of history's most prolific inventors in medicine." In 1989 Dr. Langer was elected to the Institute of Medicine of the National Academy of Sciences, and in 1992 he was elected to both the National Academy of Engineering and the National Academy of Sciences. He is one of very few people ever elected to all three United States National Academies and the youngest in history (at age 43) to ever receive this distinction.

He received his bachelor's degree from Cornell University in 1970 and his ScD from MIT in 1974, both in Chemical Engineering.

Bob Langer writes of his own career:

I was born in Albany, New York. My mother always wanted to do nice things for people, and my father often played math games with me. I had a sister named Kathy. When I was little, my parents bought me a Gilbert chemistry set. I enjoyed mixing chemical solutions together and watching them change color as reactions took place, and this started to get me interested in science.

When I went to college, Cornell University, I didn't have a very clear idea of what I wanted to do. In fact, in high school, the only things I was good at were math and science, and the guidance counselor told me that I ought to become an engineer, even though I really did not know what engineers did.

Nonetheless, I decided to go down that road, and I was fortunate that I did well in chemistry my first year at Cornell—although I didn't do that well in too much else—and I decided to major in chemical engineering. When I finished college, I still didn't have a clear idea of what I wanted to do. One of the things that impressed me about MIT's Chemical Engineering Department was that it was so diverse. You could do so many different things, ranging from BME to polymer chemistry, so I was very excited when I was accepted there. I did my doctoral thesis, entitled "Enzymatic Regeneration of adenosine triphosphate," with Clark Colton.

(*continued*)

Profiles in BME (*continued*)

But again, when I finished at MIT, I didn't have a clear idea of what I wanted to do with my life. It was 1974, and I received many job offers to join oil companies. They had many openings, and that's really where the action was in chemical engineering at that time. But I had this dream of using my background to improve people's health, and I was lucky, and this time I ended up working with Dr. Judah Folkman, Professor of Surgery at Boston Children's Hospital and Harvard Medical School. This job had a profound impact on what I ended up doing with my life. One of the great things about Dr. Folkman was that he believed that almost anything was possible, and seeing his example was terrific for me. The projects that I worked on in Dr. Folkman's lab involved developing polymer systems that might be able to slowly release molecules to study cancer. Many of these molecules were large or had a lot of electrical charges associated with them, and no one had been able to develop ways to release them in a steady, controlled manner. But when I was working on this, I made the discovery that I could modify certain types of polymers and use them to slowly release those molecules. Even though I published this work in *Nature*, the work generated considerable skepticism. Many people didn't believe large molecules could move through polymer matrices. In fact, I spent a good part of my early career understanding the mechanism of release and answering that question.

When I was done with my postdoctoral work, I applied for faculty positions in a number of chemical engineering departments. But I had trouble getting jobs there because people felt at that time that what I was doing wasn't engineering. So I ended up joining what was then the Nutrition and Food Science Department at MIT. Later when MIT reorganized their departments, I joined the new Chemical Engineering Department. I now also hold appointments in MIT's Bioengineering Division and the Harvard-MIT Division of Health Sciences and Technology.

In the early 1980s, I had some ideas on how one could create new polymers for medicine. I was very fortunate that I began to collaborate with Dr. Henry Brem, who's now the Chief of Neurosurgery at Johns Hopkins, and we came up with a new way of using these polymers to locally deliver drugs to treat tumors and other diseases. That is now being used to treat patients today.

I also had the good fortune of working with Jay Vacanti, a transplant surgeon. Jay and I came up with the idea that, if you used certain types of synthetic polymers and put cells on them, and if you did it the right way, perhaps you could make new tissues. Today, tissue-engineered skin and cartilage are available to treat certain medical conditions, and several other engineered tissues are in clinical trials.

As the years went by, many of the graduate students and postdocs in my laboratory would make other discoveries. However, many times, these findings too would be met with skepticism. But I really believe it's important to encourage students to follow their dreams and today, more than 40 different products coming out of our patents are now either FDA approved or in clinical trials, and I'd like to think that the principles we've developed are being used by many pharmaceutical and medical device companies around the world. Equally important is the fact that 110 former members of my laboratory are now professors who are training others in these fields.

On the personal side, I live in Newton, Massachusetts, with my wife Laura and our three children, Michael, Susan, and Sam.

6. Schechner JS, Nath AK, Zheng L, Kluger MS, Hughes CC, Sierra-Honiqmann MR, et al. In vivo formation of complex microvessels lined by human endothelial cells in an immunodeficient mouse. *Proc Natl Acad Sci U S A*. 2000;97(16):9191–9196.
7. Jay SM, Shepherd BR, Bertram JP, Pober JS, Saltzman WM. Engineering of multifunctional gels integrating highly efficient growth factor delivery with endothelial cell transplantation. *FASEB J*. 2008;22(8):2949–2956.
8. Enis DR, Shepherd BR, Wang Y, Qasim A, Shanahan CM, Weissberg PL, Kashgarian M, Pober JS, Schechner JS. Induction, differentiation, and remodeling of blood vessels after transplantation of Bcl-2-transduced endothelial cells. *PNAS*. 2005;102(2):425–430.
9. Poh M, Boyer M, Solan A, Dahl SL, Pedrotty D, Banik SS, et al. Blood vessels engineered from human cells. *Lancet*. 2005;365:2122–2124.
10. Luo D, Haverstick K, Belcheva N, Han E, Saltzman WM. Poly(ethylene glycol)-conjugated PAMAM dendrimer for biocompatible, high-efficiency DNA delivery. *Macromolecules*. 2002;35(9):3456–3462.
11. Fahmy TM, Fong PM, Goyal A, Saltzman WM. Targeted for drug delivery. *Materials Today*. 2005;8:18–26.
12. Shen H, Tan J, Saltzman WM. Surface-mediated gene transfer from nanocomposites of controlled texture. *Nat Mater*. 2004;3(8):569–574.

FURTHER READING

Lauffenburger DA, Linderman JJ. *Receptors: Models for Binding, Trafficking, and Signaling*. New York: Oxford University Press; 1993.

Saltzman, WM. *Drug Delivery: Engineering Principles for Drug Therapy*. New York: Oxford University Press; 2001.

Saltzman WM. *Tissue Engineering: Engineering Principles for the Design of Replacement Organs and Tissues*. New York: Oxford University Press; 2004.

Shuler ML, Kargi F. *Bioprocess Engineering*. 2nd ed. Upper Saddle River, NJ: Prentice-Hall; 2002.

KEY CONCEPTS AND DEFINITIONS

acellular not containing cells

allogenic cells cells from the same species that are genetically different but can be used for transplantation

analgesics drug agents that help to relieve pain, also referred to as "painkillers"

antibodies large Y-shaped proteins that are used in the immune system to identify and counteract invaders such as viruses, bacteria, or transplanted organs by binding to the antigen

anticoagulants a substance found in the blood that prevents clotting

artery occlusion blockage or closure of an artery

autografted cells cells that have been removed from one part of an individual's body and placed in a different location in the same individual

autologous describes a situation in which the recipient and donor of blood, organs, or tissue are the same individual

biodegradable polymer a compound of many covalently linked molecules forming a high total molecular weight with the ability to be broken down by biological agents in the body

biomolecular engineering a field of engineering that examines the changes in chemical components within a biological system and develops methods for modifying these chemicals or their interactions

bioprocess engineering a subspecialty of engineering that focuses on the conversion of materials through a process that makes them more useful to humans. Applications include biofuels, food and drug processing, and fermentation systems.

biopsy a medical procedure involving the removal of cells or tissues from a patient for examination

bioreactor an apparatus that forms a biologically active environment to grow organisms such as cells or bacteria for the production or processing of biological materials

blood transfusion the transfer of blood from one individual to another individual with a compatible blood type

catheter a hollow tube that is inserted into a cavity of the body to provide a route for drainage or insertion of fluids and access for medical instruments. For instance, a catheter can be inserted into the urethra to remove urine from the bladder.

cellular engineer a type of biomolecular engineer who uses the methods and principles of engineering to understand and manipulate the behavior of cells

chemical engineering the application of engineering and chemistry to the design and maintenance of industrial processes

chemotherapy the use of chemical agents, or drugs, in the treatment of cancer and other diseases

chondrocyte mature cells found in cartilage matrix

cross-linked molecules that have been joined by the creation of chemical covalent bonds

cul-de-sac a vessel, tube, or sac (e.g., cecum) open at only one end

degradable scaffolds a mesh framework or matrix that can be inserted into the body for cell culture or tissue replacement and can be broken down under biological conditions

deoxyribonucleic acid (DNA) molecules nucleic acids of helical structure found in the nuclei of cells that contain the genetic instructions for an organism's structure and development

drug delivery the study and administration of drugs through the use of controlled release technology and polymers that allow the drug to be encapsulated and released from the polymer in a controlled manner

elastomers plastics, either synthetic or natural, that are made up of polymerized chain units with elastic properties

endothelial cells cells found on the interior of blood vessels, forming a simple squamous layer

enzymes proteins produced by living organisms that biochemically catalyze reactions

epidermal from the outermost layer of skin

extracellular matrix any part of a tissue that is not considered part of a cell, including components such as collagen and glycoproteins

feedback control a control system that is set up to monitor itself and change the output conditions accordingly to regulate the system

fibroblast cell a type of cell (found in connective tissue) with the ability to secrete proteins and collagen

fibronectin a fibrous protein that forms as an anchor between cell membranes by attaching to integrins and other extracellular matrix components such as collagen, fibrin, and other proteins

growth factors proteins that assist in the organization, growth, and differentiation of cells and tissues

human growth hormone any natural or synthetic substance that controls the development and growth of an organism and its cellular components

hydrogel cross-linked, water-soluble polymer networks that swell in water to form a soft, but stable, material

immunosuppressive drugs drugs that inhibit the normal activity of the immune system, typically used to prevent the rejection of a transplanted organ and to treat autoimmune diseases

insulin a polypeptide hormone produced in the pancreas that helps to regulate the metabolism of carbohydrates in the body

intrauterine within the uterus

irradiated having been exposed to radiation such as x-rays or gamma rays

ligand any molecule, other than an enzyme substrate, that binds tightly and specifically to a macromolecule, usually a protein, forming a macromolecule–ligand complex

liposome an artificial vesicle with an aqueous core and a surrounding phospholipid bilayer used to transport drugs, enzymes, or vaccines to cells and organs within the body

matrix a biocompatible surface, commonly made from titanium or polymers, that can be inserted into the body to function as a mechanism for drug delivery or as a tissue replacement to provide for cell attachment and proliferation

metabolic engineering the study of the metabolism of cultured cells and the manipulation of culture conditions to control cellular function and metabolic activity

micelle a spherical aggregrate of amphipathic molecules with hydrophilic "head" groups on the outer solvent-exposed surface and hydrophobic "tails" in the protected interior

microparticulates drug-encapsulated particles of the micrometer range that can be used as a controlled drug-delivery mechanism within the body

molecular motors cellular components used in movement of the body that have the ability to convert chemical energy from adenosine-5′-triphosphate hydrolysis into mechanical work

nanominerals minerals on the nanometer scale used to create surfaces designed to influence various biological processes, such as the uptake and expression of DNA

nanotechnology a field of study focusing on the development, characterization, and application of materials of nanometer scale

nondegradable polymer a compound – one comprised of many covalently linked molecules forming a high total molecular weight – that will not be broken down by biological agents in the body

overexpression excessive expression by a gene in the body that causes an overproduction of gene products

perfuse to deliver arterial blood to the capillaries of the body

plasma the liquid component of blood that suspends the blood cells and assists in the delivery of oxygen, carbon dioxide, lipids, amino acids, and other biological components throughout the body and contains clotting agents such as fibrin

polymer a compound of many covalently linked molecules, or monomers, forming a high total-molecular-weight chemical compound

polypeptides a family of molecules consisting of many subunits of α-amino acids that have been linked together by amide bonds

recombinant protein a protein that is produced by an organism after it has been genetically modified and a new DNA sequence has been inserted

recombinant tissue plasminogen activator a tissue plasminogen activator that is produced by recombinant DNA technology; tissue plasminogen activator is an enzyme that helps to dissolve blood clots

ribosomes organelles found within cells that translate messenger ribonucleic acid (RNA) into proteins

skin grafts skin that is used in a transplant

subcellular existing or occurring within a cell

synthetic formed through chemical processing as opposed to being obtained in nature

systemically pertaining to the body as a whole

tissue engineering combines knowledge from the biological sciences with the materials and engineering sciences to quantify structure–function relationships in normal and pathological tissues, develop new approaches to repair tissues, and develop replacements for tissues

transdermal administered through application to the skin

type I collagen the primary protein of connective tissues composed of fibers in the extracellular matrix. Type I is the most abundant form in the human body.

vasculature the arrangement of blood vessels found in a particular part of the body

viruses microscopic infectious agents that are inactive outside of living cells and can only replicate within a host cell. The structure of a virus consists of a core of DNA or RNA surrounded by a protein coat.

NOMENCLATURE

c	Concentration	kg/L
D	Diffusion coefficient	cm^2/s
k	First-order elimination constant	L/s
k_P	Rate constant for cell proliferation	L/s
M	Mass	kg
M_0	Mass at time = 0	kg
N	Total number of cells	—
N_0	Initial cell number	—
S	Substrate concentration	kg/L
t	Time	s
$t_{1/2}$	Half-life	s
t_D	Doubling time	s
V_d	Volume of distribution for a compound	L
$Y_{N/S}$	Yield coefficient	kg/cell

QUESTIONS

1. Artificial skin may or may not contain living cells. What are the advantages and disadvantages to having cells in the artificial skin? Are keratinocytes immunogenic? Why? If you are going to make artificial skin with cells, where are you going to get cells?
2. You are asked to design a delivery system for local treatment of breast cancer; the system will be implanted at the site of the solid tumor and slowly release chemotherapy to the local tissue. Will you use a degradable or a nondegradable polymer? Will you use a matrix system or microspheres? Make a list of the advantages and disadvantages of each for this particular application.
3. Consider that you are designing a therapy to deliver a biomolecule or drug to a tissue over an *extended* period of time.
   a. What is the key criterion that you will look for in a polymer to provide this extended delivery time? (Be specific.) Why?
   b. Would polyanhydride (which is highly hydrophobic) or polylactic acid (which is considerably less hydrophobic) be better for this application? Why?
4. For tissue engineering:
   a. Briefly describe the three general strategies for creating new tissues mentioned in an early review article: Langer R, Vacanti JP. Tissue engineering. *Science*. 1993;260:920–932.

b. From the applications discussed in the review article, give one example of each general strategy.
c. From the article, written in 1993, would you expect these technologies to be available now? What do you think is the major impediment in developing these technologies?

5. Should nondegradable or biodegradable materials be used for bone tissue engineering? Why?
6. In tissue-engineering constructs, what are the advantages and disadvantages to using a patient's own cells to seed scaffolds? Name one alternative cell source.

PROBLEMS

1. Box 13.1 presents a simplified model that can be used to predict changes in drug concentration with time in the body. In this simplified model, the body is assumed to behave like a well-mixed vessel.
   a. Why is this model not realistic?
   b. Why does it work so well for certain kinds of drugs?
2. Norplant® is an implantable drug-delivery system for contraception. Read a brief description of the system and how it is used in people.

   http://www.helioshealth.com/birth_control/norplant/

   http://www.emory.edu/WHSC/MED/FAMPLAN/norplant.html

   a. Norplant® is a reservoir delivery system. Draw a diagram showing its important components. (You do not have to get a correct answer here, just draw a cylindrical system that you think—based on what you learned in the lecture—would work to produce 5 years of release.)
   b. Recall the well-stirred model that we developed in class. Describe in words how you would change this model to describe the changes in blood concentration that occur after implantation of a Norplant® system.
3. Write a one- to two-sentence description of a matrix drug-delivery system and a reservoir drug-delivery system. Use a diagram, if that is helpful.
4. You are asked to design a delivery system that provides a constant rate of release of insulin. You want the system to release 1 mg of insulin per day. Will you use a matrix or a reservoir system? Make a list of the advantages and disadvantages of each design for this particular application.

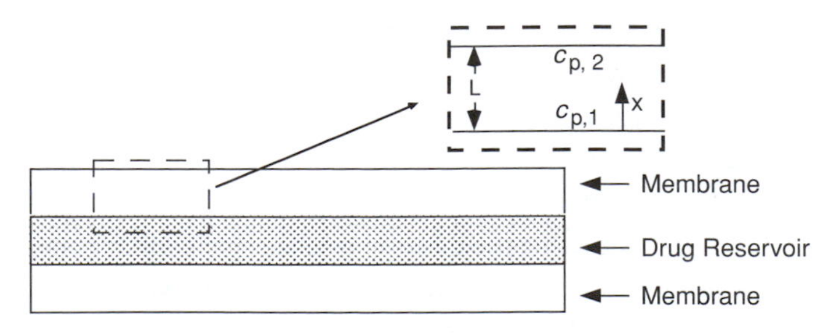

5. A schematic diagram of the Ocusert® system for delivery of pilocarpine is shown in the accompanying figure. Using your knowledge of diffusion through polymer membranes (from Chapter 2), derive an equation that will predict the rate of drug release (M, mg/h) from the following parameters: drug concentration in the reservoir (c_p, mg/mL), thickness of the membranes (cm), diffusion coefficient of drug in the membrane (D_A, cm^2/s), and surface area of the device (A, cm^2).

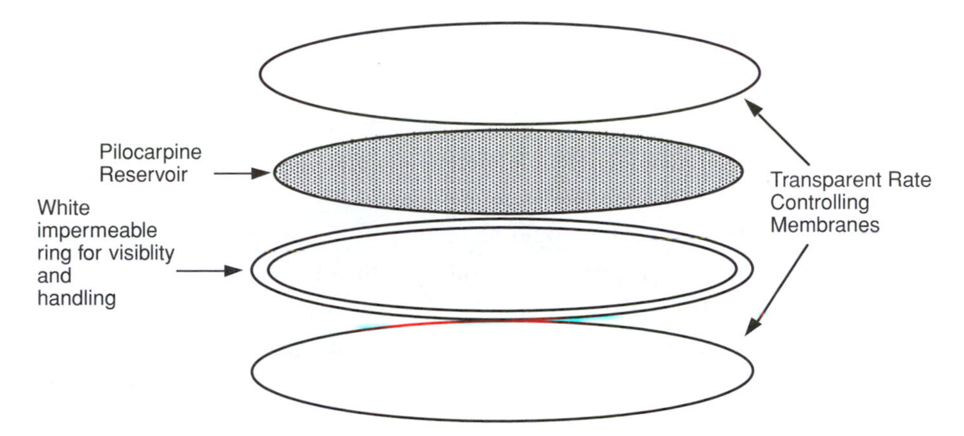

6. Consider tissue engineering of the intestine. Based on the functions of the intestine, what are the *two* primary tissue types (specific) that you would need to recreate and what are their specific roles? What type of polymer would most likely be suitable as a matrix and why?
7. You are culturing insulinoma cells (called INS-1) for in vitro experiments. These cells have a doubling time of 36 hours.
   a. What is the specific growth rate of the cells?
   b. Plot a graph describing the growth behavior of these cells over time.
   c. A T-flask is seeded with 10×10^6 cells; how many days can you wait until you have to passage the cells? [Note: Assume that a confluent T-flask contains approximately 75×10^6 cells.]
8. One of the major limitations in tissue engineering is vascularization. Without an adequate blood supply, most grafted tissues will suffer tremendously. In fact, most researchers in this field would argue that this is the single most significant hurdle that must be overcome for us to achieve overwhelming success. Even if we can establish conditions that support the proliferation and differentiation of three-dimensional tissue-engineered constructs in vitro, the survival of these constructs after implantation will be unlikely unless there is an adequate supply of blood. The liver, in particular, requires a very large supply of blood for its survival. Currently, polyglycolic acid (PGA)/hepatocyte constructs are grown in culture then implanted into the mesentery of the rat for in vivo studies. The mesentery is a highly vascularized membranous fold that attaches the small intestine to the dorsal wall of the body. Despite this highly vascularized bed for implantation, the present survival rate of

implanted hepatocytes is very low. There is research underway to improve the vascularization of implanted hepatocytes for replacement of the liver:

- In one case, researchers are impregnating the PGA scaffold with VEGF, a growth factor that induces the growth of new blood vessels (angiogenesis).
- In a second case, researchers are using advanced free-form fabrication techniques to create a three-dimensional polymer structure that looks like a capillary bed.

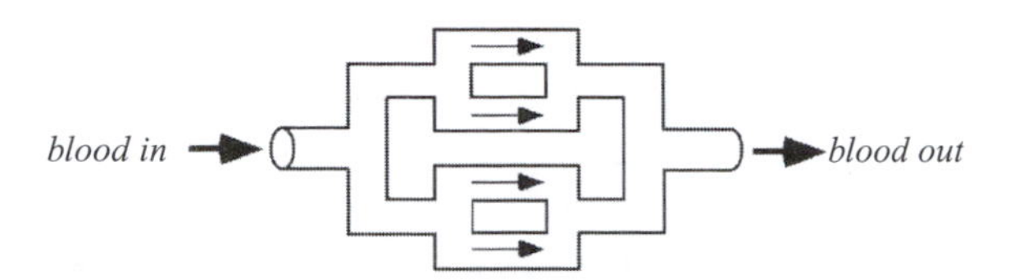

The hepatocytes would be seeded on the exterior of the capillary bed structure. The two ends of the structure can then be sutured directly in line with a blood vessel. The device would NOT be encased within any other membrane (i.e., this is an open system). From the information given here, from our discussions in class, and considering the many *functions* of the liver, thoroughly discuss the following:

a. What would be the *major concerns* with each approach and why?
b. What would be the *major advantages* to each approach and why?

14 Biomolecular Engineering II: Engineering of Immunity

LEARNING OBJECTIVES

After reading this chapter, you should:

- Understand the role of vaccines in the prevention of disease.
- Understand the role of antibodies (Abs) in the immune system, and some of the ways that Abs can be used to prevent disease in humans.
- Understand the basic elements of Ab structure, and the difference in chemical structure between Ab classes.
- Understand the difference between monoclonal and polyclonal Abs.
- Understand how monoclonal antibodies (mAbs) are manufactured.
- Understand some of the basic approaches for vaccine development.

14.1 Prelude

The previous chapter introduced three of the major subjects of interest in biomolecular engineering: drug delivery, nanobiotechnology, and tissue engineering. This chapter focuses on additional applications of biomolecular engineering, particularly approaches for enhancing the function of the **immune system**. The most familiar application of biomedical engineering (BME) in immunology is the development of **vaccines**.

The development of vaccines that are both safe and effective has been one of the great achievements of modern medicine. Because of an effective vaccine, smallpox—a frequently fatal disease that claimed thousands of lives in previous centuries—has been **eradicated**, or eliminated as a natural infectious agent. Other severe infectious diseases, such as polio and influenza, are now in control in most countries of the world. There are, however, many diseases that have proven to be difficult for vaccine makers. Acquired immune deficiency syndrome (AIDS), which is caused by infection with human immunodeficiency virus (HIV), has killed millions of people worldwide (Figure 14.1), and there is still no effective vaccine available. Most residents of developed nations, such as the United States, receive a spectrum of vaccines during childhood, but there are no vaccines for

Table 14.1 Progress in vaccine development

Vaccines available	Vaccine not available
Smallpox	AIDS (0.7 $\times$ 10^6 deaths/y)
Rabies	Diarrheal disease (5–10 $\times$ 10^6 deaths/y)
Typhoid	Malaria (1.2 $\times$ 10^6 deaths/y)
Diphtheria	Schistosomiasis (0.5–1 $\times$ 10^6 deaths/y)
Tetanus	Dengue (50 $\times$ 10^6 cases/y)
Polio	Japanese encephalitis (50,000 cases/y)
Measles	Cholera (0.2 $\times$ 10^6 cases/y)
Rubella	African trypanosomiasis (0.5 $\times$ 10^6 cases/y)
Hepatitis B	Hepatitis C (4 $\times$ 10^6 cases/y)
Varicella	Leishmaniasis (12 $\times$ 10^6 cases/y)

Notes: These lists are not inclusive: Both columns represent only a fraction of the possible diseases. AIDS: Acquired immune deficiency syndrome.

other diseases (such as malaria, schistosomiasis, and diarrheal disease) that are major causes of death in the developing world (Table 14.1).

According to the World Health Organization (WHO), vaccines and clean water are the two public health interventions that have had the most substantial impact on world health. Both of these interventions represent the work of engineers: civil and environmental engineers, who developed water sanitation systems, and bioengineers, who developed methods for the mass production of safe vaccines. Vaccines prolong the lives of millions of people each year by preventing infectious diseases. Because of the development of an effective and portable vaccine, smallpox—a disease that took the lives of millions and likely caused the fall of Aztec civilizations—has been eradicated from the world.

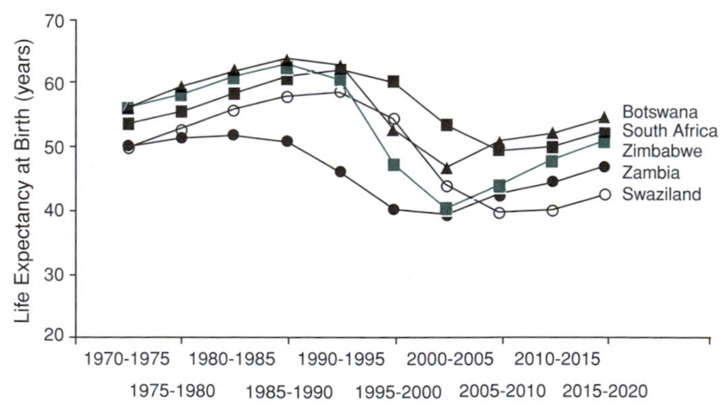

Figure 14.1 **Human immunodeficiency virus/acquired immune deficiency syndrome (HIV/AIDS) has had a dramatic impact on almost every aspect of life in African countries.** This graph shows changes in life expectancy for some selected African countries in the era of HIV/AIDS. Although there has been some progress in the development of an HIV vaccine in the past two decades, HIV has proven to be a great challenge for vaccine scientists. *Source:* United Nations Population Division. World Population Prospects: The 2006 Revision Population Database.

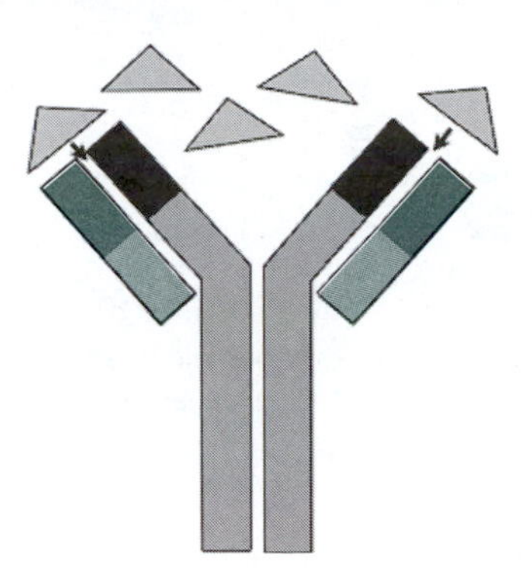

Figure 14.2 **Antibody.** Schematic diagram of an antibody—a protein formed from four connected polypeptide chains (two gray and two colored in this diagram)—and antigen (triangles), illustrating the binding interaction between antibody and antigen.

All schoolchildren in the United States and many other nations are now vaccinated for diseases that could cripple or kill them; as a result, children live longer and healthier lives. This chapter will review approaches to vaccine development and highlight the roles that biomedical engineers play in vaccine design and production.

Many vaccines work by enhancing the formation of **Abs** within the body of the vaccine recipient. Abs are specialized proteins that can inactivate pathogens directly, or mark them for destruction by other elements of the immune system (see Chapter 6). One of the great biomedical feats of the 20th century was creating methods for manufacturing Abs outside the body. This achievement has permitted the development of Abs as tools. Abs are now used as agents to prevent disease and as reagents for the detection of small amounts of materials. Because of their importance in medicine, the engineering of Abs is discussed first in this chapter.

14.2 Antigens, Abs, and mAbs

Ab molecules are proteins that carry out important functions in the immune system. The most important property of an Ab is its ability to bind to an **antigen** (Ag), or specific chemical (Figure 14.2). For their function in protecting us from disease, Ab molecules bind to chemical targets on the surface of **pathogens**, such as bacteria or viruses; this binding is often the first step in elimination of the pathogen from the body.

The target chemical to which an Ab binds is called an Ag (Figure 14.2). An Ag can be a protein, a polysaccharide, or even a small molecule that is attached to a larger carrier. Often, a whole Ag is large enough so that an Ab only binds to a small piece of it. For an individual person or animal, an Ag is a substance that is "foreign" or not ordinarily found in the body. One of the most important functions of each person's immune system is to make Abs that will bind to "foreign" Ags that happen to find their way into our bodies. The Ag may be associated with a bacterium, a virus, a transplanted organ, or some other entity that is "foreign," or not part of our body. When an Ag enters a host, it stimulates the immune system of that host. That stimulation often leads to the production of Ab molecules, which bind to the Ag.

Abs are produced by specialized cells of the immune system that are differentiated forms of circulating white blood cells called B cells. A **B cell** is stimulated to

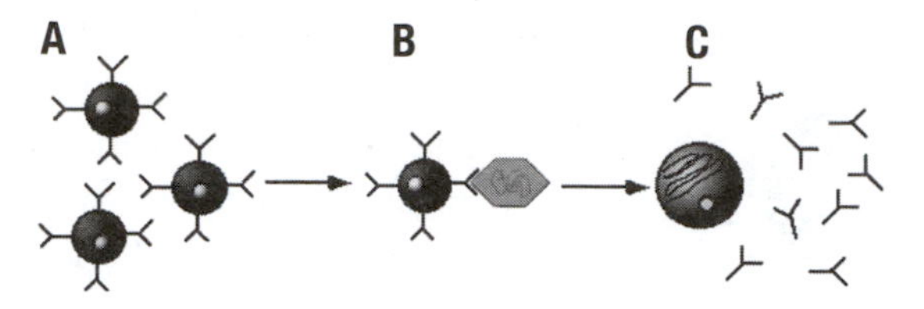

Figure 14.3 **B cells that are circulating in the body (A) can be activated by an encounter with an antigen (B).** This activation leads to the production of a mature B cell, also called a plasma cell, which secretes an antibody that binds to the antigen **(C)**.

produce an Ab by an encounter with a foreign Ag (Figure 14.3). When stimulated by an Ag, the B cell matures into an Ab-producing cell, called a **plasma cell**. Each individual plasma cell produces only one kind of Ab, but it produces many copies of that particular Ab. Every Ab molecule produced by an individual B cell is identical: It has exactly the same amino acid composition and it binds to the same target.

Abs were identified by scientists many decades ago, but medical uses of Abs did not develop until relatively recently. Their use in treating disease was limited by the inability to produce large quantities with a desired specificity. Kohler and Milstein (1) eliminated that barrier by demonstrating a method for producing Abs with a predetermined specificity from clones of an **immortalized cell line**. These special types of Abs that are produced by immortalized cells are called mAbs, and they have profoundly influenced biomedical research and medical diagnostics around the world. MAb technology has led to new therapeutic agents, which possess potent and highly specific biological activity and can be engineered to distinguish diseased cells from healthy cells. In the United States, the first mAb was approved for use in patients in 1986; this mAb is called OKT3 or Orthoclone® (Ortho-McNeil, Raritan, NJ) and it is used as an **immunosuppressive** agent in cases of **acute renal allograft rejection**. Since that time, a number of mAbs have been approved for use in treating a wide range of human diseases (Table 14.2).

Despite tremendous progress in the development of mAb and other Ag-binding reagents, the much-heralded "magic bullet" cures for cancer have not yet been realized. As scientists and clinicians have learned more about mAb-based therapeutic agents, and about their fate following introduction into the human body, expectations regarding mAb therapy have become more realistic: They are powerful tools, but they are not magic. Although much of the initial attention centered on mAb–toxin conjugates for targeting cells of solid tumors, this task has been difficult to achieve. It has proven easier to target other cells, such as the circulating cells involved in B-cell lymphoma and graft-versus-host disease. Although mAb can help in the therapy of some kinds of solid tumors, perhaps its most significant impact will be the prevention and treatment of infectious diseases. Because every human confronts many potentially infectious agents each day, new prophylactic methods could have tremendous impact worldwide.

Table 14.2 Examples of therapeutic mAbs that have been approved for use in the United States

Generic name	Trade name	Target Ag	Type	Indication	Year of FDA approval
Muromonab-CD3	Orthoclone	CD3	Murine IgG2a	Acute rejection in organ transplants	1986
Abciximab	ReoPro	GP IIb/IIIa	Fab fragment of chimeric IgG	Prevention of arterial thrombus	1994
Rituximab	Rituxan	CD20	Chimeric IgG1k	B cell NHL	1997
Daclizumab	Zenapax	CD25	Humanized IgG1	Prevention of rejection in organ transplants	1997
Basiliximab	Simulect	CD25	Chimeric IgG1k	Prevention of rejection in organ transplants	1998
Palivizumab	Synagis	RSV	Humanized IgG1k	Prevention of RSV infection	1998
Infliximab	Remicade	TNF-α	Chimeric IgG1k	Rheumatoid arthritis, Crohn's disease, and others	1998
Trastuzumab	Herceptin	Her-2/neu	Humanized IgG1k	Metastatic breast cancer	1998
Gemtuzumab ozogamicin	Mylotarg	CD33	Humanized IgG4k conjugated to calicheamicin	Relapsed AML	2000
Alemtuzumab	Campath	CD52	Humanized IgG1k	B cell CLL	2001
Ibritumomab tiuxetan	Zevalin	CD20	Murine IgG1k conjugated to 90-yttrium	Certain forms of NHL	2002
Tositumomab	Bexxar	CD20	Murine IgG2aλ	Certain forms of NHL	2003
Cetuximab	Erbitux	EGFR (Her-1)	Chimeric IgG1k	Metastatic colorectal cancer; head and neck cancers	2004
Bevacizumab	Avastin	VEGF	Humanized IgG1k	Colorectal cancer	2004
Panitumumab	Vectibix	EGFR	Human IgG2k	Metastatic colorectal cancer	2006

Notes: mAbs: Monoclonal antibodies; FDA: U.S. Food and Drug Administration; IgG, immunoglobulin G; TNF-α: tumor necrosis factor-α; AML: acute myeloid leukemia; CLL: chronic lymphocytic leukemia; EGFR: epidermal growth factor receptor; GP: glycoprotein; Her-1: human-1 epidermal growth factor receptor-1; Her-2: human-2 epidermal growth factor receptor-2; NHL: non-Hodgkin's lymphoma; RSV: respiratory syncytial virus; VEGF: vascular endothelial growth factor.

14.3 What are Abs?

A tremendous amount of information on Ab structure, and the relationship of Ab structure to Ab function, has been acquired over the last 50 years. Ab molecules, or immunoglobulins (Igs), are abundant within the blood. They were first recognized because of their common chemical properties (solubility and charge),

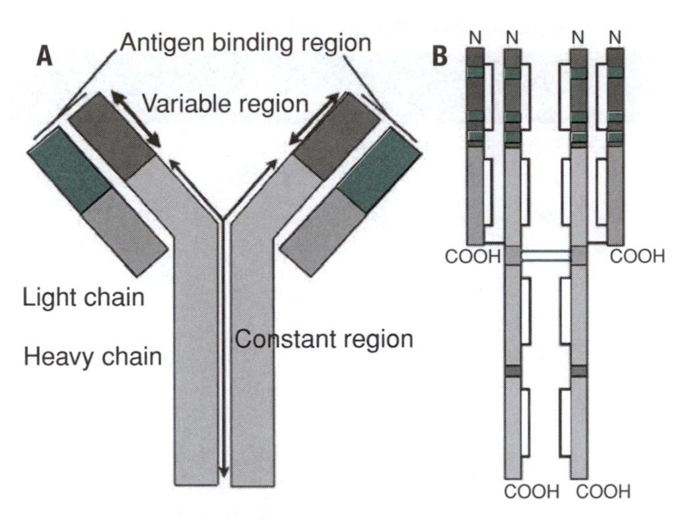

Figure 14.4

Structure of a typical antibody. A. The entire antibody consists of two light chains (colored) and two heavy chains (gray) that combine to make a functional molecule. The light shaded regions indicate the constant regions of each chain, and the darker regions indicate the variable regions. The variable regions of the heavy and light chains combine to form the antigen-binding region. **B.** A more detailed schematic of antibody structure, which shows a linear map of the amino-acid sequence of each chain. Disulfide bonds (dark lines) connect the four chains: The heavy chains are connected to each other by two disulfide bonds, in the hinge region (colored); each heavy and light chain is connected by a disulfide bond, near the carboxyl terminus (COOH) of each light chain; additional disulfide bonds stabilize the chains. The variable regions of each chain contain hypervariable regions (called complementary determining regions, or CDRs; shown in color): These CDRs form the three-dimensional structures that bind to antigens in the antigen binding regions.

suggesting a common molecular structure. Indeed, all Igs share several important characteristics: They are similar in structure, consisting of two identical light chains and two identical heavy chains that contain a series of common globular domains (see Figure 14.4a).

The basic Ab molecule is shaped like a "Y," with two Ag-binding sites at the top of the arms of the "Y" and an **effector region** within the base of the "Y" (Figure 14.4a). Therefore, the basic Ab contains: 1) two identical Ag-binding regions, which are formed by variable regions from a light and a heavy chain, and 2) one effector region, which is formed by conserved regions of the two heavy chains. By binding to cell receptors and other immune molecules, the effector region is responsible for producing class-specific functions.

The Ag-binding sites of all Ab molecules include amino-acid sequence contributions from both the heavy and light chains (Figure 14.4b). Within the variable regions of the heavy and light chains are hypervariable regions (called **complementary determining regions**, or CDRs). It is the composition of these CDRs that determine the Ag-binding specificity of each Ab.

The entire family of Igs can be divided into classes and subclasses, each containing Ab molecules with distinct characteristics. Although they share the common features discussed previously, the Ab classes—IgA, IgD, IgE, IgG, and IgM—differ from each other in the composition of the heavy-chain effector

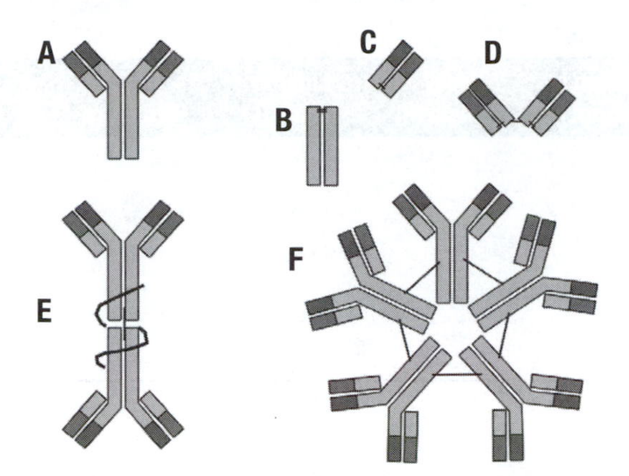

Figure 14.5 **Schematic diagram of the most important immunoglobulin (Ig) classes: IgG (A), secretory IgA (S-IgA; E), and IgM (F).** The light chains (~24 kD each) are indicated by the smaller rectangles; heavy chains (~55 or 70 kD) are indicated by longer, hinged rectangles. The diagram also shows the proteolytic fragments of IgG, Fc **(B)** and Fab **(C)**, which are produced by papain digestion, and $F(ab')_2$ **(D)**, the product of digestion with pepsin.

regions. Two of the Ab classes, IgA and IgG, can be further divided into subclasses, which also differ in heavy-chain composition.

IgG is the predominant Ab class in blood, where it circulates as a single four-chain unit (Figures 14.4 and 14.5a). IgG can be digested by enzymes into fragments: Treatment with papain produces one crystallizable fragment, Fc (Figure 14.5b) and two Ag-binding fragments, Fab (Figure 14.5c); treatment with pepsin yields a bivalent Ag-binding fragment, $F(ab')_2$ (Figure 14.5d).

IgA and IgM monomers are similar to IgG, but because they contain an extra **cysteine** residue in the C-terminal extension of their heavy chains, these **monomers** can polymerize (Figure 14.5e and f). **Polymerization** is initiated and stabilized by a 15-kD peptide, the joining (J) chain. The product of IgA-secreting cells is usually **dimeric** or **trimeric** (although other polymers also occur in small amounts), whereas IgM is usually **pentameric**. Polymerization is important for IgM, the Ig class produced on first exposure to an Ag. Although the Ag-binding sites are relatively low in affinity, the pentameric structure greatly increases the ability of IgM to bind Ag-bearing cells because **avidity** increases markedly with **valency**.

IgA is the predominant Ab class in the mucus secretions, and therefore must function within an environment that is frequently hostile, containing many microorganisms and proteolytic enzymes. IgA structure is most distinct in the hinge region of the heavy chain, where a unique amino-acid sequence and extensive **glycosylation** make the molecule more resistant to enzyme digestion. Polymerization is also required for IgA binding to the polymeric IgA (pIgA) **mucosal transport receptor** (FcαR), an essential step in the transport of functional pIgA into the mucus secretions. PIgA, produced in both peripheral and mucosal lymphoid tissue, binds to FcαR on the basal surface of epithelial cells lining the lumen of mucosal tissue. The pIgA–FcαR complex is endocytosed, transported

Table 14.3 Ab classes found in humans

Class	Concentration in serum (mg/mL)	Heavy chain (H)	Light chain (L)	Structure	Role
IgG	13	γ	κ, λ	H_2L_2	Memory
IgM	0.5–2.5	μ	κ, λ	$(H_2L_2)_5 + J$	First Abs
IgA	0.5–3.0	α	κ, λ	$(H_2L_2)_2 + J$	Secretions
IgD	0.03	δ	κ, λ	H_2L_2	?
IgE	0.0003	ε	κ, λ	H_2L_2	Allergic rxn

Notes: Ab: Antibody; Ig, immunoglobulin.

through the cell, and secreted from the luminal surface. Following secretion, a portion of the bound FcαR is cleaved, leaving a polypeptide called a **secretory** (S) **piece**, which remains bound to the newly released secretory IgA (S-IgA) molecule (see Figure 14.5e).

14.4 How can specific Abs be manufactured?

In the body, Abs are produced by the progeny of B cells (also called B lymphocytes), which differentiate into Ab-producing plasma cells after stimulation by an Ag. Each B cell produces an Ab of a certain class, specificity, and affinity for the Ag. Following synthesis and secretion by B cells, Abs become distributed throughout the body, where they accumulate in blood plasma and in secretions (Table 14.3). Abs are also found within the interstitial space of tissues, although at lower concentrations than in plasma.

Because humans naturally make Abs that are specific for some Ag, one method for obtaining Abs for clinical use is to collect the **sera** of patients known to have high concentrations, or **titers**, of Abs specific for a particular Ag. For example, **antiserum** for postexposure prophylaxis against hepatitis B is obtained by isolation of Ab from pooled plasma obtained from patients with high concentrations of Ab against hepatitis B surface Ag (HBsAg). This approach is workable with hepatitis B, hepatitis A, rabies, and measles because it is feasible to identify a population of patients producing high concentrations of specific Abs who are willing and able to donate plasma. In general, however, it is difficult to obtain sufficient quantities of a specific Ab by this method. Even the intravenous Ig preparations that are currently in clinical use, which are prepared by different procedures, may differ in effectiveness. Isolation of Abs from pooled plasma also has certain disadvantages, including the risk of transmission of other infectious agents.

Alternately, Abs can be produced in animals by intentionally exposing the animals to an Ag of interest. The procedures for doing this are now well-known; immunization of animals can lead to high titers of Abs of a predefined class and specificity within the plasma or secretions of laboratory animals. The time course

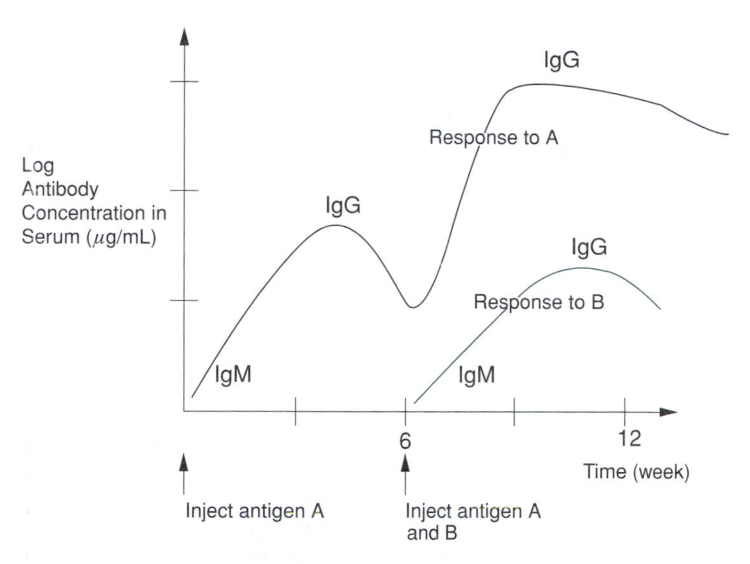

Figure 14.6 **Specific antibodies (Abs) are produced after exposure to antigens (Ags).** When a new Ag (one that the animal has not seen before) is introduced into an animal with a healthy immune system, the immune system produces Abs that are specific to (i.e., that bind to) that Ag. In this diagram, a new Ag (A) is administered at time zero. In the initial response to A, Ab production increases over time; the Ab produced over the first weeks are of class immunoglobulin (Ig) M, but class IgG is produced in abundance after several weeks. If the animal encounters Ag A again, the immune response is swifter: Class IgG molecules are produced quickly and in greater quantities. This "memory" effect is specific to the Ag. If a new Ag (B) is introduced at the same time as the old Ag (A), the response to A is swift and abundant, whereas the response to B is an initial response (colored line).

for generation of specific Abs after immunization of an animal (or a human) is shown in Figure 14.6. Upon first exposure to an Ag, Abs of class IgM are formed first; Ag-specific Abs of class IgG are formed after a few weeks. The second exposure to an Ag produces a vigorous "memory" response that is high in class IgG Abs.

The blood serum from animals that are immunized, as in Figure 14.6, contains **polyclonal Abs** against the administered Ag. Within the serum, there are Ab molecules produced by many different B-cell clones. Each of these clones was stimulated by the introduction of the Ag, but the Abs produced are different. Although these Abs all have in common the ability to bind to the Ag, they may bind to different sites on the Ag molecule, or with different affinities. Polyclonal Abs produced by this method have many valuable uses in biomedical research and diagnostics, but they are usually not appropriate for clinical use because foreign Abs cause immune reactions, thus limiting treatment to a single exposure: Only Abs produced in a human (such as the anti-hepatitis B antisera mentioned previously) can be used in humans. Antisera can be easily produced for any Ag in a mouse or goat (or other species), but these cannot be used safely in humans.

Almost two decades ago, Kohler and Milstein revolutionized immunology by developing a procedure for creating Ab-producing cell lines called **hybridomas** (1). These cell lines, created by hybridization of normal B cells with **myeloma cells**, are the progeny of a single B cell. They produce mAb molecules that are identical in chemical structure, Ag specificity, and Ag-binding affinity. The method is simple and elegant: B cells from an immunized animal are fused with myeloma cells; these cells are maintained in a selective medium that permits only the proliferation of successful fusions; and proliferating cells are cloned and screened for mAb production. To obtain large quantities of mAbs, hybridomas can be grown as **ascitic tumors** in animals or in culture. Although this cell-fusion technique has been extraordinarily successful for the production of mouse and rat mAbs, it has proven to be more difficult to obtain stable fusions of human cells.

Methods for producing mAbs are being rapidly improved by genetic engineering. Because a frequently encountered obstacle to mAb therapy is the human immune response to mouse mAbs, a major goal has been the production of Ab molecules, or Ag-binding subunits, with defined specificity that are non-immunogenic in humans and capable of eliciting human **effector mechanisms**. To create "humanized" Abs, for example, **recombinant deoxyribonucleic acid (DNA)** technology has been used to produce **chimeric mAbs**, which consist of mouse variable regions and human constant regions, and **humanized mAbs**, which are fully human with the exception of a mouse CDR. Both chimeric and humanized Igs have improved biological activity, reduced immunogenicity, and improved pharmacokinetics because of their prolonged plasma **half-life**. It is now possible to express genes for Fab fragments of the mouse Ab repertoire in bacteria. In this way, a diverse array of Ag-binding fragments can be rapidly produced. Single-domain Abs, composed of heavy-chain variable domains, can be expressed and secreted by microorganisms, such as *Escherichia coli (E. coli)*. Human Abs can also be produced in immunodeficient mice that have been repopulated with human **lymphocytes**. Human Abs can even be produced in plants, making it conceivable that Abs can be produced in large quantities using agricultural techniques.

14.5 Clinical uses of Abs

In 1890, von Behring and Kitasato demonstrated that immunity to certain diseases could be provided to non-immune individuals by the transfer of serum. This demonstration was critical in uncovering the role of Abs in the immune response; it was also the first example of **passive immunization** in animals. In passive immunization, Abs that are manufactured outside of the recipient's body are administered to an individual, usually to prevent an infectious disease. Passive immunization is a method of disease **prophylaxis**: Prophylaxis is a term that refers to a medical procedure that is intended to prevent a disease, rather than cure a disease that is already established. In certain situations, the natural effector functions of Abs (such as opsonization or complement activation, described in Section 6.5.1) provide all of the necessary biological activity for treating disease (Figure 14.7a). In other cases, the biological activity of Ab must be augmented by the addition of other toxic compounds.

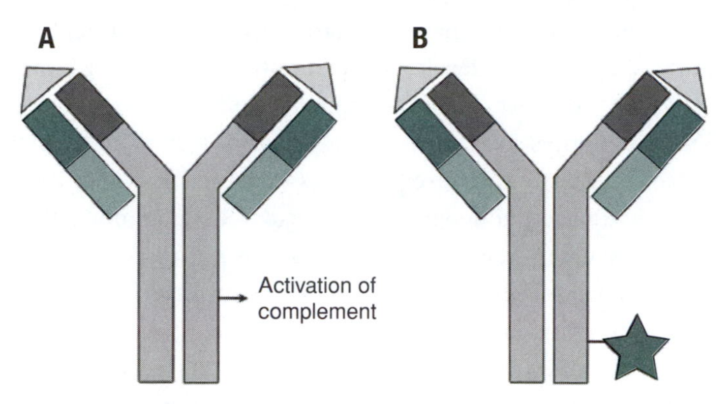

Figure 14.7 **Biological activity of antibody (Ab). A.** Abs possess a number of biological activities. Binding to an antigen (Ag; triangle) can neutralize the Ag by preventing its binding to a cell receptor or preventing its movement through tissue. Often, binding of an Ag leads to aggregation. The Fc region of the Ab can also initiate other biological reactions, primarily through the activation of molecules of the complement cascade, a sequential series of biochemical reactions that can lead to opsonization (targeting of the Ag–Ab complex for ingestion and degradation by macrophages or other cells) or formation of cytotoxic molecules (which are particularly important if the Ag is a bacterium). **B.** Additional functions can be added to Ab by chemical linking of toxins (colored star) to the Ab, which is usually done in the Fc region. This new molecule—called an immunotoxin—will still bind to Ag, but it will carry with it a toxic payload: a chemical toxin, a drug, or a radioactive species. The binding capability of the Ab allows it to accumulate in a region of interest (near tumor cells, e.g., in the event that the Ag is tumor-specific) where the toxin will act.

Also in 1890, Paul Ehrlich, a German scientist and Nobel Prize winner, who coined the phrase "magic bullet," predicted that the availability of antisera would lead to the development of agents that could provide selective activity for toxic compounds. That concept has evolved into modern **immunotoxin therapy**. Immunotoxins are chemical agents that are bound to an Ab or other molecule from the immune system that can bind with molecular specificity (Figure 14.7b). In theory, the Ab, guided by its binding specificity, carries the toxin molecule to a specific site in the body.

Unfortunately, immunotoxin therapy—which has the goal of treating a disease, such as cancer, that is already established in the patient—has not yet lived up to Ehrlich's promise. Current experience suggests that it may be limited to the treatment of certain diseases in which mAb–toxin conjugates will have ready access to the target cells. In contrast, passive immunization—using unmodified Abs and relying on their intrinsic biological activity—has been used successfully for many years as prophylaxis against infectious diseases, and is likely to increase in importance as more mAbs become available.

14.5.1 Passive immunization

Current methods for passive immunization, in which antisera or Ab preparations are provided to individuals at risk of acquiring specific diseases, followed directly from the work of von Behring. These methods became feasible with the development of simple and inexpensive techniques for separation of Ig from plasma. Today, injections of Ig (gammaglobulin) are used for prophylaxis against hepatitis A and hepatitis B, prevention and treatment of measles, and prophylaxis against the development of anti-Rh_0 Abs in Rh-negative women, as well as for treatment of chronic immunodeficiency syndromes. Travelers to developing countries sometimes get injections of gammaglobulin to decrease the chances of contracting diseases, such as hepatitis A, which can be contracted from food and water (although this practice is less common now because of the hepatitis A vaccine). These clinical indications are variations on a strategy used effectively in nature: Mother's milk contains large quantities of S-IgA class Abs, which provide immunoprotection to infants in the period following birth.

Humans are often infected through mucosal surfaces: Viruses and bacteria invade the vulnerable mucosal tissues of the respiratory, intestinal, ocular, or reproductive tracts. Therefore, several experimental therapies involve the delivery of Abs to mucosal sites, thereby interfering with pathogen entry directly at the site of invasion. For example, orally administered rabbit anti–*E. coli* serum prevents accumulation of bacteria in the blood after oral inoculation with a pathogenic *E. coli* strain in rats. Similarly, oral administration of a concentrate of Ig from milk protects humans from an oral challenge with a pathogenic strain of *E. coli*. Vaginally applied mAb can prevent both unwanted pregnancy and the transmission of sexually transmitted diseases. Passive immunization by administration of Abs to HIV can prevent disease progress in a number of settings.

14.5.2 Immunotherapy with Ab–toxin conjugates

The development of mAb–toxin conjugates as cell-selective cytotoxic agents has been vigorously pursued over the last several decades (Figure 14.7b). Immunotoxins are designed by selecting mAbs that are produced to bind to components of cells, such as receptors on the surface of cancer cells. Whereas conventional chemotherapy drugs are toxic to normal cells as well as diseased cells, **immunotoxins** should selectively bind to diseased cells, making chemotherapy safer and more effective. By conjugating Abs directed against cell-surface Ags with cytotoxic compounds, many investigators have demonstrated the selective cytotoxicity of immunotoxins in cell culture.

Immunotoxin therapy can lead to tumor regression and increased survival in animals bearing solid tumors. But human applications of immunotoxin therapy have developed slowly. In some previous clinical studies, the treatment was limited by toxicity, for example, conjugates designed to target breast and ovarian cancer cells also reacted with a brain Ag. Even when toxicity was not present, the first studies of immunotoxin therapy for breast, ovarian, colorectal, and lymphoid tumors yielded a disappointingly small number of remissions. A number of factors limited the usefulness of early immunotoxin therapy in humans—including the stability of the toxin–mAb bond, nonspecific binding to liver cells, and immune system reactions to the immunotoxin preparation.

Many of those early limitations were eliminated by the development of chimeric and humanized mAb. Today, there are a few immunotoxin preparations that are available for clinical use. Two are based on mAbs that bind to a receptor called

Table 14.4 Partial list of human vaccines and dates of introduction

Vaccine	Date introduced
Smallpox	1798
Rabies	1885
Plague	1897
Diphtheria	1923
Pertussis	1926
Tuberculosis (bacille Calmette-Guérin)	1927
Tetanus	1927
Yellow Fever	1935
Polio (injectable)	1955
Polio (oral)	1962
Measles	1964
Mumps	1967
Rubella	1970
Hepatitis B	1981

Notes: From the World Health Organization (http://www.who.int/vaccines-diseases/history/history.shtml).

CD20 that is abundant on B-cell lymphomas: These two immunotoxins both use radioactive species (^{131}I or ^{90}Y) as the toxin. Another is based on a mAb that binds to CD33, a receptor found on certain immature white blood cells and found abundantly on cells in certain acute myeloid leukemias. This immunotoxin contains a conjugated drug called calicheamicin. When this immunotoxin binds to CD33 on the tumor cells, the immunotoxin is internalized into the cell. The toxic drug is released intracellularly, causing reactions with DNA that lead to cell death.

There are many other immunotoxins in development. Improved engineering of mAbs is likely to produce more treatments for cancer, autoimmune diseases, and other indications in the future [see (2) for further reading].

14.6 Vaccines

The development of safe and effective vaccines is one of the most satisfying medical stories in modern history. Vaccines currently protect individuals around the world from debilitating or life-threatening diseases including smallpox, polio, and hepatitis (Table 14.4). Vaccines that prevent common diseases of childhood have dramatically reduced mortality during the early years of life. With many research groups around the world pursuing vaccines for AIDS, malaria, and other devastating diseases, vaccine development continues to be an active, promising area for the work of biomolecular engineers.

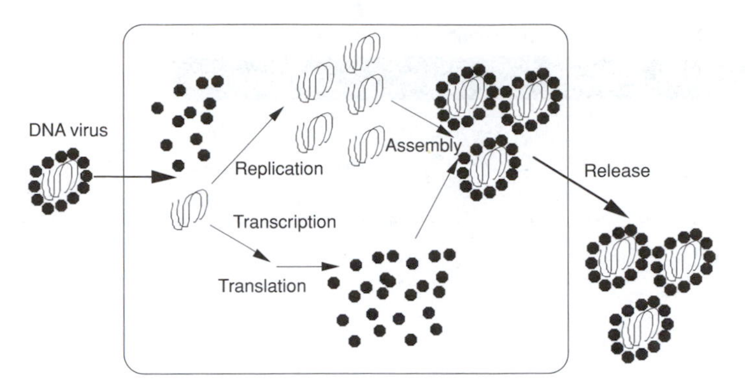

Figure 14.8 Highly simplified life cycle of a deoxyribonucleic acid (DNA) virus within a host cell.

Vaccines can be produced to treat infectious diseases that are caused by viruses, bacteria, parasites, or other infectious agents. Usually, vaccines are used as prophylaxis, but vaccines may also be useful for treatment of cancer, as described in Chapter 16, Section 16.6. Most of the successful vaccines target viral pathogens, so this section focuses on vaccines to prevent viral diseases.

14.6.1 Viral diseases

Viruses are tiny, disease-causing agents—usually on the order of 100 nm in diameter (although size varies greatly among virus families)—that can only multiply within cellular hosts. In general, a virus consists of genetic material, which can be DNA or ribonucleic acid (RNA), surrounded by a protective, **glycoprotein** coat called a **capsid**. Some viruses have complex capsid structures that include an **envelope**, or phospholipid-rich membrane.

Viruses cause disease by infecting cells within a host organism. Viruses use host mechanisms and structures for replication. The life cycles of viruses vary widely, but a typical viral life cycle involves the following steps (Figure 14.8): 1) viral entry into the host cell; 2) disassembly of the virus, which releases the genetic material; 3) replication of the viral genetic material; 4) transcription/translation of the viral genes to produce new viral proteins; 5) assembly of new viral units from the replicated genetic materials and new proteins, and 6) release of the newly assembled viral particles from the cell.

14.6.2 Examples of vaccine development

To illustrate some of the opportunities and challenges in vaccine development, this section describes approaches used to develop vaccines against smallpox, polio, and hepatitis B.

SMALLPOX

Smallpox is a physically devastating and frequently fatal infectious disease, which is caused by infection with the DNA-containing virus, **variola**. After exposure, the disease has a ~7- to 12-day incubation period, which is immediately followed by a high fever accompanied by malaise, aches, and a rash. After 3–4 days, the fever subsides and the patient seems to recover. Shortly thereafter, skin lesions develop that are characteristic of smallpox: becoming vesicular, then pustular, then developing scabs that produce deep scars across the face, arms, and legs. Death, which occurs in about 20% of patients, occurs within 1–2 weeks.

In the late 1700s, an English physician named Edward Jenner made an important observation. He noticed that farmers did not get smallpox. Cattle are infected by a virus, called **vaccinia**, which causes a disease in cows, called cowpox, which is similar to the disease caused by smallpox in humans. It was later discovered that, when humans are infected with the cowpox virus (vaccinia), they usually do not develop any disease; they do, however, develop an immune response that protects them from subsequent infection by variola. Jenner recognized the association between contact with cows and protection against the disease (even though he did not know about the similarity of the disease-causing agents; viruses were not discovered by biologists until the late 19th century). He began "vaccinating" individuals by injecting into their arms small quantities of fluid harvested from the skin lesions of cows.

The protective value of this new **vaccination** method was soon recognized, but distribution of the vaccine presented many problems. Obviously, it was not practical to bring infected cattle to every village and town. During the early 1800s, vaccination within a community was accomplished "arm to arm." When individuals were vaccinated, they developed a localized vesicular lesion. The fluid from the arm lesion of one individual was therefore used to vaccinate the arm of the next individual. This arm-to-arm procedure was not perfect: The vaccine tended to diminish in effectiveness over time, syphilis was occasionally transmitted with the vaccine fluid, and many individuals still acquired smallpox after several years.

Over the century that followed, biologists and engineers experimented with different methods to produce large quantities of effective and safe vaccine. Most of the smallpox vaccine used during the 20th century was produced by growing vaccinia in skin lesions on calves, housed under carefully controlled conditions (Figure 14.9). The virus was harvested from the skin lesions after killing the animal, purified by centrifugation, treated with phenol to kill any bacteria, and freeze-dried for transport and storage.

In 1967, with large quantities of safe and effective vaccine available (Figure 14.9), the WHO launched an effort to eradicate smallpox. This ambitious project was possible because of two important features of variola infection: 1) the absence of nonhuman reservoirs for the virus, and 2) the non-existence of asymptomatic carriers of the virus. Following a tremendous worldwide effort to distribute the vaccine to all parts of the world, the last reported case of smallpox occurred in Somalia in 1977. After two more years of continued surveillance, the WHO confirmed the eradication in 1979. At that time, all of the laboratory stocks of variola were destroyed, except for a small number that are retained for research purposes.

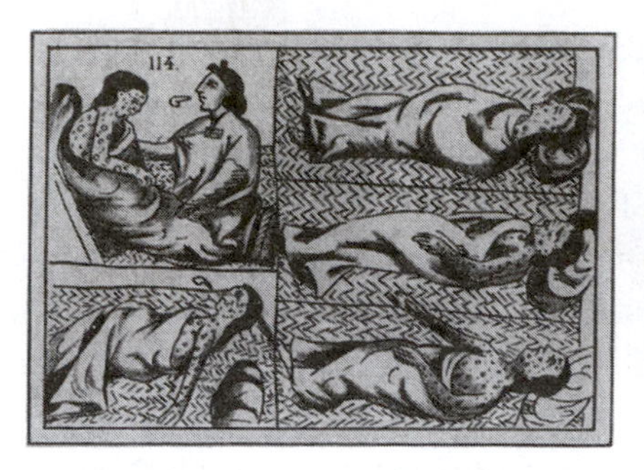

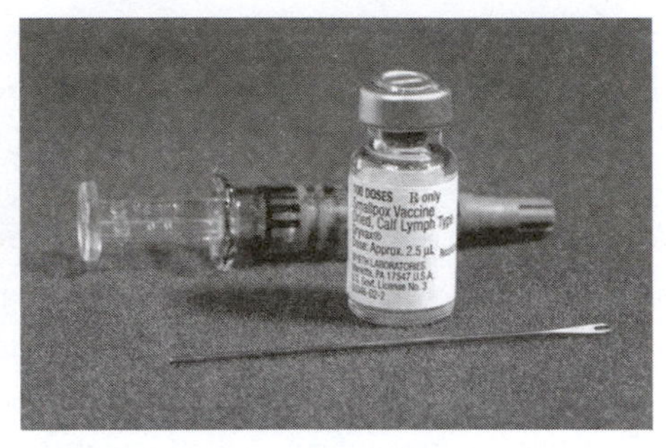

Figure 14.9

Smallpox. A. Illustration from Fray Bernardino de Sahagún's 16th century work describing Aztec culture and history known as *The Florentine Codex*. The image shows the stages of the disease, which decimated the Aztec population after the arrival of the Spanish conquistadors. **B.** The smallpox vaccine and bifurcated needle that was developed for introducing the vaccine into the skin. Photo by James Gathany and provided courtesy of the Centers for Disease Control and Prevention.

The vaccinia virus provided the basis for an extraordinarily successful vaccine for smallpox. The mechanism of action of the vaccinia-based vaccine is now understood: Vaccinia is structurally and immunologically similar to variola virus, so that the immunity to vaccinia, which is produced by intentional exposure to the vaccine, protects individuals against future contact with variola. Importantly, however, the vaccinia virus is pathologically quite different from variola: It does not cause any serious disease in humans. Because of this property, vaccinia is a naturally occurring **attenuated** variant of variola. A virus is called attenuated if it retains the immunological characteristics of another virus, without the disease-causing characteristics.

The vaccinia virus can now be grown in culture, by infection of primary cultures of rabbit kidney cells. There is a great deal of interest in using the vaccinia virus—an agent that is known to produce a good immune response in humans while being quite safe—to immunize against other diseases by using recombinant DNA techniques to modify viral properties.

POLIOVIRUS

Poliovirus is an RNA virus that causes a severe paralytic disease in humans (Figure 14.10). Poliovirus is an **enterovirus**; it infects through the intestinal tract and so is primarily contracted by the fecal–oral route. For this reason, the virus is commonly acquired by children, and transmitted through them to the rest of a family.

In the majority of cases (~95%), poliovirus infection will remain confined to the cells of the intestinal tract, where it produces few symptoms. In the other cases, it can produce a spectrum of disease: either a minor illness (where fever is the predominant symptom), meningitis, or paralytic disease. Paralytic disease, a tragic manifestation of poliovirus infection, is the result of the virus's ability to infect cells of the spinal cord and central nervous system. When these cells are infected, the disease produces irreversible paralysis. Death occurs in about 2%–5% of children and 14%–30% of adults with paralytic polio.

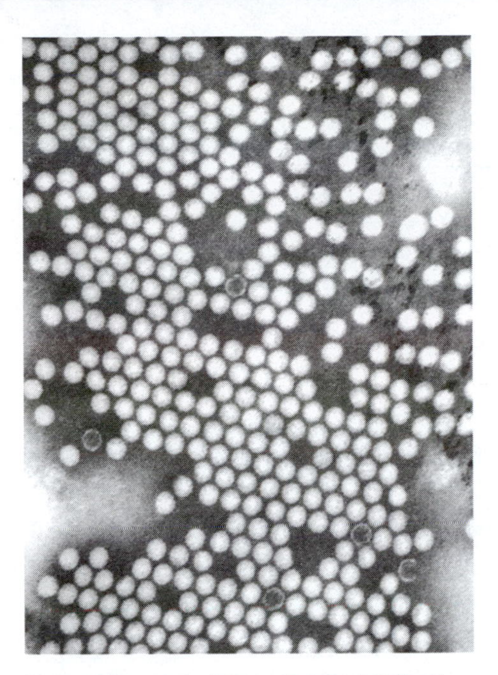

Figure 14.10 **Electron micrograph of the poliovirus.** Poliovirus is a species of *Enterovirus*, which is a genus in the family of *Picornaviridae*, and is a ribonucleic acid (RNA) virus. Photo courtesy of The Centers for Disease Control and Prevention; photo credit Dr. Fred Murphy, Sylvia Whitfield.

In the early 1950s, Jonas Salk produced a vaccine for the prevention of poliomyelitis. The Salk vaccine contains all three of the major **strains** of poliovirus (a strain is a variant of an organism—in this case a virus—that is distinct enough to be recognized, but not different enough to be classified as another species). The vaccine is prepared by killing the virus particles, using a chemical treatment, so that they will produce immunity but not disease. To obtain large quantities of the virus, it is first grown in culture by inoculating monolayer cultures of primary monkey kidney cells with live virus. The virus is purified from the culture and killed by treatment with formaldehyde and other chemicals. It is given to recipients by injection. The Salk vaccine is also called **inactivated** or **killed polio vaccine** (IPV or KPV).

After successful testing in animals and small-scale safety testing in humans, field trials for the large-scale testing of the efficacy of the Salk vaccine were initiated in 1954. For this field test, the vaccine was administered to 1,829,916 U.S. schoolchildren in the first to third grades. After initial tests had begun, it was discovered that the primary monkey kidney cells used to produce the virus were also infected with a simian tumor virus (SV-40). It was also discovered that certain lots of the virus were incompletely killed, presenting some risk of active infection following vaccination. Both of these problems were solved by switching to a human cell line for virus production and by modifying the formaldehyde treatment protocol to insure complete killing. With these problems solved, the field test was judged a success and the Salk vaccine was released to the general public.

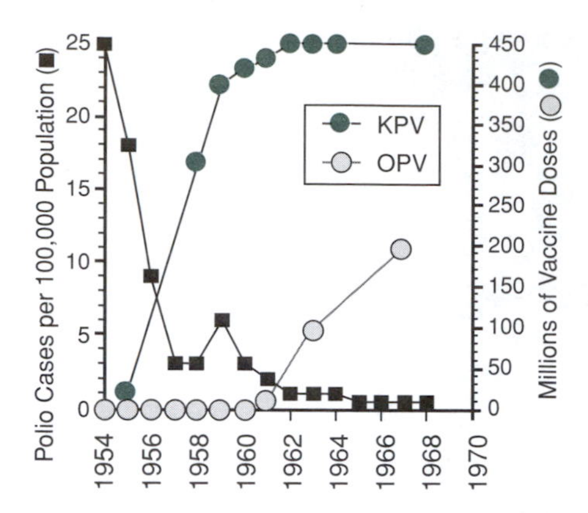

Figure 14.11 Effectiveness of the Salk killed polio vaccine (KPV) and the Sabin oral polio vaccine (OPV).

A second type of polio vaccine was developed by Albert Sabin. Unlike the Salk vaccine, which contains killed virus, Sabin used a living, attenuated virus to produce an **oral polio vaccine** (called OPV or LPV). Vaccinia, the attenuated virus used as a vaccine for smallpox, was found naturally. The Sabin vaccine was created in the laboratory, using special techniques to mutate the virus and identify new, attenuated forms.

When the Sabin vaccine is administered orally, the attenuated virus infects the cells of the intestinal tract and actively reproduces. Because of this, OPV is 100,000 times more potent than IPV. Because it can be swallowed rather than injected, it is much easier to deliver to patients than IPV is. By vaccinating one family member with OPV, which actively reproduces in the recipient, there is a high likelihood that the vaccine will be inadvertently passed to other family members: another significant advantage of OPV. However, because OPV is a mutated strain of poliovirus, there is some risk that the live virus will revert to its disease-causing form. During the years 1972–1983, 279×10^6 doses of OPV were administered in the United States; in this same period, 87 cases of vaccine-induced disease were reported.

The use of polio vaccines in the United States has led to dramatic reductions in the incidence of paralytic disease (Figure 14.11). In 1954 there were more than 18,000 new cases; in 1976 there were only 8. Unfortunately, as the likelihood of infection in the United States has decreased, so has participation in vaccination programs. Currently, about 38% of children ages 1–4 have not had primary polio vaccination; the rate of vaccination is even lower in disadvantaged urban and rural areas. Certainly, with many susceptible individuals in the population, there is a significant risk that localized polio epidemics could still occur.

Presently, killed and attenuated forms of the polio vaccine are used in different populations. The choice of the form of vaccine to use for a particular population is related to properties of the preparations (Table 14.5). Attenuated vaccine, because it contains viral particles that are capable of replication, is not appropriate for use in individuals with immune deficiencies. Killed vaccine is easier to transport to isolated communities.

Table 14.5 Typical features of attenuated versus killed virus vaccines

Attenuated	Killed
Cell-mediated immunity	Nonreplicating
Longer protection against disease	Non-infectious
Stronger immune response	Boosters required
May revert to the wild-type virus	High purity

HEPATITIS B INFECTION

Hepatitis B is probably most commonly transmitted by intimate contact with the bodily fluids of an infected individual. The most obvious route of transmission is **parenteral** through shared needles, accidental needlesticks, blood transfusions, and even shared razors and toothbrushes. It can be transmitted by sexual contact or from mother to child during birth or breast-feeding. After infection, hepatitis B Ags can be identified in virtually every fluid and tissue of an infected individual: saliva, tears, semen, cerebrospinal fluid, ascites, breast milk, synovial fluid, gastric fluid, pleural fluid, urine, and (rarely) feces.

The hepatitis B virus is a DNA-containing virus with a protein coat that is formed from many copies of a protein called the HBsAg (Figure 14.12). After entry into the body, the virus has an incubation period of 30–180 days. After infection, about 25% of individuals develop jaundice, 5% require hospitalization, and 0.1% die of a serious form of disease called fulminant hepatitis. A fraction (6%–10%) of infected individuals become carriers: 25% of these carriers eventually develop chronic hepatitis, 20% die of **cirrhosis** of the liver, and 5% die of hepatocellular carcinoma. It is a serious infectious disease, complicated by the fact that there are more than 200 million carriers of hepatitis B in the world. The lifetime risk of acquiring hepatitis B infection in the United States is ~5%.

The hepatitis virus infects, and is actively reproduced within, the cells of the human liver. Liver cells overproduce a piece of the virus, HBsAg, which assembles to form small particles (~22 nm in diameter) within the blood. Unfortunately, it is not possible to produce the virus—or these HBsAg particles—in culture. It is these particles of HBsAg that are the basis of the hepatitis B vaccine. Because the vaccine is made from these small pieces, or subunits, of the virus, it is called a **subunit vaccine**.

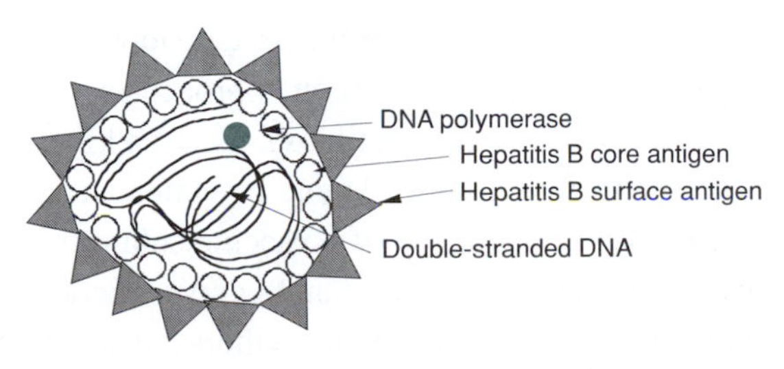

Figure 14.12 **Schematic diagram of hepatitis B virus.** DNA: deoxyribonucleic acid.

The original hepatitis B vaccine was produced from HBsAg isolated from the blood of infected patients (Heptovax®, Merck Pharmaceutical). Blood was collected from hepatitis B carriers (after carefully screening to make sure that these patients did not carry any other infectious agents in their blood), and HBsAg subunits were purified from the blood. The resulting subunit-enriched product was inactivated with chemicals and formulated into a vaccine. This initial vaccine was used only in individuals that were at high risk for hepatitis B infection, such as health care workers. It was not recommended for vaccination of the general population because, although the subunits were carefully purified and the blood donors thoroughly screened, there was still some risk from a vaccine prepared from pooled blood of infected individuals.

A safer vaccine was produced using recombinant DNA technology. To avoid the problem of purification from the blood, the gene for HBsAg was introduced into yeast. These altered yeast cells were cultivated in the laboratory, and the recombinant subunit product was isolated from the cultures. The result is a recombinant subunit vaccine (the first to be produced was Recombivax®, Merck). This vaccine is now given to children before they enter school.

Recombinant subunit vaccines do not generate an immune response that is as vigorous as whole-virus vaccine. The recombinant hepatitis B vaccine must be given in three shots (and more for some individuals) to generate protective immunity. For that reason, the vaccine ingredients must be formulated into preparations that stimulate the immune system as efficiently as possible. Usually this involves the addition of **adjuvants**, which are other ingredients added to a vaccine to enhance the immune response. Salts of aluminum (such as aluminum phosphate and aluminum hydroxide) act as adjuvants, for example. Although the mechanisms of adjuvant activity are still not completely understood, this is an active area of investigation: Immunologists are beginning to uncover some of the biological mechanisms that act in adjuvants.

For short-term protection against infection, or prophylaxis in the immediate period following a suspected exposure, passive immunization can be achieved by hepatitis B Ig (HBIG). The Abs in HBIG are obtained by fractionation of the blood from patients with known hepatitis B infections, and are injected into a recipient. This is becoming less common, as more of the population is vaccinated.

14.6.3 Approaches to vaccine development

There are several possible approaches to vaccine development. The previous sections illustrated several of the most common approaches: use of a naturally occurring attenuated virus (vaccinia for smallpox), use of a killed virus vaccine (the Salk vaccine for poliomyelitis), use of laboratory-attenuated virus vaccines (Sabin vaccine for poliomyelitis), and use of a vaccine subunit, either collected from natural sources or produced by recombinant DNA technology (hepatitis B vaccines). This section discusses some alternate methods, many of which are now being tested for new vaccines.

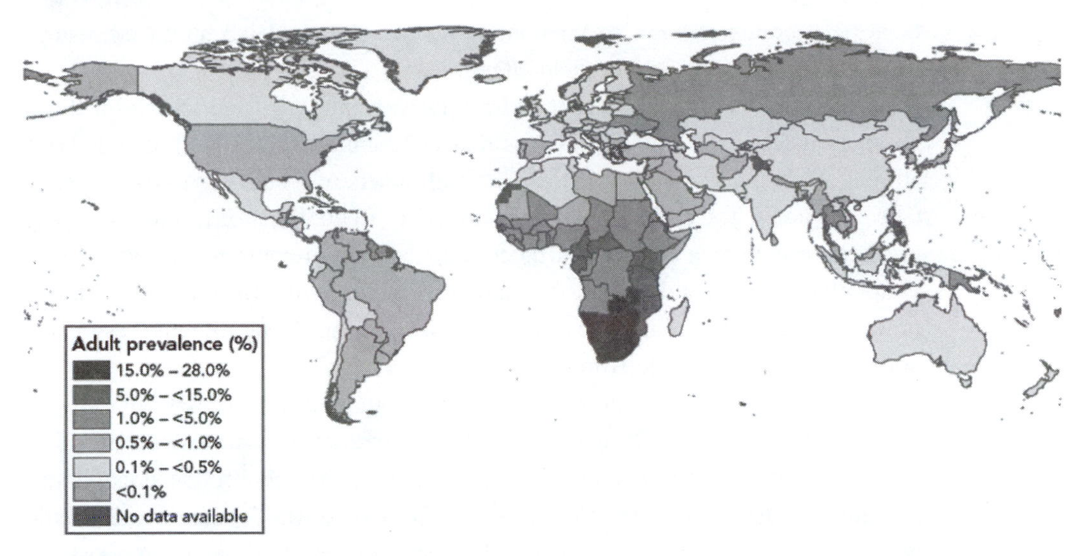

Figure 14.13 **Prevalence of human immunodeficiency virus (HIV) infection among adults in different regions of the world.** HIV infections are most prevalent in areas of Africa, but there are few regions of the world that are untouched by the virus. In 2007, it was estimated that 33 million people were living with HIV infections. Reprinted with permission from the UNAIDS 2008 "Report on the Global AIDS Epidemic." (See color plate.)

HIV VACCINES

The most significant incentives to produce new vaccines over the past decades have been the dramatic increase in the number of HIV infections and the deadly course of AIDS in humans (Figure 14.13). HIV is a retrovirus, which uses RNA as its genetic material (Figure 14.14). The search for an HIV vaccine is international, with countries from around the world developing new candidate vaccines. In countries around the globe, dozens of clinical trials of the best candidate vaccines

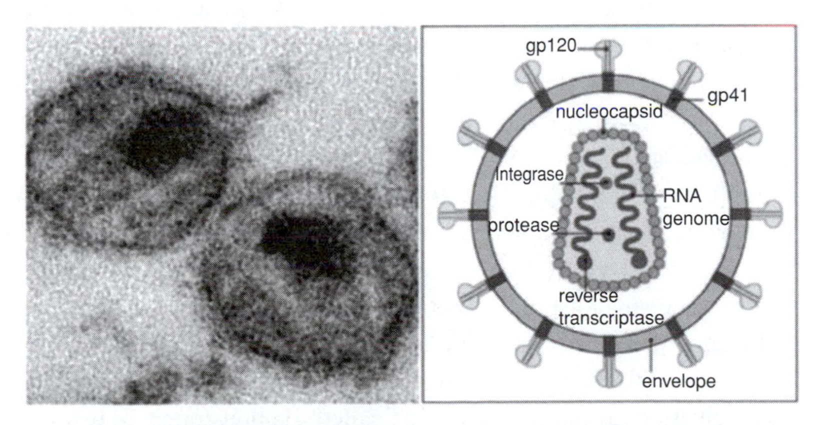

Figure 14.14 **Human immunodeficiency virus (HIV).** HIV is an example of a retrovirus (recall the discussion of retrovirus in Chapter 3). Its genetic material is ribonucleic acid (RNA). The virus, therefore, carries the reverse transcriptase enzyme so that it can synthesize double-stranded deoxyribonucleic acid (DNA) from its RNA genome after entering a host cell. Other enzymes—protease and integrase—are necessary for different steps in the HIV life cycle. The outer shell of HIV contains two proteins, called gp41 and gp120, which are involved in recognition of cells for infection and have frequently been used as targets for vaccines. The two proteins, gp120 and gp41, are derived from a precursor protein, gp160, which has also been used as a vaccine target. Photo courtesy of Dr. A. Harrison, Dr. P. Feorino, and the Centers for Disease Control and Prevention.

have been conducted, but so far none of these vaccines have produced results needed for protection against HIV infections.

Most experts agree that a successful HIV vaccine will need to exhibit three characteristics: 1) stimulation of Abs that can neutralize a broad range of HIV strains; 2) stimulation of consistent and high levels of reactive cytotoxic T cells (see Chapter 6 for a description of cytotoxic T-cell function); and 3) strong immune responses at mucosal surfaces. This last requirement reflects the observation that HIV transmission most frequently occurs across the mucosal surfaces of the reproductive and intestinal tracts and, therefore, strong immune responses at these sites are more likely to be protective.

Recombinant subunit approaches—similar to those used successfully for hepatitis B—were among the first HIV vaccine candidates to be tested clinically. Two U.S. biotechnology companies, Chiron and Genentech, produced candidate subunit vaccines based on recombinant gp120, which was produced in cell culture. In 1994, the U.S. government, which was providing financial support for vaccine trials, decided not to proceed with large-scale, phase III clinical trials to test these vaccines because the results from initial, smaller scale trials did not indicate adequate production of Ab or T-cell responses. Many different variants on the recombinant subunit approach—using different subunits, prepared by alternate methods, injected on different schedules, or mixed with adjuvants that are designed to strengthen the immune response—are still being tested in smaller scale clinical trials.

The first large-scale clinical trial of an alternate AIDS vaccine, based on an engineered virus that carries HIV genes, was initiated in South Africa in early 2007. The vaccine, called MRKAd5 HIV-1 trivalent vaccine, was developed by Merck & Co. This live-virus vaccine is based on adenovirus, which causes the common cold, that was modified using recombinant DNA techniques to include three genes from HIV called *gag*, *pol*, and *nef*. Although the virus carries these genes, it is incapable of making a live HIV virus.

DNA VACCINES

In the early 1990s, almost 200 years after Edward Jenner demonstrated the effectiveness of the smallpox vaccine, a new paradigm for vaccination emerged. The conventional method of vaccination required delivery of whole pathogens or structural subunits, but in this new approach DNA or genetic information could be administered to elicit an immunological response. Once it was observed that plasmid DNA delivered in vivo led to production of an encoded transgene, DNA vaccination (as this approach was called) demonstrated, in two ground-breaking studies, that immunological responses could be generated against antigenic transgenes delivered via plasmid DNA [see (3) for further reading]. The appearance of this new vaccination strategy coincided with many advancements in molecular biology, which provided new tools to study and manipulate the basic elements of an organism's genome that could also be applied to the design and production of DNA vaccines.

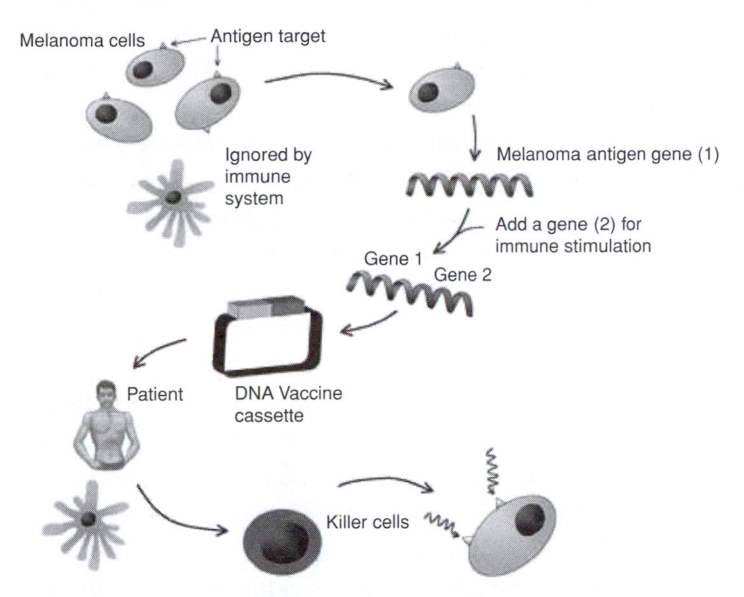

Figure 14.15 **Deoxyribonucleic acid (DNA) vaccine for treating melanoma.** Antigens are identified from melanoma cells, and the genes encoding these melanoma antigens are inserted into plasmids that are appropriate for use in humans. When the engineered plasmid is injected into a patient with melanoma, muscle cells will make the foreign protein, which will activate the immune system. The hope is that the foreign protein antigens will be expressed strongly in the patient, hopefully stimulating the patient's immune system to respond with a strong anti-melanoma response.

Why the rapid growth of interest in DNA vaccines? First, DNA vaccines represent a simple and powerful concept: The coding sequence of an antigenic pathogen gene is incorporated into plasmid DNA, which will allow its expression in host cells. Thus, DNA vaccines circumvent the need for preparation, purification, and delivery of a pathogen or antigenic protein. Instead, they use the intrinsic machinery of host cells. Second, conventional vaccination approaches have failed to yield useful vaccines for a large number of infectious diseases, and have made only modest progress in treating cancer. DNA vaccination may be able to engage immunological mechanisms that are not easily attainable with other approaches. Third, methods to produce, manipulate, and purify DNA are now standard in most biology and bioengineering laboratories, making the tools of DNA vaccine production widely accessible. Investigators rely on recombinant DNA principles—described in Chapter 3—to generate plasmids with the promoter, antigenic coding regions, and other sequences necessary to achieve optimal expression in host cells.

To date, clinical trials have investigated the utility of plasmid DNA for vaccination for infectious diseases caused by HIV, influenza, malaria, hepatitis B, and human papillomavirus, as well as neoplastic diseases such as melanoma (Figure 14.15). The primary objective in phase I clinical trials for DNA vaccines is to establish safety and tolerability of the therapy. Clinical trials for DNA vaccines to date have used a wide variety of doses (ranging from 0.25 μg to 2500 μg), delivery strategies (needle or needleless injection), administration sites, (intramuscular or epidermal), carriers (naked, polymeric microparticles, or cationic liposomes), and dosing schedules. So far, DNA vaccines appear to be safe and well tolerated by patients, with mild side effects at the site of administration that are comparable to existing vaccination strategies.

DNA vaccines also appear to be capable of eliciting both humoral and cellular immune responses, leading to clinical effectiveness. Unfortunately, in most studies the immune response to DNA vaccines has varied substantially among subjects. Intramuscular administration of naked DNA, which has been successful in generating both humoral and cellular responses in animal models, has not been as successful in humans. In one study, intramuscular injection of DNA stimulated strong cytotoxic T lymphocyte responses, but failed to induce detectable Ag-specific Abs. In contrast, intramuscular injections of DNA encapsulated in microparticles or intraepidermal delivery of DNA was able to elicit both humoral and cellular responses. More work is needed to reconcile results from these disparate studies, but it seems clear that the route of DNA vaccine delivery and the methods of vaccine preparation have strong effects on the immune response and the effectiveness of that response in preventing or treating disease.

One important clinical application of DNA vaccines may be to complement or augment traditional vaccines. For example, DNA vaccines have been tested in patients who failed to respond to conventional vaccinations for hepatitis B. The majority of subjects in one study developed a protective Ab response after DNA vaccination. It may also be advantageous in some settings to stimulate immunity with multiple vaccine preparations. To induce malaria immunity, for instance, subjects were primed with a DNA vaccine and then boosted with the recombinant Ag. This prime–boost approach was able to elicit both humoral and cellular immunity in subjects.

It is an exciting time for DNA-based vaccine technology, which has moved quite rapidly from pioneering animal studies to clinical testing. As the technology moves from the benchtop to the patient, the work of biomedical engineers is becoming more important. DNA vaccination offers a new platform technology for treatment and prevention of human disease with attributes that make it suitable for both developed and developing nations.

Summary

- One of the most important advances in medical science over the past 200 years is the use of vaccines to prevent infectious disease.
- The immune system responds to the presence of many foreign Ags by producing specific Abs. Abs vary in specificity as well as class; the ability to make a range of Abs is important in health and creates many opportunities in biotechnology.
- A variety of approaches to vaccine production have been developed, including the use of attenuated or killed viruses, protein subunits, engineered viruses, and DNA as the vaccine component.

- Despite important progress in disease prevention, there is still much to learn about how to make vaccines and how the immune system responds to different vaccine preparations.

REFERENCES

1. Kohler G, Milstein C. Continuous cultures of fused cells secreting antibody of predefined specificity. *Nature*. 1975;256:495–497.
2. Liu XY, Pop LM, Vitetta ES. Engineering therapeutic monoclonal antibodies. *Immunol Rev*. 2008;222:9–27.
3. Saltzman WM, Shen H, Brandsma JL, eds. *DNA Vaccines*. New York: Humana Press; 2006.

USEFUL LINKS ON THE WORLD WIDE WEB

http://www.aidsvaccineclearinghouse.org
A resource for information on HIV vaccines, including a listing of all of the clinical trials presently in effect throughout the world

KEY CONCEPTS AND DEFINITIONS

acute renal allograft rejection a condition that develops in some kidney transplant patients in which the recipient's immune system attacks the cells of the transplanted organ, leading to loss of kidney function

adjuvants other ingredients added to a vaccine or other drug to enhance its effects

antibodies (Abs) large, Y-shaped proteins that are used in the immune system to identify and counteract invaders such as viruses, bacteria, or transplanted organs by binding to the Ag

antigen (Ag) a substance introduced into the body that can stimulate the production of Abs

antiserum a serum composed of Abs that has been inoculated into an animal for use in a second animal to provide immunity against a particular disease

ascitic tumors tumors caused by the accumulation of fluid in the peritoneal cavity

attenuated retaining the immunological properties, in reference to a virus, without the disease-causing characteristics

avidity in immunology, the combined strength of all the Ab–Ag bonds involved in the interaction of an Ag and Ab

B cells lymphocytes that develop in the bone marrow that produce Abs against specific Ags; also called B lymphocytes

capsid the protective, glycoprotein coat of a virus

chimeric monoclonal Ab (mAb) a half-mouse, half-human mAb produced via recombinant DNA technology that merges DNA sequences from the two different species

Profiles in BME: Eliah R. Shamir

I was born in Israel and have lived in Northern Virginia since I was four years old. When I was little, I had no idea what I wanted to do when I grew up; I always told my parents I was interested in too many things. I excelled in math, but science never really grabbed me until my enthusiastic ninth-grade biology teacher got me excited about cell biology, genetics, and human physiology. I loved learning about this hidden, complex world about which we still know so little. I pursued this passion further by doing interdisciplinary research in the Mitre Corporation's Nanosystems Group for three summers. This experience taught me to problem-solve, think outside the box, integrate scientific disciplines, write and speak more precisely, and apply basic knowledge to help design workable devices.

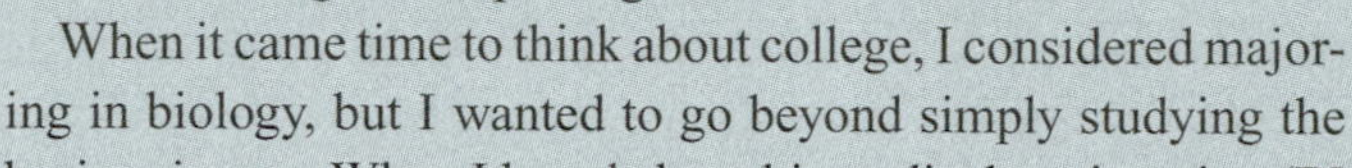

When it came time to think about college, I considered majoring in biology, but I wanted to go beyond simply studying the basic sciences. When I heard about biomedical engineering (BME) at a college information session, I thought I had found the perfect profession for myself. It would apply my love of biology to the medical and health world that I was drawn to and open doors to all kinds of careers.

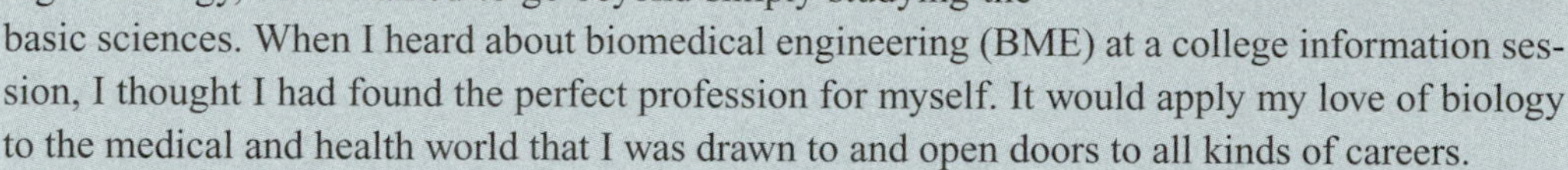

From the beginning, my BME classes took a very application-based approach to science, always with the emphasis on skill sets in computer science, math, and other analytical techniques. After my freshman year, I did an internship in an immunology lab at Memorial Sloan-Kettering Cancer Center, where I worked to better characterize the role of a transcription factor in T-cell differentiation. I loved delving into a problem yet unsolved and getting results yet undiscovered.

I enjoyed the many challenges of engineering, but I also yearned for something more—a human component; a vision of the problems I would encounter within their societal context. Starting my first year in college, I became involved with UVA's Center for Global Health and the Global Public Health Society, which aim to get students actively involved in and knowledgeable about health issues affecting both the local and the global community.

During my second year, I was inspired by a documentary called "Trading Women" that very personally captured the stories of exploited minority women in Thailand. I began to read about this and was immediately fascinated by the close relationship between HIV/AIDS and sexual and labor exploitation as applied to the vulnerable and mostly illegal migrant community from Burma, Laos, and Cambodia. I applied for and was lucky enough to receive research grants from the Center for Global Health, Pfizer Foundation, and UVA's Raven Society to fund a project that took me to Thailand for the whole summer. Throughout the year, I contacted organizations, hospitals, and clinics working in Thailand on HIV prevention and treatment. Once there, I was exposed to problems of health rooted in societal divisions of class and ethnicity, political conflict, and economic disparities in a culture with different priorities and ways of thinking and limited resources to address these issues.

Throughout this research experience, my BME background remained essential in helping me ask questions such as: What is the problem at stake? What are the factors being measured and how? Are the methods used appropriate to answering the question? Effective critical thinking was vital to properly evaluating what is being done to combat HIV/AIDS in Thailand in terms of multiple, complex causal factors, different strategies, and their outcomes.

I see engineering as a service profession. It differs from basic science research, which often explores fundamental principles without necessarily yielding a concrete product. Engineering uses scientific and mathematical tools to serve human needs. In the case of BME, these tools may be diagnostic or therapeutic. I graduated with a B.S. degree in biomedical engineering from the University of Virginia in 2008. Currently, I am pursuing an MD/PhD at the Johns Hopkins University School of Medicine, with my ultimate goal being conducting research in immunology and infectious diseases that has clinical significance and medical benefit. Although I may not be a practicing engineer, I know that wherever my career goes, my BME degree will serve as a firm foundation for how I approach my work.

cirrhosis a disease of the liver in which healthy cells are replaced by scar tissue, leading to the loss of liver function

complement cascade a sequential series of biochemical reactions in the immune system that can lead to opsonization (targeting of the Ag–Ab complex for ingestion and degradation by macrophages or other cells) or formation of cytotoxic molecules

complementary determining region (CDR) a hypervariable region located in an Ab that determines the Ag-binding specificity of the Ab

cysteine a naturally occurring amino acid that contains a thiol group

dimeric consisting of two similar molecules, or monomers, that have been linked together

effector mechanism the secondary mechanisms of action of an Ab that are initiated by Ab binding, including complement activation

effector region the region of an Ab molecule that is responsible for activating the effector functions, but not for Ag binding

enteroviruses a group of small, RNA-containing viruses that chiefly inhabit the intestinal tract and are primarily contracted by the fecal–oral route

envelope a phospholipid-rich membrane that covers the capsid of some viruses and is used to help the virus enter host cells

eradicate to completely eliminate the presence of a natural infectious agent among the global human population

glycoprotein a group of macromolecules that all contain a protein connected to a carbohydrate

glycosylation an enzyme-directed addition of a glycosyl group to a protein to form a glycoprotein

half-life the time required for a quantity to decay to exactly half of the initial amount present

humanized monoclonal Ab a monoclonal Ab produced via recombinant DNA technology that is fully human with the exception of a mouse CDR

hybridomas cells that produce large amounts of specific Abs, formed by the fusion of a normal cell with a cancer cell

immortalized cell line cells that have been altered to grow indefinitely under proper culture conditions, unlike primary and secondary cells

immune system a complex system that protects the body from foreign invaders and newly arising tumors through recognition, destruction, and removal

immunosuppressive capable of reducing the normal function of the immune system

immunotoxin a hybrid protein molecule used in immunotoxin therapy that is constructed by joining an Ab to a toxin to use the binding specificity of the Ab to home in on a target

immunotoxin therapy a treatment method for diseases, particularly cancer, that uses Abs to guide toxin molecules (which are linked to the Abs) to their specific targets

inactivated polio vaccine (IPV) a polio vaccine, first developed by Jonas Salk, that is prepared by killing the virus particles so that they will produce immunity but not disease

killed polio vaccine (KPV) another name for inactivated polio vaccine

lymphocyte a particular type of white blood cell that plays a major role in fighting infection and disease; the three major categories of lymphocytes are natural killer cells, B cells, and T cells

monoclonal antibodies (mAb) Abs that have been derived from the same clone

monomer a molecule that has the ability to combine with other molecules to form a polymer

mucosal transport receptor (FcαR) a cell receptor that is responsible for binding and transport of IgA Abs at mucosal surfaces

myeloma cell a type of cancer cell that arises from Ab-producing cells found in bone marrow

oral polio vaccine an orally administered polio vaccine that is produced by using a living, attenuated virus

parenteral a method of entering the body other than through the gastrointestinal tract

passive immunization a method of conferring immunity by giving Abs or sensitized immune cells from an immune individual to a non-immune individual

pathogen a disease-producing agent

pentameric consisting of five identical molecules, or monomers, that have been linked together

plasma cell a B cell that has matured into an Ab-producing cell

polyclonal Ab a mix of Abs produced by many different B-cell clones, thus having the ability to bind to different sites on the target Ag, or with different affinities

polymerization the process of forming a polymer

prophylaxis a medical procedure intended to prevent a disease, rather than cure a disease that is already established

recombinant DNA a DNA molecule that has been created from the insertion or combination of one or more gene segments resulting in a new genetic sequence

secretory piece a polypeptide chain with the ability to bind to secretory IgA molecules

sera plural of serum; the liquid portion of blood after all clotting factors have been removed

strain a variant of an organism that is distinct enough to be recognized, but not different enough to be classified as another species

subunit vaccine a vaccine made from small pieces, or subunits, of a virus

titer concentration of a substance in a fluid

trimeric consisting of three identical molecules, or monomers, that have been linked together

vaccination a method of producing immunity by introducing a vaccine of any type to the body

vaccine a preparation, usually of a weakened or killed pathogen, or of a part of a pathogen, that has the ability to prevent or treat infectious disease when administered to the body

vaccinia the virus that causes cowpox, a disease in cattle that is similar to smallpox in humans

valency in immunology, it is a property of Ags and Abs that describes the number of binding sites of an Ab or epitopes of an Ag

variola the virus that causes smallpox

NOMENCLATURE

[Ab]	Concentration of antibody	M
c	Concentration of antigen	M
K	Association constant	1/M
[L]	Concentration of antigen	M
n	Number of binding sites per antibody	
r	Fraction of antigen bound to sites on antibody	
[S]	Concentration of antigen-binding sites	M

QUESTIONS

1. Design an experiment to determine whether a vaccine has been effective.
2. Explain the difference between a subunit vaccine and a DNA vaccine.
3. The chapter states, regarding eradication of smallpox, that: "This ambitious project was possible because of two important features of variola infection: 1) the absence of nonhuman reservoirs for the virus, and 2) the non-existence of asymptomatic carriers of the virus." Why are these two features of smallpox important for eradication?

PROBLEMS

1. It is predicted that a certain vaccine will cause 1 death per million population when administered to humans. If the vaccine were administered to the entire U.S. population, how many deaths would be expected?
2. There are 10 million people in New York City. One person contracts smallpox, and this person subsequently spreads it to the rest of the city.

a. Derive an equation that you can use to find the number of infectious exchanges that must occur to infect 10 million people. Assume that every newly infected person, during the entire course of their sickness, will infect only two new uninfected people. With each new generation of disease propagation, every single newly infected person will also infect only two new uninfected people (see diagram).
b. Assume that each day, a newly infected person will transmit disease to two new uninfected people. Based on your answer from **a**, how many days would it take to wipe out the city?
c. Discuss why this scenario is unrealistic.

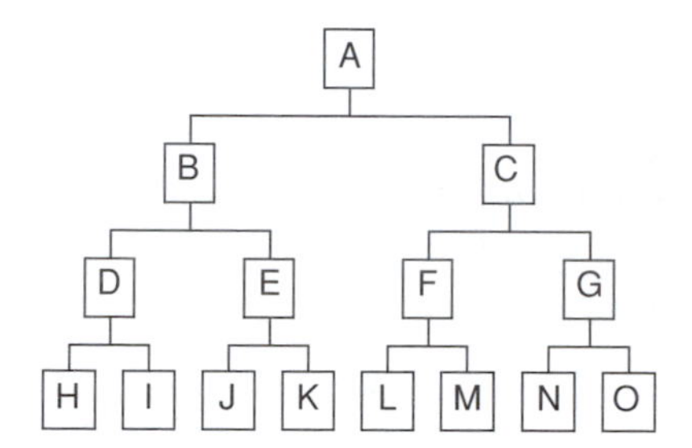

3. Imagine that a small city (similar to New Haven, Connecticut, population 124,000) is facing an outbreak of a new infectious disease. The disease is spreading quickly; each infected person is infecting three uninfected people per day. To keep track of the disease, the city's health service is trying to register all newly infected patients each day. However, resources are limited and the service can only register a maximum of 15,000 patients in one day. On what day of the outbreak will the city health service reach its limit of the number of people it can register?

15 Biomaterials and Artificial Organs

LEARNING OBJECTIVES

After reading this chapter, you should:

- Understand the various types of biomaterials that are available and their common uses.
- Understand coagulation response to biomaterials in contact with blood and the foreign body response (FBR) to implanted biomaterials.
- Understand the importance of hemodialysis in the treatment of kidney disease and the materials and methods that are used to achieve hemodialysis.
- Describe, in quantitative terms, the efficiency of hemodialysis, as well as the changes in blood and dialysate composition that occur during hemodialysis.
- Understand the functions of membrane oxygenators and their role in open heart surgery.
- Learn about the range of medical devices—from artificial hearts and valves to drug-eluting stents—that are now used to treat heart disease.
- Understand the principles of biohybrid organs, which are similar to those used for tissue engineering, but usually applied for the creation of devices that treat blood outside the body.

15.1 Prelude

The search for artificial replacements for failing human organs is long and filled with great successes. Today, hemodialysis is routinely used to replace kidney function, artificial hip prostheses allow millions of people to walk, and artificial lenses provide cataract sufferers with clear vision. There are many disappointments as well; despite decades of serious effort, there is still no proven artificial heart, liver, or pancreas.

This chapter describes the success of several artificial organs, including hemodialysis for treating kidney failure and artificial hips. It also describes the efforts to build artificial hearts, livers, and pancreases, and the challenges that remain for these artificial organs. Because all artificial organs depend on the use of materials, usually man-made materials, the chapter begins with a review of biomaterials science, and the body's response to implanted biomaterials.

15.2 Biomaterials

A **biomaterial** is often defined as "a nonviable material used in a medical device, intended to interact with biological systems" (1). Notice that, in this definition, biomaterials are "nonviable," or not living; they are nonliving materials that are introduced, for some purpose, into a living body. To be most useful, these materials should be biocompatible, or not harmful to the living body in which they are placed.

Over the years, some materials have been demonstrated to be biocompatible after implantation, although the degree of compatibility can vary significantly depending on the site of implantation and the function of the material at that site. Biomaterials are now used routinely in diverse medical applications—from contact lenses to artificial hearts. Frequently, the materials used in biomedical applications are polymers (Table 15.1) or metals (Table 15.2).

The concept of **biocompatibility** is essential for understanding how biomaterials are used in humans. Unfortunately, biocompatibility in materials is difficult to define precisely. A useful definition, which is accepted by most experts, is "the ability of a material to perform with an appropriate host response in a specific application" (2).

15.2.1 Historical uses of biomaterials

Prosthetic joints or limbs and artificial organs are among the most sophisticated tools of modern medicine. In fact, ancient surgeons used materials available to them to fashion new parts for their patients (Figure 15.1). Much of the initial progress in biomaterials was related to repair in the vascular system, primarily in an effort to close severed vessels to stop bleeding. Natural and synthetic threads, fibers, and **sutures** are still used today to repair wound sites. From ancient times until the early 1970s, only collagenous materials such as catgut found general acceptance as absorbable sutures, although a variety of innovative approaches have origins in the distant past. Sushruta, an Indian surgeon, described absorbable sutures derived from animal sinews in 600 BC. A surgeon in Newcastle-upon-Tyne in England, Richard Lambert, wrote about his suggestion to a colleague, Samuel Hallowell, for an approach for closing vessels; Hallowell executed the procedure successfully, using a wooden peg

Figure 15.1 **Ancient prosthesis.** Bioengineers in ancient Egypt crafted wooden prostheses and used them to help patients. This photograph shows an artificial toe applied in a 50- to 60-year-old woman. Reprinted from: Nerlich AG, Zink A, Szeimies U, Hagedorn HG. Ancient Egyptian prosthesis of the big toe. *Lancet.* 2000;356:2176–2179, with permission from Elsevier.

Table 15.1 Some of the polymers that might be useful in tissue engineering, based on past use in biomedical devices

Polymer	Medical applications
Polyurethanes (PEU)	Artificial hearts and ventricular assist devices Catheters Pacemaker leads
Poly(tetrafluoroethylene) (PTFE)	Heart valves Vascular grafts Facial prostheses Hydrocephalus shunts Membrane oxygenators Catheters Sutures
Polyethylene (PE)	Hip prostheses Catheters
Polysulphone (PSu)	Heart valves Penile prostheses
Poly(ethylene terephthalate) (PET)	Vascular grafts Surgical grafts and sutures
Poly(methyl methacrylate) (pMMA)	Fracture fixation Intraocular lenses Dentures
Poly(2-hydroxyethyl methacrylate) (pHEMA)	Contact lenses Catheters
Polyacrylonitrile (PAN)	Dialysis membranes
Polyamides	Dialysis membranes Sutures
Polypropylene (PP)	Plasmapheresis membranes Sutures
Poly(vinyl chloride) (PVC)	Plasmapheresis membranes Blood bags
Poly(ethylene-co-vinyl acetate) (EVAc)	Drug-delivery devices
Poly(L-lactic acid), Poly(glycolic acid), and Poly(lactide-co-glycolide) (PLA, PGA, and PLGA)	Drug-delivery devices, sutures
Polystyrene (PS)	Tissue culture flasks
Polyvinylpyrrolidone (PVP)	Blood substitutes
Polydimethylsiloxane, silicone elastomers (PDMS)	Breast, penile, and testicular prostheses Catheters Drug-delivery devices Heart valves Hydrocephalus shunts Membrane oxygenators

Note: Collected from (16) and (17).

Table 15.2 Summary of the physical and mechanical properties of various implant materials in comparison to natural bone

Properties	Natural bone	Magnesium	Ti alloy	Co–Cr alloy	Stainless steel	Synthetic hydroxyapatite
Density (g/cm^3)	1.8–2.1	1.74–2.0	4.4–4.5	8.3–9.2	7.9–8.1	3.1
Elastic modulus (Gpa)	3–20	41–45	110–117	230	189–205	73–117
Compressive yield strength (Mpa)	130–180	65–100	758–1,117	450–1,000	170–310	600
Fracture toughness (MPam$^{1/2}$)	3–6	15–40	55–115	N/A	50–200	0.7

Note: From (18).

and thread to repair a brachial artery in 1759. Others used ivory clamps or silk and needles to repair vessels in the 1880s. The first absorbable synthetic material, poly(glycolic acid), was developed by the company American Cyanamid in the 1960s; this material is still used in sutures and as scaffolds for tissue engineering.

Polymers have been used as biomaterials in dentistry for more than 100 years. Vulcanized caoutchouc was used in 1854 for dental bases, and celluloid was used in 1868 for dental prostheses. Poly(methyl methacrylate) has been used since 1930 for denture bases, artificial teeth, removable orthodontics, surgical splinting, and fillings. Many new polymers appeared during the 1930s, including polyamides, polyesters, and polyethylene. The first implanted synthetic polymeric biomaterial appears to be poly(methyl methacrylate), which was used as a hip prosthesis in 1947. Polyethylene and other polymers were used as implants in the middle ear in the early 1950s, yielding good initial results, but local inflammation limited the use of these materials.

Catheters are thin, hollow tubes formed of polymer, which are used to introduce or remove fluid from vessels far from the body surface. In less than 100 years of use, catheters have become an invaluable tool in diagnosis and disease management. Fritz Bleichroeder was the first individual to perform catheterization, when he inserted a catheter into his own femoral artery. Another brave individual, Werner Forssman, performed the first cardiac catheterization in 1929 when, as a 23-year-old urology student, he inserted a urethral catheter via the antecubital vein into his heart. With the catheter in place, Forssman reportedly ascended a flight of stairs to the x-ray room, where he documented this experiment, which eventually earned him a Nobel Prize.

Biomaterials are used frequently for diagnosis and treatment of disease in the nervous system, particularly as **shunts** to divert excess cerebrospinal fluid from the ventricular system in patients with hydrocephalus. Miculicz reported the use of a glass wool "nail" for this purpose in 1890. Autologous vessels and rubber tubes were used in the period from 1900 through 1930, with some success. Better results were achieved with poly(vinyl chloride) and silicone tubing in the 1950s. Implanted polymeric materials, as well as electronic materials, are now used routinely in sundry neurological settings, including regeneration of damaged peripheral nerves, monitoring of brain signals in patients with epilepsy, and drug

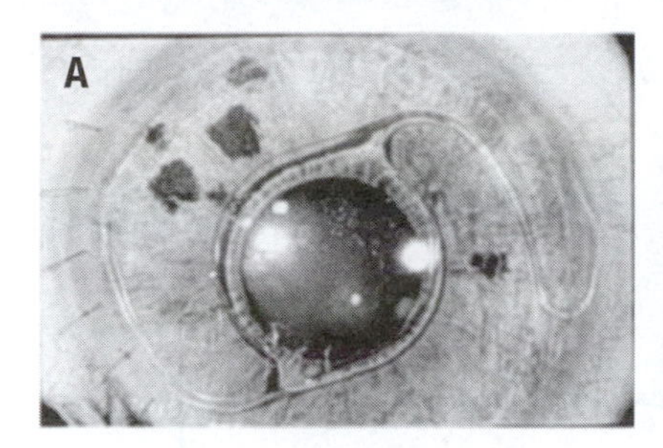

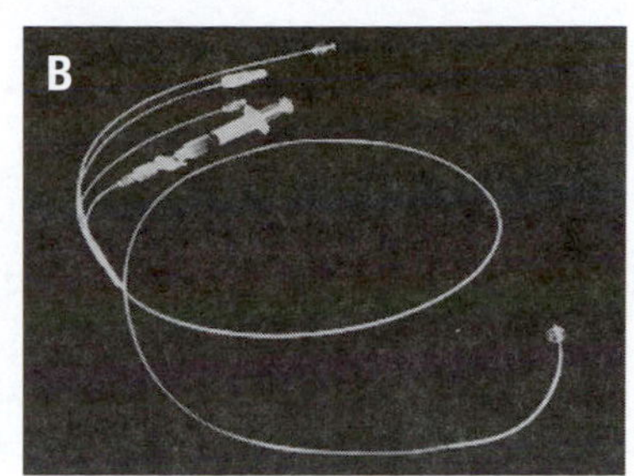

Figure 15.2

Plastic implantable lenses and catheter. A. An Artisan® phakic intraocular lens, which is implanted in front of the natural lens in extremely nearsighted patients, to allow normal vision without glasses. The artificial lens is made of polymethyl methacrylate. Image courtesy of Dr. James J. Salz and the American Eye Institute. **B.** Example of a polymer catheter. Image courtesy of Thermedical.com. (See color plate.)

delivery to the brain (e.g., see the description of the GLIADEL® drug-delivery system in Chapter 16).

15.2.2 Biomaterials in current use

Today, many medical devices involve polymeric biomaterials (see Table 15.1). Polymers of a variety of chemistries, in diverse shapes and forms, are introduced into patients around the world each day. Many of these polymer-based devices are now commonplace: catheters, coatings for pacemaker leads, and contact lenses (Figure 15.2). In some cases, synthetic polymer materials have enabled dramatic and heroic technologies, such as the artificial hip and total artificial heart. Still, there is considerable work remaining: Even the most modern artificial hips are a pale replacement of the natural material, and no patient has survived for long on a total artificial heart. Clearly, the development of better materials and smarter ways of using existing materials will improve human health care.

The creation of better biomaterials requires a clear understanding of the most common modes of failure. For materials that contact blood, the most common problem is formation of blood clots, which may be initiated by the material surface. The most common problem for non–blood-contacting applications, such as tissue implants, is the development of a FBR that isolates the device from the rest of the body. These two biological responses are reviewed in the sections that follow.

15.2.3 Biological responses to biomaterials

BIOMATERIALS THAT CONTACT BLOOD

Vascular surgeons have been manipulating arteries for wound treatment since the technique was first demonstrated by Sushruta in India and Galen in Rome. Charles A. Hufnagel used smooth, poly(methyl methacrylate) tubes as vascular grafts in the late 1940s. In the early 1950s, textile grafts were first introduced based on the work of Arthur B. Voorhees, Jr., and his collaborators; these were primarily composed of silk fibers initially and then poly(vinyl chloride-co-acrylonitrile).

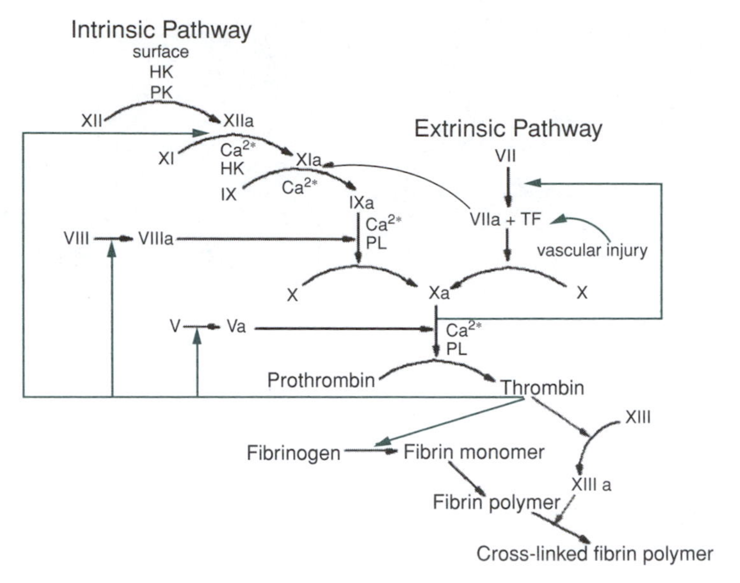

Figure 15.3 Proteins and reactions of the coagulation cascade.

Currently, polyethylene terephthalate and polytetrafluorethylene (PTFE) grafts are the most widely used. Despite this long history, however, vascular grafts are still not perfect. Only large-diameter vessels, of more than 5-mm diameter, can be replaced with synthetic grafts; replacement of smaller grafts inevitably leads to clotting and vessel failure.

The **coagulation cascade**, and the formation of a blood clot, is one of the better-known examples of physiological responses that depend on a regulated series of interrelated, enzyme-mediated reactions. The general concept was introduced in Chapter 6. In the coagulation cascade of reactions (Figure 15.3), the end point is formation of a cross-linked mesh of the protein fibrin; fibrin cross-linking produces a solid material by trapping red blood cells and platelets within a protein-rich network. Often, the clotting process is initiated by damage to a blood vessel and leakage of blood out of the vessel. Formation of this solid material closes the area of damage and stops bleeding.

Although clot formation provides life-saving protection in a damaged vessel, the unwanted formation of clots can cause serious health problems. **Clots** that form within blood that is flowing through large vessels, for example, can travel downstream, blocking smaller vessels and impeding blood flow to tissue. This process of clot formation and vessel blockade, called embolization, can lead to strokes (if the clot goes to the brain) or breathing problems (if the clot goes into the lung). Therefore, the biological process of clot formation is tightly controlled, reducing the chances of unnecessary clot formation.

The clotting cascade is mediated by a number of proteins, called clotting factors, which are assigned Roman numerals as names (e.g., factor VIII and factor X). The ultimate goal of the cascade, or sequence of chemical reactions, is the conversion of the protein prothrombin to thrombin: Thrombin can then

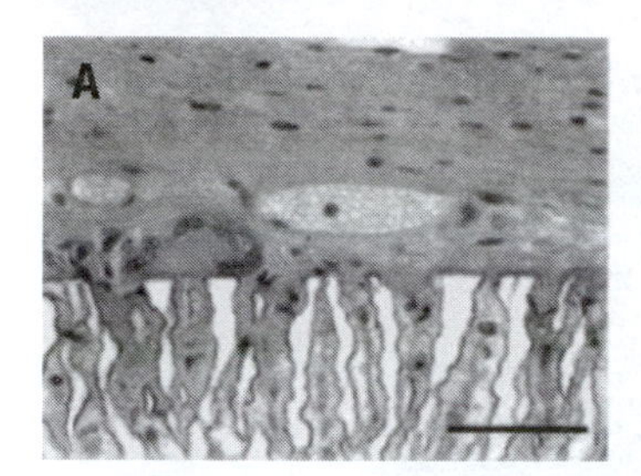

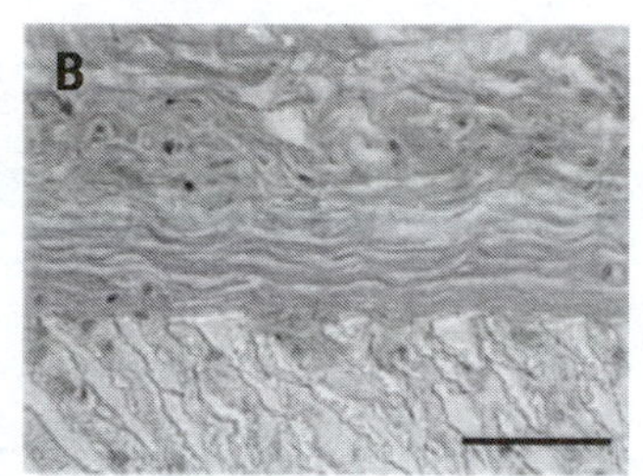

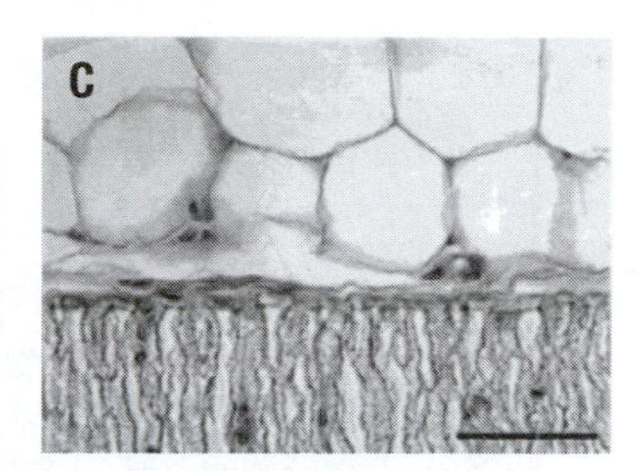

Figure 15.4 **Expanded polytetrafluoroethylene implants in various locations.** Light micrographs of hematoxylin and eosin–stained sections of expanded polytetrafluoroethylene disks after 5-week implantation in epicardial **(A)**, adipose **(B)**, and subcutaneous **(C)** tissues.The scale bar represents 50 μm in each image. After 5 weeks of implantation, the subcutaneous implant is surrounded by fibrous tissue with few cells, whereas the adipose tissue has a limited fibrous capsule and the epicardial tissue has a vascularized granulation tissue. Copyright 2002, Wiley Periodicals, Inc., reprinted from (3) with permission of John Wiley & Sons, Inc.

catalyze the conversion of the soluble molecule fibrinogen into fibrin, which polymerizes to form the insoluble clot matrix.

The clotting process can be initiated in either of two ways: through what are called the intrinsic or extrinsic pathways. The formation of a clot in response to tissue injury and vessel damage occurs by the extrinsic pathway; this pathway is initiated by tissue factor and activated factor VII (VIIa). The interaction of these proteins activates factor X. The intrinsic pathway for clot formation requires factors VIII, IX, X, XI, and XII, as well as some other proteins and the mineral Ca^{++}. This pathway is activated by the presence of charged surfaces (usually, the charged surface is extracellular matrix outside the vessel wall, but it can also be a protein-coated biomaterial within the blood vessel). These two initiation pathways work together, with the extrinsic pathway activating more quickly to augment the action of the intrinsic pathway.

IMPLANTED MATERIALS AND THE FOREIGN BODY RESPONSE

When a nonbiological material is implanted into a tissue, the body responds. For some materials, implanted in some sites, the response is mild or difficult to observe. These materials are considered to be "biocompatible" for that use. But even though the response is mild, there is a response, and the long-term consequences of that response for the patient can be difficult to predict. In some cases, such as the implantation of silicone implants for breast augmentation, the effects are still difficult to predict even after years of careful study.

The response to an implanted biomaterial is inherently complex because of the presence of blood, interstitial fluids, and multiple cell types, which may be in diverse states of activation at the implant site. Some examples are shown in Figure 15.4. A typical response is divided into stages, which develop with time (Figure 15.5). First, proteins from the blood and interstitial fluid adsorb to surfaces of the implant; proteins will absorb nonspecifically to most materials, including polymers and metals. These absorbed proteins provide a substrate for cell adhesion. Most experts believe that cells interact with this protein layer, and not directly with the material surface, which explains why similar reactions to a wide range of materials are observed. Almost all implanted materials induce

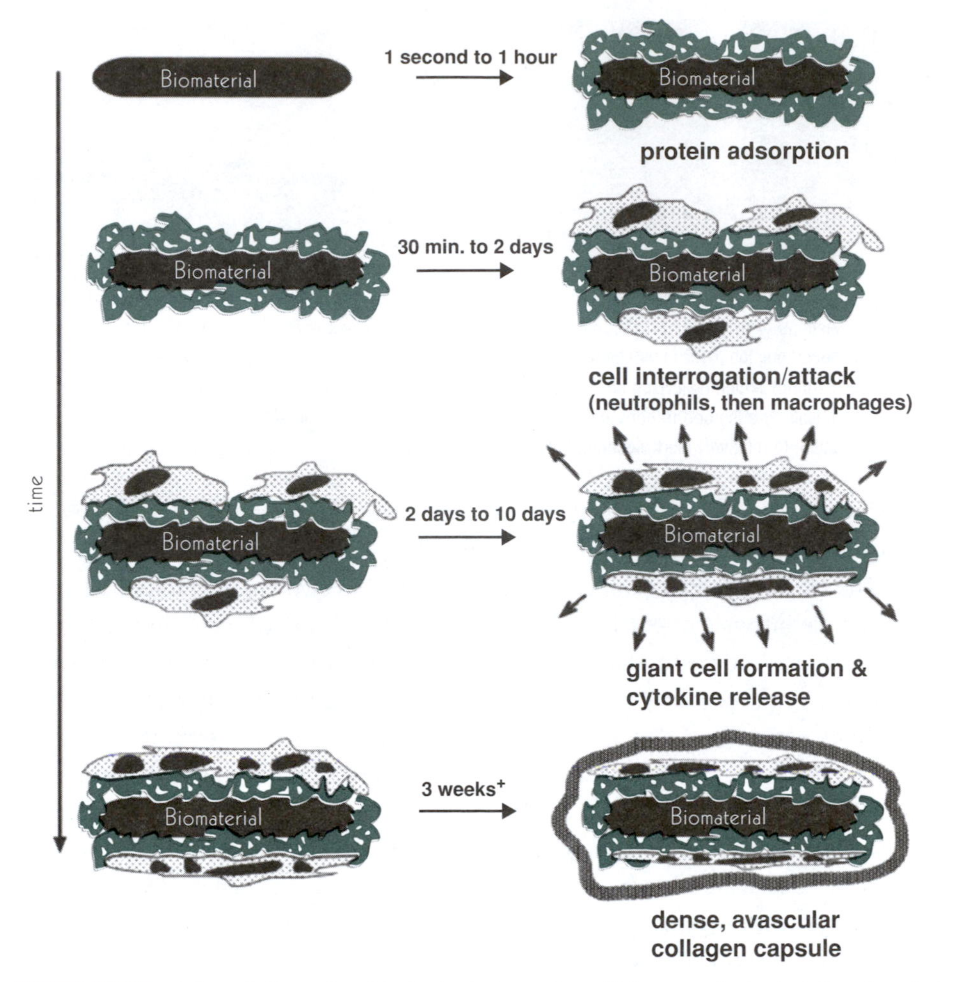

Figure 15.5 Schematic illustration of the foreign body reaction, which occurs after implantation of a material into the body. Color represents protein adsorption. Adapted from (4).

a unique inflammatory response termed the **FBR** (Figure 15.5). The FBR can be divided in overlapping phases that include protein adsorption, recruitment of cells (predominantly **neutrophils** and **macrophages** from the blood), macrophage fusion to form **foreign body giant cells**, and finally, involvement of tissue-specific fibroblasts and endothelial cells. The end result of the FBR is the formation of foreign body giant cells directly on the polymer surface and the subsequent encapsulation of the implant by a fibrous capsule, which usually contains almost no blood vessels. The composition and size of the fibrous capsule can vary significantly between materials and implantation sites (Figure 15.6).

15.3 Hemodialysis

As described in Chapter 9, the kidney provides an essential excretory function by removing water-soluble waste molecules from the blood and concentrating them

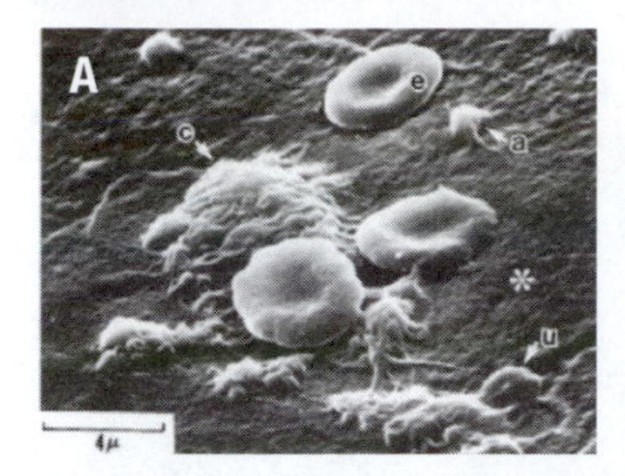

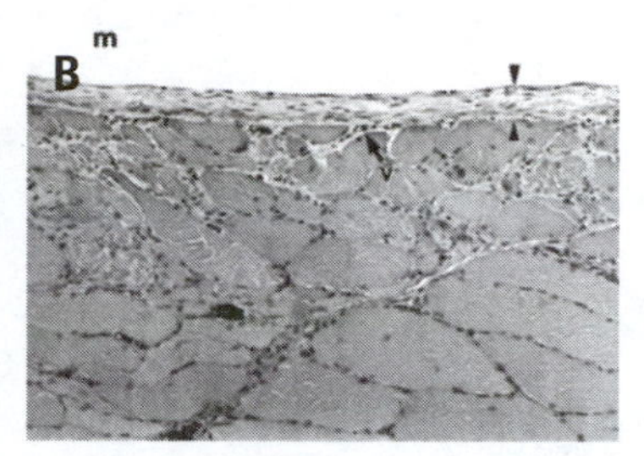

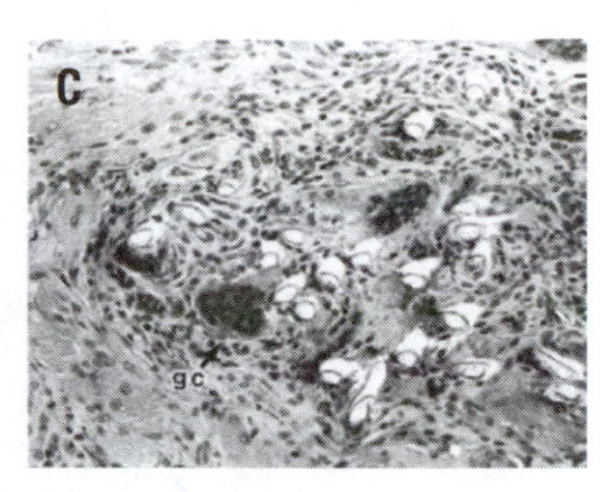

Figure 15.6 **Morphology of biomaterial–tissue interactions. A.** Scanning electron photomicrograph of early thrombotic deposits on heart valve material exposed to calf blood for a few minutes in an extracorporeal experiment. Numerous platelets in various stages of adhesion ranging from relatively unaffected (u) to adherent to activated (a) forms; some aggregated in clumps (c), are adherent to a background of protein absorption (asterisk). Erythrocytes (e) are probably passively adsorbed in early surface-induced thrombus formation. **B.** Photomicrograph of experimental tissue reaction to relatively inert biocompatible material (m), implanted into rabbit muscle. A thin discrete fibrous capsule with virtually no inflammatory cell infiltrate is adjacent to the implant (between arrowheads). Fine vascularity (v) is noted at the interface between the fibrous capsule and the surrounding muscle. Hematoxylin and eosin staining; 150× **C.** Granulomatous inflammatory response to surgically implanted Dacron mesh. This is an intense inflammatory response that includes abundant foreign body giant cells (gc). Hematoxylin and eosin staining; 200× **A:** From Schoen and Padera (5). Reproduced with permission from Schoen FJ. Cardiac valve prostheses: pathological and bioengineering considerations. *J Cardiac Surg.* 1987;2:265 with permission from Blackwell Publishing. **B** and **C:** Reproduced with permission from Schoen FJ. *Interventional and Surgical Cardiovascular Pathology: Clinical Correlations and Basic Principles.* Philadelphia: WB Saunders; 1989.

in a liquid form. Remarkably, it performs this function while maintaining the blood balance of essential ions (such as Na^+ and K^+) and with a minimal loss of total body water (~1 L/day). Our metabolic machinery is diverse and, therefore, the kidney is responsible for excretion of thousands of different chemicals. Thus, loss of kidney function results in an accumulation of many toxic molecules in the blood. The end products of nitrogen metabolism are a particularly important subset of these chemicals; for example, blood concentration of **urea** increases to toxic levels during kidney failure.

Blood treatments are often performed by extracorporeal devices. **Extracorporeal** devices—the word extracorporeal literally meaning "outside of the body"—require the removal of blood from the patient during the treatment. Usually, extracorporeal systems are designed so that a small flow of blood is removed from the patient continuously during treatment; the blood is treated and returned to the patient continuously as well (Figure 15.7). In this fashion, the patient can receive continuous treatment. Of course, there are limits on both the flow rate of blood from the patient and the total volume of blood that is outside of the patient at any given time. Flow rates of blood are usually in the range of 100–500 mL/min, and the total volume of the extracorporeal circuit is usually several hundred milliliters.

Synthetic materials are critical components in extracorporeal systems for blood purification or treatment. Willem Kolff, a Dutch physician, developed the first successful kidney dialysis unit in 1943, by using the material cellophane to remove urea from the blood of diabetic patients. The physical basis of that experiment

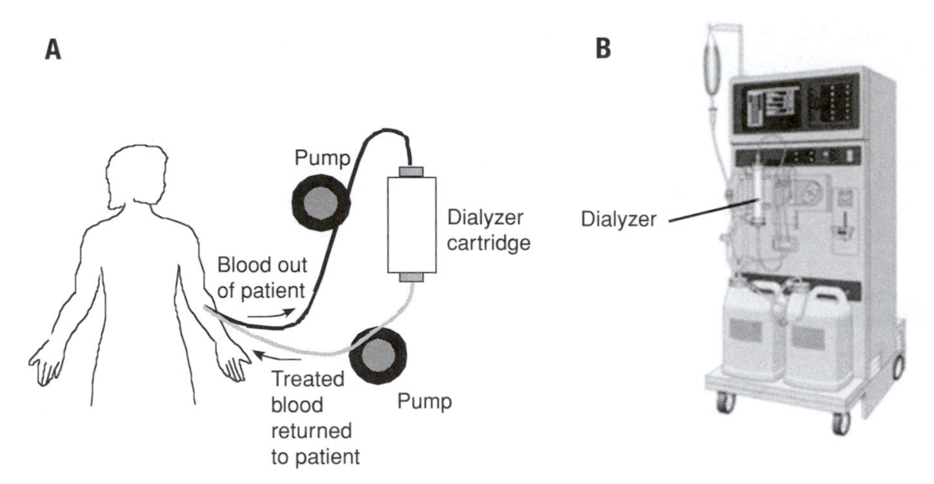

Figure 15.7 **A. Typical extracorporeal blood treatment. B. Typical dialysis machine.** The dialyzer cartridge holding the hollow-fiber dialysis membranes, shown in more detail in Figure 15.9, is only a part of the system.

was simple; by placing blood on the opposite side of a membrane from a second solution, called the **dialysate**, unwanted molecules could be removed from the blood by their diffusion through the membrane and into the dialysate (Figure 15.8).

That early experiment led to the development of modern hemodialysis. **Hemodialysis** is a life-extending procedure for ~200,000 patients with end-stage renal disease in the United States. The success of hemodialysis is a direct result of engineering analysis. The formulation of a mathematical description of solute movement during hemodialysis, which was based on classical engineering principles, has led directly to the improvements in artificial kidney design and operation that make it a safe and reliable life-saving procedure.

Current hemodialysis systems do more than remove urea. They remove a large number of waste products from the blood and they balance key ion concentrations such as sodium and potassium. Much of the work of the dialysis unit is

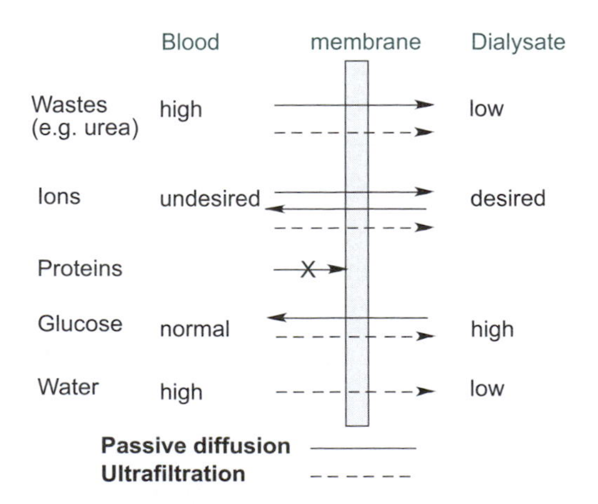

Figure 15.8 **In dialysis, waste molecules are removed from the blood by diffusion across a selective membrane into a dialysate solution.**

Table 15.3 Major components of dialysate solutions

Component*	Hemodialysate Range	Hemodialysate Typical	Peritoneal Dialysate Range
Sodium (mmol/L)	135–155	140	130–135
Potassium (mmol/L)	0–4.0	2.0	0
Calcium (mmol/L)	0–2.0	1.25	0.75–1.75
Magnesium (mmol/L)	0–0.75	0.25	0.25–0.75
Chloride (mmol/L)	87–120	105	95–102
Bicarbonate (mmol/L)	25–40	35	—
Lactate (mmol/L)	—	—	35–40
Glucose (g/dL)	0–0.20	0.20	1.36–3.86†

Notes: From (19).

* To convert values for calcium and magnesium to milliequivalents per liter, divide by 0.5; to convert values for glucose to millimoles per liter, multiply by 0.05551.

† These values correspond to 1.5–4.25 g of D-glucose monohydrate per deciliter.

accomplished by diffusion through the membrane, so membrane materials must be selected carefully. The ideal dialysis membrane should provide:

- selective separation (i.e., some molecules should be allowed to pass, whereas others are retained in the blood);
- efficient removal of waste molecules because of rapid diffusion through the membrane;
- low resistance to blood flow;
- compatible materials that exhibit low toxicity and low susceptibility to clotting;
- ease of use (pre-assembled, packaged, sterilized, and disposable); and
- low cost and reproducible operation.

Most artificial kidneys function by controlling the contact of a patient's blood with the dialysate solution (Table 15.3). Contact between the blood and dialysate is indirect and mediated by the membrane. An artificial membrane regulates the solute exchange; the membrane serves as a common boundary for the flowing liquid blood and dialysate streams (Figure 15.8). Chemical extraction from the blood is controlled via the permeability characteristics of the membrane and the flow rates of blood and dialysate (Q_B and Q_D, respectively).

The most commonly used artificial kidneys contain an array of hollow fibers (Figure 15.9). The parallel arrangement of hollow fibers provides convenient flow pathways for both blood and dialysate and an extensive surface area for solute exchange. The total surface area of the fibers, S, available for exchange of molecules is given by:

$$S = N\pi d L, \tag{15.1}$$

where N is the number of hollow fibers in the unit, d is the diameter of each fiber, and L is the length of each fiber.

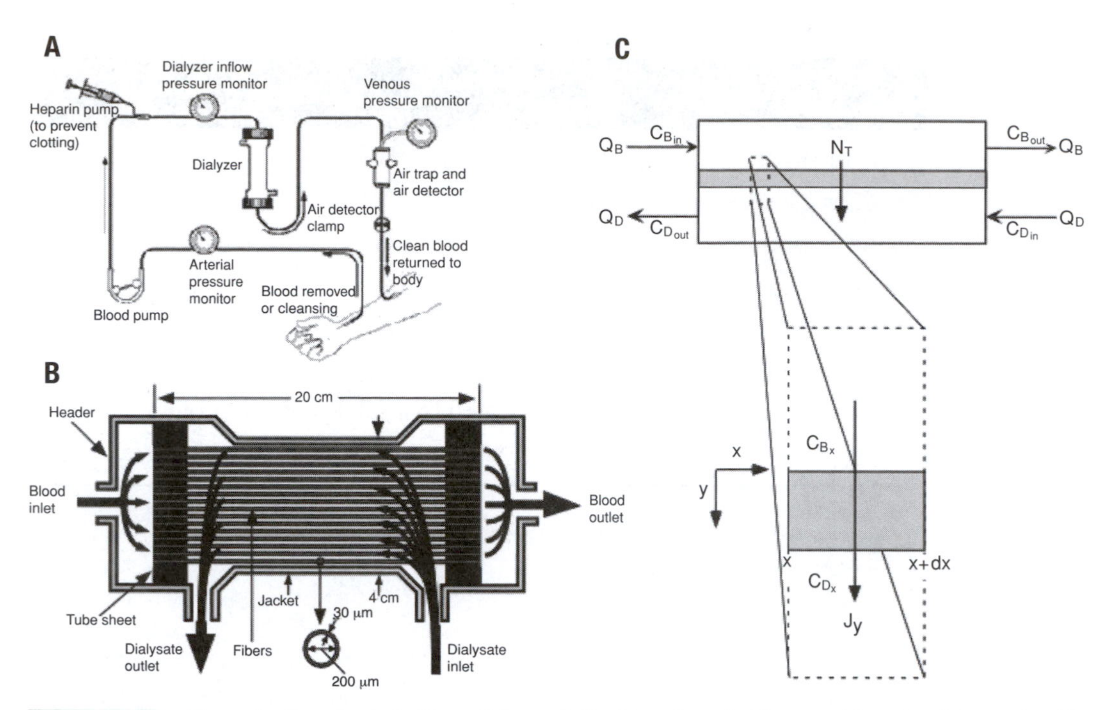

Figure 15.9 **Flow paths through a typical hemodialyzer. A.** Blood from the patient is pumped through a dialyzer unit, while the pressure is monitored. **B.** A typical dialyzer unit is composed of hollow fibers. Blood flows through the lumen of each hollow fiber, and dialysate flows around the fiber bed. **C.** Schematic diagram illustrating the indirect contact of blood and dialysate, which allows for solute exchange without mixing of the two liquid streams. The variables are described in the text.

SELF-TEST 1 What is the total surface area for contact between blood and dialysate, in a hollow-fiber unit with 10,000 fibers, in which each fiber has diameter 200 μm and length 20 cm?

ANSWER: 1.3 m^2

What is the volume of blood contained in this same hollow-fiber device?

Answer: 6.3×10^{-5} m^3 or 63 mL

How much dialysate can the device in Figure 15.9b hold, assuming that it surrounds 10,000 fibers, with the properties described above?

Answer: 1.9×10^{-4} m^3 or 190 mL (with the volume of the fibers subtracted)

The overall effect of the dialyzer is to reduce the concentration of a solute in the blood, $C_{B_{in}}$, to some lower value, $C_{B_{out}}$. It does this by transferring the solute into a dialysate fluid, with the effect of increasing solute concentration in the dialysate from $C_{D_{in}}$ to $C_{D_{out}}$. These concentration variables are defined in Figure 15.9c. If the total amount of solute that is removed from the blood is called N_T, then N_T is also equal to the amount of solute that crosses the membrane and the amount of new solute that appears in the dialysate:

$$N_T = Q_B(C_{B_{in}} - C_{B_{out}}) = Q_D(C_{D_{out}} - C_{D_{in}}). \tag{15.2}$$

The overall performance of a dialysis unit—in other words, how good it is at removing a particular solute from the blood, under its current state of operation—can be expressed in a number of ways: dialysance, clearance, or extraction ratio. **Dialysance**, D, is expressed in units of flow rate:

$$D = \frac{Q_B\left(C_{B_{in}} - C_{B_{out}}\right)}{C_{B_{in}} - C_{D_{in}}}. \tag{15.3}$$

Clearance, CL, is an extension of a concept used frequently to describe behavior of the kidneys (see Chapter 9) and is commonly used to describe the overall performance of artificial kidneys. Clearance can be interpreted as the flow rate at which the blood can be completely cleared of a solute (i.e., the volume of blood per unit time in which all of a solute is removed):

$$CL = \frac{Q_B\left(C_{B_{in}} - C_{B_{out}}\right)}{C_{B_{in}}} = \frac{N_T}{C_{B_{in}}} \tag{15.4}$$

The **extraction ratio**, E, is the solute concentration change in the blood compared to the theoretical solute concentration change that would occur if the blood and dialysate came to equilibrium:

$$E = \frac{C_{B_{in}} - C_{B_{out}}}{C_{B_{in}} - C_{D_{in}}}. \tag{15.5}$$

Equations 15.3 through 15.5 are defined for countercurrent flow operation. In countercurrent flow, the blood and dialysate streams flow in opposite directions; each is moving in a different direction with respect to the membrane between them. In cocurrent flow, the blood and dialysate streams flow in the same direction.

SELF-TEST 2 A dialyzer is operating with a blood flow rate of 300 mL/min, so that the outlet concentration of urea is 5 mg/dL when the inlet concentration is 50 mg/dL. What are the dialysance, clearance, and extraction ratio for this state of operation?

ANSWER: $D = 270$ mL/min; $CL = 270$ mL/min; and $E = 0.9$ (all assuming the inlet dialysate urea concentration is zero).

Engineering analysis allows us to predict the efficiency of a dialysis operation, when the operating variables are known. The major variables are the flow rate of blood, the flow rate of dialysate, the permeability of the membrane, and the surface area of the membrane. If the inlet dialysate stream contains no solute ($C_{D_{in}} = 0$, which is the usual situation for most solutes that are to be removed), the extraction ratio can be written in terms of characteristics of the dialysis unit:

$$E^* = \frac{1 - e^{\gamma(1-z)}}{z - e^{\gamma(1-z)}}, \tag{15.6}$$

where γ is the relative rate of permeation through the membrane to blood flow (KS/Q_B) and z is the relative rate of blood flow to dialysate flow (Q_B/Q_D).

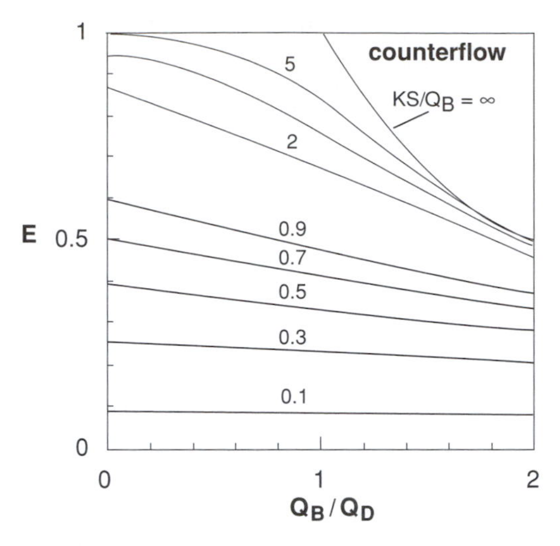

Figure 15.10 **Mathematical model for the efficiency of a dialysis unit.** The extraction ratio, E, is shown for different flow rates (Q_B/Q_D) and values of relative membrane permeability (KS/Q_B).

Equation 15.6, which applies to a unit in countercurrent operation, was obtained by calculating concentrations within a dialysis unit as a function of device design: that is, by calculating the inlet and outlet solute concentrations for known values of K, S, Q_D, and Q_B. Those calculations are beyond the scope of this text (see Box 15.1), but the result of this calculation is shown in Figure 15.10.

SELF-TEST 3 What is the extraction ratio for urea in a dialyzer in which the blood and dialysate flow rates are 300 and 500 mL/min, respectively, the total surface area is 1 m^2, and the membrane permeability to urea is 90 mL/min-m^2.

ANSWER: 0.23 (because $Q_B/Q_D = 0.6$ and $KS/Q_B = 0.3$)

The analysis just presented assumes that diffusion is the only mechanism for molecular exchange across the membrane. This assumption is not always true; hemodialysis units are sometimes operated with a pressure drop across the membrane to provide an additional **ultrafiltration** flux of water from the blood to the dialysate. Ultrafiltration is the movement of water from the blood side of the membrane to the dialysis side by means of a pressure gradient. Ultrafiltration may serve an important clinical function, by allowing for regulation of the fluid volume of the patient. By performing dialysis together with ultrafiltration, removal of fluid from the patient occurs simultaneously with molecular exchange to balance salts and other solutes. The more advanced operation with ultrafiltration is now possible because materials with good separation characteristics that also permit the convective movement of water are now available.

Peritoneal dialysis is achieved by using the patient's own peritoneal membrane (the lining of the abdominal cavity) as the dialysis membrane. To accomplish this, a large volume of dialysis fluid (usually about 2 L of dialysate) is injected into the abdominal cavity. Wastes and other excess electrolytes then diffuse from the

Box 15.1 Example: Use of the extraction ratio in design of a dialyzer

Consider the design of a counterflow, hollow-fiber hemodialysis unit with the structure shown in Figure 15.9 containing 10,000 fibers. Assume that you are asked to design a unit with this general composition, and you would like for the blood urea concentration to drop by 90% from the inlet to the outlet. If $C_{B_{out}}$ is equal to 10% of $C_{B_{in}}$, then:

$$C_{B_{out}} = 0.1C_{B_{in}} \qquad \text{(Equation 1)}$$

or, the extraction ratio E is:

$$E = \frac{C_{B_{in}} - 0.1C_{B_{in}}}{C_{B_{in}} - 0} = 0.9. \qquad \text{(Equation 2)}$$

To calculate this extraction ratio, you assumed that there is no urea in the dialysate solution that enters the device ($C_{D_{in}} = 0$).

When you examine Figure 15.10, it should become clear that an extraction ratio of 0.9 can be achieved only for certain values of the relative permeability. Of the values selected for plotting on Figure 15.10, only KA/Q_B values of 3, 5, and ∞ allow an extraction ratio E of 0.9 or greater, as shown in the plot in this box. These values should make sense to you: To achieve a high extraction ratio, the permeability of the membrane must be large. For a fixed extraction ratio of 0.9, as the membrane becomes more permeable, a slower rate of dialysate flow is needed.

This information can be used to design a dialysis unit. Assume, for example, that you elect to build a dialyzer that has a relative permeability of 3. To achieve an E of 0.9, that dialyzer would need to be operated so that Q_B/Q_D is 0.4. If the flow rate of blood that was to be treated was known, then the permeability and the dialysate flow rate could be calculated. These calculations are shown for the situation in which $Q_B = 100$ mL/min in the table below.

KA/Q_B	Q_B/Q_D	KA (if $Q_B = 100$ mL/min)	Q_D(if $Q_B = 100$ mL/min)
3	0.4	300 mL/min	250 mL/min
5	0.8	500 mL/min	125 mL/min

The critical decision to make in design of your dialyzer is now clear. You must pick the material that you will use to construct the hollow fibers. What materials will provide the highest permeability, K? Table B.3 in Appendix B lists some permeability properties for typical dialyzer membranes. Try selecting one of these materials and, using the permeability of the material, calculate the hollow-fiber surface area, A, that is needed to achieve the dialyzer performance that we are seeking. Does this area sound reasonable? What does this tell you about the design that you have been pursuing? Will it be physically possible to construct a dialyzer that meets these design specifications?

You might be wondering how Equation 15.5, which was used to construct Figure 15.10 and Figure Box 15.1, was derived. It was derived by applying the principles of conservation of mass to the solute as it moves from the blood to the dialysate. Although that derivation is beyond the scope of this book, you can find it in other biomedical engineering textbooks, such as (14) and (15).

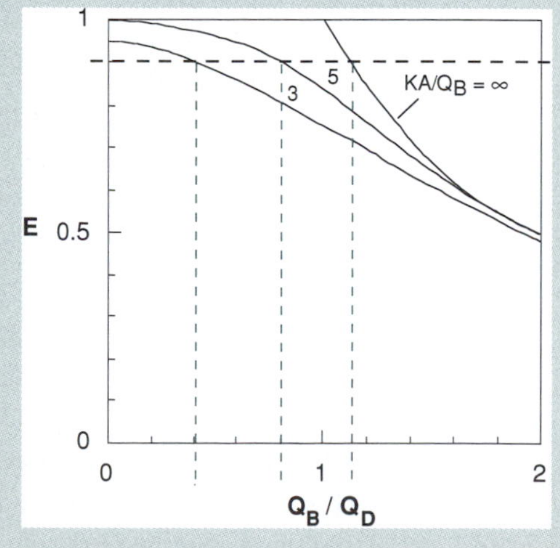

Figure Box 15.1

plasma through the peritoneal membrane into the fluid, which is drained off and replaced several times a day. The method permits patients to carry out normal activities while dialysis is being performed. Some systems permit the peritoneal dialysis to be performed automatically at night, while the patient is sleeping.

The movement of solutes across a membrane—either the peritoneal membrane or an artificial dialysis membrane—depends on several factors (recall the description of diffusion in Box 2.5). The rate of passive diffusion across the membrane depends on the concentration gradient, the type of membrane, the thickness of the membrane, and the size of solute. The rate of diffusion can be predicted by an extension of the equations presented in Box 2.5:

$$J_{\mathrm{y}} = -D\frac{\mathrm{d}c}{\mathrm{d}y}. \tag{15.7}$$

Note that Equation 15.7 is written with reference to the coordinate system shown in Figure 15.9c, in which the x direction measures distance along the membrane from the blood inlet to the outlet, y measures distance across the membrane in the direction from the blood to the dialysate and, therefore, J_{y} is the flux of solute from the blood to the dialysate, which will be a positive number if the concentration of solute in the blood is greater than the concentration of solute in the dialysate. For diffusion across a membrane of fixed thickness, Δy, the diffusion gradient, dc/dy, can be approximated:

$$J_{\mathrm{y}} \approx -D\frac{(C_{\mathrm{D,x}} - C_{\mathrm{B,x}})}{\Delta y}. \tag{15.8}$$

Equation 15.8 allows estimation of the membrane permeability, K, which is defined as:

$$J_{\mathrm{y}} = K\left(C_{\mathrm{B,x}} - C_{\mathrm{D,x}}\right). \tag{15.9}$$

Therefore, the membrane permeability, K, is related to the diffusion coefficient for the solute in the membrane, D, and the thickness of the membrane, Δy:

$$K = D/\Delta y. \tag{15.10}$$

The simple analysis in Equations 15.8 through 15.10 does not incorporate the effect of solute solubility in the dialysis membrane. Most solutes will dissolve more readily either in blood or in the membrane material, but rarely are the solubilities equal. When the difference in solubility is accounted for, the permeability is given by:

$$K = \alpha D/\Delta y. \tag{15.11}$$

where α is the ratio of solubility in the membrane to solubility in blood. For solutes that are less soluble in the membrane material, permeation through the membrane occurs more slowly.

This analysis applies to membranes in which there is no overall flow of water. If the membrane material permits ultrafiltration, the hydrostatic pressure drop across the membrane and resulting water flow can also contribute to solute flux.

A

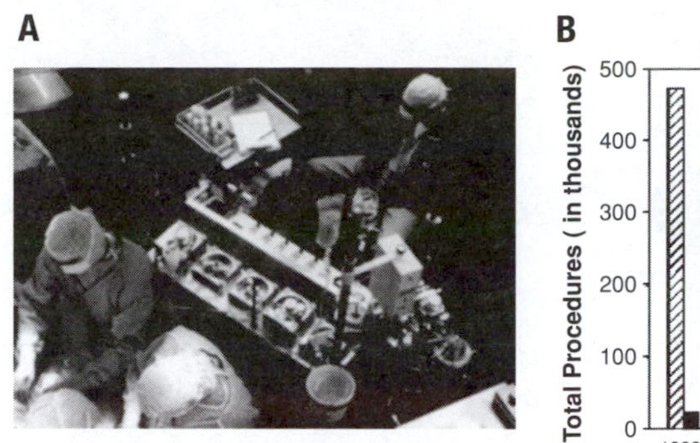

B

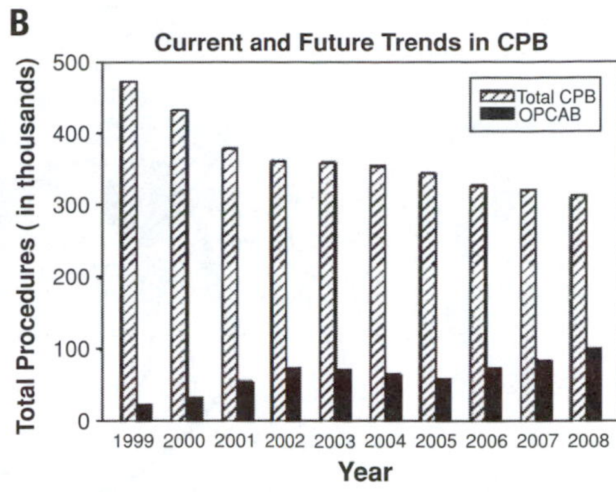

Figure 15.11 **Heart–lung machines. A.** Image of a perfusionist operating a heart–lung machine. Photo courtesy of National Institutes of Health. (See color plate.) **B.** Number of cardiopulmonary bypass procedures (CBP) performed in the United States. The number of off-pump coronary artery bypass (OPCAB) procedures, which is growing, is also shown.

15.4 Membrane oxygenators

Open heart surgery is now commonplace, having grown from origins early in the middle of the last century to a major enterprise today. In the United States, there are more than 1,000 hospitals that routinely perform **open heart surgery**. In open heart surgery, the patient's chest is opened to fully expose the heart for operations that include heart transplantation, transplantation of a coronary artery bypass, or replacement of a heart valve.

Open heart surgery was made possible by another set of extracorporeal devices, called **heart–lung machines** or **pump oxygenators** (Figure 15.11). These machines are used during surgical procedures called **cardiopulmonary bypass procedures**. In these procedures, the heart–lung machine replaces the function of both the heart and the lungs, by providing for the mechanical force to move blood through the body and providing O_2/CO_2 exchange within the flowing blood. Gentle, reliable mechanical pumps provide blood flow, and membrane oxygenators provide for gas exchange. Together, these machines keep the patient alive during the short period in which the surgeon must stop the heart to allow for direct viewing in an open, motionless, blood-free chest. Dr. John H. Gibbon accomplished the first successful surgical procedure with a heart–lung machine in 1953; this event demonstrated that a person could be kept alive, at least for the 26 minutes required for this procedure, with a man-made machine providing all of the functions of the human heart and lungs.

Chapters 7 and 8 provided background on the role of the lung and the heart in blood circulation and delivery of oxygen to cells deep within the body. The overall function of a heart–lung machine to replace the function of these normal, vital organs is shown in Figure 15.12.

Why is the number of cases of cardiopulmonary bypass procedures declining in recent years, as shown in Figure 15.11? New surgical techniques, which permit

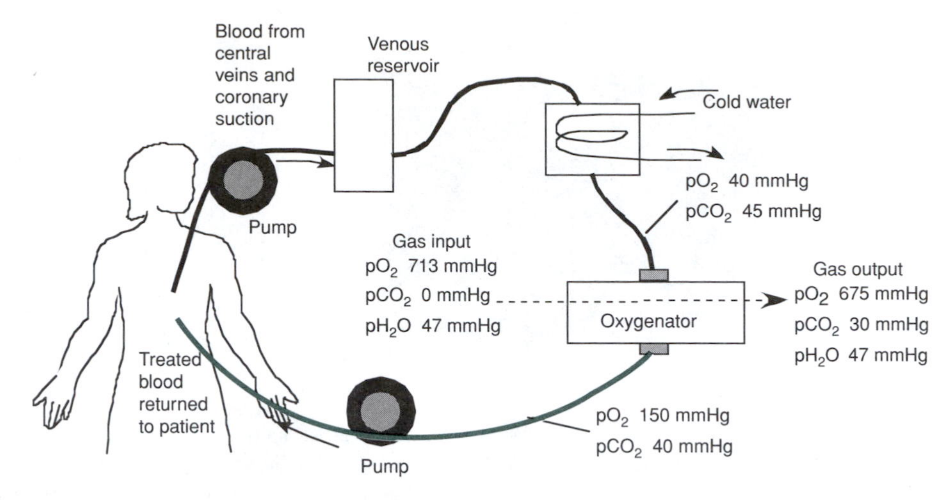

Figure 15.12 Cardiopulmonary bypass with a heart–lung machine.

operations without stopping the beating heart, are now available. These new techniques allow for "off-pump" coronary artery bypass graft (CABG) surgery, for example. Because the use of the heart–lung machine adds some elements of risk to the patient and complexity to the operation, there is currently great interest in learning whether "on-pump" or "off-pump" is more beneficial to the patient. Another factor in the decrease in cardiopulmonary bypass device procedures is probably the success of coronary artery **stents**, which decreases the number of patients who need CABG operations.

15.5 Artificial heart

To many people, design of a synthetic replacement for the human heart is the most heroic goal in biomedical engineering. Teams of scientists, engineers, and physicians have long collaborated on the development of an artificial heart. Although there is still no implantable replacement heart, substantial progress has been made (Figure 15.13).

Many of the first attempts to develop artificial hearts focused on implantable devices to supplement, rather than totally replace, heart function. An artificial ventricle, the DeBakey tube pump, which was composed of polyethylene terephthalate–reinforced silicone, was implanted into a patient in 1966. In its first human use, the device was used in a woman who had been placed on a heart–lung machine after valve replacement. In the period after the operation, her heart was unable to provide sufficient function to allow her to be removed from (or weaned from) the heart–lung machine. The **ventricular assist device**, which provided flow up to 1200 mL/min, allowed her to be removed from the heart–lung machine and, after 10 days of support from the assist device, her own heart resumed function. Today, artificial hearts and ventricular assist devices (sometimes called left ventricular assist devices or LVADs) are still used in this mode: as temporary

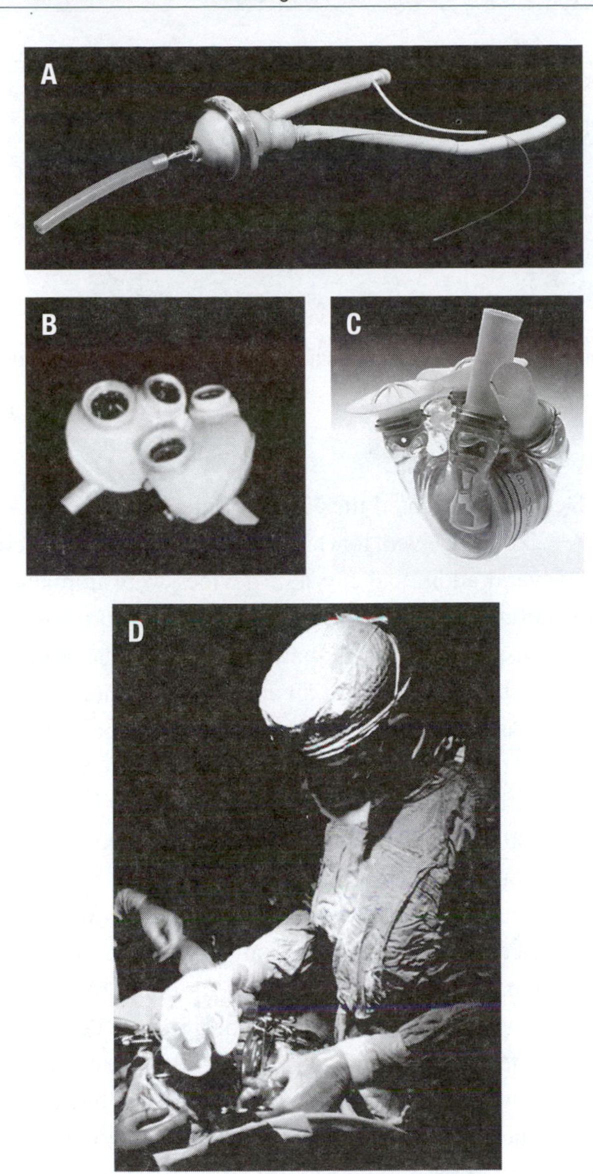

Figure 15.13 **Artificial hearts and ventricular assist devices. A.** A model of the DeBakey ventricular assist device, circa 1966. Used with permission from the Howard Hughes Medical Institute, Copyright (1998). **B.** The Jarvik 7 totally implantable heart, which was powered by an external air pump. Photo courtesy of National Institutes of Health. **C.** The implantable AbioCor. Image courtesy of Abiomed. **D.** The Liotta-Cooley heart. Reprinted with permission from Macmillan Publishers Ltd: Cooley, DA. The total artificial heart. *Nature Medicine.* 2003;9:108–111.

measures to support heart function, with the intent that the patient's own heart will eventually recover and take over.

There have been major efforts to make artificial units that will provide all of the functions of the human heart. A large, bedside artificial heart machine was first used on a human in Houston in 1969; it was designed by Domingo Liotta and used to keep Haskell Karp alive, without a heart, and awaiting a donor for 63 hours. The first permanent total artificial heart was the Jarvik 7, which was composed of polyethylene terephthalate with a polyurethane diaphragm.

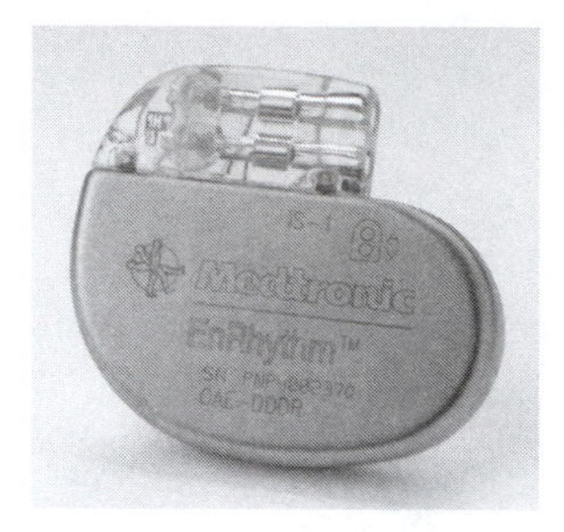

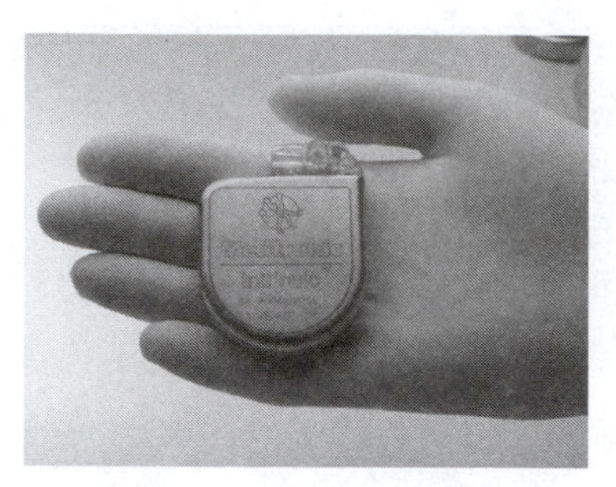

Figure 15.14 **A. Medtronic EnRhythm® pacemaker B. Medtronic Intrinsic® cardioverter-defibrillator (ICD).** Images courtesy of Medtronic, Inc.

Dr. William C. DeVries implanted the Jarvik 7 into a human in 1982; that first recipient, Barney Clark, survived for 112 days before dying of complications from the device. Several additional Jarvik 7 devices were implanted during the 1980s; the longest duration of survival was 620 days. Versions of the Jarvik 7 are still used in some transplant centers for patients with advanced heart disease, as a life support bridge while they wait for an appropriate donor.

Some newer designs are raising the hope that a permanent artificial heart replacement may be possible. The AbioCor totally implantable heart was first tested in a human patient in 2001. That patient, Robert Tools, lived for 151 days with the artificial heart. The AbioCor has now been implanted in more patients, and may eventually get U.S. Food and Drug Administration (FDA) approval as a temporary measure to keep patients alive while they wait for transplant. These devices are expensive, and patients require extensive in-hospital care after implantation, making widespread use unlikely. In addition, there are still major technical hurdles in artificial heart design. All designs tested so far require continuous external power, which requires the patient to be tethered to a power supply or frequently recharging heavy batteries. The AbioCor is totally implanted, which reduces the risk of infections that have plagued patients with devices in which tubes or wires must exit the body. Even totally implanted devices, however, have synthetic materials in extensive contact with blood; coagulation of blood inside the artificial heart chambers can lead to strokes.

In addition to LVADs and artificial hearts, there are a variety of other artificial devices that are now used routinely to extend the lives and enhance the quality of life for patients with heart disease. In 1952, Charles A. Hufnagel implanted an artificial heart valve composed of a plastic ball in a metal socket. Artificial heart valves are now routinely used to replace failing valves in humans. Although the technology is not perfect, there are now devices that can provide decades of service. Implantable cardiac pacemakers were also developed in the 1950s. Pacemakers, and now implantable cardioverter-defibrillators (ICDs), are complex medical devices that allow for modulation of the electrical stimulation of the heart (Figure 15.14). Designs of pacemakers and ICDs have improved continuously over the years, and both are proven to prolong life in certain patients. Millions of these

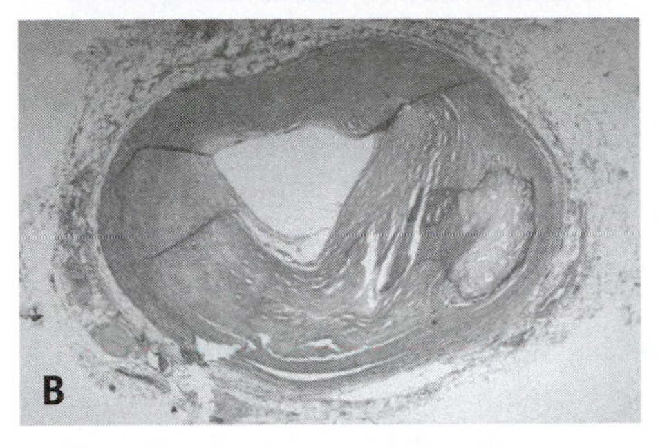

Figure 15.15 **Images of normal (A) and diseased (B) coronary artery.** Note loss of lumen and increased thickness of the muscular layer in the diseased artery. Image from WebPath, courtesy of Edward C. Klatt MD, Mercer University School of Medicine.

devices are used around the world; in the United States alone, more than 2 million pacemakers and 400,000 ICDs were implanted from 1990 through 2002 (6).

Loss of patency in coronary vessels (Figure 15.15) can lead to a loss of sufficient blood flow to heart muscle and myocardial infarction (i.e., heart attack). Coronary vascular disease is one of the world's most frequent causes of death. Vascular stents are tubular devices that provide mechanical support for a blood vessel. Metal stents can be placed into the coronary artery by catheters. Stents are now used together with angioplasty—the inflation of a balloon within a narrowed, or stenotic, vessel to increase the diameter of the lumen (Figure 15.16). Placement of

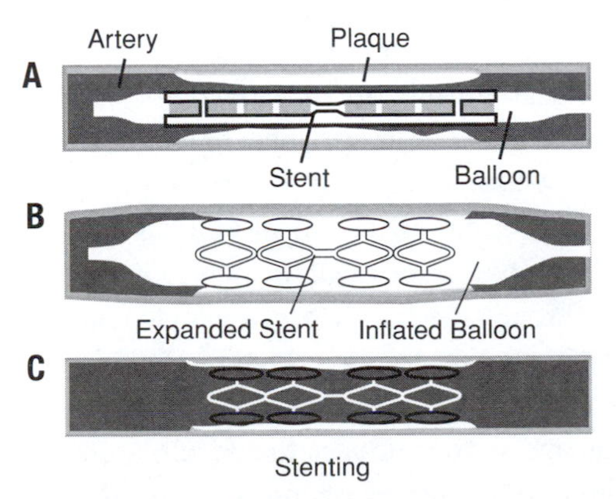

Figure 15.16 **Drug-eluting stent placement during angioplasty. A.** A catheter with the unexpanded stent is inserted into the artery. **B.** The balloon on the stent is inflated, causing the stent to expand into the vessel wall. **C.** The balloon is deflated and the catheter withdrawn, leaving the stent in place.

a stent during angioplasty keeps the lumen open for substantially longer periods, providing the patient with longer duration of enhanced blood flow through the vessel. One problem with bare metal stents is that they are frequently subject to restenosis, or the regrowth of scar tissue around the stent, which can re-narrow the lumen. Drug-eluting stents are metal stents that are equipped with a drug-releasing polymer coating to continuously administer drugs that block restenosis. These combination devices—combination because they provide the mechanical device function as well as the drug-delivery function—provide long-lasting blood flow and prevention of restenosis. Drug-eluting stents represent a major advance in medical device design and patient care.

15.6 Biohybrid artificial organs

In the preceding sections, the artificial organs were composed of largely man-made ingredients: polymers, electronics, and metals. There is another approach to artificial organs, in which synthetic components are used to support the activity of living biological components, particularly cells. The devices that result from this approach are often called **biohybrid artificial organs**, to designate the role of the biological component.

Biohybrid artificial organs share many characteristics with tissue engineering (Chapter 13), and the literature on these areas is merging. There are important distinctions that can be made between the two fields, however. The goal of biohybrid artificial organ design is often the creation of an extracorporeal device, akin to hemodialysis. This approach is different from tissue engineering, in which the goal is to create implantable systems that will perhaps even merge with native tissue. The similarity in approach between biohybrid artificial organs and hemodialysis makes it natural for them to be considered in this chapter.

15.6.1 Cell-based treatments for diabetes

The role of the pancreas in human health was described in Chapter 7. The pancreas is an extended organ that resides in the rear section of the upper abdomen, behind the stomach. One of the essential functions of the pancreas is the production and release of the hormone insulin into the blood. Insulin is produced by beta cells, which reside within cellular aggregates called islets (Figure 15.17). Insulin, a protein hormone, travels through the circulatory system to reach cells throughout the body; insulin regulates glucose levels in blood, and in cells, by binding to insulin receptors on the cell surface (see Chapter 6).

Diabetes is a common and serious disease that results from the failure of the pancreas to produce adequate quantities of insulin (type 1 diabetes) or failure of the body to use insulin properly (type 2 diabetes); type 2 diabetes is usually

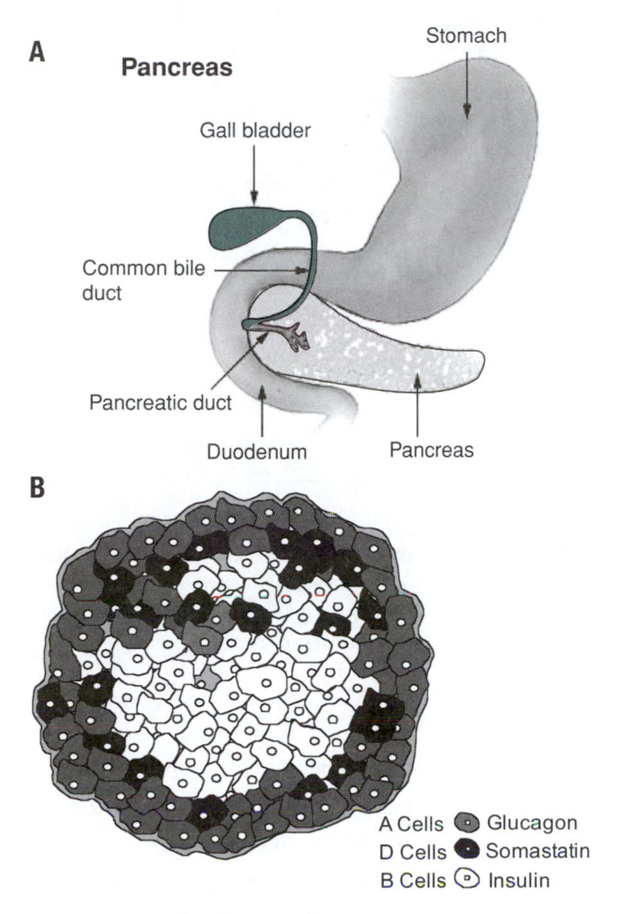

Figure 15.17 **Location and structure of the pancreas. A.** The pancreas is located along the back wall of the abdomen, behind the stomach. It secretes pancreatic digestive juices into the intestine through the pancreatic duct. The exocrine pancreas contains ~1,000,000 cell aggregates, called islets. **B.** Each islet contains cells that secrete glucagons (**A** cells) and insulin (**B** cells or beta cells).

coupled with low insulin production. Diabetes affects 7% of the U.S. population, and the prevalence is growing each year; many diabetics need to self-administer daily injections of insulin to control their blood glucose levels. It is now known that diabetics who achieve the best control of their blood glucose levels live longer, with fewer of the serious side effects of diabetes, such as heart disease, blindness, kidney disease, and nervous system disease.

Because of the prevalence and severity of the disease, and the difficulty that many diabetics experience in controlling their blood glucose levels with insulin injections, many biomedical engineers have worked to develop new methods for treating diabetes. Some of the approaches have been intended to improve the delivery of the drug insulin, such as the wearable or implantable insulin pumps that are described in Chapter 13.

The physiology of insulin release from the pancreas involves a network of cellular interactions and molecular signals. Cells within the pancreas secrete

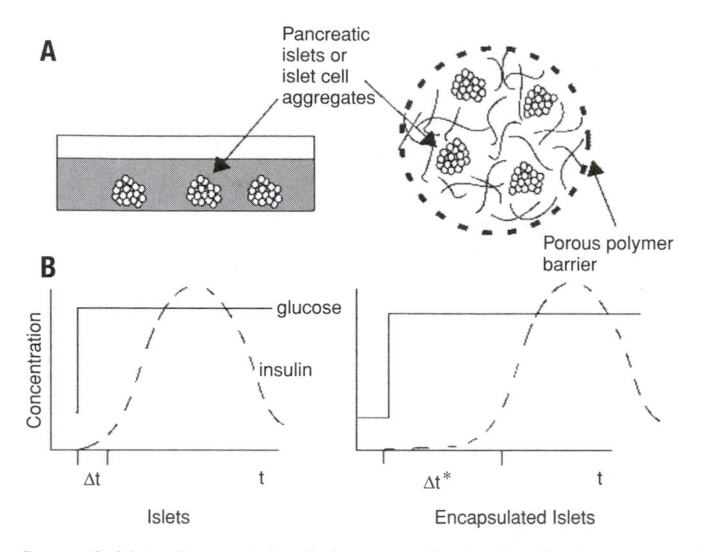

Figure 15.18 **Pancreatic islet cells secrete insulin in response to elevations in glucose concentration. A.** Pancreatic islets can be removed from the pancreas and maintained in cell culture or further encapsulated in a polymer membrane. **B.** When maintained in normal culture conditions, the time course of the insulin release response is normal, but there is an additional lag time in release when the islets are encapsulated.

insulin; the rate of insulin secretion depends on the glucose concentration in the blood. Other cells in the body store glucose; the rate of glucose removal from the blood depends on insulin concentration. Because this function is conceptually simple, many patients with diabetes can control their blood glucose levels by periodically injecting doses of insulin; the timing and quantity of the injected dose is based on estimated or measured levels of glucose in the blood. In this manner, one of the functions of the pancreas is replaced by adding appropriate amounts of insulin to the blood as needed to maintain glucose levels.

Insulin injection therapy is effective in many patients, but it is imperfect. The normal pancreas continuously secretes insulin in response to continuous sensing of glucose concentration; the injection of insulin mimics this function without the high level of regulation (i.e., without continuous sensing and evaluation of the need for insulin). Patients with diabetes have long-term side effects of their illness that correlate with the precision of glucose control. In addition, the injection of insulin does not provide other functions naturally performed by a normal pancreas.

Certain cells of the pancreas can sense glucose concentration and release insulin in response. In the intact pancreas, these cells are located within clusters called pancreatic islets (or Islets of Langerhans) (Figure 15.17). When islets are removed from the pancreas, and maintained in the right conditions in culture, they can be maintained in a viable, functional state. When islet cells from the pancreas are exposed to elevated levels of glucose, they respond by secreting insulin (Figure 15.18).

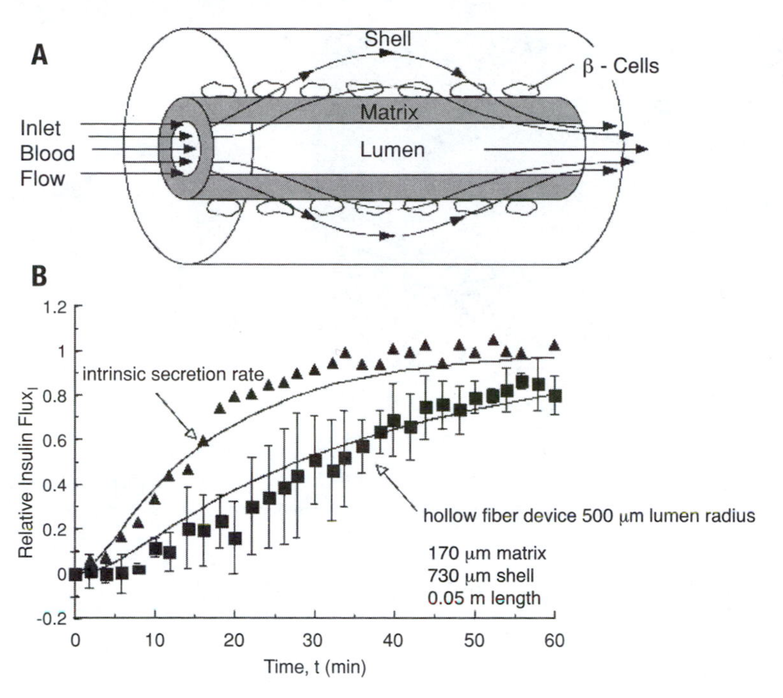

Figure 15.19

Design of a hollow-fiber biohybrid artificial pancreas. A. Islets are maintained in the shell of the hollow-fiber device, and blood flows through the lumen of the fibers. **B.** Islets will release insulin continuously in response to glucose as blood flows through the device, although the rate of insulin secretion is lower than observed when the islets contact blood directly.

An artificial pancreas can use normal pancreatic cells as functional units. For example, cells from the pancreatic Islets of Langerhans can be added to an artificial device that is conceptually similar to the hollow-fiber hemodialyzer (Figure 15.19). Because of their position within the device, islet cells are exposed to glucose concentrations from the flowing blood, but they are protected from immunological reactions caused by blood cells and proteins.

There has been considerable progress in the translation of this concept into a working system for clinical use. One design for an implantable biohybrid artificial pancreas uses an annular shaped acrylic housing 9 cm in diameter, 2 cm high, weighing 50 g, containing 30–35 cm of coiled tubular membrane with an inner diameter of 5–6 mm and a wall thickness of 120–140 μm (Figure 15.20). This membrane material was connected to standard 6-mm PTFE graft for connection to the vasculature. This design provides a total membrane surface area of 60 cm^2 and a total cell compartment volume of 5–6 mL; islets are seeded within the device as a suspension in agar. The ability of these devices to produce insulin was tested in vitro; when seeded with $1–2 \times 10^5$ canine islets, islets within the device produced insulin for many months in response to changes in glucose concentration with a time lag of ~21 minutes. When the devices were implanted into dogs that had their pancreases removed, the need for insulin in these animals was decreased

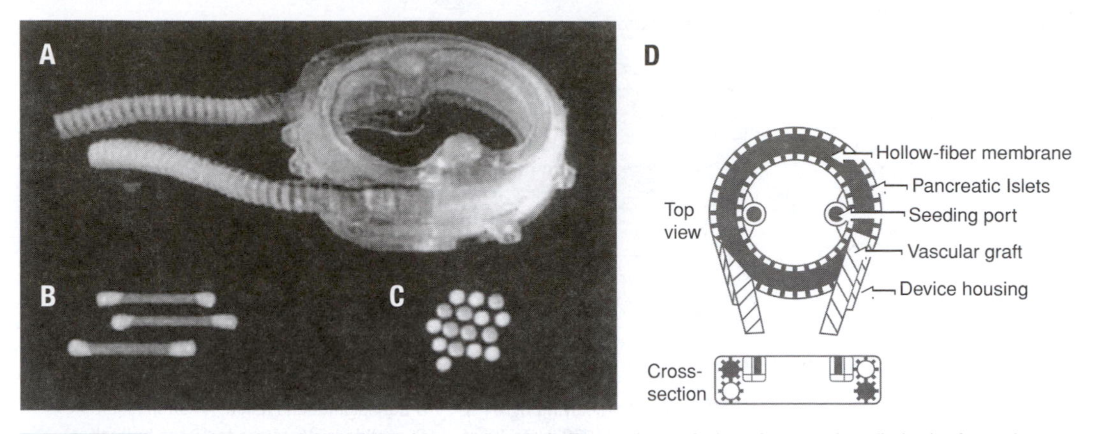

Figure 15.20 **Design of a clinical biohybrid pancreas.** Left: Encapsulation devices that contain cells in the form of a vascular implant **(A)**, membrane diffusion chambers **(B)**, or microcapsules **(C)**. Reprinted with permission from (7). **D**. A schematic of the vascular implant device. Adapted from (8) with permission from AAAS.

to ~0 while the fasting blood glucose level was decreased from ~250 mg/dL to 150 mg/dL.

It has, however, been difficult to develop similar devices of an adequate size to use in humans because of problems with coagulation within the fiber lumen and surgical connection of the synthetic device to the patient's vessels, as well as difficulty in identifying the appropriate source for islets.

As some biomedical engineers have been laboring on the use of islet cells and synthetic materials to build a biohybrid pancreas, others have been refining techniques for direct transplantation of islet cells into diabetic patients. Sometimes these transplanted cells have been encapsulated within polymeric materials, to isolate the cells and protect them from rejection by the immune system (as shown in Figures 15.18 and 15.20). It appears, however, that if they are transplanted in sufficient number correctly, naked islets can survive and function in people. A technique that has become known as the Edmonton protocol (because it was developed in Edmonton, Canada) is to inject the harvested islets into the liver through the portal vein (Figure 15.21). Table 15.4 provides a summary of the results of islet transplantation in human patients. Although a technique that is acceptable for widespread use has not been developed, the progress in this area is encouraging. It is likely that the work of interdisciplinary teams of biomedical engineers, cell biologists, immunologists, and physicians will bring these approaches into routine clinical practice for the next generation.

15.6.2 Artificial liver

Fulminant liver failure presents a dramatic and life-threatening situation. Patients experience rapid progression of jaundice, clotting disorders, and encephalopathy, which leads to death in 50%–80% of cases. The liver can recover from pathological changes, and frequently does. Unfortunately, in many cases, the progression of disease is more rapid than the process of regeneration; therefore, the patient

Table 15.4 Review of islet transplantation

Study	Year of report	No. of recipients and size of transplant	Outcome
Largiader et al.	1980	1 recipient of pancreas microfragments containing 200,000 islets	Insulin independent with normal glucose level at 9.5 mo
Scharp et al.	1990	1 recipient of 800,000 islets	Insulin independent at 22 days
Tzakis et al.	1990	9 patients with cancer and abdominal exenteration without diabetes received 205,000–746,000 islets	Normal glycosylated hemoglobin values in 5 patients, with some receiving insulin supplementation
Warnock et al.	1991	1 recipient of 611,000 islets	Insulin independent with normal glucose levels at 3 mo
Scharp et al.	1991	First 9 patients receiving 6,161 $\pm$ 911 to 13,916 $\pm$ 556 islets-kg of body weight	3 transplantations failed; 4 had measurable C-peptide levels for up to 10 mo but not insulin independent; 2 with normal glucose levels and insulin independent for 1–5 mo
Warnock et al.	1992	4 recipients of 261,000–896,000 fresh and cryopreserved islets	3 had measurable C-peptide levels for 1–8 mo, but not insulin independent; 1 insulin independent for 1 y
Gores et al.	1993	2 recipients of 502,000–528,000 islets	1 had measurable C-peptide levels but not insulin independent at 9 mo; 1 with normal glucose levels and insulin independent at 8 mo
Soon-Shiong et al.	1994	1 recipient of 678,000 encapsulated islets	Insulin independent with normal glucose levels at 9 mo
Carroll et al.	1995	1 patient with cancer and abdominal exenteration without diabetes	Insulin independent with normal glycosylated hemoglobin values at 3 y
Luzi et al.	1996	15 recipients of 98,587–1,294,125 islets	8 had C-peptide levels $>$1.4 ng/L; 4 insulin independent with glycosylated hemoglobin values of 5.6%–7.2% at 1–8 mo
Alejandro et al.	1997	8 recipients of 478,000–1,271,000 islet equivalents	2 insulin independent at 1 mo and 2 insulin-independent at 6 y with normal to near-normal glycosylated hemoglobin values
Secchi et al.	1997	20 recipients of 3,461–14,488 islet equivalents-kg	9 had measurable C-peptide levels with decreased need for insulin; 6 insulin independent at 3–11 mo; 1 insulin independent at 48 mo; all with normal or near-normal glycosylated hemoglobin values
Keymeulen et al.	1998	7 recipients of 2,100–5,300 islet equivalents-kg	3 had measurable C-peptide levels for $>$1 y; 2 insulin independent with normal to near-normal glycosylated hemoglobin values for 1 y
Oberholzer et al.	2000	13 recipients of 199,000–863,000 islets	All had measurable C-peptide levels for $>$3 mo; 5 of 8 had normal C-peptide levels $>$1 y; 2 patients insulin independent at 4 and 36 mo
Shapiro et al.	2000	7 recipients of 11,546 $\pm$ 1,604 islets	All insulin independent at 4–15 mo with 6-month glycosylated hemoglobin values of 5.7 $\pm$ 0.2%

Note: Adapted from (10); complete references are available in the source. Original article on islet transplantation by the Edmonton protocol (20).

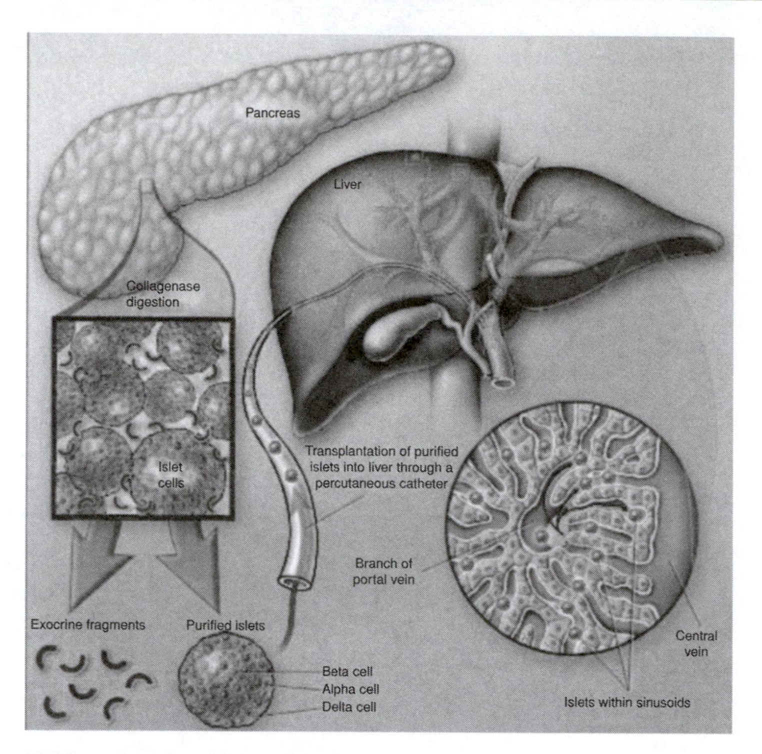

Figure 15.21

Islet transplantation. A pancreas is obtained from a donor. The pancreas is digested with collagenase to free islets from the surrounding exocrine tissue. The freed islets, containing mostly beta and alpha cells, are purified to remove remaining exocrine cellular debris. The purified islets are infused into a catheter that has been placed percutaneously through the liver into the portal vein, whence they travel to the liver sinusoids. Drawing and caption from (17) and used with permission. Copyright © 2004 Massachusetts Medical Society. All rights reserved. (See color plate.)

succumbs prior to regeneration. The loss of metabolic, synthetic, and regulatory functions of the liver creates multisystem failures. The only current therapy is liver transplantation, with 1-year survivals now exceeding 70%. Donors are severely limited, however, and many patients die awaiting a donor liver. Artificial measures to support the patients with rapidly progressive disease would be extremely useful.

Extraordinary efforts are frequently enlisted to provide temporary support for patients awaiting liver transplants, including ex vivo perfusion through animal livers. Xenoperfusion was first used to support patients in liver failure in 1965. Studies of pig-liver perfusion suggest that this treatment can reduce serum bilirubin and ammonia levels, while reducing the symptoms of hepatic encephalopathy. This approach is limited, however, because each perfused liver retains essential functions for only a few hours. Other approaches for liver support include charcoal hemoperfusion, plasmapheresis, hemodialysis, and cross-circulation with humans and animals [see (9)].

Approaches similar to those used to develop an artificial biohybrid pancreas have been used to create artificial livers, as well. At least four different artificial liver-assist devices have been tested clinically using hepatocytes from either a

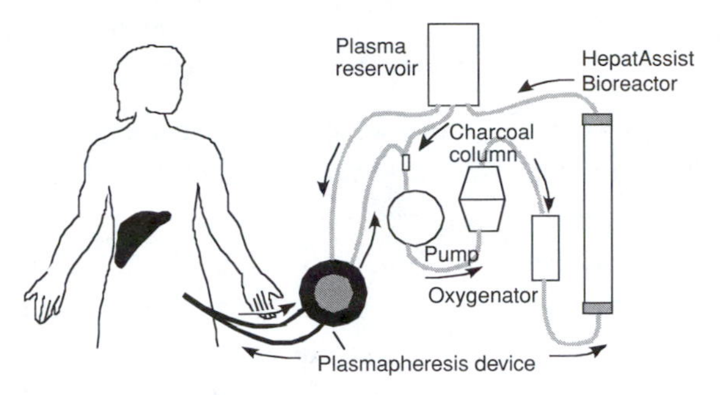

Figure 15.22 **Biohybrid artificial liver.**

well-differentiated human hepatoblastoma (12) or hepatocytes from pigs (13). In one system, porcine hepatocytes are used in a hollow fiber bioreactor, two charcoal columns, a membrane oxygenator, and a pump (see Figure 15.22). Blood is first separated by plasmapheresis prior to HepatAssist^{TM} (Arbios) treatment of the plasma fraction. In its earliest clinical trial, which included 171 patients treated at 11 U.S. and 9 European hospitals, the HepatAssist^{TM} device did not provide encouraging results for a broad range of liver disease. It was discovered, however, that the device was successful in patients with certain types of liver failure and, importantly, that the procedure is safe in these very sick patients.

Transplantation of hepatocytes has also been tested, but it has not progressed as rapidly as has islet transplantation. One interesting early result, suggesting that hepatocyte transplantation may work in certain situations, is shown in Figure 15.23. In this case, the hepatocytes were attached to the outside of polymer microparticles prior to transplantation.

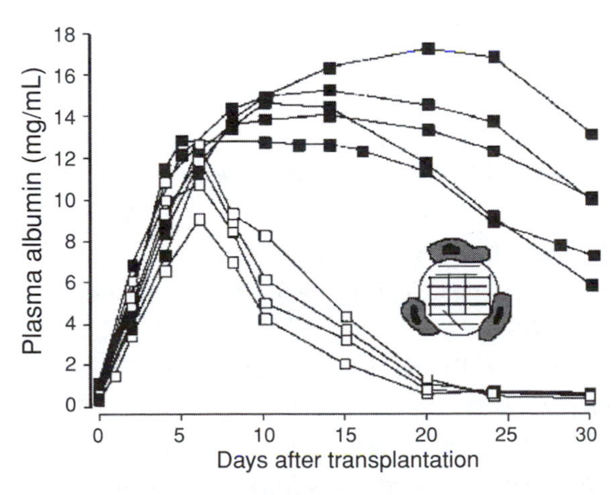

Figure 15.23 **Albumin production after transplantation of hepatocytes on microspheres.** Adapted from Figure 1 of (11). Graph shows albumin production by transplanted cells (into rats that are naturally deficient in albumin). Open symbols represent animals that did not receive immunosuppression; closed symbols represent animals that did receive immunosuppression, which prolonged the life of the transplanted cells.

Summary

- Scientists have long searched for methods to replace organs that are lost to disease or failing, and the last half of the 20th century saw major progress in artificial kidneys, hearts, and pancreases.
- Biomaterials are a critical part of the design of artificial organs; although the perfectly biocompatible material is not yet known, there are many materials that function in the body without many side effects.
- The most important biological response to materials in contact with blood is coagulation, whereas the most important response to implanted materials is the FBR; the basic mechanisms of these responses are now understood.
- Hemodialysis, which is based solidly on engineering principles of biomaterials and molecular separations, is now well integrated into medical care, and it prolongs the lives of many people.
- The rate of movement of a solute across a synthetic membrane is defined by its permeability, which allows for quantitative analysis of concentrations and fluxes.
- Membrane oxygenators revolutionized heart surgery, allowing surgeons to have a direct view of the surgical field, by temporarily taking over all of the functions of the patient's heart and lungs.
- Permanent artificial hearts do not yet exist, but major progress has been made in heart design; other devices—such as valves, pacemakers, stents, and ventricular assist devices—have changed dramatically the life expectancy for patients with many types of heart disease.
- Cells can be used as the basis for artificial organ function, and have been incorporated into the design of the artificial pancreas and artificial liver.

REFERENCES

1. Williams, DF. Definitions in biomaterials. In: *Progress in Biomedical Engineering*. Amsterdam: Elsevier; 1987:72.
2. Williams DF. Definitions in biomaterials. In: *Proceedings of a Consensus Conference of the European Society of Biomaterials*. New York: Elsevier; 1987.
3. Kellar RS, Kleinert LB, Williams SK. Characterization of angiogenesis and inflammation surrounding ePTFE implanted on the epicardium. *J Biomed Mater Res*. 2002;61:226–233.
4. Ratner BD, Bryant SJ. Biomaterials: Where we have been and where we are going. *Annu Rev Biomed Eng*. 2004;6:41–75.
5. Schoen FJ, Padera RF. Cardiac surgical pathology. In: Cohn LH, Edmunds LHJ, eds. *Cardiac Surgery in the Adult*. New York: McGraw-Hill; 2003:119–185.
6. Maisel WH, Moynahan M, Zuckerman BD, Gross TP, Tovar OH, Tillman DB, et al. Pacemaker and ICD generator malfunctions: Analysis of Food and Drug Administration annual reports. *JAMA*. 2006;295(16):1901–1906.

7. Lanza RP, Hayes JL, Chick WL. Encapsulated cell technology. *Nat Biotechnol.* 1996;14:1107–1111.
8. Sullivan SJ, Maki T, Borland KM, Mahoney MD, Solomon BA, Muller TE, et al. Biohybrid artificial pancreas: Long-term implantation studies in diabetic, pancreatectomized dogs. *Science.* 1991;252:718–721.
9. Munoz SJ. Difficult management problems in fulminant hepatic failure. *Semin Liver Dis.* 1993;13(4):395–413.
10. Robertson RP. Islet transplantation as a treatment for diabetes—A work in progress. *N Engl J Med.* 2004;350:694–705.
11. Demetriou AA, Whiting JF, Feldman D, Levenson SM, Chowdhury NR, Moscioni AD, et al. Replacement of liver function in rats by transplantation of microcarrier-attached hepatocytes. *Science.* 1986;23:1190–1192.
12. Sussman NL, Chong MG, Koussayer T, He DE, Shang TA, Whisennand HH, et al. Reversal of fulminant hepatic failure using an extracorporeal liver assist device. *Hepatology.* 1992;16:60–65.
13. Rozga J, Podesta L, LePage E, Morsiani E, Moscioni AD, Hoffman A, et al. A bioartificial liver to treat severe acute liver failure. *Ann Surg.* 1994;219:538–546.
14. Cooney DO. *Biomedical Engineering Principles: An Introduction to Fluid, Heat and Mass Transport Processes.* New York: Marcel Dekker; 1976.
15. Saltzman WM. *Tissue Engineering: Engineering Principles for the Design of Replacement Organs and Tissues.* New York: Oxford University Press; 2004.
16. Peppas NA, Langer R. New challenges in biomaterials. *Science.* 1994;263:1715–1720.
17. Marchant RE, Wang I. Physical and chemical aspects of biomaterials used in humans. In: Greco RS, ed. *Implantation Biology.* Boca Raton, FL: CRC Press; 1994:13–53.
18. Staiger MP, Pietak AM, Huadmai J, Dias G. Magnesium and its alloys as orthopedic biomaterials: a review. *Biomaterials.* 2006;27(9):1728–1734.
19. Pastan S, Bailey J. Medical progress: Dialysis therapy. *N Engl J Med.* 1998;338:1428–1437.
20. Shapiro AM, Lakey JR, Ryan EA, Korbutt GS, Toth E, Warnock GL, et al. Islet transplantation in seven patients with type 1 diabetes mellitus using a glucocorticoid-free immunosuppressive regimen. *N Engl J Med.* 2000;343:230–238.

FURTHER READING

DeBakey ME. Development of mechanical heart devices. *Ann Thorac Surg.* 2005;79(6):S2228–S2231.

Galletti PM, Colton CK. Artificial lungs and blood-gas exchange devices. In: Bronzino J, ed. *Tissue Engineering and Artificial Organs.* Boca Raton, FL: CRC Press; 2006:1–19.

Gravelee GP, Davis RF, Kurusz M, Utley JR. *Cardiopulmonary Bypass: Principles and Practice.* 2nd ed. Philadelphia: Lippincott Williams & Wilkins; 2000.

KEY CONCEPTS AND DEFINITIONS

biocompatibility "the ability of a material to perform with an appropriate host response in a specific application" (2)

biohybrid artificial organs organs created from synthetic components that are used to support the activity of living biological components, such as cells

biomaterial "a nonviable material used in a medical device, intended to interact with biological systems" (1)

cardiopulmonary bypass procedure a procedure used during open heart surgery during which a heart–lung machine replaces the function of both the heart and the lungs, by providing the mechanical force to move blood through the body and providing O_2/CO_2 exchange within the flowing blood

catheter a hollow tube that is inserted into a cavity of the body to provide a route for drainage or insertion of fluids and access for medical instruments. For instance, a catheter can be inserted into the urethra to remove urine from the bladder

clearance the rate at which the blood can be completely cleared of a solute, measured in volume per unit of time; usually used to describe the ability of the kidneys to remove waste products from the blood

clot a clump that forms as a result of coagulation of the blood

coagulation cascade a series of interrelated, enzyme-mediated reactions that result in the formation of a blood clot

dialysance a value, D, that represents the overall performance of a dialysis unit in removing a particular solute from the blood under the unit's current state of operation

dialysate a balanced solution of salts and other solutes that is used as an extraction medium during dialysis

extracorporeal literally "outside of the body"; in medicine, it is used to refer to procedures that are performed on tissues, such as blood, when the tissue is first taken outside the patient's body

extraction ratio a value, E, which stands for the solute concentration change in the blood compared to the theoretical solute concentration change that would occur if blood and dialysate came to equilibrium during dialysis

foreign body giant cells cells formed by the fusion of macrophages around a foreign object in the body, such as during the foreign body response (FBR)

foreign body response (FBR) the long-term inflammatory response to a foreign object or material within a tissue

heart–lung machine a machine that replaces the function of both the heart and the lungs, by providing for the mechanical force to move blood through the body and providing O_2/CO_2 exchange within the flowing blood during a cardiopulmonary bypass procedure

hemodialysis the process of removing wastes and other unwanted materials from the bloodstream during dialysis

macrophages cells, originating from white blood cells, that are responsible for the digestion, or phagocytosis, of foreign invaders into the body

neutrophils a kind of white blood cell, the most numerous kind in the blood, which is involved in the early process of inflammation

open heart surgery a surgical procedure during which a patient's chest is opened to fully expose the heart for operations that include heart transplantation, transplantation of a coronary artery bypass, or replacement of a heart valve

peritoneal dialysis the process of using a patient's own peritoneal membrane (the lining of the abdominal cavity) as a dialysis membrane during kidney dialysis

pump oxygenator another name for a heart–lung machine

shunt a passageway that allows the movement of fluids from one area of the body to another

stent a tubular device that provides mechanical support for blood vessels

suture a material, usually a thread or a fiber, that is used to sew a wound, to bring layers of tissue into contact with one another for healing

ultrafiltration the filter of solutions through a membrane that allows only the permeation of small molecules

urea a product of protein metabolism containing carbon, nitrogen, hydrogen, and oxygen that is most commonly found in urine

ventricular assist device an instrument used during heart failure to pump blood through the body as a temporary measure to support heart function, with the intent that the patient's own heart will eventually recover and take over

NOMENCLATURE

C	Concentration	Moles/L or kg/L
CL	Clearance	mL/min
D	Dialysance	L/s
E	Extraction ratio	Dimensionless
J	Flux	kg/m^2-s
K	Permeability	
L	Fiber length	m
N	Solute flow across a dialysis membrane	Moles/s or kg/s
Q	Volumetric flow rate	L/s
x, y	Length	m

GREEK

γ	Relative rate of permeation through a dialysis membrane to blood flow

SUBSCRIPTS

B	Related to the blood flow through a dialyzer
D	Related to the dialysate flow through a dialyzer
In	Related to an inlet stream
Out	Related to an outlet stream
T	Related to a total, as in the total flow of solute across a membrane

QUESTIONS

1. The blood coming out from a dialyzer should not go directly back to the patient. The FDA requires dialysis machines to include a set of monitors and controls to ensure safety. One of these is a pH sensor to monitor the pH of the blood. Think of three other monitors or detectors you would install into a dialysis device.
2. Countercurrent flow and cocurrent flow patterns give different efficiencies in dialysis. Why?

PROBLEMS

1. Baxter Healthcare Corporation has long been a pioneer in the design and manufacture of dialyzers. In 2006, they produced a high-flux dialyzer system that they call EXELTRA. Their product literature contains the following information:
   The hollow fibers have an inner diameter of 200 microns (10^{-6} m). How many fibers are in each of the units?

	EXELTRA 150	EXELTRA 170	EXELTRA 190	EXELTRA Plus 210
Surface area (m^2)	1.5	1.7	1.9	2.1
Priming volume (mL)	95	105	115	125

2. Dialyzer clearance data for the Baxter EXELTRA—obtained from their product information—is shown in the following table:

EXELTRA Dialyzer Clearance Data

	Urea				Creatinine				Phosphate				Vitamin B_{12}			
Qb (mL/min)	200	300	400	500	200	300	400	500	200	300	400	500	200	300	400	500
Qd (mUmin)	500				500				500				500			
EXELTRA 150	193	262	305	332	186	242	274	297	179	227	255	274	132	152	163	170
EXELTRA 170	196	268	310	341	190	252	286	307	179	232	261	280	138	160	172	180
EXELTRA 190	197	273	323	354	190	251	289	313	186	242	276	296	143	168	183	193
EXELTRA Plus 210	199	287	350	384	198	277	328	363	191	252	292	318	164	202	222	232

   a. From this information, use Figure 15.10 to determine the total membrane permeation rate (*KS*) for urea, creatinine, phosphate, and vitamin B_{12}, for each system (i.e., for EXELTRA 150, 170, 190, and Plus 210).
   b. Use the additional data from problem 1 to answer this question: Does the permeability for each solute differ in each of the units (i.e., in EXELTRA 150 vs. 170 vs. 190 vs. Plus 210)? Explain your answer.
3. The following setup is a simplified version of hemodialysis, where Q = Flow rate of fluids (mL/min); B = Concentration of urea in blood;

D = Concentration of urea in dialysate; and the subscripts i = in and o = out.

a. Identify the system of the dialyzer and draw boundaries around it.
b. Write a mass balance equation for urea in this system.
c. Knowing that this is an actual system with a patient connected to the dialysis machine, which two variables would you have control over?
d. Because urea is a waste product toxic to the body if accumulated, what would you set D_i, the concentration of urea in the dialysate, to be?
e. At a particular moment, suppose the patient's blood flow rate out to the dialysis machine is 300 mL/min, with a urea concentration of 300 mg/L. You set the dialysate flow rate to be 1400 mL/min, and you measured the urea concentration of the exiting dialysate to be 54 mg/L. What concentration of urea does the blood going back to the patient contain?

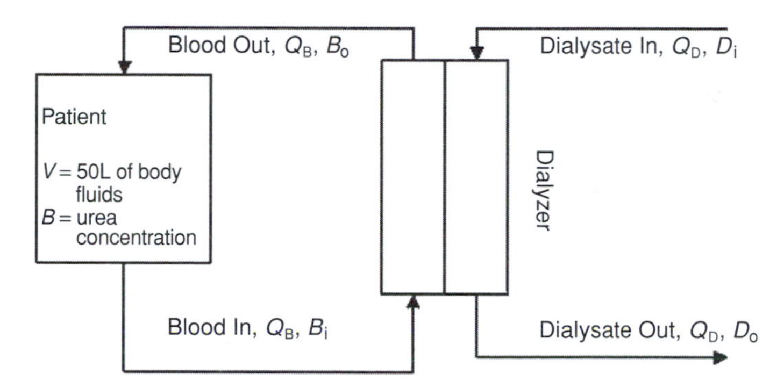

f. We have analyzed the system of the dialyzer, but we are more interested in knowing the concentration of urea within the patient over time. Set up a mass balance equation for the system of the patient.

4. In the cardiopulmonary bypass procedure shown in Figure 15.12:
   a. How much O_2 (in cm^3 at standard temperature and pressure) is returned to the patient if the blood flow rate is 5 L/min?
   b. Assuming that the blood flow rate is 5 L/min, what gas flow rate is needed to accomplish the indicated changes in O_2 and CO_2 composition?

16 Biomedical Engineering and Cancer

LEARNING OBJECTIVES

After reading this chapter, you should:

- Understand the magnitude of the problem of cancer in modern society.
- Develop an elementary understanding of the biology of cancer cells and be able to describe some of the methods for characterizing the progression of tumors in cancer patients.
- Know some of the ways that ionizing radiation interacts with biological tissues and understand the use of radiation in treatment of solid tumors.
- Understand the role of surgery in diagnosis and treatment of tumors and be able to predict some of the ways that surgical treatments for cancer will develop in the future.
- Understand the value and limitations of chemotherapy in the treatment of cancer.
- Know about some of the new approaches, based on our understanding of the molecular and cellular biology of cancer, for creating biological treatments.

16.1 Prelude

Cancer is a common, often life-threatening, disease involving the uncontrolled growth and spread of abnormal cells. Cancer is one of the leading causes of death in the world, particularly in developed nations such as the United States. Cancer is really a group of diseases; it can arise in any organ of the body and has differing characteristics that depend on the site of the cancer, the degree of spread, and other factors. Mutations in certain genes within cells—called proto-oncogenes and tumor suppressor genes—are the primary cause of cancer (1).

Sadly, almost every college student has some knowledge of cancer, gained through experience with classmates, family members, or friends. Cancer occurs in people from all segments of society and people of all ages, although it is relatively rare in childhood. Some cancers can be inherited, but most cancers are strongly linked to environmental factors such as exposure to radiation, tobacco, viruses, or chemicals. Cancer can occur many years after exposure, making it difficult to find precise links between factors and disease. However, much is now known about the risk of exposure to certain toxins and infectious agents. For example, cigarette smoking increases the risk of lung cancer by 23 times in males and by a similar factor in females (2).

Biomedical engineers have played important roles in the development of methods for cancer treatment, and many biomedical engineers are working on new approaches that they hope will prevent or treat cancers in the future. This chapter reviews some of the methods that are used to treat cancer.

16.2 Introduction to cancer

Cancer cells can arise in almost any location in the body, and tumors—or collections of cancer cells—can be vastly different from organ to organ and person to person. A cell that is capable of forming a tumor is different from normal cells within an organ; it has undergone a malignant transformation. The molecular events that occur during this transformation are not completely understood, and there is more than one set of molecular changes that will cause a malignant transformation. All cancer cells share a number of characteristics, however. They proliferate or divide rapidly compared to normal cells. Importantly, they do not halt proliferation in response to normal signals. They generally do not respond normally to signals that are provided by neighboring cells. They do not differentiate normally, but tend to remain as immature, dedifferentiated cells. They do not adhere as readily to other cells and extracellular matrix, and they do not become specialized or die, even when they are moved to a part of the body that is different from their normal environment.

Pathologists classify cancer cells by examining them microscopically and assigning them a "grade." The grade is a measure of how abnormal the cancer cell is compared to the normal cell from that organ. Cancer cells that look similar to normal cells receive a low grade. Cancer cells that appear abnormal, usually less well differentiated than normal cells, receive a high grade. The grade of a cancer is often a determinant of the optimal treatment.

Cancer begins as a cell or a few cells that develop these abnormal properties. As these cells divide, a tumor develops, grows, and becomes more difficult to treat. To help guide treatments, cancers are graded according to their stage, which is defined for each type of tumor. Here, the **Tumor, Node, and Metastases (TNM)** staging system is described briefly, to provide a general sense of how cancers spread and the issues that are considered important for their treatment.

The T is used to classify the tumor on a scale of 0 to 4, or T0 to T4. The classification is specific for each tumor, and is shown for bladder cancer in Figure 16.1. T0 indicates a tumor that has not yet invaded the local environment, whereas T4 indicates a tumor that has spread into an adjacent organ.

Cancer tends to spread through the body's lymphatic system. The N is used to classify the extent of **lymph node** involvement (lymph nodes are small organs found throughout the body that contain many cells, particularly white blood

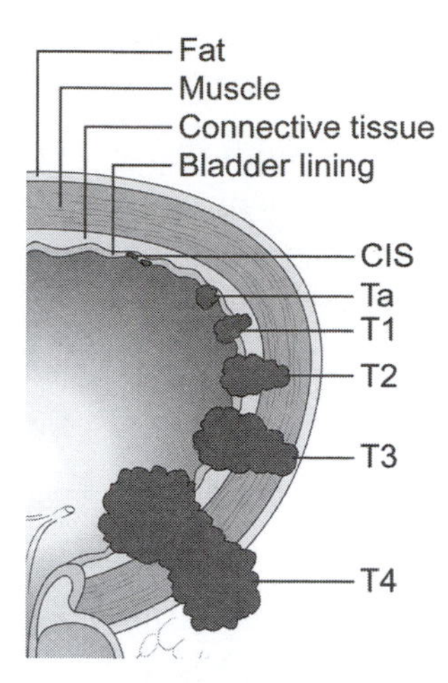

Figure 16.1 **Tumor (T) stages of bladder cancer.** The T of the Tumor, Node, and Metastases (TNM) staging system indicates the extent of local growth. CIS: early-stage cancer cells are detected in the innermost layer of the bladder lining; Ta: the cancer is just in the innermost layer of the bladder lining; T1: the cancer has started to grow into the connective tissue beneath the bladder lining; T2: the cancer has started to grow into the muscle under the connective tissue layer; T3: the cancer has grown all the way through the muscle layer into the fat layer; T4: the cancer has spread outside the bladder to the prostate, vagina, or other organs in the pelvis. Image courtesy of CancerHelp UK, the patient information Web site of Cancer Research UK (www.cancerhelp.org.uk).

cells, and that act as filters and sites for immune activity). N0 indicates that no tumor cells have entered the local lymph nodes, whereas N4 indicates extensive involvement. Again, the grade from 0 to 4 is defined for the particular tumor, and reflects the clinical experience with tumors of this type.

Spread of a tumor to a distant site is called **metastasis**, and it is a sign that the tumor is at an advanced stage. The M is used to classify the extent of metastasis, where M0 indicates no metastasis and M1 indicates that metastasis is present. An example of the application of TNM staging to melanoma is provided in Table 16.1.

Treatments for cancer vary depending on site, stage, and many other factors, but the available treatments fall into a few general categories: surgery, radiation therapy, chemotherapy, and—in a few cases—biological therapies and hormonal therapies. Many patients receive treatment by a combination of approaches, called multimodality treatment.

16.3 Surgery

Surgery is often used in the diagnosis of cancers. Small samples of tissue, called **biopsies**, are surgically removed so that the type and stage of the tumor can be determined by microscopic analysis (Figures 16.2–16.4). The use of biopsies to define the properties of a tumor—and therefore the proper course of treatment for the patient—has been tremendously important in the development of treatments for cancer.

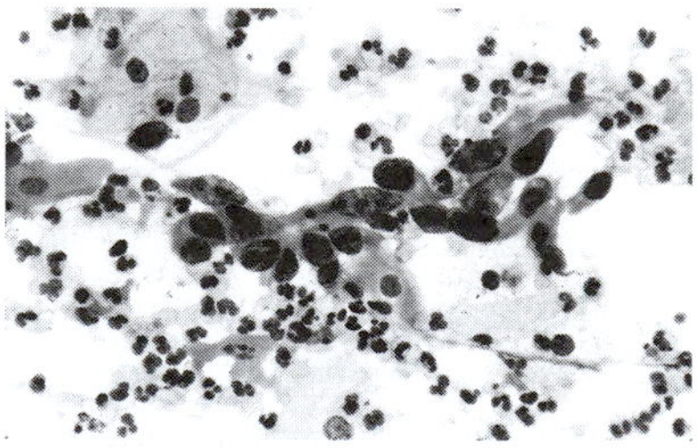

Figure 16.2 **Histological section showing cervical cancer, specifically, squamous cell carcinoma in the cervix.** Tissue is stained with Pap stain and magnified 200×. Photo courtesy of National Cancer Institute. (See color plate.)

As more is learned about the biology of cancer, and as this new information is applied to the characterization of surgical biopsies, treatment of cancer should improve. It is now known that most tumors are **heterogeneous**, containing cells that differ widely in their biology and therefore in their response to different treatments.

Table 16.1 Classification system for TNM staging of melanoma

T classification	Thickness	Ulceration status
T1	<1.0 mm	a: w/o ulceration and Clark level II/III b: with ulceration or Clark level IV/V
T2	1.01–2.0 mm	a: w/o ulceration b: with ulceration
T3	2.01–4.0 mm	a: w/o ulceration b: with ulceration
T4	>4.0 mm	a: w/o ulceration b: with ulceration
N classification	**No. of metastatic nodes**	**Nodal metastatic mass**
N0	No evidence	
N1	1 node	a: micrometastasis b: macrometastasis
N2	2–3 nodes	a: micrometastasis b: macrometastasis
N3	4 or more metastatic nodes, or or matted nodes, or in-transit metastases/satellites and metastatic nodes	
M classification	**Site**	**Serum LDH**
M0	No evidence of metastasis to distant tissues or organs	
M1a	Distant skin, subcutaneous, or nodal metastases	Normal
M1b	Lung metastases	Normal
M1c	All other visceral metastases or any distant metastases	Normal Elevated

Notes: Adapted from publications of the American Joint Committee on Cancer (2002). TNM: Tumor, Node, and Metastases; LDH, lactate dehydrogenase.

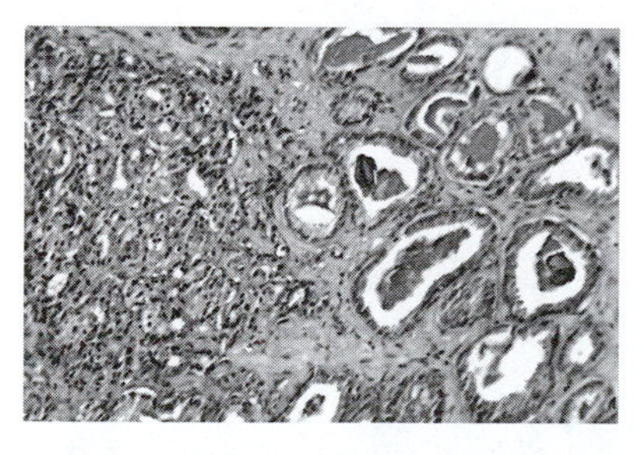

Figure 16.3 **Histological slide (hematoxylin and eosin stain at $300\times$) showing prostate cancer.** Right: Somewhat normal Gleason value of 3 (of 5) with moderately differentiated cancer. Left: Less normal tissue with a Gleason value of 4 (of 5) that is highly undifferentiated. The Gleason score is the sum of the two worst areas of the histological slide. Photo credit: Otis Brawley (See color plate.)

Improved methods for measuring this heterogeneity—perhaps by using new **microarray** technologies (see Section 16.7)—and application of treatment methods that account for heterogeneity should improve cancer care (3).

Surgical treatments for cancer are conceptually simple: A surgeon removes the tissue containing the malignant cells. A surgical attempt to remove a tumor from the body is called a **resection**. Resection of early skin cancers often leads to a complete cure. Depending on the location and stage of the cancer, surgery can be technically

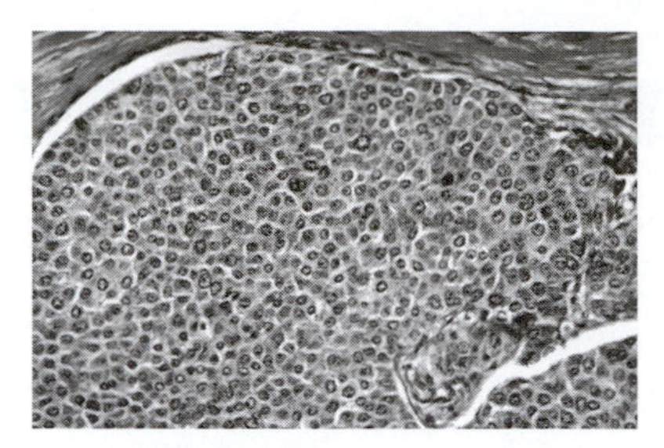

Figure 16.4 **Breast cancer invades normal tissues and establishes new centers of growth.** Note that the duct (demarcated by the two long, white areas, which are the walls of the duct) on inside of breast is completely filled with tumor cells. This histological slide was stained with hematoxylin and eosin and magnified to 100×. Photo by Dr. Cecil Fox and provided courtesy of the National Cancer Institute Visuals Online. (See color plate.)

complex. Pancreatic cancer is difficult to treat surgically because of the location of the pancreas, in the retroperitoneal space beneath the abdominal cavity, and because pancreatic cancer is difficult to detect in the earliest stages. Tumors in the brain are difficult to treat surgically because of the risk that resection will destroy brain tissue that provides essential functions.

Biomedical engineers have made important contributions to surgery, by design of many of the instruments and tools that are used by surgeons in the operating room. **Endoscopes** (see Chapter 12) are used for minimally invasive surgery, allowing physicians to look inside the body, to take pictures, and, frequently, to take small pieces of tissues from internal organs for biopsy or other analysis (Figure 16.5). Endoscopes have different names—and different designs—depending on the organ that they are intended to observe: colonoscopy (for the colon), bronchoscopy (for the lung), cystoscopy (for the bladder), colposcopy (for the cervix), laparoscopy (for the abdomen and pelvis), and arthroscopy (for joints).

16.4 Radiation therapy

Radiation therapy for cancer involves the use of high-energy **electromagnetic rays**, which are focused at the site of cancer to damage and kill malignant cells.

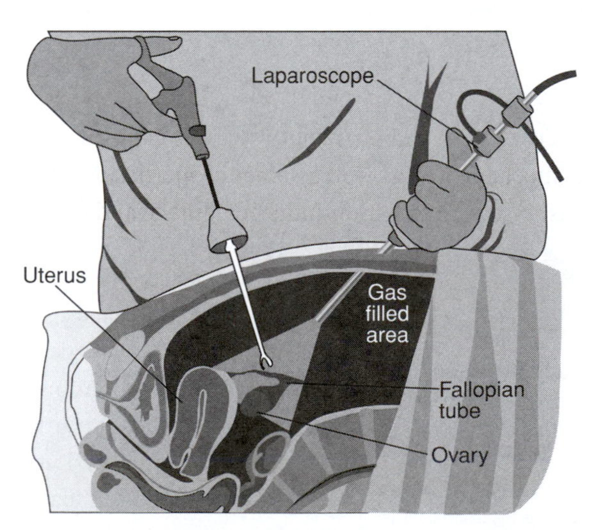

Figure 16.5 **Instruments for laparoscopic surgery.** In this image, a surgeon is using a laparoscope to see inside of the abdomen, with only a small incision. Other instruments allow surgeons to perform surgical procedures as well.

Because the radiation is physically focused on a specific location in the body, radiation therapy (like surgery) is a local therapy.

Ionizing radiation was defined earlier (see Chapter 12) as radiation with enough energy to eject electrons from electrically neutral atoms, leaving behind charged atoms or ions. Recall that there are four basic types of ionizing radiation: **alpha particles** (helium nuclei), **beta particles** (electrons), **neutrons**, and **gamma rays**. High frequency electromagnetic waves, **x-rays**, are generally identical to gamma rays except for their place of origin: X-ray photons are generated by energetic electron processes, whereas gamma rays are produced by transitions within atomic nuclei. Neutrons are not ionizing, but their collisions with nuclei lead to the ejection of other charged particles that do cause ionization.

When ionizing radiation hits a biological tissue, it has effects on the living cells within the tissue. Because of its highly energetic nature, ionizing radiation can eject or deviate electrons from molecules in its path; these free electrons can damage the cell, usually by causing chemical changes in deoxyribonucleic acid (DNA) in the cell nucleus. The dose of ionizing radiation that is deposited in a medium (such as a cell or tissue exposed to radiation) is measured in units of energy per mass, usually in units of J/kg or grays (Gy) (another unit that is sometimes used is the rad—the Roentgen Absorbed Dose: 1 Gy is also equal to 100 rad).

We are continuously exposed to ionizing radiation, but it usually has no noticeable effect because our body has efficient repair mechanisms, including the ability to produce new cells. Sometimes, however, the damage caused by radiation exposure is irreparable: Cells will die (an early effect of radiation) or suffer mutation (a delayed effect of radiation). Most cells are efficient at repairing DNA damage from ionizing radiation (4). It is estimated that a cell exposed to 1 Gy of ionizing radiation experiences $\sim 2 \times 10^5$ ion pairs within its nucleus: ~2,000 of these pairs are generated directly within the DNA. This high level of active ions probably results in a significant amount of damage, which is estimated as ~1,000 single-strand DNA breaks and ~40 double-strand breaks. Normal cells will repair most of this damage; cell survival curves suggest that only a small number of lethal changes (estimated to be between 0.3 and 10) result from this exposure. Because they differ in their repair mechanisms (as well as other properties), cells within the body have different sensitivities to radiation-induced damage and killing (Figure 16.6).

SELF-TEST 1 What percentage of ion pairs produced by ionizing radiation cause a lethal lesion in mammalian cells?

ANSWER: From the data above, for 1 Gy of exposure, the range is between:
$(0.3 \text{ lethal changes}/2 \times 10^5 \text{ ion pairs}) \times 100 = 0.00015\%$
and $(10 \text{ lethal changes}/2 \times 10^5 \text{ ion pairs}) \times 100 = 0.005\%$

The absorbed dose is not the only feature important in determining the likelihood of biological damage, as some forms of radiation are more efficient at

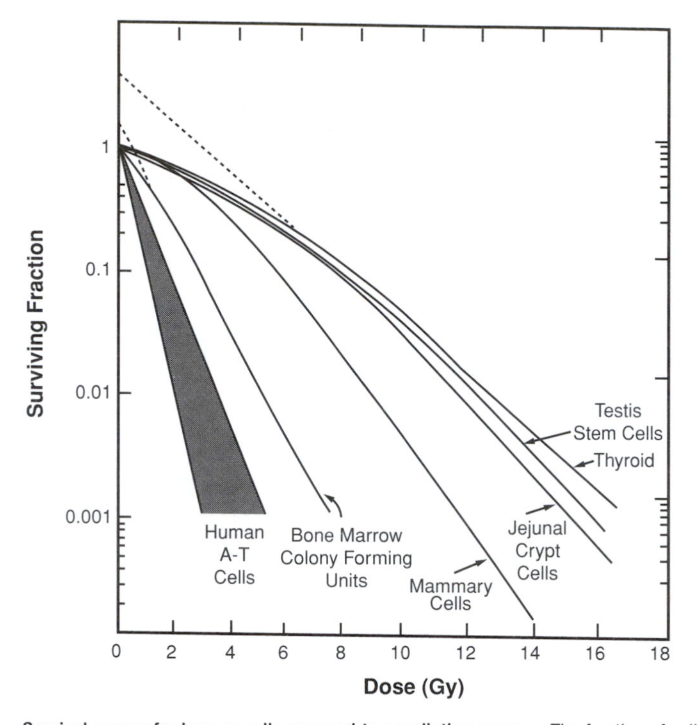

Figure 16.6 **Survival curves for human cells exposed to a radiation source.** The fraction of cells that will survive a dose of radiation is plotted as a function of dose for cells from the testis, thyroid, intestine (jejunum), breast, and bone marrow. Stem cells are more resistant to lethal damage, whereas bone marrow cells are sensitive. Human A-T cells were obtained from a patient with ataxia-telangiectasia, where DNA repair mechanisms are lacking. Adapted from Hall EJ. *Radiobiology for the Radiologist.* 4th ed. Philadelphia, PA: Lippincott; 1994:60.

causing damage than are others, and some tissues are more sensitive to radiation damage than are others. When the efficiency of biological damage of the radiation is taken into account, another unit is used to measure the biological dose: This unit, which also is measured in J/kg, is called the sievert or Sv (another unit that is sometimes used is the rem—from Roentgen Equivalent Man). For a particular type of radiation, the dose equivalent is equal to the absorbed dose multiplied by a weighting (W_R) factor:

$$\text{Dose Equivalent[Sv]} = \text{Absorbed Dose [Gy]} \times W_R. \qquad (16.1)$$

The weighting factor, W_R, is determined by two factors: $W_R = QN$, where Q is the quality of the radiation source and N is a function of the tissue that is irradiated (Table 16.2).

Because ionizing radiation can kill cells, exposure of the body to radiation can cause disease, including cancer. A large dose of radiation can kill so many cells that the body cannot replace them rapidly enough. This sudden, massive cell death results in serious effects, such as skin burns, vomiting, migraine headaches, and internal bleeding. An exceptionally high dose can kill a person within days or weeks. When growth-regulating genes become damaged by radiation, cancer can occur (but not immediately). This cancer may remain latently present for several decades before becoming apparent. Studies of populations exposed to an

Table 16.2 Q values for the quality of different radiation sources; N values for sensitivity of different tissues

Photons	Q = 1
Electrons	Q = 1
Neutrons	Q = 5, 10, 20
Protons	Q = 5
Alpha particles	Q = 20
Gonads	N = 0.20
Bone marrow, colon, lung, stomach	N = 0.12
Bladder, brain, breast, kidney, liver, muscles, pancreas, small intestine, spleen, thyroid, uterus	N = 0.05
Bone surface, skin	N = 0.01

exceptionally large dose of radiation, notably the survivors of the atomic bombing of Hiroshima and Nagasaki, have shown that very large doses can increase the risk of cancer and, possibly, cause genetic damage. Effects such as these cannot usually be confirmed in any particular individual exposed; they occur at random in the irradiated population.

Radiation can also be used to treat diseases; the intentional exposure of tumors to radiation can produce powerful effects on the growth of tumors and often adds years of life to patients with cancer. In radiation therapy, the biological effects of radiation are harnessed and controlled in an attempt to destroy tumor cells without damaging normal tissues and cells. The objective of radiation therapy is to kill enough of the cancer cells to maximize the chances for cure of the cancer, without substantial side effects to normal tissues. Often, the objective is to shrink the cancer with radiation, so that surgery or chemotherapy can be more effective.

Because the effects of radiation-induced DNA damage do not usually appear until cells divide, tissue responses to radiation differ. Some tissue responses to radiation are early, occurring in the first few hours or days, whereas other responses are late, occurring over weeks or months. Early responding tissues include most cancers, skin, testis, and intestine. Late responding tissues (LRTs) include kidney, lung, bladder, and spinal cord. LRTs might be more sensitive to radiation, however: In Figure 16.7b, the responses of a tumor and an LRT to a single dose of radiation (curves labeled $n = 1$) are shown. At a dose of 9 Gy, most of the LRT cells will be killed (only 1 in 10^6 survive), whereas many more tumor cells survive (1 in 100).

SELF-TEST 2 Using the information in Figure 16.7b, with a single dose of radiation ($n = 1$), what dose of radiation must be delivered to kill 99.99% of tumor cells in a tissue?

ANSWER: For a survival fraction of 0.0001, the dose is ~15 Gy

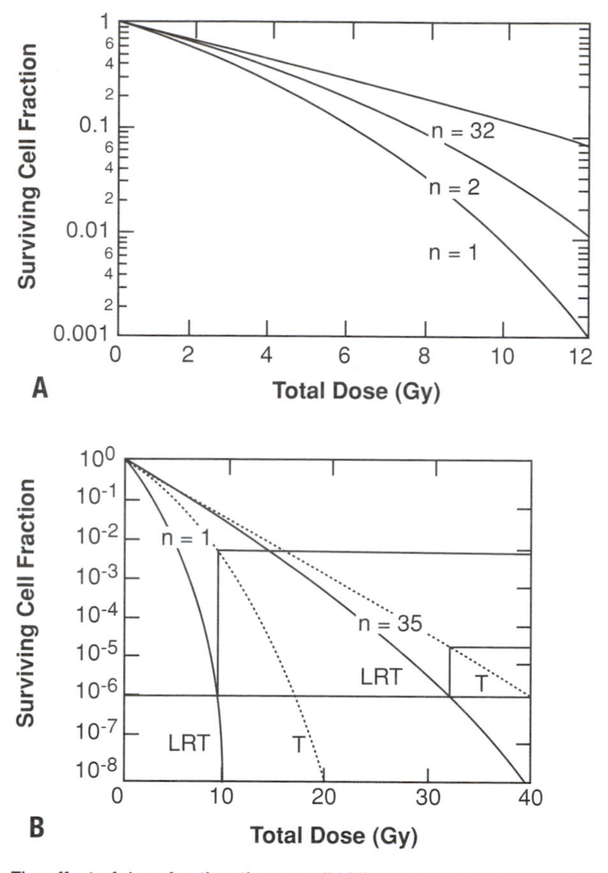

Figure 16.7

The effect of dose fractionation on cell killing. A. The fraction of cells that survive a given dose of radiation is plotted for doses divided into 1, 2, or 32 fractions. **B.** Estimated survival curves for a late responding tissue (LRT) and tumor (T), showing the improvement in tumor cell killing with dose fractionation. Adapted from Hobbie RK. *Intermediate Physics for Medicine and Biology.* 3rd ed. New York: Springer AIP Press: 1997:443.

The major problem faced by radiation oncologists (physicians who specialize in the administering of radiation therapy to cancer patients) is to deliver the radiation in a way that kills as many cancer cells as possible, while having a minimal effect on other normal tissues. They address this problem in several ways: first, by adjusting the timing of the doses and, second, by focusing the dose on the tumor tissue.

The first important method for narrowing the difference between tissue responses is by **dose fractionation**. Here, the total amount of delivered radiation is split in a number of doses: This splitting allows more sensitive tissue (such as the LRT described in the last paragraph) to recover from reversible damage. Figure 16.7a shows the effect of dose fractionation on a sensitive tissue: By splitting a dose of 12 Gy into 32 fractions, and delivering these fractions over time, the fraction of tissue surviving the radiation increases from 1 in 1,000 to 1 in 10, a substantial increase, which could translate to a considerable difference in terms of survival of tissue function. Figure 16.7b shows the effect of dose

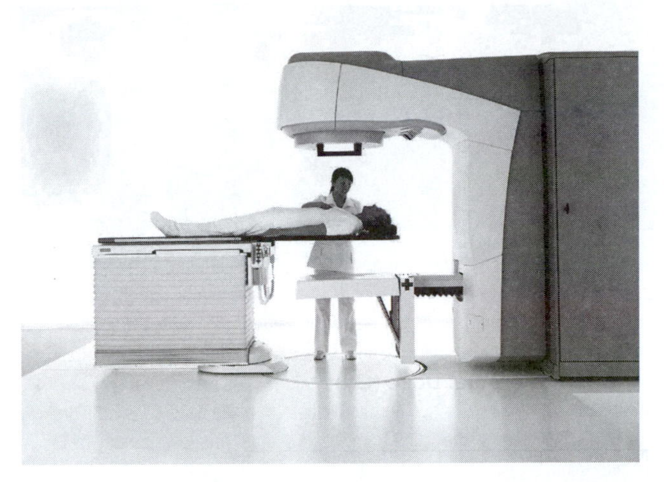

Figure 16.8 **Linear accelerator.** The ONCOR linear accelerator, manufactured by Siemens, is capable of intensity-modulated radiation therapy, stereotactic radiation therapy and surgery, and image-guided radiation therapy. Photo courtesy of Siemens Medical Solutions USA, Inc.

fractionation: By dividing the dose into 35 fractions, the LRT can accept a dose of 30 Gy, which is sufficient to provide substantial damage to the tumor.

The second important method for optimizing radiation therapy is to focus the dose, so that delivery of radiation to normal tissue is avoided to the maximum extent possible. Radiation therapy can be administered externally, from radiation beams that are produced outside the patient and aimed inward at the tumor, or internally. Internal radiation therapy is often given as **brachytherapy**, which involves the implantation of radioactive materials (called seeds) into or near the tumor. In a few cases, as in the delivery of radioactive iodine for the treatment of thyroid cancer, chemically targeted sources of radiation can be injected intravenously. **External beam radiation therapy (EBRT)** is produced from external radiation beams generated by machines called linear accelerators (Figure 16.8). Effective therapy is provided by a combination of modern equipment, which provides a high degree of focus for the beams, and careful planning of how the radiation will be applied in the patient. The pattern of delivery of radioactivity, also called the spatial field, is optimized by considering clinical examination, biopsy results, diagnostic imaging studies, and mathematical models of energy deposition in tissue. This technique can be made even more precise by the use of specialized computed tomography (CT) scans, which are taken just prior to the delivery of radiation, in an approach known as **three-dimensional conformal radiation therapy (3DCRT)**. Further specialized rotating devices provide radiation delivery from a greater number of angular positions, in an approach called **intensity-modulated radiation therapy (IMRT)** (Figure 16.9).

Brachytherapy, sometimes called interstitial brachytherapy, involves the placement of radiation sources, or seeds, into the patient at sites selected to maximize the exposure of the tumor to radiation. This mode of internal delivery of radiation

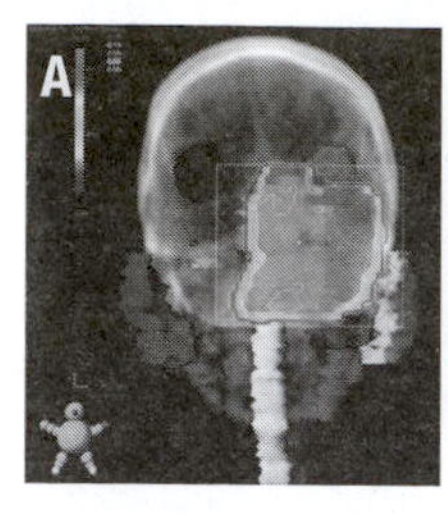

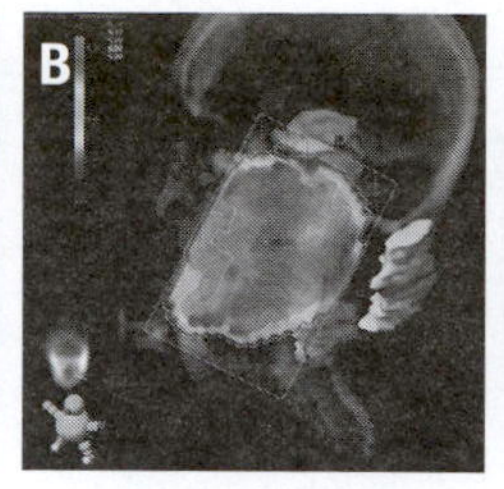

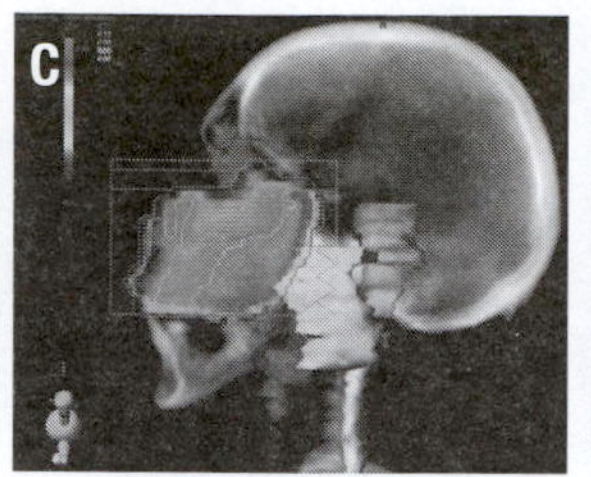

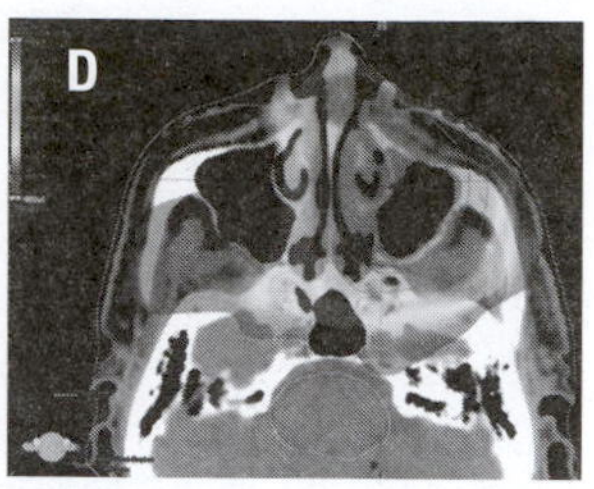

Figure 16.9 **Example of intensity-modulated radiation therapy (IMRT).** These images show the planning of radiation therapy for a head and neck cancer using IMRT. Anterior **(A)**, left superior oblique **(B)**, and left lateral **(C)** views are shown. Each image shows the pattern of the beam intensities, the surrounding bony anatomy, and manually segmented important radiosensitive anatomical structures such as the eyes, lenses, optic nerves, inner ear, brain stem, and spinal cord. The beam intensity patterns were determined from constraints on dose delivery to the tumor and avoidance of dose delivery to sensitive normal tissues. A single axial CT slice **(D)** shows the radiation dose distribution that was calculated for that slice from these beams. Images courtesy of Dr. Jonathan Knisely, Yale University. (See color plate.)

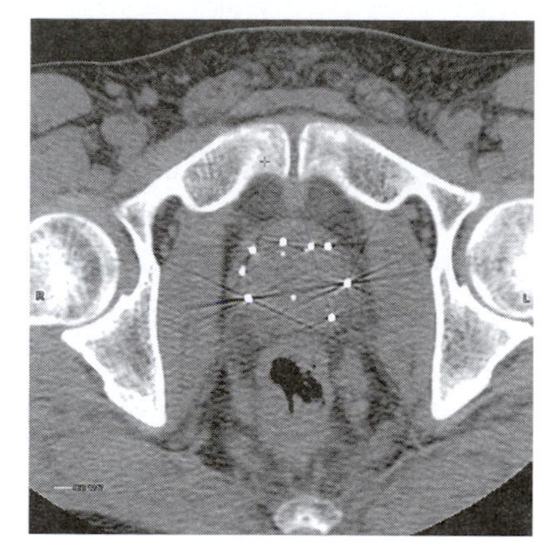

Figure 16.10 **Computed tomography (CT) image showing radioactive seeds implanted into a prostate tumor.** Image courtesy of Dr. Jonathan Knisely, Yale University.

is used most often in treatment of prostate cancer (Figure 16.10). The seeds for brachytherapy are implanted into the prostate while an ultrasound imaging probe, inserted into the rectum, is used to observe the procedure.

Brachytherapy is sometimes used, as an alternative to EBRT, to treat breast cancer. In one form of breast brachytherapy, which is used for early-stage breast cancer, a balloon on the end of a catheter is inserted into the tumor resection cavity (Figure 16.11). The brachytherapy treatment is applied by inserting a radioactive seed through the catheter and into the balloon.

Brachytherapy is aided by engineering design of radioactive seeds (Figure 16.12) and by engineering calculations to determine the distribution of radioactive dose within the tumor and the patient (Figure 16.13). Interstitial brachytherapy generates highly conformal dose distributions in a target volume because radioactive seeds are implanted directly inside the target tissue. In a typical interstitial brachytherapy procedure, 50–100 radioactive seeds, each about the

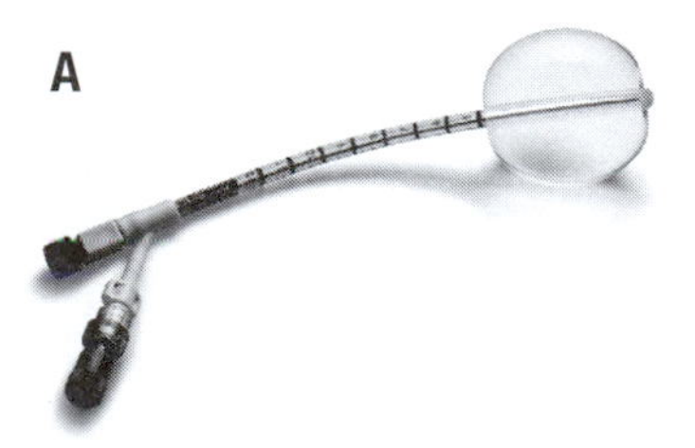

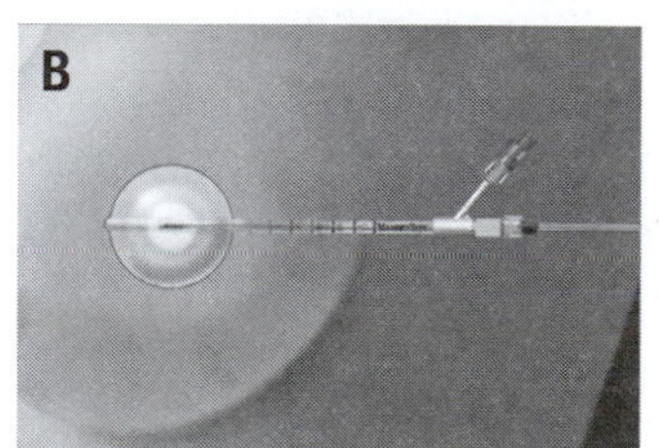

Figure 16.11 **Brachytherapy of breast cancer.** Photos courtesy of Hologic.

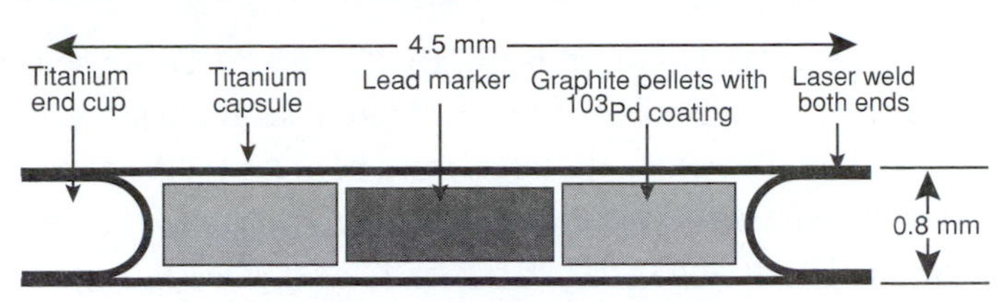

Figure 16.12 **Schematic diagram of the design of a brachytherapy seed device.** This diagram is modeled after the Theragenics Theraseed® model 200 ^{103}Pd brachytherapy seed.

size of a rice grain, are implanted in the tumor using image-guided implantation techniques such as ultrasound, CT, or fluoroscopy, which allow the physician to place radioactive seeds precisely at desired locations within a three-dimensional volume with minimal invasiveness. A typical seed such as the example of a ^{103}Pd seed (Figure 16.12) has outer dimensions of 4.5 mm (length) $\times$ 0.8 mm (diameter). Low energy photon-emitting radionuclides such as ^{125}I and ^{103}Pd are preferred because they 1) provide adequate coverage of the tumor when used in a grid of about 1-cm spacing and 2) produce minimal exposures to distant organs in the patient, to the hospital personnel performing the procedure, and to the family members and friends of the patient after he/she is released from the hospital with the radioactive seeds in place.

Most of the brachytherapy procedures today are performed in one-day surgery suites without the need for hospitalization. This and the depth dose characteristics (Figure 16.13) make brachytherapy a very cost-effective and patient-friendly procedure compared to 3DCRT or IMRT, which also produce highly conformal dose distributions. A key advantage of 3DCRT or IMRT over brachytherapy is

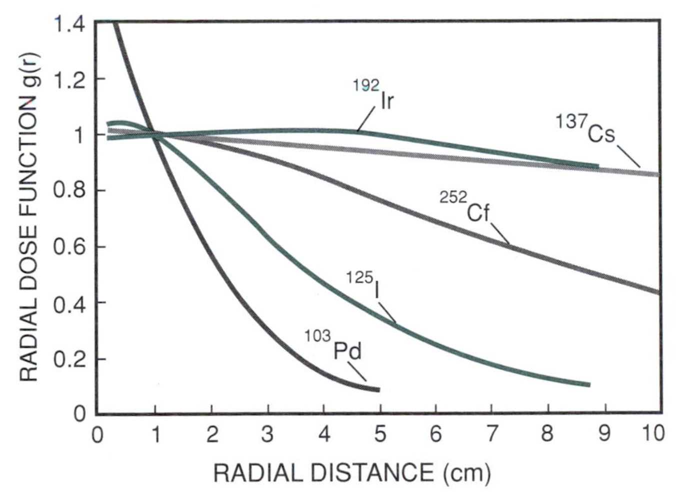

Figure 16.13 **Radial dose functions for various brachytherapy sources.** Radioactive isotopes differ in the characteristics of the radiation they emit. One of the major differences between isotopes is the distance over which the energy they emit is deposited, or delivered, to tissues. This graph shows the distance of energy deposition in tissue for some typical isotopes: Pd deposits most of its dose within 5 cm of the source; Cs has a much broader distribution. Adapted from Coursey BM and Nath R. Radionuclide therapy. *Physics Today* 2000;53:25–30.

that it is non-invasive. However, both 3DCRT and IMRT are sensitive to patient localization and setup errors because of high dose gradients at the periphery of the target volume. Therefore, the target must be placed at the correct position with a precision of about a millimeter relative to the linear accelerator over a course of 5–6 weeks of 3DCRT and IMRT. In contrast, brachytherapy requires a single visit to a one-day surgery unit. Unlike 3DCRT and IMRT, brachytherapy is far more forgiving of localization and target motion errors because the implanted sources of radiation in an interstitial implant move with the target. Thus brachytherapy solves a critical problem of 3DCRT and IMRT. This high-precision requirement makes 3DCRT and IMRT expensive, labor intensive, and technically complex. For these practical reasons and other important radiobiological reasons related to continuous low-dose-rate irradiation, brachytherapy remains a valuable treatment modality for selected cancers.

16.5 Chemotherapy

The term **chemotherapy** refers to the use of chemical agents, or drugs, for their effect in treating any disease, but it is most often used to refer to the use of drugs to treat cancer. Unlike surgery and radiation therapy, which are local treatments, chemotherapy drugs are typically given systemically, rendering them able to act on any cells within the body. Drugs that are used to treat cancer usually work by killing cells and, therefore, produce side effects when they kill normal, nontumor cells. Chemotherapy of cancer is common; more than 50% of people diagnosed with cancer will receive chemotherapy. Chemotherapy of cancer has typically involved the willingness of patients to endure side effects including anemia, depression, fatigue, hair loss, infection, pain, nausea, vomiting, and mouth sores. Because side effects are a well-known problem that often limits the dose of drug that can be administered to patients with fatal diseases, current research is attempting to produce chemotherapy agents that have fewer side effects, or are better targeted to cancer cells.

Chemotherapy to treat cancer is an ancient practice. There is evidence that ginger root was used to treat skin cancer as early as 300 BC; that the Romans used red clover, as well as mastectomy, to treat certain cancers; and that the 11th-century Arab physician Ibn Sina used arsenic compounds for cancer. Nitrogen mustard for chemotherapy was developed during World War II; this derivative of the chemical warfare agent, mustard gas, was used as a treatment for lymphoma in the 1940s.

Methotrexate is often considered to be the first modern chemotherapy agent because it is a chemically defined agent that acts specifically on a defined biological target, in this case an enzyme. Methotrexate inhibits the action of the enzyme dihydrofolate reductase, which disrupts DNA replication for all cells, but especially rapidly growing cancer cells. Many other chemotherapy agents have now been characterized. These drugs act by a range of mechanisms (Table 16.3): Most

Table 16.3 Examples of commonly used classes of chemotherapy compounds

Class of agents	Examples of the class	Mechanism of action
Alkylating agents	Nitrogen mustard; cyclophosphamide; ifosphamide, Melphalan, Chlorambucil, carmustine (BCNU), lomustine (CCNU), Procarbazine, Busulfan, and thiotepa	Bind to DNA and interfere with replication and transcription
Antimetabolites	Methotrexate; 5-fluorouracil; cytarabine, gemcitabine (Gemzar®), 6-mercaptopurine, 6-thioguanine, fludarabine, and cladribine	Interfere with normal metabolic pathways, including those necessary for making new DNA
Anthracyclines	Daunorubicin; doxorubicin	Form free radicals, which leads to strand breaks in DNA, and inhibition of topoisomerases
Antibiotics	Bleomycin	Generate free oxygen radicals that result in DNA damage
Camptothecins	Camptothecin, irinotecan, and topotecan	Inhibit topoisomerase
Vinca alkaloids	Vincristine, vinblastine, and vinorelbine	Bind to tubulin and disrupt mitotic spindle function
Taxanes	Paclitaxel, docetaxel	Bind to microtubules and inhibit function
Platinums	Cisplatin, carboplatin	Cross-link DNA

drugs interfer with the function of DNA in replication; therefore, these drugs are selectively active against cells undergoing cell division.

Chemotherapy drugs (including the ones listed in Table 16.3) are often administered intravenously, although sometimes they can be injected into a body cavity, such as intraperitoneal injection into the abdominal cavity for treatment of ovarian cancer or intrathecal injection into the space around the spinal cord for treatment of meningeal tumors. A few chemotherapy drugs can be administered orally (Table 16.4), which allows the patient to receive therapy more comfortably and easily.

New methods for delivery of chemotherapy drugs have been introduced recently. One system involves a controlled-release drug-delivery system for treatment of malignant glioblastoma, a particularly aggressive form of brain cancer. The treatment of **malignant glioma** with chemotherapy requires special consideration because of the location of the **neoplasm**. The tight capillary **cellular junctions** of the **blood–brain barrier (BBB)** restrict the entry of drugs from systemic circulation into the brain, with the result that only a few chemotherapy drugs are capable of reaching cytotoxic concentrations at the tumor target when delivered intravenously. Several approaches have been applied to overcome this physiological limitation and improve the pharmacokinetic profile of

Table 16.4 Chemotherapy drugs that can be administered orally

Cancer type	Name of oral chemotherapy agent
Breast cancer	Capecitabine Vinorelbine Oral cyclophosphamide Idarubicin
Colon and colorectal cancer	Capecitabine Tegafur with uracil + leucovorin
Leukemia	Oral cyclophosphamide
Chronic myeloid leukemia (CML)	Imatinib mesylate
Chronic lymphocytic leukemia (CLL), palliative therapy	Chlorambucil
Acute promyelocytic leukemia (APL)	Tretinoin
Acute non-lymphocytic leukemia (ANLL)	Idarubicin
Lymphoma	Etoposide
Cutaneous T cell lymphoma	Bexarotene Oral cyclophosphamide
Small cell lung cancer	Etoposide
Non-small cell lung cancer (NSCLC)	Vinorelbine
Lung cancer	Oral cyclophosphamide
Kaposi's sarcoma	Etoposide
Prostate cancer	Etoposide
Multiple myeloma	Oral cyclophosphamide
Ovarian cancer	Oral cyclophosphamide
Brain tumor	Temozolomide

intravenously administered drugs, including high dose delivery, the use of drug-loaded liposomes, and the disruption of the BBB with biochemical or osmotic agents. Given that primary gliomas rarely metastasize, that they often recur within centimeters of the original tumor location, and that systemic exposure to chemotherapy can result in a variety of toxicities, attempts have also been made to limit the volume distribution of chemotherapy drugs in the body. Limitation of distribution has been achieved to some extent by using intra-arterial administration of **nitrosoureas**, but at the expense of central neurotoxicity. Direct infusion using the **Ommaya reservoir** in conjunction with a catheter, implantable pumps, and drug-loaded controlled-release polymers are methods to achieve delivery intracranially while minimizing the exposure of normal tissues to drugs. Of these devices, controlled-release polymers offer the theoretical advantage of continuous delivery that is not subject to clogging by tissue debris and protection of labile drugs prior to their release. In 1996, the **carmustine** implant (GLIADEL®)—a biodegradable, chemotherapy-loaded polymer matrix—was approved by the

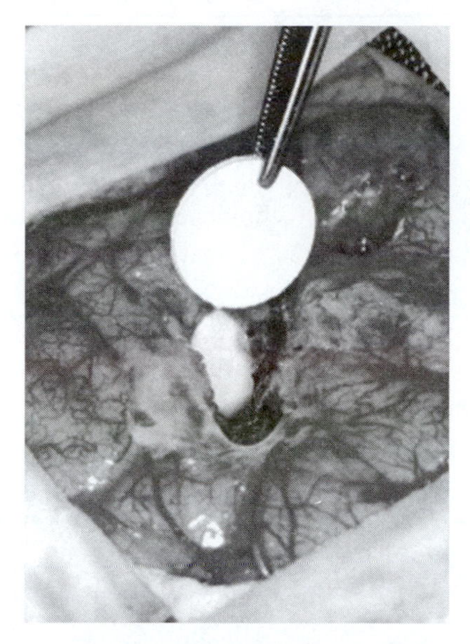

Figure 16.14 **Photograph showing GLIADEL® wafer placement into a human brain following surgical resection of a malignant glioma.** The walls of the tumor cavity are lined with up to eight such wafers. Each wafer is 14.5 mm in diameter and 1 mm in thickness, and contains 7.7 mg of carmustine. Reproduced with permission from Fleming AB, Saltzman WM. Pharmacokinetics of the carmustine implant. *Clin Pharmacokinet.* 2002;41:403–419. (See color plate.)

U.S. Food and Drug Administration (FDA) after it was demonstrated to be a safe and effective treatment for recurrent malignant glioma. This was the first new treatment to be approved for malignant gliomas in 20 years, since the introduction of systemic chemotherapy. Biomedical engineers contributed to this effort by designing the materials and by mathematical modeling, which was essential for understanding the fate of drugs in the brain after local delivery (Box 16.1).

GLIADEL® is composed of the agent carmustine (BCNU) homogeneously distributed in a solid copolymer disk, and is currently approved for the treatment of recurrent **glioblastoma multiforme**. After removal of the bulk tumor mass, 200-mg wafers containing 3.85% carmustine are placed in the resection cavity (Figure 16.14). The GLIADEL® implant then degrades in the brain over time, releasing carmustine. Most patients receive seven or eight wafers at the time of surgical resection, amounting to a total dose of about 60 mg of carmustine. However, patients have received up to 100 mg intracranially in clinical trials. Patients in clinical trials did not experience a significant reduction in blood cell counts, nor was there evidence of renal or hepatic injury, demonstrating that local delivery at this dose can obviate the systemic toxicities encountered with other methods of administration.

One of the most exciting advances in chemotherapy is the development of drugs that target specific molecular pathways that regulate cancer cell growth. Chapter 6 describes the biological importance of enzymes with tyrosine kinase activity. About 30 years ago, it was discovered that leukemic cells from patients with **chronic myelogenous leukemia (CML)** had a change in their chromosomes: Two of the chromosomes (9 and 22) were reciprocally translocated. This abnormal change brought two gene sequences together, *bcr* and *c-abl*, which encodes a nonreceptor tyrosine kinase. The fusion of these two genes created a new protein in the cell, the oncoprotein BCR-ABL, which is able to drive the cell to divide rapidly. The drug imatinib (or Gleevec®, Novartis) was designed to specifically inhibit this tyrosine kinase. In human clinical trials, imatinib produced dramatically positive responses in patients with CML with few side effects.

Box 16.1 Penetration of chemotherapy into tissues

Consider the consequences of implanting a delivery system, such as GLIADEL®, within the brain. Molecules released into the interstitial fluid in the brain extracellular space must penetrate into the brain tissue to reach tumor cells distant from the implanted device. Before these drug molecules can reach the target site, however, they might be eliminated from the interstitium by partitioning into brain capillaries or cells, entering the cerebrospinal fluid, or being inactivated by enzymes present in the extracellular space. Elimination always accompanies dispersion; therefore, regardless of the design of the delivery system, one must understand the dynamics of both processes to predict the spatial pattern of drug distribution after delivery. Although this box focuses on drug transport in the context of polymeric controlled release to the brain, many of these issues apply to other novel drug-delivery strategies.

Assume that the drug diffuses through, and is simultaneously eliminated from, tissue in the vicinity of an implant releasing active molecules. The polymer implant is surrounded by biological tissue, composed of cells and an extracellular space filled with extracellular fluid (ECF). Immediately following implantation, drug molecules escape from the polymer and penetrate the tissue. Once in the brain tissue, a number of processes influence the movement of drug molecules in the brain (Figure Box 16.1a). All of these events influence drug therapy, but a few seem to be most important: Diffusion is the primary mechanism of drug distribution in brain tissue; elimination of the drug occurs when it is removed from the ECF or transformed; and binding or internalization may slow the progress of the drug through the tissue.

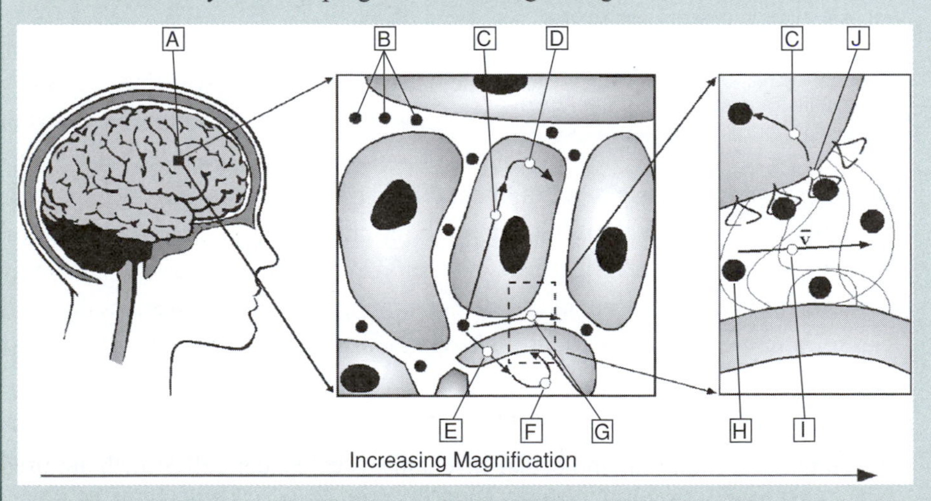

Figure Box 16.1A The fate of compounds delivered to the interstitial space of the brain. Following direct interstitial delivery to the brain, by a polymer implant, for example, compounds can move through the extracellular space. Compounds are eliminated from the extracellular space by permeation through the capillary wall, making them available for clearance by hepatic and renal mechanisms, or by internalization by cells at the local sites. Rates of migration through the extracellular space may be influenced by a variety of factors, including the presence of receptors on the cell surface. A. Site of administration. B. Molecules of agent in extracellular space. C. Internalization. D. Metabolism. E. Permeation through capillary. F. Elimination and biotransformation. G. Transport in the extracellular fluid. H. Diffusion through extracellular matrix gel phase. I. Fluid movement through extracellular space (with velocity, v). J. Association with receptor or transport proteins.

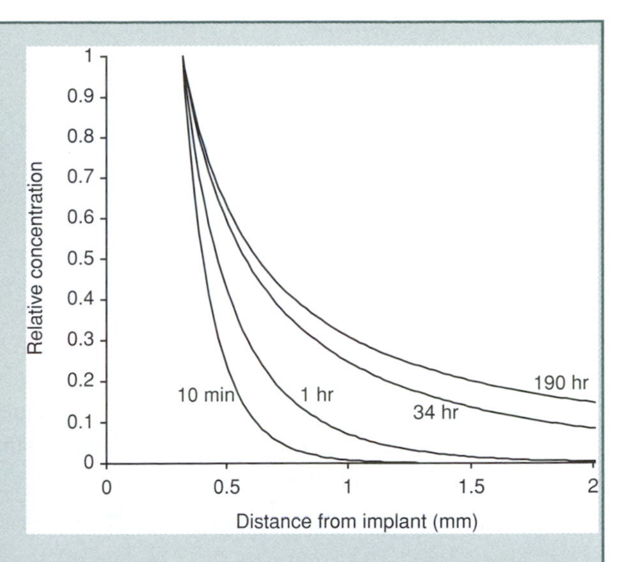

Figure Box 16.1B Concentration profiles after implantation of a spherical drug-releasing implant. Solid lines in this plot were obtained from the equation with the following parameters: $D^* = 4 \times 10^{-7}$ cm^2/s; $R = 0.032$ cm. Each curve represents the steady-state concentration profile for drugs with different elimination half-lives in the brain, corresponding to different values $t_{1/2} = (\ln 2/k)$: $t_{1/2} =$ 10 min; 1 h (0.7); 34 h (0.12); and 190 h (0.016). As the half-life increases, the concentration profile is shifted farther to the right, indicating better penetration of drug.

Mathematical models of these events can be constructed. One of the major results is that simple solutions to the diffusion equation (recall Box 2.5) can explain drug movement in many situations:

$$\frac{c_t}{c_i} = \frac{R}{r} \exp\left[-R\sqrt{\frac{k}{D}}\left(\frac{r}{R} - 1\right)\right], \qquad \text{(Equation 1)}$$

where c_t is the concentration of drug in the tissue, c_i is the maximum concentration (e.g., at the surface of the GLIADEL® implant), R is the radius of the device, r is position measured from the center of the device, k is the rate constant for elimination of the drug from the brain, and D is the diffusion coefficient for drug in the brain. See Saltzman WM. *Drug Delivery: Engineering Principles for Drug Therapy*. New York: Oxford University Press; 2001, for more details.

Imatinib is only the first of the new wave of tyrosine kinase inhibitors to be used for chemotherapy (Figure 16.15). It is likely that work in this area, in which understanding the molecular mechanisms of cancer cell growth are used in rational design of drugs, will continue to yield progress in cancer therapy.

16.6 Hormonal and biological therapies

Hormonal therapy is similar to conventional chemotherapy, in that a drug—in this case, a hormone—is introduced systemically to influence tumor growth. **Hormones**, which are described in Chapter 6, occur naturally in the body, where they are used as signals for biological events including cell growth and differentiation. Some tissues in the body, such as the breast and the prostate, are naturally responsive to hormones. Cancer that arises in these tissues is sometimes

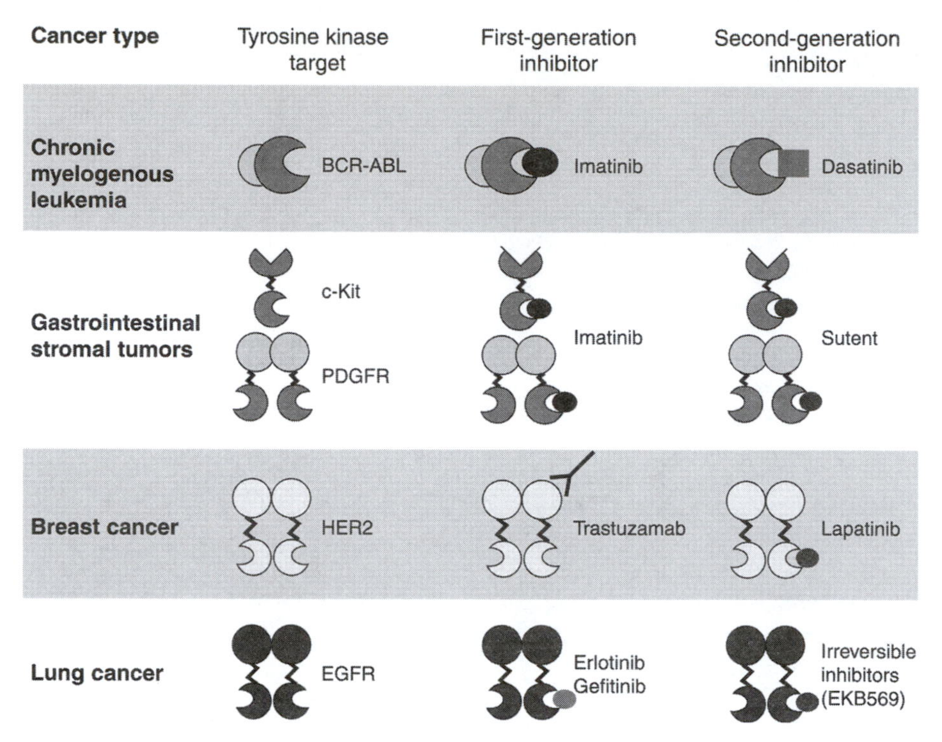

Figure 16.15 **Tyrosine kinase inhibitors for treating cancer.** From Baselga J. Targeting tyrosine kinases in cancer: The second wave. *Science.* 2006;312:1175–1178. Reprinted with permission from the American Association for the Advancement of Science (AAAS).

sensitive to hormones as well. For example, the testicles secrete hormones including **testosterone** that stimulate prostate cell growth; therefore, blockade of this natural hormone can be useful for decreasing the rate of growth of prostate tumors. As an alternative to orchiectomy (surgical removal of the testes), a hormone called **luteinizing hormone-releasing hormone (LHRH)** can be administered to block the natural production of testosterone. This is another example in which a controlled-release drug-delivery system can be useful for treating cancer: Controlled-release implants are commonly used to deliver LHRH for up to several months. Many breast cancers are also sensitive to hormones, particularly estrogen and progesterone. Hormone therapy with drugs such as tamoxifen—a selective estrogen-receptor modulator—is therefore used with chemotherapy to treat breast cancer.

A variety of other approaches for treatment of cancer are in development. Many of these treatments fall under the general category of **biological therapy**, and most of them involve approaches to stimulate the natural function of the immune system to eliminate malignant cells from the body (Table 16.5). It is now commonly believed that malignant cells develop throughout human life and that the immune system recognizes these abnormal cells and eliminates most of them before they become tumors (recall Figure 6.19). A major goal of biological therapy is to enhance or supplement this natural response.

Table 16.5 Examples of biological therapy for cancer

Form of therapy	Rationale	Example
Microorganisms	Historically, it had been noticed that cancer patients who developed severe bacterial infections would sometimes have remissions of their tumors	Bacillus Calmette-Guérin (BCG) for treatment of bladder cancer by instillation into the bladder
Cytokines	Cytokines are naturally occurring mediators of immune system performance; therefore, administration of cytokines might enhance immune responses to tumors	Interleukin-2 (IL-2) infusion to treat metastatic melanoma
Monoclonal antibodies	Antibodies that bind specifically to molecules on the surface of cancer cells would target them for treatment, either by a drug that is attached to the antibody or by a natural biological mechanism	Trastuzumab (Herceptin) for treatment of breast cancer
Adoptive immunotherapy	Transfer of large numbers of pre-activated T cells into the body would enhance the T cell–mediated immune killing of tumor cells	Photopheresis for treatment of cutaneous T cell lymphoma
Cancer vaccines	Introduction of antigens derived from tumor cells would stimulate an immune response directed against the tumor	None have yet been proven

Some of these approaches are not new. For example, **bacillus Calmette-Guérin (BCG)** is an attenuated live tuberculosis bacillus that has been used since the 1880s as a vaccine for tuberculosis. Several decades ago, it was found that instillation of this organism into the bladder was an effective approach for treatment of superficial tumors of the bladder wall (recall Figure 16.1). The mechanism of BCG action in bladder cancer is not known, but it is assumed that the organism enhances the immune response to the tumor.

Great progress has been made over the past two decades in the development of therapies that use adoptive transfer of activated or engineered T cells (5). In **adoptive immunotherapy**, the patient's own immune system is stimulated to produce active T cells, hopefully T cells that will help control the tumor.

The first evidence that the immune system could be manipulated to treat solid tumors in humans was achieved by high-dose delivery of a **cytokine**. **Interleukin-2 (IL-2)** has diverse activities in the immune system, including stimulating the proliferation of activated lymphocytes. High-dose IL-2 led to tumor regression in a fraction of patients with metastatic melanoma (15% of patients experienced regression) and metastatic renal cell cancer (19%). IL-2 has no direct effect on tumor cell growth; therefore, the regression must be caused by an influence of the cytokine on cells of the immune system. In this case, the IL-2 is acting on cells that already reside in the patient; responses, therefore, are limited only to patients who harbor a sufficient number of cells that are capable of responding to the cytokine.

In adoptive cell transfer therapy, large numbers of cells are administered to the patient. Often, the cells are isolated from the patient and then activated ex vivo, allowing for control of the activation process. In this way, the exact cells that are required for therapy can be identified and the host can be manipulated prior to transfer to optimize therapy. For example, IL-2 has also been administered together with an additional population of activated lymphocytes. In this approach, the activity of IL-2 on the patient's immune system is augmented by the large number of new cells that are also provided. Success of this cell transfer therapy depends on 1) the ability to identify cell populations that are effective at mediating tumor regression and 2) the ability to amplify the cell population ex vivo. Often, the precursor cells are collected from within the tumor itself; these cells are called **tumor-infiltrating lymphocytes**.

Other T-cell therapies are already available for clinical use. For example, extracorporeal photopheresis is used to treat patients with cutaneous T cell lymphoma (6). In photopheresis, ultraviolet light is focused on the blood of lymphoma patients as it flows through a serpentine-shaped plastic chamber outside the body. Exposure to the light causes the tumorigenic blood-borne T lymphocytes to enter apoptosis; debris resulting from cell death is taken up in part by circulating monocytes. The monocytes are re-introduced to the blood of the patient; it is believed that immune activation accounts for full or partial recovery in a substantial percentage of cases. It is suspected, but not proven, that this therapy results in the activation of antitumor killer T cells that attack the tumorigenic T cells at the site of the skin lesion.

Cancer vaccines take advantage of the appealing idea that vaccines, which have been so effective at reducing infectious diseases, can be used to treat cancer. The immune system has an important role in the elimination of cancer; therefore, a cancer vaccine would potentially harness this natural activity, by stimulating cells of the immune system to enhance their effectiveness at recognizing and killing cancer cells. Although there are many variations, the approach is simple: Antigens from tumor cells are administered to a patient with cancer to stimulate an immune response to cells that display that antigen. Many such approaches have been successful in experimental animals, but none have yet been demonstrated in humans.

There is, however, a vaccine that prevents an infectious disease that will probably prevent the formation of cancers in the future. **Human papillomavirus (HPV)** is a common sexually transmitted disease: The Centers for Disease Control and Prevention in the United States estimate that 80% of women will have contracted at least one (of the more than 100) strains of HPV. HPV leads to cervical dysplasia (a precursor of cervical cancer) in a fraction of women. Two vaccines for HPV were introduced in the United States in 2006: These vaccines prevent

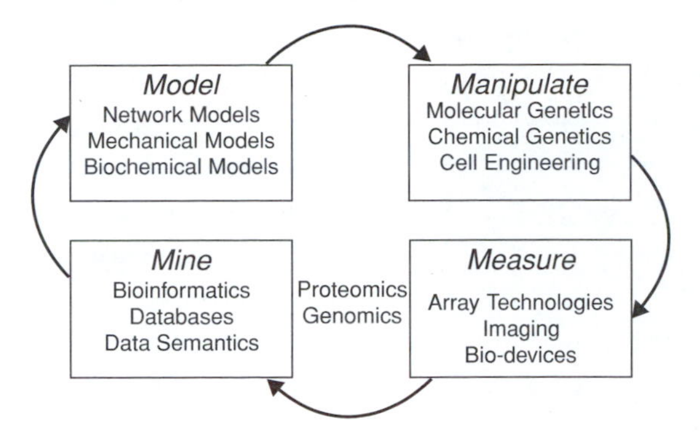

Figure 16.16 **Systems biology.** This diagram shows an organizational framework for systems biology, an emerging field at the intersection of biology and engineering. In systems biology, tools of mathematical modeling are coupled with high-throughput experimental approaches to generate and test novel hypotheses about complex biological systems. These approaches are useful in all application areas of biomedical engineering, but are particularly useful in the study of cancer. Diagram adapted from Ideker T, Winslow LR, Lauffenburger DA. Bioengineering and systems biology. *Ann Biomed Eng.* 2006;34:1226–1233.

transmission of several strains of HPV. Therefore, it is now recommended that young women receive the HPV vaccine, to prevent the formation of precancerous lesions that might lead to cancer.

16.7 Systems biology, biomedical engineering, and cancer

Systems biology is a relatively new area of study for biomedical engineers, but it is one of the most important fields of inquiry for making future progress. The relationship of systems biology to biomedical engineering was presented earlier (Chapter 1). Here, systems biology is illustrated by describing its use in an important application, the understanding of cancer, a complex disease process.

Systems biologists are interested in interactions between the diverse elements that make up a complex biological system. For example, the diversity of function in human cells results from the expression of genes: Different cells express different sets of genes. There are ~30,000 genes in the human genome: Each cell expresses each of these genes at some level (although for many of the genes, that level is zero). How is the overall behavior of the cell at any point in time related to the genes that are being expressed? How does a change in the level of expression of any one of the genes influence the overall behavior of the cell? If the expression of two, or three, or ten different genes changes when a cell undergoes the transformation from normal to malignant, which of these changes is the most important?

To answer these kinds of questions, new technologies are needed. A helpful framework for thinking about the new tools that are needed—and the relationship between the new tools—is shown in Figure 16.16. Systems biology requires innovation in measurement, mining, modeling, and manipulation.

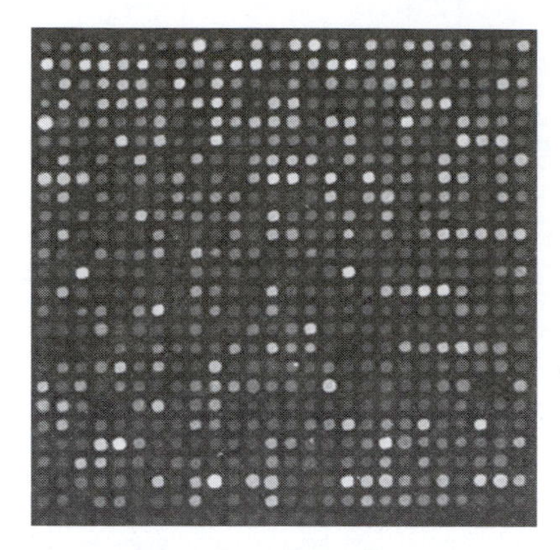

Figure 16.17 **Deoxyribonucleic acid (DNA) microarray.** In this array, each of the colored spots corresponds to a tiny island filled with single-stranded DNA encoding a particular gene (this 25 × 25 array contains 625 genes). An experiment is performed by exposing this array simultaneously to samples obtained from two different messenger ribonucleic acids (mRNAs): one mRNA sample (sample 1) is reverse transcribed to complementary DNA (cDNA) with red fluorescence, and the other (sample 2) is reverse transcribed with green fluorescence. The two samples compete for hybridization to the array: On spots that appear red, sample 1 had more mRNA; on spots that appear green, sample 2 had more mRNA; and on spots that appear yellow, both samples had substantial mRNA. Image courtesy of Dr. Valerie Reinke, Yale University. (See color plate.)

Advances in measurement—particularly in the development of microarray technologies—make systems biology possible. To test a hypothesis about the expression of many genes in a cell population, there must be a method for measuring gene expression simultaneously in a cell population. Microarray techniques permit the simultaneous measurement of multiple events. Microarray techniques are built on a simple concept, the placement of multiple, highly controlled, small "spots" of material onto microscope slides.

In DNA microarrays, tiny spots of DNA are attached to a solid substrate, usually made of glass or silicon (Figure 16.17). Often these DNA sequences are covalently attached to the surface. By making very small islands, or spots, of DNA, many samples can be arranged into a small space. Figure 16.17 shows 625 spots on a substrate, in which each spot contains a concentrated collection of DNA, but each spot has a different DNA sequence, representing a separate gene. Because different DNA sequences have the same general chemistry (sequences of nucleotides that differ only in the ordering of the nucleotides), the same chemical method can be used to bind the DNA to each spot: The challenge is to place the right sample of DNA at each location on the substrate. Because the spots are small, and because tiny volumes of liquid are used to deliver the DNA samples to each location, robots are often used to make the microarrays (Figure 16.18).

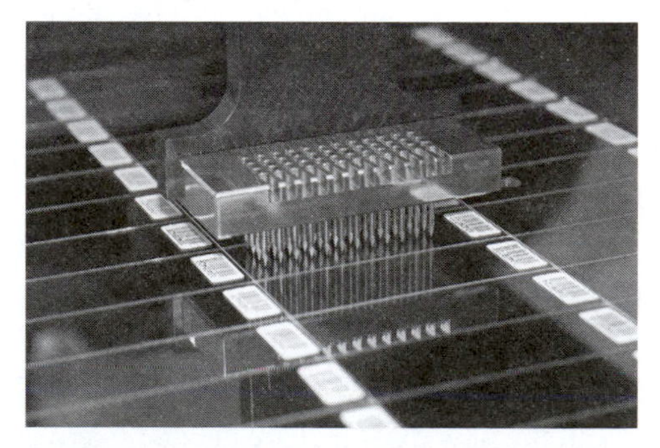

Figure 16.18 **Deoxyribonucleic acid (DNA) microarray robot.** Close-up view of the printing head on a microarray robot. This robot has 48 separate pins—or mini-pipettes—that pick up fluid samples and place them onto the surface. Each pin places a fluid drop of 0.7 nL. Photo courtesy of Andre Nantel, www.digitalapoptosis.com.

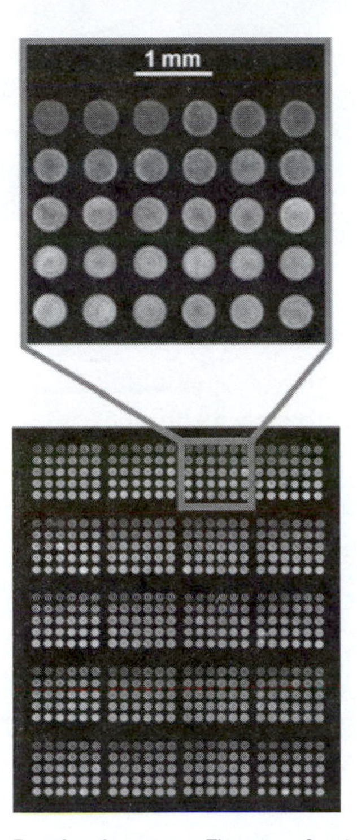

Figure 16.19 **Protein microarray.** The array of proteins is created by dispensing very small volumes of protein-rich solution onto specially designed surfaces, which hold the proteins in place. In this particular example, each spot of the array contains a dilution of human immunoglobulin G (IgG) (the brightness of the spot is related to the concentration of IgG within the spot). The microarray can be used to perform immunoassays (such as enzyme-linked immunosorbent assay [ELISA]). By creating arrays in which each spot is a different protein, the protein microarray can be used to screen for interactions of a test protein with many other proteins simultaneously.

Scientists have developed a variety of approaches to use DNA microarrays. One common approach is to use two colors of fluorescence to represent two different states of a cell population (e.g., normal vs. malignant). The differences between the genes expressed in these two states are revealed by color differences in the completed microarray, with red representing genes expressed in one state, green representing genes expressed in the other, and yellow representing genes expressed in both (Figure 16.17). The experiment can also be performed with only one sample, in which the intensity of gene binding on each spot represents the relative level of expression of that particular gene in that sample. All microarray experiments have a feature in common: They rapidly generate a lot of information on gene expression.

Microarrays for proteins are also available (Figure 16.19). These arrays contain small concentrated spots of protein. Production of the array can be more challenging because covalent bonding to the surface might change the three-dimensional structure of the protein and alter its biological activity. Scientists are finding many interesting applications for protein microarrays: performing high-throughput immunological measurements (enzyme-linked immunosorbent assays [ELISAs]), testing for protein–protein interactions, testing for biological activity of cells on proteins, and designing tissue-engineering substrates, to name a few.

The importance of microscopic examination of tissue samples in assessing the stage of a tumor was described earlier in the chapter. Pathologists—physicians who are experts in identifying disease stages from microscopic specimens—have developed microarray approaches for examining tissue samples (7). Tissue microarrays are microscope slides that contain many unique "spots" of tissue (Figure 16.20): Each spot might represent a tissue from an individual patient, so that the microarray contains information on the spectrum of tissue anatomies that represent a large population of patients (Figure 16.21). The tissue microarrays can be stained using different techniques, to detect features of interest on the samples. Conventional hematoxylin and eosin staining uses a blue stain (hematoxylin) and

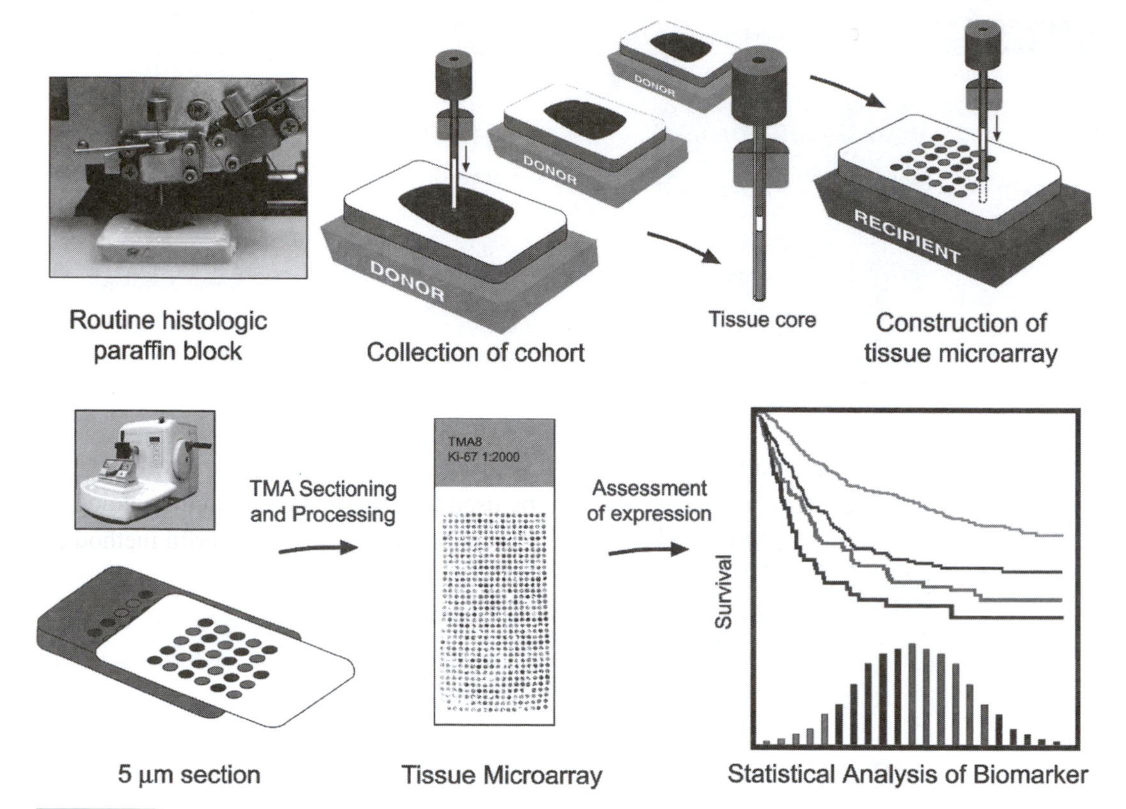

Figure 16.20 **Tissue microarray.** Samples of tissue (biopsies) are often taken from patients with cancer. These samples are used to assess the state of their disease. The tissue microarray technique takes small pieces of tissue from each individual in a large set of patient samples. Usually these patient samples have something in common: They might all be samples for one kind of cancer, such as breast cancer. These small pieces are then assembled onto a single microscopic slide, so that a large sampling of manifestations of one underlying human disease can be studied simultaneously. In the tissue microarray technique, small pieces are cut from a paraffin-mounted tissue sample using a needle, and are mounted onto a new block. The new block (recipient) contains needle-collected samples from many different specimens, and it is cut into sections and mounted on a slide (a spot size of 0.6 mm allows 600 specimens). These slides can then be analyzed for the presence of specific proteins or other biomarkers. Images courtesy of Dr. David Rimm, Yale University and reprinted from (7) with permission from Nature Publishing Group.

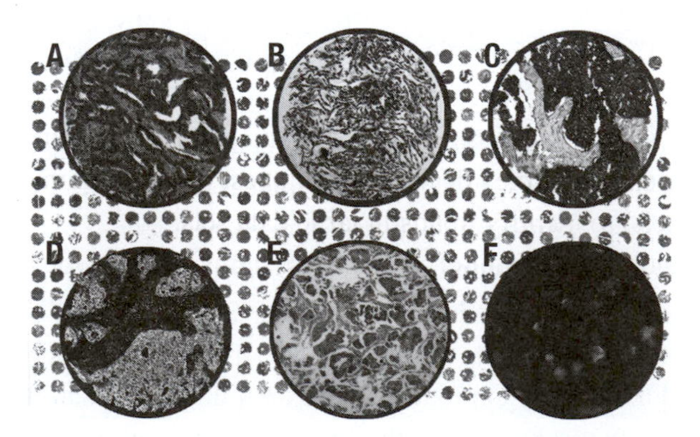

Figure 16.21 **Tissues prepared in a variety of formats can be arrayed.** **A**. Breast carcinoma stained with hematoxylin and eosin (H&E). **B**. Cell line A431 stained with H&E. **C**. Breast carcinoma stained for HER-2/*neu*. **D**. Breast carcinoma stained for cytokeratin (green), DNA (blue), and HER-2/*neu* (red). **E**. Breast carcinoma with F-actin detected by in situ hybridization. **F**. Breast carcinoma with *ERBB2* amplification by fluorescence in situ hybridization. Images courtesy of Dr. David Rimm, Yale University and reprinted from (7) with permission from Nature Publishing Group. (See color plate.)

a red stain (eosin) to detect nucleic acids and cytoplasm, respectively (Figure 16.21, a and b). Immunohistochemical staining, in which an antibody is used to detect a particular protein or other substance in the cell, can also be used: Figure 16.21c shows an example of this technique, using an antibody to HER-2/*neu*, a receptor that is found on some breast cancer cells. In situ hybridization uses nucleic acid probes to locate the complementary DNA or ribonucleic acid (RNA) sequence in a tissue sample (Figure 16.21, e and f).

Tissue microarrays permit the testing of hypotheses that would have been cumbersome in the past. For example, with a microarray containing 600 colon cancer samples, a scientist can quickly test to see what fraction of the tumors are expressing a particular protein. Substances—such as proteins, other molecules, or molecular complexes—that can be used to distinguish a particular biological state are called **biomarkers**. Using tissue microarrays is a powerful method for searching for biomarkers of cancer.

After a reliable biomarker of a disease is found, it can be used to screen individuals for the presence of that disease. An example is the use of prostate specific antigen (PSA), which is used to screen for prostate cancer in men. It is now routine for men to get checked for PSA, as an early indicator of prostate disease. PSA was identified the old-fashioned way, by decades of study of the role of this particular protein in patients with cancer. With tissue microarrays (and all of the microarray approaches), scientists should be able to find biomarkers that are reliable much more rapidly.

DNA, protein, and tissue microarrays differ in their construction, and they are used to make different kinds of measurements. They are similar, however, in their potential to generate large amounts of data. How do you get information from these microarrays? The experiment illustrated in Figure 16.17 reveals many green spots, many red spots, and many yellow spots. What does this mean in terms of the function of each of the genes that are represented by the spots? If there are five red spots—indicating five genes that are highly expressed in one of the two samples—are the functions of these genes within the cell related? To answer these questions, and the many others that arise when microarray data are collected, systems biologists do more than make a lot of measurements quickly. They **mine** the data (i.e., they develop techniques, often using mathematical and statistical models) to identify trends in large data sets.

Systems biologists also **model** biological systems—that is, they develop mathematical models of complex biological processes that they can compare to their microarray data. Models of gene regulatory networks are particularly useful (Figure 16.22). The results of a microarray experiment often reveal only the steady-state level of expression of genes under a certain experimental condition. It is far more interesting to know how gene expression changes over time or, more likely, how the expression of a set of genes changes over time. Mathematical models, which incorporate the relationships between a set of genes (as represented by the arrows in Figure 16.22d), can be used to predict how gene networks will behave; often, these predictions can be tested with additional microarray measurements.

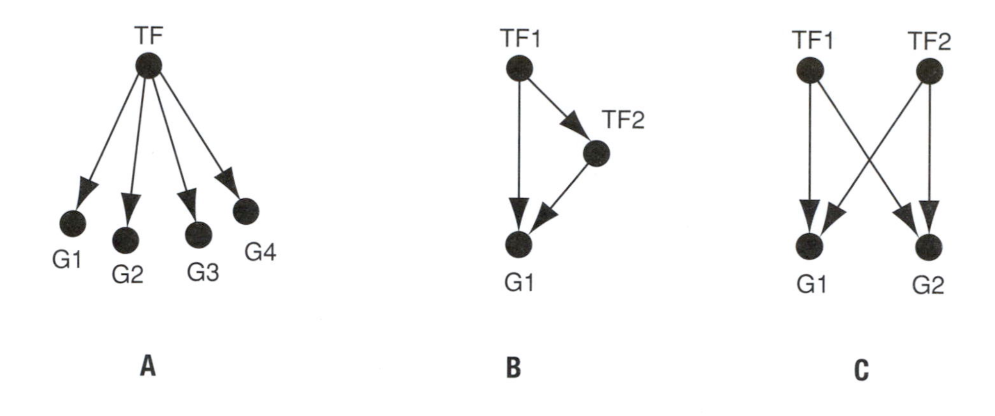

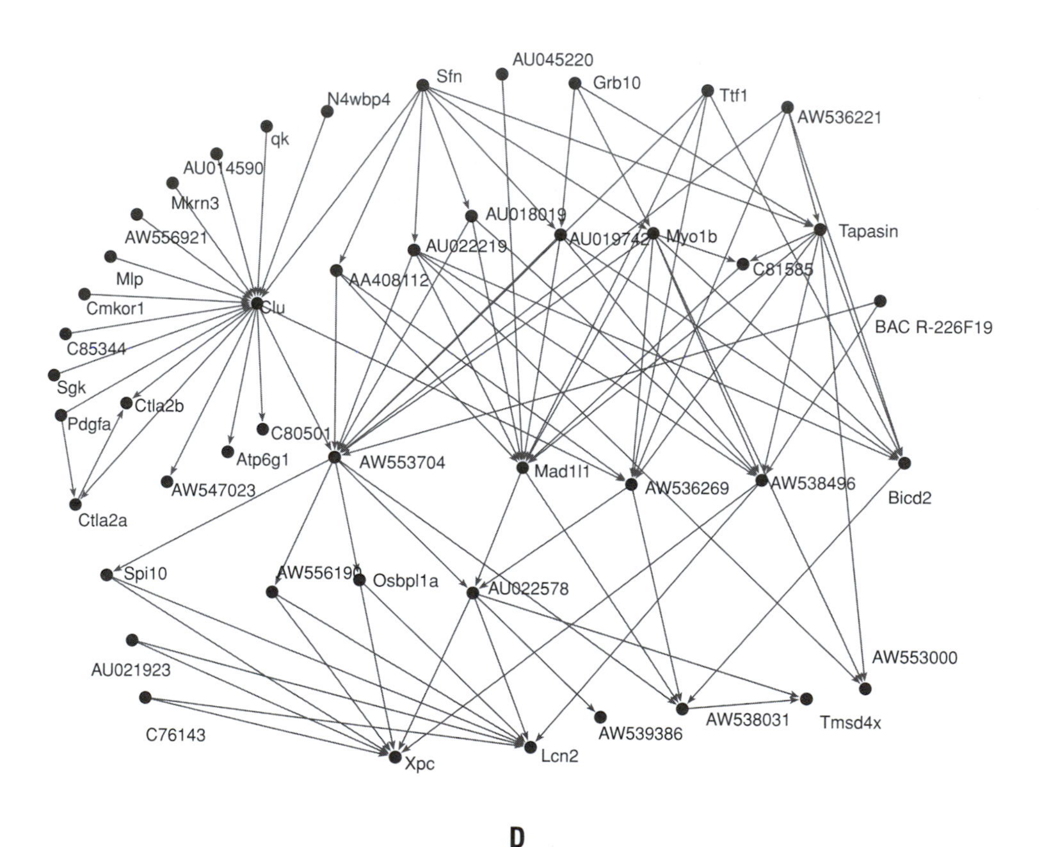

Figure 16.22 **Elements of a gene regulatory network.** The circles, or nodes, in each diagram represent genes, and the arrows represent relationships between the genes. In **(A)**, a transcription factor (TF) has a positive effect on four other genes (G1, G2, G3, and G4). In **(B)**, a first transcription factor (TF1) has a positive effect on a gene (G1), but also on a second transcription factor (TF2), which also has a positive effect on G1. This is an example of feed-forward feedback in a gene network. Because microarray techniques permit the simultaneous measurement of many different quantities, the models can become complex, as in **(D)**. Reprinted with permission from Wang E, Lenferink A, O'Connor-McCourt M. Cancer systems biology: Exploring cancer-associated genes on cellular networks. *Cell Mol Life Sci.* 2007;64:1752–1762.

These models are different from the kinds of models seen earlier in this book—such as the models for blood flow through a capillary (Chapter 8) or diffusion of drugs in the brain (Box 16.1)—but they often involve the same tools that all engineers use to describe complex systems, such as differential equations and linear algebra.

Finally, systems biologists **manipulate** biological systems using approaches from molecular biology, genetics, chemistry, and bioengineering. Work with microarrays, mining, and modeling often leads to predictions regarding cell, tissue, or animal behavior. For example, all of the results might predict that if gene X is expressed at a high level for 10 days in these cells within a tumor, then a tumor will regress. By manipulating the experimental system to create an elevation of gene X (without changing anything else!), this prediction can be tested.

Through new techniques in measurement and modeling, systems biology promises to give us a more detailed understanding of human biology. Systems biology is integrative, in that it proposes to build—from the individual elements that molecular and cellular biology have revealed—an understanding of how the individual parts of a cell or tissue (the proteins, genes, and complexes within cells) contribute to the whole. It is likely to be an extremely valuable approach for understanding diseases such as cancer.

Summary

- Cancer, one of the leading causes of death in the world, is caused by the uncontrolled proliferation of cells, as a result of mutations in the genes that control cell growth.
- Surgery is one of the most successful treatments for cancer: Biomedical engineers have contributed to surgery for cancer by designing tools for use in the operating room.
- Ionizing radiation can be used to kill tumor cells in the body: Optimal application of radiation for therapy requires an understanding of radiation physics and the biological effects of radiation on normal and tumor cells.
- Chemotherapy, or the administration of toxic drugs to kill tumor cells, is another common approach for cancer treatment: Rational design of drugs and new drug-delivery systems promise to make chemotherapy safer and more effective.
- Biological therapies—involving monoclonal antibodies, cytokines, and adoptive immunotherapy—now work in certain kinds of cancer: A better understanding of the basic mechanisms of action, as well as improvements in our ability to manufacture these agents inexpensively, will improve treatment for more patients.
- Systems biology is allowing integration of molecular information on cell function into models for the behavior of cells and tissues: These approaches will improve—perhaps dramatically—our understanding of cancer and our ability to design better therapy for cancer.

REFERENCES

1. Varmus H. The new era in cancer research. *Science*. 2006;312:1162–1165.
2. American Cancer Society. *Cancer Facts and Figures 2006*. 2006, Atlanta, GA: American Cancer Society.
3. Dalton WS, Friend SH. Cancer biomarkers—an invitation to the table. *Science*. 2006;312:1165–1168.
4. Steele GG. From targets to genes: A brief history of radiosensitivity. *Phys Med Biol*. 1996;41:205–222.
5. Rosenberg SA. Progress in the development of immunotherapy for the treatment of patients with cancer. *J Intern Med*. 2001;250:462–475.
6. Wolfe JT, Lessin SR, Singh AH, Rook AH. Review of immunomodulation by photopheresis: Treatment of cutaneous T-cell lymphoma, autoimmune disease, and allograft rejection. *Artif Organs*. 1994;18(1):888–897.
7. Giltnane JM, Rimm DL. Technology insight: Identification of biomarkers with tissue microarray technology. *Nat Clin Pract Oncol*. 2004;1:104–111.

FURTHER READING

Giltnane JM, Rimm DL. Technology insight: Identification of biomarkers with tissue microarray technology. *Nat Clin Pract Oncol*. 2004;1:104–111.

Hobbie RK. Chapter 15: Medical use of x-rays. In: *Intermediate Physics for Medicine and Biology*. 3rd ed. New York: Springer AIP Press; 1997.

Ideker T, Winslow LR, Lauffenburger DA. Bioengineering and systems biology. *Ann Biomed Eng*. 2006;34:1226–1233.

Saltzman WM. *Drug Delivery: Engineering Principles for Drug Therapy*. New York: Oxford University Press; 2001.

KEY CONCEPTS AND DEFINITIONS

adoptive immunotherapy a cancer treatment in which the patient's immune system is stimulated to produce active T cells that may help control the tumor

alpha particle a helium nucleus composed of two protons and two neutrons that is the product of radioactive decay

bacillus Calmette-Guérin (BCG) an attenuated live tuberculosis bacillus that is used to treat superficial tumors of the bladder wall

beta particle an electron emitted from an atom during beta decay, a form of radioactive decay

biological therapy a mode of cancer treatment that harnesses the body's immune system to either attack cancer cells or to fight the side effects of other cancer treatments

biomarkers substances that can be used to distinguish a particular biological state; in oncology, their presence or absence is used as an indicator of whether the cells producing them are malignant

biopsies tissue samples that are surgically removed to determine if cancer is present

blood–brain barrier (BBB) a layer of cells covering the brain that prevents the passage of blood into the brain tissue

brachytherapy the implantation of radioactive materials, called seeds, into or near a tumor

cancer vaccines vaccines that are designed to strengthen the immune system's ability to fight existing cancer cells, or to prevent infections by viruses known to cause cancer

carmustine an alkylating agent, $C_5H_9Cl_2N_3O_2$, used in the treatment of tumors

cellular junctions spaces between adjacent cells where cell-to-cell communication can occur either allowing or preventing the flow of ions and minerals between the cells. There are three types of cellular junctions that serve different purposes: adherens, gap, and tight.

chemotherapy the use of chemical agents, or drugs in the treatment of cancer and other diseases

chronic myelogenous leukemia (CML) a form of cancer characterized by the abundant production of myeloid cells in the bone marrow

cytokine a class of signaling proteins and glycoproteins that is crucial to the regulation of the immune system

dose fractionation a method of optimizing radiation therapy by splitting the total amount of delivered radiation into smaller doses, allowing more sensitive healthy tissue time to recover from reversible damage

electromagnetic rays waves with both electric and magnetic characteristics

endoscope a surgical instrument that can be used to look inside the body, take pictures, and take tissue samples in a minimally invasive manner

external beam radiation therapy (EBRT) a cancer treatment method produced by the use of external radiation beams generated by machines called linear accelerators that provide a high degree of focus for the beams, and careful planning of the radiation field

gamma rays electromagnetic radiation emitted from the nucleus of an atom; they possess the highest frequency and energy as well as the shortest wavelength of all the types of radiation within the electromagnetic spectrum

glioblastoma multiforme an invasive form of primary brain tumor arising from the glial cells of the brain

heterogeneous containing cells that differ widely in their biology and therefore in their response to different treatments

hormonal therapy the use of hormones for medical treatment

hormone a ligand that induces specific responses in target cells especially in the endocrine system; hormones regulate the growth, differentiation, and metabolic activities of various cells, tissues, and organs

human papillomavirus (HPV) a common sexually transmitted disease that causes cervical dysplasia, which is a precursor of cervical cancer

intensity-modulated radiation therapy (IMRT) a treatment for cancer that uses computer-controlled x-ray accelerators to provide radiation delivery from different angular positions to target a tumor and provide minimal exposure to surrounding tissues

interleukin-2 (IL-2) a cytokine used in biological cancer therapy whose function in the immune system includes stimulating the proliferation of activated lymphocytes

luteinizing hormone-releasing hormone (LHRH) a hormone that is responsible for controlling the sex hormones of humans

lymph node a small, kidney-shaped organ found throughout the body that stores cells, particularly white blood cells, and acts as a filter and site for immune activity

malignant glioma an invasive brain tumor

manipulate in systems biology, to use approaches from molecular biology, genetics, chemistry, or bioengineering to arrange factors within a biological system to produce a desired outcome

metastasis spread of a tumor to a distant site

methotrexate a chemotherapy agent that specifically targets the enzyme dihydrofolate reductase and inhibits the action of DNA replication in cells. It is often considered the first modern chemotherapy agent.

microarray a method of analyzing a large number of samples (e.g., DNA, protein) in a small space by attaching tiny quantities of each sample to a surface and surveying them for the presence of a particular gene or substance

mine in systems biology, to develop techniques to identify trends in large data sets, often with the use of mathematical and statistical models

model in systems biology, to develop mathematical models of complex biological process that can be used to compare with microarray data

neoplasm cell with uncontrolled growth and division, also called a tumor

neutron a subatomic particle with no net charge

nitrosoureas a class of chemicals containing nitroso groups and urea, commonly used in the treatment of cancer

Ommaya reservoir a device that is implanted under the scalp with the ability to deliver chemotherapeutic drugs to the cerebrospinal fluid

radiation therapy the treatment of a disease or cancer by the administration of radioactive materials or rays

resection surgical removal of a tissue from the body

testosterone a hormone, derived from cholesterol and with the ability to affect libido and energy, that is secreted in the sex organs of males and females

three-dimensional conformal radiation therapy (3DCRT) a treatment for cancer that uses computer-controlled x-rays to provide radiation delivery through a beam with the ability to conform to the shape of a tumor

tumor-infiltrating lymphocytes white blood cells that have infiltrated a cancer tumor; they are often collected from within the tumor, cultured ex vivo, and re-introduced into the patient together with a cytokine in hopes of boosting the immune system's ability to fight the tumor

Tumor, Node, and Metastases (TNM) a system used in the classification of cancer, indicating the rate and extent of growth within the body

x-ray a type of electromagnetic radiation with wavelengths in the range of 10–0.01 nanometers

NOMENCLATURE

N	A functional value of a tissue that is being irradiated
Q	Quality of a radiation source
W_R	Weighting factor

QUESTIONS

1. Biological dyes are used to prepare a Pap smear, such as the one shown in Figure 16.2. How do you think these dyes work? Why are different cells, and different regions of cells, different colors?

PROBLEMS

1. A drug has a diffusion coefficient in the brain of 10^{-8} cm^2/s and it is eliminated from the brain with a half-life of 45 minutes. If you design an implantable drug-delivery system—a sphere with a diameter of 1 cm—how large a region of the brain can this delivery system treat?
2. A standard microscope slide has dimensions of 1 in × 3 in. If each spot on a tissue microarray is 0.6 mm in diameter, what is the maximum number of spots that can be placed on a microscope slide? Are there other practical limitations that determine the number of spots per slide?
3. The isotopes ^{103}Pd and ^{137}Cs have different radial dose functions when used in brachytherapy (Figure 16.13). Why might you use one versus the other in interstitial radiation therapy?
4. Which of the following leads to the highest biological dose, per Gy of absorbed radiation?
   a. Photons to the bone marrow
   b. Protons to the brain
   c. Electrons to the gonads
   d. Alpha particles to the intestine
   e. Neutrons to the skin

Appendix A

Physiological Parameters

Table A.1 Standard data for a man

Age	30 y
Height	1.73 m
Weight	68 kg
Surface area	1.80 m^2
Normal body core temperature	37.0°C
Normal mean skin temperature	34.2°C
Heat capacity	0.86 kcal/kg °C
Percent body fat	12% (8.2 kg)
Subcutaneous fat layer	5 mm
Body fluids	41 L (60 wt % of body)
Intracellular	28 L
Interstitial	10.0 L
Transcellular	—
Plasma	3.0 L
Basal metabolism	40 kcal/m^2-h, 72 kcal/h
O_2 consumption	250 mL/min[a]
CO_2 production	200 mL/min[a]
Respiratory quotient	0.80
Blood volume	5 L
Resting cardiac output	5 L/min
Systemic blood pressure (systolic/diastolic)	120/80 mmHg
Mean arterial pressure	93 mmHg (at 120/80 mmHg)
Heart rate at rest	65/min
General cardiac output	3.0 + 8M L/min, where M = liters O_2 consumed/min at STP
Total lung capacity	6 L[b]
Vital capacity	4.2 L[b]
Pulmonary ventilation rate	6 L/min[b]
Alveolar ventilation rate	4 L/min[b]
Tidal volume	500 mL[b]
Dead space	150 mL[b]
Respiratory rate at rest	12/min
Pulmonary capillary blood volume	75 mL
Arterial O_2 content	0.195 mL O_2/mL blood[a]
Arterial CO_2 content	0.492 mL CO_2/mL blood[a]
Venous O_2 content	0.145 mL O_2/mL blood[a]
Venous CO_2 content	0.532 mL CO_2/mL blood[a]

Notes: Adapted from Seagrave RG. *Biomedical Applications of Heat and Mass Transfer* Ames, IA: Iowa State University Press, 1971, p. 66.

[a] At standard temperature and pressure (STP). [b] At body temperature and pressure (37°C, 1 atm).

Table A.2 Normal clinical values for blood and urine

Chemical	Blood	CSF	Urine
Water content	93%	99%	99%
Protein content	7,000 mg/dL	35 mg/dL	~0
Osmolarity	295 mOsm/L	295 mOsm/L	
Inorganic substances			
Ammonia	12–55 μM		
Bicarbonate	22-26 mEq/L		
Calcium	4.8 mEq/L	2.1 mEq/L	0–300 mg/d
Carbon dioxide	24–30 mEq/L		
Chloride	100–106 mEq/L	119 mEq/L	
Copper	100–200 μg/dL		0–60 μg/d
Iron	50–150 μg/dL		
Lead	<10 μg/dL		<120 μg/d
Magnesium	1.5–2.0 mEq/L	0.3 mEq/L	
P_{CO2}	35–45 mmHg 4.7–6.0 kPa		
pH	7.35–7.45	7.33	
Phosphorous	3.0–4.5 mg/dL		
P_{O2}	75–100 mmHg 10–13.3 kPa		
Potassium	3.5–5.0 mEq/L	2.8 mEq/L	
Sodium	135–145 mEq/L	135–145 mEq/L	
Organic molecules			
Acetoacetate	Negative		0
Ascorbic acid	0.4–15 mg/dL		
Bilirubin	Direct: 0–0.4 mg/dL Indirect: 0.6 mg/dL		
Carotenoids	0.8–4.0 μg/mL		
Creatinine	0.6–1.5 mg/dL		15–25 mg/kg body weight
Glucose	70–110 mg/dL	60 mg/dL	0
Lactic acid	0.5–2.2 mEq/L		
Lipids	Total: 450–1000 mg/dL Cholesterol: 120–220 mg/dL Phospholipids: 9–16 mg/dL (as lipid P)		
Fatty acids	190–420 mg/dL		
Triglycerides	40–150 mg/dL		
Phenylalanine	0–2 mg/dL		
Pyruvic acid	0–0.11 mEq/L		
Blood urea nitrogen (BUN)	8–25 mg/dL		
Uric acid	3–7 mg/dL		
Vitamin A	0.15–0.6 μg/mL		

Notes: Collected from Devlin TM, ed. *Textbook of Biochemistry with clinical correlations.* New York: Wiley-Liss; 1997, and *New England Journal of Medicine,* 1992;327:718. Deciliter (dL) is 100 mL. CSF: cerebrospinal fluid.

Table A.3 pH range of body fluids

Blood	7.2–7.8
Colon	7.0–7.5
Conjunctival sac	7.8–8.0
Duodenum	4.8–8.2
Milk	8.5–8.7
Mouth	6.2–7.2
Stomach	1–3
Sweat	4.7–4.8
Urethra	5–7
Vagina	3.4–4.2

Table A.4 Protein composition of human blood

Category	Protein	M_w	c_p (mg/100 mL)
	Albumin	69,000	3,500–4,500
	Prealbumin	61,000	28–35
	Insulin	5,000	0–29 μUnits/mL
	Fibrinogen	341,000	200–600
	Erythropoietin	34,000	0.1–0.5 ng/mL
α_1-globulins	α_1-lipoprotein		
	HDL2	435,000	37–117
	HDL3	195,000	217–270
	α_1-acid glycoprotein (orosomucoid)	44,100	75–100
	α_1-antitrypsin (α_1-glycoprotein)	45,000	210–500
	Transcortin		7
	α_{1x}-glycoprotein		14–35
	Haptoglobin	100,000	30–190
α_2-globulins:	Ceruloplasmin	160,000	27–39
	α_2-macroglobulin	820,000	220–380
	α_2-lipoprotein (low density)	5,000,000–20,000,000	150–230
	α_2–HS-glycoprotein	49,000	8(6)
	Z_n-α_2-glycoprotein	41,000	4(6)
	Prothrombin	62,700	9(6)
β-globulins:	β–lipoprotein	3,200,000	280–440
	Transferrin	90,000	200–320
	β_{1C}-globulin		35
	Hemoplexin (β_{1B}-globulin)	80,000	80–100
	β_2-glycoprotein		20–25
γ-globulins:	γG-immunoglobulin	160,000	1,200–1,800
	γM-immunoglobulin	1,000,000	75
	γA-immunoglobulin	350,000	100
Enzymes:	Aldolase		0–7 U/mL
	Amylase		4–25 U/mL
	Cholinesterase		0.5 pH U or more/h
	Creatinine kinase		40–150 U/L
	Lactate dehydrogenase		110–210 U/L
	Lipase		<2 U/mL
	Nucleotidase		1–11 U/L
	Phosphatase (acid)		0.1–0.63 Sigma U/mL
	Phosphatase (alkaline)		13–39 U/L
	Transaminase (serum glutamic oxaloacetic transaminase [SGOT])		9–40 U/mL
	Total		**6,726–9,819**

Note: Information in this table was adapted from Colton C, Merrill EW, Reece JM, Smith KA. Diffusion of organic solutes in stagnant plasma and red cell suspensions. *Chemical Engineering Progress, Symposium Series.* 1970;66:85–99.

Table A.5 Cellular composition of human blood

Cell type	Cell count (10^3 cells/mm^3)	Percentage of WBCs	Percentage of lymphocytes		
			Blood	Lymph	Spleen
Red cells	5,000 ± 350				
Platelets	248 ± 50				
WBCs	7.25 ± 1.7				
Neutrophils		55			
Eosinophils		3			
Basophils		0.5			
Monocytes		6.5			
Lymphocytes		35			
B cells (class II MHC)			10–15	20–25	40–45
T cells (CD3+)			50–60	50–60	50–60
Helper (CD4+,CD8–)					
Cytolytic (CD4–,CD8+)			20–25	15–20	10–15
Natural killer cells (CD16)			~10	Rare	~10

Notes: Information in this table was collected from Abbas AK. Lichtman AH, and Pober JS. *Cellular and Molecular Immunology.* 2nd ed., Philadelphia, PA: W.B. Saunders Company 1994. WBC: White blood cell; MHC: major histocompatibility complex.

Table A.6 Blood distribution for hypothetical man

Pulmonary	Volume (mL)	Systemic	Volume (mL)
Pulmonary arteries	400	Aorta	100
Pulmonary capillaries	60	Systemic arteries	450
Venules	140	Systemic capillaries	300
Pulmonary veins	700	Venules	200
		Systemic veins	2,050
Total pulmonary system	1,300	Total systemic vessels	3,100
Heart	250	Unaccounted	550[a]

Notes: Age 30, weight 63 kg, height 1.78 m, blood volume (assume 5.2 L). From Burton AC. *Physiology and Biophysics of the Circulation.* Chicago: Yearbook Med. Publ: 1965, p. 64.
[a] Probably represents extra blood in reservoirs of the liver and spleen.

Table A.7 Systemic circulation of man

Structure	Diameter (cm)	Blood velocity (cm/s)	Tube Reynolds number[a]
Ascending aorta	2.0–3.2	63[b]	3,600–5,800
Descending aorta	1.6–2.0	27[b]	1,200–1,500
Large arteries	0.2–0.6	20–50[b]	110–850
Capillaries	0.0005–0.001	0.05–0.1[c]	0.0007–0.003
Large veins	0.5–1.0	15–20[c]	210–570
Vena cavae	2.0	11–16[c]	630–900

Notes: Information in this table was obtained from Whitmore RL. *Rheology of the Circulation.* Oxford: Pergamon; 1968.
[a] Assuming viscosity of blood is 0.035 P.
[b] Mean peak value.
[c] Mean velocity over indefinite period of time.

Table A.8 Size and composition of vessels in the human circulatory system

	Aorta	Medium artery	Arteriole	Precapillary sphincter	True capillary	Venule	Vein	Vena cava
Internal radius:	12 mm	2 mm	15 μm	15 μm	3 μm	10 μm	2.5 mm	15 mm
Wall thickness:	2 mm	1 mm	20 μm	30 μm	1 μm	2 μm	0.5 mm	1.5 mm

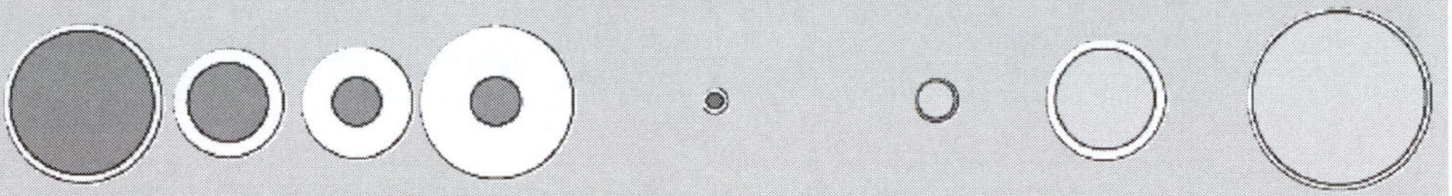

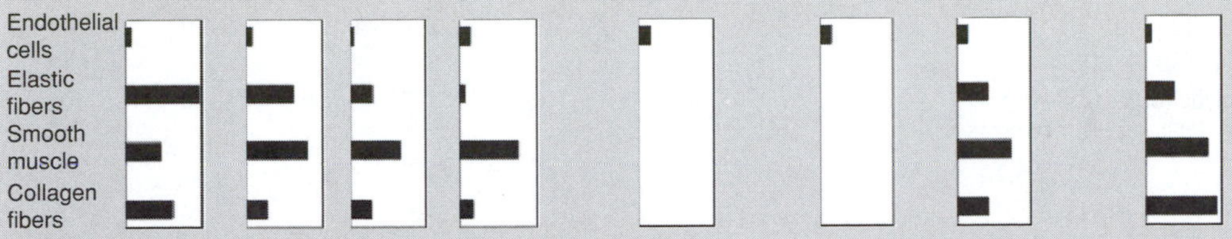

Endothelial cells
Elastic fibers
Smooth muscle
Collagen fibers

Note. From Boron and Boulpaep, *Textbook of Medical Physiology.* Philadelphia, PA: W.B. Saunders Company; 2002.

Table A.9 Systemic circulation of a dog

Structure	Diameter (cm)	Number	Total cross-sectional area (cm^2)	Length (cm)	Total volume (cm^3)	Blood velocity (cm/s)	Tube Reynolds number[a]
Left atrium					25		
Left ventricle					25		
Aorta	1.0	1	0.8	40	30	50	1,670
Large arteries	0.3	40	3.0	20	60	23	230
Main arterial branches	0.1	600	5.0	10	50	8	27
Terminal branches	0.06	1,800	5.0	1	5	6	12
Arterioles	0.002	40×10^6	125	0.2	25	0.3	0.02
Capillaries	0.0008	12×10^8	600	0.1	60	0.07	0.002
Venules	0.003	80×10^6	570	0.2	110	0.07	0.007
Terminal veins	0.15	1800	30	1	30	1.3	6.5
Main venous branches	0.24	600	27	10	270	1.5	12
Large veins	0.6	40	11	20	220	3.6	72
Vena cavae	1.25	1	1.2	40	50	33.0	1,375
Right atrium					25		
Right ventricle					25		
Main pulmonary artery	1.2	1	1.1	2.4	24[b]	36.4	2,090
Lobar pulmonary artery branches	0.4	9	1.19	17.9		33.6	670
Smaller arteries and arterioles	—	—	—	—	18		
Pulmonary capillaries	0.0008	—	300	0.05	16	0.14	0.006
Pulmonary veins	—	600	—	—	52[c]		
Large pulmonary veins	—	4	—	—			

Notes: Information in this table was adapted from Whitmore RL. *Rheology of the Circulation.* Oxford: Pergamon; 1968, p. 92.

[a] Assumes viscosity of blood is 0.03 P.

[b] Includes both main pulmonary artery and lobar pulmonary artery branches.

[c] Includes both pulmonary veins and large pulmonary veins.

Table A.10 Distribution of blood flow at rest

Tissue	Blood flow at rest (mL/min)
Brain	650 (13%)
Heart	214 (4%)
Muscle	1,030 (20%)
Skin	430 (9%)
Kidney	950 (20%)
Abdominal organs	1,200 (24%)
Other	525 (10%)
Total	5,000

Note: Adapted from Sherwood L. *Human Physiology: From Cells to Systems.* 5th ed., Belmont, CA: Thomson Brooks/Cole; 2004, p. 384.

Appendix B

Chemical Parameters

Table B.1 Structures and K_a values of the selected compounds and amino acids

Compound	Structure	pKa	Conjugate base
Ethane	C_2H_6	50	$C_2H_5^-$
Ethene	C_2H_4	44	$C_2H_3^-$
Ammonia	NH_3	38	NH_2^-
Hydrogen	H_2	35	H^-
Toluene	$C_6H_5CH_3$	35	$C_6H_5CH_2^-$
Ethyne	C_2H_2	25	C_2H^-
Acetone	CH_3COCH_3	19	$CH_3COCH_2^-$
Water	H_2O	16	OH^-
Methanol	CH_3OH	16	CH_3O^-
Phenol	C_6H_5OH	10	$C_6H_5O^-$
Hydrogen cyanide	HCN	9	CN^-
Carbonic acid	H_2CO_3	6.1	HCO_3^-
Acetic acid	CH_3COOH	5	CH_3COO^-
Hydrogen fluoride	HF	3	F^-
Hydrogen phosphate	H_3PO_4	2.1	$H_2PO_4^-$
Hydronium ion	H_3O^+	−2	H_2O
Hydrogen chloride	HCl	−7	Cl^-
Hydrogen bromide	HBr	−9	Br^-
Sulfuric acid	H_2SO_4	−9	HSO_4^-
Hydrogen iodide	HI	−10	I^-

Name	pK			pI at 25°C
	α-COOH	NH_3	R group	
Alanine	2.35	9.87		6.11
Arginine	2.18	9.09	13.2	10.76
Asparagine	2.18	9.09	13.2	10.76
Aspartic acid	1.88	9.60	3.65	2.98
Cysteine	1.71	10.78	8.33	5.02
Glutamic acid	2.19	9.67	4.25	3.08
Glutamine	2.17	9.13		5.65
Glycine	2.34	9.60		6.06
Histidine	1.78	8.97	5.97	7.64
Isoleucine	2.32	9.76		6.04
Leucine	2.36	9.60		6.04
Lysine	2.20	8.90	10.28	9.47
Methionine	2.28	9.21		5.74
Phenylalanine	2.58	9.24		5.91
Proline	1.99	10.60		6.30
Serine	2.21	9.15		5.68
Threonine	2.15	9.12		5.60
Tryptophan	2.38	9.39		5.88
Tyrosine	2.20	9.11	10.07	5.63
Valine	2.29	9.74		6.02

Table B.2 Heats of formation for some compounds

Compound	ΔH$_f$ (kJ/mol)	Compound	ΔH$_f$ (kJ/mol)
$Ag_2O(s)$	−30.6	$HgO(s)$	−90.7
$Ag_2S(s)$	−31.8	$HgS(s)$	−58.2
$AgBr(s)$	−99.5	$HI(g)$	+25.9
$AgCl(s)$	−127.0	$HNO_3(l)$	−173.2
$AgI(s)$	−62.4	$KBr(s)$	−392.2
Al^{3+} (aq)	−524.7	$KCl(s)$	−435.9
$Al_2O_3(s)$	−1,669.8	$KClO_3(s)$	−391.4
$BaCl_2(s)$	−860.1	$KF(s)$	−562.6
$BaCO_3(s)$	−1,218.8	$Mg(OH)_2(s)$	−924.7
$BaO(s)$	−558.1	$MgCl_2(s)$	−641.8
$BaSO_4(s)$	−1,465.2	$MgCO_3(s)$	−1,113
$C_2H_2(g)$	+226.7	$MgO(s)$	−601.8
$C_2H_4(g)$	+52.3	$MgSO_4(s)$	−1,278.2
$C_2H_5OH(l)$	−277.6	$MnO(s)$	−384.9
$C_2H_6(g)$	−84.7	$MnO_2(s)$	−519.7
$C_3H_8(g)$	−103.8	$NaCl(s)$	−411.0
$Ca(OH)_2(s)$	−986.6	$NaF(s)$	−569.0
$CaCl_2(s)$	−795.0	$NaOH(s)$	−426.7
$CaCO_3$	−1,207.0	n-$C_4H_{10}(g)$	−124.7
$CaO(s)$	−635.5	n-$C_5H_{12}(l)$	−173.1
$CaSO_4(s)$	−1,432.7	$NH_3(g)$	−46.2
$CCl_4(l)$	−139.5	$NH_4Cl(s)$	−315.4
$CH_3OH(l)$	−238.6	$NH_4NO_3(s)$	−365.1
$CH_4(g)$	−74.8	$NiO(s)$	−244.3
$CHCl_3(l)$	−131.8	$NO(g)$	+90.4
$CO(g)$	−110.5	$NO_2(g)$	+33.9
$CO_2(g)$	−393.5	$Pb_3O_4(s)$	−734.7
CO_2 (aq)	−412.9	$PbBr_2(s)$	−277.0
$CoO(s)$	−239.3	$PbCl_2(s)$	−359.2
$Cr_2O_3(s)$	−1,128.4	$PbO(s)$	−217.9
$Cu_2O(s)$	−166.7	$PbO_2(s)$	−276.6
$CuO(s)$	−155.2	$PCl_3(g)$	−306.4
$CuS(s)$	−48.5	$PCl_5(g)$	−398.9
$CuSO_4(s)$	−769.9	$SiO_2(s)$	−859.4
$Fe_2O_3(s)$	−822.2	$SnCl_2(s)$	−349.8
$Fe_3O_4(s)$	−1,120.9	$SnCl_4(l)$	−545.2
H_2CO_3 (aq)	−699.7	$SnO(s)$	−286.2
$H_2O(g)$	−241.8	$SnO_2(s)$	−580.7
$H_2O(l)$	−285.8	$SO_2(g)$	−296.1
$H_2O_2(l)$	−187.6	$So_3(g)$	−395.2
$H_2S(g)$	−20.1	$ZnO(s)$	−348.0
$H_2SO_4(l)$	−811.3	$ZnS(s)$	−202.9
$HBr(g)$	−36.2	Ethanol(l)	−276.98
$HCl(g)$	−92.3	Glucose (s)	−1,274.5
$HF(g)$	−268.6	Sucrose (s)	−2,221.7

Note: Enthalpy change (ΔH) when 1 mole of compound is formed at 25°C and 1 atm from elements in their stable form.

Table B.3 Types of dialysis membranes

Membrane type	Example membrane name	High or low flux	Biocompatibility*
Cellulose	Cuprophane	Low	−
Semisynthetic cellulose derivatives			
Cellulose diacetate	Cellulose acetate	High and low	+
Cellulose triacetate	Cellulose triacetate	High	++
Diethylaminoethyl-substituted cellulose	HEMOPHAN®	High	+
Synthetic polymers			
Polyacrylonitrile methallyl sulfonate copolymer	PAN/AN-69	High	++
Polyacrylonitrile methacrylate copolymer	PAN	High	++
Poly(methyl methacrylate)	PMMA	High and low	++
Polysulfone	Polysulfone	High	++

Notes: From Pastan S and Bailey J. Medical Progress: Dialysis Therapy. *The New England Journal of Medicine* 1998; 338:1428–1437.

* Biocompatibility is based on complement activation (indicated by in vivo plasma levels of C3a). The minus sign indicates that the membrane is not biocompatible, and the plus signs (+ to ++) increasing degrees of biocompatibility.

Appendix C

Units and Conversion Factors

Table C.1 Prefixes of SI units

Factor	Prefix	Symbol	Factor	Prefix	Symbol
10^{24}	yotta	Y	10^{-1}	deci	d
10^{21}	zetta	Z	10^{-2}	centi	c
10^{18}	exa	E	10^{-3}	milli	m
10^{15}	peta	P	10^{-6}	micro	μ
10^{12}	tera	T	10^{-9}	nano	n
10^{9}	giga	G	10^{-12}	pico	p
10^{6}	mega	M	10^{-15}	femto	f
10^{3}	kilo	k	10^{-18}	atto	a
10^{2}	hector	h	10^{-21}	zepto	z
10^{1}	deka	da	10^{-24}	yocto	y

Table C.2 Gas constant values

8.314 $m^3 \cdot Pa/(mol \cdot K)$
0.08314 $L \cdot bar/(mol \cdot K)$
0.08206 $L \cdot atm/(mol \cdot K)$
62.36 $L \cdot mm\ Hg/(mol \cdot K)$
0.7302 $ft^3 \cdot atm/(lb\text{-}mole \cdot {}^\circ R)$
10.73 $ft^3 \cdot atm/(lb\text{-}mole \cdot {}^\circ R)$
8.314 $J/(mol \cdot K)$
1.987 $cal/(mol \cdot K)$
1.987 $Btu/(lb\text{-}mole \cdot {}^\circ R$

Table C.3 Conversion factors

Quantity	Equivalent values
Mass	1 kg = 10^{-3} metric ton = 2.20462 lb_m = 35.27392 oz. 1 lb_m = 16 oz. = 5×10^{-4} ton = 453.593 g
Length	1 m = 10^{10} angstroms (Å) = 39.37 in. = 3.2808 ft = 1.0936 yds = 0.006214 mile 1 ft = 12 in. = 1/3 yd = 0.3048 m
Volume	1 m^3 = 1,000 L = 35.3145 ft^3 = 264.17 gal = 1056.68 quarts 1 ft^3 = 1728 $in.^3$ = 7.4805 gallons = 0.028317 m^3 = 28.317 L
Force	1 N = 1 kg·m/s^2 = 10^5 dynes = 10^5 g·cm/s^2 = 0.22481 lb_f 1 lb_f = 32.174 lb_m·ft/s^2 = 4.4482 N = 4.4482×10^5 dynes
Pressure	1 Pa = 1 N/m^2 = 10 $dyne/cm^2$ = 1.45×10^{-4} $lb_f/in.^2$ 1 atm = 1.01325×10^5 Pa = 14.7 $lb/in.^2$ = 760 mmHg (at 0°C) = 1,033 cm-H^2O (at 4°C)
Energy	1 J = 1 N·m = 10^7 ergs = 10^7 dyne·cm = 0.23901 cal = 0.7376 ft·lb_f = 9.486×10^{-4} Btu 1 kW·h = 3.6×10^6 J 1 cal = 4.19 J
Power	1 W = 1 J/s = 0.23901 cal/s = 0.7376 ft·lb_f/s = 9.486×10^{-4} Btu/s = 1.341×10^{-3} hp 1 hp = 746 W = 550 ft·lb/s
Temperature	Temperature (°F) = $[\frac{9}{5} \times$ Temperature (°C)] + 32 Temperature (°C) = $\frac{5}{9} \times$ [Temperature (°F) − 32] Temperature (K) = Temperature (°C) + 273.15
Viscosity	1 poise = 100 centipoise = 1 g/(cm-s) = 0.1 Pa-s = 100 mPa-s
Acceleration due to gravity, g	9.8 m/s^2 = 32.174 ft/s^2

Table C.4 Geometric formulas

Surface area of sphere	$4\pi r^2$	r is the radius of the sphere
Surface area of cylinder	$2\pi r^2 + 2\pi r h$	r is the radius of the cylinder, and h is the height
Volume of sphere	$(4/3)\pi r^3$	r is the radius of the sphere
Volume of cylinder	$\pi r^2 h$	r is the radius of the cylinder, and h is the height

Index